Advances in Intelligent Systems and Computing

Volume 249

Series Editor

Janusz Kacprzyk, Warsaw, Poland

For further volumes:
http://www.springer.com/series/11156

Suresh Chandra Satapathy · P.S. Avadhani
Siba K. Udgata · Sadasivuni Lakshminarayana
Editors

ICT and Critical Infrastructure: Proceedings of the 48th Annual Convention of Computer Society of India - Volume II

Hosted by CSI Vishakapatnam Chapter

Springer

Editors
Suresh Chandra Satapathy
Anil Neerukonda Institute of Technology
and Sciences, Sangivalasa
(Affilated to Andhra University)
Vishakapatnam
Andhra Pradesh
India

P.S. Avadhani
College of Engineering (A)
Andhra University
Vishakapatnam
India

Siba K. Udgata
University of Hyderabad
Hyderabad
Andhra Pradesh
India

Sadasivuni Lakshminarayana
CSIR-National Institute of Oceanography
Vishakapatnam
India

ISSN 2194-5357 ISSN 2194-5365 (electronic)
ISBN 978-3-319-03094-4 ISBN 978-3-319-03095-1 (eBook)
DOI 10.1007/978-3-319-03095-1
Springer Cham Heidelberg New York Dordrecht London

Printed on acid-free paper

Springer is part of Springer Science+Business Media (www.springer.com)

Preface

This AISC volume contains the papers presented at the 48th Annual Convention of Computer Society of India (CSI 2013) with theme 'ICT and Critical Infrastructure' held during 13th –15th December 2013 at Hotel Novotel Varun Beach, Visakhapatnam and hosted by Computer Society of India, Vishakhapatnam Chapter in association with Vishakhapatnam Steel Plant, the flagship company of RINL, India.

Computer society of India (CSI) was established in 1965 with a view to increase information and technological awareness among Indian society, and to make forum to exchange and share the IT- related issues. The headquarters of the CSI is situated in Mumbai with a full-fledged office setup and is coordinating the individual chapter activities. It has 70 chapters and 418 students' branches operating in different cities of India. The total strength of CSI is above 90000 members.

CSI Vishakhapatnam Chapter deems it a big pride to host this prestigious 48th Anuual Convention after successfully organizing various events like INDIA-2012, eCOG-2011, 28th National Student convention, and AP State Student Convention in the past.

CSI 2013 is targeted to bring researchers and practitioners from academia and industry to report, deliberate and review the latest progresses in the cutting-edge research pertaining to emerging technologies.

Research submissions in various advanced technology areas were received and after a rigorous peer-review process with the help of program committee members and external reviewer, 173 (Vol-I: 88, Vol-II: 85) papers were accepted with an acceptance ratio of 0.43.

The conference featured many distinguished personalities like Dr. V.K. Saraswat, Former Director General, DRDO, Prof. Rajeev Sangal, Director, IIT-BHU, Mr. Ajit Balakrishnan, Founder & CEO Rediff.com, Prof. L.M. Patnaik, Former Vice Chancellor, IISc, Bangalore, Prof. Kesav Nori, IIIT-H & IIT-H, Mr. Rajesh Uppal, Executive Director & CIO, Maruti Suzuki-India, Prof. D. Krishna Sundar, IISc, Bangalore, Dr. Dejan Milojicic, Senior Researcher and Director of the. Open Cirrus Cloud Computing, HP Labs, USA & President Elect 2013, IEEE Computer Society, Dr. San Murugesan, Director, BRITE Professional Services, Sydney, Australia, Dr. Gautam Shroff, VP and Chief Scientist, Tata Consultancy Services, Mr. P. Krishna Sastry, TCS, Ms. Angela R. Burgess Executive Director, IEEE Computer Society, USA, Mr. Sriram Raghavan,

Security & Digital Forensics Consultant, Secure Cyber Space, and Dr. P. Bhanu Prasad, Vision Specialist, Matrix Vision GmbH, Germany among many others.

Four special sessions were offered respectively by Dr. Vipin Tyagi, Jaypee University of Engg. & Tech., Prof. J.K. Mandal, University of Kalyani, Dr. Dharam Singh, CTAE, Udaipur, Dr. Suma V., Dean, Research, Dayananda Sagar Institutions, Bengaluru. Separate Invited talks were organized in industrial and academia tracks in both days. The conference also hosted few tutorials and workshops for the benefit of participants.

We are indebted to Andhra University, JNTU-Kakinada and Visakhapatnam Steel plant for their immense support to make this convention possible in such a grand scale. CSI 2013 is proud to be hosted by Visakhapatnam Steel Plant (VSP), which is a Govt. of India Undertaking under the corporate entity of Rashtriya Ispat Nigam Ltd. It is the first shore-based integrated steel plant in India. The plant with a capacity of 3 mtpa was established in the early nineties and is a market leader in long steel products. The Plant is almost doubling its capacity to a level of 6.3 mtpa of liquid steel at a cost of around 2500 million USD. RINL-VSP is the first integrated steel plant in India to be accredited with all four international standards, viz. ISO 9001, ISO 14001, ISO 50001 and OHSAS 18001. It is also the first Steel Plant to be certified with CMMI level-3 certificate and BS EN 16001 standard.

Our special thanks to Fellows, President, Vice President, Secretary, Treasurer, Regional VPs and Chairmen of Different Divisions, Heads of SIG Groups, National Student Coordinator, Regional and State Student Coordinators, OBs of different Chapters and Administration staff of CSI-India. Thanks to all CSI Student Branch coordinators, Administration & Management of Engineering Colleges under Visakhapatnam chapter for their continuous support to our chapter activities. Sincere thanks to CSI-Vizag members and other Chapter Members across India those who have supported CSI-Vizag activities directly or indirectly.

We take this opportunity to thank authors of all submitted papers for their hard work, adherence to the deadlines and patience with the review process. We express our thanks to all reviewers from India and abroad who have taken enormous pain to review the papers on time.

Our sincere thanks to all the chairs who have guided and supported us from the beginning. Our sincere thanks to senior life members, life members, associate life members and student members of CSI-India for their cooperation and support for all activities.

Our sincere thanks to all Sponsors, press, print & electronic media for their excellent coverage of this convention.

December 2013

Dr. Suresh Chandra Satapathy
Dr. P.S. Avadhani
Dr. Siba K. Udgata
Dr. Sadasivuni Lakshminarayana

Organization

Chief Patrons

Shri A.P. Choudhary, CMD, RINL
Prof. G.S.N. Raju, VC, Andhra University
Prof. G. Tulasi Ram Das, VC, JNTU-Kakinada

Patrons

Sri Umesh Chandra, Director (Operations), RINL
Sri P. Madhusudan, Director (Finance), RINL
Sri Y.R. Reddy, Director (Personnel), RINL
Sri N.S. Rao, Director (Projects), RINL

Apex (CSI) Committee

Prof. S.V. Raghavan, President
Shri H.R. Mohan, Vice President
Dr S. Ramanathan, Hon. Secretary
Shri Ranga Rajagopal, Hon. Treasurer
Sri Satish Babu, Immd. Past President
Shri Raju L. Kanchibhotla, RVP, Region-V

Chief Advisor

Prof D.B.V. Sarma, Fellow, CSI

Advisory Committee

Sri Anil Srivastav, IAS, Jt. Director General of Civil Aviation, GOI
Sri G.V.L. Satya Kumar, IRTS, Chairman I/C, VPT, Visakhapatnam

Sri N.K. Mishra, Rear Admiral IN (Retd), CMD, HSL, Visakhapatnam
Capt.D.K. Mohanty, CMD, DCIL, Visakhapatnam
Sri S.V. Ranga Rajan, Outstanding Scientist, NSTL, Visakhapatnam
Sri Balaji Iyengar, GM I/C, NTPC-Simhadri, Visakhapatnam
Prof P. Trimurthy, Past President, CSI
Sri M.D. Agarwal, Past President, CSI
Prof D.D. Sharma, Fellow, CSI
Sri Saurabh Sonawala, Hindtron, Mumbai
Sri R.V.S. Raju, President, RHI Clasil Ltd.
Prof. Gollapudi S.R., GITAM University & Convener, Advisory Board, CSI-2013

International Technical Advisory Committee:

Dr. L.M. Patnaik, IISc, India
Dr. C. Krishna Mohan, IIT-H, India
Dr. K. RajaSekhar Rao, Dean, KLU, India
Dr. Chaoyang Zhang, USM, USA
Dr. Wei Ding, USA
Dr. Cheng-Chi Lee, Taiwan
Dr. Pramod Kumar Singh, ABV-IIITM, India
Dr. B. Biswal, GMRIT, India
Dr. G. Pradhan, BBSR, India
Dr. B.K. Panigrahi, IITD, India
Dr. S. Yenduri, USA
Dr. D. Janikiram, IIT-M
Dr. H.R. Vishwakarma, VIT, India
Dr. Deepak Garg, TU, Patiala
Dr. Hamid Arabnia, USA
Dr. Azah Kamilah Muda, Malaysia
Dr. Yun-Huoy Choo, Malaysia
Dr. Hongbo Liu, Dalian Maritime
Dr. B.N. Biswal, BEC, India
Dr. Saihanuman, GRIET, India
Dr. L. Perkin, USA
Dr. V. Shenbagraman, SRM, India
and many others

Organizing Committee

Chair
Sri T.K. Chand, Director (Commercial), RINL & Chairman, CSI-Vizag

Co-Chairmen
Sri P.C. Mohapatra, ED (Projects), Vizag Steel
Sri P. Ramudu, ED (Auto & IT), Vizag Steel

Vice-Chairman
Sri K.V.S.S. Rajeswara Rao, GM (IT), Vizag Steel

Addl. Vice-Chairman:
Sri Suman Das, DGM (IT) & Secretary, CSI-Vizag

Convener
Sri Paramata Satyanarayana, Sr. Manager (IT), Vizag Steel

Co-Convener
Sri C.K. Padhi, AGM (IT), Vizag Steel

Web Portal Convener
Sri S.S. Choudhary, AGM(IT)

Advisor (Press & Publicity)
Sri B.S. Satyendra, AGM (CC)

Convener (Press & Publicity)
Sri Dwaram Swamy, AGM (Con)

Co-Convener (Press & Publicity)
Sri A.P. Sahu, AGM (IT)

Program Committee

Chairman:
Prof P.S. Avadhani, Vice Principal, AUCE (A)

Co Chairmen
Prof D.V.L.N. Somayajulu, NIT-W
Dr S. Lakshmi Narayana, Scientist E, NIO-Vizag

Conveners:
Sri Pulle Chandra Sekhar, DGM (IT), Vizag Steel – Industry
Prof. S.C. Satapathy, HOD (CSE), ANITS – Academics
Dr. S.K. Udgata, UoH, Hyderabad

Finance Committee

Chairman
Sri G.N. Murthy, ED (F&A), Vizag Steel

Co-Chairmen
Sri G.J. Rao, GM (Marketing)-Project Sales
Sri Y. Sudhakar Rao, GM (Marketing)-Retail Sales

Convener
Sri J.V. Rao, AGM (Con)

Co-Conveners
Sri P. Srinivasulu, DGM (RMD)
Sri D.V.G.A.R.G. Varma, AGM (Con)

Sri T.N. Sanyasi Rao,
Sr.Mgr (IT)

Members Sri V.R. Sanyasi, Rao
Sri P. Sesha Srinivas

Convention Committee

Chair Sri D.N. Rao, ED(Operations), Vizag Steel

Vice-Chairs

Sri Y. Arjun Kumar
Sri S. Raja
Dr. B. Govardhana Reddy
Sri ThyagaRaju Guturu
Sri Bulusu Gopi Kumar
Sri Narla Anand

Conveners

Sri S.K. Mishra
Sri V.D. Awasthi
Sri M. Srinivasa Babu
Sri G.V. Ramesh
Sri A. Bapuji
Sri B. Ranganath
Sri D. Satayanarayana
Sri P.M. Divecha
Sri Y.N. Reddy
Sri J.P. Dash
Sri K. Muralikrishna

Co-Conveners

Sri Y. Madhusudan Rao
Sri S. Gopal
Sri Phani Gopal
Sri M.K. Chakravarty
Sri P. Krishna Rao
Sri P. Balaramu
Sri B.V. Vijay Kumar
Mrs M. Madhu Bindu
Sri P. Janardhana
Sri G.V. Saradhi

Members:

Sri Y. Satyanarayana
Sri Shailendra Kumar
Sri V.H. Sundara Rao
Mrs K. Sarala
Sri V. Srinivas
Sri G. Vijay Kumar
Mrs. V.V. Vijaya Lakshmi
Sri D. Ramesh
Shri K. Pratap
Sri P. Srinivasa Rao
Sri S. Adinarayana
Sri B. Ganesh
Sri Hanumantha Naik
Sri D.G.V. Saya

Sri V.L.P. Lal
Sri U.V.V. Janardhana
Sri S. Arun Kumar
Sri K. Raviram
Sri N. Prabhakar Ram
Sri BH.B.V.K. Raju
Sri K. Srinivasa Rao
Mrs T. Kalavathi
Mrs A. Lakshmi
Sri N. Pradeep
Sri K. Dilip
Sri K.S.S. Chandra Rao
Sri Vamshee Ram
Ms Sriya Basumallik
Sri Kunche Satyanarayana
Sri Shrirama Murthy

Contents

Session 2: Software Engineering and Bio-Informatics

Content Based Retrieval of Malaria Positive Images from a Clinical Database VIA Recognition in RGB Colour Space

Somen Ghosh and Ajay Ghosh

Department of Applied Optics and Photonics, University of Calcutta
92, APC Road, Kolkata 700009
ghosh_somen@hotmail.com, aghosh.cu@gmail.com

Abstract. Modern hospitals are trying to create a database of patients' diagnostic history that also contains multiple images taken during different clinical tests on a patient. This has lead to a demand for easy retrieval of images matching a query condition, so that this database can be used as a clinical decision support system. This paper presents a technique for retrieval of malaria positive images, matching a specific query condition, from a clinical image database. The candidate image is segmented in RGB colour space, and a pseudo-colour is imparted to the non-region of interest pixels. The technique additionally retains the full features of the chromosomes, and hence the modified image can be used for further studies on the chromosomes. The algorithm utilizes 4-connected labeled region map property of images to analyze and modify the image, i.e., delete unwanted artifacts, etc. This property is also used to count the number of RBCs.

Keywords: Malaria, Segmentation, RGB space, cell counting, labeled regions.

1 Introduction

Modern hospitals are attempting to create a database of patients' diagnostic history. More and more patient records in this database contain data in form of images. This has lead to a demand for an application for easy retrieval of images and associated diagnostic details so that it can be used as a clinical decision support system.

While it is easy to do a query on structured information available in a database, image data being random and statistical in nature, query on them is not trivial. In the past many content based image retrieval systems have been proposed using different visual features. Shape is one key feature proposed in literature [1,4,6,9,14] . However, the shape of real objects varies in images depending on the pose, illumination, etc. Hence such techniques have limited applications. Recently a few alternative methods have been proposed that use texture, colour, etc. [2,3,8,10,11,13,15]. However these methods suffer from poor segmentation quality. None of these methods can however be directly applied to biomedical images. Such images are characterized by the occurrence of multiple objects in the same image, and are usually stained with

S.C. Satapathy et al. (eds.), *ICT and Critical Infrastructure: Proceedings of the 48th Annual Convention of CSI - Volume II*, Advances in Intelligent Systems and Computing 249,
DOI: 10.1007/978-3-319-03095-1_1,

objective specific stains. Hence such images need segmentation techniques that search for specific signatures of the object being searched.

In this paper we attempt to create an algorithm for identifying the presence of malaria, the type of smear (thick or thin) whose image is being studied, and the degree of disease. The method segments the chromosomes of the parasites from the rest of the image, and, at the same time retains its complete colour (rgb) details. A pseudo-coloring technique is used to impart a specific colour to the pixels that do not represent the chromosomes. The method utilizes 4-connected labeled region properties of digital images to count the RBCs within the image so that the degree of the disease can be estimated.

Clinical databases usually have the following types of images: (1)x-ray, (2)ultra-sonographs and echo-cardiographs, (3)ECG (4)CT and MRI (5)pathological and histo-pathological and (6) nuclear medicine (bone density/vessel maps, etc.). The application is expected to successfully identify malaria positive images from a database containing these diverse types of images.

2 Signature of Malaria

The presence of the colored chromatin dots is a sufficient signature to indicate the presence of parasite in the digital images of stained blood smears. Both thin and thick smears carry this signature. The manual clinical test process involves visual discrimination between the colours of chromatin dots and the RBCs under a microscope. These chromatin dots are usually located within the RBCs but in advanced disease they can exit and RBC. Besides, a diseased RBC can have more than one chromatin dots, but they are considered as single infection. Thus we can say that the presence of chromatin dots is a sufficient signature to indicate the presence of malaria parasite in the digital image of stained blood smears.

Extracts from malaria positive images are shown in figure 1.

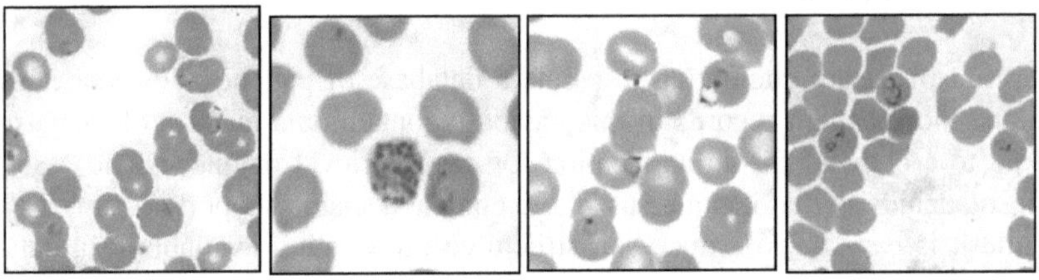

Fig. 1. Extracts from malaria-positive images

3 Algorithm for Recognition of Malaria

To meet the objective of the segmentation, our approach would be: 1)work in rgb colour space, 2) pixels other than those representing chromosomes should be assigned a specific colour, and then, 3)the presence of the chromosomes should be tested in the modified image.

In the first step, some knowledge about images found in a clinical database is used to eliminate certain category of images from consideration. Since malaria positive images are RGB images, hence gray images (single layered or 3-layered) are eliminated from consideration.

The colour that is to be assigned to the non-Region of Interest (non-RoI) pixels must satisfy the following conditions: 1)have a colour distinct from that rendered by the stain to the chromatin dots, and 2)have an intensity that is less than the intensity of the RBCs. The color is defined in the HSI space(H and I can be defined separately) and then the equivalent r, g and b values calculated. A hue of green was chosen to be rendered to the non-RoI pixels, since it has maximum contrast with the hue of the stained chromatin dots. A hue value of 227^0 was selected that is located within the range of green hues (210^0 – 269^0) in the hue-scale of HSI space. To arrive at the intensity value, the candidate RGB image was converted to a gray scale image, and then the average intensity of the pixels representing the RBCs was measured. An intensity value that was 50% of the average intensity existing within the RBCs was selected. The saturation value (S) was taken at 0.2. These HSI values (H=227^0,S=0.2 and I=as calculated) were converted to the corresponding r, g, and b values for RGB space. This calculated r, g and b values were the pseudo-colour that was assigned to the non-RoI pixels.

Identification and pseudo-coloring of non-RoI pixels is a two step process. In the first step we make use of the fact that in the gray scale version of the image (fig. 2a), the background has an intensity distinct from that of the pixels representing RBCs. Hence this image is thresholded at an intensity value given by Otsu's formula. This results in a binary image that has the background pixels identified as 1(white). This binary image is used as a template image to identify the location of the background pixels in the original RGB image, and these pixels are assigned the new defined pseudo-colour (fig. 4). The modified image still has the RBCs and the chromosomes retained in their original colour.

Additionally, the count of the background pixels in the binary image is used to decide whether the smear was thick or thin. Only images of thin smears are analysed.

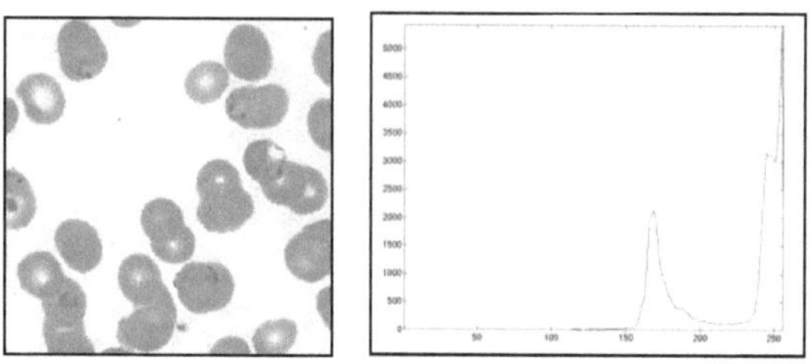

Fig. 2. (a) Gray scale image of 1(a) and (b) its histogram

A histogram of the gray scale version of the modified image shows three distinct regions (fig. 5): 1) the intensity assigned to the background pixels, this is also the lowest intensity, 2) an intermediate intensity possessed by a low number of pixels, and 3) a higher intensity possessed by pixels representing the RBCs. Otsu's formula returns an intensity value that is between regions 2 and 3 indicated above. However, for this to be successful, the volume of background pixels should be high. The gray scale image is thresholded at this threshold value to create another binary mask image. This new template image is used to identify the foreground pixels and they are also assigned the pseudo-colour defined.

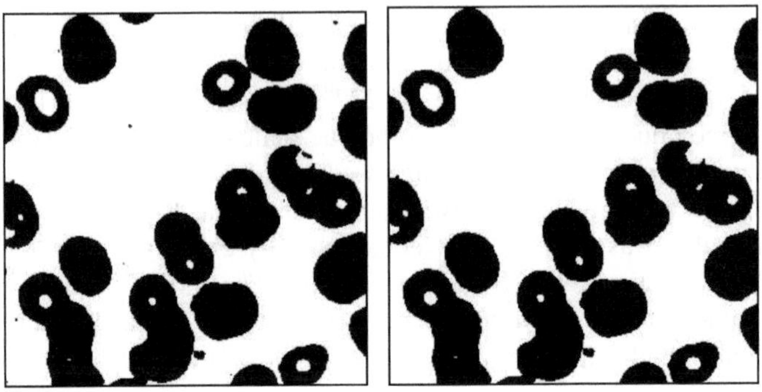

Fig. 3. (a),(b) Binary mask images of fig 1(a), before and after removing artifacts from background

Note that the first binary mask image (fig. 3a) clearly shows artifacts in the background region. These are black regions on a white background. To remove the artifacts, a digital negative of the binary image is converted into a 4-connected region labeled image. The population of each labeled region is an indication of area occupied by each region. Any region with an area less than 0.3% of the total image size was erased and the corresponding pixels in the binary mask image were marked as background pixels.

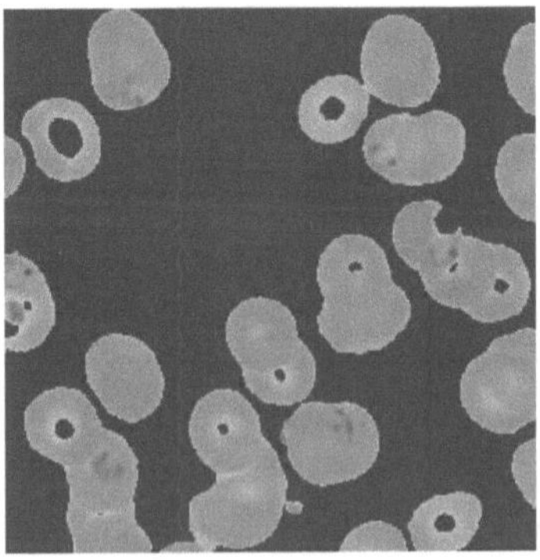

Fig. 4. Image of 1(a) with background pixels pseudo-colored

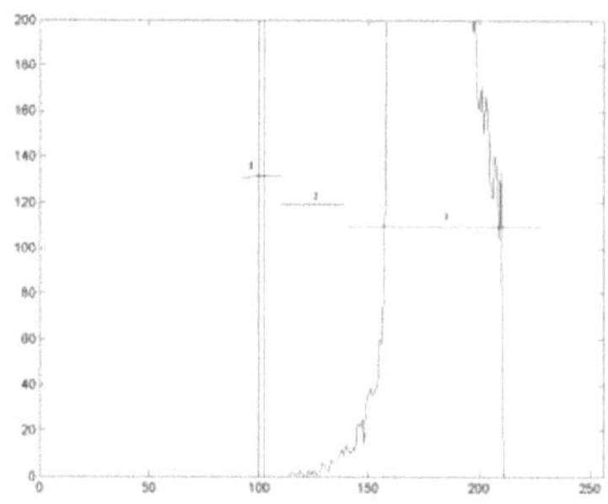

Fig. 5. Zoomed histogram of modified image

Fig. 6. Chromatin dots identified for 1(a) – binary image used for easy demonstration

4 Algorithm for Counting of RBCs

The degree of disease is estimated for thin smears only. The degree of the disease is given by the formula

[No. of affected RBCs / Total number of RBCs in view] x 100

Hence we need to count the number of RBCs and affected RBCs in the field of view. For this we have used the final binary mask prepared previously. The mask image is converted into a 4-connected, labeled region map array. The centroids, and major and minor axis of all disjoint regions are measured. The number of disjoint regions is essentially the number of cells in the field of view. However the count needs to be corrected to account for two factors: 1) partially visible cells located at the boundary of the image and 2) overlapped cells.

The ratio of the major axis to minor axis is used to decide whether the cells are free standing or are overlapped. Free standing cells have a ratio ~1. The minor axis of cells with ratio ≈ 1 is taken as the diameter of a RBC. If the major or the minor axis of any region is less than the radius of a free standing cell, then the cell is partially visible. For cells having ratio > 1, if their major or minor axis is $\geq$ diameter of the cell, then cells are partially overlapped.

Fig. 7. Binary template with holes removed

Partially visible cells are identified as those cells having their centroids at a distance less than the radius of the cell from the boundary of the image. These cells are removed from the count. Overlapped cells are identified by their ratio. All regions having ratio greater than a threshold were considered for resolving. The major axis was divided by the diameter of the cells to find out how many cells were overlapped. A similar study was done on minor axis. However, for cases where the ratio was higher than the threshold, but the minor axis was less than 80% of the diameter of an RBC, the resolution was not done. These additional numbers of cells were added to arrive at the total number of cells.

5 Experimental Results

The training set of images was sourced from [5]. The prototype was developed using MATLAB.

The program fetches one image at a time and rejects images that do not match the filtering condition. A candidate image is analysed and if found to be malaria- positive, the information obtained (thick or thin smear, degree of infection) is written to the database.

The image in figure 1(a) exhibited an average intensity of 227 within the RBCs. Hence the I value was taken as 110 for calculation of R,G and B values. The HSI values of $H=227^0$, $S=0.2$ and $I=110$ translates to R=85, G=99 and B=146 in RGB space. However, since MATLAB was used for the initial study, it calculates intensity using the formula $0.299R+0.587G+0.114B$ and not $1/3[R+G+B]$. Thus, the resultant histogram shows the intensity value as 100 for the pseudo-colored pixels, and not 110 as expected.

The following table gives details of the calculations to arrive at the number of RBCs.

Table 1. Calculation of number of RBCs

Reg.	Centroid 1	Centroid 2	Axes 1	Axes 2	Ratio	RBC
1	5.52	86.80	33.55	12.72	2.64	0
2	12.52	163.29	56.16	29.15	1.93	0
3	33.47	71.96	49.27	39.95	1.23	1
4	47.82	255.71	109.94	48.03	2.29	3
5	68.11	203.18	51.59	40.69	1.27	1
6	69.65	29.17	57.14	45.83	1.25	1
7	129.56	259.08	97.33	58.52	1.66	3
8	150.40	190.05	65.77	41.59	1.58	2
9	184.17	239.85	55.53	47.23	1.18	1
10	208.85	40.92	99.00	46.17	2.14	2
11	200.47	159.80	68.37	50.14	1.36	1

Table 1. (*continued*)

12	208.61	293.39	37.92	17.88	2.12	0
13	232.52	77.74	59.72	41.15	1.45	1
14	258.23	143.24	93.01	44.27	2.10	2
15	248.86	284.61	43.97	31.52	1.39	0
16	281.67	208.17	58.92	43.33	1.36	0
17	290.78	93.40	49.14	24.01	2.05	0
18	293.19	34.54	50.15	18.51	2.71	0
19	294.86	256.08	37.23	14.50	2.57	0

Diameter of free standing cell = 40.
Following regions are partially visible and are not counted: 1,2,12,15,16,17,18,19.
Partially overlapped cells that needed to be resolved are : 2,5,12.
Number of regions = 19
Partially visible cells to be removed = 8
Overlapped cells to be added = 7
Total cells = 19 - 8 + 7 = 18.

An analysis of the results obtained shows: (1) Malaria positive images available in the database in gray scale are not retrieved. These images are however not normal for clinical databases. (2) Images of malaria positive slides taken under very high magnification could not be processed. This is because a distinct background could not be identified in such images. These images too are not normal for a clinical database. (3) Images of thick smears were retrieved irrespective of the degree of the disease. This is because the algorithm does not do counting of cells in images of thick smears.

6 Conclusion

Using the algorithm described, we were able to successfully demonstrate that it is possible to build a system to identify malaria positive images in a clinical database, especially for images of thin smears. Thus we are able to convert an image database to a 'fetch-able' information repository. However, to develop a fully functional decision support system, separate query services need to be written for different categories of images : X-rays, MRI, ECG, etc. A key advantage of the algorithm is that is automatically adapts to the different stains used in different laboratories. However the algorithm does not yet identify the type of parasite. This will be included in a future development.

Acknowledgement. Financial assistance from the Department of Applied Optics and Photonics, University of Calcutta, under the Technical Education Quality Improvement Program is acknowledged by the first author. Opinions expressed and conclusions drawn are of the authors and are not to be attributed to the Department of Applied Optics and Photonics.

References

1. Adoram, M., Lew, M.S.: IRUS: Image Retrieval using Shape. In: IEEE International Conference on Multimedia Computing and Systems, vol. 2, pp. 597–602 (1999)
2. Avula, S.R., Tang, J., Acton, S.T.: An object based image retrieval system for digital libraries. Multimedia Systems Journal 11(3), 260–270 (2006)
3. Avula, S.R., Tang, J., Acton, S.T.: Image retrieval using Segmentation. In: Proceedings of IEEE 2003 Systems and Information Engineering Design Symposium, pp. 289–294 (2003)
4. Bartolini, I., Ciaccia, P., Patella, M.: WARP: Accurate Retrieval of shapes using phase of Fourier descriptors and time warping distance. IEEE Transactions on Pattern Analysis and Machine Intelligence 27(1), 142–147 (2005)
5. Center for Disease Control and Prevention - malaria image library, `http://www.dpd.cdc.gov/dpdx/HTML/ImageLibrary/Malaria_il.htm` (accessed July 2013)
6. Folkers, A., Samet, H.: Content based image retrieval using Fourier descriptors on a Logo database. In: IEEE Proceedings of 16th International Conference on Pattern Recognition 2002, vol. 3, pp. 521–524 (2002)
7. Ghosh, S., Kundu, S., Ghosh, A.: Rapid detection of malaria via digital processing of images of thin smears of peripheral Blood. In: Proceedings of International Conference on Trends in Optics and Photonics II, pp. 181–187 (2011)
8. Kutics, A., Nakajima, M., Ieki, T., Mukawa, N.: An object-based image retrieval system using an inhomogeneous diffusion model. In: Proceedings of International Conference on Image Processing 1999, ICIP 1999, vol. 2, pp. 590–594 (1999)
9. Milios, E., Petrakis, E.G.M.: Shape retrieval based on dynamic programming. IEEE Transactions on Image Processing 9(1), 141–147 (2000)
10. Natsev, A., Rastogi, R., Shim, K.: WALRUS: a similarity retrieval algorithm for image databases. IEEE Transactions on Knowledge and Data Engineering 16(3), 301–316 (2004)
11. Rajakumar, K., Muttan, S.: Medical image retrieval using energy efficient wavelet transform. In: International Conference on Computing Communication and Networking Technologies (ICCCNT), pp. 1–5 (2010)
12. Ross, N.E., Pritchard, C.J., Rubin, D.M., Duse, A.G.: Automatic Image processing method for the diagnosis and classification of malaria on thin blood smears. Med. Biol. Eng. Comput. 44, 427–436 (2006)
13. Sheikh, A.R., Lye, M.H., Mansor, S., Fauzi, M.F.A.: A content based image retrieval system for marine life images. IEEE 15th International Symposium on Consumer Electronics (ISCE), pp. 29–33 (2011)
14. Wang, Z., Chi, Z., Feng, D.: Shape based leaf image retrieval. In: IEEE Proceedings on Vision, Image and Signal Processing, vol. 150, pp. 34–43 (2003)
15. Huang, Y., Zhang, J., Zhao, Y., Ma, D.: Medical Image Retrieval with Query-Dependent Feature Fusion Based on One-Class SVM. In: IEEE 13th International Conference on Computational Science and Engineering (CSE), pp. 176–183 (2010)

Modified Clustered Approach for Performance Escalation of Distributed Real-Time System

Urmani Kaushal[1] and Avanish Kumar[2]

[1] Department of Physical & Computer Sciences, FASC,
MITS Deemed University, Lakshmangarh-332311, Sikar, Raj., India
[2] Department of Maths, Stats & Computer Applications, Bundelkhand University,
Jhansi, U.P., India
{urmani10kaushal,dravanishkumar}@gmail.com

Abstract. In Distributed Real-Time Systems the execution of computational tasks must be completed within time else catastrophe may ensue. This constraint can be achieved by optimal task allocation in DRTS. In the process of task allocation, an appropriate and efficient load sharing strategies can improve the performance of distributed system. In this paper an efficient, two level clustering approach has been proposed. The initial level of clustering will be done on the basis of execution cost and inter task communication cost and second at the stage of task allocation. In the allocation of tasks a more scalable active approach is being followed which iteratively refines the performance of DRTS. The proposed model has been simulated in MATLAB 7.11.0. Two examples have been demonstrated to illustrate the model and algorithm for performance improvement in distributed system.

Keywords: Distributed Real-Time System, Task Allocation, Load Sharing, Clustering, Execution Cost.

1 Introduction

The availability of inexpensive high performance processors and memory chips has made it attractive to use Distributed Computing Systems for real time applications [1]. Distributed Real-Time Systems are characterized by their requirement that the execution of their computational tasks must be not only logically correct but also completed within time [2]. The task allocation in DRTS finds extensive applications in the faculties where large amount of the data is to be processed in relatively short periods of time, or real time computations are required [3]. Fields of research applications are: Meteorology, Cryptography, Image Analysis, Signal Processing, Solar and Radar Surveillance, simulation of VLSI circuits, Industrial Process Monitoring. All these applications require not only very fast computing speeds but different strategies involving DRTS, because in such application the quality of the output is directly proportional to the amount of the real time computations.

Appropriate and efficient load sharing strategies can improve the performance of distributed system. Under load sharing scheme, the workload is to be distributed

S.C. Satapathy et al. (eds.), *ICT and Critical Infrastructure: Proceedings of the 48th Annual Convention of CSI - Volume II*, Advances in Intelligent Systems and Computing 249,
DOI: 10.1007/978-3-319-03095-1_2,

among the nodes available to receive the workload in the system. In this paper static load sharing policy has been considered. In this policy system state information is not required to make load distribution decisions [4]. Once the load sharing decisions has been made, all the tasks should be placed to the appropriate node in the system. Task allocation over the distributed system must be performed in such a manner that the total system cost is minimized. Task allocation in distributed real time systems is an open research area. In task allocation time and space dimension and information requirement for decision making, makes it very complex and difficult. Task allocation can be viewed as a strategic factor which if not managed properly will affect the performance of the system [5]. Task allocation is a NP-complete problem [6][7][8]. Various models for task allocation have been proposed in literature [6][7][9][10][11][12][13][14][8]. Effective clustering technique for tasks of distributed application reduces the allocation search space [8]. In [12][14][6] clustering has been done only once at initial level.

Here in this model, clustering is done at two level, first at initial level on the basis of execution cost and inter task communication cost and second at the stage of task allocation. In the allocation of tasks a more scalable active approach is being followed which iteratively refines the performance of DRTS. The heterogeneity of the system has been considered at the time of allocation. Organization of the paper is as follows. In the next section problem formulation has been done. In the third section the problem has been stated. Fourth section describes the model and its components in detail. In the fifth section, the technique has been described and sixth section provides the algorithm for clustering and allocation of tasks. In next section the model has been simulated using MATLAB7.11.0. Here two examples have been taken to illustrate the proposed model. Results of the simulated experiments are collectively presented in the section 8.

2 Problem Formulation

DRTS is a useful platform for huge and complex real-time parallel application. The execution of a parallel application can be seen as the execution of multiple tasks (parallel application divided into number of tasks) over different processors in the system concurrently. The performance of parallel applications on DRTS basically depends on the arrangement of the tasks on various processors available in the system [15]. Methodical resource management is required to properly allocate tasks to achieve the constrained performance. Task allocation should be made in such a way that it can minimize the inter task communication and processor's capabilities must suit to the execution requirements of the task. Proposed model offer an optimal solution by assigning a set of "m" tasks of the parallel application to a set of "n" processors (*where,* $m > n$) in a DRTS with the goal to enhance the performance of DRTS. The objective of this problem is to enhance the performance of the distributed system by making optimal utilization of its processors and suitable allocation of tasks.

2.1 Notations

T : the set of tasks of a parallel program to be executed.

P : the set of processors in distributed system.

n : the number of processors.
m: the number of tasks formed by parallel application .
k: the number of clusters.
t_i : i^{th} task of the given parallel application.
P_l: l^{th} processor in P.
ec_{il} : incurred execution cost (EC), if i^{th} task is executed on l^{th} processor.
cc_{ij}: incurred inter task communication cost between task t_i and t_j , if they are executed on separate processors.
X: an allocation matrix of order $m*n$, where the entry $x_{il} = 1$; *if* i^{th} task is allocated to l^{th} processor and 0; otherwise.
CI : cluster information vector.
ECM (,) : execution cost matrix.
ITCCM (,) : inter task communication cost matrix.

2.2 Definitions

2.2.1 Execution Cost (EC)

The execution cost ec_{il} of a task t_i , running on a processor P_l is the amount of the total cost needed for the execution of t_i on that processor during process execution. If a task is not executable on a particular processor, the corresponding execution cost is taken to be infinite (∞).

2.2.2 Communication Cost (CC)

The communication cost (cc_{ij}) incurred due to the inter task communication is the amount of total cost needed for exchanging data between t_i and t_j residing at separate processor during the execution process. If two tasks executed on the same processor then $cc_{ij} = 0$.

2.3 Assumptions

To allocate the tasks of a parallel program to processors in DRTS, following assumptions has been made:

2.3.1. The processors involved in the distributed system are heterogeneous and do not have any particular interconnection structure.

2.3.2. The parallel program is assumed to be the collection of m- tasks that are free in general, which are to be executed on a set of n- processors having different processor attributes.

2.3.3. Once the tasks are allocated to the processors they reside on those processors until the execution of the program has completed. At whatever time a cluster of tasks is assigned to the processor, the inter task communication cost (ITCC) between those tasks will be zero.

2.3.4. Data points for *k-mean clustering* will be collection of vectors which represents the execution cost of the task t_m on each processor.

2.3.5. Number of tasks to be allocated is more than the number of processors (m>>n) as in real life situation.

3 Problem Statement

For the evaluation of the proposed model, the problem has been chosen where a set $P = \{p_1, p_2, p_3, \ldots\ldots.p_n\}$ of '*n*' processors and a set $T = \{t1, t_2, t_3, \ldots\ldots.t_m\}$ of '*m*' tasks. The processing time of each task to each and every processor is known and it is mentioned in the Execution Cost Matrix of order *m x n*. The inter task communication cost is also known and is mentioned *ECM (,)*in Inter Task Communication Cost Matrix *ITCCM (,)* of order *m x m*.

There are *m* number of tasks which are to be allocated on *n* processors (where *m*>*n*). Here*m* tasks will be clustered into *k* clusters which equal to *n* in the case. Here we have *m* data points in the form of task vectors to be clustered in *k* clusters. Cluster information will be stored in *CI (,)* vector of *m x 1*.

According to the cluster information the *ECM (,)* will be recalculated. Now at the second level *ECM (,)* will be clustered again it required and the final assignment will be made. With the final task assignment *ECM (,)* and *ITCCM (,)* will be calculated and the total system cost will be calculated.

4 Proposed Model

In this section, a model for two level clustering and task allocation has been proposed, which is used to get an optimal system cost for system's performance enhancement. This objective can be achieved by efficient clustering and allocating tasks in suitable manner.

In order to cluster and allocate the tasks of parallel application to processors in DRTS, we should know the information about tasks attributes like execution cost, inter task communication cost. While obtaining such information is beyond the scope of this paper therefore, a deterministic model that the required information is available before the execution of the program is assumed.

For preparing the clusters of tasks *k-mean* clustering has been used. In *k-mean* algorithm *k* centroids should be defined, one for each cluster. Defined centroids should be positioned in a calculating way because of different location causes different result. Pick each point from the given data set and associate it to the nearest centroid. At this level early grouping is done. Now re-calculate *k* new centroids as centers of the clusters resulting from the previous step. With *k* new centroids binding will be done with the same data set and *k* new centroids. It should be iterated till the centroids do not move any more.

Clusters obtained by *k-mean* clustering again an iterative method will be enforced on it. In the process of minimizing system cost further iterative clustering may decrease the number of clusters and mean while the assignment will be made according to the lowest execution cost of task on the processor.

4.1 Execution Cost (EC)

The task allocation given as: X: T→P, X $(i) = l$ (1)

The execution cost ec_{il} represents the execution of task t_i on processor P_l and it is used to control the corresponding processor allocation. Therefore, under task allocation X, the execution of all the tasks assigned to l^{th} processor can be computed as:

$$\mathrm{EC(X)} = \sum_{i=1}^{n}\sum_{l=1}^{m} ec_{il}x_{il} \tag{2}$$

$$where, x_{il} = \begin{cases} 1\ if\ i^{th} task is assigned to l^{th} processor \\ 0, otherwise \end{cases}$$

4.2 Task Clustering

Evaluation of cluster compactness as the total distance of each point (task vector of n dimension) of a cluster from the cluster mean which is given by [15] [16], Z_{ki}

$$\sum_{x_i I C_k} \|X_i\text{-}\overline{X}_k\|^2 = \sum_{i=1}^{m} Z_{ki}\,\|X_i\text{-}\overline{X}_k\|^2 \tag{3}$$

Where the cluster mean is defined as $\overline{X}_k = \frac{1}{m_k}\sum_{x_i I C_k} X_i$ and $m_k = \sum_{i=1}^{m} Z_{ki}$ is the total number of points allocated to cluster k. The parameter Z_{ki} is an indicator variable indicating the suitability of the ith data point X_i to be a part of the kth cluster.

The total goodness of the clustering will then be based on the sum of the cluster compactness measures for each of the k clusters. Using the indicator variables Z_{ki}, we can define the overall cluster goodness as $\varepsilon_k = \sum_{i=1}^{m}\sum_{k=1}^{k} Z_{ki}\,\|X_i\text{-}\overline{X}_k\|^2$ (4). Here $\overline{X}_k$ should be found in such a manner that the value of ε_k can be minimized.

5 Technique

Model proposed in this paper consists of multiple components. A set $P = \{p_1, p_2, p_3, ...p_n\}$ of 'n' processors and a set $T = \{t_1, t_2, t_3, ...t_m\}$ of 'm' tasks. Execution time of each task to each and every processor is known and it is mentioned in the Execution Cost Matrix *ECM (,)* of order m x n. The communication cost of task is also known and is mentioned in Inter Task Communication Cost Matrix *ITCCM (,)* of order m x m.

Minimization of total system cost will boost the performance. For this purpose tasks is to be grouped according to the similarity in their attributes. Since m tasks are to be processed over n processors ($m>n$), so n clusters should be formed. This condition gives the number of clusters to be formed which is represented by k and it is equal to n in the case. For this level *k-means* clustering is being used to form k clusters. Here m vectors of tasks are to be placed onto k clusters. For grouping the m vectors, k initial points is calculated for each cluster. These initial points are called centroids. Each task vector is to be assigned that has the closest centroid. This assignment is to be performed for all tasks in *ECM (,)*[15].Repeat the work of assignment and recalculation of centroid's positions until the centroids no longer move[17]. This produces a separation of the task vectors into clusters from which the metric to be minimized will be calculated using eq (3).

Modify the *ECM(,)* according the *k* clusters by adding the processing time of those tasks that occurs in the same cluster. Modify the *ITCCM(,)* by putting the communication zero amongst those tasks that are in same cluster.

In the process of assigning *k* clusters to *n* processors, next level of clustering will be done on the basis of ec_{il} (execution cost) constraint. If there is any change in the clusters the *ECM (,)* and *ITCCM (,)* should be recalculated accordingly. Once the final assignments are in hand optimal cost of assignment is to be computed using eq. (2). The objective function to calculate total system cost is as follows:

$$\text{Total Cost} = \text{EC} + \text{CC} \tag{7}$$

6 Algorithm

The algorithm consists of following steps:

Step-0: Read the number of processors in *n*
Step-1: Read the number of tasks in *m*
Step-2: Read number of clusters in *k(in this case it equal to number of processors)*
Step-3: Read the Execution Cost Matrix *ECM (,)* of order *m x n*
Step-4: Read the Inter Task Communication Cost Matrix *ITCCM (,)* of order *m x m*
Step-5: Apply *k-mean* clustering algorithm on ECM (,)
Step-6: Cluster information is stored in vector *CI*
Step-7: Modify the ECM (,) by adding the processing time of tasks in each cluster
Step-8: Modify the ITCCM (,) by putting communication zero amongst those tasks which are in the same cluster
Step-9: Divide Modified *ECM (,)* in n column matrix (which includes the execution cost of each task on processor P_l)
Step-10: Sort each column matrix
Step-11: For each column matrix *i*=1 to *n*

```
    For j=1 to m
        If t_j is not assigned
            Assign t_j on P_i processor
        Else
            If cost on pre assignment > cost of current assignment
                Assign t_j on P_i processor
            Else
                Skip current assignment
            End if
        End if
    End for
End for
```

Step-12: Calculate the total execution cost using equation 6
Step-13: Calculate Inter Task Communication Cost
Step-14: Optimal Cost = Execution Cost + Inter Task Communication Cost
Step-15: End

7 Implementation

Two examples are considered to illustrate the performance of proposed method for better allocation, as well as to test the proposed model on these examples.

Example 1. The efficacy of the proposed algorithm has been illustrated by solving the same running example as in [13]. The results obtained for this example, with the proposed algorithm, [13], [11] and [9] have been given below in Table 1.

Table 1. Comparative results with proposed algorithm, [13], [11] and [9]

	System Cost
Proposed algorithm	195
Bora et al. algorithm [13]	270
Lo's et al. algorithm [11]	275
Kopidakis et al. algorithm[9]	285

In comparison with [13], [11] and [9] the proposed model has minimized the total system cost by 27.78%, 29.09% and 31.58% respectively.

Example 2. Distributed system considered in this example, consisting of three processors $P = \{P_1, P_2, P_3\}$, a parallel application with four executable tasks $T = \{t_1, t_2, t_3, t_4\}$, ECM and ITCCM as in [14]. The results obtained with the proposed algorithm, [14] and [10]for this example has been given below in Table: 2.

Table 2. Comparative results with proposed algorithm, [14] and [10]

Proposed Algorithm			Yadav et.al. algorithm [14]			Kumar et. al. algorithm [10]		
Tasks	Processors	System's Cost	Tasks	Processors	System's Cost	Tasks	Processors	System's Cost
t_2, t_4	P_1		t_2	P_1		t_2	P_1	
t_1	P_2	18	Nil	P_2	24	t_1, t_4	P_2	27
t_3	P_3		t_1, t_4, t_3	P_3		t_3	P_3	

In comparison with [14] and [10] the proposed model has minimized the total system cost by 25% and 33.33% respectively.

8 Conclusion

The critical phase in distributed system is to minimize the system cost. In this paper the proposed model has considered the task allocation under static load sharing scheme. Here the *ITCC* has been reduced by multilevel clustering and the system cost has been minimized by efficient and straightforward algorithm for optimal task allocation. The proposed model has been implemented the example presented in [13] [11] [9] [14] and [10] to illustrate the performance intensification and the proposed model has reduced total system cost by 27.78%, 29.09%, 31.58%, 25% and 33.33%respectively. So by using this model the optimal solution can be achieved at all the times.

References

[1] Hou, C., Shin, K.: Load shaing wiht consideration of future task arrivals in heterogeneous real time system. In: Real Time Systems Symposiun (1992)

[2] Shin, K., Krishna, C., Lee, Y.: A unified method for evaluating real time computer controllers and its applications. IEEE Trans. Automat. Contr. (1985)

[3] Gu, D.: Resource Management for Dynamic, Distributed Real-Time Systems. Russ College of Engineering and Technology (2005)

[4] Lo, M., Dandamudi, S.P.: Performance of Hierarchical Load Sharing in Heterogeneous Distributed Systems. Appears in Proc. Int. Conf. Parallel and Distributed Computing Systems, Dijon, France (1996)

[5] Imtiaz, S.: Architectural Task Allocation in Distributed Environment: A Traceability, pp. 1515–1518. IEEE (2012)

[6] Govil, K., Kumar, A.: A Modified and Efficient Algorithm for Static Task Assignment in Distributed Processing Environment. International Journal of Computer Applications 23(8), 1–5 (2011)

[7] Barroso, A., Torreão, J., Leite, J., Loques, O.: A Heuristic Approach to Task Allocation in Real-Time Distributed Systems. Brazilian Research Funding Agencies CNPq, Finep and Faperj, Brazil

[8] Vidyarth, D.P., Tripathi, A.K., Sarker, B.K., Dhawan, A., Yang, L.T.: Cluster-Based Multiple Task Allocation in Distributed Computing System. In: International Parallel and Distributed Processing Symposium (2004)

[9] Kopidakis, Y., Lamari, M., Zissimopoulos, V.: Task assignment problem:two new heuristic algorithms. J. Parallel Distrib. Comput. 42(1), 21–29 (1997)

[10] Kumar, H.: A Task Allocation Model for Distributed Data Network. Journal of Mathematical Sciences 1(4), 379–392 (2006)

[11] Lo, V.: Heuristic algorithms for task assighment in distributed system. IEEE Trans. Comput. 37(11), 1384–1397 (1988)

[12] Saxena, P., Govil, K.: An Optimized Algorithm for Enhancement of Performance of Distributed Computing System. International Journal of Computer Applications 64(2), 37–42 (2013)

[13] Ucara, B., Aykanata, C., Kayaa, K., Ikincib, M.: Task Assignment in heterogeneous computing system. J. Parallel Distrib. Comput. 66, 32–46 (2006)

[14] Yadav, P., Singh, M., Sharma, K.: An Optimal Task Allocation Model for System Cost Analysis in Hetrogeneous Distributed Computing Systems: A Heuristic Approach. International Journal of Computer Applications 28(4), 30–37 (2011)

[15] Kaushal, U., Kumar, A.: Performance Intensification of DRTS under Static Load Sharing Scheme. International Journal of Computer Applications 71(16), 55–59 (2013)

[16] Kaushal, U., Kumar, A.: Performance Intensification of DRTS under Static Load Sharing Scheme. International Journal of Computer Applications 71(16), 55–59 (2013)

[17] Aravind, H., Rajgopal, C., Soman, K.P.: A simple Approach to Clustering in Excel. International Journal of Computer Application 11(7) (December 2010)

[18] Masera, M., Fovino, I.N.: Methodology for Experimental ICT Industrial and Critical Infrastructure Security Tests. In: International Conference on Availability, Reliability and Security (2009)

Indian Stock Market Predictor System

C.H. Vanipriya[1] and K. Thammi Reddy[2]

[1] Department of CSE, Sir M. Visvesvarya Institute of Technology, Bangalore, Karnataka
[2] Department of CSE, GIT, GITAMUniversity, Vishakhapatnam, Andhra Pradesh
vani_hm72@yahoo.co.in, thammireddy@yahoo.com

Abstract. Stock market prediction models are one the most challenging fields in computer science. The existing models are predicting stock market prices either by using statistical data or by analyzing the sentiments on the internet. Our proposed model combines both of these methods to develop a hybrid machine learning Stock Market Predictor based on Neural Networks, with intent of improving the accuracy.

Keywords: Machine Learning, Neural networks, Sentiment analysis.

1 Introduction

People generally want to invest their money in stock markets and expect high returns in short period of time. All of these investors have one common goal, which is to maximize their profits .They need to know the right time to buy or sell their investment. Only with a deep understanding of the working principles of the stock markets can one make the right decisions.

The investors try to predict whether the stock price will go high or low by various methods, so that they can sell or buy the stock.

For time series prediction, traditional statistical models are widely used in economics. These statistical models are capable of modeling linear relationships between factors that influence the market and the value of the market. In economics, there are two basic types of time series forecasting simple regression and multivariate regression. Historical pricing models involve analysis of historical prices of a particular stock to identify patterns in the past and are extended to predict the future prices. Market sentiment is monitored with a variety of technical and statistical methods such as the number of advancing versus declining stocks and new highs versus new lows comparison. A major share of overall movement of an individual stock has been attributed to the market sentiment. Various prediction techniques used for stock prediction do far are traditional time series prediction, Neural networks, State Vector Machines.

Earlier people use to take advices from experts and, they use to go through newspapers to monitor the stocks they invested in. But recently they use internet for these activities. As of June 2012 India has the world's third-largest Internet user-base

S.C. Satapathy et al. (eds.), *ICT and Critical Infrastructure: Proceedings of the 48th Annual Convention of CSI - Volume II*, Advances in Intelligent Systems and Computing 249,
DOI: 10.1007/978-3-319-03095-1_3, © Springer International Publishing Switzerland 2014

with over 137 million. The major Indian stock markets introduced Internet trading (online-trading) in February 2002.

In the last decade, investors are also known to measure market sentiment through the use of news analytics, which include sentiment analysis on textual stories about companies and sectors. For instance, in October 2008, there was an online attack on the ICICI bank, which made the stock value of that to come down from 634.45 to 493.30 within 7 days of span. It clearly shows that the rumors which were spreading through the online media make an impact on the stock price. This clearly shows that the news about a particular financial firm on various on line media also plays an important role in stock price prediction.

With the exponential growth of online trading in India, large amount of information is available on the net about stock related data. There are two kindsof data available, numerical data in the form of historical statistics and textual data in the form of news feeds provided by online media. Most of the earlier work on stock prediction was based on the numeric data like historical stock prices .They used to analyze the technical predictors to expect the rise or fall of stock. Now the online media is also playing an important role in the stock market.

The investors are looking in to various news items and expert advices etc. to predict the price of the stock they invest in. There is an emotion attached to this which is called a sentiment .An expert may say that there are some chances of a particular stock price goes up. Then the sentiment is positive. Sentiment can be defined as a thought, view or attitude. Evaluation of large volume of text is tedious and time consuming. Sentiment analysis aims to automatically identify the feeling, emotion or intent behind the given text using various text mining techniques. These sentiment analysis tools have the ability to evaluate large quantities of text without manual intervention. These tools were used for analysis of sentiment about a product or company; most recently tools to measure sentiments have been successfully applied to stock market domain.

In the literature, it has been shown that Neural Networks offer the ability to predict market directions more accurately than other existing techniques. The ability of NNs to discover non-linear relationships between the training input/output pairs makes them ideal for modeling nonlinear dynamic systems such as stock markets.

Support Vector Machine (SVM) based on the statistical learning theory, was developed by Vapnik and his colleagues in the late 1970s. It has become a hot topic of intensive study due to its successful application in classification and regression tasks, especially in time series prediction and financial related applications.

This research is aimed at improving the efficiency of stock market prediction models by combining historical pricing models with sentimental analysis by developing a hybrid neural network to which historical prices and sentimental values are fed as inputs.

The paper is organized as follows. Section 2 discusses related work of our study. Section 3 presents proposed system architecture. Section 4 describes about data collection. Section 5 describes experiments used. Section 5 includes results, section 6 concludes the paper.

2 Literature Review

Several authors have attempted to analyze the stock market. They used quantitative and qualitative information on the net for predicting the movement of the stock.

In [1], the authors proposed a system for quantifying text sentiment based on Neural Networks predictor. By using the methodology from empirical finance, they proved statistically significant relation between text sentiment of published news and future daily returns.

In [2], the author work used only volume of posted internet stock news to train neural network and predict changes in stock prices.

In [3], the authors employed natural language processing techniques and hand crafted dictionary to predict stock returns. They used feed forward neural network with five neurons in the input layer, 27 in the hidden layer, and one output neuron. Since only 500 news items was used for the analysis, no statistical significance of the results could be found.

In [4], the authors proposed a system learns the correlation between the sentiments and the stock values. The learned model can then be used to make future predictions about stock values. They showed that their method is able to predict the sentiment with high precision and also showed that the stock performance and its recent web sentiments are also closely correlated.

In [5],the author developed a system called E-Analyst which collects two types of data, the financial time series and time stamp news stories .It generates trend from time series and align them with relevant news stories and build language models for trend type. In their work they treated the news articles as bag of words.

3 Proposed System Architecture

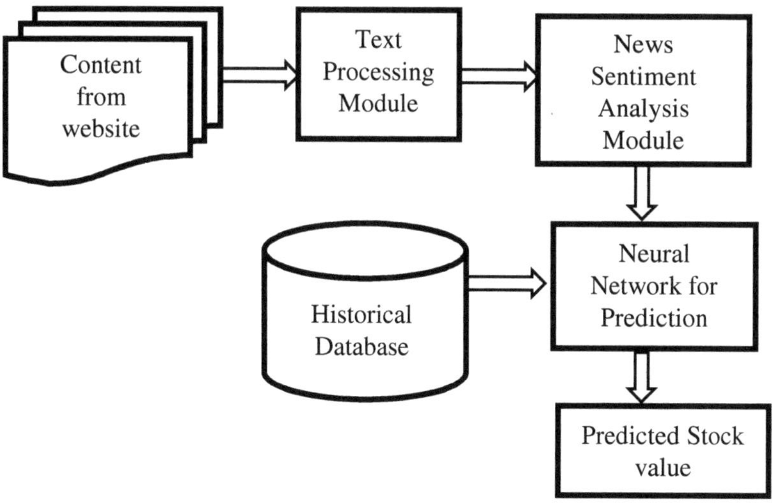

Fig. 1. System Model

Steps involved in financial sentiment extraction:

3.1 Extraction of Data

There are two kinds of data to be extracted historical data about a firm and news articles regarding that company. Gathering data from internet is solely based on the (SOR) Subject of Reference (e.g. ICICI bank).Some web mining techniques (ex. crawler) are used to gather all web pages. Extracting historical stock values is easy, but extracting news articles is tricky, since different websites have different structures. Various sources of financial related data are:

1. 20 to 30 mainstream digital newspapers or online news channels.
2. Authorized sources: financial newsletter
3. Expert commentary: moneycontrol, traderG
4. Social media: Face book, Twitter
5. Alerts &feeds: Bloomberg, Google alerts.

3.2 Text Preprocessing

It is mostly heuristic based and case specific. By this we mean is to identify the unwanted portions in the extracted contents with respect to different kinds of web documents (e.g. News article, Blogs, Review, Micro Blogs etc.) Simple cleanup codes are written to remove such unwanted portions with high accuracy.

3.3 Extract the Sentiment

There are mainly two methods for extraction of sentiments: Lexicon methods, machine learning methods.

Lexicon Methods The basic paradigm of the lexical approach is :

1. Pre-process each post or news item.
2. Initialize the total polarity score: s= 0.
3. Tokenize each post. For each token, check if it is present in a dictionary.
   (a) If token is present in dictionary,
      i. If token is positive, then s =s + w.
      ii. If token is negative, then s= s- w.
4. Look at total polarity score s,
   (a) If s > threshold, then classify the post as positive.
   (b) If s < threshold, then classify the post as negative.

Machine Learning. It is a subfield of Artificial Intelligence dealing with algorithms that allow computers to learn. This means that an algorithm is given a set of data and subsequently infers information about the properties of the data; that information allows it to make predictions about other data that it might come across in the future. Machine learning usually distinguishes between three learning methods: supervised, weakly supervised and unsupervised learning. Supervised machine learning techniques implicate the use of a labelled training corpus to learn a certain classification function and involve learning a function from examples of its inputs and outputs. The output of this function is either a continuous value or can predict a category or label of the input object.Most of the researchers have used five supervised learning classifiers: Naïve Bayes, Maximum Entropy, Decision Trees, TiMBL and Support Vector Machine.

3.4 Predict Future Stock Price

Neural Networks**.** There are many tools available for classification. Neural networks are one amongst those. An Artificial Neural Network is an information processing paradigm. It is inspired by the way biological nervous systems process information. Novel structure of the information processing system is the key element. It is composed of a large number of highly interconnected neurons working together to solve specific problems. These systems learn by example.

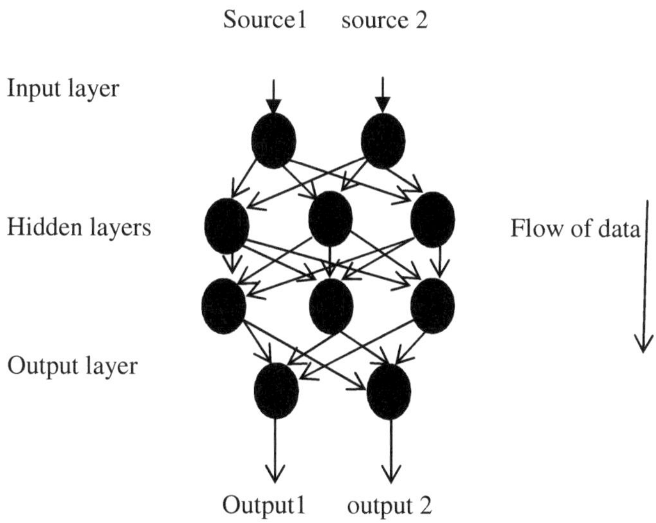

Fig. 2. A typical feed forward neural network

An ANN is configured for a specific application, such as pattern recognition or data classification, through a learning process. Learning in biological systems involves adjustments to the synaptic connections that exist between the neurons. This is true of ANNs as well. As computers become more powerful, Neural Networks are gradually taking over from simpler Machine Learning methods

Individual nodes in a neural network emulate biological neurons by taking input data and performing simple operations on the data, selectively passing the results on to other neurons. The output of each node is called its "activation". Weight values are associated with each vector and node in the network, and these values constrain how input data are related to output data. Weight values associated with individual nodes are also known as biases. Weight values are determined by the iterative flow of training data through the network. Weight values are established during a training phase in which the network learns how to identify particular classes by their typical input data characteristics. Once trained, the neural network can be applied toward the classification of new data. Classifications are performed by trained networks through 1) the activation of network input nodes by relevant data sources [these data sources must directly match those used in the training of the network], 2) the forward flow of this data through the network, and 3) the ultimate activation of the output nodes.

There are three types of artificial neural networks: 1.Single layer feed forward neural network 2.Multi neural feed forward neural network 3.Recurrent network. A neural network in which the input layer of source nodes projects into an output layer of neurons but not vice-versa is known as single feed-forward or acyclic network. This type of network consists of one or more hidden layers, whose computation nodes are called hidden neurons or hidden units. The function of hidden neurons is to interact between the external input and network output in some useful manner and to extract higher order statistics. The source nodes in input layer of network supply the input signal to neurons in the second layer (1st hidden layer). The output signals of 2nd layer are used as inputs to the third layer and so on. The set of output signals of the neurons in the output layer of network constitutes the overall response of network to the activation pattern supplied by source nodes in the input first layer. A feed forward neural network having one or more hidden layers with at least one feedback loop is known as recurrent network.

Once a network has been structured for a particular application, it is ready for training. At the beginning, the initial weights are chosen randomly and then the training or learning begins. There are two approaches to training; supervised and unsupervised. In supervised training, both the inputs and the outputs are provided. The network then processes the inputs and compares its resulting outputs against the desired outputs. Errors are then propagated back through the system, causing the system to adjust the weights, which control the network. This process occurs over and over as the weights are continually tweaked. The set of data, which enables the training, is called the "training set." During the training of a network, the same set of data is processed many times, as the connection weights are ever refined. The other type is the unsupervised training (learning). In this type, the network is provided with inputs but not with desired outputs. The system itself must then decide what features it will use to group the input data.

Multi-layer Networks and Back Propagation. The back propagation algorithm is the most commonly used training method for feed forward networks. Consider a multi-layer perception with ‘k’ hidden layers. Together with the layer of input units and the layer of output units this gives k+2 layers of unit's altogether, which are numbered by 0... k+1. Let the number of input units be K, output units be L and of units in hidden layer m be N^m. The weight of jth unit in layer m and the ith unit in layer m+1 is denoted by w_{ij}^m. The activation of the ith unit in layer m is x_i^m (for m = 0 this is an input value, for m = k+1 an output value). The training data for a feed forward network training task consist of T input-output (vector-valued) data pairs

$$u(n) = (x_1^0(n),\ldots, x_K^0(n))^t, d(n) = (d_1^{k+1}(n),\ldots,d_L^{k+1}(n))t. \tag{1}$$

Where 'n' denotes training instance. The activation of non-input units is computed according to

$$x_i^{m+1}(n) = f(\sum_{j=1,\ldots,N^m} w_{ij}^m x_j(n)) \tag{2}$$

Presented with training input u(t), the previous update equation is used to compute activations of units in subsequent hidden layers, until a network response $y(n) = (x_1^{k+1}(n),\ldots, X_L^{k+1}(n))^t$ is obtained in the output layer. The objective of training is to find a set of network weights such that the summed squared error

$$E = \sum_{n=1,\ldots,r} |d(n) - y(n)|^2 = \sum_{n=1,\ldots,r} E(n) \tag{3}$$

is minimized. This is done by incrementally changing the weights along the direction of the error gradient with respect to weights

$$\frac{\partial E}{\partial w_{ij}^m} = \sum_{i=1,\ldots,r} \frac{\partial E(n)}{\partial w_{ij}^m} \tag{4}$$

using a (small) learning rate γ:

$$\text{New } w_{ij}^m = w_{ij}^m - \frac{\partial E}{\partial w_{ij}^m} \tag{5}$$

This is the formula used in batch learning mode, where new weights are computed after presenting all training samples. One such pass through all samples is called an epoch. Before the first epoch, weights are initialized, typically to small random numbers. A variant is incremental learning, where weights are changed after presentation of individual training samples.

4 Data Collection

We merge two sources of data: a corpus of news articles, a dataset of historical data. The first source is extraction of news articles from moneycontrol [7] website; it contains important news for individual stocks. Moneycontrol is India's leading financial information source. It manages our finance with their online Investment Portfolio, Live Stock Price, Stock Trading news etc. Many of the investors in India go through this website for manipulating their stocks. The corpus is taken for INFOSYS for one year from Jan 2012 to Jan 2013. A web scraper in R language is written to get all the news articles and calculate the sentiments.Historical DataWe have extracted historical values from http://ichart.finance.yahoo.com for the year of 2012 for the stock of Infosys in NIFTY.

5 Experiments

Lexicon method has been used for extracting the sentiment from the news articles [13].For extraction of news articles from money control, we coded a scraper in R to get the news articles for the year 2012 of INFY. The news articles have been cleaned andscaled from -1 score to +1 score and -1 being the most negative article.

The score is calculated according to the following formula:

$$SCORE = \frac{\sum(\text{positivematches}) + \sum(\text{negativematches})}{\sum(\text{positivematches}) + \sum(\text{negativematches})} \tag{6}$$

The proposed system uses multi-Layer perceptron. It is a feed forward neural network with1 input layer, 2 hidden layers and 1 output layer. Feed forward means that data flows in one direction from input to output layer (forward). This network is trained with the back propagation learning algorithm. We have chosen MLP because it is widely used for pattern classification, recognition, prediction and approximation. Back propagation has been used as the learning rule.

The main steps using the learning algorithm as follows:

Step 1: Take some sample input data (training set) and, compute the corresponding output.
Step 2: Compute the error between the output(s) and the actual target(s)
Step 3: The connection weights and membership functions are adjusted
Step 4: At a fixed number of epochs, delete useless rule and membership function nodes, and add in new ones;
Step 5: IF Error > Tolerance THEN go to Step 1 ELSE stop.

Stock values of stock for 4 consecutive days were taken as inputs .The fifth input will be the previous day's news article score. These values are fed into neural networks .We have used a java framework, Neuroph [8] is used to train and predict the values. The model is tested with different number of nodes in the hidden layers .The model predicts the rise or fall in the stock price based on which the investor will take a decision to buy or sell on a day to day basis that is for intraday trading. The model is predicting for the day i using the previous four days values for the company and the fifth input is the Sentiment value of i-1 day. If there is no news article for the previous day, the sentiment value is 0.

The proposed system has 5 input neurons (three for previous 4 consecutive days' stock prices and 1 for sentiment), 1 neuron in output layer. Different number of neurons is taken in the hidden layers and the system is tested. The number of neurons (nodes) in both the hidden layers and its training dataset and test data set accuracy is shown in table 1. All the neurons had Tanh transfer function. The system uses supervised training set with maximum 1000iterations, learning rate 0.7 and Max error as 0.001.Training stops after some iteration with total net error under 0.001.After training the system is tested with test data. The system's accuracy is calculated as follows:

Training set accuracy=no. of news articles correctly predicted/Total no. of articles
Test set accuracy=no. of news articles correctly predicted/Total no. of articles.

6 Results

Based on the analysis of various neural network structures developed by changing the number of hidden layers and input layer nodes, the following results as shown in table I were obtained.

Table 1. The accuracy of the predicted values

No of nodes in the 1st Hidden layer	No of nodes in the 2nd Hidden layer	Training set accuracy	Test set accuracy
20	10	80.12%	79%
5	4	80.12%	71%
10	10	78%	74.756%

The correlation between the actual stock values and predicted stock values are shown in the graph below in figure.

Fig. 3. Plotting of Stock values

7 Conclusion and Future Work

The main contribution of this paper is to suggest a new method for automatically predicting the stock price. We have shown that stock prices predicted from historical prices and sentiments are significantly correlated with actual stock prices of a particular company. Future work would be extending these results by using Deep Multilayer Neural Networks with more than two hidden layers for determining text sentiment. One can go even further and use information of more companies. And also use complex algorithms like SVM, Naïve Bayes theorem to classify the sentiments.

References

1. Bozi, C., Seese, D.: Neural Networks for Sentiment Detection in Financial. JEL Codes: C45, D83, an G17, Institute of Applied Informatics and Formal Description Methods, Karlsruhe Institute of Technology (KIT)
2. Liang, X.: Impacts of Internet Stock News on Stock Markets Based on Neural Networks. In: Wang, J., Liao, X.-F., Yi, Z. (eds.) ISNN 2005. LNCS, vol. 3497, pp. 897–903. Springer, Heidelberg (2005)
3. Liang, X., Chen, R.: Mining Stock News in Cyber world Based on Natural Language Processing and Neural Networks. In: ICNN &B 2005, pp. 13–15 (2005)
4. Sehgal, V., Son, C.: SOPS: Stock Prediction using Web Sentiment. IEEE, doi:10.1109, 2007
5. Ahmad, K., Almas, Y.: Visualizing Sentiments in Financial Texts
6. Gao, Y., Zhou, L., Zhang, Y., Xing, C.: Sentiment Classification for Stock News. IEEE, 978-1-4244-9142-1/10/2010
7. Glissman, S., Terrizzano, I., Lelescu, A., Sanz, J.: Systematic Web Data Mining with Business Architecture to Enhance Business Assessment Services. In: Annual SRII Global Conference, IEEE (2011), doi:10.1109, 978-0-7695-4371-0/11
8. Chen, H., De, P., Hu, Y(J.), Hwang, B.-H.: Sentiment Revealed in Social Media and Effect on the Stock Market. In: Statistical Signal Processing Workshop. IEEE
9. Li, Y., Wang, J.: Factors on IPO under-pricing based on Behavioral Finance Theory: Evidence from China. IEEE, 978-1-4577-0536-6/11/2011
10. Zhang, K., Li, L., Li, P., Wenda: Stock Trend Forecasting Method Based on Sentiment Analysis and System Similarity Model
11. Nair, B.B., Mohandas, V.P., Sakthivel, N.R.: A Decision Tree- Rough Set Hybrid System for Stock Market Trend Prediction. International Journal of Computer Applications (0975 – 8887) 6(9) (September 2010)
12. Yoo., P.D., Kim, M.H., Jan, T.: Machine Learning Techniques and Use of Event Information for Stock Market Prediction: A Survey and Evaluation. IEEE (2005)
13. `http://www.cs.uic.edu/~liub/FBS/sentiment-analysis.html`

Text Clustering Using Reference Centered Similarity Measure

Ch.S. Narayana[1], P. Ramesh Babu[1], M. Nagabushana Rao[2], and Ch. Pramod Chaithanya[1]

[1] CSE Department, Malla Reddy Engineering College (Autonomous), Hyderabad
{satiesh.ch,prameshbabu526,pramod.mrec}@gmail.com
[2] CSE Department, Swarnandra Engineering College, Narsapur
mnraosir@gmail.com

Abstract. The majority clustering skill must presume some cluster relationship relating to the data set. Similarity among the items is usually defined sometimes clearly or even absolutely. With this paper, we introduced some sort of novel numerous reference centered similarity measure and two related clustering approaches. The significant difference between a traditional dissimilarity/ similarity measure and our's is to compared the performance of the former method using single viewpoint, which may be the source, the number of mention sources. Using several reference points, more useful assessment of similarity could possibly be achieved. Two qualification functions with regard to document clustering are proposed determined by this novel measure. We examine them with well-known clustering algorithm cosine similarity and exposed the development. Performance Analysis is conducted and compared.

Keywords: Document Clustering, Similarity Measure, Cosine Similarity, Multi View Point Similarity Measure.

1 Introduction

Clustering is among the most useful and essential topics within data mining. The aim of clustering is always to find implicit structures within data, and organize all of them into important subgroups intended for further study and analysis. according to some recent study, more compared to half a hundred years after it had been introduced; the uncomplicated algorithm k-means still remains among the top 10 information mining algorithms these days. It could be the most commonly used partitioned clustering algorithm used. Another recent scientific conversation states of which k-means could be the favourite algorithm that practitioners from the related fields go for. Needless to cover, k-means has many basic cons, such because sensitiveness to be able to initialization and to cluster sizing, and it's performance is usually worse compared to other state-of-theart algorithms in many domains. Despite that, it's simplicity

S.C. Satapathy et al. (eds.), *ICT and Critical Infrastructure: Proceedings of the 48th Annual Convention of CSI - Volume II*, Advances in Intelligent Systems and Computing 249,
DOI: 10.1007/978-3-319-03095-1_4, © Springer International Publishing Switzerland 2014

understand capability and scalability will be the reasons to its tremendous recognition. An algorithm with enough performance and usability practically in most of application scenarios may very well be preferable to a single with much better performance sometimes but minimal usage on account of high complexness. While supplying reasonable outcomes, k-means is easily to combine with other techniques in more substantial systems. A common approach to the clustering problem is usually to treat it just as one optimization procedure. A best partition is available by optimizing a unique function connected with similarity (or distance) amongst data. That's why effectiveness involving clustering algorithms under this approach depends around the appropriateness of the similarity measure to the data at hand. For illustration, the initial k-means features sum-of-squared-error aim function in which uses Euclidean distance. In an exceptionally sparse as well as high dimensional area like text documents, spherical k implies, which utilizes cosine similarity rather than Euclidean distance since the measure, is deemed for being more suited.

2 Previous Work

Document clustering has long been studied as a post retrieval document visualization technique to provide an intuitive navigation and browsing mechanism by organizing documents into groups, where each group represents a different topic. In general, the clustering techniques are based on four concepts: data representation model, similarity measure, clustering model, and clustering algorithm. Most of the current documents clustering methods are based on the Vector Space Document (VSD) model. The common framework of this data model starts with a representation of any document as a feature vector of the words that appear in the documents of a data set. A distinct word appearing in the documents is usually considered to be an atomic feature term in the VSD model, because words are the basic units in most natural languages (including English) to represent semantic concepts. In particular, the term weights (usually tf-idf, term-frequencies and inverse document-frequencies) of the words are also contained in each feature vector. The similarity between two documents is computed with one of the several similarity measures based on the two corresponding feature vectors, e.g., cosine measure, Jaccard measure, and euclidean distance. To achieve a more accurate document clustering, a more informative feature term phrase has been considered in recent research work and literature. A phrase of a document is an ordered sequence of one or more words.. Reference proposed a phrase based document index model namely Document Index Graph (DIG), which allows for the incremental construction of a phrase-based index for a document set. The quality of clustering achieved based on this model significantly surpassed the traditional VSD modelbased approaches in the experiments of clustering Web documents. [2]

Particularly, the Suffix Tree Document (STD) model and Suffix Tree Clustering (STC) algorithm were proposed by Zamir et al. and Zamir and Etzioni. The STC algorithm was used in their Meta searching engine to real time cluster the document snippets returned from other search engines. The STC algorithm got poor results in clustering the documents in their experimental data sets of RCV1 corpus. [2]

Text document clustering has been traditionally investigated as a means of improving the performance of search engines by preclustering the entire corpus, and a postretrieval document browsing technique as well.. K-Nearest Neighbor (K-NN) algorithm is well known for classification. It has also been used for document clustering. In the traditional document models such as the VSD model, words or characters are considered to be the basic terms in statistical feature analysis and extraction. To achieve a more accurate document clustering, developing more informative features has become more and more important in information retrieval literature.

The *k*-Means Algorithm

The k-means algorithm is a simple iterative method to partition a given dataset into a user specified number of clusters, *k*. This algorithm has been discovered by several researchers across different disciplines, most notably Lloyd (1957, 1982), Forgey (1965), Friedman and Rubin (1967), and McQueen (1967). A detailed history of k means alongwith descriptions of several variations are given. Gray and Neuhoff provide a nice historical background for k-means placed in the larger context of hill climbing algorithms. The algorithm operates on a set of *d*-dimensional vectors, $D = \{\mathbf{xi} \mid i = 1, \ldots, N\}$, where $\mathbf{xi} \in \_d$ denotes the *i*th data point. The algorithm is initialized by picking *k* points in $\_d$ as the initial *k* cluster representatives or centroids||. Techniques for selecting these initial seeds include sampling at random from the dataset, setting them as the solution of clustering a small subset of the data or perturbing the global mean of the data *k* times. Then the algorithm iterates between two steps till convergence: [1]

Step 1: Data Assignment. Each data point is assigned to its *closest* centroid, with ties broken arbitrarily. This results in a partitioning of the data.

Step 2: Relocation of "means". Each cluster representative is relocated to the center (mean) of all data points assigned to it. If the data points come with a probability measure (weights), then the relocation is to the expectations (weighted mean) of the data partitions. The algorithm converges when the assignments (and hence the **cj** values) no longer change. The algorithm execution is visually depicted. Note that each iteration needs $N \times k$ comparisons, which determines the time complexity of one iteration. The number of iterations required for convergence varies and may depend on *N*, but as a first cut, this algorithm can be considered linear in the dataset size. One issue to resolve is how to quantify —closest|| in the assignment step.

The greedy-descent nature of k-means on a non-convex cost also implies that the convergence is only to a local optimum, and indeed the algorithm is typically quite sensitive to the initial centroid locations. Figure below illustrates how a poorer result is obtained for the same dataset as in Fig. below for a different choice of the three initial centroids. [1]

3 Proposed System

3.1 Similarity Measure

The particular cosine similarity might be expressed inside form applying Sim(di, dj) in which vector which represents the origin point are employed. According to the present formula, your measure requires reference stage. The similarity between a couple of documents di in addition to dj is decided w. r. t. the angle between two items when looking from the origin. To build a new thought of similarity, it's possible to use a lot more than just 1 point involving reference. We may have a additional accurate assessment of exactly how close or perhaps distant a couple of points are, if we look at them from numerous reference items are suggested respectively because of the difference vectors (di – dh) in addition to (dj – dh). The similarity of a couple of documents di in addition to dj simply because they are inside same cluster pertains to the common of similarities measured relatively from the references of other papers outside which cluster.

3.2 Multi-Reference Point Similarity

We call this module this Multi-Reference stage based Likeness, or MRCS. Out of this point onwards, we all will denote the suggested similarity calculate between 2 document vectors di in addition to dj simply by MRCS (di, dj). The MRCS form is dependent upon particular formulation on the individual similarities from the sum. If the relative likeness is identified by dot-product on the difference vectors, we now have: The likeness between 2 points di in addition to dj interior cluster Sr, reference from a point dh external this chaos, is equal to the product on the cosine on the angle between di in addition to dj shopping from dh and also the Euclidean distances from dh to these two points. This definition will be based upon the assumption that dh just isn't in the same cluster together with di in addition to dj. Small the distances di–dh in addition to dj –dh tend to be, the higher the chance that dh is in fact in the same cluster together with di in addition to dj, and also the similarity according to dh should also be little to echo this potential. Therefore, via these distances, we also supplies a measure of inter chaos dissimilarity, since points di in addition to dj fit in with cluster Sr, whereas dh belongs to an alternative cluster. The overall similarity between di in addition to dj depends on taking average over all the research points not belonging to cluster Sr. It's possible to argue that while a large number of reference points are useful, there may be a variety of them giving mistaken information just as it can happen with the origin point. Nonetheless, given a sizable enough amount of reference points and their particular variety, it is usually reasonable to help assume that most them will be useful. For this reason, the result of mistaken reference points is constrained and reduced through the averaging move. It sometimes appears that this kind of offers more informative assessment of similarity than the single foundation point.

Procedure. BUILD MRCSMatrix (A)

for $r \leftarrow 1 : c$ **do**
$D_{S \setminus S_r} \leftarrow \sum_{di \notin S_r} di$
$n_{S \setminus S_r} \leftarrow |S \setminus S_r|$
end for
for $i \leftarrow 1 : n$ **do**
$r \leftarrow$ class of di
for $j \leftarrow 1 : n$ **do**
if $dj \in Sr$ **then**
$a_{ii} \leftarrow d^t_i d_i - d^t_i - d^t_j + 1$
end if
end for
end for
return $A = \{ a_{ij} \}_{n \times n}$
end procedure

3.3 Validity Computation

The particular validity computation was created as pursuing. For every sort of likeness measure, a likeness matrix A is established. For CS, this is simple, as aij = dti dj. The process for developing MRCS matrix is Firstly, the particular outer blend w. r. t. each class is established. Then, for every single row ai of an, i = 1,..., n, if the set of documents di and dj, t = 1,..., n will be in the same class, aij is calculated. Normally, dj is assumed to stay di's type, and aij is calculated. After matrix A is shaped, the procedure is employed to receive its validity report. For each and every document di matching to line ai of an, we pick out qr files closest for you to di. The worth of qr is chosen reasonably as percentage of the length of the type r that contains di, exactly where percentage ∈ (0, 1]. And then, validity t. r. testosterone levels. di is calculated with the fraction of those qr documents getting the same type label with di, The closing validity depends on averaging over all the rows of an. It is clear that validity report is bounded within just 0 and 1. The more expensive validity report a likeness measure offers, the a lot better it must be for the particular clustering activity.

3.4 Clustering Criteria's

Having defined our similarity evaluate, we now formulate our clustering qualifying measure functions. The first function, called IR, would be the cluster size-weighted amount of average pair wise parallels of documents from the same cluster. We would want to transform this objective functionality into several suitable form in ways that it might facilitate this optimization procedure for being performed in a simple, fast

When comparing F with all the min-max lower, both functions secure the two terms an intra-cluster similarity measure and also inter-cluster similarity measure. On the other hand, while the intention of min-max cut is to minimize this inverse relation between the two of these terms, our aim the following is to maximize their weighted difference. This difference term is resolute for every single cluster. They're weighted by the inverse from the cluster's dimensions, before summed up overall the groups. One issue is that this formulation is supposed to be very sensitive in order to cluster dimensions. It shows up that IR' s performance dependency on the value connected with α is not very important. The qualifying measure function yields relatively excellent clustering effects for α ∈ (0, 1). Inside formulation connected with IR, a cluster quality is actually measured by the average pair wise similarity between papers within which cluster. On the other hand, such an approach can result in sensitiveness towards size and also tightness from the clusters. Using CS, for example, pair wise similarity of documents in a sparse cluster is generally smaller as compared to those in a dense cluster. To reduce this, a different approach is to consider similarity between every single document vector and its particular cluster's centroid alternatively.

4 Results

The concept of this paper is implemented and different results are shown below, The proposed paper is implemented in Java technology on a Pentium-IV PC with 20 GB hard-disk and 256 MB RAM. The propose paper's concepts shows efficient results and has been efficiently tested on different Reuters Datasets. The Fig 1, Fig 2, Fig 3, Fig 4 and Fig 5 shows the evaluation of the results.

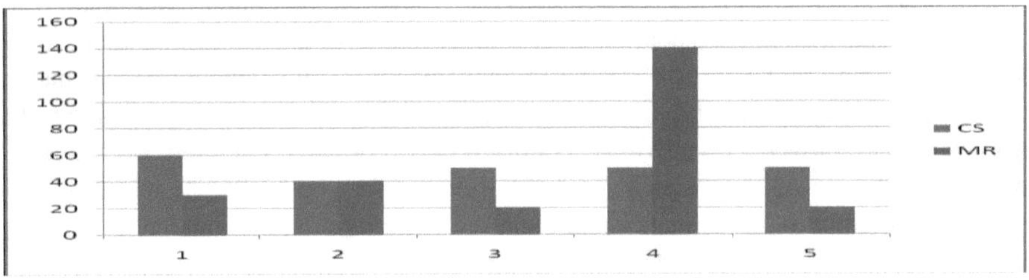

Fig. 1. Comparison of 200 documents in clusters

The above graph represents the number of documents clustered, we have taken 200 documents from reuters dataset and clustered them in 5 clusters. we can observe the documents in clusters changes with respect to cosine similarity and multi reference similarity.

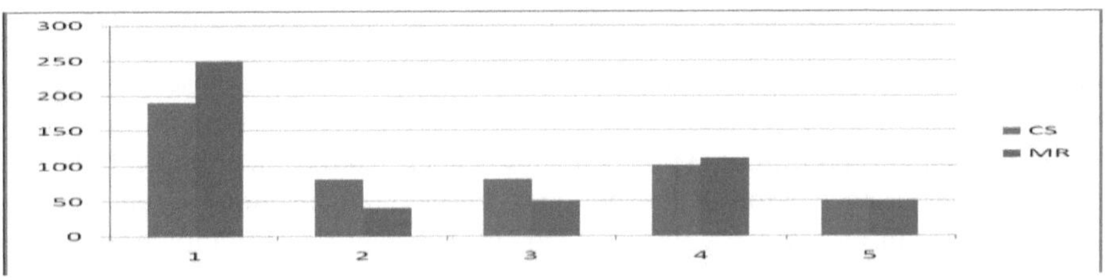

Fig. 2. Comparison of 500 documents in clusters

The above graph represents the number of documents clustered, we have taken 500 documents from reuters dataset and clustered them in 5 cvlusters. we can observe the documents in clusters changes with respect to cosine similarity and multi reference similarity.

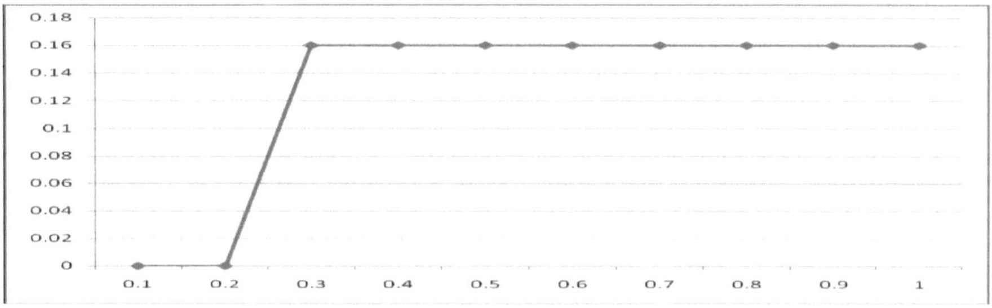

Fig. 3. Validity graph for 25 documents

In the above graph we compare the validity score for various percentages ranging from 0.1 to 1.0 in steps of 0.1. The above graph is plotted for 200 documents. we observe that the validity score is null till 0.2 percentage, but it increases to 0.16 and remains constant.

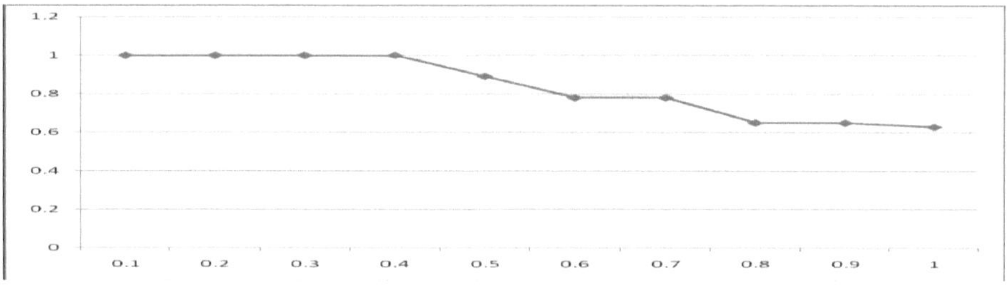

Fig. 4. Validity graph for 50 documents

In the above graph we compare the validity score for various percentages ranging from 0.1 to 1.0 in steps of 0.1. The above graph is plotted for 500 documents. we observe that the validity score is 0.1 till 0.4 percentage, but increases as the percentage increases.

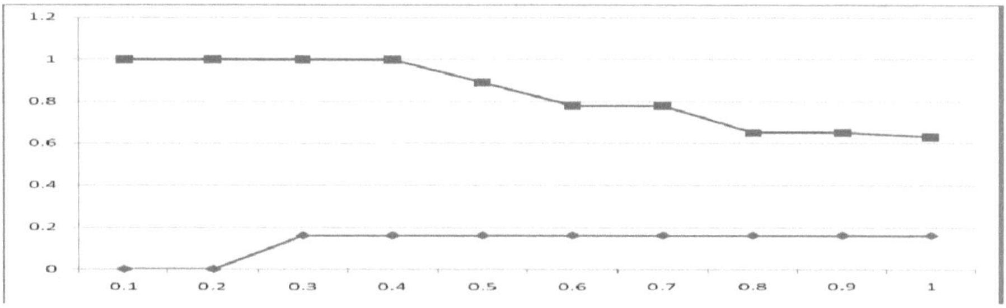

Fig. 5. Comparison of Validity graph

In the above graph we show the comparison of validity scores for 200 and 500 documents respectively for various percentages.

5 Conclusions

Within this paper, we propose a Multi-Reference stage based Similarity measuring procedure, named MRCS. Theoretical research and empirical cases show which MRCS is actually potentially far better for text message documents than the popular cosine likeness. Based about MRCS, a couple criterion characteristics, IR and IV, and their respective clustering algorithms, MRCS-IR and MRCS-IV, happen to be introduced. Compared along with other state-of-the-art clustering techniques that use unique variations of similarity gauge, on a large number of document datasets and under different evaluation metrics, the proposed algorithms show them to could produce significantly enhanced clustering effectiveness.

The key contribution of the paper is the fundamental concept of similarity measure from several reference factors. The comparison graph demonstrates the effectiveness of the MRCS.

Acknowledgement. We would like to take this opportunity to thank Sri Ch.Malla Reddy, Chairman , Malla Reddy Group of Institutions, Dr.S.Sudhakara Reddy, Principal of Malla Reddy Engineering College(Autonomous), Dr.P.V.Ramana Murthy, Head Of Department of CSE, for allowing and encouraging me to complete the work, Dr. M.Nagabushana Rao, Professor, CSE Department, Swarnandra Engineering College, Narsapur for his valuable guidance.

References

[1] Wu, X., Kumar, V., Ross Quinlan, J., Ghosh, J., Yang, Q., Motoda, H., McLachlan, G.J., Ng, A., Liu, B., Yu, P.S., Zhou, Z.-H., Steinbach, M., Hand, D.J., Steinberg, D.: Top 10 algorithms in data mining. Knowl. Inf. Syst. 14(1), 1–37 (2007)

[2] Chim, H., Deng, X.: Efficient phrase-based document similarity for clustering. IEEE Trans. on Knowl. and Data Eng. 20(9), 1217–1229 (2008)

[3] Lee, D., Lee, J.: Dynamic dissimilarity measure for support based clustering. IEEE Trans. on Knowl. and Data Eng. 22(6), 900–905 (2010)

[4] Lakkaraju, P., Gauch, S., Speretta, M.: Document similarity based on concept tree distance. In: Proc. of the 19th ACM conf. on Hypertext and Hypermedia, pp. 127–132 (2008)

[5] Ienco, D., Pensa, R.G., Meo, R.: Context-based distance learning for categorical data clustering. In: Proc. of the 8th Int. Symp. IDA, pp. 83–94 (2009)

[6] Guyon, I., von Luxburg, U., Williamson, R.C.: Clustering: Science or Art? In: NIPS 2009 Workshop on Clustering Theory (2009)

[7] Pękalska, E., Harol, A., Duin, R.P.W., Spillmann, B., Bunke, H.: Non-euclidean or non-metric measures can be informative. In: Yeung, D.-Y., Kwok, J.T., Fred, A., Roli, F., de Ridder, D. (eds.) SSPR & SPR 2006. LNCS, vol. 4109, pp. 871–880. Springer, Heidelberg (2006)

[8] Pelillo, M.: What is a cluster? Perspectives from game theory. In: Proc. of the NIPS Workshop on Clustering Theory (2009)
[9] Dhillon, I., Modha, D.: Concept decompositions for large sparse text data using clustering. Mach. Learn. 42(1-2), 143–175 (2001)
[10] Zhong, S.: Efficient online spherical K-means clustering. In: IEEE IJCNN, pp. 3180–3185 (2005)
[11] Banerjee, A., Merugu, S., Dhillon, I., Ghosh, J.: Clustering with Bregman divergences. J. Mach. Learn. Res. 6, 1705–1749 (2005)
[12] Banerjee, A., Dhillon, I., Ghosh, J., Sra, S.: Clustering on the unit hypersphere using von Mises-Fisher distributions. J. Mach. Learn. Res. 6, 1345–1382 (2005)
[13] Xu, W., Liu, X., Gong, Y.: Document clustering based on nonnegative matrix factorization. In: SIGIR, pp. 267–273 (2003)

A Comparative Analysis of New Approach with an Existing Algorithm to Detect Cycles in a Directed Graph

Shubham Rungta, Samiksha Srivastava, Uday Shankar Yadav, and Rohit Rastogi

CSE Department, ABES Engineering College, Ghaziabad (U.P.), India
{shubhamrungta93,samikshasrivastava607,uday4792}@gmail.com,
rohit.rastogi@abes.ac.in

Abstract. In various applications such as discovering infinite loops in computer programs, periodic scheduling, communication systems etc. there are always requirement for cycle detection. Graph theories and algorithms are very helpful for this type of problems. In this paper, we proposed our new SUS_CycleDetection algorithm for detecting cycle in any directed graph, with the help of linked list. This algorithm has the ability to count total number of cycles in the graph along with displaying the set of vertices responsible for the formation of each cycle. A comparison is also made between the proposed algorithm and an existing algorithm in terms of their modes of execution. Informer, space is allocated during runtime and nodes are stored using linked list which is more efficient in terms of memory utilization while in the latter, space is allocated before execution and nodes are stored using queue.

Keywords: Directed graph, Cycle, Linked list, Graph theory, Data structure.

1 Introduction

The analysis of cycles in network has different application in the design and development in communication systems such as the investigation of topological features and consideration of reliability and fault tolerance. There are various problems related to the analysis of cycles in network among which the most important one is the detection of cycles in graph. A Graph is a data structure comprises of vertices and edges. In Fig.1. 1, 2, 3 to 6 are the vertices and the lines joining any two vertices are the edges. And cycle is a repeating part during graph traversal in a sequence. In computer science, cycle detection is the algorithmic problem of finding a cycle in a sequence of iterated function values [1]. Suppose in a function F(x), if x repeats the same sequence of values once again, then there exist a cycle. In Fig.1. F(x) = [x: x is 1, 2, 3, 4, 5, 6, y… in sequence, where y is 2, 3, 4, 5, and 6 in sequence and is repeated], here cycle exists because x repeats the value 2, 3, 4, 5, 6 repeatedly.

To overcome this problem, we introduced an algorithm SUS_CycleDetection algorithm which will not only detect the cycles but also will provide total no. of cycles and the set of vertices that are responsible for cycle formation. Although there are various cycle detection algorithm, we are comparing the proposed algorithm with

S.C. Satapathy et al. (eds.), *ICT and Critical Infrastructure: Proceedings of the 48th Annual Convention of CSI - Volume II*, Advances in Intelligent Systems and Computing 249,
DOI: 10.1007/978-3-319-03095-1_5, © Springer International Publishing Switzerland 2014

Hongbo Liu'scycle detection algorithm in order to analyse the benefits of proposed over existing. The existing algorithm is more complex than proposed because it is going through higher number of comparisons. The existing algorithm is dependent on queue data structure so memory utilization is not as efficient as proposed algorithm because this algorithm uses Linked list Data structure where nodes are formed during runtime. A linked list is a data structure used to implement several abstract data types such as Stack, Queue, D-Queue, associative arrays etc. In the singly linked list, there are nodes linked with other nodes and each node contains two fields. First field resembles the information or data and other contains the address of next node, known as link field. Firstly, the proposed algorithm will take memory for entering the vertices in linked lists (during a walk) and when the traversal is completed (i.e. cycle detects or null node appears), the particular list will be deleted and memory becomes free. As far as complexity is concerned both are using same amount of time to execute. But proposed algorithm is more simple and effective than existing algorithm.

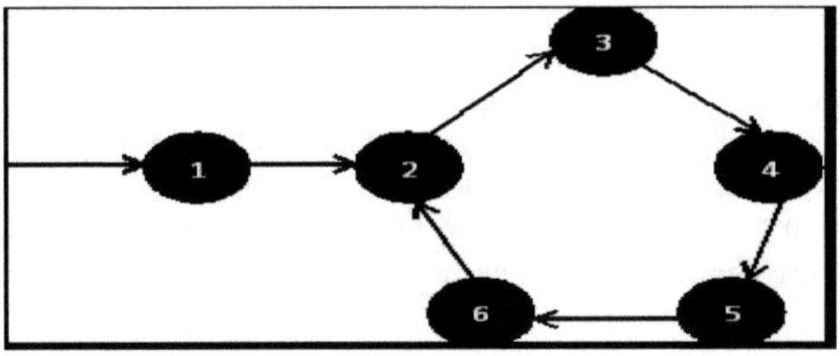

Fig. 1. Example of a cycle with 1, 2, 3, 4, 5, 6 as Vertices of the Graph

2 Existing Algorithm

2.1 Hongbo Liu'sCycle Detection Algorithm

Step 1. Put all the vertices $v_1,v_2....v_n$ in a queue.
Step 2. Initialize the value of register with 0
Step 3. Fetch an open path P from the queue Q, whose length is k.
Initialize the value of register with k.
Step 4. Check if there is any edge connecting the tail (Let v_t be the tail node) of the chosen open path to its head (Let v_h be the head node).
If the edge is not found GOTO Step6.
Step 5.If the above condition is true, a cycle is detected, and then output the cycle.
OUTPUT (P+e) where, e is the edge connecting the tail of the open path to its head.
When the register is 0 in such case, it means the cycle is a self-loop.
Step 6. Get an adjacent edge e of the tail whose end does not occur in the open path and the order of its end is greater than the order of the head.
This edge and the k length open path construct a new k + 1 length open path.
Put (P+e) into the queue Q.

After having generated all the k + 1 length open paths from the k length open path, if this open path is the last k
length open path of the queue, set register to k + 1.

Step 7. IF all adjacent nodes of the vertex v_i is not handled
THEN Repeat Step 6.
ELSE
Check if the queue is empty
IF Queue is empty
THEN END
Else Repeat from Step 1.

2.2 Comparisons between Our New SUS_CycleDetection Algorithm and Existing Algorithm

In this existing algorithm, data structure queue is used to store the nodes of the graph due to which traversal become less efficient as compared to proposed algorithm in which linked list is used to perform the same task, thus making more efficient traversal and better memory utilization. A large amount of comparisons are made in the existing algorithm which makes it more complex.The amount of computation seems complex in existing algorithm, because there are so many combinations of edges, most of which could not construct a cycle. In existing method, duplicate cycles may be generated from different vertices of the same cycle. To avoid duplicate cycles integer designators (1, 2, 3,…..N) and some constraints are added in existing but in proposed there are no chances to form any duplicate cycle as it does not allow to traverse through same path.As far as complexity is concerned both are approximately using same amount of time to execute.

3 Proposed Idea

3.1 Proposed SUS_CycleDetection Algorithm

SUS_CycleDetection
(INFO, LINK, START1, START2, LIST1, LIST2, PTR1, PTR2, PTR3, ITEM, Counter, FREE(x), TEMP)

INFO. It stores the information field of the node in the linked list.
LINK. It is the address field of the node that contains the address of the next node in the linked list.
LIST1. It is the linked list, which will store the base address of all the linked lists formed during runtime.
LIST2. It is a linked list, which stores the first node of digraph.
START1. It is the pointer that points to the first node of linked LIST1.
START2. It is the pointer, which points to the first node of linked LIST2.
Counter. It is a global variable to count the number of cycle.
SAVE, PTR, PTR2, PTR3 and PTR4, TEMP. These are the pointer variables.
ITEM. It contains the info character.
FREE(x). The function will remove the node x.

Step 1. Insert the first node in linked list LIST2.(Let the first node be A.)

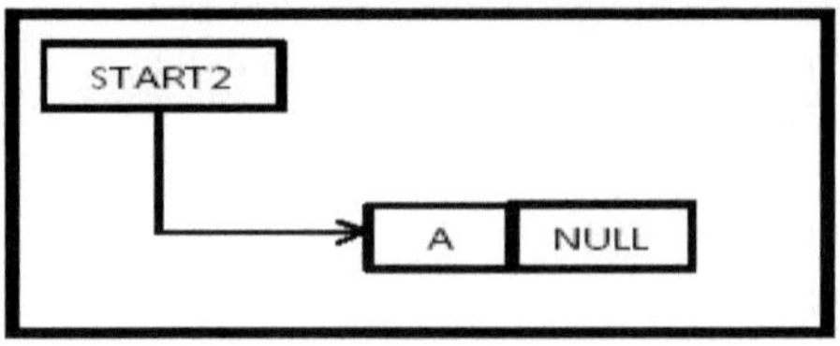

Fig. 2. Insertion of first node in LIST2

Step 2. Put the base address of the LIST2 in LIST1.

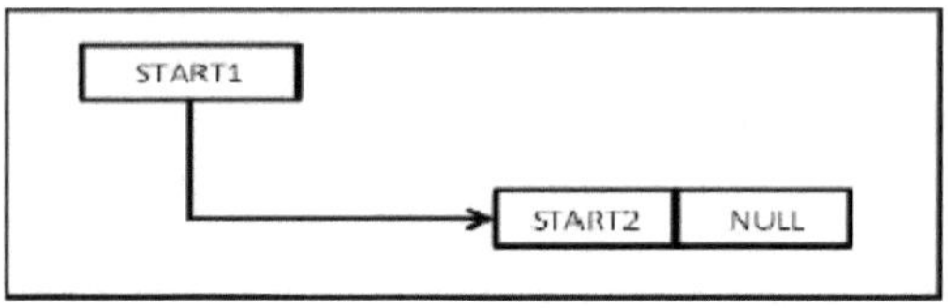

Fig. 3. Insertion of the address of first node of LIST2 in the info field ofLIST1

Step 3. Now, if there are n directed paths from A, then total number of new linked list will be n-1 and they all will be duplicate of LIST2.

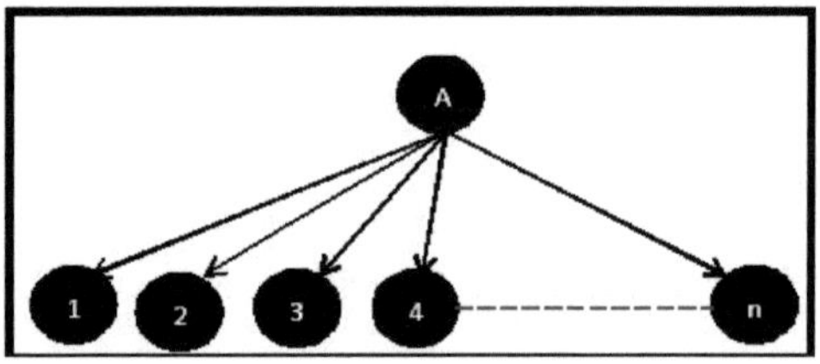

Fig. 4. Example of n directed path from A

Step 4.Now put each next node of the graph directed from the last node into separate linked lists.

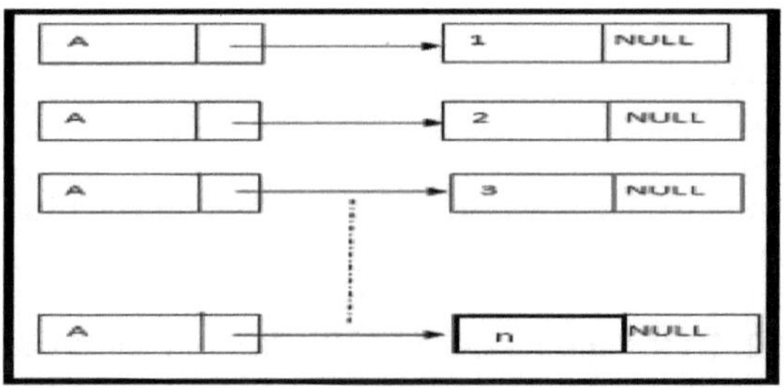

Fig. 5. Insert 1 in LIST2 and others in the copies in LIST2

Step 5. Put the base address of each newly created list into the LIST1 in a consecutive manner.

// Iteration of outer loop
Step 6. Repeat Step (7) to (10) while (START1! = NULL)

/*Compare the element at last node of LIST2 with all its previous nodes starting from starting node.*/
Step 7. [Initialize the value of pointer PTR and SAVE with the value of pointer START2.]
PTR<-START2
SAVE<-START2

Step 8. [Repeat Steps (a) to (b) until (PTR! =NULL)]
a) PTR<-LINK [PTR]
b) TEMP<-INFO [PTR]

Step 9. [Initialize PTR2 with START2.]
PTR2<-START2

/*Compare the last element with all the elements of LIST2.*/
Step 10. [Repeat the following steps (a) to (b) and 11 to 13until (PTR2! =NULL)]
SAVE=PTR2
IF (TEMP=INFO [SAVE])
THEN
i. Counter<- Counter+1

/*If the counter is incremented then drop the wholeLIST2 list if cycle exist in addition, the node with the base address of the dropped list is removed from List1 and START1 points the next node to the deleted node in List1*/
ii. [Display the detected cycle by repeating the following step until PTR3! =NULL]
ITEM=INFO [START2]
DISPLAY [ITEM]
PTR3=LINK [START2]
[In addition freeing the displayed nodes]
FREE (START2)
START2=PTR3

iii. [Remove the node containing the base address ofLIST2 list in which cycle occurs or null node appears]
PTR4=LINK [START1]
FREE (START1)
START1=PTR4
ELSE
PTR2=LINK [PTR2]
ENDIF

Step 11. [Enter the next node directed by the last node in LIST2 using Step (3) to (10) until NULL node encountered.

IF (NULL is encountered then GOTO STEP 11-(ii))]

/* the final value of counter will result in total number of cycles in the test graph.*/
Step 12. DISPLAY [Counter]

Step 13.END SUS_CycleDetection.

3.2 Example

Let's take a directed graph with n (V) =6

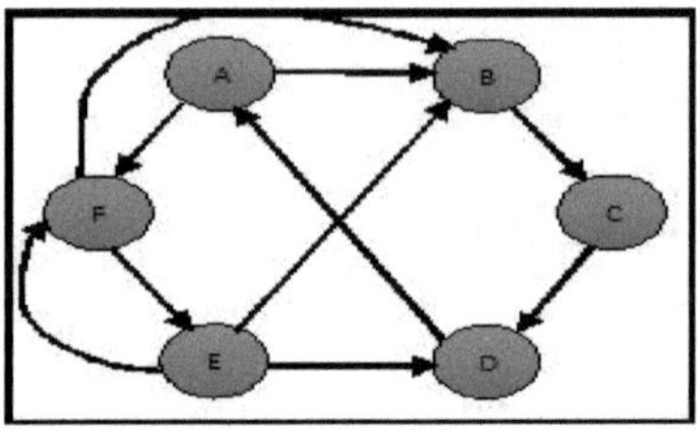

Fig. 6. A directed graph with A, B, C, D, E and F as its vertices

Let us take two-linked list LIST1, LIST2 withSTART1, START2 as their pointers to their base address respectively.
Step 1. Insert the first node in linked list LIST2 (let it be node A).

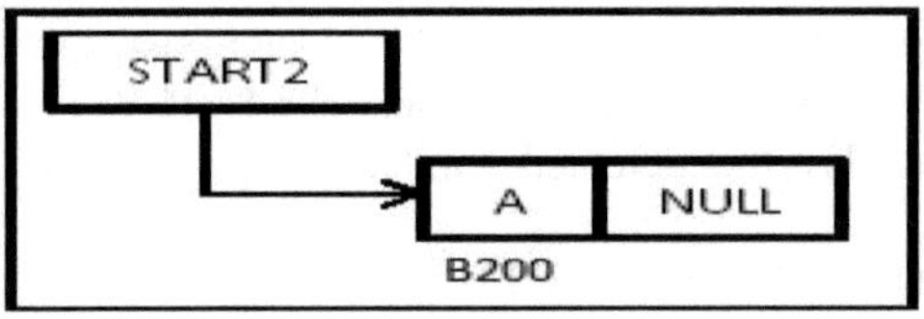

Fig. 7. Linked list LIST2with vertex A in its first info field

In addition, put the base address of this linked list LIST2 in LIST1 in its info field.

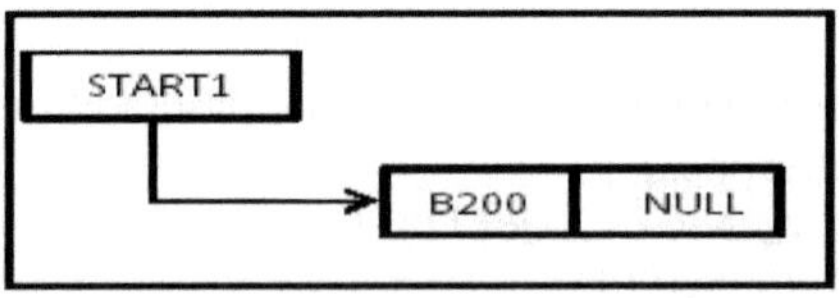

Fig. 8. START1 points to the base address of LIST1 with base address of LIST2 in its info field

Step 2. Since, there are two directed paths from A i.e. B and F so, total number of new linked lists will be one (2-1=1).And the new lists will be duplicate of present

LIST2. Now put each next node of the graph directed from the last node (B &F) into separate linked lists and put the base address of each newly created linked lists into the LIST1 in a consecutive manner.

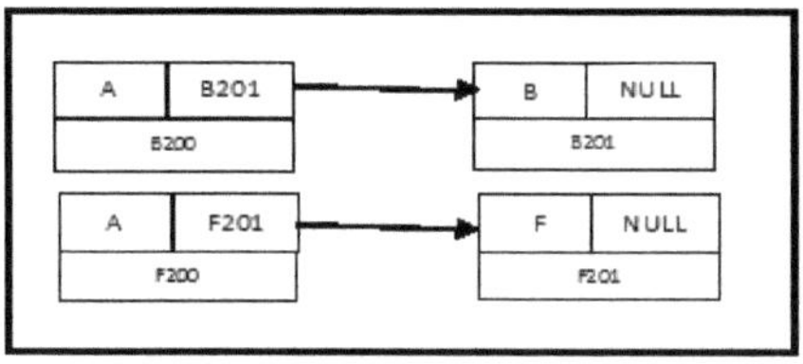

Fig. 9. Insert B in LIST2 and F in the newly created copies of LIST2

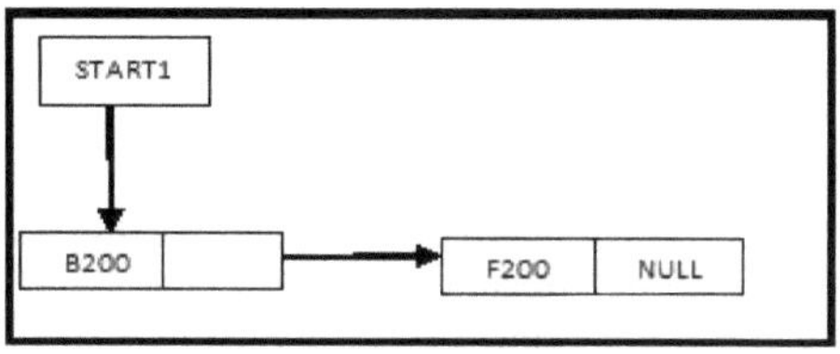

Fig. 10. Insertion of base address of the newly created lists in LIST1

Step 3. Repeat step (4) to (9) until START1! = NULL

Step 4.//Take PTR as a temporary variable.
PTR=INFO [START1] (Here, PTR = B200)

Step 5. Now as PTR points to the LIST2, compare the element at last node of LIST2 with all its previous nodes from starting.

Step 6. Since cycle is not detected, insertion of node directed by B in LIST2 will take place. Since B directs only one node (C) so there is no need to discover new lists, if there will be two path say towards x &towards y then new nodes will be formed with A linked with B and B links with x and second list will be A linked with B and B links with y.

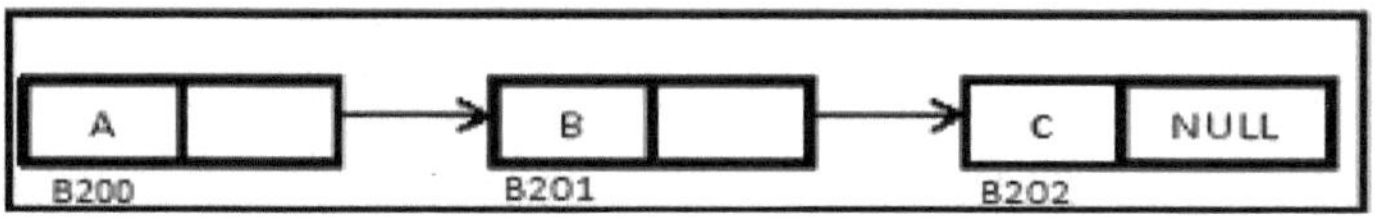

Fig. 11. Insertion of C node in the LIST2

Step 7. Again check for the appearance of cycle, here cycle doesn't appear so further traversal. Node C only directs to node D, so put node D next to C in LIST2.

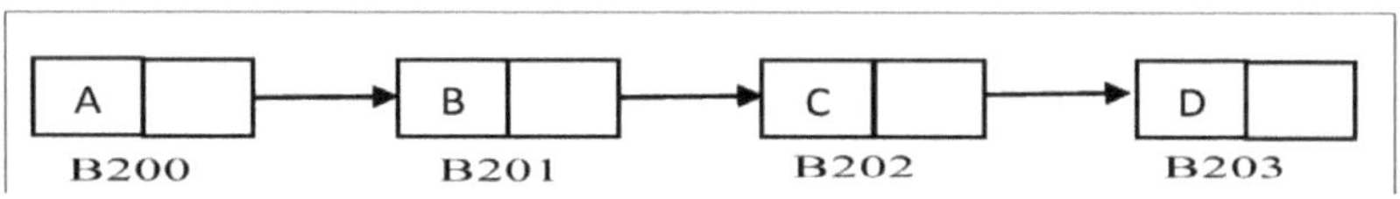

Fig. 12. Insertion of node D in LIST2

Step 8. Again check for the appearance of cycle, here cycle doesn't appear so further traversal. Node D only directs to node A, so put node A next to C in LIST2.

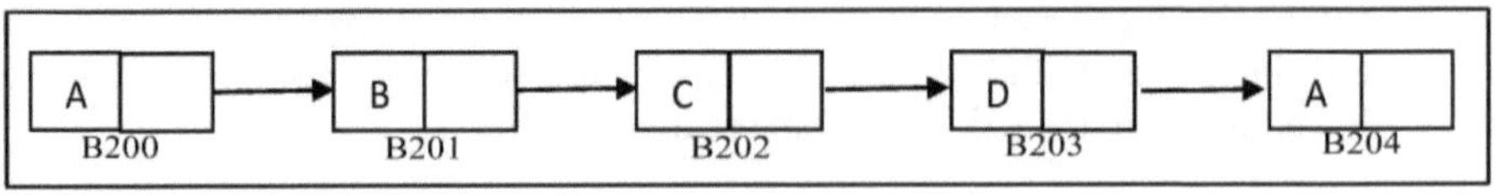

Fig. 13. Insertion of node A in LIST2

Step 9. Again check for the appearance of cycle, here cycle is detected hence counter incremented and the list displayed and then deleted. In addition, if NULL node appears then there is no cycle, in this case counter does not incremented but list deleted.

Step 10. Now, PTR should points to next node of LIST1.
PTR =LINK [PTR]

Step 11. Display the counter to get total number of cycles in the graph.

Hence, there are 2 cycles in the graph (Fig.6.).Therefore, using this algorithm we are able to find the set of nodes that forms the cycle and number of cycles in any digraph.

3.3 Complexity Analysis

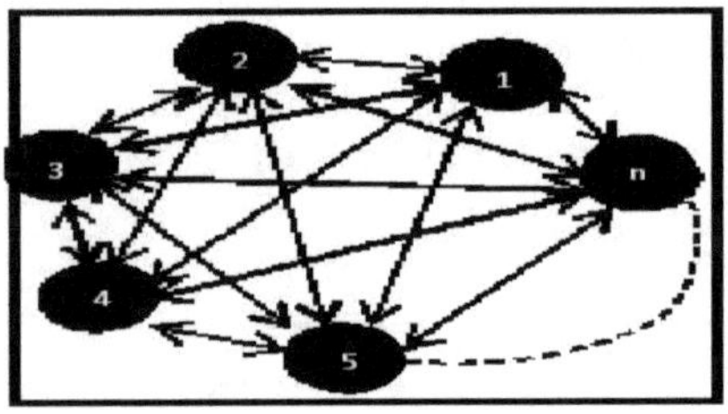

Fig. 14. An example to calculate the worst case of algorithm

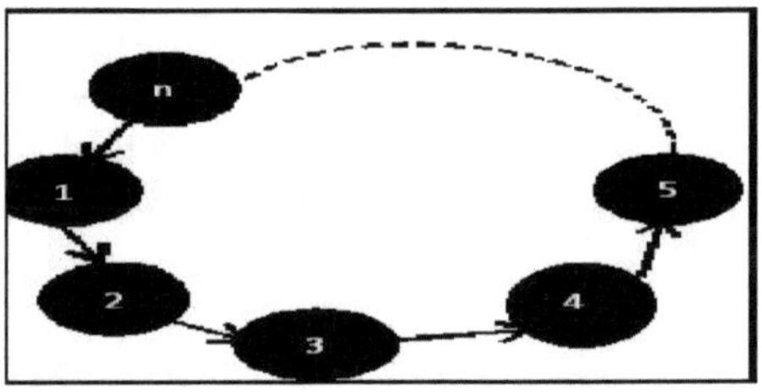

Fig. 15. Example to calculate the best case of algorithm

Let n be the number of vertices in the directed complete graph.

3.3. a. Worst Case (T_w)

3.3. a. i. Proposed SUS_CycleDetection Algorithm

=> T_{1w} (n) =Pointers assigning+ New lists+ Cycle detection
=> T_{1w} (n) =O (n-1) +O $((n-1)^{(n-1)})$ +O (n-1)
=> T_{1w} (n) =O (n^n)

3.3. a.ii. Existing Hongbo Liu'sCycle Detection

=> T_{2w} (n) =Nodes Order Assigning + Comparison of nodes+ Insertion of detected path in queue.

=> T_{2w} (n) =O (n) +O ((n-1) $^{(n-1)}$) +O (n-1)
=> T_{2w} (n) =O (n^n)

3.3. b. Best Case (T_b)

3.3. b. i. Proposed SUS_CycleDetection Algorithm

=> T_{1b} (n) = Pointers assigning+ Cycle detection
=> T_{1b} (n) =O (1) +O (n-1)
=> T_{1b} (n) =O (n)

3.3. b. ii. Existing Hongbo Liu'sCycle Detection Algorithm

=> T_{2b} (n) =Order Assigning + Comparison of nodes+ Insertion of detected path in queue.
=> T_{2b} (n) =O(n)+{O(n-1) + O(n-2)+ - - - - - - - - + 3 +2+1}+O(n-1) = O(∑(n))+O(n-1)
=> T_{2b} (n) =O [{n (n-1)/2} + n-1]
=> T_{2b} (n) =O (n^2)

4 Recommendations

The proposed SUS_CycleDetection algorithm by authors is not only an easier technique to detect cycle in any digraph but also very helpful in finding number of cycles in the graph. Moreover, since once the cycle is detected in a graph then it is displayed and deleted, that reduces the space complexity. In addition, it does not have the same walk to detect cycle, which lowers the time complexity. Therefore, this algorithm is efficient and can be used in several applications such as detecting infinite loops in various computer programs, analysis of electrical networks, periodic scheduling and in many more places where there is a need to detect cycle.

5 Limitations

In this paper, the cycle detection is done with the help of linked list. However, this method is easier to implement, in worst case its complexity reaches to O (n^n), which is much higher and because of the formation of new lists in run time it needs large space to act upon. Although, this algorithm removes those linked lists in which traversal is completed, computers with large space will be required to execute the proposed algorithm.

6 Conclusion

After all comparisons with the already existing algorithms, we can conclude that our algorithm is not much efficient as detecting cycle in any digraph, takes much space to execute and its complexity is much higher in case of worst case but it has its own advantages. It can detect each and every cycle and will also display the nodes responsible for its formation. It can be applied in any type of Directed graphs. The main benefit of this algorithm is that it is making all the nodes free after the comparisons. So it requires only temporary memory during execution and saves memory space. The generality of this algorithm is good as procedure of this algorithm is simple and its realization is efficient. The procedure of this algorithm is much easier to implement and execute for digraph and directed multigraph. This algorithm can be implemented directly thus avoiding some errors and wrong results in programming. Some amendments to this algorithm such as reduction in time complexity might be implemented in future to provide better result and further related work can be done to make it more efficient. This algorithm is expected to be of great interest in theory and practice alike.

7 Novelties in This Paper

In this paper, we have not only presented the new way to detect the cycle in any simple or strongly connected digraph but also presented the new way to count number of cycles in the graph. We used here an efficient data structure named linked list to form new nodes in run-time and used in storing the base address of the newly form linked list in run-time. This algorithm also reduces the time complexity by covering the individual walk in the graph only once, i.e. does not spend time in detecting same cycle repeatedly. We have also compared the proposed SUS_CycleDetection algorithm with Hongbo Liu's cycle detection algorithm in order to analyse the benefits of proposed over existing. Proposed algorithm is better than existing in terms of memory utilization, less number of comparisons, no extra constraints or integer designators, no duplicate cycles hence less cycle generation time, also the algorithm is efficient in its generality and simplicity.

Acknowledgement. The completion of this paper would not have been possible without the help and guidance of our rev. HOD-CSE, Prof. Dr. R. Radhakrishnan and MANAGEMENT of ABES Engineering College whose support was always there with us to correct at every step. We would like to thank our friends, family and seniors for their motivation and encouragement. Last but definitely not the least we would thank the almighty god without whose grace this paper would not have achieved success.

References

1. Puczynski, P.: The cycle detection algorithms, Wroclaw University of Technology, Faculty of Management
2. Van Gelder, A.: Efficient loop detection in Prolog using the tortoise-and-hare technique. Journal of Logic Programming 4(1), 23–31 (1987), doi:10.1016/0743-1066(87)90020-3.
3. Silberschatz, A., Galvin, P., Gagne, G.: Operating System Concepts, p. 260. John Wiley & Sons, Inc. (2003) ISBN 0-471-25060-0

4. Nivasch, G.: Cycle detection using a stack. Information Processing Letters 90(3), 135–140 (2004), doi:10.1016/j.ipl.2004.01.016.
5. Rozenfeld, H.D., et al.: Statistics of cycles: how loopy is your network? J.Phys. A: Math.Gen. 38, 4589 (2005)
6. Medard, M., Lumetta, S.S.: Network reliability and fault tolerance. In: Proakis, J. (ed.) Wiley Encyclopaedia of Engineering
7. Liu, H., Wang, J.: A new way to enumerate cycles in graph., Tsinghua University, State Key Lab of Intelligent Technology and System Department of Computer Science and Technology

[illegible]

Super Resolution of Quality Images through Sparse Representation

A. Bhaskara Rao and J. Vasudeva Rao

Department of Computer Science and Engineering,
GMR Institute of Technology, Rajam, AP, India
{bhaskararao.anem,jvr.vasu}@gmail.com

Abstract. This paper addresses the problem of generating the super resolution (SR) image from a single low resolution input image. Image patches can be represented as a sparse linear combination of elements from an over-complete dictionary. The low resolution image is viewed as down sampled version of a high resolution image. We look for a sparse representation for each patch of the low resolution image, and then use the coefficients of this representation to generate high resolution. Theoretically the sparse representation can be correctly recovered from the down sampled signals. The low and high resolution image patches are mutually training two dictionaries. We can look for the similarity between low and high resolution image patch pair of sparse representations with respect to their own dictionaries. Hence the high resolution image patch is applied to sparse representation of a low resolution image patch. This approach is more compact representation of the patch pairs compared to previous approaches. The earlier approaches simply sample a large amount of image patch pairs. The effectiveness of sparsity prior is demonstrated for general image super resolution. In this case, our algorithm generates high resolution images that are competitive or even superior in quality to images produced by other similar SR methods. This algorithm is practically developed and tested and it is generating high resolution image patches. The results are compared and analyzed with other similar methods.

Keywords: SR, sparse representation, dictionary, noise, sparsity.

1 Introduction

Super- Resolution is the process of combining multiple low resolution images to form a higher resolution one**.** Image representation using sparse approximation and learned over complete dictionaries has been given some attention in recent years. Several theoretical works have been obtained and probable applications are given like face compression [9], general image compression [10] and image denoising. Traditional approaches to generating a super resolution (SR) image require multiple low-resolution images of the same scene, typically aligned with sub-pixel exactness. The SR task is cast as the inverse problem [1] of recovering the original high-resolution picture by fusing the low-resolution pictures, based on prior knowledge about the

S.C. Satapathy et al. (eds.), *ICT and Critical Infrastructure: Proceedings of the 48th Annual Convention of CSI - Volume II*, Advances in Intelligent Systems and Computing 249,
DOI: 10.1007/978-3-319-03095-1_6,

generation model from the high-resolution image to the low-resolution images. The individual low resolution images have sub-pixel displacements relative to each other; it is possible to take out high frequency details of the observation well beyond the Nyquist [2] limit of the individual Source images. The concepts of sparsity and over completeness, together or separately, in representation of signals were proved to be highly effective. Applications that can benefit from them include compression, regularization in inverse problems, feature extraction, and more. Sparse signal representation has proven to be an extremely powerful device for acquiring, representing, and compressing high-dimensional signals. The fundamental reconstruction limitation for SR is that the recovered image, after applying the same generation model, should reproduce the observed low-resolution images. However, SR image reconstruction is generally a severely ill-posed problem because of the insufficient number of low-resolution images, ill-conditioned registration and unknown blurring operators and the solution from the reconstruction constraint is not unique.

The basic reconstruction constraint is that applying the image formation model to the improved image should produce the same low-resolution images. The reconstruction problem is severely underdetermined, and the solution is not unique because much information is lost in the high-to-low generation process. But to further normalize the problem various methods have been proposed. This paper mainly focuses on the problem of improving the super-resolution version of a given low-resolution image. Even though this method can be extended to handle multiple input images, we mostly deal with a single input image only. Like the above mentioned learning-based methods, we will rely on patches from example images. This method does not need any knowledge on the high-resolution patches, instead working directly with the low-resolution training patches. Super resolution is the process of combining a sequence of low-resolution (LR) noisy blurred images to produce a higher resolution image or sequence. An underlying assumption during this paper is that natural signals are represented well using a sparse linear composition of atoms from redundant dictionaries. We refer to this hypothesis as the "Sparse land" model, and provide an initial study of its geometrical behavior. More specially, we provide bounds on the estimated ratio of signals that can be represented by this dictionary with fixed sparsity constraints.

This algorithm requires a much smaller database. The online recovery [1] of the sparse representation uses the low resolution dictionary only; the high-resolution dictionary is used only to calculate the closing high-resolution image. The computed sparse representation adaptively selects the most important patches in the dictionary to best represent each patch of the given low-resolution image. This leads to superior performance, both qualitatively and quantitatively, compared to methods that use an unchanging number of nearest neighbors, generating sharper edges and clearer textures.

This paper focuses on the problem of recovering the SR version of a given low-resolution image. Similar to the aforesaid learning-based methods, we will rely on patches from the input image. However, instead of working directly with the image patch pairs sampled from high- and low-resolution images, we learn a compact symbol for these patch pairs to confine the co occurrence prior, significantly improving the speed of the algorithm. Our approach is guided by recent results in sparse signal representation, which suggest that the linear associations among high-resolution signals can be truly recovered from their low-dimensional projections. Although the SR problem is very ill-posed [3], resulting into non-recovery of such signals, the image patch sparse representation demonstrates both effectiveness and robustness in regularizing the inverse problem.

Notations: *H*igh- and low-resolution images are represented with X and Y whereas high- and low-resolution image patches with x and y respectively. The bold uppercase D is used to denote the dictionary for sparse coding and we use D_h and D_l to denote the dictionaries for high- and low-resolution image patches, respectively. Bold lowercase letters denote vectors. Plain uppercase letters denote regular matrices, i.e; S is used as a down sampling operation in matrix form. Plain lowercase letters are used as scalars.

In the super-resolution context, x is a high-resolution image (patch), while y is its low-resolution version (or features extracted from it). If the dictionary D is over complete, the equation $p = D\alpha$ is underdetermined for the unknown coefficients α. The equation $q = LD\alpha$ is even more radically underdetermined. However, under gentle situation, the sparsest solution α_0 to this equation is unique. Furthermore, if D satisfies an appropriate near-isometric situation, then for a wide variety of matrices L, any sufficiently sparse linear representation of a high-resolution image x in terms of the D can be recovered (almost) perfectly from the low-resolution image. Even for such a complicated texture, sparse symbol recovers a visually attractive upgrading of the original signal. Recently sparse representation has been useful to many other related inverse problems in image processing, such as compression, denoising, and restoration, often improving on the state-of-the-art. The K-SVD algorithm can be used to learn an over complete dictionary from natural image patches and successfully apply it to the image denoising problem. In our setting, we do not directly compute the sparse representation of the high-resolution patch. Instead, we will work with two coupled dictionaries, D_h for high-resolution patches, and $D_l = LD_h$ for low-resolution patches. The sparse representation of a low-resolution patch in terms of D_l will be directly used to recover the corresponding high-resolution patch from Dh. We obtain a locally consistent [3] solution by allowing patches to overlap and demanding that the reconstructed high-resolution patches agree on the overlapped areas. Finally, we apply global optimization to eliminate the reconstruction errors [1] in the recovered high-resolution image from local sparse representation, suppressing noise and ensuring consistency with the low-resolution input. The leftovers of this paper are organized as follows. The Section 2 gives details of our formulation and solution to the image super-resolution problem based on sparse representation. In Section 3, we discuss how to prepare a dictionary from sample images and what features to use. Various experimental results in Section 4 demonstrate the efficacy of sparsity as a prior for image super-resolution.

2 Image SR Using Sparsity

Here the Image SR Using Sparsity is the proposed scheme to convert the low resolution image to high resolution image. There are three steps to implement the constituent:

- Super-resolution using Sparsity
- Home Model using sparse Representation
- Universal optimization analysis

2.1 Super-Resolution Using Sparsity

The super resolution single image is having a problem that is, to recover a high resolution image P on the same screen from low resolution image Q. The basic constraint is that recovered image P should be consistent with the input Q. To solve the above problem two constraints are introduced in this paper. 1. Modernization constraint, in which the recovered constraint P should be consistent with input Q with respect to the image observation model. 2. Sparse representation prior, which assumes that the high resolution patches can be sparsely represented in appropriate chosen over complete dictionary. Sparse representations can be recovered from the low resolution observations.

Modernization Constraint. From the low resolution observation model the input Q is a down sampled blurred version of high resolution image P.

$$Q = DHP \tag{1}$$

Here, H represents a blurring filter, and D the down sampling operator. Super-resolution remains particularly ill-posed, since for a given low-resolution input Q, infinitely many high resolution images P satisfy the above modernization constraint. We regularize the problem via the following previous on small patches p of P:

Sparse Representation Prior. In this constraint the patches are represented as a sparse linear combination in a dictionary. The patches of high resolution image P are represented in the dictionary named D_h. The patches of low resolution image Q are represented in the dictionary D_l.

$$p \approx D_h \alpha \text{ for some } \alpha \in R^k \text{ with } \|\alpha\|_0 \leq K \tag{2}$$

The sparse representation α will be recovered from patches q of the input is Q, with respect to the low resolution dictionary D_l consistent with D_h.

Using the sparse prior, we find the sparse representation for each local patch, with regard to spatial compatibility between neighbors. Next, using the result from this local sparse representation, we further regularize and refine the entire image using the reconstruction constraint. In this approach, a local model from the sparse prior is used to recover lost high-frequency for local details. The global model from the reconstruction constraint is then applied to remove possible artifacts from the first step and make the image more dependable and natural.

2.2 Home Model Using Sparse Representation

From the previous patch methods, we are trying to get high resolution patch for each low resolution patch of the input. For this home model we have two dictionaries D_l and D_h, D_h is a linear combination of high resolution patches, and D_l is a linear combination of low resolution patches. We deduct the mean pixel value for each patch; therefore dictionary represents image textures rather than absolute intensities.

For each low resolution patch q we find the sparse representation in $D_{l.}$ The corresponding high resolution patches D_h will be combined according to these coefficients to get the high resolution out patch p.

$$\mathbf{min}\|\alpha\|_0 \ \ \mathbf{s.t.}\ \|\mathbf{F}\, D_l\, \alpha\text{-}Fq\|^2_2 <= t; \tag{3}$$

where F is a feature extraction operator. The main role of F in (3) is to provide a perceptually meaningful constraint on how closely the coefficients α must approximate q.

$$\mathbf{min}\|\alpha\|_1 \ \ \mathbf{s.t.}\ \|\mathbf{F}\, D_l\, \alpha\text{-}Fq\|^2_2 <= t; \tag{4}$$

Solving (4) separately for each patch does not guarantee the compatibility between adjacent patches. We enforce compatibility between adjacent patches using a one-pass algorithm parallel to that of [6]. The patches are processed in raster-scan order in the image, from left to right and top to bottom. We change so that the super resolution reconstruction $D_h\alpha$ of patch y is constrained to closely agree with the previously computed adjacent high-resolution patches.

2.3 SR Algorithm

The Super Resolution algorithm is given below:

1. **Name of Algorithm**: Super Resolution via Sparse Representation
2. **Input**: training dictionaries D_h and D_l, a low-resolution image Q.
3. **Process**: we have two dictionaries D_h and D_l, which are trained to have the same sparse representations for each high-resolution and low-resolution image patch pair. In the recovery process for each input low-resolution patch q, we find a sparse representation with respect D_l to the corresponding high resolution patch bases D_h, which will be combined according to these coefficients to generate the output high-resolution patch p.
4. **Outpu**t: SR image P*

2.4 Universal Optimization Analysis

The general Sparse Representation frame work of the SR algorithm discussed above can be viewed as a special case for inverse problems in image processing. Similar techniques have been applied in image compression, denoising [7], and restoration [8]. These applications provide context for understanding our paper and also gives suggestions to further improve the performance.

Using the resources available, one could in principle solve for coefficients associated with all patches simultaneously with a given sufficient computational resources. The high resolution image P is treated as a variable. Rather than demanding that P be perfectly reproduced by the sparse coefficients α, we can reprimand the difference between P and high resolution image given by these coefficients, allowing solutions that are not perfectly sparse but satisfy the reconstruction constraints. This results in to a large optimization pr0oblem.

3 Dictionary Preparations

Gaining knowledge of an over-complete dictionary accomplished of optimally indicating broad classes of image patches is a difficult problem. In place of trying to learn such a dictionary or using a generic set of basis vectors, we create dictionaries by simply randomly sampling raw patches from training images of similar statistical nature. We will demonstrate that so simply prepared dictionaries are already capable of generating high-quality reconstructions, when used together with the sparse representation prior. Sparse coding is the problem of discovery sparse representations of the signals with regard to an over complete dictionary D.

The dictionary is naturally learned from a set of training examples $P=\{p_1,p_2,p_{3...}p_t\}$. Generally, it is hard to learn a compact dictionary which guarantees that sparse representation can be obtained from l_1 minimization. Fortunately, many sparse coding algorithms proposed previously are sufficient for practical applications.

Given the sampled instruction image patch pairs $P=\{P^h,Q^l\}$, where $P^h=\{p_1,p_2,p_3....p_n\}$ are the set of sampled high-resolution image patches and $Q^l=\{q_1,q_2,q_3.....q_n$ are the corresponding low-resolution image patches (or features), our goal is to learn dictionaries for high-resolution and low-resolution image patches, so that the sparse representation of the high-resolution patch is the same as the sparse representation of the parallel low-resolution patch.

4 Experimental Results

The SR algorithm which is discussed previously is implemented in dot net. We first demonstrate the SR results obtained by applying the SR method via sparse representation on generic images. We then move on to discuss various in influential factors for the proposed SR algorithm including dictionary size and input images with noise and the global re-construction constraints.

In our experiments with dot net, we magnify the input low resolution image by a factor of 3 for generic images, which is common place in the literature of single frame SR. In SR generic image, for the low resolution images we usually taken 3×3 low resolution patches, with over lap of one pixel between adjacent patches, corresponding to 9×9 patches with over lap of 3 pixels for high resolution patches. In dot net this is achieved with bit map class. The bit map class stores an image in pixel format and it is having clone method, it takes two parameters one is the shape area which we want convert low resolution to high resolution, and the second parameter is pixel format. This clone method takes the α value of the bit map image and uses for loop, repeats for all 3×3 patches in the given shape area by finding the mean value.

We applied our method to generic images such as flowers, human faces and architectures. The two dictionaries for high resolution and low resolution image patches are trained using patch pairs randomly sampled from natural images collected from the internet. We pre process these images by cropping out textured regions and discard the smooth cards. We always fix the image size 273×227 pixels and the dictionary size is 1024 in all our experiments. The Fig.1 shows a low resolution image and its corresponding high resolution image. The Fig.2 shows original images and the super resolution image. The Fig.3 shows another low resolution image and its high resolution Image. The Fig.4 shows original images and super resolution Image.

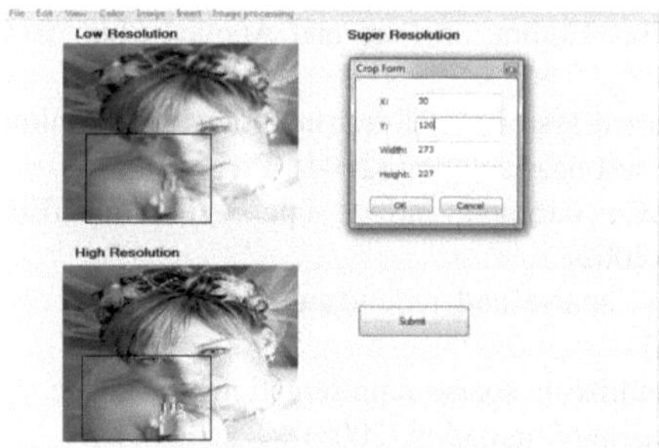

Fig. 1. Crop Images

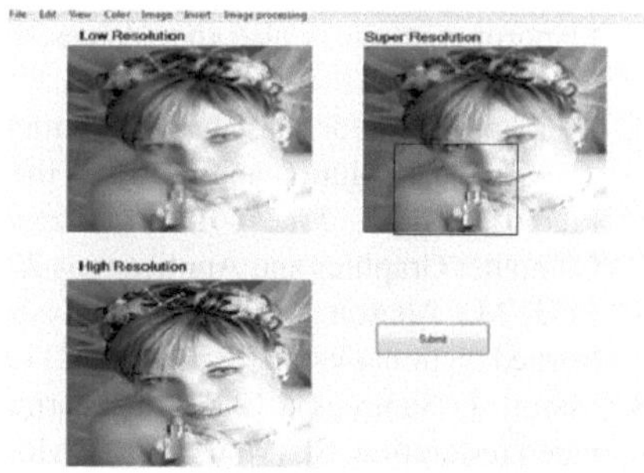

Fig. 2. Super Resolution Image

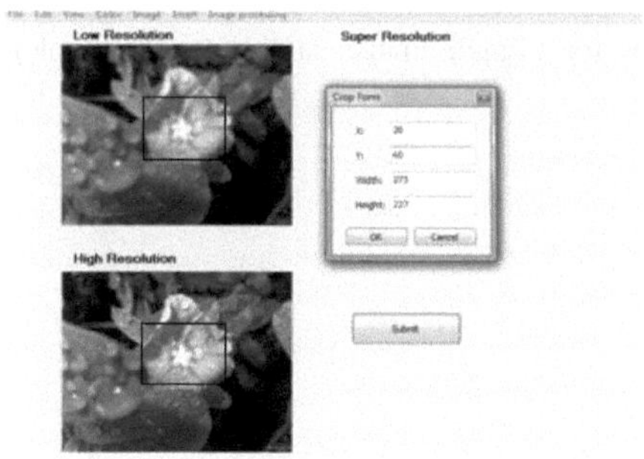

Fig. 3. Crop Images

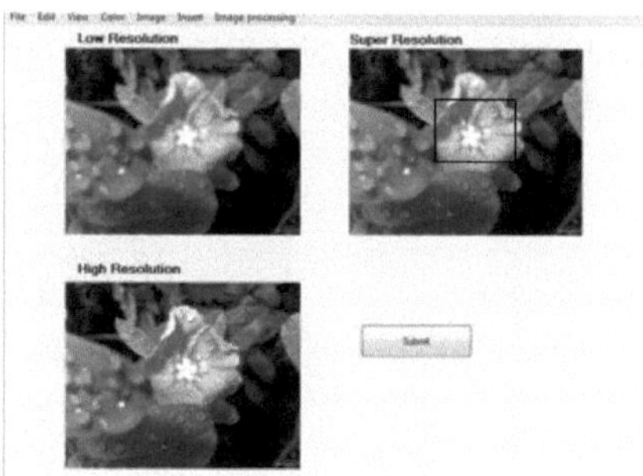

Fig. 4. Super Resolution Image

5 Conclusion

This paper presents a new approach for single image super-resolution based on sparse representations in terms of coupled dictionaries jointly trained from high- and low resolution image patch pairs. The compatibilities among adjacent patches are enforced. Experimental results demonstrate the effectiveness of the sparsity as a prior for patch-based super-resolution for images. However, one of the most important questions for future investigation is to determine the optimal dictionary size for natural image patches in terms of SR tasks.

References

1. Yang, J., Wright, J., Huang, T., Ma, Y.: Image super-resolution as sparse representation of raw image patches. In: IEEE Conference on Computer Vision and Pattern Recognition (CVPR), pp. 1–8 (2008)
2. Tipping, M.E., Bishop, C.M.: Bayesian image super-resolution. In: Proc. Advances in Neural Information and Processing Systems 16, NIPS (2003)
3. Yang, J., Wright, J., Huang, T., Ma, Y.: Image Super-Resolution via Sparse Representation. IEEE Transactions on Image Processing 19(11) (2010)
4. Donoho, D.L.: For most large underdetermined systems of linear equations, the minimal ℓ1-norm solution is also the sparsest solution. Comm. on Pure and Applied Math 59(6) (2006)
5. Donoho, D.L.: For most large underdetermined systems of linear equations, the minimal ℓ1-norm near-solution approximates the sparsest near-solution (2004) (Preprint)
6. Freeman, W.T., Jones, T.R., Pasztor, E.C.: Example based super-resolution. IEEE Computer Graphics and Applications 22(2) (2002)
7. Elad, M., Aharon, M.: Image denoising via sparse and redundant representations over learned dictionaries. IEEE TIP 15(12) (2006)
8. Mairal, J., Sapiro, G., Elad, M.: Learning multiscale sparse representations for image and video restoration. SIAM Multiscale Modeling and Simulation (2008)
9. Aharon, M., Elad, M., Bruckstein, A.: K-SVD: An algorithm for designing over complete dictionaries for sparse representation. IEEE Transactions on Signal Processing 54(11), 4311–4322 (2006)
10. Murray, J.F., Kreutz-Delgado, K.: Learning sparse over complete codes for images. The Journal of VLSI Signal Processing Systems for Signal, Image, and Video Technology 46(1), 1–13 (2007)

An Interactive Rule Based Approach to Generate Strength Assessment Report: Graduate Student Perspective

P. Ajith[1], K. Rajasekhara Rao[1], and M.S.S. Sai[2]

[1] KL University
ajipenu@yahoo.com,
rajasekhar.kurra@klce.ac.in
[2] KKR & KSR Institute of Technology and Sciences
msssai@gmail.com

Abstract. Data and Information or Knowledge has a significant role on human activities. Data mining means extracting or discovering knowledge from large volume of data. Now-a-days, data mining process is applying in educational field also; it is called as educational data mining. Educational Data Mining used many techniques such as Decision Trees, Neural Networks, Naïve Bayes, K- Nearest neighbor, and many others. By using these techniques, the discovered knowledge can be used for prediction and analysis purposes of student patterns. Existing techniques like tree classification and some clustering techniques are suffering with decision-making problems. To solve this problem, in this paper, an interactive approach is used to prune and filter discovered rules. In proposed system, an integrate user knowledge is used in the post processing task. A set of rules or measures are given as input to proposed system in order to evaluate the student performance. Furthermore, an interactive framework is designed to assist the user throughout the analyzing task. By applying proposed approach to discover the likelihood of student's deviations / requiring special attention is organized and efficient providing more insight by using Strength Assessment Report. After analyzing the student performance, strength assessment reports are generated which lists the career skills or competencies that are strong/good in.

Keywords: Educational Mining, Association Rules, Post Processing, Strength Assessment Report(SAR), Competency Skills, Student Performance.

1 Introduction

Data mining is a type of sorting technique, which is actually used to extract hidden patterns from large databases. The major objectives for data mining are fast retrieval of data or information, Knowledge Discovery from the databases, to detect hidden patterns and those patterns which are previously unknown.

S.C. Satapathy et al. (eds.), *ICT and Critical Infrastructure: Proceedings of the 48th Annual Convention of CSI - Volume II*, Advances in Intelligent Systems and Computing 249,
DOI: 10.1007/978-3-319-03095-1_7, © Springer International Publishing Switzerland 2014

Now a days, the topic of explanation and prediction of academic performance is widely researched. The main objective of higher education institutes is to provide quality education to its students and to improve the quality of managerial decisions. Data Mining Techniques is the promising methodology to extract valuable information in this objective. The data collected from different applications require proper method of extracting knowledge from large repositories for better decision making. Knowledge discovery in databases (KDD) [1], often called data mining, aims at the discovery of useful information from large collections of data. In this perspective, Data Mining can analyze relevant information results and produce different perspectives to understand more about the students' activities to customize the course for student learning [2].

In existing system, for optimally analyzing the student performance, the classification task is used on student database to predict the students division on the basis of previous database. As there are many approaches that are used for data classification, the decision tree method is used here. Information is like Attendance, Class test, Seminar and Assignment marks were collected from the student's previous database, to predict the performance at the end of the semester.

In this paper, an interactive approach is used to prune and filter discovered rules. In proposed system, an integrate user knowledge is used in the post processing task. A set of rules or measures are given as input to proposed system in order to evaluate the student performance. Furthermore, an interactive framework is designed to assist the user throughout the analyzing task. Proposed approach is used to discover the likelihood of student's performance, behavior and requiring special attention is organized. The discovered knowledge is more efficiently provided by using Strength Assessment Report. According to gained knowledge, analysis is performed; after analyzing each and every attribute of student, performance of the student is generated using strength assessment reports. The generated report lists the career skills or competencies that are strong/good in. According to the reports of dataset, Association rules are mined which are used for enhancing the student performance.

In this paper, Section-2 describes various data mining techniques. Related work is specified in section-3, Section-4 describes the background work of proposed system, proposed data mining process is specified in section-5 and finally performance, result and discussions are discussed in section-6 & 7.

2 Data Mining Techniques

2.1 Decision Tree

Decision tree is a predictive model that, as its name implies, can be viewed as a tree. Specifically each branch of the tree is a classification question and the leaves of the tree are a partition of the dataset with their classification. A decision tree is a classifier expressed as a recursive partition of the instance space. The decision tree consists of nodes that form a rooted tree, meaning it is a directed tree with a node called "root" that has no incoming edges. All other nodes have exactly one incoming edge. A node with outgoing edges is called an internal or test node. All other nodes are called leaves (also known as terminal or decision nodes). In a decision tree, each internal node splits the instance space into two or more sub-spaces according to a certain discrete function of the input attributes values.

Interesting things about decision trees are:

- It divides up the data on each branch point without losing any of the data (the number of total records in a given parent node is equal to the sum of the records contained in its two children).
- The number of churners and non-churners is conserved as you move up or down the tree.
- It is easy to understand how the model is being built.

2.2 kNN: k-Nearest Neighbor Classification

A more sophisticated approach, *k*-nearest neighbor (*k*NN) classification, finds a group of *k* objects in the training set that is closest to the test object, and bases the assignment of a label on the predominance of a particular class in this neighborhood. There are three key elements of this approach: a set of labeled objects, e.g., a set of stored records, a distance or similarity metric to compute distance between objects, and the value of *k*, the number of nearest neighbors. To classify an unlabeled object, the distance of this object to the labeled objects is computed, its *k*-nearest neighbors are identified, and the class labels of these nearest neighbors are then used to determine the class label of the object.

2.3 Naïve Bayes

Naïve Bayes method is important for several reasons. It is very easy to construct, not needing any complicated iterative parameter estimation schemes. This means it may be readily applied to huge data sets. It is easy to interpret, so users unskilled in classifier technology can understand why it is making the classification it makes. And finally, it often does surprisingly well: it may not be the best possible classifier in any particular application, but it can usually be relied on to be robust and to do quite well. Here, the initial set of objects are used with known class memberships (the training set) to construct a score such that larger scores are associated with class 1 objects (say) and smaller scores with class 0 objects. Classification is then achieved by comparing this score with a threshold, *t*.

2.4 CART

The CART decision tree is a binary recursive partitioning procedure capable of processing continuous and nominal attributes as both targets and predictors. Data are handled in their raw form; no binning is required or recommended. Trees are grown to a maximal size without the use of a stopping rule and then pruned back to the root via cost-complexity pruning. The next split to be pruned is the one contributing least to the overall performance of the tree on training data. The procedure produces trees that are invariant under any order preserving transformation of the predictor attributes. The CART mechanism is intended to produce not one, but a sequence of nested pruned trees, all of which are candidate optimal trees.

3 Related Work

Modeling student performance at various levels and comparing different data mining algorithms are discussed in many recently published research papers. Kovacicin [3] uses data mining techniques (feature selection and classification trees) to explore the socio-demographic variables (age, gender, education, and disability) and study environment that may influence persistence or dropout of students, identifying the most important factors for student success and developing a profile of the typical successful and unsuccessful students. Vandamme et al. [4] use decision trees, neural networks and linear discriminant analysis for the early identification of three categories of students: low, medium and high-risk students.

Kotsiantis et al. [5] apply five classification algorithms (Decision Tree, Perceptron-based Learning, Bayesian Net and Instance-Based Learning) to predict the performance of computer science students from distance learning. . Y u et al. [6] explores student retention by using classification trees, Multivariate Adaptive Regression Splines (MARS), and neural networks. Cortez and Silva [7] attempt to predict student failure by applying and comparing four data mining algorithms –Decision Tree, Random Forest, Neural Network and Support Vector Machine. Ramaswami and Bhaskaran [8] focus on developing predictive data mining model to identify the slow learners and study the influence of the dominant factors on their academic performance, using the popular CHAID decision tree algorithm.

Al-Radaideh, Al-Shawakfa & Al-Najjar et al. [9] was using the classification trees to predict the final grade among undergraduate students of the Information Technology & Computer Science Faculty, at Yarmouk University in Jordan. Han and Kamber[10] describes data mining software that allow the users to analyze data from different dimensions, categorize it and summarize the relationships which are identified during the mining process. Galit[11] gave a case study that use students data to analyze their learning behavior to predict the results and to warn students at risk before their final exams. Ayesha, Mustafa, Sattar and Khan[12] describe the use of k-means clustering algorithm to predict students learning activities.

Romero et al.[13] compared different methods of data mining in order to predict final assessment based on the data obtained from the system of e-learning. Minaei-Bidgolim, et al.[15] was among the first authors who classified students by using genetic algorithms to predict their final grade. Zekić-Sušac, Frajman-Jakšić and Drvenkar et al.[16] created a model for predicting students' performance using neural networks and classification trees decision-making, and with the analysis of factors which influence students' success. Kumar and Vijayalakshmi et al. [17] using the decision tree predicted the result of the final exam to help professors identify students who needed help, in order to improve their performance and pass the exam.

Hongjie Sun et al.[18] conducts a research on student learning result based on data mining. It is aimed at putting forward a rule-discovery approach suitable for the student learning result evaluation and applying it into practice so as to improve learning evaluation skills and finally better serve learning practicing. Hua-long Zhao et al. has done Multidimensional cube analysis by taking use of OLAP technology and has shown that the curriculum chosen by the students can depend upon many angles like teacher, semester and student. Fadzilah Siraj and Mansour Ali Abdoulha.et al.[20] have used data mining techniques for understanding student enrolment data. They have done comparative study of three predictive data mining techniques namely Neural Network, Logistic regression and Decision tree.

4 Back Ground Work

Existing as specified in section-3, various authors proposed various techniques and various measures like behaviors, age, economical background, regularity, exams and soon. This paper consider following career skills or competencies for modern corporate world, which are used for analyzing students performance. They are:

- F1: Gregariousness: It means feels comfortable while conversing with strangers and wants to be centre of attention.
- F2: Adaptability: It means easily accepts and adapts to changing ideas, technology, situations and conditions. Handles crises appropriately functioning effectively under stress.
- F3: Follow-up: It means pursues activities to reinforce the earlier ones without being fed up and sticks to schedules.
- F4: Ability to meet deadlines: It means utilizes time in the best possible way and identifies important and unimportant things in life.
- F5: Competitiveness: It means strives for the best when making comparisons with some standards of excellence and compares own potential with others which make efforts to pursue goals.
- F6: Logical Reasoning: It has the ability to analyze information, make inferences, and draw logical conclusions. Logical reasoning critically evaluates all aspects and the possible consequences of each action.
- F7: Proactive: It means control key events in one's life and feels responsible for it. Makes decisions based on own conscious choice.
- F8: Updating Knowledge: It means spends time and resources to know what is happening in professional field in a continuous manner by taking membership of various bodies, by reading magazines and indulging in self-study and formal and informal discussions etc.
- F9: Numerical Ability: It has the capability to deal with numbers or qualitative values quickly and accurately.
- F10: Ability to work in a team: It has the willingness to work closely with others towards a common goal as opposed to working in competition with others.
- F11: Organization: It means arranges groups, things and activities in order and allocates resources required to carry out a specific plan.

- F12: Calmness: It means rarely gets excited or agitated, accepts people as they are and handles tasks smoothly.
- F13: Problem Solving Skills: It means systematically addresses and resolves problems.
- F14: Politeness: It means behaves in a way that is socially correct and creates an environment of comfort and respect for others.
- F15: Cheerfulness: It means expresses child like joy and smiles a lot.
- F16: Resourcefulness: It means able to meet situations and capable of devising ways and means and resources appropriately to solve complex problems.
- F17: Mind quickness: It means applies knowledge and experiences to understand tricky situations and have an attentive mind.
- F18: Empathy: It has deep sympathy, sorrow for others and tends to forgive others easily.
- F19: Patience: It means shows capacity to endure hardships of delay or incompetence without complaining and has rich imagination.

5 Data Mining Process

Now-a-days, performance of individual student in educational system is evaluated based on the various factors like internal exams, regularity, uniformity, intelligence and soon. However, in this paper, career skills and competencies are considered as major factor, which are discussed in above section. According to discussed factors, a basic framework and an interactive post mining process is used to analyze student performance.

5.1 Basic Framework

Firstly, a basic mining process is applied over data and a set of association rules. Secondly, the knowledge base allows formalizing user knowledge and goals. Domain knowledge allows a general view over user knowledge in database domain, and user expectations express prior knowledge over the discovered rules. Finally, the post-processing step consists in applying several operators (i.e. pruning) over user expectations in order to extract the interesting rules.

The novelty of this approach resides in supervising the knowledge discovery process using two different conceptual structures for user knowledge representation: integrate user knowledge in the post-processing task and several rule schemas generalizing general impressions, and proposing an iterative process.

5.2 Interactive Post Mining Process

The framework proposes to the user an interactive process of rule discovery. Taking into account his/her feedbacks, the user is able to revise his/her expectations in function of intermediate results. Several steps are suggested to the user in the framework as follows:

1. Knowledge Gained—starting from the database, and eventually, from existing Knowledge, the user develops knowledge on database items;
2. Defining Measures or Skills —the user expresses his/her local goals and expectations concerning the association rules that he/she wants to find;
3. Choosing the right operators to be applied over the measures created, and then, applying the operators;
4. Visualizing the results— results generated in reports (SAR) format. From the reports, association rules are proposed to the user;

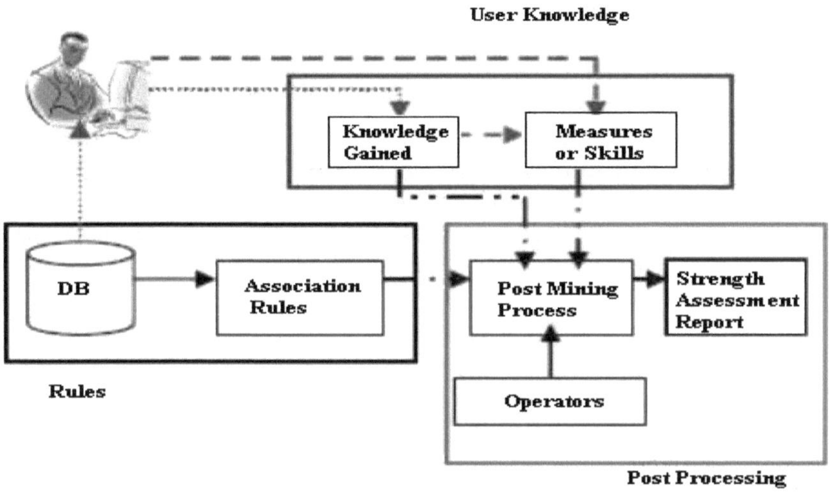

Fig. 1. Framework description

5. Selection/validation—starting from these preliminary results, the user can validate the results or he/she can revise his/her information;
6. The interactive loop permits to the user to revise the information that he/she proposed. Thus, he/she can return to step 2 in order to modify the rule schemas, or he/she can return to step 3 in order to change the operators. Moreover, in the interactive loop, the user could decide to apply one of the two predefined filters discussed in step 6.

6 Implementation

The performance of proposed system is analyzed by generating the Strength Assessment Report (SAR), which talks about your personality, attitudes, and talents —under 19 factors. Those factors are already discussed in section 4. The purpose of this report is to identify and make full use of student strengths and help to develop an awareness of areas that could be limiting your effectiveness.

Generally, there are two directions in which people or students focus their energy. These two directions are: extroversion and introversion. These (extroversion and introversion) are two tendencies that define people to a certain extent. Now, introversion in psychology refers to a disposition that shows inward concern — with one's own thoughts and feelings. Introverts derive energy from going within themselves. Extroversion refers to a disposition that reflects concern with what is outside the self. Extroverts derive their energy from the external environment.

There are two more factors that decide our tendencies (of dealing with the world): Perceiving(F21) and Judging(F20). Perceiving means spending time in taking in information and being open to new experiences. Judging is the cognitive process of reaching a decision or drawing conclusions. It implies organizing, taking decisions, and completing things on time.

6.1 Data Preparation

The data set used in this study was obtained from SRIJI College of various branches like B.Sc, B.Com, and BCA of various academic years. Initially size of the data set is 192. Let consider 30 out of 192 students, those student performance are analyzed under 21 factors, which are related to career skills of modern corporate world; factors are discussed in section 4. The analyzed results are displayed in table 1.

Table 1. Analyzed results of data sets

S No	StudentID	Batch	F1	F2	F3	F4	F5	F6	F7	F8	F9	F10	F11	F12	F13	F14	F15	F16	F17	F18	F19	F20	F21
1	SID001	BCA-II	12	18	14	24	16	13	16	18	14	9	17	11	13	15	14	18	20	18	15	18	14
2	SID002	BCOM-III	12	14	13	20	12	18	17	17	19	14	25	12	18	18	11	18	15	19	13	18	13
3	SID003	BCOM-III	12	12	16	13	18	14	17	18	12	15	15	13	14	11	15	19	20	16	17	17	16
4	SID004	BSC-III	12	10	13	17	15	9	20	14	20	11	17	11	9	14	19	13	9	8	13	14	13
5	SID005	BSC-II	11	12	11	16	20	16	17	19	18	16	18	12	16	15	14	19	17	17	20	12	11
6	SID006	BSC-III	10	20	13	21	16	20	14	13	13	13	17	15	20	14	16	15	20	12	15	19	13
7	SID007	BSC-II	16	18	16	15	18	12	20	16	13	17	20	13	12	18	19	19	16	19	19	20	16
8	SID008	BSC-III	15	14	12	17	14	15	13	16	14	11	18	20	15	12	18	14	16	15	14	15	12
9	SID009	BCA-II	15	9	17	12	18	8	11	13	13	9	16	13	8	10	16	15	17	19	23	12	17
10	SID010	BCA-II	25	10	16	24	9	12	21	13	8	21	17	14	12	13	17	16	25	17	20	25	16
11	SID011	BSC - II	11	17	18	24	22	18	22	19	19	11	21	14	18	22	16	18	22	18	19	19	18
12	SID012	BCA - III	16	20	16	13	14	10	9	13	16	11	11	10	13	12	17	14	12	17	18	12	16
13	SID013	BCA - III	18	13	16	19	21	18	11	13	16	13	16	14	18	16	16	22	11	18	16	11	16
14	SID014	B.Com - II	17	13	14	20	13	17	12	21	13	21	10	12	17	14	13	18	23	13	15	15	14
15	SID015	BCA - I	15	16	25	15	20	17	12	17	15	16	21	17	14	12	20	20	25	19	12	23	25
16	SID016	BSC - III	12	15	16	15	13	19	11	14	12	11	18	17	19	10	18	16	14	12	14	12	16
17	SID017	BCA - II	15	15	17	16	16	11	14	12	24	12	11	11	15	13	17	16	19	16	12	15	17
18	SID018	B.Com - III	14	14	11	20	22	17	19	22	15	11	16	15	17	14	17	19	12	14	17	15	11
19	SID019	BSc - II	22	20	20	20	17	19	21	14	17	11	18	19	19	22	23	19	20	21	21	17	20
20	SID020	BSc-III	12	17	16	14	20	19	13	16	15	15	11	13	19	12	20	18	16	17	18	14	16
21	SID021	B.Com - III	15	16	16	15	16	14	13	16	19	16	16	16	16	18	18	18	18	18	14	13	16
22	SID022	BCA - III	16	18	15	9	14	18	18	18	13	16	16	14	13	16	15	14	12	16	18	13	15
23	SID023	BSc - III	13	20	16	13	14	23	17	14	12	14	15	23	18	20	19	22	20	24	15	14	16
24	SID024	BCA - III	25	22	7	20	22	11	19	17	16	15	16	11	14	20	12	17	24	23	23	5	7
25	SID025	BCA - II	11	13	12	14	16	15	20	21	19	12	22	13	15	20	23	16	14	13	22	23	12
26	SID026	BCA - II	12	12	16	13	18	14	17	18	12	15	15	13	14	11	15	18	2	16	17	17	16
27	SID027	B.Com - III	16	22	16	10	12	16	17	19	19	17	18	16	16	17	16	13	20	17	12	17	16
28	SID028	B.Com - III	15	19	14	16	15	16	12	12	13	15	14	16	15	14	7	8	11	9	6	4	14
29	SID029	BSc - III	11	17	13	18	15	14	17	21	15	17	21	8	14	19	23	11	17	17	12	16	13
30	SID030	BCA - III	10	18	13	24	21	15	19	19	16	14	18	13	15	20	17	15	20	24	13	14	13

7 Results and Discussions

The results generated after applying the proposed system are given below. Now, consider the some students which are presented in the table.

Consider student "SID001" studying BCA-II year. The results are generated using Strength Assessment Report (SAR) as shown in Fig 2. It lists the competencies that are strong/good in. Before taking any career-related decision about career, these strengths will be used.

According to the report, first student was,(i) "Resourceful", which means to able to meet situations or capable of devising ways and means and using resources appropriately for solving complex problems.

(ii) "Facing Deadlines". People with this competencies denotes a high self-motivation toward work which includes working effectively even under material constraints or if conditions are not supporting well to perform a job well without getting frustrated. These people generally meet the deadlines.

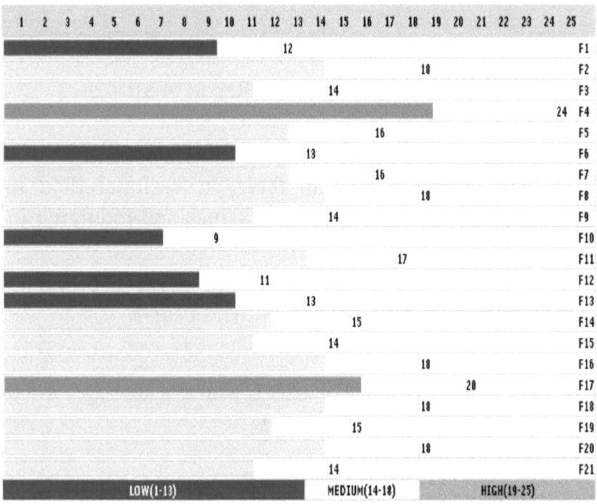

Fig. 2. Strength Assessment Report of SID001

Consider student "SID029" studying BSC-III year. The results are generated using Strength Assessment Report (SAR) as shown in fig 3.

According to the report, student was (i) "Organizing": It is defined, as arranging and grouping the things & activities, establishing authority and allocating resources required carrying out a specific plan.

(ii) "Polite to others". Person with this characteristic behave in a way that is socially correct and shows awareness and caring for other people's feelings. It means creating an environment of comfort and respect of others and not annoying by others behavior and knowing the art of winning people.

(iii) "Problem Solving Ability". It is the ability to systematically address and resolve problems by Understanding the problem, devising a plan, implementing a solution and reflecting on the problem.

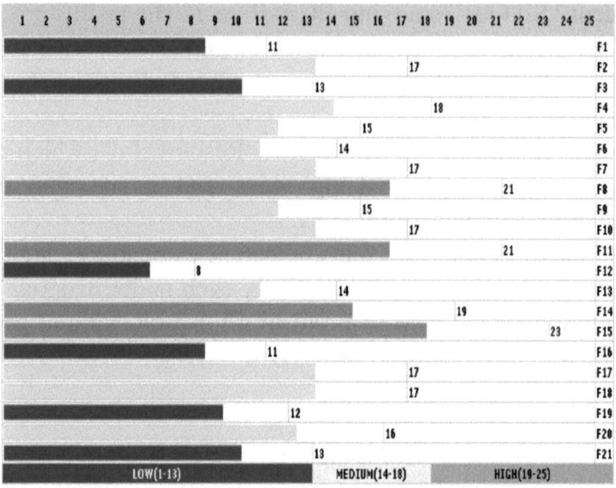

Fig. 3. Strength Assessment Report of SID029

(iv) Tendency to "Update Knowledge". Person of this attitude like to Spend time and resources to know what is happening in their professional field in a continuous manner, by membership of various bodies, magazines and indulging in self-study and formal or informal discussions etc.

Consider student “SID018” studying B.Com-III year. The results are generated using Strength Assessment Report (SAR) as shown in fig 4.

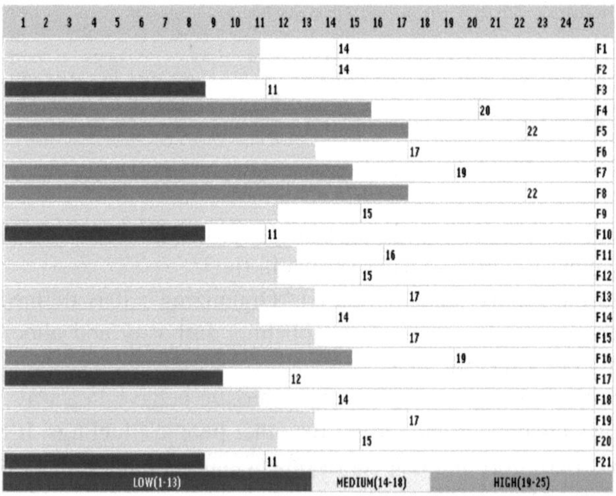

Fig. 4. Strength Assessment Report of SID018

According to the report, student was, (i) "Cheerful". Cheerful person carries a happy disposition; expressing childish joy and smiling a lot. These people see the beauty in things, which other might not notice. They have rich imagination, tell entertaining stories, and enjoy romance and fashion.
(ii) "Facing Deadlines". People with this competencies denotes a high self-motivation toward work which includes working effectively even under material constraints or if conditions are not supporting well to perform a job well without getting frustrated. These people generally meet the deadlines.
(iii) "Proactive Attitude". It is the belief that one controls key events and consequences in one's life, feeling responsible for one's own life. People govern from it do more than required, do not blame circumstances and their decisions are based on their own conscious choice.
(iv) An attitude of "Competitiveness". People govern from this attitude strive for best at the time of comparing with some standard of excellence. It involves comparison of own potential with others, efforts to pursue goals, persistence and intensity of efforts. Competitiveness means doing better and perhaps faster than others, and hopefully at lower cost.
(v) Tendency to "Update Knowledge". Person of this attitude like to Spend time and resources to know what is happening in their professional field in a continuous.

8 Analysis

In the above sections, only 30 students information is considered out of 192 students. In section 7, some student's reports are presented. But here, in the same manner, reports are generated for 192 student based on considered measures or rules.

According to the proposed approach, initially the information means details about the student are gained from database and other resources. Along with the student information, some set of rules or measures are framed which are described in section-4. According to framed rules, the student information is analyzed based on post mining process. After analyzing, the final strength assessment report (SRA) is generated. Based on each and every report of the student, some association rules are framed. Some of them are:

1. If student have "Adaptability=HIGH" and "Ability to meet deadlines=HIGH" then "Mind Quickness=HIGH".
2. If student have "Follow-up="HIGH" then "perceiving=HIGH".
3. If Student have "Problem solving skills = HIGH" then "Logical reasoning=HIGH".
4. If student have "proactive=HIGH" and "Patience=HIGH" then "organization=HIGH".
5. If student have "Resourcefulness=HIGH" then "Competitiveness=HIGH".
6. If student have "Update Knowledge=HIGH" then "Competitiveness=HIGH" and "Resourcefulness=High".
7. If student have "Politeness=HIGH" then "Empathy=HIGH".

8. If student have “Ability to work in a team=HIGH” then “Ability to meet deadline=HIGH”.
9. If student have “Cheerfulness=HIGH” then “Proactive=HIGH”.
10. If student have “Competitiveness=HIGH” then “Patience=HIGH”.

The above specified are subset of association rules which are collected from set of association rules generated according to student assessment reports.

9 Conclusion

Now-a-days, data mining process is applying in educational field also; it is called as educational data mining. The main objective of higher education institutes is to provide quality education to its students and to improve the quality of managerial decisions. Existing techniques like tree classification and some clustering techniques are suffering with decision-making problems. In this paper to solve this problem, an interactive approach is specified for pruning and filtering discovered rules. Strength Assessment report are generated in step-by-step process. Initially, discover the likelihood of student’s in different aspects i.e., behavior, attitude, requiring special attention and soon. After discovery the basic requirement, analyze the student performance according to attributes then final reports are generated. The generated reports sort the career skills or competencies that are strong/good in. According to report, student performance is enhanced by mining the association rules.

References

1. Heikki, M.: Data mining: machine learning, statistics, and databases. IEEE (1996)
2. Cios, K.J., Pedrycz, W., Swiniarski, R.W., Kurgan, L.A.: Data Mining: A Knowledge Discovery Approach. Springer, New York (2007)
3. Kovačić, Z.: Early Prediction of Student Success: Mining Students Enrolment Data. In: Proceedings of Informing Science & IT Education Conference (InSITE 2010), pp. 647–665 (2010)
4. Vandamme, J., Meskens, N.: Predicting Academic Performance by Data Mining Methods. Education Economics 15(4), 405–419 (2007)
5. Kotsiantis, S., Pierrakeas, C., Pintelas, P.: Prediction of Student’s Performance in Distance Learning Using Machine Learning Techniques. Applied Artificial Intelligence 18(5), 411–426 (2004)
6. Yu, C., DiGangi, S., Jannasch-Pennell, A., Kaprolet, C.: A Data Mining Approach for Identifying Predictors of Student Retention from Sophomore to Junior Year. Journal of Data Science 8, 307–325 (2010)
7. Cortez, P., Silva, A.: Using Data Mining to Predict Secondary School Student Performance. In: Brito, A., Teixeira, J. (eds.) EUROSIS, pp. 5–12 (2008)
8. Ramaswami, M., Bhaskaran, R.: A CHAID Based Performance Prediction Model in Educational Data Mining. IJCSI International Journal of Computer Science Issues 7(1(1)) (January 2010)
9. Al-Radaideh, Q.A., Al-Shawakfa, E.M., Al-Najjar, M.I.: Mining student data using decision trees. In: The Proceedings of the 2006 International Arab Conference on Information Technology (2006)

10. Han, J., Kamber, M.: Data Mining: Concepts and Techniques. Morgan Kaufmann (2000)
11. Galit, A., et al.: Examining online learning processes based on log files analysis: a case study. Research, Reflection and Innovations in Integrating ICT in Education (2007)
12. Ayesha, S., Mustafa, T., Sattar, A.R., Inayat Khan, M.: Data mining model for higher education system. Europen Journal of Scientific Research 43(1), 24–29 (2010)
13. Romero, C., Ventura, S., Espejo, P.G., Hervás, C.: Data mining algorithms to classify students. In: International Conference on Educational Data Mining (EDM), Montreal, pp. 8–17
14. Romero, C., Ventura, S.: Educational Data Mining: a Survey from 1995 to 2005. In: Expert Systems with Applications, pp. 135–146. Elsevier
15. Minaei-Bidgoli, B., Kashy, D.A., Kortemeyer, G., Punch, W.F.: Predicting student performance: an application of data mining methods with the educational web-based systems. In: LONCAPA, 33rd ASEE/IEEE Frontiers in Education Conference, Boulder, pp. 13–18
16. Zekić-Sušac, M., Frajman-Jakšić, A., Drvenkar, N.: Neuron Networks and Trees of Decision-making for Prediction of Eficiency in Studies. Ekonomski Vjesnik (2), 314–327
17. Kumar, S.A., Vijayalakshmi, M.N.: Efficiency of Decision Trees in Predicting Student's Academic Performance. In: First International Conference on Computer Science, Engineering and Applications, CS and IT 2002, Dubai, pp. 335–343 (2002)
18. Sun, H.: Research on Student Learning Result System based on Data Mining. IJCSNS International Journal of Computer Science and Network Security 10(4) (April 2010)
19. Khan, Z.N.: Scholastic achievement of higher secondary students in science stream. Journal of Social Sciences 1(2), 84–87 (2005)
20. Siraj, F., Abdoulha, M.A.: Uncovering hidden Information within University's Student Enrollment Data using Data Mining. In: Third Asia International Conference on Modeling and Simulation (2009)
21. Klosgen, W., Zytkow, J.: Handbook of data mining and knowledge discovery. Oxford University Press, New York
22. Wu, X., Kumar, V.: The Top Ten Algorithms in Data Mining. Chapman and Hall, Boca Raton
23. Fayadd, U., Piatesky, Shapiro, G., Smyth, P.: From data mining to knowledge discovery in databases. AAAI Press / The MIT Press, Massachusetts Institute Of Technology (1996) ISBN –262 56097–6
24. Hijazi, S.T., Naqvi, R.S.M.M.: Factors affecting student's performance: A Case of Private Colleges. Bangladesh e-Journal of Sociology 3(1) (2006)
25. Bean, J.P., Metzner, B.S.: A conceptual model of nontraditional undergraduate student attrition. Review of Educational Research (1985)
26. Murtaugh, P., Burns, L., Schuster, J.: Predicting the retention of university students. Research in Higher Education 40(3), 355–337
27. Rokach, L., Maimon, O.: Data mining with decision trees – Theory and applications. World Scientific Publishing, New Jersey

Analysis of Stutter Signals with Subsequent Filtering and Smoothening

Mithila Harish and M. Monica Subashini

School of Electrical Engineering, VIT University, Vellore, Tamil Nadu
{mitziyer,monicasubashini.m}@gmail.com

Abstract. The problems of communication disorders are many. The individual who suffers from such a disorder, such as stuttering, faces difficulty in getting his or her point across. Communication is a holistic process which spans multiple levels. An error in any one level can lead to misunderstanding and may even result in severe repercussions. People who stutter have a disadvantage. The time lag between what a person without this condition says and what a person with this condition says is appreciable, as stuttering causes many words, vowels and fillers to be repeated. This paper suggests a method for improving communicability of stutter signals obtained from audio recording. Under the method suggested, audio signals are read and spliced into different portions depending on the length of the given signal. Presence of stutter type repetitions are assessed by applying loops. Repeated signals, if present, are eliminated using windowing techniques. In totality this results in the smoothening out of the signal and removing disfluency-inducing repetitions.

Keywords: Audio Processing, Stuttering, PSD, Splicing, Windowing.

1 Introduction

Communication is a complex process which involves steps including the sender, encoding, decoding, media and receiver. A misstep in one process can cause serious misunderstanding of the message passed.

Stuttering is a serious medical condition that causes great difficulty to the speaker. The causes of stuttering are not known, but there have been studies linking it to brain activity [1] [2]. No known cure is available for this disorder. Progress is possible through speech therapy and in some cases, passing of time. There are cases of people overcoming this problem, but there are also people who suffer grievously without improvement.

In this work, a novel technique for removing stutter words and repetitions is found. This cannot be a replacement, but it is hoped that further analysis may lead to further smoothening of these signals which may be a possible aid for these people who can use this technique for recording their speeches which are, for example, intended for an audience.

S.C. Satapathy et al. (eds.), *ICT and Critical Infrastructure: Proceedings of the 48th Annual Convention of CSI - Volume II*, Advances in Intelligent Systems and Computing 249,
DOI: 10.1007/978-3-319-03095-1_8,

1.1 Existing Difficulty

Speech filtration is a complicated process, which includes various steps. Currently, barring a few novel discoveries, there is no significant technique which filters out stutter signals in an efficient manner. The problem lies in identifying the repetitions- the stutter words being repeated by the speaker. What word is being repeated, and how does one analyse it, was the main question.

Methods like windowing must be done carefully. If the wrong signal is windowed, the whole process may collapse. The technique presented in the work uses windowing in a continuous loop, adjusted so that the signal is filtered out carefully.

2 Methodology

Step One: The first step is in obtaining the signal from the sender. This is done where the speaker directly speaks into an audio recorder program created in MATLAB [3]. Alternatively, this is derived from audio processing software such as Audacity, which can be then input into the MATLAB program.

A snippet of MATLAB code is given below [4]:

```
%sample for Audio Processing
recObj = audiorecorder(44100, 16, 2);
get(recObj)
% Record your voice for 5 seconds.
recObj = audiorecorder;
disp('Start speaking.')
recordblocking(recObj, 5);
```

This can be adjusted accordingly for the usage. If time is user defined, then an input from the user can be given so that the recording can be done for the appropriate amount of time.

Step Two: The input signal, after being read, is stored as a wav file in a temporary location.

Step Three: Now the signal is ready for manipulation. The signal is spliced into smaller segments based on time. This is based on the length of the desired segments that can be determined by the user for more accuracy. The sampling frequency is set at 44100 Hz.

Step Four: The signal is continuously scanned.

2.1 Background Theory

Power Spectral Density: This is a highly useful computation that gives the power representation of a spectrum. In this method, the PSD [5] was taken for each spliced section of the signal. Since the time frames were small, the peak frequency was taken to be the representative frequency, which was then compared to the previous spliced

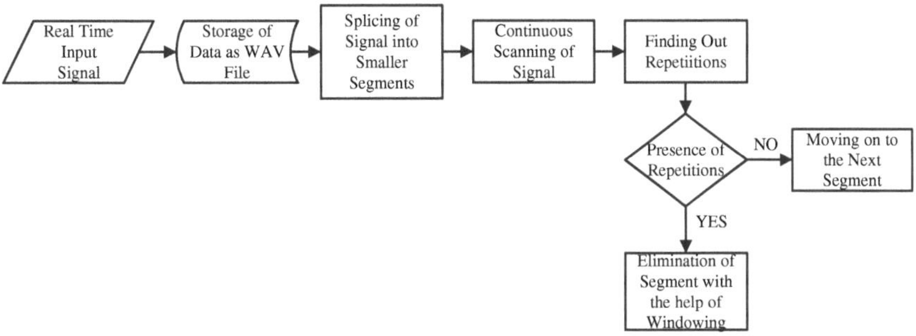

Fig. 1. The flowchart represents the methodology

section of the signal. The loop proceeds in this direction. The following equation represents the mathematical expression for the total power of a sequence x(t).

$$P = \lim \frac{1}{2T} \int_{-T}^{T} x(t)^2 dt \tag{1}$$

where P= power and T=Time Period. The PSD can be represented as in equation (2):

$$S_{xx}(w) = \lim_{T \to \infty} E[|z_T(w)|^2] \tag{2}$$

$$z_T(w) = \frac{1}{\sqrt{T}} \int_0^T x(t) e^{-iwt} dt \tag{3}$$

$z_T(w)$ is the Truncated Fourier Transform depicted in equation (3). Peak frequency is obtained after this, again in a loop.

2.2 Windowing Techniques

Windows in signal processing can be of various types. These include, but are not restricted to, the Rectangular window, the B-Spline Window, the Triangular Window, the Hamming Window and the Hanning Window.

In this method, we have used the Rectangular Window [6].

Rectangular Window: This is one type of window, which is defined as being 1 or 0 depending on the specifications of the signal to be windowed. It makes the data look turned on and turned off.

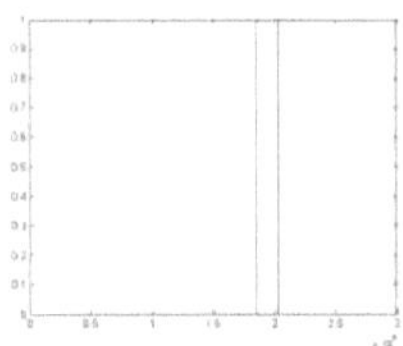

Fig. 2. This represents a version of a Rectangular Window

MATLAB Code: (Sample)

```
if (n<=ms2)||(n>=ms20)
    w(n)=0;
else w(n)=1;
```

where ms2 and ms20 are frequencies associated with the start and end points of the spliced signal, respectively.

The procedure for identifying repetitions is as follows:

(a) The Power Spectral Density (PSD) of each segment is calculated and the peak frequency of the segment is derived using the calculated PSD. Since the samples are small, the peak frequency is the distinguishing characteristic of each sample.

(b) The peak frequency is stored in a temporary location.

(c) The same procedure is repeated for the next segment, where its peak frequency is found out. This is stored in another temporary location.

(d) The values of the two peak frequencies are compared.

(e) If they are found to be present within a given margin, 25 %, the segment containing the repetition is eliminated. This is done with the help of windowing.

Windowing: The windows used are rectangular windows.

W(n)=1 throughout the length of the sample being scanned;

W(n)=0 before and after this sample;

The window and the original signal are multiplied on an element-by-element basis. The resulting signal will be the original signal without the repetition. If they are not found to be within this given margin, the analysis for the next sample starts. This is carried out throughout the length of the original sample.

3 Observation

The following is the speech signal obtained consisting of stuttering of two words – *lucky* and *me*. The speech signal is *'Lll..lll...lll....llll...lucky mm..mmm...mmm... mmm...Me'*

The wav signal of the speech input is highlighted in Fig.3:

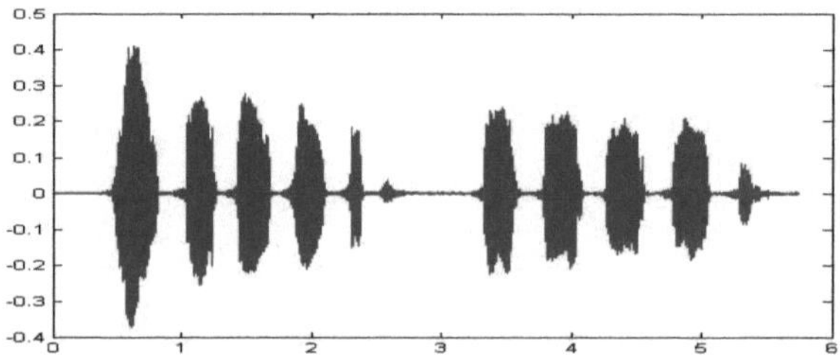

Fig. 3. This displays the input wav signal used

The spliced signals generated after each subsequent loop are shown in Table 1, where Figures 4(a) to 4(j) represent each spliced segment. The variations in the signals can be noted. For instance, Figure 4(g) is mostly uninterrupted, depicting a silence segment, whereas 4(i) is clearly a speech segment.

Table 1. Spliced Signals Generated

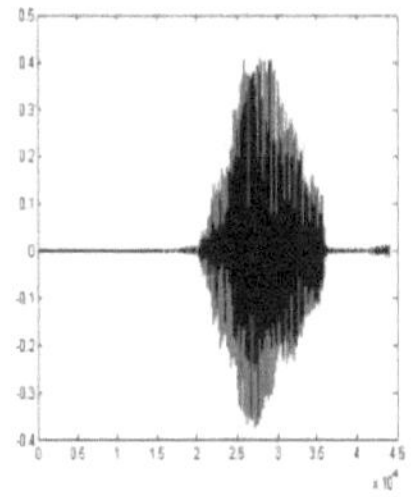

Fig. 4(a)

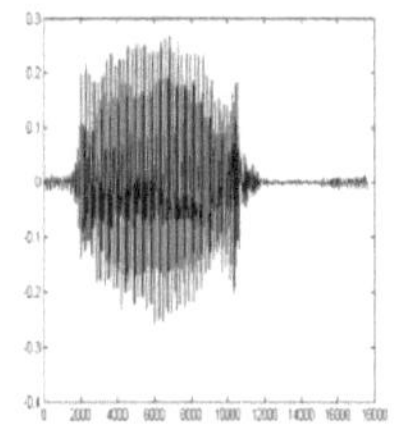

Fig. 4(b)

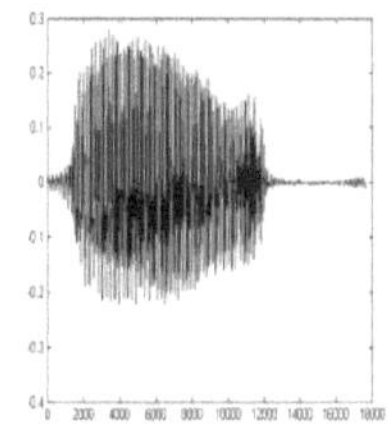

Fig. 4(c)

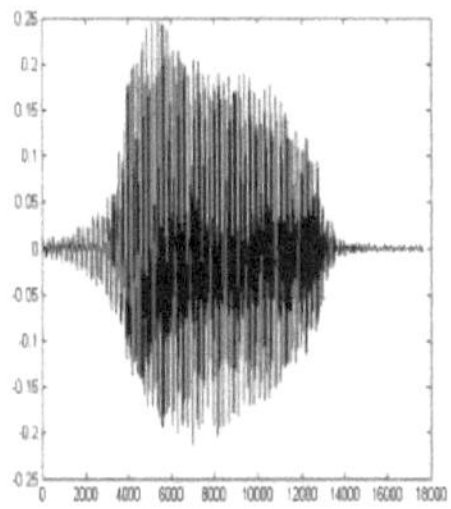

Fig. 4(d)

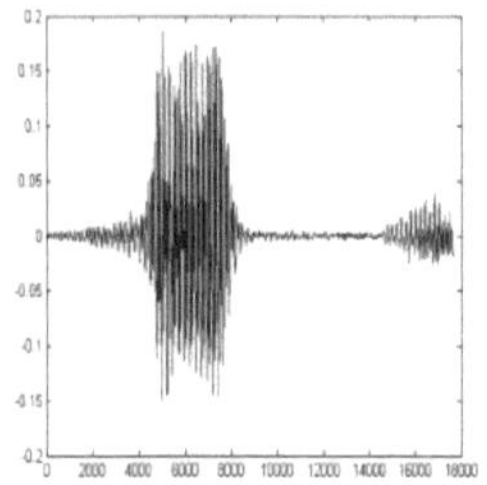

Fig. 4(e)

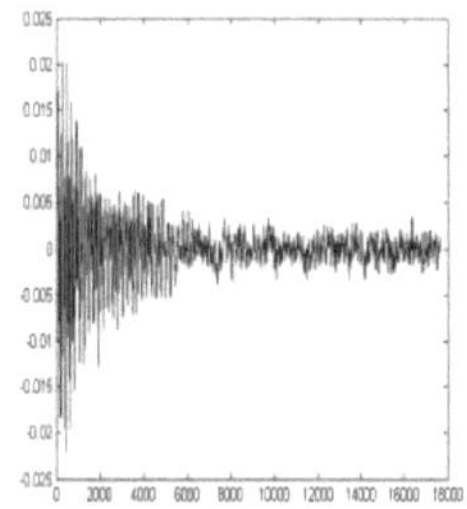

Fig. 4(f)

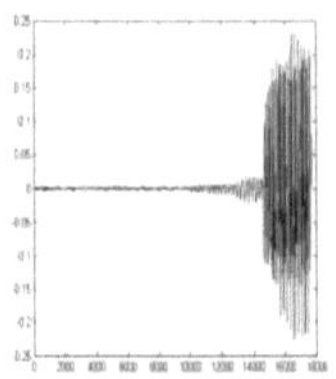

Fig. 4(g)

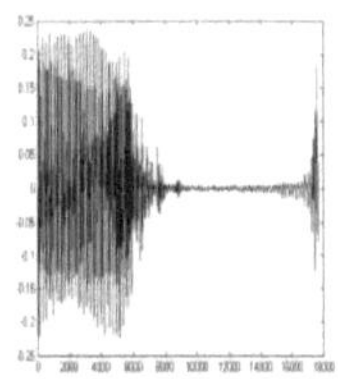

Fig. 4(h)

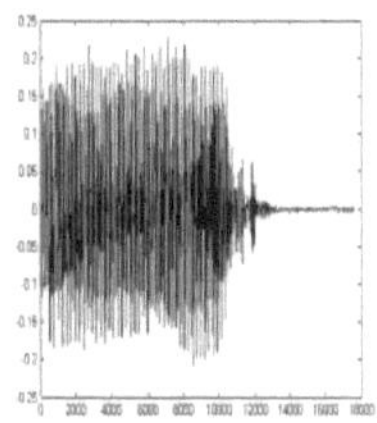

Fig. 4(i)

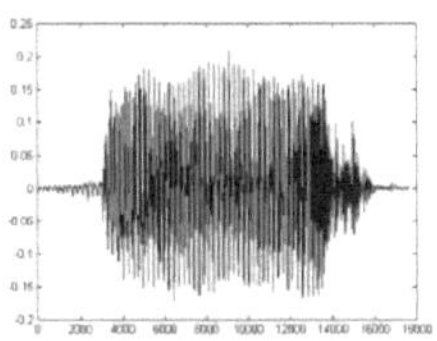

Fig. 4.(j)

Table 2, with its Figures 5(a) to 5(g), depicts the rectangular windows used. Each window has its value as either '1' or '0', depending upon the area of the speech signal in which it is to act. The duration is continuously altered, based on the loop. After each iteration, the 't' and 't1' values, which represent the start and end time, respectively, are incremented by the chosen amount so that the entire portion of the speech signal can be covered. If it is not required, the signal simply moves on to the next iteration, disregarding the current window generated. Extreme care was taken in the windowing section, as even slight shifts can change the output signal considerably. Excessive shifting can result in large errors being generated. The start and end times of the window are clearly defined so that ambiguity is removed.

Table 2. Rectangular Window Functions Generated

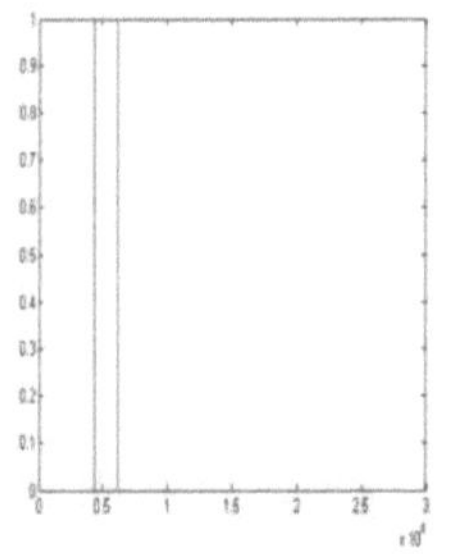 **Fig.** 5(a)	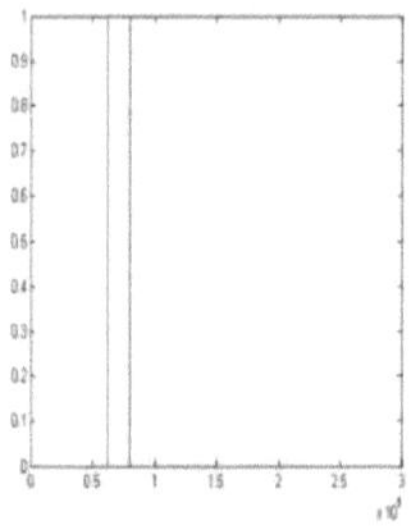 **Fig.** 5(b)	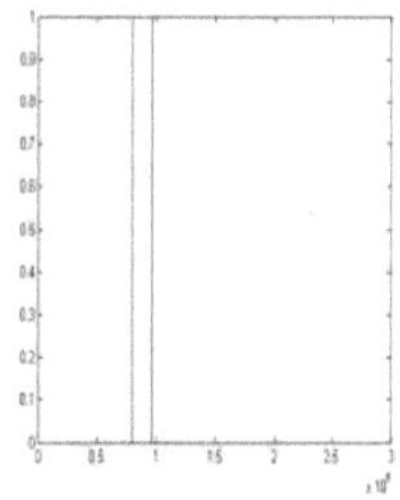 **Fig.** 5(c)
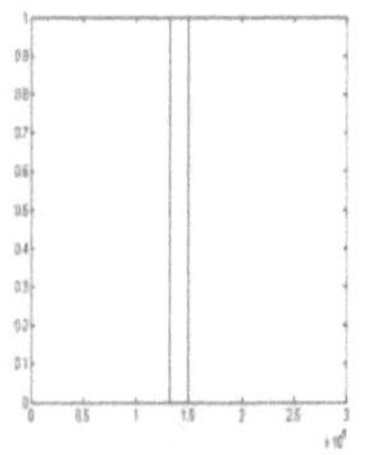 **Fig.** 5(d)	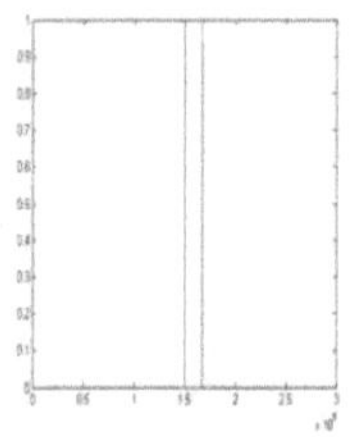 **Fig.** 5(e)	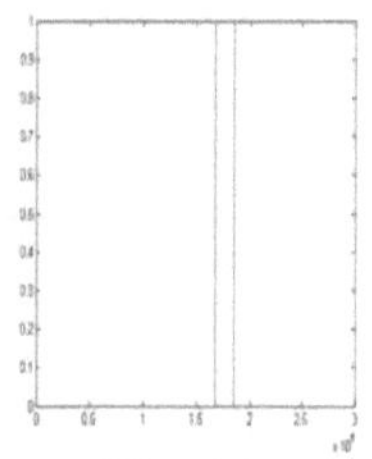 **Fig.** 5(f)
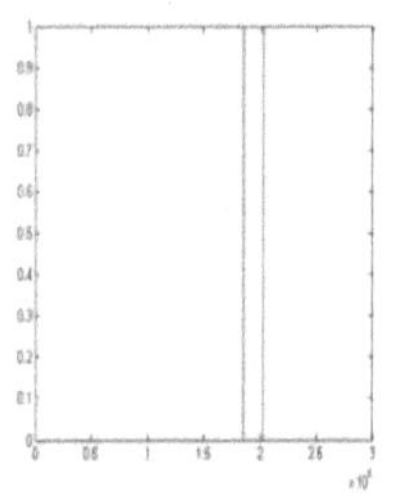 **Fig.** 5(g)		

Table 3 depicts the signals after windowing. Figures 6(a) to 6(e) show the results of the signal changing at each stage after windows are used. As can be noted, the signal is progressively smoothened. Each window helps in eliminating one stutter repetition. The more the signal is segmented, with the time duration of each spliced portion being reduced, the better will be the result.

Table 3. Results after Windowing at each stage

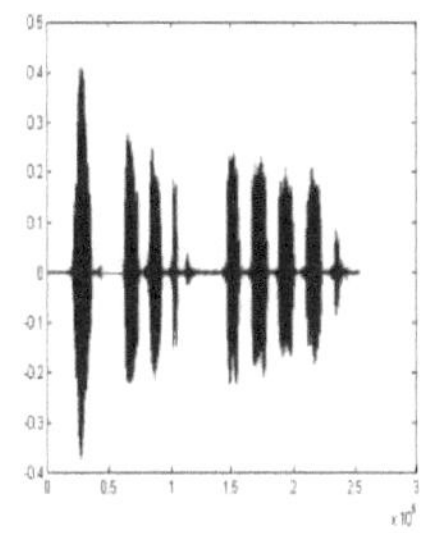

Fig. 6(a)

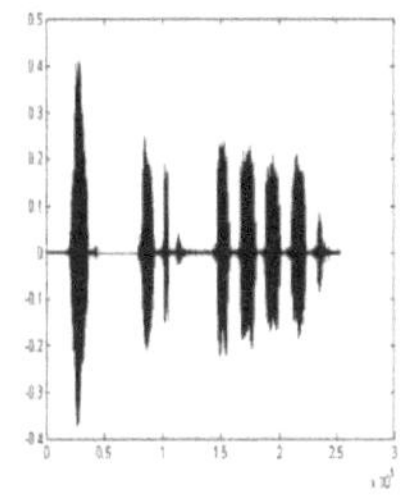

Fig. 6(b)

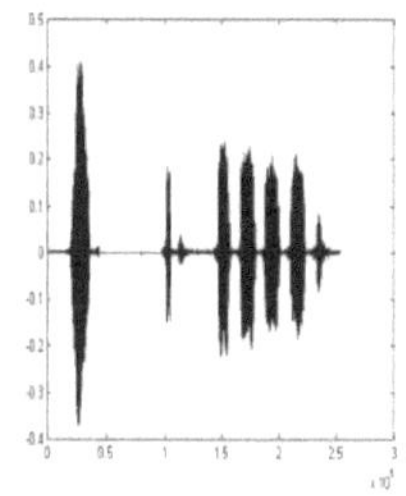

Fig. 6(c)

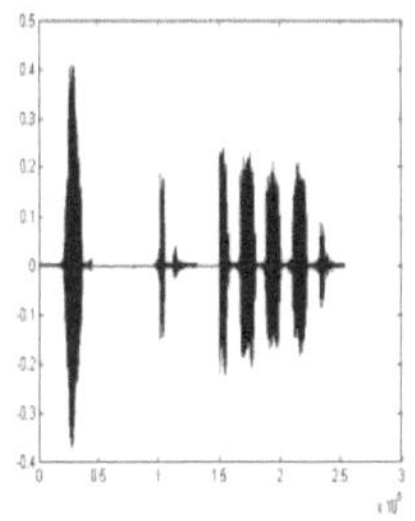

Fig. 6(d)

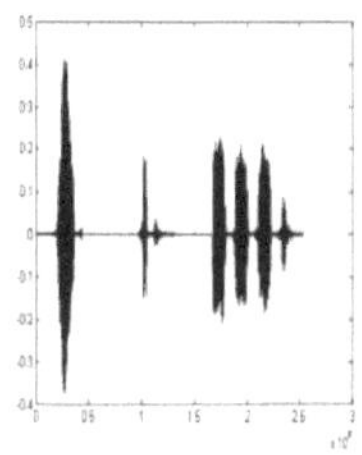

Fig. 6(e)

Finally, Figures (7) and (8) display the initial input and the final output, respectively, for comparison purposes.

Initial versus Final Output:

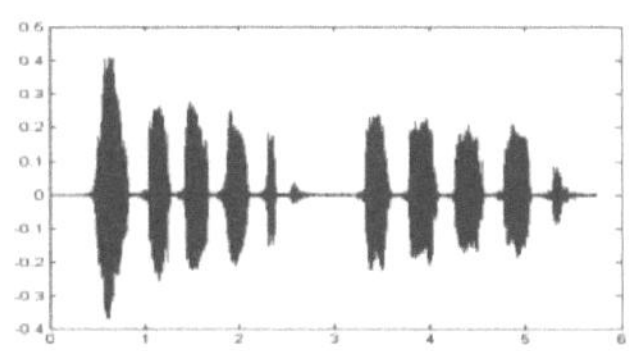

Fig. 7. Initial Input Wav Signal

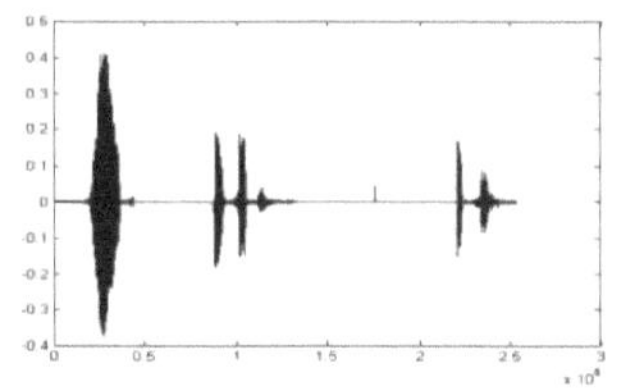

Fig. 8. Final Output Wav Signal

It is apparent from Figures 7 and 8 that the method suggested in this paper for removal of repeated sounds of the stutter type results in a fair amount of smoothening and significant reduction in repeated [stutter] signals.

The exercise described herein is able to obtain a final output signal of '*Lll...lucky...m...me*', which is a considerable improvement from the original input signal - '*Lll..lll...lll....llll...lucky mm..mmm...mmm...mmm...Me*' - which had multiple repetitions. The final result is noticeably smoother than the input signal which consisted of many more "jerks".

4 Conclusion

Smoothening of signals with multiple, stutter-like repetitions was performed. This was accomplished by multiple splicing of the signal and comparison of the spliced signals on the basis of peak values in their respective Power Spectral Densities (PSD).

Communication is an integral part of our lives. The method described herein hopes to be an aid for people with severe stuttering to smoothen their speech. It can also be a device that helps in speech therapy of people with this disorder, by continuously replaying the smoothened signal so they can practise talking without stuttering.

This method successfully removes the manual component, i.e. apart from obtaining an input feed from an external source [software such as Audacity, recorded speech from MATLAB or from other audio recorder programs], everything else is done by the program. This especially can be used to aid individuals without a technical background, who can feed their input into the program and generate a smoothened output that can help them. It is sincerely hoped that individuals with severe communication disorders can benefit from methods like these.

5 Future Work

An extension of the process would be in using the PSD obtained using the Yule-Walker algorithm, which uses autoregression. This could provide improved efficiency by efficient comparison between the spliced signal, in addition to the method presented in this method. This can be further improved by adding parameters such as Hidden Markov Models(HMMs).

Acknowledgements. The work is supported by School of Electrical Engineering, VIT University, Vellore, India.

References

1. Webster, W.G.: Neural mechanisms underlying stuttering: Evidence from bimanual handwriting performance. Brain and Language 33(2), 226–244 (1988)
2. Ingham, R.J., Ingham, J.C., Finns, P., Fox, P.T.: Towards a functional neural systems model of developmental stuttering. Journal of Fluency Disorders 28(4), 297–318 (2003)
3. Kuo, S.M., Lee, B.H., Tian, W.: Real-Time Digital Signal Processing: Implementations and Applications. Wiley, New Jersey (2006)
4. Mathworks Audiorecorder, `http://www.mathworks.in/help/matlab/ref/audiorecorder.html`
5. Stoica, P., Moses, R.L.: Introduction to Spectral Analysis. Prentice Hall, New Jersey
6. Semmlow, J.L.: Biosignal and Biomedical Image Processing: Matlab-Based Applications. Taylor & Francis, London (2004)

Fingerprint Reconstruction: From Minutiae

B. Amminaidu and V. Sreerama Murithy

Department of Computer Science and Engineering
GMR Institute of Technology, Rajam, AP, India
ammicse06@gmail.com,
sreeramamurthy.v@gmrit.org

Abstract. Fingerprint is one of the very important biometric features used to identify humans across the globe. Minutiae based representation is the most widely used fingerprint representation scheme among other schemes available. Since minutiae representation is a compacted one, there has been an impression that the minutiae template does not contain sufficient information to reconstruct the original grayscale fingerprint image. This misconception has been proven to be incorrect, several algorithms have been proposed that can reconstruct fingerprint images from minutiae templates. But all these algorithms have one common drawback that many spurious minutiae, which are not included in the original minutiae template are generated in the reconstructed image. Moreover, some of these techniques can only reconstruct a partial fingerprint. In this paper, a novel fingerprint reconstruction algorithm is proposed, which not only reconstructs the whole fingerprint, but the reconstructed fingerprint contains very few spurious minutiae. The proposed algorithm reconstructs the continuous phase from minutiae. Experimental results have shown that the proposed algorithm reconstructs whole fingerprint. It also contains very few spurious minutiae.

Keywords: Fingerprint, fingerprint reconstruction, phase image, minutiae, and orientation field.

1 Introduction

Fingerprint recognition systems play a vital role in several situations where a person needs to be verified or identified with high confidence. As a result of the interaction of genetic factors and embryonic conditions, the friction ridge pattern on fingertips is unique to each finger. Fingerprint features are generally categorized into three levels (as shown in Fig. 1, level 3 is not shown because this paper does not deal with it):

1. Level 1 feature mainly refers to ridge orientation field and features derived from it, i.e., singular points and pattern type.
2. Level 2 features refer to ridge skeleton and features derived from it, i.e., ridge bifurcations and endings.
3. Level 3 features include ridge contours, position, and shape of sweat pores and incipient ridges.

S.C. Satapathy et al. (eds.), *ICT and Critical Infrastructure: Proceedings of the 48th Annual Convention of CSI - Volume II*, Advances in Intelligent Systems and Computing 249,
DOI: 10.1007/978-3-319-03095-1_9, © Springer International Publishing Switzerland 2014

Most fingerprint matching systems are based on four types of fingerprint representation schemes (Fig. 2): grayscale image [1], phase image [2], skeleton image.

[3], [4], and minutiae [5], [6]. Due to its distinctiveness, compactness, and compatibility with features used by human fingerprint experts, minutiae-based representation has become the most widely adopted fingerprint representation scheme.

(a) (b) (c)

Fig. 1. Features at three levels in a fingerprint. (a) Grayscale image, (b) Level 1 feature (orientation field), (c) Level 2 feature (ridge skeleton).

Fingerprints are the graphical flow-like ridges present on human fingers. Finger ridge configurations do not change throughout the life of an individual except due to accidents such as damages and cuts on the fingertips. This unique property makes fingerprints a very eye-catching biometric identifier. Fingerprint-based personal identification has been used for a very long time [7]. Fingerprint reconstruction from minutiae (hereinafter simply referred to as fingerprint reconstruction) is very similar to fingerprint synthesis [8] except that the goals and the inputs of the two techniques are different. The goal of fingerprint reconstruction is to acquire an artificial fingerprint that resembles the original fingerprint as close as possible, while the goal of fingerprint synthesis is to generate artificial fingerprints that are as realistic as possible. For fingerprint reconstruction, the minutiae from a given fingerprint must be provided, while for fingerprint synthesis, no input is needed (except for statistical models learned from many real fingerprints). In this paper, a novel approach to fingerprint reconstruction from minutiae template is proposed, which uses enhancement and adaptive thresholding. Authors have excerpted the major conceptions from [reference 9 and 10].

2 Fingerprint Representation

Larkin and Fletcher [11] proposed to represent a fingerprint image as 2D amplitude and frequency modulated (AM-FM) signal:

$$I(x,y) = a(x,y) + b(x,y)\cos\left(\Psi(x,y)\right) + n(x,y) \tag{1}$$

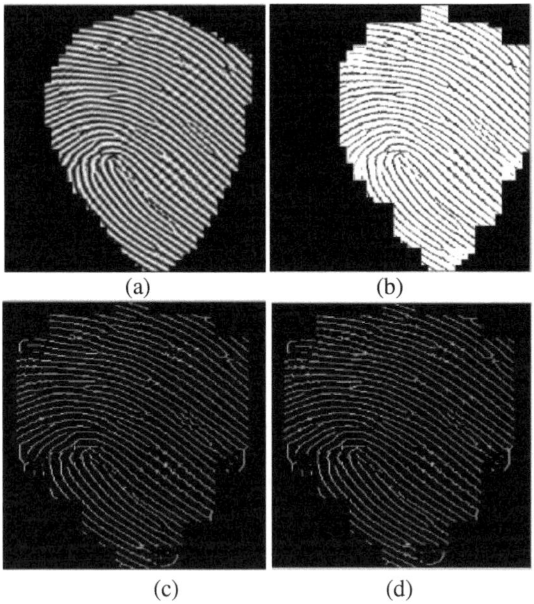

Fig. 2. Fingerprint representation schemes. (a) Grayscale image (b) Phase image (c) Skeleton image and (d) Minutiae.

which is composed of four components: the intensity offset $a(x,y)$, the amplitude $b(x,y)$, the phase $\Psi(x,y)$, and the noise $n(x,y)$. Here we are only interested in the phase $\Psi(x,y)$, since ridges and minutiae are totally determined by the phase; the other three components just make the fingerprint appear realistic. Therefore, an ideal fingerprint is represented as a 2D FM signal:

$$I(x,y) = \cos(\Psi(x,y)). \tag{2}$$

According to the Helmholtz Decomposition Theorem [12], the phase can be uniquely decomposed into two parts: the continuous phase and the spiral phase:

$$\Psi(x,y) = \Psi c(x,y) + \Psi s(x,y) \tag{3}$$

The gradient of the continuous phase $\Psi c(x,y)$ is termed as instantaneous frequency $G(x,y)$. The direction of instantaneous frequency is normal to ridge orientation. The amplitude of instantaneous frequency is equal to the ridge frequency.The spiral phase $\Psi s(x,y)$ corresponds to minutiae:

$$\psi_s(x,y) = \sum_{n=1}^{N} p_n \arctan\left(\frac{y-y_n}{x-x_n}\right) \tag{4}$$

Where x_n And y_n denote the coordinates of the *n*th minutia, and *N* denotes the total number of minutiae. The direction of a minutia is determined by its polarity $p_n \in \{1,-1\}$ and the local ridge orientation O (x_n, y_n), which is defined in the continuous phase. Assume the ridge orientation is in the range [0, π]. The direction of a minutia is equal to O (x_n, y_n) when it has positive polarity, or O (x_n, y_n)+π when it has negative polarity. Adding spiral to a continuous phase generates minutiae.

3 Fingerprint Reconstruction

3.1 Problem Statement

A set of *N* minutiae $\{x_n, y_{n,} \alpha_n\}$, $1 \le n \le N$ of a fingerprint is given, where (x_n, y_n) an α_n denote the location and direction of the nth minutia, respectively. FM model requires the spiral phase and the direction of instantaneous frequency of the continuous phase, which is known at the locations of the *N* minutiae. The real problem is to reconstruct the original fingerprint image as given in Eq. (1). This is clearly an ill-posed problem, because ridge frequency is not known which is the important information required to reconstruct the continuous phase of fingerprints. Moreover, information needed to reconstruct realistic fingerprints, such as brightness, contrast, the background noise of fingerprint sensor, and detailed ridge features (pores, contours) and so on are also not available. Therefore, a more practical way is to estimate the FM representation of the original fingerprint, $\cos(\psi(x,y))$. To obtain the phase $\Psi(x, y)$, the following three steps are to be performed: orientation field reconstruction, continuous phase reconstruction, and combination of the spiral phase and the continuous phase. The flow chart of the proposed fingerprint reconstruction algorithm is depicted in Fig. 3.

3.2 Orientation Field Reconstruction

Ross et al. [13] used selected minutiae triplets to estimate the orientation field in triangles. Cappelli et al. [14] estimated orientation field by fitting an orientation field model to the orientations at minutiae. Both these methods have a minimum requirement on the number of minutiae. We propose a novel orientation field reconstruction algorithm that can work even when only one minutia is available. Figure 4 shows the reconstructed orientation field for the four main types of fingerprints

The image is divided into non-overlapping blocks of 8*8 pixels. Foreground mask for the fingerprint image is obtained by dilating the convex hull of minutiae using a diskshaped structuring element of 8 * 8 pixels. The local ridge orientation at block (*m,n*) is predicted by using the nearest minutia in each of the 8 sectors. The minutia direction α_k is doubled to make α_k equivalent to α_k +π The cosine and sine components of 2α_k of all the *K* selected minutiae are summed:

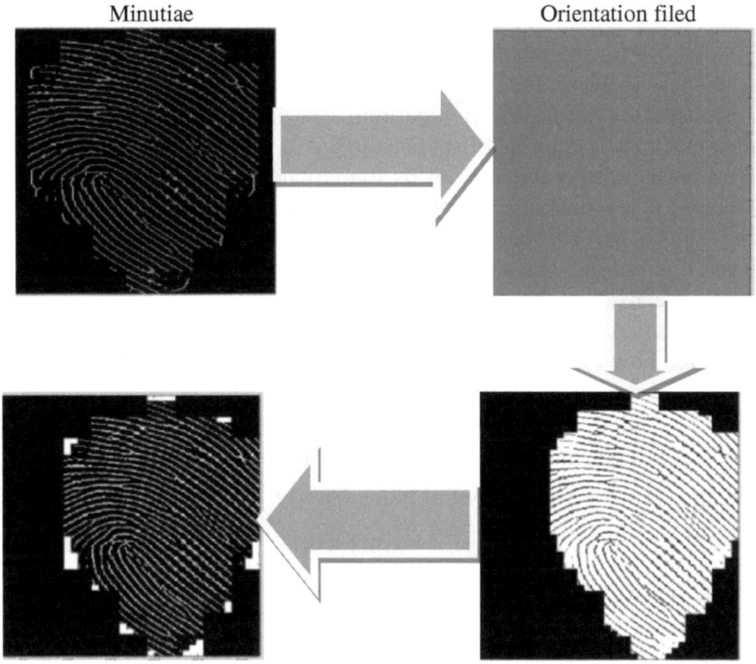

Fig. 3. Flow chart of the proposed fingerprint reconstruction algorithm

$$u = \sum_{K=1}^{K} \frac{\cos(2\alpha_k)}{d_k}, v = \sum_{K=1}^{K} \frac{\sin(2\alpha_k)}{d_k} \tag{5}$$

Where d_k denotes the Euclidean distance between the block center and the k th minutia. Then the orientation at block (m, n) is computed as: $O(m,n) = \frac{1}{2}\arctan(\frac{v}{u})$

3.3 Continuous Phase Reconstruction

The continuous phase of a fingerprint is modeled by piecewise planes at each foreground block (m, n) of 8 * 8 pixels:

$$\begin{aligned} &\varphi_c(x, y) = 2\pi.F(m,n).(\cos(O(m,n))x + \\ &\sin(O(m,n))y) + P(m,n), 8(m-1) \le x < 8m, \\ &8(n-1) \le y < 8n, \end{aligned} \tag{6}$$

where $F(m, n)$, $O(m, n)$ and $P(m, n)$ denote the ridge frequency, the ridge orientation and the phase offset at block (m, n), respectively. Since it is not possible to estimate the ridge frequency from minutiae (if the ridge count information between minutiae is provided, then it is possible to estimate the ridge frequency), we have used a constant frequency value 0.1 for the whole image, which corresponds to a ridge period of 10 pixels in 500 ppi images. The only unknown value in Eq. (6), the phase offset *P(m,n)*,is estimated by the following algorithm .Starting with a queue containing the top left-most block (whose phase offset is assumed to be 0), in each iteration, a block is obtained from the queue and each of its four-connected neighbors is checked if it has been reconstructed (namely, the phase offset has been estimated). If one of the neighboring blocks has not been reconstructed, the phase offset of this block is estimated and it is put into the queue. This procedure is performed until the queue is empty (which means that the continuous phase has been reconstructed at all the foreground blocks). An ancillary image is used to record the reconstructed blocks. Here we describe how to estimate the phase offset of a block using all of the already reconstructed four-connected neighbors. Consider one of the neighbors, say block (*m ,-1, n*), of block (*m, n*). The phase images of these two blocks are required to be continuous at the border pixels $\{(x, y): x = 8(m-1), 8(n-1) \le y < 8n\}$ For each border pixel (*x, y*), a phase offset of block (*m, n*) is estimated:

$$\begin{aligned} \varphi = 0.2\pi.(\cos(O(m-1,n)).x + \sin(O(m-1,n)).y) + P(m-1,n) \\ -0.2\pi.(\cos(O(m,n)).x + \sin(O(m,n)\,y) \end{aligned} \quad (7)$$

The mean value is used as the phase offset of block (*m, n*).

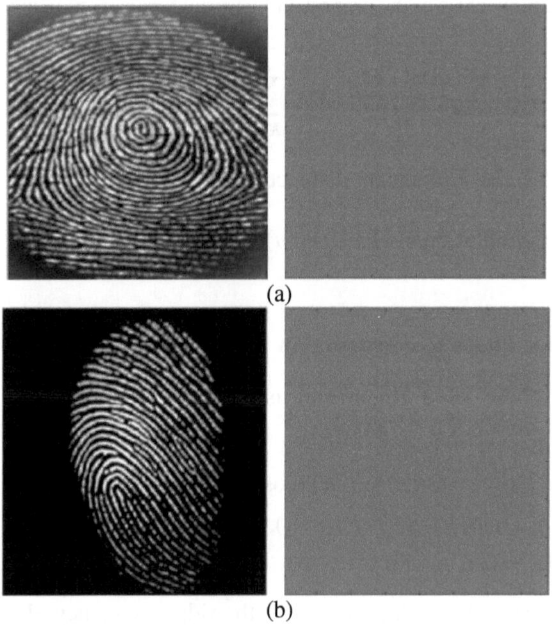

Fig. 4. Reconstructed orientation field for the two main types of fingerprints: (a) whorl and (b) right loop

4 Experimental Results

The environment used to build the application which implements the proposed algorithm is a PC with 4 GB of RAM and Core 2 dual processor. MATLAB image processing API is used to achieve fingerprint reconstruction. Reconstruction image of the whorl fingerprint shown in figure 5.

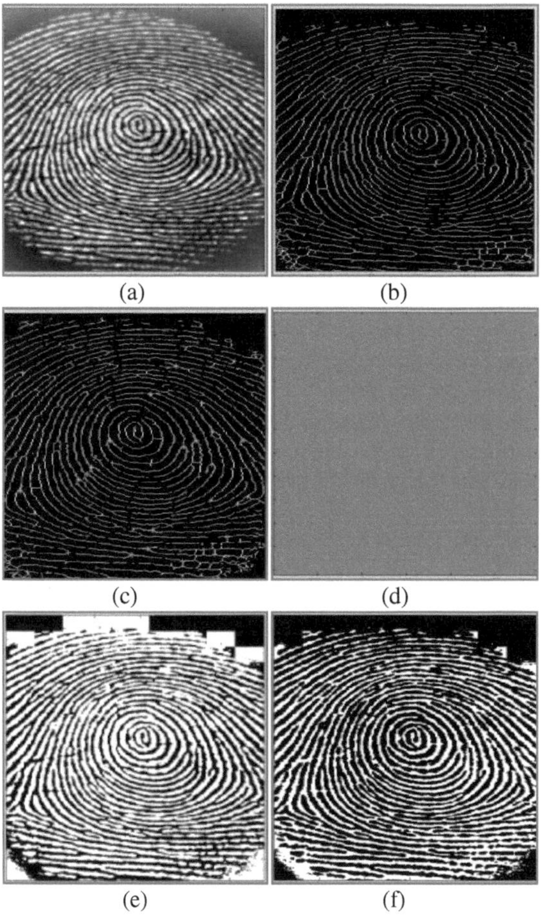

Fig. 5. Expermental results for whorl fingerprint image,(a)Grayscal image, (b)Thinning image,(c)Extraction of Minutiae,(d)Orientation Flow,(e)Phase Image, (f)Reconstructed image

5 Conclusions and Future Work

In this paper, we proposed a novel fingerprint reconstruction algorithm can be used to reconstruct fingerprint images from minutiae representation scheme. Out of the representation schemes such as grayscale image, phase image, skeleton image and minutiae, the minutiae is widely used representation in fingerprint recognition systems. However, its compactness led the people believe that it has no sufficient information to reconstruct the whole fingerprint image. Many algorithms in the literature and the proposed algorithm in this paper proved that the minutiae representation can be used to accurately reconstruct original fingerprint image. The experiments revealed that it is very effective.

References

[1] Bazen, A.M., Verwaaijen, G.T.B., Gerez, S.H., Veelenturf, P.J., van der Zwaag, B.J.: A Correlation-Based Fingerprint Verification System. In: Proc. 11th Ann. Workshop Circuits Systems and Signal Processing, pp. 205–213 (November 2000)

[2] Thebaud, L.R.: Systems and Methods with Identity Verification by Comparison and Interpretation of Skin Patterns Such as Fingerprints. US Patent No. 5,909,501 (1999)

[3] Feng, J., Ouyang, Z., Cai, A.: Fingerprint Matching Using Ridges. Pattern Recognition 39(11), 2131–2140 (2006)

[4] Hara, M., Toyama, H.: Method and Apparatus for Matching Streaked Pattern Image. US Patent No. 7,295,688 (2007)

[5] Ratha, N.K., Bolle, R.M., Pandit, V.D., Vaish, V.: Robust Fingerprint Authentication Using Local Structural Similarity. In: Proc. Fifth IEEE Workshop Applications of Computer Vision, pp. 29–34 (2000)

[6] Bazen, A.M., Gerez, S.H.: Fingerprint Matching by Thin-Plate Spline Modelling of Elastic Deformations. Pattern Recognition 36(8), 1859–1867 (2003)

[7] Lee, H.C., Gaensslen, R.E. (eds.): Advances in Fingerprint Technology. Elsevier, New York (1999)

[8] Maltoni, D., Maio, D., Jain, A.K., Prabhakar, S.: Hand book of Fingerprint Recognition. Springer (2003)

[9] Feng, J., Jain, A.K.: Fingerprint Reconstruction: From Minutiae to Phase. IEEE Transactions on Pattern Analysis And Machine Intelligence 33(2) (February 2011)

[10] Feng, J., Jain, A.K.: FM Model Based Fingerprint Reconstruction from Minutiae Template. In: Proc. Second Int'l Conf. Biometrics, pp. 544–553 (June 2009)

[11] Larkin, K.G., Fletcher, P.A.: A Coherent Framework for Fingerprint Analysis: are Fingerprints Holograms? Optics Express 15(14), 8667–8677 (2007)

[12] Ghiglia, D.C., Pritt, M.D.: Two-Dimensional Phase Unwrapping: Theory, Algorithms, and Software. John Wiley and Sons, NewYork (1998)

[13] Ross, A., Shah, J., Jain, A.K.: From template to image: Reconstructing Fingerprints from Minutiae Points. IEEE Trans. Pattern Analysis and Machine Intelligence 29(4), 544–560 (2007)

[14] Cappelli, R., Lumini, A., Maio, D., Maltoni, D.: Fingerprint Image Reconstruction from Standard Template. IEEE Trans. Pattern Analysis and Machine Intelligence 29(9), 1489–1503 (2007)

Performance Analysis of Asynchronous Periodic Pattern Mining Algorithms

G.N.V.G. Sirisha[1], Shashi Mogalla[2], and G.V. Padma Raju[1]

[1] Department of CSE, S.R.K.R. Engineering College, Bhimavaram
{sirishagadiraju,gvpadmaraju}@gmail.com
[2] Department of CS & SE, A.U. College of Engineering, Visakhapatnam
smogalla2000@yahoo.com

Abstract. Periodic pattern is a pattern that repeats itself with a specific period in a given sequence. Patterns that occur frequently with strict periodicity in one or more subsequences separated by tolerable disturbance are called asynchronous periodic patterns. Longest Subsequence Identification (LSI) is the pioneering algorithm to mine asynchronous periodic patterns. For each asynchronous periodic pattern the algorithm detects the longest subsequence containing it. Simple Multiple Complex and Asynchronous periodic pattern miner (SMCA) is a four phase algorithm that detects all the subsequences containing asynchronous periodic patterns. One Event One Pattern (OEOP) algorithm uses a linked list structure to detect single event one patterns in a single scan of a sequence. OEOP can be used to replace the first phase of SMCA for data sets like data streams. When compared to SMCA, E-MAP can efficiently mine all patterns in a single step and single scan of the sequence in the presence of large primary memory.

Keywords: asynchronous periodic pattern, periodic pattern, sequence mining, data streams, temporal databases.

1 Introduction

Sequence is an ordered list of elements from any domain. The order among the elements of a sequence may be implied by time order as in stock market data or by physical position as in DNA or protein sequences [1]. If the order is implied by time order the sequences are called event sequences. Sequences where the order is implied by physical position are called biological sequences. Frequently occurring subsequences are referred to as sequential patterns. Sequential patterns with high support extracted from sequence databases are called frequent sequential patterns while the repeating patterns found in a lengthy sequence are called periodic patterns. Periodic analysis is often performed over time-series data which consists of sequences of values or events typically measured at equal time intervals [5]. There are many events in our lives which occur periodically. For example power consumption, daily traffic patterns, telecommunication traffic, maintenance of vehicles, stock market price change, web click streams, sales histories in super market, seasonal changes in climate, data sent

S.C. Satapathy et al. (eds.), *ICT and Critical Infrastructure: Proceedings of the 48th Annual Convention of CSI - Volume II*, Advances in Intelligent Systems and Computing 249,
DOI: 10.1007/978-3-319-03095-1_10,

by sensors, biological sequences. Generalized Sequential Patterns (GSP) [2], Sequential PAttern Discovery using Equivalence Class(SPADE) [3], Prefix and Suffix Projection (PrefixSpan) [4] are a few algorithms for finding frequent sequential patterns. This paper presents a comparative analysis of state of art techniques for asynchronous periodic pattern mining.

Periodic patterns can be classified as full periodic or partial periodic. Full periodic pattern is a pattern where every position in the pattern exhibits the periodicity [6]. Periodic patterns in which one or more elements do not exhibit the periodicity are called partial periodic patterns. If $\{a\}\{b\}\{c\}\{b\}\{c\}\{a\}\{c\}\{d\}$ is an input sequence $\{b\}\{c\}$ is a full periodic pattern with period 2. It is all also called as full periodic pattern because every position in the pattern exhibits the periodicity. The sequence $\{a\}\{b\}\{c\}\{a\}\{d\}\{c\}\{a\}\{c\}\{c\}$ contains a partial periodic pattern $\{a\}\{*\}\{c\}$ with period 3 where the second element is not exhibiting the periodic behavior.

Periodic patterns can also be classified as perfect and imperfect periodic patterns. A pattern X is said to satisfy perfect periodicity in sequence S with period p if starting from the first occurrence of X until the end of S every next occurrence of X exists p positions away from the current occurrence of X. $\{a\}\{b\}\{*\}$ is perfect periodic pattern with period 3 in the sequence $\{a\}\{b\}\{d\}\{a\}\{b\}\{v\}\{a\}\{b\}\{f\}\{a\}\{b\}\{c\}$. $\{a\}\{b\}\{*\}$ has occurred for 4 times starting from its first occurrence till the end of the sequence. If some of the expected occurrences of X miss it is called imperfect periodicity. $\{a\}\{b\}\{*\}$ is an imperfect periodic pattern in the sequence $\{a\}\{b\}\{c\}\{d\}\{b\}\{f\}\{a\}\{b\}\{g\}\{a\}\{b\}\{v\}$. The occurrences of $\{a\}\{b\}\{*\}$ is missed in one of its expected positions.

A pattern which occurs periodically without any misalignment is called as synchronous periodic pattern. In the sequence $\{a\}\{b\}\{c\}\{a\}\{d\}\{c\}\{a\}\{c\}\{c\}$, $\{a\}\{*\}\{c\}$ is the synchronous partial periodic pattern. The pattern has repeated for three times consecutively in the sequence with a period 3. Asynchronous periodic patterns are patterns with some disturbance between the repetitions of the pattern. Disturbance is allowed not only in terms of missing occurrences but also as insertion of random noise events [7]. $\{a\}\{*\}\{c\}$ is an asynchronous periodic pattern in the sequence $\{a\}\{b\}\{c\}\{a\}\{c\}\{c\}\{c\}\{b\}\{a\}\{b\}\{c\}\{a\}\{d\}\{c\}$. The above pattern has appeared for four times in the sequence where there is a disturbance between second and third occurrences of the pattern.

2 Problem Definition

Asynchronous periodic patterns can be mined from event sequences or eventset sequences. An event sequence consists of a single event at every time instant. Event set sequences consist of one or more events at every time instant. A brief discussion about asynchronous periodic pattern mining problem [7][8] is given here.

Given a single long sequence of events/event sets, *min_rep*, $L_{\max}$, *global_rep*, and *max_dis* as input, the asynchronous periodic pattern mining problem is to mine periodic patterns that are significant within a subsequence of event/eventset sequence. A pattern is said to appear significantly in a subsequence if the number of repetitions of pattern is greater than or equal to *global_rep*. This subsequence is called valid subsequence and it is a collection of valid segments with tolerable disturbance between the valid segments. A valid segment is a subsequence of the input sequence where the pattern has appeared for atleast *min_rep* number of times consecutively without any overlaps. *Max_dis* gives the maximum allowed disturbance between valid segments of a pattern. Asynchronous periodic pattern mining aims to discover all the patterns that appear with a periodicity that is less than or equal to maximum period bound $L_{\max}$ along with their valid segments and valid subsequences.

3 Algorithms For Mining Asynchronous Periodic Patterns

LSI [7], SMCA [8], Progressive Time List based Verification (PTV) [9], Linked List structure based OEOP algorithm [10] and E-MAP [11] were accepted as successful algorithms for asynchronous periodic pattern mining.

3.1 LSI

Longest Subsequence Identification (LSI) is the pioneering algorithm to mine asynchronous periodic patterns. LSI algorithm was proposed by J. Yang el. al. in [7] to detect asynchronous periodic patterns from event sequences. Event sequences are sequences where every position in the sequence consists of only one event. For every significant pattern, the algorithm mines the longest subsequence containing it.

LSI is an iterative level wise search algorithm. It works in three phases. In the first phase it detects all potential periods for all events. This requires single scan of the sequence. In the second phase all candidate *1-patterns* are validated. A *1-pattern* is a pattern where one position in the pattern is defined and rest of them are *'s. For example {a}{*}{*},{b}{*} are *1-patterns* with periodicity 3 and 2 respectively. An *i-pattern* is a pattern where *i* positions in the pattern are defined for e.g. {a}{b}{*} is a 2-pattern with periodicity 3 where 2 positions out of 3 are defined. In the third phase an iterative level wise approach is used where in the *i th* iteration candidate *i-patterns* are formed from *(i-1)-patterns*. Validation of these *i-patterns* in *i th iteration* requires single scan of the sequence. The second and third phases of LSI require multiple scans of the sequence.

For every asynchronous periodic pattern, LSI algorithm identifies only the longest subsequence containing it. LSI algorithm also does multiple scans of the sequence for detecting all asynchronous periodic patterns and their valid subsequences .SMCA, OEOP algorithm and E-MAP mine all subsequences containing the asynchronous periodic patterns instead of confining to longest subsequence identification. So these algorithms provide more knowledge to the end user. All these algorithms do fewer scans of the sequence when compared to LSI. These algorithms can handle simultaneously occurring events (event sets). LSI mines patterns from single event sequences only.

3.2 Single, Multi-Event, Complex and Asynchronous Periodic Pattern Miner (SMCA)

K. Y. Huang, C.H. Chang proposed SMCA in [8]. It is a four phase algorithm used to find asynchronous periodic patterns from a sequence of event sets. SMCA model detects all valid subsequences of an asynchronous periodic pattern.

Three parameters namely *min_rep*, *global_rep* and *max_dis* are used to find valid patterns and subsequences containing them. Each valid segment should have at least *min_rep* contiguous matches of the pattern. Every valid segment should be maximal. A segment is maximal if there are no other contiguous matches at either end. A valid subsequence is a list of valid segments interleaved by a disturbance. The overall number of repetitions of a pattern in a valid subsequence should be greater than *global_rep*.

SPMiner, MPMiner, CPMiner and APMiner are the algorithms that comprise the SMCA model. SPMiner mines all significant single event *1-patterns* and their corresponding valid segments. We need to scan the sequence $2 * L_{\max}$ times to accomplish this step. The single event *1-patterns* and their respective valid segments that are found by SPMiner are used as input to MPMiner. MPMiner is used to mine multi-event *1-patterns* and their valid segments using depth first enumeration. The outputs of SPMiner and MPMiner were used as input by CPMiner to generate complex patterns. Complex patterns are *i-patterns* with $i \geq 2$. The first three algorithms mine one event *1-patterns*, multi-event *1-patterns*, complex patterns (*i-patterns*) and their corresponding valid segments. The last algorithm APMiner uses the outputs of the first three algorithms to mine the valid subsequences of the valid patterns. A valid subsequence is a group of valid segments where distance between every pair of consecutive valid segments is less than or equal to *max_dis* and the number of repetitions of a pattern in the subsequence is greater than or equal to user specified threshold called *global_rep*.

K. Y. Huang, C.H. Chang has also proposed Progressive Time List Based Verification (PTV) [9] algorithm for mining single event one patterns and multi event one patterns. Its performance was found to be better than LSI. K. Y. Huang and C.H. Chang declare that the scalability of PTV is better than SPMiner and MPMiner.

In order to deal with data streams like stock market data, sensor inputs the algorithm must be able to find patterns in a single scan of sequence. For this purpose, we can use linked list structure based OEOP algorithm and E-MAP algorithm.

3.3 One Event One Pattern (OEOP) Algorithm

An efficient linked list structure based One Event One Pattern (OEOP) algorithm was proposed by J.S. Yeh, S.C. Lin [10] to improve the speed of mining single event *1-patterns* from event sets sequence. Given a sequence of event sets *D*, for each event *e*, a list of time instants of *e* called time list of *e* is generated. This algorithm can mine all valid *1- patterns* at all periods by a single scan of time lists using the efficient linked list structure. Three node structures called Start structure, end structure and valid structure are used for this purpose. Once the single event *1-patterns* are found, algorithms similar to MPMiner, CPMiner and APMiner as proposed in SMCA model can be used to find the multi-event *1-patterns*, complex patterns and asynchronous patterns.

We have compared the performances of SPMiner and OEOP using human genome sequence collected from National centre for Biotechnology information [12]. Fig 1. shows the running time taken by both the algorithms when *min_rep* is 3 and L_{max} is 9. We can see that SPMiner performs well when compared to OEOP algorithm. But OEOP has the advantage of mining single event *1-patterns* in a single scan of sequence. As we scan the sequence at any point of time we can find valid *1-patterns* and the valid subsequences containing them from the linked list structure. So this algorithm can be used on datasets like datastreams which require single scan to generate patterns. OEOP generates redundant patterns like $(ab,1,7,2,4)$, $(ab,3,7,2,3)$ where the first pattern indicates that the multievent *1-pattern "ab"* repeated for four times from time instant 1 to time instant 7 with periodicity 2. The second pattern shows that *"ab"* has repeated for 3 times from time instant 3 to 7 with periodicity 2. The second pattern is redundant.

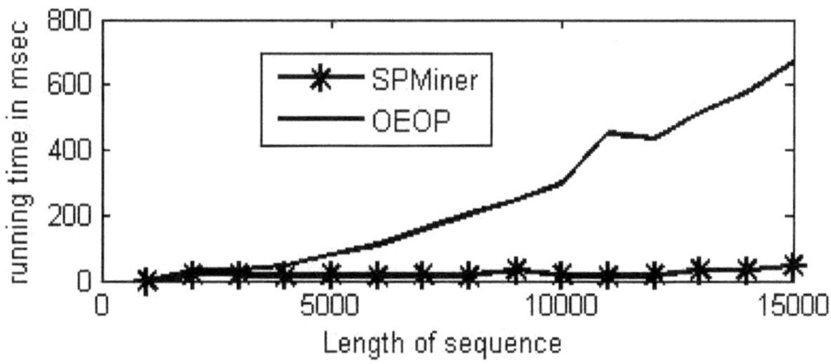

Fig. 1. Performance analysis of SPMiner and OEOP algorithms when different lengths of genetic sequence are considered

3.4 E-MAP: Efficiently Mining Asynchronous Periodic Patterns

F. Maqbool, S. Bashir, and A.R. Baig proposed E-MAP [11] which mines all single event *1-patterns*, multievent *1-patterns*, maximal complex patterns and asynchronous periodic patterns with a single scan of the sequence. All these are done in a single phase. In E-MAP each event in the event set sequence is associated with memory called event block memory (EBMem). Each EBMem is in turn made up of L_{max} period block memories (PBMem). There is a period block memory associated with each period p in the range $1 \le p \le L_{max}$. Each PBMem corresponding to period p is in turn made up of p offset segment memories (OSMem) with respect to each offset for offsets in the range $0 \le offset < p$. Each OSMem stores *start_occurrence*, *repetition* and *last_occurence* corresponding to a extendable segment.

Using EBMems, the E-MAP algorithm is able to maintain information corresponding to all possible extendable segments of all events at all periods and starting at any offset o where $0 \leq o \leq p$ in main memory parallely. If a segment corresponding to an event cannot be extended then it is removed from EBMem. If the number of repetitions of an event in that segment is greater than *min_rep* then that segment is considered as a valid segment of that event. If a valid segment is detected and being removed from EBMem then the other OSMem's of same event and other events are searched to form multi-event *1-patterns* and complex patterns. The details of how valid segments are used to form valid sequences were not clearly given in [11]. In order to implement the E-MAP algorithm for finding both the valid segments and valid sequences of patterns, we have maintained extendable sequence lists corresponding to every pattern and period (p,l) pair. The valid segments that were removed from EBMem in case of single event *1-patterns* and that are constructed from EBMem in case of multi-event *1-patterns* and complex patterns were added to the corresponding (p,l) pair's extendable sequence list. If any sequence cannot be extended, it is removed from extendable sequence list. If the pattern has repeated for atleast *global_rep* number of times in such a sequence then it is called a valid sequence .All the valid sequences were stored in a file for future processing.

The performances of SMCA and E-MAP are compared on stock market data. The daily closing prices of five stock indices namely SENSEX, BSE100, BSE200, BSE500 and TECK were collected for 17 years i.e. from 2nd Jan 1991 to 16th Oct 2007 from BSE [13]. The data consisted of 4000 time stamps. The data is preprocessed by applying the following transformation. The original data is transformed into sequence of event sets in the following way. Saturdays and Sundays and all public holidays will not have any trading done, so they are ignored. Time gap between every two consecutive trading days is taken as 1 day. The closing price of every index on a given day was compared with its average closing price in the last five days (considering only the days when trading took place). Depending on the percentage change of closing price value when compared with average closing price, its price is replaced with one of the five different events (symbols) for e.g. A1,A2,A3,A4,A5. $<-3\%$,-3% to -1%, -1% to 1%, 1% to 3% and >3% were the ranges considered for percentage change of value of stock index price. Figure 2 shows the performances of SMCA and E-MAP for different values of L_{max} with *min_rep*=5, *global_rep*=10, *max_dis*=5. *Max_dis* is chosen as 5 because the number of working days of stock market is 5 in a week. So we allow maximum gap of one week between valid segments to form valid subsequences. We can see that E-MAP outperforms SMCA. But if *min_rep* was set to 4 or less value the E-MAP algorithm has failed in finding asynchronous periodic patterns. This was because E-MAP requires a very large primary memory to store all valid segments and extendable sequences simultaneously in memory. The primary memory requirement crossed 2GB for *min_rep*=4 itself. So in the presence of a very large primary memory E-MAP can be used for asynchronous periodic pattern mining. Fig.3. shows the running time requirement of both the algorithms for different values of *min_rep* with Lmax=35.As the value of *min_rep* increases the performance of both

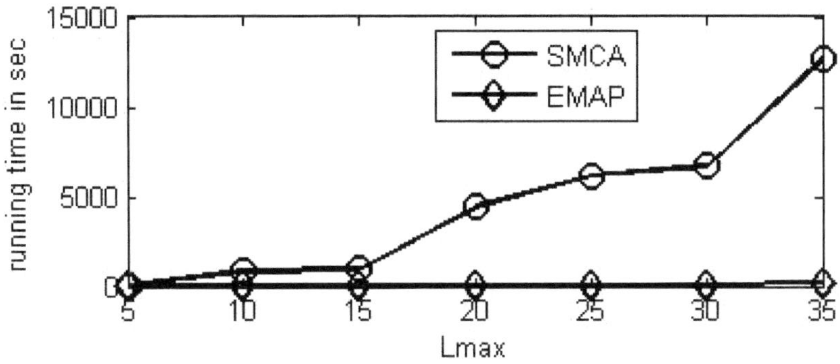

Fig. 2. Performance of SMCA and EMAP for different values of L_{max} with *min_rep*=5

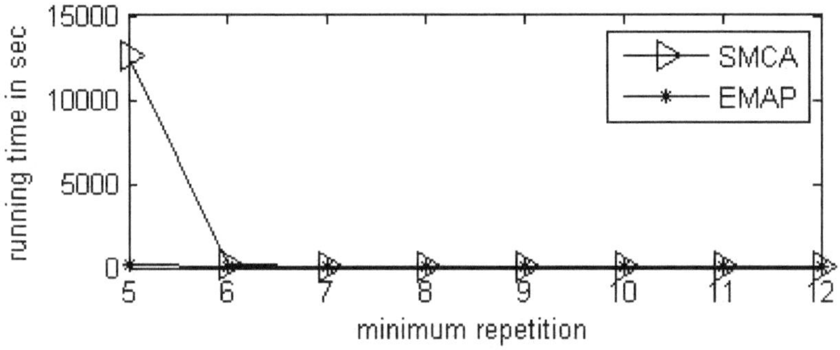

Fig. 3. Performance of SMCA and EMAP at different values of *min_rep* with L_{max} =35

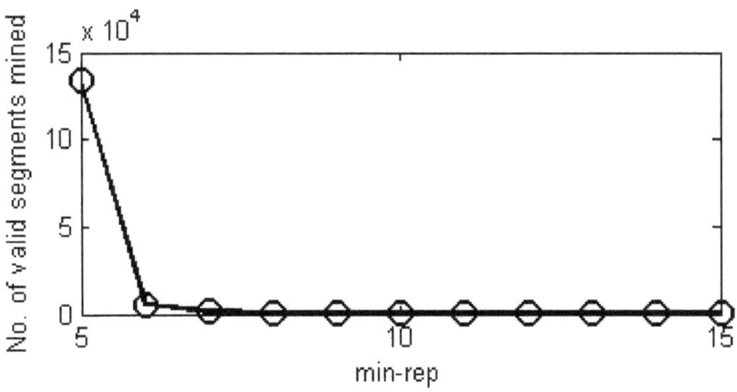

Fig. 4. Number of valid segments mined by both SMCA and E-MAP for different values of *min_rep* with L_{max}=*35* and sequence length 4000

the algorithms is found to be the same. Fig. 4. shows the number of valid segments mined by both SMCA and E-MAP for different values of *min_rep* and *Lmax*=35.. In all the cases *global_rep* is taken as twice that of *min_rep* 140000 segments were detected for *min_rep*=5 and *Lmax* =35. For all the different inputs the valid segments generated by both the algorithms were exactly the same.

4 Conclusions

Four different algorithms for mining asynchronous periodic patterns were discussed in this paper. Performances of OEOP algorithm and SPMiner are compared using genetic sequence. SPMiner can mine single event one patterns in less time when compared to OEOP algorithm. But OEOP algorithm has the advantage of mining the patterns in a single scan of sequence. The performance comparison of SMCA and E-MAP shows that E-MAP algorithm outperforms SMCA in the presence of large primary memory. For large values of *min_rep* both the algorithms have performed equally well. In our implementation of E-MAP algorithm it is also found that it is consuming more than 2GB of primary memory and is terminating abruptly when *min_rep* is set to 4 or less than 4. E-MAP's advantage is that it can mine all the patterns with a single scan of sequence. At any point of time as we scan the sequence we can get all valid segments and valid subsequences of any event at any periodicity from the event block memories and hence E-MAP can be applied for data sets like data streams. Because of the large main memory requirement for maintaining the extendable sequence lists we can make best use of the E-MAP algorithm if it can be made as a distributed algorithm. One of the drawbacks of all the above algorithms is that they mine redundant patterns like $((A, B),2,6,10)$, $((A,*),2,6,10)$, $((*, B),2,6,10)$. The first pattern shows that *(A, B)* repeats with periodicity 2 for 6 times starting from time instant 10. The knowledge conveyed by second and third patterns is already present in first pattern. To improve the time and space utilization of algorithms used for asynchronous periodic pattern mining, algorithms that mine only the closed, maximal periodic patterns need to be developed. It also leads to good interpretability. As the size of datasets is growing tremendously, algorithms that can mine the patterns in an incremental manner need to be developed.

References

1. Dong, G., Pei, J.: Sequence Data Mining. Advances in Database Systems. Springer science (2007)
2. Srikant, R., Agarwal, R.: Mining sequential patterns: Generalizations and Performance Improvements. In: 5th International Conference on Extending Database Technology, Avignon, France, pp. 3–17 (1996)
3. Zaki, M.J.: SPADE: An Efficient Algorithm for Mining Frequent Sequences. J. Machine Learning 42, 31–60 (2001)
4. Pei, J., Han, J., Mortazavi-Asl, B., Wang, J., Pinto, H., Chen, Q., Dayal, U., Hsu, M.-C.: Mining sequential patterns by pattern growth: The prefixspan approach. J. IEEE Transactions on Knowledge and Data Engineering 16, 1424–1440 (2004)

5. Han, J., Kamber, M.: Data Mining: Concepts and Techniques, 2nd edn. Morgan Kaufmann (2006)
6. Wang, W., Yang, J.: Mining Sequential Patterns from Large Data Sets. Advances in Database Systems, vol. 28. Springer Science (2005)
7. Yang, J., Wang, W., Yu, P.S.: Mining Asynchronous Periodic Patterns in Time Series Data. J. IEEE Transactions on Knowledge and Data Engineering 15, 613–628 (2003)
8. Huang, K.Y., Chang, C.H.: SMCA: A General Model for Mining Asynchronous Periodic Patterns in Temporal Databases. J. IEEE Transactions on Knowledge and Data Engineering 17, 774–785 (2005)
9. Huang, K.-Y., Chang, C.-H.: Mining Periodic Patterns in Sequence Data. In: Kambayashi, Y., Mohania, M., Wöß, W. (eds.) DaWaK 2004. LNCS, vol. 3181, pp. 401–410. Springer, Heidelberg (2004)
10. Yeh, J.S., Lin, S.C.: A New Data Structure for Asynchronous Periodic Pattern Mining. In: 3rd International Conerence on Ubiquitous Information Management and Communication, New York, pp. 426–431 (2009)
11. Maqbool, F., Bashir, S., Baig, A.R.: E-MAP: Efficiently Mining Asynchronous Periodic Patterns. International Journal Computer Science and Network Security 6, 174–179 (2006)
12. National Center for Biotechnology Information, http://www.ncbi.nlm.nih.gov/
13. BSE, http://www.bseindia.com/indices/indexarchivedata.aspx

A Comparative Study of the Classification Algorithms Based on Feature Selection

A. Sravani, D.N.D. Harini, and D. Lalitha Bhaskari

Dept. of CS&SE,
Andhra University

Abstract. Image classification is a large and growing research field with its applications in the areas of CBIR (Content Based Image Retrieval), image mining and automatic image annotation. In this digital era there is a huge voluminous multimedia data available and the challenge lies in retrieving and classifying the most similar images based upon an input query. Images can be classified according to their nature, content or domain and Feature extraction is the key process to classify the images accordingly. In this paper, an attempt is made to calculate all the possible features of an image based on color, texture, shape, and statistical. Based up on the features the images are further classified, studied and compared with four Classification algorithms namely Naïve Bayes, Instance Based Learning, J48 and Random forest Classification. Further the classification is applied on a prescribed set of features, so as to test the best feature set for the query image to be classified. An image database of 1150 images divided into 17 categories are considered for Classification and a brief comparative study is done.

Keywords: Classification, Image Classification, Feature Extraction, Classification Algorithms, Naïve Bayes, Random Forest Classification, J48, Instance based learning.

1 Introduction

In this modern world, advances in multimedia technologies such as image digitization, storage and transmission along with the growth of the World Wide Web, mobile device, cameras have lead to the proliferation of online digital images. Content-based image classification has been an interesting subject of many researchers in recent years. There are many great efforts in developing the classification approaches and techniques to improve the classification accuracy [1].Image Classification aims to find a description that can best describe the images in one class and to distinguish these images from all other classes [2]. To classify an image into a certain class, its feature vectors must be constructed from the image. There are many possible ways to construct features. But no theoretical guidelines suggest the appropriate features to use in classification of data [5,6]. The general criteria for choosing a feature is they should be independent and posses maximum information about the image and another important property of a feature is it should

S.C. Satapathy et al. (eds.), *ICT and Critical Infrastructure: Proceedings of the 48th Annual Convention of CSI - Volume II*, Advances in Intelligent Systems and Computing 249,
DOI: 10.1007/978-3-319-03095-1_11, © Springer International Publishing Switzerland 2014

be easily computable even for a large database. Usually the features considered with respect to images are color, shape, texture, edge and so on. In this paper the set of features computed are: Hue, Saturation, Value, Color percentage (Red, Blue, Green, Yellow, Magenta and cyan), GLCM features, entropy, singular value decomposition, wavelets, fast Fourier transform. Based on these features the images are classified.

The classification techniques used in this paper are: Naïve Bayes Classification Algorithm, Random forest Classification Algorithm, J48 classification Algorithm, Instance Based Learning Algorithm. After Classification, the image retrieval is done based on the set of classified images.

The work in this paper is divided into sections where section2 deals with Classification and different types of image classification methods used. Section 3 deals with Image retrieval based on Classification. Section 4 compares various classification techniques for the given image data. Section 5 analyses the Comparison results and concludes. Section 6 and 7 deals with future work and references

2 Classification

Classification is used in every field of our life. Classification is an important data mining technique with broad applications and it is also a challenging task with many applications in computer vision. It classifies data of various kinds. Classification is used to classify each item in a set of data into one of predefined set of classes or groups [2,8]. The problem of Classification is defined as a set of training records D = {X1,...,XN}, such that each record is labeled with a class value drawn from a set of k different discrete values indexed by {1...k}. The training data is used in order to construct a Classification model, which relates the features in the underlying record to one of the class labels. For a given test instance for which the class is unknown, the training model is used to predict a class label for this instance.

2.1 Classification Algorithms

The problem of Classification has been widely studied in the database, data mining, and information retrieval communities. As the problems and applications are numerous in this area, there is no single algorithm that is better than all the others on all the problems. Therefore, for each problem, right algorithm should be selected. A Classification algorithm is a procedure for selecting a hypothesis from a set of alternatives that best fits a set of observations [9].

In this paper a Comparative study is performed on a huge data set of 1150 images. The Classification algorithms used for comparison are: Naïve Baye's Classification, J48, Random Forest Classification and Instance based classification algorithm.

2.1.1 Naïve Bayes Classification. In Bayesian classifiers, it attempts to build a probabilistic classifier based on modeling the underlying features in different classes. The idea is then to classify based on the posterior probability belonging to the different classes. A naïve Bayes classifier models a join distribution over a label Y and a set of features $\{F_1, F_2, ..Fn\}$, using the assumption that the full join distribution can be factored as follows [3,9]:

$$P(F_1, .. F_n, Y) = P(Y) \prod_i P(F_i|Y)$$

To classify a datum, we can find the most probable class given the feature values for each pixel:

$$P(y|f_1, ..., f_m) = \frac{P(f_1, ..., f_m|y)P(y)}{P(f_1, ..., f_m)}$$

$$= \frac{P(y) \prod_{i=1}^{m} P(f_i|y)}{P(f_1, ... f_m)}$$

$$\arg max_y \; P(y|f_1, ... f_m) = \arg max_y \frac{P(y) \prod_{i=1}^{m} P(f_i|y)}{P(f_1, ... f_m)}$$

$$= argmax_y \; P(y) \prod_{i=1}^{m} P(f_i|y)$$

Because multiplying many probabilities together often results in underflow, we will instead compute log probability which will have the same argmax:

$$argmax_y \log(P(y|f_1, ..., f_m) = argmax_y(\log(P(y)) + \sum_{i=1}^{m} \log(P(f_i|y)))$$

2.1.2 J48 (C4.5). J48 is an open source Java implementation of the C4.5 algorithm in the Weka data mining tool. C4.5 is a program that creates a decision tree based on a set of labeled input data. This algorithm was developed by Ross Quinlan. The decision trees generated byC4.5 can be used for classification, and for this reason, C4.5 is often referred to as a statistical classifier [3,11].

2.1.3 Instance Based Learning Classification Algorithm. Instance Based Learning (which puts efforts in classification) is also referred to as Lazy learning as opposed to eager learning. Eager learning(which invests its efforts in Learning phase) algorithms put significant effort in abstracting from the training instances by creating condensed representations like Decision trees, rule sets, and hyper planes etc. during the learning phase. The classification phase of an eager learner reduces to a relatively effortless application of the abstracted representation to new instances [7,10].

2.1.4 Random Forest Algorithm. The Random forest is an ensemble learning method that grows many classification trees. Each tree gives a classification. The forest selects the classification that has the most votes [4]. The term came from random decision forests that were first proposed by Tin Kam Ho of Bell Labs in 1995. This method combines Breiman's "bagging" idea and the random selection of features [12].

3 Methodology

A training image data set of 1150 images are considered which belong to several categories like Ant(5), Barrel(36), kangaroo(115), Buildings(100),People(100), Beach(100), Bus(100), Elephant(100), Rose(114), Horse(100), Mountain(100), Food(100), Tiger(20), Bear(20),Apple(6), Banana(15),Jelly Fish(15) where the Number in the brackets represents the number of images in that category. The method followed is explained as below:

Step 1: Feature Calculation: A total of 53 features corresponding to Color, Texture, Shape and Statistical features are calculated and maintained in a database.

Step2: Classification: Classification is done using Weka tool for Naïve Bayes, J48, Random Forest and IB1 Classification algorithms.

Step 3: Comparison between Classification Techniques: The results are computed from each algorithm for the train set and test sets. A brief comparison is done and checks how many correctly classified instances and incorrectly classified instances. Along with that True Positive Rate, False Positive Rate, Precision, Recall, F- Measure and ROC area are also computed and compared for each Classification method. The Methodology is explained in detail in the following diagram.

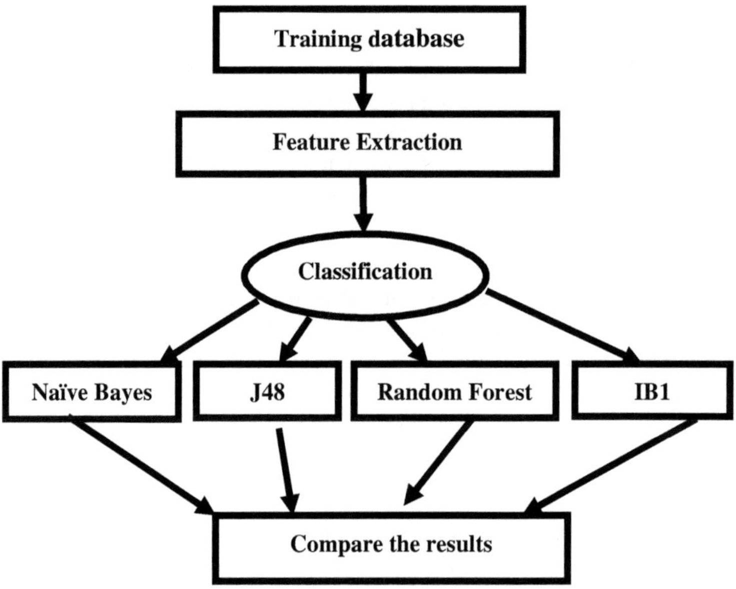

Fig. 1. Process flow of Methodology

4 Test Cases and Results

A total of 53 features have been computed for each image in a dataset of 1150 images which is termed as Trained Data.

Test Cases

Test Case 1: In test case1 a subset of 12 features are considered which are color components using HSV, fast Fourier transform, correlation, contrast, homogeneity, energy and entropy.

Test Case 2: A total of 39 features considered in test case2 are color components using HSV, Wavelets, GLCM features.

Test Case 3: In this test case3, a total of 14 features corresponding to color components using HSV, Wavelets, SVD and texture components are considered.

All the four classification algorithms are applied on each test cases and the results are observed in Weka and the results are as shown below.

Table 1. The Number of Correctly Classified and Incorrectly Clasiified instances for all classification algorithms for the trained data and 3 test cases

Instance	Number Of Correctly Classified Instances (%)				Number of Incorrectly Classified Instances (%)			
	NaïveBayes	**J48**	**Random Forest**	**IB1**	**Naïve bayes**	**J48**	**Random Forest**	**IB1**
Trained Data	400 (34.81%)	938(81.6362 %)	1127(98.0853 %)	1129(98.2594 %)	749(65.1871%)	211(18.363 8%)	22(1.9147%)	20(1.740%)
Test Case 1	313(27.2411 %)	860(74.8477 %)	1128(98.1723 %)	1129(98.2594 %)	836(72.7589%)	289(25.152 3%)	21(1.8277%)	20(1.740%)
Test Case 2	310(26.98%)	918(79.8956 %)	1126(97.9983 %)	1129(98.2594 %)	839(73.02%)	231(20.104 4%)	23(2.0017%)	20(1.740%)
Test Case 3	431(37.5109 %)	921(80.1567 %)	1127(98.0853 %)	1129(98.2594 %)	789(62.4891%)	228(19.843 3%)	22(1.9147%)	20(1.740%)

```
=== Detailed Accuracy By Class ===

               TP Rate  FP Rate  Precision  Recall  F-Measure  ROC Area  Class
                 0.2     0.007     0.467     0.2      0.28      0.786    Barrel
                 0.25    0.003     0.25      0.25     0.25      0.924    Ant
                 0.505   0.064     0.432     0.505    0.466     0.828    People
                 0.152   0.035     0.288     0.152    0.199     0.804    Beach
                 0.67    0.145     0.306     0.67     0.42      0.871    Buildings
                 0.01    0.004     0.2       0.01     0.019     0.675    Bus
                 0.033   0.021     0.154     0.033    0.055     0.676    Kangaroo
                 0.05    0.025     0.161     0.05     0.076     0.738    Elephant
                 0.5     0.069     0.445     0.5      0.471     0.809    Rose
                 0.57    0.039     0.582     0.57     0.576     0.858    Horse
                 0.39    0.087     0.3       0.39     0.339     0.803    Mountain
                 0.55    0.077     0.404     0.55     0.466     0.877    Food
                 0.2     0.002     0.667     0.2      0.308     0.924    Tiger
                 0.5     0.01      0.476     0.5      0.488     0.92     Bear
                 0.5     0.002     0.6       0.5      0.545     0.861    Apple
                 0.8     0.021     0.333     0.8      0.471     0.958    Banana
                 0.8     0.094     0.101     0.8      0.179     0.925    Jelly Fish
Weighted Avg.    0.348   0.052     0.337     0.348    0.31      0.801
```

```
=== Confusion Matrix ===

  a  b  c  d  e  f  g  h  i  j  k  l  m  n  o  p  q   <-- classified as
  7  1  1  1  4  3  0  4  3  2  2  5  0  0  0  1  1 |  a = Barrel
  0  1  0  1  0  0  0  0  0  0  1  0  0  0  0  0  1 |  b = Ant
  0  0 51  1 14  0  2  3  3  7 12  5  1  1  1  0  0 |  c = People
  3  1 16 15 37  0  1  1  4  5  8  7  0  1  0  0  0 |  d = Beach
  2  0  3  5 67  0  0  2  1  4 12  4  0  0  0  0  0 |  e = Buildings
  0  0  0  1 28  1  4  1  3  0  5  2  0  1  0  3 51 |  f = Bus
  0  0  5  8  9  0  4  1  1  3 13 17  0  1  0 13 45 |  g = Kangaroo
  1  0  3  6 10  0  8  5 39  3 14 11  0  0  0  0  0 |  h = Elephant
  0  0 10  4  6  0  0  1 57 12  7  8  1  3  1  2  2 |  i = Rose
  0  0  6  0 10  0  1  3  3 57  7 10  0  0  0  0  3 |  j = Horse
  1  0 17  4 20  0  1  5  2  2 39  7  0  0  0  2  0 |  k = Mountain
  0  1  5  5 13  1  4  2  6  1  7 55  0  0  0  0  0 |  l = Food
  0  0  0  0  0  0  0  2  3  1  1  2  4  3  0  1  3 |  m = Tiger
  1  0  1  1  0  0  0  1  1  0  1  3  0 10  0  1  0 |  n = Bear
```

Fig. 2. Detail Accuracy Of each Class, Confusion matrix using **Naïve Baye's algorithm**

=== Detailed Accuracy By Class ===

	TP Rate	FP Rate	Precision	Recall	F-Measure	ROC Area	Class
	0.771	0.013	0.643	0.771	0.701	0.991	Barrel
	0.5	0.002	0.5	0.5	0.5	0.998	Ant
	0.792	0.023	0.769	0.792	0.78	0.984	People
	0.859	0.015	0.842	0.859	0.85	0.992	Beach
	0.85	0.011	0.876	0.85	0.863	0.994	Buildings
	0.85	0.03	0.733	0.85	0.787	0.984	Bus
	0.792	0.02	0.819	0.792	0.805	0.984	Kangaroo
	0.82	0.01	0.891	0.82	0.854	0.993	Elephant
	0.86	0.017	0.845	0.86	0.852	0.992	Rose
	0.89	0.016	0.84	0.89	0.864	0.994	Horse
	0.82	0.016	0.828	0.82	0.824	0.989	Mountain
	0.88	0.02	0.807	0.88	0.842	0.992	Food
	0.7	0.003	0.824	0.7	0.757	0.995	Tiger
	0.65	0.002	0.867	0.65	0.743	0.997	Bear
	0	0	0	0	0	0.994	Apple
	0.867	0.002	0.867	0.867	0.867	0.999	Banana
	0	0	0	0	0	0.98	Jelly Fish
Weighted Avg.	0.816	0.017	0.805	0.816	0.809	0.99	

=== Confusion Matrix ===

a	b	c	d	e	f	g	h	i	j	k	l	m	n	o	p	q	<-- classified as
27	0	0	0	1	0	0	0	2	0	2	2	0	0	0	1	0	a = Barrel
1	2	0	1	0	0	0	0	0	0	0	0	0	0	0	0	0	b = Ant
2	0	80	2	0	2	3	2	0	1	5	4	0	0	0	0	0	c = People
4	0	4	85	1	0	0	1	0	1	2	1	0	0	0	0	0	d = Beach
0	1	5	3	85	1	0	1	2	0	1	1	0	0	0	0	0	e = Buildings
2	0	1	0	1	85	6	1	1	0	2	1	0	0	0	0	0	f = Bus
3	0	2	1	1	12	95	1	0	0	1	2	1	0	0	1	0	g = Kangaroo
1	0	2	3	2	1	4	82	3	1	0	1	0	0	0	0	0	h = Elephant
1	0	1	1	0	1	1	1	98	9	1	0	0	0	0	0	0	i = Rose
0	0	2	1	1	0	1	1	2	89	0	2	1	0	0	0	0	j = Horse
0	0	6	2	3	1	0	0	2	1	82	3	0	0	0	0	0	k = Mountain
0	1	1	2	1	1	1	0	2	2	1	88	0	0	0	0	0	l = Food
0	0	0	0	0	1	0	0	1	1	1	1	14	1	0	0	0	m = Tiger
0	0	0	0	1	1	1	1	1	0	0	2	0	13	0	0	0	n = Bear
1	0	0	0	0	0	1	1	0	1	1	0	1	0	0	0	0	o = Apple
0	0	0	0	0	1	0	0	0	0	0	0	0	1	0	13	0	p = Banana
0	0	0	0	0	9	3	0	2	0	0	1	0	0	0	0	0	q = Jelly Fish

Fig. 3. Detail Accuracy Of each Class and Confusion matrix using **J48 algorithm**

=== Detailed Accuracy By Class ===

	TP Rate	FP Rate	Precision	Recall	F-Measure	ROC Area	Class
	1	0	1	1	1	1	Barrel
	1	0	1	1	1	1	Ant
	1	0	1	1	1	1	People
	1	0	1	1	1	1	Beach
	1	0	1	1	1	1	Buildings
	1	0.016	0.855	1	0.922	0.992	Bus
	0.925	0.002	0.982	0.925	0.953	0.962	Kangaroo
	1	0	1	1	1	1	Elephant
	1	0	1	1	1	1	Rose
	1	0	1	1	1	1	Horse
	1	0	1	1	1	1	Mountain
	1	0	1	1	1	1	Food
	1	0	1	1	1	1	Tiger
	1	0	1	1	1	1	Bear
	1	0	1	1	1	1	Apple
	1	0.001	0.938	1	0.968	1	Banana
	0.267	0	1	0.267	0.421	0.633	Jelly Fish
Weighted Avg.	0.983	0.002	0.985	0.983	0.98	0.99	

=== Confusion Matrix ===

a	b	c	d	e	f	g	h	i	j	k	l	m	n	o	p	q	<-- classified as
35	0	0	0	0	0	0	0	0	0	0	0	0	0	0	0	0	a = Barrel
0	4	0	0	0	0	0	0	0	0	0	0	0	0	0	0	0	b = Ant
0	0	101	0	0	0	0	0	0	0	0	0	0	0	0	0	0	c = People
0	0	0	99	0	0	0	0	0	0	0	0	0	0	0	0	0	d = Beach
0	0	0	0	100	0	0	0	0	0	0	0	0	0	0	0	0	e = Buildings
0	0	0	0	0	100	0	0	0	0	0	0	0	0	0	0	0	f = Bus
0	0	0	0	0	8	111	0	0	0	0	0	0	0	0	1	0	g = Kangaroo
0	0	0	0	0	0	0	100	0	0	0	0	0	0	0	0	0	h = Elephant
0	0	0	0	0	0	0	0	114	0	0	0	0	0	0	0	0	i = Rose
0	0	0	0	0	0	0	0	0	100	0	0	0	0	0	0	0	j = Horse
0	0	0	0	0	0	0	0	0	0	100	0	0	0	0	0	0	k = Mountain
0	0	0	0	0	0	0	0	0	0	0	100	0	0	0	0	0	l = Food
0	0	0	0	0	0	0	0	0	0	0	0	20	0	0	0	0	m = Tiger
0	0	0	0	0	0	0	0	0	0	0	0	0	20	0	0	0	n = Bear
0	0	0	0	0	0	0	0	0	0	0	0	0	0	6	0	0	o = Apple
0	0	0	0	0	0	0	0	0	0	0	0	0	0	0	15	0	p = Banana
0	0	0	0	0	9	2	0	0	0	0	0	0	0	0	0	4	q = Jelly Fish

Fig. 4. Detail Accuracy Of each Class and Confusion matrix using **IB1** algorithm

=== Detailed Accuracy By Class ===

	TP Rate	FP Rate	Precision	Recall	F-Measure	ROC Area	Class
	1	0	1	1	1	1	Barrel
	1	0	1	1	1	1	Ant
	0.99	0	1	0.99	0.995	1	People
	1	0	1	1	1	1	Beach
	1	0	1	1	1	1	Buildings
	0.95	0.011	0.888	0.95	0.918	0.998	Bus
	0.95	0.006	0.95	0.95	0.95	0.999	Kangaroo
	1	0	1	1	1	1	Elephant
	1	0	1	1	1	1	Rose
	1	0.001	0.99	1	0.995	1	Horse
	1	0	1	1	1	1	Mountain
	1	0	1	1	1	1	Food
	1	0	1	1	1	1	Tiger
	1	0	1	1	1	1	Bear
	1	0	1	1	1	1	Apple
	0.867	0	1	0.867	0.929	1	Banana
	0.467	0.003	0.7	0.467	0.56	0.996	Jelly Fish
Weighted Avg.	0.981	0.002	0.98	0.981	0.98	1	

=== Confusion Matrix ===

a	b	c	d	e	f	g	h	i	j	k	l	m	n	o	p	q	<-- classified as
35	0	0	0	0	0	0	0	0	0	0	0	0	0	0	0	0	a = Barrel
0	4	0	0	0	0	0	0	0	0	0	0	0	0	0	0	0	b = Ant
0	0	100	0	0	0	0	0	0	1	0	0	0	0	0	0	0	c = People
0	0	0	99	0	0	0	0	0	0	0	0	0	0	0	0	0	d = Beach
0	0	0	0	100	0	0	0	0	0	0	0	0	0	0	0	0	e = Buildings
0	0	0	0	0	95	3	0	0	0	0	0	0	0	0	0	2	f = Bus
0	0	0	0	0	5	114	0	0	0	0	0	0	0	0	0	1	g = Kangaroo
0	0	0	0	0	0	0	100	0	0	0	0	0	0	0	0	0	h = Elephant
0	0	0	0	0	0	0	0	114	0	0	0	0	0	0	0	0	i = Rose
0	0	0	0	0	0	0	0	0	100	0	0	0	0	0	0	0	j = Horse
0	0	0	0	0	0	0	0	0	0	100	0	0	0	0	0	0	k = Mountain
0	0	0	0	0	0	0	0	0	0	0	100	0	0	0	0	0	l = Food
0	0	0	0	0	0	0	0	0	0	0	0	20	0	0	0	0	m = Tiger
0	0	0	0	0	0	0	0	0	0	0	0	0	20	0	0	0	n = Bear
0	0	0	0	0	0	0	0	0	0	0	0	0	0	6	0	0	o = Apple
0	0	0	0	0	0	2	0	0	0	0	0	0	0	0	13	0	p = Banana
0	0	0	0	0	7	1	0	0	0	0	0	0	0	0	0	7	q = Jelly Fish

Fig. 5. Bar Chart for Correctly Classified Instances for all algorithms

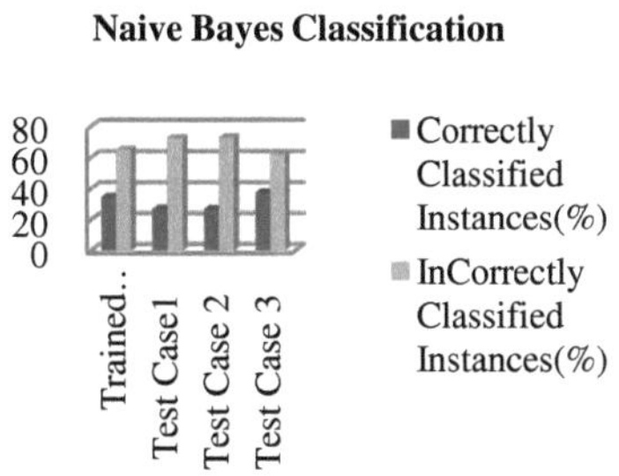

Fig. 6. Bar chart of Naïve Baye's

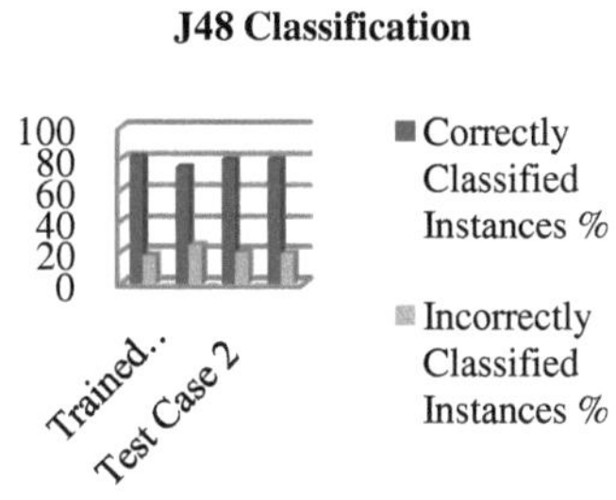

Fig. 7. Bar Chart for J48 algorithm

Classification Algorithm

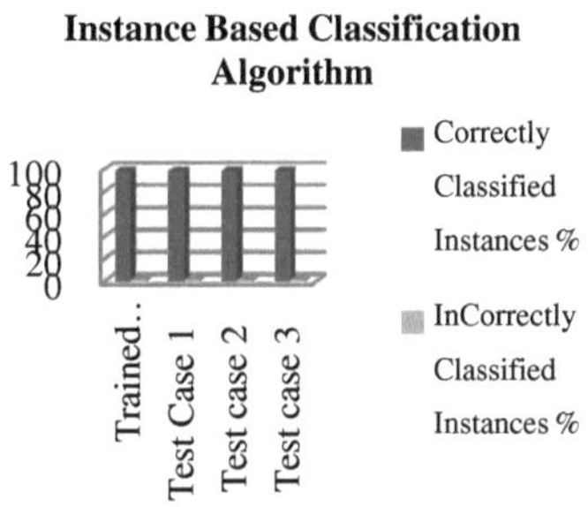

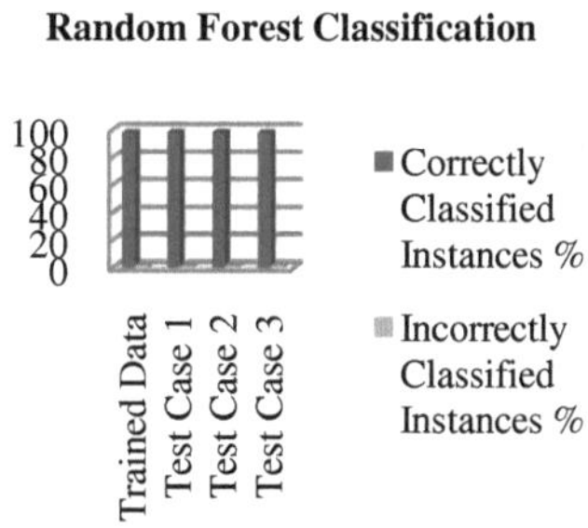

Fig. 8. Bar Chart for **IB1** Algorithm **Fig. 9.** Bar Chart for Random Forest Classification

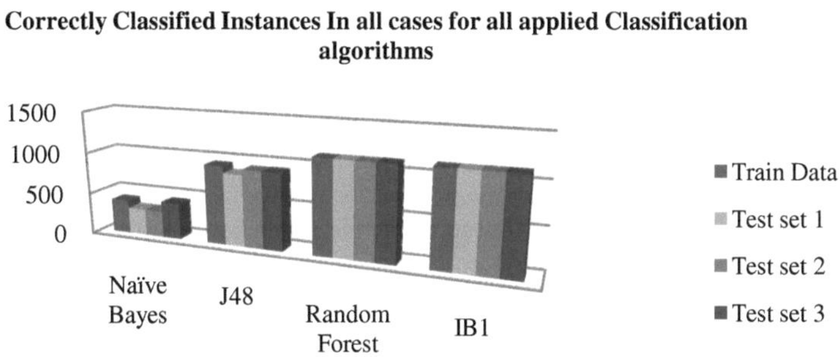

Fig. 10. Bar Chart for Correctly Classified Instances for all algorithms

5 Conclusion

From the above results it is observed that IB1 algorithm has produced the best results followed by Random forest classification algorithm. The results obtained by Naives Bayes and J48 are not satisfactory for the test cases which are considered. Another interesting observation in test case1, test case2, test case 3 only a set of features are considered have produced the same results as compared to trained data which is a collection of all the 53 features. IB1 has proved to be the best even in the test cases followed by Random forest. This observation can be further used in image retrieval and annotation techniques which have gained a lot of interest in the research community.

References

[1] Le Saux, B., Amato, G.: Image Classifiers For Scene Analysis (2003)

[2] Harini, D.N.D., Lalitha Bhaskari, D.: Image Mining Issues and Methods Related to Image Retrieval System. International Journal of Advanced Research in Computer Science 2(4) (July-August 2011) ISSN No. 0976-5697

[3] Gholap, J.: Performance Tuning Of J48 Algorithm for Prediction of Soil Fertility

[4] Kouzani, A.Z., Nahavandi, S., Khoshmanesh: Face classification by a random forest. In: K. TENCON 2007 - 2007 IEEE Region 10 Conference. Deakin Univ., Geelong (October 30-November 2, 2007)

[5] Harini, D.N.D., Lalitha Bhaskari, D.: Image Retrieval System Based on Feature Extraction and Relevance Feedback. ACM (2012)

[6] Harini, D.N.D., Lalitha Bhaskari, D.: Identification of Leaf Diseases in TomatoPlant Based on Wavelets and PCA. IEEE (2011), doi:978-1-4673-0125-1_c

[7] Hendrickx, I., van den Bosch, A.: Hybrid Algorithms with Instance-Based Classification. In: Gama, J., Camacho, R., Brazdil, P.B., Jorge, A.M., Torgo, L. (eds.) ECML 2005. LNCS (LNAI), vol. 3720, pp. 158–169. Springer, Heidelberg (2005)

[8] Kim, S., Jin, X., Han, J.: DislClass: discriminative frequent pattern based image classification. In: MDMKDD 2010, July 25. ACM (2010), doi:978-1-4503-0220-3

[9] Philippe, M.: A comparison of active classification methods for content- Based Image Retrieval. In: CVDB 2004. ACM, Paris (2004), doi:1-58113-917-9/04/06

[10] Aha, D.W., Kibler, D., Albert, M.K.: Instance based learning algorithms. In: Machine Learning, vol. 6, pp. 37–66. Kluwer Academic Publishers, Boston (1991), Manufactured in The Netherlands

[11] Pitchumani Angayarkanni, A.S., Kamal, B.N.B.: Automatic Classification Of Mammogram MRI using Dendograms. AJCSIT 2(4), 78–81 (2012) ISSN 2249-5126

[12] Bosch, Univ. of Girona, Zisserman, M.: Image classification using Random Forests and Ferns. In: IEEE 11th International Conference on Computer Vision, ICCV 2007, pp. 1–8 (2007) ISSN:1550-5499

Analysis and Classification of Plant MicroRNAs Using Decision Tree Based Approach

A.K. Mishra and H. Chandrasekharan

AKMU, Indian Agricultural Research Institute, New Delhi

Abstract. MicroRNA (miRNA) analysis have progressed tremendously in recent past, but further indepth computational study is required to know the complete potential of these RNAs. Due to its short length (~20 nucleotides), it is difficult to use the conventional lab techniques for microRNA prediction and analysis. This has led to this work in the domain of computational biology. These are the non coding small RNAs which are responsible for the gene regulation at the post translational level by binding to the mRNAs and thereby stopping the translation activities. Therefore,the effect of microRNAs on the various proteins is important. In this paper we have studied 1010 microRNA and precursor microRNA sequences from monocots . Our study in this paper is on the microRNA classification using decision trees and determining dominating attributes. We have used WEKA, a data mining tool which helps us to study the large data and classify it. The decision trees based classification was best suited for the miRNA study and the dominating attributes derived are biologically significant.

Keywords: miRBase, precursor, RNAFold, WEKA, J48, Decision Trees.

1 Introduction

MicroRNAs are small non coding RNA molecules which are 18-25bp long sequences. They are responsible for gene regulation in plants as well as in animals. Present in the nucleus, microRNA genes are transcribed into primary transcript or pri-miRNA with the help of RNA polymerase-2. The dsRNA specific ribonuclease Drosha digests the long primary miRNA transcript in the nucleus and releases hairpin precursor microRNA (pre-miRNA). Exportin-5(Exp5) and RAN- GTP transports the pre-miRNA into the cytoplasm where an enzyme called Dicer, processes the pre-miRNA into mature microRNA. Dicer (endonuclease) is a member of Rnase-3 superfamily and cleaves the pre-miRNA approximately 19bp from the Drosha cut site. Only one of the two strands is the miRNA. The double stranded RNA produced by Dicer separate and associate with RISC (RNA-induced silencing complex), on the basis of the stability of the 5'end. In plants as Drosha is lacking so Dicer performs the processing [1, 2].

Mature miRNA are partially complimentary to messenger RNA and they play an important role in gene regulation through mRNA cleavage or translational repression by associating with the RISC. Prediction of miRNA helps us to understand its structure and thus its function and role in organism. miRNA function in cell death,

S.C. Satapathy et al. (eds.), *ICT and Critical Infrastructure: Proceedings of the 48th Annual Convention of CSI - Volume II*, Advances in Intelligent Systems and Computing 249,
DOI: 10.1007/978-3-319-03095-1_12,

proliferation and fat metabolism in *Drosophila melanogaster*[3]. In plants, they regulate the development of leaves and flower. Thus intense study is required to find out the regulation of most fundamental biological processes in the organisms.Need for computational prediction: The short length of microRNA makes it difficult to analyse it with the help of conventional genetic techniques. Some microRNAs have low expression levels and some are expressed in specific conditions only, due to this reason their cloning is difficult. Also Deep-sequencing techniques require intense computational analysis to differentiate the miRNAs from other non-coding miRNAs [4]. Therefore we look up to the computational approaches to predict miRNA sequences and do their analysis.

Due to the short length of miRNA sequences, tools like BLAST give a large number of irrelevant hits. Hence only nearly perfect matches are to be found. Also the pre-miRNA sequences are less conserved which makes it difficult to use the conventional sequence alignment methods to find the homologous. Unlike the sequences, the secondary structures are more conserved which is helpful in predicting new miRNAs. Therefore more sensitive methods which consider both sequence and structure conservation are needed [5].

2 Review of Literature

To carry out the computational prediction and their analysis there are some tools which are based on the following techniques.

- Filter based- this approach uses different features and conservation criteria to restrict the presursor candidates.
- Machine learning- it uses the concept of learning through previously known miRNAs.
- Mixed approach- in this a combination of computational tools and high-throughput experimental procedures are used.
- Target centered approach- from conservation analysis a putative set of miRNA targets are developed which helps to find out new miRNAs
- Homology based- identifies the miRNAs similar to previously known pre-miRNAs.
- Rule based- it is based on some rules by studying the features of the sequences.

3 Materials and Methods

3.1 Reference miRNAs

The set of miRNAs and precursor miRNAs we used were downloaded from miRBase (version15, http://www.mirbase.org/). It has 1010 known mature miRNA sequences from 5 species; Oryza sativa(447), Arabidopsis thaliana(199), Zea mays(170), Sorghum bicolor(148) and Brassica napus(46). Oryza sativa, Arabidopsis thaliana are more in number as their genome sequence information is available.

Table 1. Tools for computational prediction of miRNA sequences are:

Name	Type	URL	Techniques
Mir Scan	W	http://genes.mit.edu/mirscan/	Filter-based
MiRFinder	D	http://www.bioinformatics.org/mirfinder/	Filter-based
ProMIR	W	http://cbit.snu.ac.kr/_ProMiR2/	Machine learning
TripletSVM	D	http://bioinfo.au.tsinghua.edu.cn/mirnasvm/	Machine learning
RNAMicro	D	http://www.bioinf.uni-leipzig.de/_jana/software/RNAmicro.html	Machine learning
MiPred	W	http://www.bioinf.seu.edu.cn/miRNA/	Machine learning
Mireval	W	http://tagc.univ-mrs.fr/mireval/	Mixed approaches
findMiRNA	D	http://sundarlab.ucdavis.edu/mirna/download.html	Target based
MirAlign	W	http://bioinfo.au.tsinghua.edu.cn/miralign/	Homology-based
BayesmiRNAfind	W	http://wotan.wistar.upenn.edu/miRNA	Rule based

3.2 Preparation of Dataset

Using the PERL scripts we automated the retrieval of 1010 sequences from miRBase. These sequences were put into the RNAfold, a software developed by M. Zuker and P. Stiegler [6, 7, 8, 9] for the secondary structure of our sequences and their MFE. Coding in PERL was done to calculate the values of the set attributes from their secondary structures.

3.3 Computational Analysis

WEKA (Waikato Environment for Knowledge Analysis) is a JAVA based software developed at the University of Waikato, New Zealand. WEKA version 3.6.2 was used to do our research.

It is a data mining tool written in java language which is a collection of machine learning algorithms. We are using the J48 classifier to classify our data as is the easiest and simplest way to interpret the results.

3.4 Data Curation

The 1010 miRNA sequences were downloaded from miRBase (15 release) with the help of Perl script. RNAfold was run on all the sequences and a secondary structure was generated along with the MFE of the sequences. The secondary structure is in the form of dot bracket format. Each bracket represents a base pairing and each dot a non paired base.

>osa-MIR395s MI0001037
GUAUCACCGUGAGUUCCCUUCAAGCACUUCACGUGGCACUAUUUCAAU
GCCUAUU GUGAAGUGUUUGGGGAACUCUCGAUGUUCC

>osa-miR395s MIMAT0000968
GUGAAGUGUUUGGGGAACUC

3.5 Sequences from miRBase and Their Ids

>osa-MIR395s MI0001037

GUAUCACCGUGAGUUCCCUUCAAGCACUUCACGUGGCACUAUUUCAAU
GCCUAUU GUGAAGUGUUUGGGGAACUCUCGAUGUUCC

....((.((.(((((((((.(((((((((((((.((((.........))))...))))))))))))))))))))))).)).))....

RNA fold dot bracket secondary structure

3.6 Identification of Attributes and Calculating the Values

From the sequences, 9 and 14 attributes were considered for mature microRNA sequences and precursor sequences respectively. Attributes for mature microRNA are ARM sequence on first or second arm of the hairpin structure, DFL distance of the mature sequence from loop sequence, BPN base pair per nucleotide, LNM length of the mature miRNA sequence, POP percentage of pairing, GCC Guanine and Cytosine nucleotide content in the sequence, MFE minimum free energy to fold the mature microRNA, DAS dominating nucleotide at start of the sequence and DAE dominating nucleotide at end of the sequence.

Attributes for precursor sequences are LEN length of the precursor sequence, NBP number of base pairs in the sequence, BLR base length ratio, NHP number of hairpins, HPL hairpin length, FRE free energy(minimum) to fold the sequence, FEN free energy(minimum) per nucleotide, AUC Adenine and Uracil nucleotide content in the sequence, MSK maximum stack in the sequence, SDI symmetric difference, MBL maximum length of the bulge, MBS maximum bulge symmetry, MTL maximum number of tails and NTL number of tails [10].

Table 2. Attributes for mature/precursor microRNA

Attributes for mature microRNA	Attributes for precursor microRNA
ARM, DFL, BPN, LNM, POP, GCC, MFE, DAS, DAE	LEN, NBP, BLR, NHP , HPL, FRE, FEN AUC, MSK, SDI, MBL MBS, MTL, NTL

Based on these attributes the values were calculated with the help of Perl scripts and sets of datasets were prepared for different species.

3.7 Attributes

A ={LEN,NBP,BLR,NHP,HPL,FRE,FEN,AUC,MSK,SDI,MBL,MBS,MTL,NTL}

3.8 Attribute Values

87, 33, 0.37, 1, 7, -37.20, 0.42, 56, 17, 2, 3, 6, 0, 0

To feed in the calculated data of the attributes of miRNAs, it was converted into ARFF format. Shuffle DNA was used to shuffle the precursor sequences of all the species in such a way that we generated sequences having hairpins. Using Perl script, randomly mature sequences were picked from the new randomised sequences and a negative dataset was created.

4 Result and Discussion

In this study, we find out the dominating attributes of the existing microRNAs of the plant species. In addition, decision trees building of the plant species using a classifier.

The graphical view represents some patterns found in the microRNA sequences. In the sequences the dominating nucleotide at the beginning of the sequence is Uracil whereas Adenine has the lowest percentage. Near the end of the sequences Cytosine percentage is highest and Adenine percentage is lowest. The data shows that the mature microRNA sequences are mostly found on the first arm on the hairpin loop. Maximum number of mature miRNA has minimum distance from the hairpin thus restating the fact that the mature microRNA sequences are to be found near the hairpin loop. Maximum precursor sequence shows no tail in the secondary structures. The AU content is 60% in maximum number of precursor sequences and GC content is 62% in maximum number of mature miRNA sequences. The free energy to fold the precursor sequences was found mostly between -52Kcal/mol and -37Kcal/mol. The hairpin length was between 4 to 6 nucleotides long in maximum sequences. Presence of mostly single hairpin was found though there were cases of more than one hairpin loop in precursor sequences. There were sequences found having six hairpins which were considered under special occurrences.

We tested the data as a training set and generated the decision trees for the species *Oryza sativa*(447), *Arabidopsis thaliana*(199), *Zea mays*(170), *Sorghum bicolor*(148) and *Brassica napus*(46).

4.1 Decision Tree Construction

The datasets were fed into the software and run. A graphical representation of the dataset was displayed. This graphical view indicates various patterns in out dataset values. WEKA revealed that there are some attributes which are dominating than rest of the attributes. Weka provides us with attribute evaluators and search methods which help us find the dominating attributes. These attributes vary with different search methods and the evaluators. A tabular view of selected attributes is shown below.

Table 3. Selected attributes

	Oryza sativa (20)	*Arabidopsis thaliana* (20)	*Zea mays*(20)	*Sorghum bicolor*(20)	*Brassica napus*((20)
BestFirst+CfsSubset Ev al	BLR, NHP, FRE, FEN, MSK, NTL, DFL	NBP, BLR, NHP, FEN, MTL, DFL	BLR, NHP, FRE, FEN, SBR, DFL, DAE	BLR, FRE, FEN, MSK, SDI, MBA, MBS,	BLR, FEN, SBR, MBA, MBS, DFL
Ranker+ChiSquared At tributeEval	FEN, NHP, FRE, DFL, SDI, MSK	FEN, BLR, DFL, NHP, FRE, MTL, MSK, SDI	BLR, FEN, DFL, NHP, MSK, FRE, SDI, MTL	MBS, BLR, FEN, SDI, MBA, DFL, FRE	MBS, FEN, BLR, SDI, MBA, FRE, NHP, DFL
GreedyStepwise+Co ns istencySubsetEval	BLR, NHP, FRE, FEN, MSK, SDI, MBS, DFL	NBP, BLR, NHP, FRE, FEN, MSK	BLR, NHP, FRE, FEN, MSK, MBL, BS, MTL, DFL	BLR, MBS	BLR, MBS
Ranker+ SVMAttribute Eval	NHP, DFL, FEN, AUC, FRE,	FEN, BLR, FRE, NHP, AUC, MBS	BLR, NHP, DFL, FEN, AUC, FRE, SDI	MBS, MBA, BLR, MSK, SDI, FEN, AUC	FEN, BLR, MBS, MBA, SDI, AUC, DFL, NHP
Ranker+InfoGainAt tri buteEval	NHP, DFL, FEN, AUC, FRE, MBS, MTL, SDI	FEN, BLR, DFL, NHP, FRE, MTL, MSK	BLR, FEN, DFL, NHP, MSK, FRE, SDI,MBL	MBS, BLR, FEN, SDI, MBA, DFL, FRE	MBS, BLR, FEN, MBA, SDI, DFL, NHP

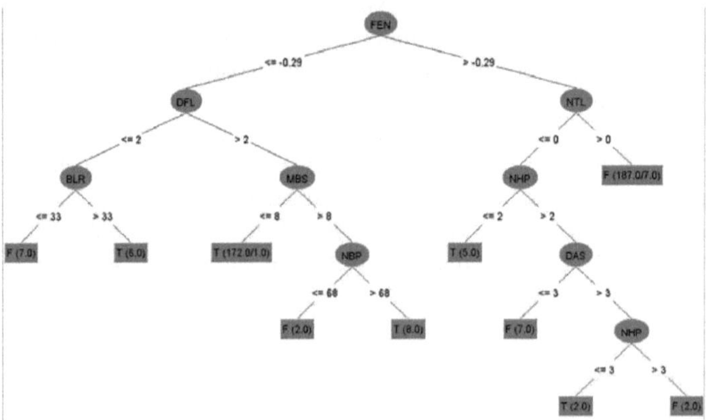

Fig. 1. Decision tree for with 20 attributes *Arabidopsis thaliana*

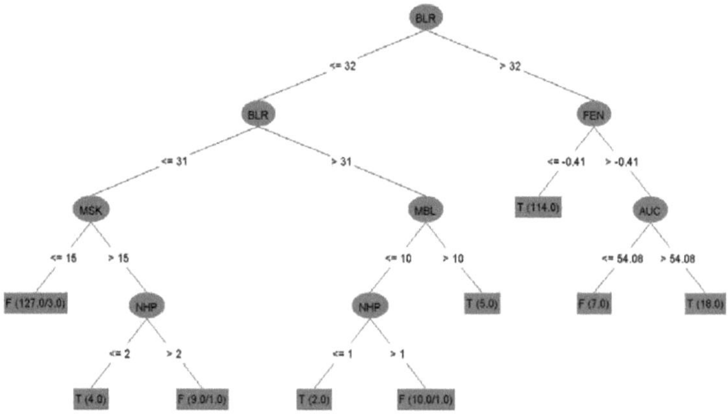

Fig. 2. Decision tree with 14 attributes for *Sorghum bicolor*

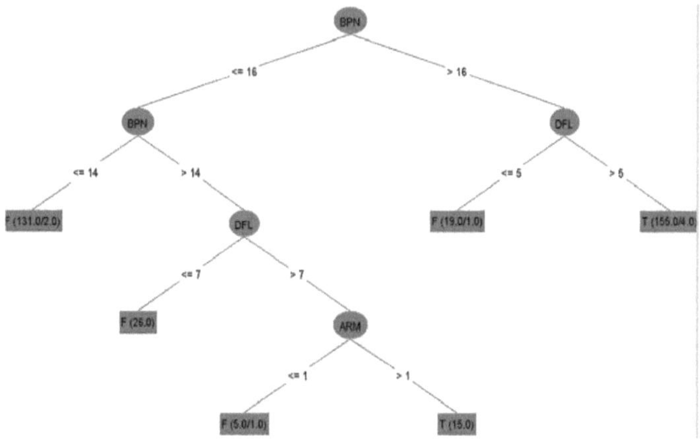

Fig. 3. Decision tree with 9 attributes for *Zea mays*

By observing the graphs certain patters are recorded of the different species. The classifier J48 which is an implementation of C4.5 algorithm works on the dataset. Analysis of the data generates a descriptive format in the form of decision trees. The decision tree checks an attribute at each node and the decision is made to classify the data. They are easy to interpret thus the decision trees of various species of plant are compared and the relevance of attribute is calculated.

4.2 Performance Evaluation Tables

The predictive performance was calculated by WEKA software. The TP rates, FP rates, precision (specificity) and recall (sensitivity) values. The values which were near one were considered good for classification. F-measure is the harmonic mean of the precision and recall. It is the threshold of precision and recall as they both cannot be increased together.

Table 4. Classification results with reference to *Oryza sativa*(9)

	TP Rate	FP Rate	Precision	Recall	F-Measure	Class
Training set	0.951	0.018	0.982	0.951	0.966	T
	0.982	0.049	0.952	0.982	0.967	F
Cross-validation with fold 10	0.94	0.065	0.935	0.94	0.938	T
	0.935	0.06	0.939	0.935	0.937	F
Percentage split (66%)	0.956	0.024	0.97	0.956	0.963	T
	0.976	0.044	0.965	0.976	0.971	F
Test against *Arabidopsis thaliana*	0.869	0.06	0.935	0.869	0.901	T
	0.94	0.131	0.878	0.94	0.908	F
Test against *Zea mays*	0.953	0.088	0.91	0.953	0.931	T
	0.912	0.047	0.954	0.912	0.932	F
Test against *Sorghum bicolor*	0.892	0.061	0.936	0.892	0.913	T
	0.939	0.108	0.897	0.939	0.917	F
Test against *Brassica napus*	0.957	0.065	0.936	0.957	0.946	T
	0.935	0.043	0.956	0.935	0.945	F

Table 5. Classification results with reference to *Arabidopsis thaliana* (14)

	TP Rate	FP Rate	Precision	Recall	F-Measure	Class
Training set	0.96	0	1	0.96	0.979	T
	1	0.04	0.961	1	0.98	F
Cross-validation with fold 10	0.925	0.05	0.948	0.925	0.936	T
	0.95	0.075	0.926	0.95	0.938	F
Percentage split (66%)	0.924	0	1	0.924	0.961	T
	1	0.076	0.903	1	0.949	F
Test against *Oryza thaliana*	0.949	0.105	0.9	0.949	0.924	T
	0.895	0.051	0.946	0.895	0.92	F
Test against *Zea mays*	0.924	0.188	0.831	0.924	0.875	T
	0.812	0.076	0.914	0.812	0.86	F
Test against *Sorghum bicolor*	0.885	0.101	0.897	0.885	0.891	T
	0.899	0.115	0.887	0.899	0.893	F
Test against *Brassica napus*	0.957	0.043	0.957	0.957	0.957	T
	0.957	0.043	0.957	0.957	0.957	F

Table 6. Classification results with reference to *Zea mays*(20)

	TP Rate	FP Rate	Precision	Recall	F-Measure	Class
Training set	0.971	0	1	0.971	0.985	T
	1	0.029	0.971	1	0.986	F
Cross-validation with fold 10	0.924	0.053	0.946	0.924	0.935	T
	0.947	0.076	0.925	0.947	0.936	F
Percentage split (66%)	0.929	0.05	0.945	0.929	0.937	T
	0.95	0.071	0.934	0.95	0.942	F
Test against *Oryza sativa*	0.919	0.085	0.915	0.919	0.917	T
	0.915	0.081	0.919	0.915	0.917	F
Test against *Arabidopsis thaliana*	0.839	0.045	0.949	0.839	0.891	T
	0.955	0.161	0.856	0.955	0.903	F
Test against *Sorghum bicolor*	0.946	0.203	0.824	0.946	0.881	T
	0.797	0.054	0.937	0.797	0.861	F
Test against *Brassica napus*	0.913	0.283	0.764	0.913	0.832	T
	0.717	0.087	0.892	0.717	0.795	F

5 Conclusion

The classification yielded good results based on the decision trees which are best suited for the classification of miRNAs. In our studies we predicted the dominating attributes which are the basis o f classification of miRNAs of related species. These attributes hold biological significance. Clustering was not required as our data was based on two classes.

References

1. He, L., Hannon, G.J.: MicroRNAs: small RNAs with a big role in gene regulation. Nature Genetics (2004)
2. Lee, Y., Jeon, K., Lee, J.-T., Kim, S., Kim, V.N.: MicroRNA maturation: stepwise processing and subcellular localization. EMBO J. (2002)
3. Alvarez-Garcia, I., Miska, E.A.: MicroRNA functions in animal development and human disease. Development. The Company of Biologists (2005)
4. Mendes, N.D., Freitas, A.T., Sagot, M.-F.: Current tools for the identification of miRNA genes and their targets. Nucleic Acids Research (May 2009)
5. Wang, X., Zhang, J., Li, F., Gu, J., He, T., Zhang, X., Li, Y.: MicroRNA identification based on sequence and structure alignment. Bioinformatics (2005)
6. Hofacker, I.L., Fontana, W., Stadler, P.F., Bonhoeffer, S., Tacker, M., Schuster, P.: Fast-Folding and Comparison of RNA Secondary Structures. Monatshefte F. Chemie 125, 167–188 (1994)

7. Zuker, M., Stiegler, P.: Optimal computer folding of large RNA sequences using thermodynamic and auxiliary information. Nucl. Acid Res. (1981)
8. Hofacker, I.L., Stadler, P.F.: Memory Efficient Folding Algorithms for Circular RNA Secondary Structures. Bioinformatics (2006)
9. Bompfunewerer, A.F., Backofen, R., Bernhart, S.H., Hertel, J., Hofacker, I.L., Stadler, P.F., Will, S.: Variations on Folding and Alignment: Lessons from Benasque. J. Math. Biol. (2007)
10. Mishra, A.K., Lobiyal, D.K.: Exploring Dominating Features from Apis Mellifera Pre-miRNA. IEEE (2009)

Features Selection Method for Automatic Text Categorization: A Comparative Study with WEKA and RapidMiner Tools

Suneetha Manne[1], Supriya Muddana[1], Aamir Sohail[1], and Sameen Fatima[2]

[1] Department of IT, VRSEC, Vijayawada, India
[2] Depratment of CSE, Osmania University, Hyderabad

Abstract. The advent of Internet over the past few decades has totally revolutionised the fields of Science and Technology. Enormous increase in data on internet has raised the need of effective representation of textual information. The organizers of technical conferences and journals have to place the research papers in various session tracks, for which they need to spend a lot of time. The investigation provides a solution for this problem by automatic document categorization approach with the help of features selection method. Researchers and students constantly face a problem that, it is almost impossible to read most of the newly published papers to be informed of the latest progress. The time spent on reading literature review seems endless. The goal of this research is to design a domain independent automatic text categorization system to alleviate, if not totally solve, this problem. Text categorization is the task of assigning predefined categories to natural language text. This paper explores the effect of word and other values of word in the document, which express the features of a word in the document. The proposed features are exploited by a tf-itf, position of the word, compactness and these features are combined. Experiments show that the feature selection method has been effective for text categorization. The proposed text categorization approach is validated with Naïve Bayesian, Decision Tree Induction, Nearest Neighbour and SVM approaches. The results of the experiment have shown comparatively good accuracy (above 95%), precision and recall, ensuring that the system is more effective and efficient. The experimental results revealed that text categorization had a significant improvement with the help of combination of these features.

1 Introduction

The people involved in research need to analyze the research papers, e-books and other resources available on the web. Even in the Human Resources department of Multinational companies, it is difficult to categorize the Curriculum Vitae received from hundreds or often thousands of applicants. In such applications, Text Categorization becomes the key tool for automatic handling and organizing of text information. According to Fabrizio Sebastiani "Text categorization (also known as text classification) is the task of automatically sorting a set of documents into

S.C. Satapathy et al. (eds.), *ICT and Critical Infrastructure: Proceedings of the 48th Annual Convention of CSI - Volume II*, Advances in Intelligent Systems and Computing 249,
DOI: 10.1007/978-3-319-03095-1_13, © Springer International Publishing Switzerland 2014

categories from a predefined set" [1]. Text categorization is a prime research area in information retrieval and machine learning. Today, text classification is necessary due to the very large amount of text documents that are to be dealt with day by day. In general, text classification plays an important role in information extraction, summarization, text retrieval, and question answering.

Text categorization is the task of assigning predefined categories to natural language text [2]. Text Categorization may be formalized as the task of approximating the unknown target function: $\Phi: D \times C \rightarrow \{T, F\}$ that describes how documents ought to be classified, according to a supposedly authoritative expert by means of a function, where $C = \{c1 \ldots c|C|\}$ is a pre-defined set of categories and D is a (possibly infinite) set of documents. The paper is organized as follows. In section 2, related research work on Text Categorization is discussed. The proposed approach is elaborated in section 3. The experimental results with observations are shown in section 4. Finally in section 5 conclusions are presented.

2 Text Categorization Approaches

The automated categorization of texts into topical categories has a long history, dating back at least to 1960. Until the late '80s, the dominant approach to the problem involved knowledge-engineering automatic categorizers that manually built a set of rules encoding expert knowledge on how to classify documents. Over the years, automated categorization technologies have used a succession of different algorithms. Some of the most longstanding technologies we could mention are semantics-based approaches, which had the disadvantage of high costs in human and financial terms if the classification system required updating. To counter this problem, several machine learning approaches are introduced.

2.1 Machine Learning Approaches

This section describes three Machine Learning techniques that are common for Text Categorization: Naive Bayes categorizers, Decision Table and Support Vector Machines. Machine Learning algorithms required to provide a set of examples from which the rules defining the machine behaviour are extracted. Automatic Text Categorization systems attempt to label documents according to ontology of classes defined by the user. The problem is a supervised task, that is the machine is tuned using a training set of labelled documents in order to minimize the error between the real target and the predict label [6]. A naive Bayes classifier assumes that the presence or absence of a particular feature is unrelated to the presence or absence of any other feature, given the class variable. Naive Bayes classifiers can be trained very efficiently in a supervised learning setting [3] [4]. However there are a few demerits which are to be stressed although. First, the naive Bayes classifier requires a very large number of records to obtain good results. Second, where a predictor category is not present in the training data, naive Bayes assumes that a new record with that category of the predictor has zero probability. This can be a problem if this rare predictor value is important [5].

Support Vector Machines are a new learning method introduced by V.Vapnik et al in the year 1979. One remarkable property of SVMs is that their ability to learn can be independent of the dimensionality of the feature space. SVMs measure the complexity of hypotheses based on the margin with which they separate the data, not the number of features [7]. Perhaps the biggest limitation of the support vector approach lies in choice of the kernel. Second limitation is speed and size, both in training and testing and the discrete data presents another problem. Another factor is that it lacks transparency of results [8].

2.2 N-Gram Based Text Categorization

At present, text categorization techniques are predominantly keyword-based. Many researchers in the field have used different classifiers, but most of them treat a document as a bag of words (BOW), that is, identify terms with all the words occurring in the document and perform categorizations based mainly on the presence or absence of keywords. The main drawback of the BOW representation is in destruction of semantic relations between words. In early 90s, Bag-Of-Bigrams (pairs of consequent words) was proposed by Lewis as a competitive representation [9]. While some of the researchers report significant improvement in text categorization results (Mladeni´c and Grobelnik), many of them show only marginal improvement or even a certain decrease [10].

The reviewed literature resulted that, most of the researchers till now relied highly on training dataset or corpus to classify a test file to do Text Categorization. Each training document of such undertaken corpus is usually of large in size due to which, most approximations, computations and analyses are time consuming. To overcome these drawbacks architecture is proposed which does text categorization. Section 3 gives the detailed description for this architecture.

3 Features Selection Method

The proposed method comes up with a new procedure of combining two or more features for text categorization which gives results better as compared to existing classifiers. It does not require trained data unlike existing classifiers. Also has advantage of reduced time and space complexity. The following are the algorithm steps that are implemented as per the proposed method.

3.1 Preprocessing

Textification: The process of converting given input data into Unicode text only format is known as Textification. The textified file is now free from codes, tags, images, graphics, etc. This plain encoded text is now considered for the Information Analysis.

Table 1. Depicts the algorithmic steps of feature selection method

Step 1: Input a random document.
Step 2: Filtering of document is done i.e. in other words preprocessing steps are applied.
- (a) Textification
- (b) Case folding and Lemmatization
- (c) Stop word removal
- (d) Tokenization
- (e) Stemming using Porters algorithm

Step 3: The filtered document is passed to various features for categorization of document.
- (a) tf- itf
- (b) Compactness
- (c) First Appearance

Step 4: The feature combinations are applied in order to categorize documents rather than using them solely.
Step 5: The input document is mapped with the existing database in order to detect the document through these features.
Step 6: If the input document is more relevant to existing field in database then the document is said to be belonging to the respective field. Hence there forth the document is said to be categorized.

Case Folding and Lemmatization: The whole document is converted to a unified font i.e. either capital case or lower case of alphabets. Also, the plural words are converted to singular. This is to get the uniqueness of all words in document.

Tokenization: The processing of text often consists of parsing a formatted input string. The sentences in the document are segmented into tokens.

Stop Word Removal: Some words are extremely common and occur in a large majority of documents. Categorization is based on the featured terms not on commas, full stops, colons, semicolons etc.. To reduce search space and processing time, these stop words are dealt separately by recognizing them in the stemmed file and remove it from document.

Stemming: In most cases, morphological variants of words have similar semantic interpretations and can be considered as equivalent for the purpose of IR applications. For this reason a number of so called stemming algorithms or stemmers have been developed which attempts to reduce a word to its stem or root form. This method uses porter's algorithm for stemming the word to its root [11].

3.2 Features Selection

Term Frequency and Inverse Term Frequency (tf-itf): The importance of a word can be measured by its term frequency which means it counts the number of occurrences of a particular term in the whole document.

$$\text{TF (t, d)} = \frac{Count\ (t,d)}{Size\ (d)}$$

Considering only terms which have occurred many times will not be sufficient to have effective categorization. There are also terms which occur less times in the document but have significant part in categorizing the document. The terms which are less frequent are determined by using following formula:

$$\text{ITF (t, d)} = \frac{\log |d|}{|\{d \in D : t \in d\}|}$$

Now by combining both term and inverse term frequency to get the terms with combined values which gives the terms of most related to the document with high values.

$$\text{TF-ITF (t, d)} = \text{importance of (t, d)} * \text{ITF (t, d)}$$

Compactness of Appearance of the Word: A word is compact if its appearances concentrating specific part of a document and less compact if its appearances spread over the whole document. This consideration is motivated by the following facts. A document usually contains several parts. If the appearances of a word are less compact, the word is more likely to appear in different parts and more likely to be related to the theme of the document [12]. The more frequent, the more important, the compactness of the appearances of a word shows that the less compact, the more important and the position of the first appearance of a word shows that the earlier, the more important. These features are calculated with the following equations.

$$\text{Count (t, d)} = \sum_{i=0}^{n-1} c_i$$

$$\text{Centroid (t, d)} = \frac{\sum_{i=0}^{n-1} c_i \times \text{i}}{\text{count(t,d)}}$$

$$\text{Compactness (t, d)} = \frac{\sum_{i=0}^{n-1} c_i \times |i - \text{centroid(t,d)}|}{\text{count(t,d)}}$$

Position of First Appearance of the Word: This consideration is based on an intuition that the author naturally mentions the important contents in the earlier parts of a document. This feature is influenced by the fact that any author specifies a word which is important in the document mentions in the early part of the document.

$$\text{First appearance (t, d)} = min_{i \in \{0 \dots n-1\}} \times c_i > 0 ?\ \text{i} : \text{n}$$

Combinations: As mentioned above each feature is capable of categorizing text documents, the proposed method of using the combinations of these features have resulted in better and accurate results. The combinations that are used here are a) tf-itf and compactness b) compactness and first appearance c) first appearance and tf-itf d) combination of three (tf-itf, compactness and first appearance).

4 Experimental Results and Observations

Samples of 10 random documents are used to represent the effectiveness of each feature to categorize the document correctly and accurately. The extent to which each document differed from the exact document is also scaled and has been mismatched with the relevant field. The calculation is done with the following equation. The document is termed as most relevant if it falls under "low" level of irrelevancy and vice versa.

Difference {matched words of document with irrelevant field and matched words of document of relevant field}

Low [0-100] ----> represents the range of words by which a particular document differed from original document.
Medium [100-500] ----> represents the range of words by which a particular document differed from original document.
High [500-1000] ----> represents the range of words by which a particular document differed from original document.

In the similar fashion the combinations are tried for documents of respective fields and the results attained are depicted to show the effectiveness of using combinations rather than using each feature alone to categorize documents. The sample graph for two different fields comprising of 30 sample documents are shown in Fig.1.

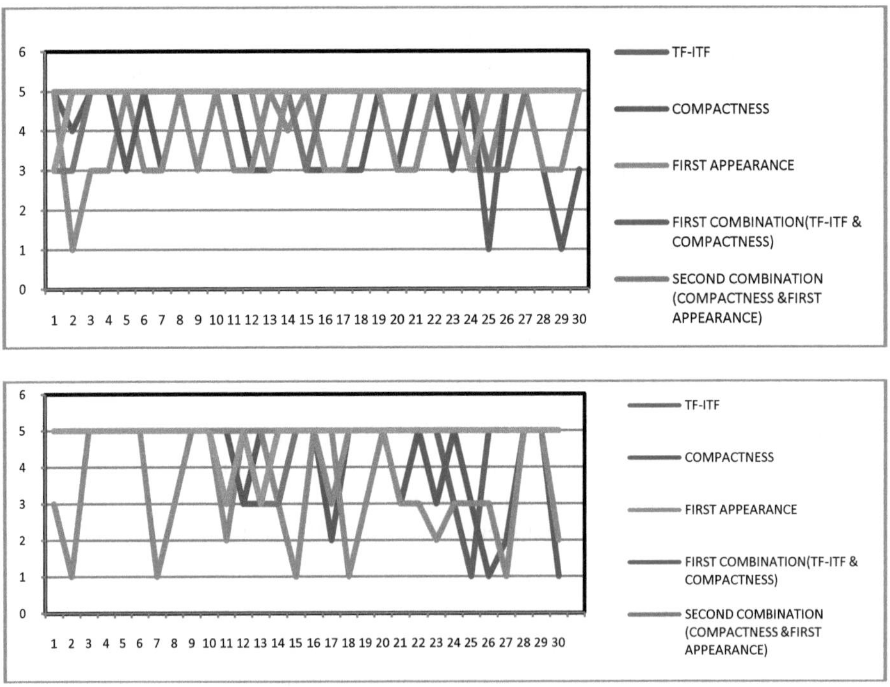

Fig. 1. Result graphs for Computer Graphics and Bio- informatics fields

A total of 204 random text files of six different fields are taken and is categorized using the combinations of different features i.e using the “all combination feature” as stated above.

The categorization of documents is also evaluated using various classification techniques with WEKA and RapidMiner tools. The level of accuracy is obtained by using Decision table, Naïve Bayes, KNN and SVM are shown in Table 2 and Table 3.

Table 2. Experimental results with WEKA tool

Parameters	Decision table	Naïve	SVM
Mean absolute error	0.1037	0.085	0.0408
Root mean squared error	0.1689	0.2914	0.2021
Precision	0.936	0.83	0.895
Recall	0.931	0.745	0.877
F-Measure	0.932	0.759	0.025
Accuracy	**87.1373%**	**74.5098 %**	**82.7451 %**

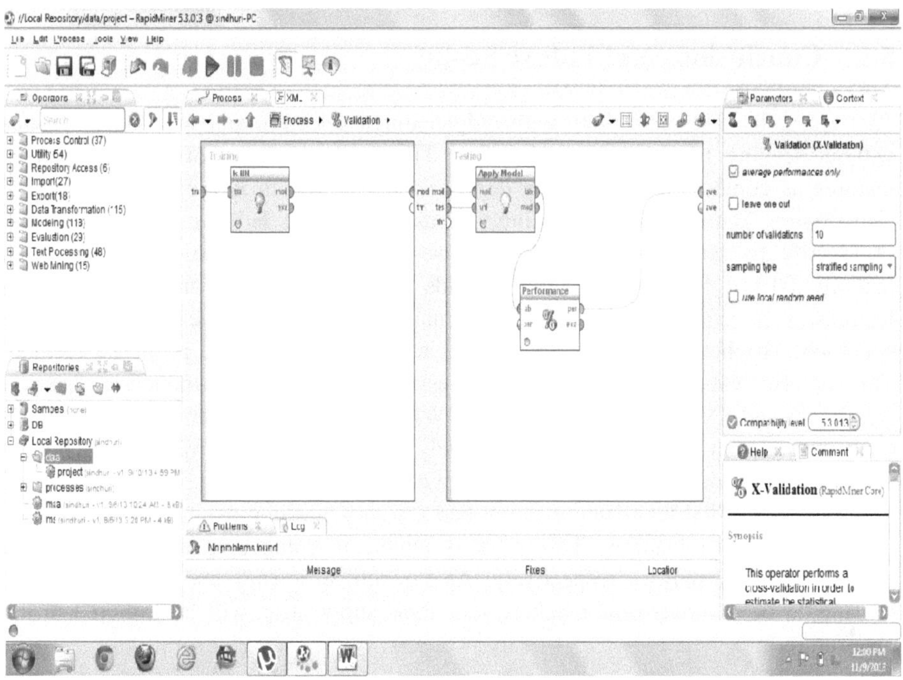

Fig. 2. Cross validation applied on k-NN classification

To begin with rapid miner, the input files are first loaded into the repository. The documents are processed from the fed input files. The processed documents are then tokenized and filtered. Now cross validation is performed on the documents. It consists of training phase and testing phase. In training phase a model is trained by passing the data set and in the testing phase the model is tested and its performance is measured. The Fig.2.depicts cross validation applied on k-NN classification.

Table 3. Experimental results with RapidMiner tool

Parameters	**k- NN**	**Naïve Bayes**
Absolute error	0.086	0.090
Relative error	8.59%	9.02%
Normalized absolute error	0.057	0.060
Squared error	0.086	0.090
Accuracy	**91.41%**	**90.98%**

5 Conclusions and Future Scope

This research was solely done with only features selection and their combinations without seeking the help of a classifier. The implementation of the selection of features as combinations has been done and it resulted in improved efficiency of categorising. The result of combining various features has led to more efficient way of identifying various documents and also has decreased time complexity. The random 204 test documents of various fields were taken as input and tested upon all features, their combinations and have finally compared with existing classifiers approach. The better accuracy above 95% was reached with the proposed method. This is also validated with Naïve Bayesian, Decision Tree Induction, Nearest Neighbour and SVM approaches by using WEKA and RapidMiner tools. The results of the experiment have shown comparatively good accuracy (above 95%), precision and recall, ensuring that the system is more effective and efficient. The experimental results revealed that text categorization had a significant improvement with the help of combination of features. This work is unable to categorize the data based on phrases and Para-phrases respectively. As a result, efficiency and time-complexity remains to be a concern on a lighter vein. This supposition will be explored in the near future.

References

1. Sebastiani's, F.: Machine Learning in Automated Text Categorization. ACM Computing Surveys 34(1), 1–47 (2002)
2. Xue, X.-B., Zhou, Z.-H.: Distributional Features for Text Categorization. In: Fürnkranz, J., Scheffer, T., Spiliopoulou, M. (eds.) ECML 2006. LNCS (LNAI), vol. 4212, pp. 497–508. Springer, Heidelberg (2006)
3. Pattern Recognition and Machine Learning. Christopher Bishop. Springer (2006)
4. Pattern Classification by Duda, R.O., Hart, P.E., Stork, D.: Wiley and Sons
5. Ng, A.Y., Jordan, M.I.: On Discriminative vs. Generative Classifiers: A comparison of Logistic Regression and Naive Bayes. In: Neural Information Processing Systems (2002)
6. Baoli, L., Shiwen, Y., Qin, L.: An Improved k-Nearest Neighbor Algorithm for Text Categorization Institute of Computational Linguistics Department of Computer Science and Technology Peking University, Beijing, P.R. China, p. 100871
7. Auria, L.: Rouslan: Support Vector Machines (SVM) as a Technique for Solvency Analysis
8. Joachims, T.: Text categorization with support vector machines: Learning with many relevant features. In: Nédellec, C., Rouveirol, C. (eds.) ECML 1998. LNCS, vol. 1398, pp. 137–142. Springer, Heidelberg (1998)
9. Lewis, D.D.: An Evaluation of Phrasal and Clustered Representations on a Text Categorization Task. In: Proceedings of the 15th Annual International ACM SIGIR Conference on Research and Development in Information Retrieval, pp. 37–50 (1992)
10. MladeniL ́C, D., Grobelink, M.: Word Sequences as Features in Text Learning. In: Proceedings of the 17th Electro technical and Computer Science Conference (ERK 1998), Ljubljana, Slovenia. IEEE section (1998)
11. Xue, X.-B., Zhou, Z.-H.: Distributional features for text categorization. IEEE Trans. Ensembles, IEEE Trans. Knowledge and Data Eng. 21(3) (2009)
12. Yang, Y., Pedersen, J.O.: A Comparative Study on Feature Selection in Text Categorization. In: Proc. of Int'l Conf. on Machine Learning, pp. 412–420 (1997)

References

Outliers Detection in Regression Analysis Using Partial Least Square Approach

Nagaraju Devarakonda[1], Shaik Subhani[1], and Shaik Althaf Hussain Basha[2]

[1] Department of Computer Science & Engg.,
Acharya Nagarjuna University,
Nagarjuna Nagar
[2] Department of MCA,
Gokaraju Rangaraju Institute of Eng. & Tech., Hyderabad
dnagaraj_dnr@yahoo.co.in,
subbu_buddu@ymail.co.in,
althafbashacse@gmail.com

Abstract. Identifying abnormal behavior in the chosen dataset is essential for improving the quality of the given dataset and decreasing the impact of abnormal values/patterns in the knowledge discovery process. Outlier detection may be established in many data mining techniques. In this paper Regression analysis have been used to detect the outliers. Partial Least Square approach is mainly used in regression analysis. Laser dataset has been used to find out the outliers. The main objective is used for constructing predictive models. The Mahalanobis distance, Jackknife distance and T^2 distance were calculated for finding the outliers.

Keywords: Outliers, Regression, classification, correlation, least squares.

1 Introduction

An outlier is a data point which is significantly different from the residual data [1].The concept of an outlier formally defined as follows: "outlier is an observation which deviates so much from the other observations as to arouse suspicions that it was generated by a different mechanism" [2]. In data mining and statistics literature Outliers are also called as abnormalities, irregularities, defects, discordant, deviants, or anomalies. In many applications, the data is created by one or more generating processes, which could either reflect activity in the system or observations collected about entities. When the generating process behaves in an unusual way, it results in the creation of outliers. Thus, an outlier contains useful information about abnormal characteristics of the systems and entities, which impact the data generation process. In data mining, one of the fundamental issue is a Outlier detection, specially it has been used to detect and remove anomalous objects from given data. Outlier occurs due to mechanical faults, changes in system performance, fraudulent behavior, network intrusions or human errors [3, 4, 5]. Most of such applications are high

S.C. Satapathy et al. (eds.), *ICT and Critical Infrastructure: Proceedings of the 48th Annual Convention of CSI - Volume II*, Advances in Intelligent Systems and Computing 249,
DOI: 10.1007/978-3-319-03095-1_14, © Springer International Publishing Switzerland 2014

dimensional domains in which the data contain hundreds of dimensions. [6, 7]. Outlier plays a major role in regression. It is important to differentiate between two types of outliers. Outliers in the response variable represent model failure. Such observations are called outliers. Outliers with respect to the predictors are called leverage points. They can affect the regression model, too. Their response variables need not be outliers [8, 9]. A well-known supervised learning technique is Regression analysis, which deals with estimation of an output value based on input value. It can be used to solve classification problems, and application such as forecasting Regression can be performed using many different types of techniques including neural networks. In actuality, regression takes a set of data and fits the data to a formula. This paper will provide an overview of regression methods and illustrate the use of the procedure to fit regression models and display outliers and leverage points. Partial least squares is a popular method for soft modeling in industrial applications. Partial least squares (PLS) is a method for constructing predictive models when the factors are many and highly collinear. The X- and Y-scores are chosen so that the relationship between successive pairs of scores is as strong as possible. In principle, this is like a robust form of redundancy analysis, seeking directions in the factor space that are associated with high variation in the responses but biasing them toward directions that are accurately predicted. In the rest of the paper, Existing work is presented in section 2, classification of regression of analysis are described in section 3, proposed work and experimental analysis are described in section 4 and 5. Conclusion of the paper is given in section 6.

2 Existing Work

Regression is an important method for analyzing data which contaminated with outliers. It can be used to detect outliers and to provide constant results in the presence of outliers. To estimate and justify an effective model from regression analysis, it is necessary to check and preprocess the data set. Without outliers , it is impossible to get a real data. Regression analysis is usually applied in many statistical aspects, science and engineering applications [10]. There are many regression techniques, out of which the least squares (LS) method has been generally adopted because of simplicity and easy computation. However, there is presently a widespread awareness of the dangers posed by the occurrence of outliers, which may be a result of keypunch errors, misplaced decimal points, recording or transmission errors, exceptions, different population slipping into the sample. Outliers occur very frequently in real data, and they often go unnoticed because now days much data is processed by computers, without careful inspection or screening. Not only the response variable can be outlying, but also the explanatory part, leading to so-called leverage points. Both types of outliers may totally spoil an ordinary LS analysis. Often, such influential points remain hidden to the user, because they do not always show up in the usual LS residual plots.

3 Classification of Regression Analysis

The Regression analysis can be classified as Linear Regression, Multiple Regression, Curve Linear Regression and Polynomial Regression. Curve Linear Regression can be divided into Power Curves, Reciprocal Curves Exponential Curves and Gas Equation etc which can be shown in figure 1.

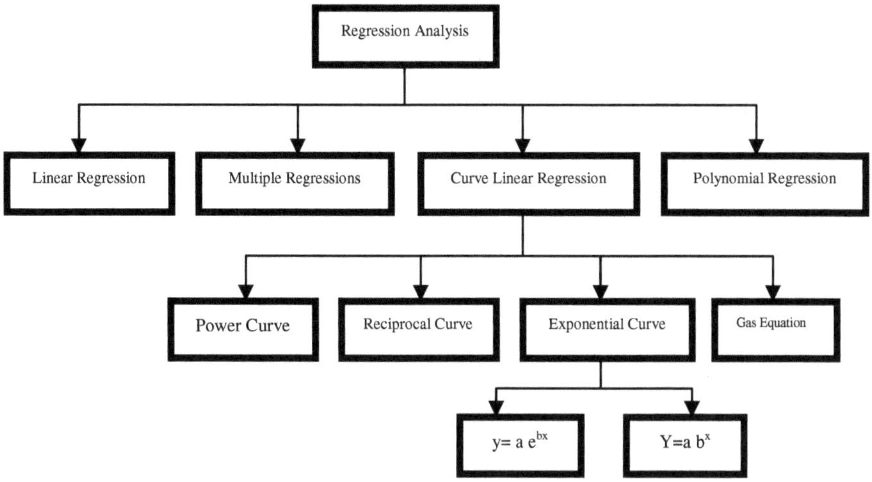

Fig. 1. Classification of Regression Analysis

3.1 Linear Regression

Linear Regression model which models the data into lower dimensional embedded subspaces with the use of linear correlations. The optimal line which passes through these points is determined with the use of regression analysis. Typically, a least squares fit is used to determine the optimal lower dimensional subspace. The distances of the data points from this plane are determined. Extreme values analysis can be applied on these deviations in order to determine the outliers. The simple linear regression in two dimensional spaces is given by

$$y = a + b\,x \tag{1}$$

Here, a and b are called regression coefficients or Y-intercept and slope of the line. x and y are called two points represent the training data

For determination of regression coefficients, we need to consider the following normal equations.

$$\sum y = n\,a + b\ \sum x \tag{2}$$

$$\sum xy = a\ \sum x\ + b\ \sum x\,2 \tag{3}$$

3.2 Multiple Regression

Multiple regressions are the most effective at identifying relationship between a dependent variable and a combination of independent variables. Each of the metric variables in multiple regression is normally distributed, and the relationships between metric variables are linear. Outliers can distort the regression results. When an outlier is included in the analysis, it pulls the regression line towards itself. This can result in a solution that is more accurate for the outlier, but less accurate for all of the other cases in the data set. We will check for uni-variate outliers on the dependent variable and multivariate outliers on the independent variables.

The general equation for multiple regressions is given by

$$Y= a_0 + a_1 x_1 + a_2 x_2 + \dots\dots + a_n x_n \quad (4)$$

By determining the regression coefficients a_0, a_1,…………, a_n ,the relationship between the output parameter 'y' and the input parameters x_1,x_2,……x_n can be estimated. For that purpose, consider the following normal equation for n = 2 are given

$$\sum y = n a_0 + a_1 \sum x_1 + a_2 \sum x_2$$
$$\sum x_1 y = a_0 \sum x_1 + a_1 \sum x_1^2 + a_2 \sum x_1 x_2$$
$$\sum x2y = a_0 \sum x2 + a_1 \sum x1x2 + a_2 \sum x22$$

3.3 Polynomial Regression

This function fits a polynomial regression model to powers of a single predictor by the method of linear least squares. Interpolation and calculation of areas under the curve are also given. The general principles which include in polynomial regression model as:

i) The fitted model is more reliable when it is built on large numbers of observations.
ii) Do not extrapolate beyond the limits of observed values.
iii) Choose values for the predictor (x) that are not too large as they will cause overflow with higher degree polynomials; scale x down if necessary.
iv) Do not draw false confidence from low P values; use these to support your model only if the plot looks reasonable.

In general, we can model the expected value of *y* as an *n*th order polynomial, yielding the general polynomial regression model.

$$Y=a_0+a_1x+a_2x^2+\dots\dots+a_nx^n \quad (5)$$

Conveniently, these models are all linear from the point of view of estimation, since the regression function is linear in terms of the unknown parameters a_0, a_1, ...a_n Therefore, for least squares analysis, the computational and inferential problems of polynomial regression can be completely addressed using the techniques of multiple regression. This is done by treating x, x^2, ...x_n as being distinct independent variables in a multiple regression model.

In general, the normal equations for polynomial model is given by

$$\sum y = n\,a0 + a1 \sum x + \dots\dots\dots\dots\dots\dots\dots + an \sum x n$$
$$\sum xy = a_0 \sum x + a_1 \sum x\,2 + \dots\dots\dots\dots + a_n \sum x\,n+1$$
$$\vdots$$
$$\sum xpy = a_0 \sum xp + a_1 \sum x\,p+1 + \dots\dots\dots\dots + a_n \sum x2\,n$$

For n=2, then it becomes a quadratic model $y=a + b\,x + c\,x^2$ and its normal equations are

$$\sum y = n\,a + b \sum x + c \sum x\,2$$
$$\sum xy = a \sum x + b \sum x2 + c \sum x\,3$$
$$\sum x2y = a \sum x2 + b \sum x\,3 + c \sum x4$$

Polynomial models are useful in situations where the analyst knows that curvilinear effects are present in the true response function. These models are also useful as approximating functions to unknown and possible very complex nonlinear relationship. Polynomial model is the Taylor series expansion of the unknown function.

4 Proposed Work

An outlier detection method is proposed and analyzed using regression algorithm. It provides efficient outlier detection. The proposed outlier detection method is divided into two stages. The first stage provides calculation of regression coefficients. The main objective of the second stage is an iterative removal of objects. The removal occurs according to a chosen threshold. Finally, we provide experimental results using the proposed approach for the benchmark dataset such as LASER dataset which shows its effectiveness and usefulness. The empirical results indicate that the proposed method was successful in detecting intrusions and promising in practice. We also compare regression algorithm with other available methods such as WEKA to show its important advantage against existing algorithms in outlier detection. The Correlation coefficient is 0.8365, Mean absolute error is 14.9227, Root mean squared error is 25.6897, Relative absolute error is 39.9791 % and Root relative squared error is 54.7486 %.

5 Experimental Analysis

While we are considering partial least squares, the model launch can be specified by two methods such as NIPALS, SIMPLS and the validation method as Leave-one-out-method can be chosen. For the Factor Search Range, the initial number of factors is taken as 1. The sample laser dataset has been chosen for outlier detection in regression, which is shown in table 1. The correlations were shown in table 2. The scatter matrix of the given dataset is shown in figure 2.

Table 1. Sample Laser dataset

Input1	Input2	Input3	Input4
95	32	138	111
41	72	111	48
21	111	23	19
32	48	19	27
72	23	27	59
138	19	59	129
48	59	129	58
23	129	58	27
27	58	19	24
59	27	24	46
129	19	46	112
129	24	112	144
58	46	144	73
27	112	73	30
19	144	30	20
24	73	20	19
46	30	19	37

Table 2. Correlations

Correlations	Input 1	Input 2	Input 3	Input 4
Input 1	1.0000	- 0.6360	0.0962	0.6917
Input 2	- 0.6360	1.0000	- 0.1987	- 0.5817
Input 3	0.0962	- 0.1987	1.0000	0.5307
Input 4	0.6917	- 0.5817	0.5307	1.0000

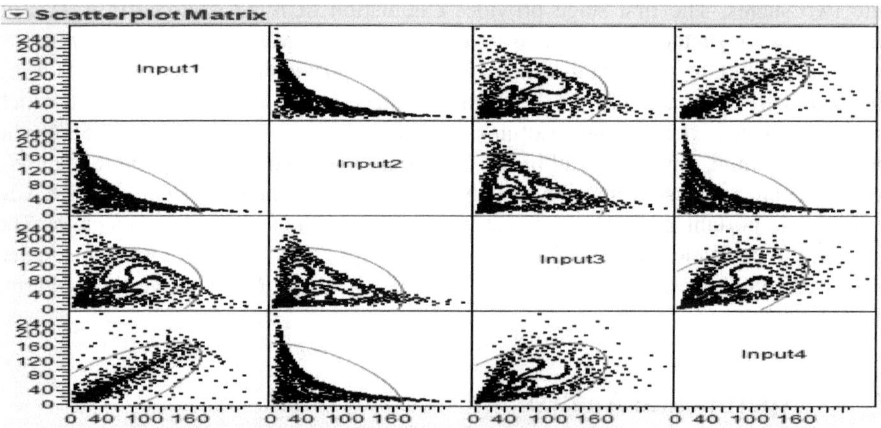

Fig. 2. Scatter Matrix of the Laser dataset

To find out the outliers for the chosen dataset, Mahalanobis distance, Jackknife distance and T^2 distance were calculated and shown in the figures 3, 4, and 5 respectively.

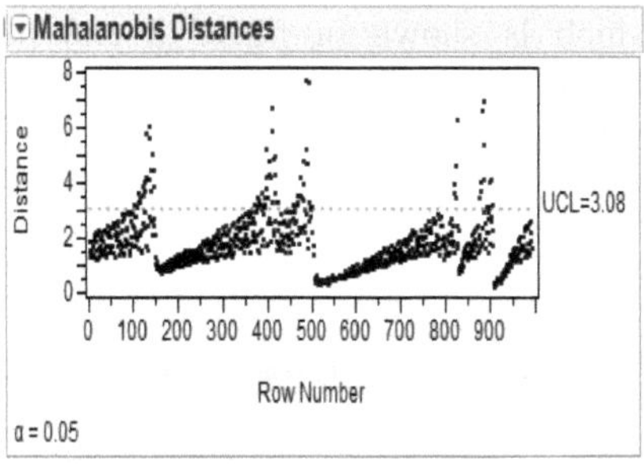

Fig. 3. Mahalanobis Distance for the given dataset

Model Comparison Summary

Method	Number of rows	Number of factors	Percent Variation Explained for Cumulative X	Percent Variation Explained for Cumulative Y	Number of VIP > 0.8
NIPALS	993	1	100	36.615659	1
SIMPLS	993	1	100	36.615659	1

Fig. 6. Model Comparison Summary

Cross Validation with Method=NIPALS

Number of factors	Root Mean PRESS	van der Voet T²	Prob > van der Voet T²
0	1.001008	296.83277	<.0001*
1	0.798182	0.000000	1.0000

Fig. 7. Cross Validation with NIPALS method

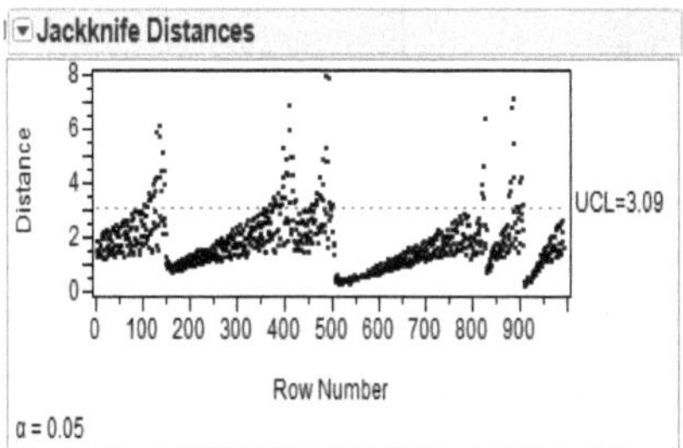

Fig. 4. Jackknife Distance for the given dataset

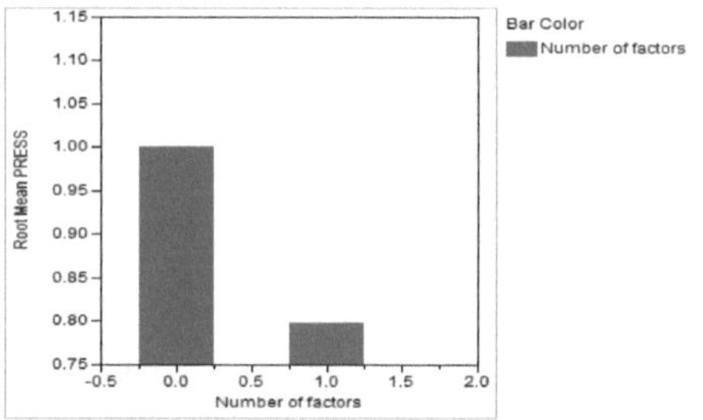

Fig. 8. The minimum root mean PRESS is 0.79818

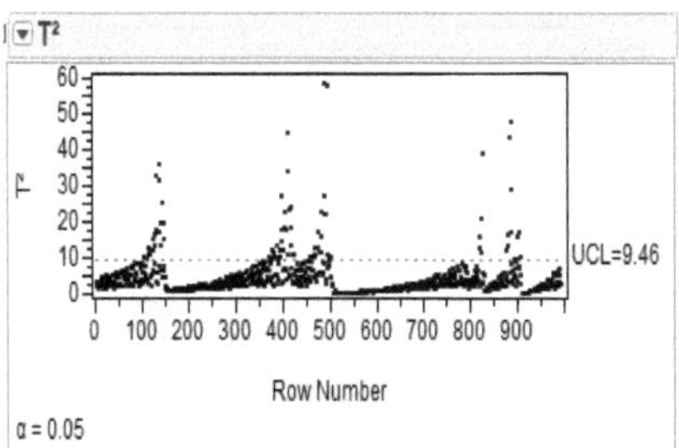

Fig. 5. T2 Distance for the given dataset

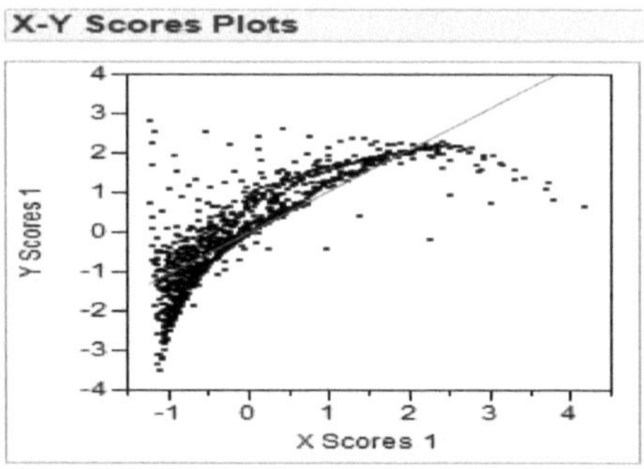

Fig. 9. X-Y Scores Plots

The Non Linear Iterative Partial Least Squares (NIPALS) and SIMPLS methods were calculated for the number of rows 993 and the number of factors is 1 and the percent variation for cumulative x and y are 100 and 36.61565 respectively were shown in figure 6. Cross Validation with NIPALS method is shown in figure 7. The minimum root mean PRESS is 0.79818, which is shown in figure 8 and the minimizing number of factors is 1. The X-Y scores plot was shown in figure 9. The Percent Variation was Explained and shown in figure 10 the chart in figure 11. Percent Variation for Number of factors 1 and cumulative X and Y is shown in figure 12.

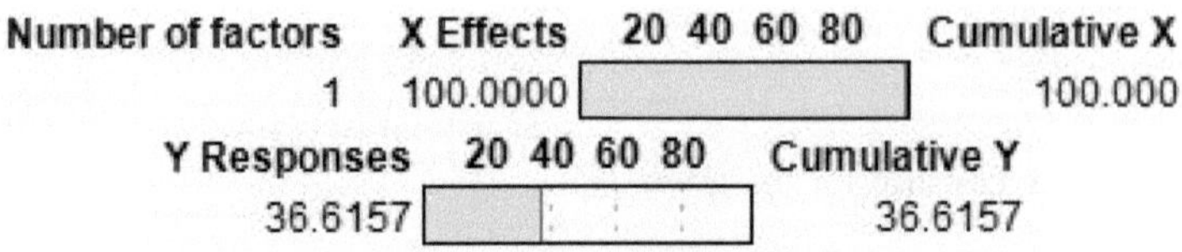

Fig. 10. Percent Variations

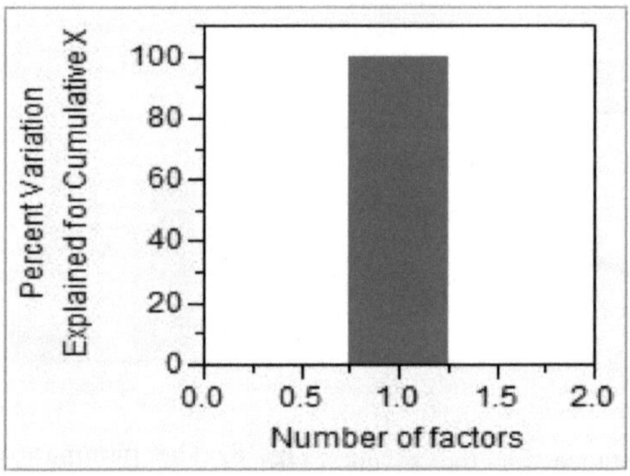

Fig. 11. Percent Variation Chart

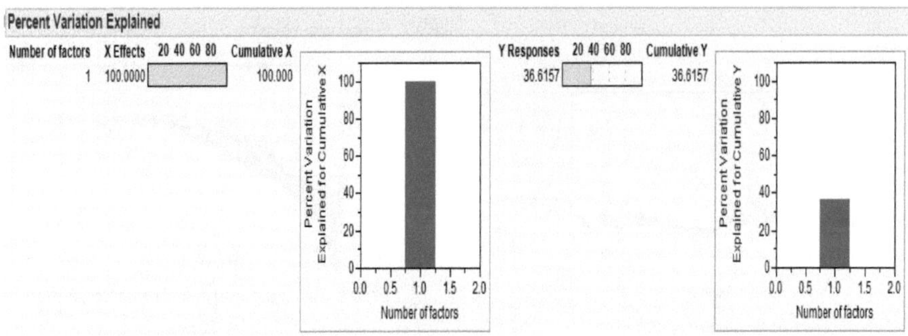

Fig. 12. Percent Variation for Number of factors 1 and cumulative X and Y

The distribution for the Laser dataset of the four attributes Input1, Input2, Input3, and Input4 is shown in figure 13. The quantiles for the Laser dataset of mean, standard deviation and standard error mean were calculated and were shown in figure 14. The Bivariate platform is the continuous by continuous personality of the

Fit Y by X platform. Bivariate Analysis shows the relationship between two continuous variables. The bivariate analysis results of Input1 by Input4, Input2 by Input4 and Input3 by Inpu4 appear in 15, 16 and 17 scatter plot.

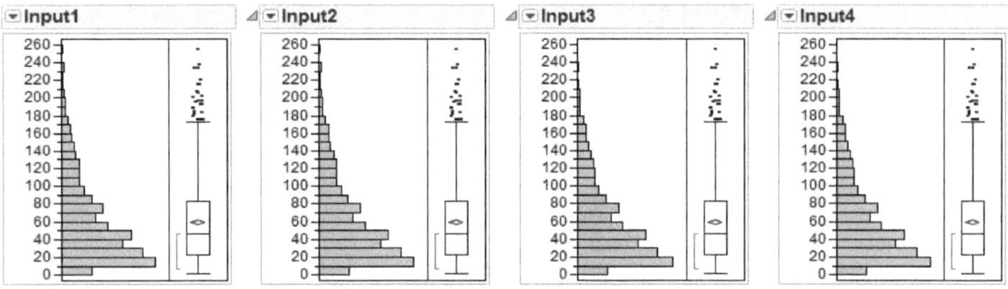

Fig. 13. The distribution for the Laser dataset

Quantiles		
100.0%	maximum	255
99.5%		220.39
97.5%		176
90.0%		130.6
75.0%	quartile	83
50.0%	median	46
25.0%	quartile	23
10.0%		13
2.5%		7
0.5%		3.97
0.0%	minimum	2

Summary Statistics	
Mean	59.882175
Std Dev	46.777304
Std Err Mean	1.4844329
Upper 95% Mean	62.795164
Lower 95% Mean	56.969186
N	993

Quantiles		
100.0%	maximum	255
99.5%		220.39
97.5%		176
90.0%		131
75.0%	quartile	83
50.0%	median	46
25.0%	quartile	23
10.0%		13
2.5%		7
0.5%		3.97
0.0%	minimum	2

Summary Statistics	
Mean	59.902316
Std Dev	46.88272
Std Err Mean	1.4877782
Upper 95% Mean	62.82187
Lower 95% Mean	56.982762
N	993

Quantiles		
100.0%	maximum	255
99.5%		220.39
97.5%		176
90.0%		131
75.0%	quartile	83
50.0%	median	46
25.0%	quartile	23
10.0%		13
2.5%		7
0.5%		3.97
0.0%	minimum	2

Summary Statistics	
Mean	59.884189
Std Dev	46.899318
Std Err Mean	1.4883049
Upper 95% Mean	62.804777
Lower 95% Mean	56.963602
N	993

Quantiles		
100.0%	maximum	255
99.5%		220.39
97.5%		176
90.0%		131
75.0%	quartile	83
50.0%	median	46
25.0%	quartile	23
10.0%		13
2.5%		7
0.5%		3.97
0.0%	minimum	2

Summary Statistics	
Mean	59.875126
Std Dev	46.905582
Std Err Mean	1.4885036
Upper 95% Mean	62.796103
Lower 95% Mean	56.954148
N	993

Fig. 14. The Quantiles for the Laser dataset

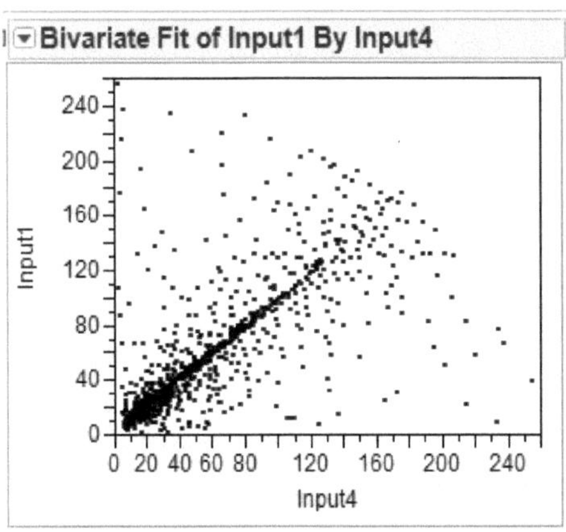

Fig. 15. Bivariate Fit of Input1 by Input4

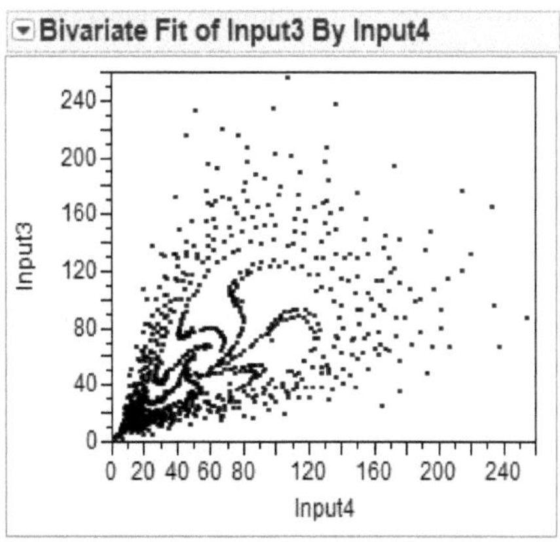

Fig. 16. Bivariate Fit of Input3 by Input4

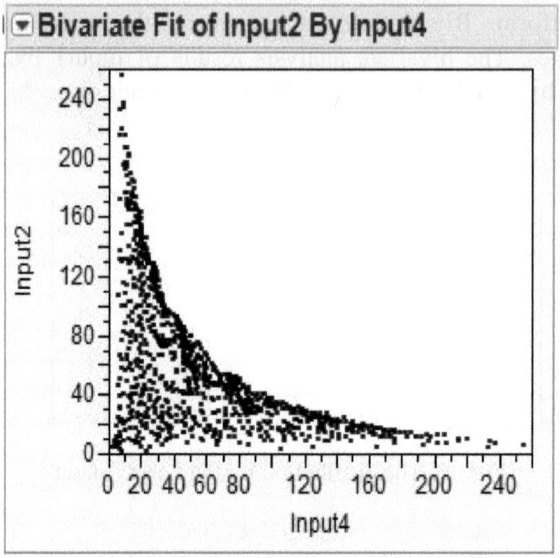

Fig. 17. Bivariate Fit of Input2 by Input4

6 Conclusion

The outlier detection problem finds applications in frequent domains, where it is desirable to determine interesting and unusual events in the activity which generates such data. The core of all outlier detection methods is the creation of a probabilistic, statistical or algorithmic model which characterizes the normal behavior of the data. In this paper we have calculated distances and calculated Cross Validation with NIPALS method. SIMPLS method is more accurate than NIPALS, but the calculations can be shown as both the methods have same result. The quantiles for the Laser dataset of mean, standard deviation and standard error mean were calculated. Outlier analysis has tremendous scope for research, especially in the area of organizational and progressive analysis.

References

1. Aggarwal, C.C.: An Introduction to Outlier Analysis
2. Hawkins, D.: Identification of outliers. Chapman and Hall, Reading (1980)
3. PrasantaGogoi, D.K., Bhattacharyya, B.B., Kalita, J.K.: A Survey of Outlier Detection Methods in Network Anomaly Identification (2011)
4. Yang, P., Zhu, Q., Zhong, X.: Subtractive Clustering Based RBF Neural Network Model for Outlier Detection (2009)
5. Ferdousi, Z., Maeda, A.: Anomaly Detection Using Unsupervised Profiling Method in Time Series Dataciteseerx (2013)
6. Bao, Z.: A Novel Proposal for Outlier Detection in High Dimensional Space. In: Proceeding SIGMOD 2001, Proceedings of the 2001 ACM SIGMOD International Conference on Management of Data, pp. 37–46 (2013)
7. Dallal, G.E.: Regression Diagnostics (2009)

8. Universidad Carlos III de Madrid,Regression Methods: An introduction to R Language (2012)
9. Salem, A.M.: An Efficient Estimator For Regression Median. International Journal Of Pure And Applied Mathematics 37(4), 541–553 (2007)
10. Rousseeuw, P.J., Leroy, A.M.: Robust Regression And Outlier Detection. John Wiley & Sons (2005)

[illegible]

Cluster Analysis on Different Data Sets Using K-Modes and K-Prototype Algorithms

R. Madhuri[1], M. Ramakrishna Murty[1], J.V.R. Murthy[2], P.V.G.D. Prasad Reddy[3], and Suresh C. Satapathy[4]

[1] Dept. of CSE, GMR Institute of Technology, Rajam, Srikakulam(Dist.) A.P., India
{ravi.madhuri5,ramakrishna.malla}@gmail.com
[2] Dept. of CSE, JNTUK-Kakinada, A.P., India
mjonnalagedda@gmail.com
[3] Dept. of CS&SE, Andhra University, Visakhapatnam, A.P., India
prasadreddy.vizag@gmail.com
[4] Dept. of CSE, ANITS, Visakhapatnam, A.P., India
sureshsatapathy@gmail.com

Abstract. The k-means algorithm is well-known for its efficiency in clustering large data sets and it is restricted to the numerical data types. But the real world is a mixture of various data typed objects. In this paper we implemented algorithms which extend the k-means algorithm to categorical domains by using Modified k-modes algorithm and domains with mixed categorical and numerical values by using k-prototypes algorithm. The Modified k-modes algorithm will replace the means with the modes of the clusters by following three measures like "using a simple matching dissimilarity measure for categorical data", "replacing means of clusters by modes" and "using a frequency-based method to find the modes of a problem used by the k-means algorithm". The other algorithm used in this paper is the k-prototypes algorithm which is implemented by integrating the Incremental k-means and the Modified k-modes partition clustering algorithms. All these algorithms reduce the cost function value.

Keywords: Cluster, K-means, K-modes, K-prototypes, mixed data.

1 Introduction

The most diverse characteristic of data mining is that it deals with very large and complex data sets.[2] The datasets to be mined often contain millions of objects described by tens, hundreds or even thousands of various types of attributes or variables. This requires the data mining operations and algorithms to be scalable and capable of dealing with different types of attributes. In terms of clustering, we are interested in algorithms which can efficiently cluster large data sets containing both numeric and categorical values because such data sets are frequently encountered in data mining applications.

S.C. Satapathy et al. (eds.), *ICT and Critical Infrastructure: Proceedings of the 48th Annual Convention of CSI - Volume II*, Advances in Intelligent Systems and Computing 249,
DOI: 10.1007/978-3-319-03095-1_15, © Springer International Publishing Switzerland 2014

In this paper, we presented three new algorithms that uses the incremental k-means paradigm to cluster data having numerical data, Modified k-modes paradigm to cluster data having categorical data and k-prototypes paradigm to cluster mixed data i.e., categorical and the numerical data. [8]The Modified k-modes algorithm extended the k-means paradigm to cluster categorical data by using (1) a new matching dissimilarity measure for categorical objects, (2) modes instead of means for clusters as centroids and (3) a frequency-based method to provide initial modes to minimize the clustering cost function. The k-prototypes algorithm in general integrates the k-means and k-modes to cluster data with mixed numeric and categorical values. So here in our paper, we used Incremental k-means and Modified k-modes paradigms to get integrated for implementing the k-prototypes algorithm.

The clustering process [4] of the k-prototypes algorithm is similar to the k-means algorithm except that it uses k-modes approach to provide initial modes for the categorical attributes of cluster prototypes. [7]Because these algorithms use the same clustering process as k-means and they preserve the efficiency of the k-means algorithm which is highly desirable for data mining. In this paper we deal with the following partitioning clustering methods, namely K-means, K-modes, K-mediods and K-prototype.

2 Our Proposed Work

We compared and implemented three algorithms in this paper, namely incremental k-means, Modified k-modes and K-prototype algorithms with different combinations of real world data sets and found that incremental K-means provide better results than simple K-means for numeric data, Modified K-modes is better than K-modes for categorical data and K-prototype is useful for mixed data clustering. We also observed K-prototypes paradigm is the combination of the K-means and the K-modes paradigms.

The number of iterations required to obtain the effective clustering results gets reduced. The cost function or the dissimilarity rate of the clustering ultimately obtained is comparatively low. Since the number of iterations converges, the time complexity is also reduced.

2.1 Incremental K-Means Paradigm

In the incremental k-means, after assigning each object to any cluster, the mean of that cluster is immediately updated.[5] So the next object comparison does with all the updated means. Thus incremental k-means is more appropriate and does clustering effectively with less number of iterations.

Algorithm

Step.1: Specify the number of clusters

Step.2: Select initial centroids randomly based on number of clusters specified.

Step.3: The centoids can be updated incrementally after each assignment of a data object to a cluster.

Step.4: Update that particular cluster's mean as:

$$mean_{jf} = mean_{jf} + a_{if} \quad (a_{if} \text{ is the data object })$$

Step.5: Repeat the Step.3to Step.4 with the updated means until all the instances' are assigned to clusters.

2.2 Modified K-Modes Paradigm

Clustering and other data mining applications frequently involve categorical data. The traditional approach of converting categorical data into numerical ones does not necessarily produce meaningful results. Thus, handling such data is a very important research topic in data mining. The simple k-modes algorithm proposed by Haung uses a simple dissimilarity measure [4] only. So, in this paper, we are going to use the "Modified k-modes algorithm" by avoiding too many iterations using frequency methodology [2] to select the initial centroids (modes) for clustering.

Algorithm

Step.1: Start

Step.2: Select the dataset used for clustering.

Step.3: Choose attributes in the dataset for clustering.

Step.4: Sort in descending order each and every field from the data set according to the most frequent number of values present in the dataset.

Step.5: Select the required number of clusters and choose the appropriate initial centroids(modes).

Step.6: Perform the concordance-discordance test [3]. Here the difference between each object y_i $(i \in n)$ and mode x_j $(j \in k)$ for each attribute using the formula:

$$D(x_{j,f}, y_{i,f}) = \frac{(m_{x_{j,f}} + m_{y_{i,f}})}{(m_{x_{j,f}} \times m_{y_{i,f}})} \times \delta(x_{j,f}, y_{i,f}) \tag{1}$$

where, $x_{j,f}$ =value of mode x_j on attribute a_f

$y_{i,f}$= value of object y_i on attribute a_f

$m_{x_{j,f}}$=number of times $x_{j,f}$appears in the set of modes on attributes a_f

$m_{y_{i,f}}$=number of times $y_{i,f}$ appears in the set of modes on attributes a_f

and $\delta(x_{j,f}, y_{i,f}) = \begin{cases} 0 \; if \; x_{j,f} = y_{i,f} \\ 1 \; if \; x_{j,f} \neq y_{i,f} \end{cases}$

Step.7: Assign instance or object to the cluster to which the above dissimilarity difference measure is low.

Step.8: Repeat the same process from step.7 and step.8 until the entire object's assignments is completed.

Step.9: Calculate the cost function [6] for each such iteration to find the dissimilarity rate obtained after the clustering process is finished. It is calculated using the formula as shown in Eqn. (2):

$$C(Q) = \sum_{j=1}^{k} \sum_{i=1}^{n} \sum_{f=1}^{F} \delta(x_{j,f}, y_{i,f}) \tag{2}$$

Where, k is number of clusters, n is number of elements present in each cluster, F is number of attributes and $\delta(x_{j,f}, y_{i,f}) = \begin{cases} 0 \; if \; x_{j,f} = y_{i,f} \\ 1 \; if \; x_{j,f} \neq y_{i,f} \end{cases}$

2.3 K-Prototypes Paradigm

K-prototypes algorithm is a combined approach of the k-means and the k-modes algorithm. Here, we are incorporating "Incremental k-means" to cluster numerical data and the "Modified k-modes" algorithm to cluster categorical data in the mixed datasets.

Algorithm

Step.1: Select the dataset containing both numerical and the categorical data for clustering process to start.

Step.2: Choose or take the required fields or attributes to start the clustering as per requirements.

Step.3: Sort each and every taken field as:

a) The numerical data fields in the dataset are sorted in the ascending order.

b) The categorical data fields present in the dataset should be taken an be sorted by taking the most frequent values in each field and should be arranged in the descending order and those values taken should be distinct.

Step.4: Choose the number of clusters to perform.

Step.5: Choose the centroids for those clusters i.e., means [as chosen in the incremental k-means procedure] and the modes [as chosen in the Modified k-modes procedure] for the clusters taken.

Step.6: Take each object or instance and perform the assignment to the appropriate cluster based on the difference and dissimilarity measures as:

a) For the numerical data typed object, we use the Euclidean distance measure i.e.,

$$d(i,j) = \sqrt{\sum_{f=1}^{F} (a_{if} - mean_{jf})^2} \tag{3}$$

where, j∈ k (k=number of clusters)

i∈ number (number=total number of instances in the dataset)

f∈ F (F=number of attributes)

b) For the categorical data typed attributes, we use the dissimilarity difference measure as:

$$\mathrm{D}(x_{j,f}, y_{i,f}) = \frac{(m_{x_{i,f}} + m_{x_{i,f}})}{(m_{x_{i,f}} \times m_{x_{i,f}})} \times \delta(x_{j,f}, y_{i,f}) \text{ and } \delta(x_{j,f}, y_{i,f}) = \begin{cases} 0 \; if \; x_{j,f} = y_{i,f} \\ 1 \; if \; x_{j,f} \neq y_{i,f} \end{cases}$$

Step.7: Measure or calculate $d(i,j) + D(x_{j,f}, y_{i,f})$ with every cluster (k) and assign that object to whichever cluster the overall difference is low.

Step.8: Repeat the same process for step.6 and step.7 until all the objects' assignments is completed.

Step.9: Calculate the dissimilarity rate obtained after the clustering process is finished for each iteration. It is calculated using the formulae as shown in Eqn. (4):

Cost function for categorical attributes:

$$C(Q) = \sum_{j=1}^{k} \sum_{i=1}^{n} \sum_{f=1}^{F} \delta(x_{j,f}, y_{i,f}) \tag{4}$$

$$\delta(x_{j,f}, y_{i,f}) = \begin{cases} 0 \; if \; x_{j,f} = y_{i,f} \\ 1 \; if \; x_{j,f} \neq y_{i,f} \end{cases}$$

Sum of squared errors calculation for numerical attributes as shown in Eqn. (5):

$$D(i,j) = \sqrt{\sum_{f=1}^{F} (a_{if} - mean_{jf})^2} \tag{5}$$

Step.10: Stop

3 Data Set Analysis

The data sets for performing clustering have been taken from the UCI machine repository. Three types of data sets have been taken to apply for Incremental k-means, modified k-modes and k-prototypes paradigms.

3.1 Data Sets for Incremental k-Means

The data sets taken for implementing Incremental k-means algorithm are given below.

***Iris Data Set*:** Iris data set consists of 4 numerical attributes and 155 instances. The attributes are "petal length", "sepal length", "petal width" and the "sepal width".

***Cholesterol Data Set*:** Cholesterol data set consists of 2 numerical attributes and 250 instances. The attributes are "Item Number" and the "Fat content". Using this data, we are going to group the persons having the similar cholesterol levels.

3.2 Data Sets for Modified k-Modes

The data sets taken for implementing Modified k-modes algorithm are given below.

Contact-Lens Data Set: Contact-lens data set consists of 5 categorical attributes and 24 instances. The attributes are "age", "spectacle", "astigmatism", "tearrate" and the "contact lenses". Using this data, we are going to group the persons having the similar eye sights and their presence of contact lenses.

***Post-operative Data Set*:** Post-operative data set consists of 7 categorical attributes and 190 instances. The attributes are "lcore", "lsurf", "lo2", "lbp", "surface stability", "core stability" and the "BP stability". Using this data, we are going to group the persons having the similar body temperatures and reactions.

3.3 Data Sets for k-Prototypes

The data sets taken for implementing k-prototypes algorithm are the mixed data sets (both numeric and categorical data sets) are discussed below.

***Blood Information Data Set*:** Blood Information data set consists of 5 attributes wherein 3 are of numerical attributes and two are of categorical attributes and 200 instances. The attributes are "name", "blood content", "plasma content", "hemoglobin in cc" and the "color". Using this data, we are going to group the persons having the similar blood structures, levels and the groups.

***Weather Data Set*:** Weather data set consists of 4 attributes in total wherein 2 are of categorical attributes and the rest are of numerical attributes and there are 350 instances. The attributes are "outlook", "temperature", "humidity" and the "windy nature". Using this data, we are going to group the similar weather reports.

4 Experimental Results

A common data set named "post operation" is taken to analyze the results obtained for all the three algorithms. "Post operation" data set consists of 8 attributes wherein seven are of numerical attributes and one is of categorical attribute.

4.1 Analysis for Incremental K-Means

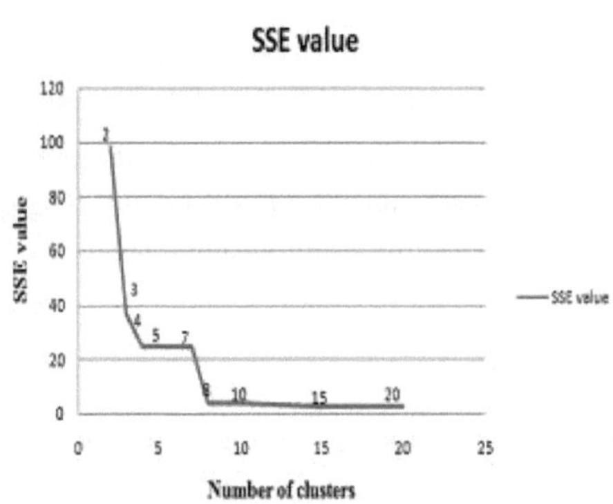

Number of clusters	SSE value
2	98
3	37
4	25
5	25
6	25
7	25
8	4
9	4
10	4
15	3
20	3
25	3

Fig. 1. Dissimilarity in incremental k-means to the number of clusters when clustering 90 records of the post operation data set

4.2 Analysis for Modified K-Modes

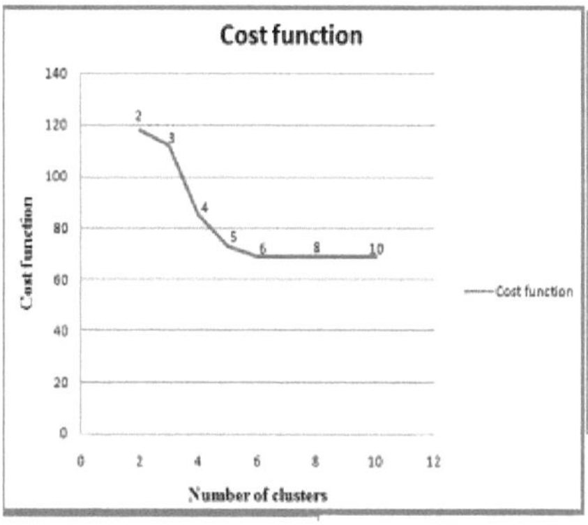

Number of clusters	Cost function
2	118
3	112
4	85
5	73
6	69
8	69
10	69

Fig. 2. Dissimilarity of Modified k-modes to the number of clusters when clustering 90 records of the post operation data set

4.3 Analysis for K-Prototypes

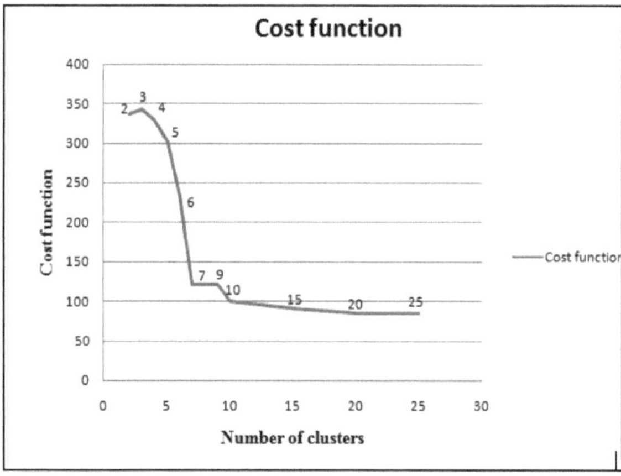

Number of clusters	Cost function
2	336
3	342
4	328
5	302
6	233
7	121
9	121
10	100
15	91
20	85
25	85

Fig. 3. Dissimilarity of k-prototypes to the number of clusters when clustering 90 records of the post operation data set

5 Conclusion

The real world data is becoming huge day-by-day with even growing data typed objects. The different data types included in the real world are categorical, numerical, scaled, Boolean etc and Sometimes there will be mixed data (combination of numerical and categorical). Clustering such different data sets as per requirements is a difficult task. To make the task easier and effective, the above three partition clustering algorithms, namely "Incremental k-means", "Modified k-modes" and "k-prototypes" are implemented.

By using the "Incremental k-means" and "Modified k-modes" independently, we have reduced the number of iterations. The dissimilarity rate i.e., the SSE value (Sum of Squared Errors) in case of Incremental k-means and the Cost function value in case of Modified k-modes paradigm can also be reduced.

References

1. Haung, Z.: Extensions to the k-Means Algorithm for Clustering Large Data Sets with Categorical Values. Canberra, ACT 2601, Australia (1998)
2. He, Z., Deng, S., Xu, X.: Improving K-Modes Algorithm Considering Frequencies of Attribute Values in Mode. Harbin Institute of Technology, China (2005)
3. Haung, Z.: A Fast Clustering Algorithm to Cluster Very Large Categorical Data Sets in Data Mining
4. Sayal, R., Vijay Kumar, V.: A Novel Similarity Measure for Clustering Categorical Data Sets. International Journal of Computer Applications (2011)
5. Witten, I., Frank, E.: Data Mining: Practical Machine Learning Tools and Techniques (2011)
6. Mastrogiannis, N., Giannikos, I., Boutsinas, B., Antzoulatos, G.: CL.E.KMODES: A modified k-modes clustering algorithm. University of Patras, Greece (2009)
7. Khan, S.S., Kant, S.: Computation of Initial Modes for K-modes Clustering Algorithm using Evidence Accumulation (2007)
8. Tan, P.-N., Steinbach, M., Kumar, V.: Introduction to Data Mining. Pearson education (2006)
9. He, Z.: Approximation Algorithms for K-Modes Clustering. Harbin Institute of Technology, China (2006)
10. Han, J., Kamber, M.: Data Mining: Concepts and Techniques, 2nd edn. Elsevier (2006)

Content Based Image Retrieval Using Radon Projections Approach

Nilam N. Ghuge[1] and Bhushan D. Patil[2]

[1] J.J.T. University, JSPM's Bhivarabai Sawant Institute of Tech. & Research(W),Wagholi, Pune, India 412207
ghuge1974@gmail.com
[2] Samsung Research and Development Institute, Bangalore, India
bhushandpatil@gmail.com

Abstract. CBIR systems are mainly used to retrieve images from huge database; the effectiveness of the CBIR system depends on the algorithm that is implemented for indexing. The way or method is used to determine similarities of available visual data by considering minute detailed low level features. Effectiveness of retrieval method depends on how the image is retrieved with maximum details and how much memory space is saved during retrieval process. Implementation of effective content-based image retrieval (CBIR) systems involves the combination of image creation, storage, security, transmission, analysis, evaluation feature extraction, and feature combination in order to store and retrieve images effectively. The goal of CBIR systems is to support image retrieval based on content i.e. shape, color, texture. In this paper we have implemented CBIR techniques using conventional Histogram and Radon Transform. Radon transform is based on projection of image intensity along a radial line oriented at a specific angle. We have test results on COREL1000 database. We have used Euclidean distance as a measure to calculate distance between two images and plot precision Vs Recall curve to show the effectiveness of the system.

Keywords: Content based Image Retrieval, Histogram, Randon transform, texture, Pattern recognition system, Euclidean distance, Precision, Recall.

1 Introduction

As a lot of visual information (image and Video) is available due improvement in communication technology and processing industry, there is vast scope for researchers to develop various methods / algorithms for sorting the visual database, archive the minute details of images and retrieve the data based on its contents. These algorithms are nothing but well known Content Based Image Retrieval System (CBIR) [1]. CBIR systems include methods of feature extraction like color, texture and shape, defining indices for various databases. So CBIR is one of the application of pattern recognition system. Computing similarities between query image and available database and retrieving the similar images [2]. CBIR system found its

S.C. Satapathy et al. (eds.), *ICT and Critical Infrastructure: Proceedings of the 48th Annual Convention of CSI - Volume II*, Advances in Intelligent Systems and Computing 249,
DOI: 10.1007/978-3-319-03095-1_16, © Springer International Publishing Switzerland 2014

applications in numerous fields like from medical imaging, commercial advertisements, scientific database management system, military purpose, remote sensing, copyright management system, criminal investigation and geographic information system [3]. There is huge demand for pictorial database as amount of digital video is generated. Effective classification, retrieval, summarization of information in the video from huge database of digital video is one of the challenging tasks. Lot of successful paradigms has emerged for video parsing, indexing, summarization, classification and retrieval.

We need a system which can effectively retrieve the desired image even if the database is not annotated. Imaging is a major factor in areas such as art galleries, interior design and weather forecasting. It is important for those areas to be able to retrieve the stored image quickly and accurately. The more effective the images are being stored, the more efficient the images can be retrieved later; this is where Content-Based Image Retrieval (CBIR) indexing comes in. Several existing applications, such as Query By Image (QBIC) which handles image databases and allows user to insert queries or interact with provided interfaces have focus on CBIR. Some of the applications have even used new algorithms or methods that help bring better result in retrieval process. However, there is potential to improve the existing algorithms, which increases the effectiveness of the retrieval process. Those existing algorithms have their own advantages and disadvantages, but we can use them by trying to combine and come up with a new algorithm that reduces the limitation of existing algorithms [4].

There are various approaches and methods for content based image retrieval. One of the simplest and easiest method is color histogram, which is based on visual features of image like color, shape and texture [5]. Another techniques have been developed for extraction of textural features. Textural features include periodicity, contrast, directionality and randomness [6]. The direction dependent Gabor filter is also used to extract the image features for image retrieval. The accuracy of filter depends on the angle chosen. To get rid of from angle dependency Radial basis function Gabor filer is used [7] and different wavelet techniques like wavelet and complex wavelet [8].

In this paper we proposed two methods of content based image retrieval system. In the next section we review CBIR using color Histogram technique. In section 3 we introduce a new algorithm of CBIR using Radon transform. We discuss and compare results in section 4. And results are compared using suitable measures like Precision and Recall. We finally conclude in Section 5.

2 CBIR Using Color Histogram

Comparing two images and deciding if they are similar or not is a relatively easy thing to do for a human. Getting a computer to do the same thing effectively is however a different matter. Many different approaches to CBIR have been tried and many of these have one thing in common, the use of color histograms.

For content based image retrieval to work, we have to find some features of the image that can be used when comparing it with another. One of the features most popular for image indexing and retrieval is color. Comparing the color distribution of two images will often say something about their similarity.

When computing a color histogram for an image, the different color axes are divided into a number called bins. A three dimensional 8X8X8 RGB histogram would therefore contain a total of 512 such bins. When indexing the image, the color of each pixel is found, and the corresponding bin's count is incremented by one [9].

An image histogram refers to the probability mass function of the image intensities. This is extended for color images to capture the joint probabilities of the intensities of the three color channels. More formally, the colour histogram is defined by,

$$h_{A,B,C}\ (a,b,c) = N \ . \ \mathrm{Prob}(A=a, B=b, C=c) \tag{1}$$

where A , B and C represent the three color channels (R,G,B or H,S,V) and N is the number of pixels in the image. Computationally, the color histogram is formed by discretizing the colors within an image and counting the number of pixels of each color.

There are several distance formulas for measuring the similarity of color histograms. Three distance formulas that have been used for image retrieval including histogram Euclidean distance, histogram intersection and histogram quadratic (cross) distance [10, 11].

In this work we had calculate Histogram Euclidean distance: Let h and g represent two color histograms. The Euclidean distance between the color histograms h and g can be computed as:

$$d^2\ (h,g) = \sum_A \sum_B \sum_C (h(a,b,c) - g(a,b,c))^2 \tag{2}$$

In this distance formula, there is only comparison between the identical bins in the respective histograms. Two different bins may represent perceptually similar colors but are not compared crosswise. All bins contribute equally to the distance. The minimum distance value signifies an exact match with the query.

3 Radon Transform in CBIR

Reconstruction of cross section of an image from its various projections is the challenging task. Projection of image can be determined by illuminating image by penetrating radiations. The purpose of reconstruction of image from various projection is to get the cross section view of image.

Radon transform follow the basic concept of CT scanning [12, 13]. This transform is able to transform two dimensional images with lines into a domain of possible line parameters, where each line in the image will give a peak positioned at the corresponding line parameters.

Radon transform is a mathematical tool developed by J. Radon in 1917.Radon transform finds its application in radar imaging, geophysical imaging, nondestructive testing medical imaging for tomography, seismic data processing, image processing and computer vision [14, 15].

The Radon transform computes projections of an image matrix along specified directions. A projection of a two-dimensional function f(x, y) is a set of line integrals. The Radon function computes the line integrals from multiple sources along parallel paths, or beams, in a certain direction. The beams are spaced 1 pixel unit apart.

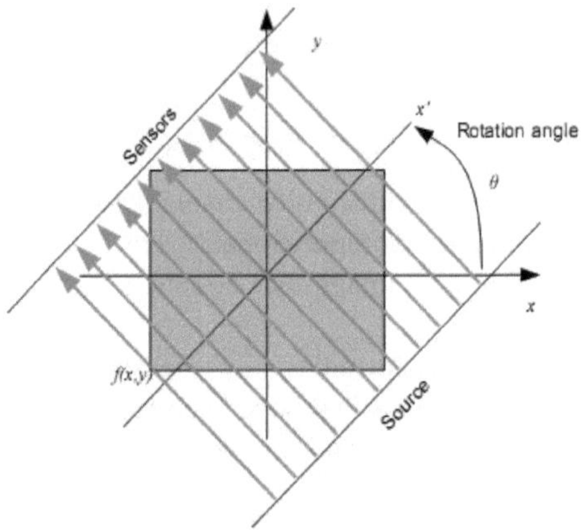

Fig. 1. Single projection at a specified rotation angle

To represent an image radon function takes multiple, parallel-beam projections of the image from different angles by rotating the source around the centre of the image. The Fig.1 shows a single projection at a specified rotation angle. The Radon transform is the projection of the image intensity along a radial line oriented at a specific angle. The radial coordinates are the values along the x'-axis, which is oriented at θ degrees counter clockwise from the x-axis.

The origin of both axes is the center pixel of the image. For example, the line integral of f(x, y) in the vertical direction is the projection of f(x, y) onto the x-axis; the line integral in the horizontal direction is the projection of f(x, y) onto the y axis.

Projections can be computed along any angle θ, by use general equation of the Radon transformation [16, 17, 18]

$$R_{\theta}(x') = \int_{-\infty}^{\infty}\int_{-\infty}^{\infty} f(x,y)\delta(x\cos\theta + y\sin\theta - x')dxdy \qquad (3)$$

where δ (xcosθ + ysinθ) is the delta function with value zero not equal zero for every argument except 0, and

$$x' = x\cos\theta + y\sin\theta \qquad (4)$$

x' is the perpendicular distance of the beam from the origin and θ is the angle of incidence of the beams.

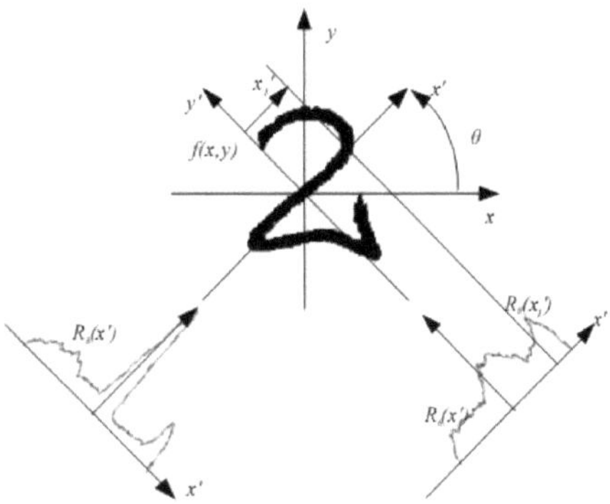

Fig. 2. Geometry of the Radon Transformation

Fig.2 illustrates the geometry of the Radon Transformation. The very strong property of the Radon transform is the ability to extract lines (curves in general) from very noise images. We can compute the Radon transform of any translated, rotated or scaled image, knowing the Radon transform of the original image and the parameters of the affine transformation applied to it [19, 20].

The 2D discrete Radon transform is defined by

$$R(k,\theta) = \sum_{x=0}^{N-1}\sum_{y=0}^{N-1} I(x,y)\delta(k - xy_\theta + yx_\theta) \tag{5}$$

Where, $\theta = \tan^{-1}\left({x_\theta}/{y_\theta}\right),\quad x_\theta \in Z,\quad y_\theta \in Z$

where I(x, y) is the image function, N×N is the image size, and N is assumed to be a prime number; δ(x) is the delta function,

$$k \in \{0,1,2 \ldots\ldots N_\theta - 1\}, N_\theta = N(|x_\theta| + y_\theta) \tag{6}$$

x_θ and y_θ are respectively the vertical and horizontal distance with the nearest pixels.

The discrete Radon transformation is obtained as a successive columns sums, designated for the image rotated by an angle Δθ. The obtained vectors are transposed, and formed the matrix with accumulator elements.

4 Results and Discussion

The performance or evaluation of the image retrieval algorithm is measured by Precision and Recall curve[21, 22].

$$\text{Precision} = \frac{\text{Number of relavent Images Retrieved}}{\text{Total Number of Images Retrieved}} \tag{7}$$

$$\text{Recall} = \frac{\text{Number of Relavant Images Retrieved}}{\text{Total Number of Relavant Images}} \tag{8}$$

We have computed precision – recall values for queries. Figure 8 shows precision and recall curve for above techniques.

Fig. 3. Query Image and retrieved images using Color Histogram

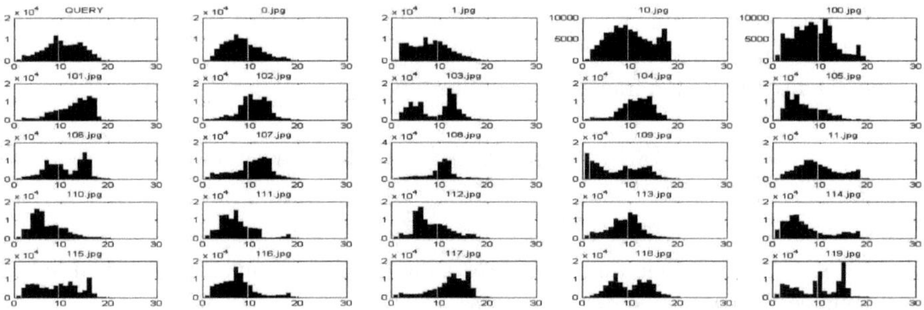

Fig. 4. Histogram of Query Image and retrieved images

Fig. 5. Query Image and retrieved images using Radon Transform

The algorithms have been tested on COREL database of 1000 images. Fig. 3 shows query image and retrieved images using Color Histogram technique and Fig. 4 shows histogram of query and retrieved images. Fig.5 shows query image and retrieved images using Radon transform. Fig. 6 shows energy diagram of query image and retrieved images using Radon transform. Fig. 7 shows Radon projection of first retrieved image.

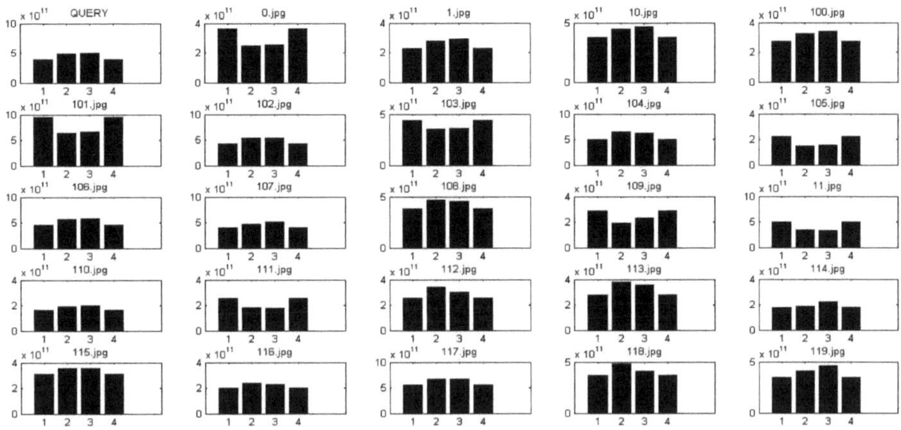

Fig. 6. Energy diagram Query Image and retrieved images using Radon Transform

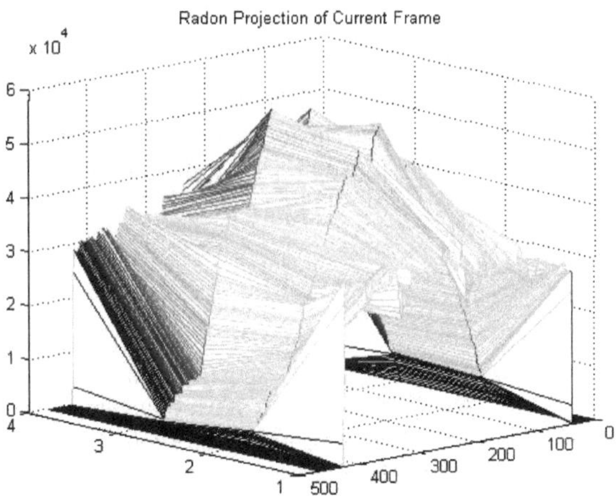

Fig. 7. Radon Projection of retrieved image

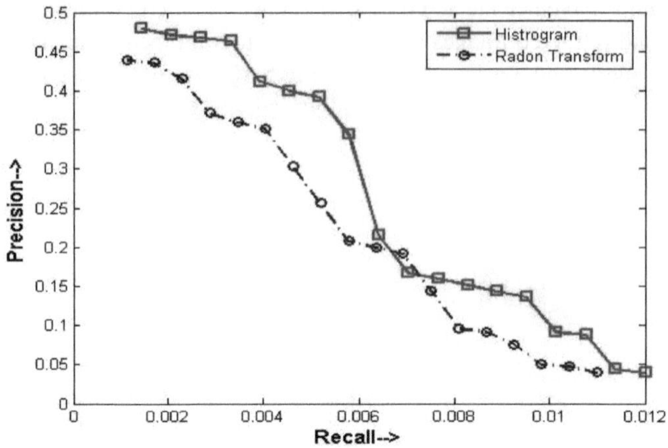

Fig. 8. Precision Recall curve

5 Conclusion

In this paper, we have presented comparative analysis of on feature extraction using Color Histogram and Radon Transform techniques. We have taken Euclidean distance as measures for retrieving the similar images from the data base. From this experiment we conclude that Color Histogram technique is based on matching of histogram of query image and retrieved images and gives result based on exact match. In second part Radon projection technique is used for image retrieval. Radon projection is a technique which is based on projection from different angle of salient points for extraction and quantization. Precision and Recall values are more precise for radon algorithm compared to color histogram technique. Radon algorithm gives more detailed information of image during retrieval process as we can use the features from different angles.

References

[1] Gupta, A., Jain, R.: Visual information retrieval. ACM Commun. 40(5), 70–79 (1997)

[2] Rui, Y., Huang, T.S., Change, S.F.: Image retrieval: current techniques, promising directions and open issues. J. Visual Commun. Image Representation 10(1), 39–62 (1999)

[3] Youssef, S.M.: ICTEDCT-CBIR: Integrating curvelet transform with enhanced dominant colors extraction and texture analysis for efficient content-based image retrieval. Elsevier, Computers and Electrical Engineering 38, 1358–1376 (2012)

[4] Datta, R., Joshi, D., Li, J., Wang, J.Z.: Image retrieval: Ideas, inuences & trends of the new age. ACM Computer Surv. 40(2), 160–173 (2008)

[5] Safar, M., Shahabi, C., Sun, X.: Image retrieval by Shape: A comparative study. In: Proceedings of IEEE International Conference on Multimedia and Expo (ICME 2000), pp. 141–144 (2000)

[6] Manjunath, B.S., et al.: Color and texture descriptors. IEEE Trans. CSVT 11(6), 703–715 (2001)

[7] Sastry, C.S., Ravindranath, M., Pujari, A.K., Deekshatulu, B.L.: A modified Gabor function for content based image retrieval. Pattern Recognition Letters 28, 293–300 (2007)
[8] Bhagavathy, S., Chhabra, K.: A wavelet-based image retrieval system. -Technical report – ECE278A, Vision Research Laboratory, University of California (2007)
[9] Suhasini, P.S., Sri Rama Krishna, K., Muralikrushna, I.V.: CBIR using color histogram processing. Journal of Theoretical and Applied Information Technology 6(1), 116–122 (2008)
[10] Smith, J.R., Chang, S.-F.: Automated image retrieval using color and texture. Technical Report CU/CTR 408-95-14, Columbia University (1995)
[11] Smith, J.R., Chang, S.-F.: Tools and techniques for color image retrieval. In: Symposium on Electronic Imaging: Science and Technology - Storage & Retrieval for Image and Video Databases IV, IS&T/SPIE, vol. 2670 (1996)
[12] Natterer, F., Wubbeling, F.: Mathematical Methods in Image Reconstruction. SIAM (2001)
[13] Averbuch, A., Shkolnisky, Y.: 3D Fourier based discrete Radon transform. Elsevier, Appl. Comput. Harmon. Anal. 15, 33–69 (2003)
[14] Peter, T.: The Radon Transform-Theory and Implementation. PhD thesis, Dept. of Mathematical Modelling Section for Digital Signal Processing of Technical University of Denmark (1996)
[15] Hoilund, C.: The Radon Transform. Aalborg University, VGIS (2007)
[16] Asano, A.: Radon transformation and projection theorem. Topic 5, Lecture notes of Subject Pattern Information Processing (2002)
[17] Averbuch, A., Coifman, R.R.: Fast Slant Stack: A notion of Radon Transform for Data in a Cartesian Grid which is Rapidly Computible, Algebraically Exact, Geometrically Faithful and Invertible. SIAM J. Scientific Computing (2001)
[18] Kupce, E., Freeman, R.: The Radon Transform: A New Scheme for Fast Multidimensional NMR. Concepts in Magnetic Resonance, Wiley Periodicals 22, 4–11 (2004)
[19] Bracewell, R.N.: Two-Dimensional Imaging. Englewood Cliffs, pp. 505–537. Prentice Hall (1995)
[20] Lim, J.S.: Two-Dimensional Signal and Image Processing. Englewood Cliffs, pp. 42–45. Prentice Hall (1990)
[21] Muller, H., Muller, W., Squire, D.M., Maillet, S., Pun, T.: Performance evaluation in content based image retrieval: overview and proposals. Pattern Recognition Letters 22, 593–601 (2001)
[22] Manjunath, B.S., Ma, W.Y.: Texture features for browsing and retrieval of image data. IEEE Trans. Pattern Anal. Machine Intell. 18(8), 837–842 (1996)

A Novel Approach for Facial Feature Extraction in Face Recognition

A. Srinivasan[1] and V. Balamurugan[2]

[1] Department of Information Technology
Misrimal Navajee Munoth Jain Engineering College
Anna University, Chennai, Tamilnadu, India
asrini30@gmail.com
[2] Department of Computer Science and Engineering
Chandy College of Engineering, Thoothukudi
Anna University, Tamilnadu, India

Abstract. Face recognition is the process of identifying a person by comparing his facial image with the existing image in a trained database. The crucial step in face recognition system is the extraction of facial feature. We propose an efficient facial feature representation by using Dual Tree Complex Wavelet Transform (DT-CWT). The Complex WT face characterizes the geometrical structure of facial images by using the properties of DT-CWT such as approximate shift in-variance and good directional selectivity. Since the efficiency retained with DT-CWT is inadequate, a new block design using Dual Tree Complex Wavelet Transform along with efficient normalization and noise reduction techniques is developed and using that design a face recognition system is developed.

1 Introduction

Face recognition recognizes the face images by extracting the facial features from a test image and compares it with trained facial images. The intensity variations due to illumination, shift, pose, and occlusion in human faces result in a highly complex distribution. Generally, the solution to this drawback is to extract the facial features before discriminant analysis which brings robustness against these variations. Face recognition system uses several techniques for feature extraction, examples for the feature extraction methods include i) Discrete Wavelet Transform, ii) Gabor Wavelet Transform and iii) Dual Tree Complex Wavelet Transform. The properties of DT-CWT such as approximate magnitude shift-invariance, good directional selectivity, limited redundancy and efficient linear computation can be harnessed to compute the accurate estimates of the geometrical structure in images.

Organization of this paper is done as, Section 2, elaborates the related works done in this field and approaches available. Section 3 reveals the existing system. Section 4 details the proposed system design. Section 5 deals with implementation, experiments done using public database and are explained, it also discusses experimental results. Section 6 concludes the paper.

S.C. Satapathy et al. (eds.), *ICT and Critical Infrastructure: Proceedings of the 48th Annual Convention of CSI - Volume II*, Advances in Intelligent Systems and Computing 249,
DOI: 10.1007/978-3-319-03095-1_17,

2 Related Work

Anudeep Gandam et al. proposed a post processing algorithm [1] for detection and removal of corner outliers. This method uses signal adaptive filtering technique to remove outliers. Heng Fui Liau et al. implemented a method for illumination invariant face recognition based on discrete cosine transform (DCT) [2]. This is done to address the effect of varying illumination on the performance of appearance based face recognition systems. Chao-Chun Liu et al. proposed [3] a novel facial representation based on the Dual Tree Complex Wavelet Transform for face recognition. It is experimentally verified that the proposed method is more powerful to extract facial features robust against the variations of shift and illumination than the discrete wavelet transform and Gabor wavelet transform. Srinivasan et al. [4] proposed and designed a Face Recognition System using HGPP & Adaptive Binning. This method overcomes the drawback of high dimensional histogram features by using adaptive binning technique.

3 Existing System

The existing system (Figure 1) is a novel feature representation based on DT-CWT for face recognition, referred as complex-WT-face. It is different from the existing DT-CWT based techniques since it uses only the single scale information and it approximately reduces the dimensionality of image to its one eighth. The performance of DT-CWT is studied under the variations of shift and illumination for facial image using DWT and GWT for comparison. Moreover, there are some large artificial singularities in border generated due to finite-support of facial image, which can affect the accuracy of representation. So, a clip method is proposed to reduce their effects on normal intrinsic singularity extraction. The efficiency in the current trend of face recognition system is reduced owing to

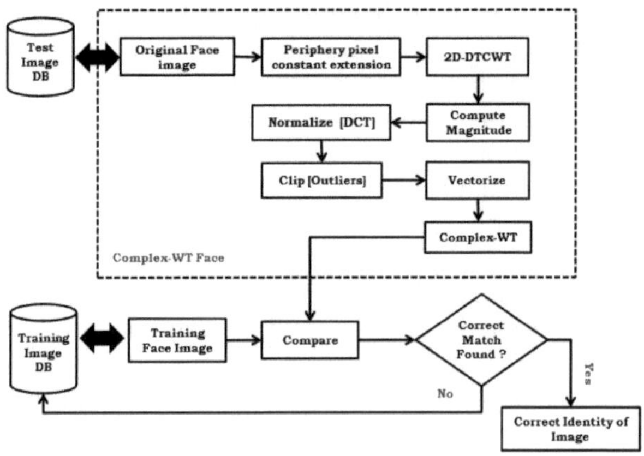

Fig. 1. Existing System

factors such as illumination variation, outliers and noises. The current trend of face recognition utilizes Zero Mean and Unit Variance normalization to overcome illumination variation. But the efficiency retained using ZMUV normalization is not adequate. The images consist of corner outliers that result in degradation at block boundaries.

4 Proposed System

By virtue of the good properties of DT-CWT, such as approximate shift invariance and good directional selectivity, the complex-WT face can characterize the geometrical structure in facial image. The proposed system improves the efficiency of face recognition, by reducing the impact caused flaws in the existing trend. The proposed design mainly concentrates on solution for i) Normalization, ii) Outlier reduction [5] and iii) Noise Reduction. Figure 2 shows the architectural flow of the proposed system. Given a facial image of size $n_r x n_c$, we extend its size to a critical size $n_r^e = ([n_r/2^L]+1)*2^L$ and $n_c^e = ([n_c/2^L]+1)*2^L$ by periphery-pixel constant extension, so that the decomposition can proceed till the given level L. Then the DT-CWT is performed on the extended facial image to generate a series of different-scale sub band constitute of complex coefficients. After that, only the sub bands of scale L are considered. The high-frequency sub bands whose scales are smaller than L are considered as noises caused by environmental variations and hence are discarded. The reason for using magnitudes of complex coefficient as features is, it provides an accurate measure of spectral energy, and approximately it becomes insensitive to small image shifts. For each L-scale sub band, we compute the magnitudes of its complex coefficients. In each scale, 2-D dual-tree complex produces two low-pass sub-images and six high-pass sub-images. Below algorithm 1 shows the steps involved in the implementation of the proposed work.

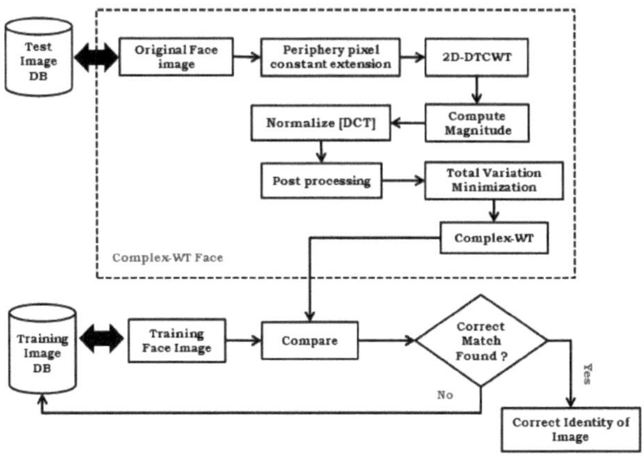

Fig. 2. Proposed System Design

Input: Input Image
Output: Complex Wavelet Face Representation
1.Extenstion
1a. Let the size of input image be $n_r x n_c$.
1b. The size of the extended image will be $n_{re} x n_{ce}$
$n_r^e = ([n_r/2^L]+1)*2^L$ and $n_c^e = ([n_c/2^L]+1)*2^L$ where L is the level of decomposition of DTCWT.
2. Dual Tree Complex Wavelet Transform
2a. DT-CWT is performed for feature extraction
2b. The input is image is decomposed to six high frequency and two low frequency sub bands.
3. Normalization
3a. Normalization is done to overcome illumination variation.
3b. The normalization is done using Discrete Cosine Transform.
4. Outlier reduction
4a. The images may consist of corner outliers.
4b. A post processing algorithm is implemented[corner outliers].
5. Noise reduction
5a. TVM approach is implemented to reduce blocky and mosquito noises.
6. Vectorization
6a. The eight sub images are vectorized and joined together to form a large vector.
6b. This represents complex-WT-face representation.

Algorithm 1. Implementation Steps for DT-CWT Processing

A post processing is used to eliminate corner outliers. A corner outlier is visible at the corner point of the block, where the corner point is either much larger or much smaller than the neighbouring pixels. A post processing is applied and based on signal adaptive filtering is proposed to reduce outliers. For smooth regions, the continuity of original pixel levels in the same block and the correlation between the neighbouring blocks is used to reduce the discontinuity of the pixels across the boundaries. For texture and edge regions, an edge preserving smoothing filter is applied. In this approach the image is divided into $8x8$ DCT blocks. The blocks are named $A, B, CandD$. The corner outliers are detected first and then they are eliminated. A global edge map is obtained by thresholding with global threshold value T_g, which is given by:

$$T_g = 10Qf + 8 \tag{1}$$

where Qf is the quantization factor of JPEG compression. The detection procedure is done by using a threshold value m, which is 20% of T_g. From this detection procedure, the return point is a corner outlier. A detected corner outlier and adjacent pixels are replaced by the weighted average. If a pixel of A to D is detected as a corner outlier, the pixels of $A, A1, andA2$ will be replaced with the proposed values of $A, A1, andA2$, respectively, as follows:

$$\begin{aligned} a &= int[(5 * A + B + C + D)/8] \\ a1 &= int[(2 * A1 + A2 + a)/4] \\ a2 &= int[(2 * A2 + A1 + a)/4] \end{aligned} \tag{2}$$

The pixels of $B, B1, andB2$ will be replaced with the proposed values of $b, b1, and$ $b2$ respectively, as follows:

$$\begin{aligned} b &= int[(5 * B + C + A + D)/8] \\ b1 &= int[(2 * B1 + B2 + b)/4] \\ b2 &= int[(2 * B2 + B1 + b)/4] \end{aligned} \tag{3}$$

The pixels of $C, C1, andC2$ will be replaced with the proposed values of $c, c1, and$ $c2$, respectively, as follows:

$$\begin{aligned} c &= int[(5 * C + B + A + D)/8] \\ c1 &= int[(2 * C1 + C2 + c)/4] \\ c2 &= int[(2 * C2 + C1 + c)/4] \end{aligned} \tag{4}$$

The pixels of $D, D1, andD2$ will be replaced with the proposed values of $d, d1, and$ $d2$, respectively, as follows:

$$\begin{aligned} d &= int[(5 * D + C + A + B)/8] \\ d1 &= int[(2 * D1 + D2 + d)/4] \\ d2 &= int[(2 * D2 + D1 + d)/4] \end{aligned} \tag{5}$$

Thus the corner outlier values are detected and are replaced with appropriate value. TVM method [6] is used to reduce blocky noise. The reconstructed images include the blocky noise and the mosquito noise. A new method for reducing the blocky noise and the mosquito noise using total variation minimization approach is proposed and used. In this method, by using the total variation filter, an image is decomposed to a skeleton component, which consists of smooth luminance and edges, and a texture component, which consists of small signals and noise. The Sobel filter is used for edge detection from the skeleton component, and the texture component corresponding to around the edges is filtered by using the Modified Adaptive Centre Weighted Median filter. As a result, the blocky noise and mosquito noise in the reconstructed images are reduced, and fine images are obtained.

5 Results and Discussions

For implementation we have used MATLAB, which is an interactive software system for numerical computations and graphics. The results are verified using FRI CVL face database [7] and Yale database [8]. The training database consists of 100 subjects (person) with seven face images (for each subject) in different directions. The experiment is verified for different sizes such as $128x128$ and $256x256$ pixels. The distance value obtained by the nearest neighbour approach is used for comparison purpose. The distance value is reduced by using the normalization and outlier reduction techniques. In the proposed method, the distance values are smaller than that of the existing method. The recognition rate is improved for the candidates with minimum distance value. The distance values by using the normalization and noise reduction techniques are compared with the results obtained without using the normalization and outlier reduction techniques for both the existing and the proposed systems.

Table 1 shows the results of the distance measure with and without using the normalization and outlier reduction in the existing system. By using the zero mean and unit variance normalization and clip method outlier reduction the distance value is minimized and the plotted version is shown in figure 3.

Table 1. Distance metric results [Existing vs Proposed for image size($128X128$)]

Test	Existing System		Proposed System	
sample	With ZMUV and Clip method	Without ZMUV and Clip method	With DCT and post processing	Without DCT and post processing
1	150.1055	150.8426	8	49
2	91.1823	98.0295	6	38
3	39.7284	44.058	4	16
4	105.2331	119.2366	6	34
5	115.5132	133.6785	0	33
6	97.0818	109.0456	0	30

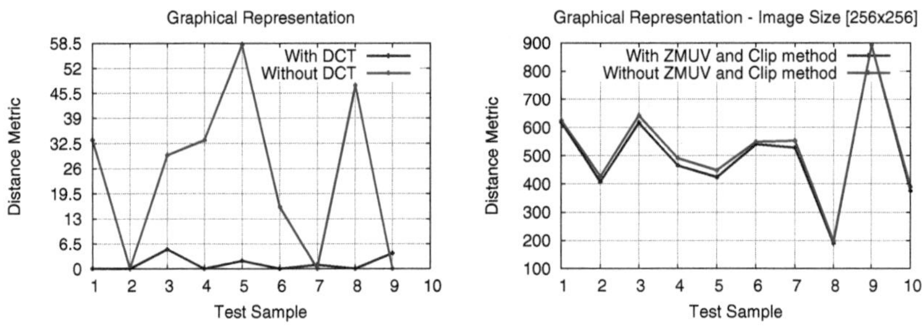

(a) Existing (b) Proposed

Fig. 3. Distance Metric Results Existing and Proposed ($128X128$)

Table 2 shows the results of the distance measure with and without using the normalization and outlier reduction in the existing system. By using the zero mean and unit variance normalization and clip method outlier reduction the distance value is minimized and the plotted results are shown in figure 4.

Table 2. Distance metric results [Existing vs Proposed for image size(256X256)]

Test	Existing System		Proposed System	
sample	With ZMUV and Clip method	Without ZMUV and Clip method	With DCT and post processing	Without DCT and post processing
1	619.218	623.7625	14	200
2	407.8989	426.9476	25	149
3	616.3582	642.8839	3	146
4	465.1737	491.0688	17	112
5	423.4235	447.4157	5	190
6	540.5289	549.2032	11	62

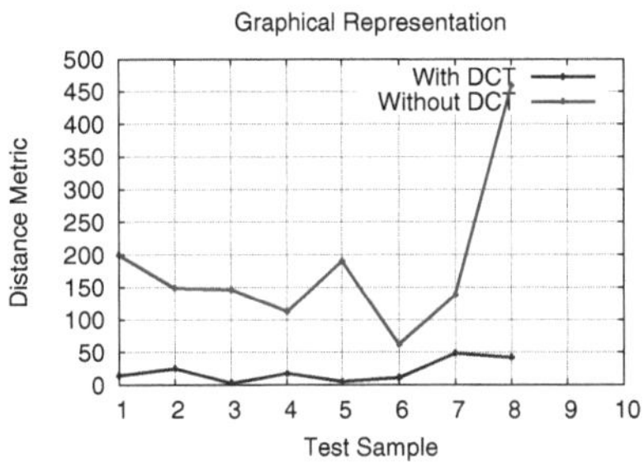

(a) Existing

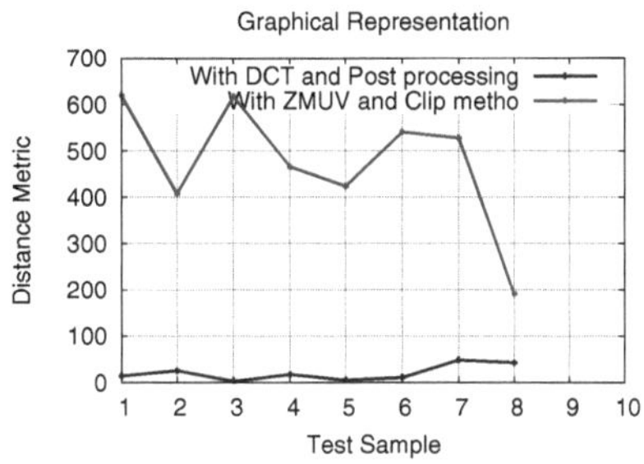

(b) Proposed

Fig. 4. Distance Metric Results Existing and Proposed (256X256)

The experiment for the recognition rate is verified for 700 images with 100 test images. It is verified for two image sizes such as 128x128 and 256x256. Table 3 shows the recognition rate for two image sizes. An example is provided where the training folder has 15 subjects with seven images for each subject. The existing system (E) recognizes thirteen images out of the fifteen for the size 128x128. The recognition rate of the existing system is 86.66%. The proposed system (P) recognizes fourteen images out of fifteen test samples. The recognition rate of the proposed system is 93.33%. The existing system recognizes ten images out of the fifteen for the size 256x256. The recognition rate of the existing system is 66.66%. The proposed system recognizes eleven images out of fifteen test samples. The recognition rate of the proposed system is 73.33%.

Table 3. PSNR table: Recognition Rate

TEST	128x128		256x256	
sample	E	P	E	P
Recognition Rate(%)	86.66	93.33	66.66	73.33

6 Conclusion

An advanced algorithm for face recognition using Dual Tree Complex Wavelet Transform is developed. The system uses Discrete Cosine Transform for normalization to overcome the effect of illumination variation. A post processing algorithm is used to reduce corner outliers. Total Variation Minimization is used to reduce blocky noises. The proposed system improves the efficiency of the Face Recognition System. The work is assessed with FRI CVL and Yale face databases.

References

1. Gandam, A., Sidhu, J.S.: A Post-Processing Algorithm for Detection & Removal of Corner Outlier. International Journal of Computer Applications 4(2) (2010) ISSN:09758887
2. Liau, H.F., Isa, D.: New Illumination Compensation Method for Face Recognition. International Journal of Computer and Network Security 2(3), 5–12 (2010)
3. Liu, C.-C., Dai, D.-Q.: Face Recognition Using Dual-Tree Complex Wavelet Features. IEEE Transactions on Image Processing 18(11), 2593–2599 (2009)
4. Srinivasan, A., Bhuvaneswaran, R.S.: A New Design for Face Recognition Using HGPP and Adaptive Binning Method. In: International Conference on Foundations of Computer Science, pp. 80–85. Monte Carlo Resort, Las Vegas (2008)
5. Cevikalp, H., Triggs, B.: Face recognition based on image sets. In: 2010 IEEE Conference on CVPR, Computer Vision and Pattern Recognition (CVPR), pp. 2567–2573 (June 2010)
6. Chambolle, A.: An Algorithm for Total Variation Minimization and Applications. Journal of Mathematical Imaging and Vision 20(1-2), 89–97 (2004) ISSN:0924-9907, doi:10.1023/B:JMIV.0000011325.36760.1e
7. http://www.lrv.fri.uni-lj.si/facedb.html
8. http://cvc.yale.edu/projects/yalefaces/yalefaces.html

Video Shot Boundary Detection Using Finite Ridgelet Transform Method

Parul S. Arora Bhalotra[1] and Bhushan D. Patil[2]

[1] J.J.T. University, G.H.R.C.E.M, Pune, India
parulsarora@gmail.com
[2] Samsung Research and Development Institute, Bangalore, India
bhushandpatil@gmail.com

Abstract. Video shot transition identification constitutes an important computer vision research field, being applied, as an essential step, in many digital video analysis domains: video scene detection, video compression, video indexing, video content retrieval and video object tracking. In this paper we propose a novel technique for shot boundary detection using finite ridgelet transform aiming to obtain fast and accurate boundary detection. We devise new two step algorithm for automatic shot boundary detection. Firstly effect of illumination change is removed using DCT and DWT. Then shot boundary is detected using finite ridgelet transform. The ridgelet transform was introduced as a sparse expansion for functions on continuous spaces that are smooth away from discontinuities along lines. This transform is a new directional resolution transform and it is more suitable for describing the signals with line or super-plane singularities. Finite ridgelet transform is a discrete orthonormal version of ridgelet transform. Experimental result indicates that finite ridgelet transform offers an efficient representation for frames that are smooth away from line discontinuities or straight edges.

Keywords: Shot boundary detection, DCT, DWT, finite ridgelet transform, radon transform.

1 Introduction

In our daily lives huge amount of digital video is generated. Effective classification, retrieval, summarization of information in the video from huge database of digital video is one of the challenging task. Lot of successful paradigms have emerged for video parsing, indexing, summarization, classification and retrieval. There is a need to organize large collections of digital videos for efficient access and retrieval. Such a task is known as video content analysis, which refers to understanding the meaning of a video document. Shot boundary detection is the most basic temporal video segmentation task. The detection of shot boundaries provides a base for all video segmentation approaches. Therefore solving the problem of shot boundary detection is one of the major prerequisites for revealing higher level video content structure. The existing methods on shot boundary detection are discussed below.

S.C. Satapathy et al. (eds.), *ICT and Critical Infrastructure: Proceedings of the 48th Annual Convention of CSI - Volume II*, Advances in Intelligent Systems and Computing 249,
DOI: 10.1007/978-3-319-03095-1_18, © Springer International Publishing Switzerland 2014

Likelihood ratio, pair-wise comparison and histogram comparison have been used as a different metric for shot boundary detection by Zhang et al. [1].Object motion and camera motion have been observed as major source of false positives by Boreczky and Lawrence [2]. They presented a comparison of several shot boundary detection classification techniques and their variations including pixel difference, statistical difference, compression difference Histogram, Edge tracking, discrete cosine transform, motion vector and block matching methods. It was seen that algorithm features that seemed to produce good results were region based comparisons, running differences and motion vector analysis. According to Boreczky combination of these three features may perform well to produce better results than either the region histogram or running histogram algorithm. Lienhart [3] has used color histogram differences. Standard deviation of pixel intensities and edge based contrast as a metric to find shot boundaries and tested results on diverse set of video sequences. Henjalic [4] have identified and analyzed the major issues related to shot boundary detection in detail. Knowledge relevant to shot boundary detection, shot length distribution, visual discontinuity pattern at shot boundaries and characteristic temporal changes of visual features around a boundary are needed to be considered for the study.

Gargi et al. [5] have evaluated and characterized the performance of number of shot detection methods using color histogram, moving picture expert group compression parameter information and image block motion matching. Ford et al. [6] have reported results on various histogram test statistics, pixel difference. Yuan et al. [7] have presented a comprehensive review of existing approaches and identified major challenges to shot boundary detection, according to them elimination of disturbances due to motion of large object and camera is a challenge in shot boundary detection. Sethi and Patel [8] have tested statistical test for changes in scene. Jinhui yuan et al. [9] employed three critical techniques i.e representation of visual content, construction of continuity signal and classification of continuity values are identified and formulated in the perspective of pattern recognition. In the proposed algorithm illumination change is removed using discrete cosine transform and discrete wavelet transform followed by detection of shot boundaries by finite ridgelet transform.

The outline of this paper is as follows. In the next section we review the concept and motivation of Ridgelet in continuous domain. In section 3 we introduce orthonormal finite ridgelet transform and its relation with radon transform. We discuss results in section 4. Finally we conclude in section 5 with some outlook,

2 Continuous Ridgelet Transform

Initially we begin with brief review of ridgelet transform and explaining its relation with other transform in continuous domain. Given an integrable bivariate function f(x), its continuous ridgelet transform (CRT) in R2 is defined by [13],[14].

$$CRT_f(a,b,\theta) = \int_{R^2} \psi_{a,b,\theta}(x) f(x)dx, \quad (1)$$

Where the ridgelets $\psi_{a,b,\theta}(x)$ in 2-D are defined from a wavelet-type function in 1-D $\psi(x)$ as

$$\psi_{a,b,\theta}(x) = a^{-\frac{1}{2}}\psi(x_1 cos\theta + x_2 sin\theta - b)/a) \quad (2)$$

Fig.1 shows an example ridgelet function, which is oriented at an angle and is constant along the lines x

$$x_1 cos\theta + x_2 sin\theta = constant$$

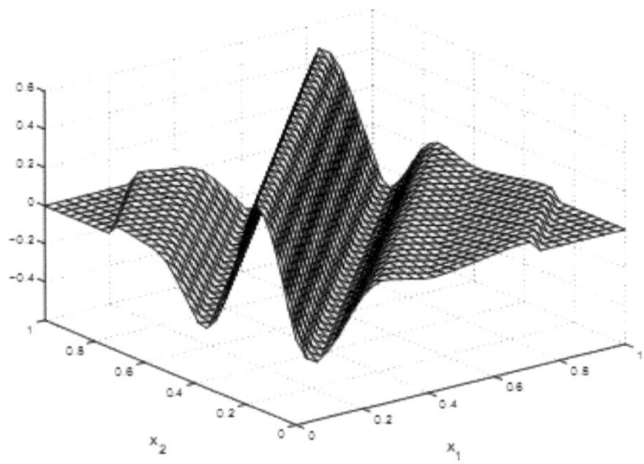

Fig. 1. An example ridgelet function $\psi_{a,b,\theta}(x_{1,}x_2)$

For comparison, the (separable) continuous wavelet transform (CWT) in R2 Of f(x) can be written as

$$CWT_f(a_{1,}a_{2,}b_1, b_2) = \int_{R^2} \psi_{(a_1,a_2,b_1,b_2)}(x) f(x)dx \quad (3)$$

Where the wavelets in 2-D are tensor products

$$\psi_{a_1,a_{2,}b_{1,}b_2}(x) = \psi_{a_{1,}b_1}(x_1)\, \psi_{a_{2,}b_2}(x_2) \quad (4)$$

$$\text{Of 1-D wavelets, } \psi a, b(t) = a^{1/2}\psi(\tfrac{t-b}{a})$$

As can be seen, the CRT is similar to the 2-D CWT except that the point parameters (b1,b2) are replaced by the line parameters (b, θ).

In 2-D, points and lines are related via the Radon transform, thus the wavelet and ridgelet transforms are linked via the Radon transform. More precisely, denote the Radon transform is given as

$$R_f(\theta, t) = \int_{R^2} f(x)\delta(x_1 cos\theta + x_2 sin\theta - t)dx \quad (5)$$

Then the ridgelet transform is the application of a 1-D wavelet transform to the slices (also referred to as projections) of the Radon transform

$$CRT_f(a,b,\theta) = \int_{R^2} \psi_{b,\theta}(t) R_f(\theta,t) dt \tag{6}$$

It is to be noted that if in (6) instead of taking 1-D wavelet transform, the application of a 1-D Fourier transform along would result in the 2-D Fourier transform. More specifically, let Ef (ω) be the 2-D Fourier transform of f(x), then we have

$$F_f(\xi cos\theta, \xi sin\theta) = \int_R e^{-j\xi t} R_f(\theta,t) dt \tag{7}$$

This is the well known projection-slice theorem and is commonly used in image reconstruction from projection methods [16],[17]. The ridgelet transform is the application of 1-D wavelet transform to the slices of the Radon transform, while the 2-D Fourier transform is the application of 1-D Fourier transform to those Radon slices [11].

3 Orthonormal Finite Ridgelet Transform

With an invertible finite radon transform (FRAT) and applying (6), we can obtain an invertible discrete ridgelet transform by taking the discrete wavelet transform (DWT) one each FRAT projection sequence, $r_k[0], r_k[1] \dots \dots . r_k[p-1]$. Where the direction k is fixed. The overall result is called the finite ridgelet transform (FRIT). Fig. 3 depicts the steps.

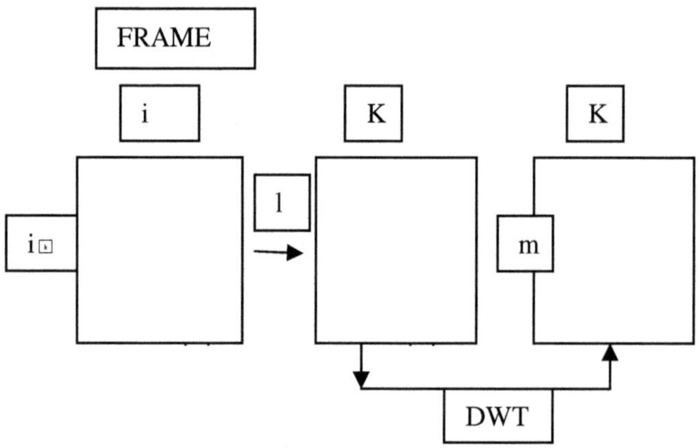

Fig. 2. Diagram of Finite Ridgelet Transform

After taking the FRAT, a DWT is applied on each of the FRAT slices where k is fixed. It is assumed that DWT is implemented by an orthogonal tree structured filter bank with J levels, where G0 and G1 are low and high pass synthesis filters, respectively. Then the family of functions

$$\{g_0^{(J)}[.-2^J m], g_1^{(J)}[.-2^J m]: j = 1, \dots \dots J; m\epsilon Z\} \tag{8}$$

is the orthogonal basis of the discrete-time wavelet series [18]. Here G(j) denote the equivalent synthesis filters at level j, or more specifically

$$G_0^{(J)}(z) = \prod_{k=0}^{J-1} G_0\left(z^{2^k}\right), \tag{9}$$

$$G_1^{(j)}(z) = G_1(z^{2j-1}) \prod_{k=0}^{J-2} G_0\left(z^{2^k}\right), \; j = 1 \dots J \tag{10}$$

The basis functions from $G_0^{(J)}$ are called the scaling functions, while all the others functions in the wavelet basis are called wavelet functions. Typically, the filter G1 is designed to satisfy the high pass condition.

$$G_1(z)|_{z=1},=0 \tag{11}$$

so that the corresponding wavelet has at least one vanishing moment.

$$G_1^{(j)}(z)|_{z=1} =0, \forall j = 1, \dots \dots \dots, J.$$

For a more general setting, let us assume that we have a collection of (p + 1) 1-D orthonormal transforms on R^p, one for each projection k of FRAT, that have bases as

$$\left\{\omega_m^{(k)} : m \epsilon z_p\right\}; k = 0,1, \dots \dots \dots, p.$$

By definition, the FRIT can be written as

$$FRIT_F[k,m] = \langle FRAT_F[k,\cdot], \omega_m^{(k)}[\cdot]\rangle \tag{12}$$

$$= \textstyle\sum_{l \epsilon Z_p} \omega_m^{(k)}[l] \langle f, \psi_{k,l}\rangle$$

$$= \langle f, \textstyle\sum_{l \epsilon Z_p} \omega_m^{(k)}[l]\, \psi_{k,l}\rangle \tag{13}$$

Here $\psi_{k,l}$ is the FRAT frame which is defined as

$$\psi_{k,l} = p^{-1/2} \delta_{L_{k,l}}$$

Hence we can write the basis functions for the FRIT as follows

$$\rho_{k,m} = \textstyle\sum_{l \epsilon Z_p} \omega_m^{(k)}[l]\, \psi_{k,l} \tag{14}$$

Let us consider inner products between any two FRIT basis function

$$\langle \rho_{k,m}, \rho_{k',m'}\rangle = \textstyle\sum_{l,l' \epsilon Z_p} \omega_m^{(k)}[l]\, \omega_{m'}^{k'}[l'](\psi_{k,l}, \psi_{k',l'}\rangle \tag{15}$$

Using properties of lines in the finite geometry Z_p^2 it is easy to verify that

$$\langle \psi_{k,l.} \psi_{k',\psi_{l'}}\rangle = 1 \quad if\ k = k'$$

$$= 0 \qquad if\ k = k', l \neq l'$$

$$= {}^1/_p \qquad if\ k \neq k'$$

Thus when any two FRIT basis function have same direction, K=k', then

$$\langle \rho_{k,m}, \rho_{k',m'} \rangle = \sum_{l,\epsilon Z_p} \omega_m^{(k)}[l]\, \omega_{m'}^{(k)}[l] \;= \delta[m - m'] \tag{16}$$

So the orthogonality of these FRIT basis function comes from the orthogonality of the basis $\{\omega^k = m \in Z_p\}$, when two FRIT basis functions have different directions, k=k′,

$$\begin{aligned}(\rho_{k,m}, \rho_{k',m'}) &= \frac{1}{p}\sum_{l,l'\epsilon Z_p} \omega_m^{(k)}[l]\, \omega_{m'}^{k'}[l'] \\ &= \tfrac{1}{p}\left(\sum_{l\epsilon Z_p} \omega_m^{(k)}[l]\right)\left(\sum_{l'\epsilon Z_p} \omega_{m'}^{k'}[l']\right)\end{aligned} \tag{17}$$

4 Results and Discussion

As depicted in Fig 3. is the original frame which is affected by illumination change disturbance. This disturbance is often mistaken as shot boundary. An algorithm is proposed using DCT and DWT, which effectively removes illumination change effect as shown in Fig. 4.

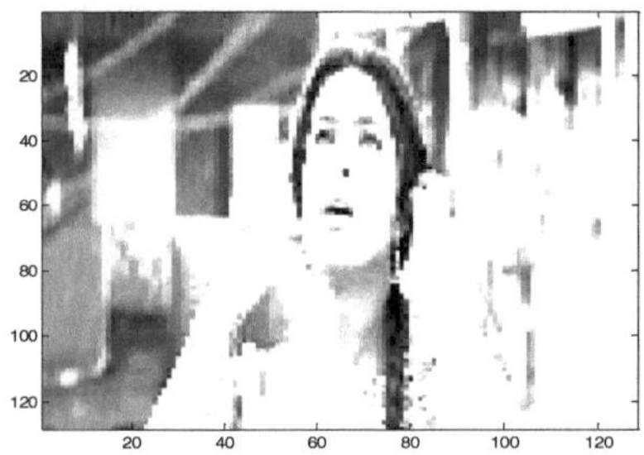

Fig. 3. Original Frame

Fig. 4. Frame after illumination removal

In the second part we employed Finite Ridgelet transform as our base for detection of shot boundaries. As shown in Fig. 5 graph depicts, boundary is detected between span of 25 to 45 frame number. It can be seen that shot boundary is detected between

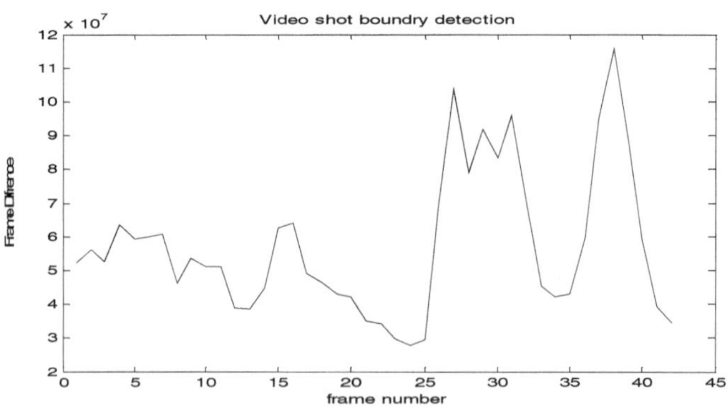

Fig. 5. Shot boundary detection

this span. To simplify we can employ adaptive thresholding technique in which highest peak above the threshold is considered for shot boundary detection. Now the FRIT of the current frame and previous frame is taken. As shown in Fig. 6 and Fig. 7. We can see difference in pattern for both the frames. As we know that ridgelet transform is more suitable at discontinuities along the straight line. So we can see the considerable change in two patterns.

A family of discrete orthonormal transforms for frames based on the ridgelet concept is presented. Owning to orthonormality, the proposed finite ridgelet transform is Self inverting, the inverse transform uses the same algorithm as the forward transform and has excellent numerical stability. Where edges are mainly along curves and there are texture regions (which generate point discontinuities), the ridgelet transform is not optimal. Therefore, a more practical scheme in employing the ridgelet transform as the building block in a more localized construction such as the curvelet transform.

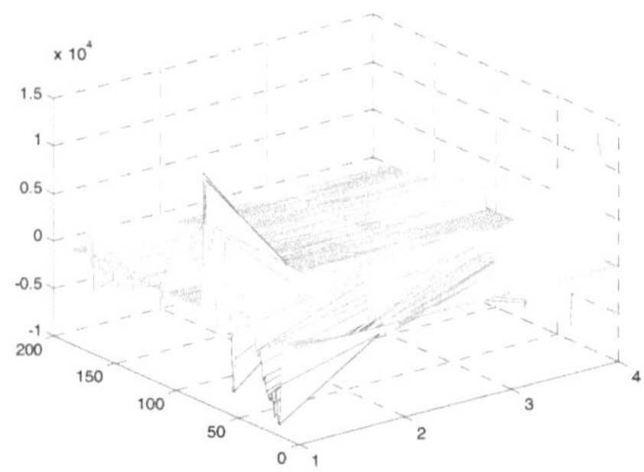

Fig. 6. Finite Ridgelet transform of current frame

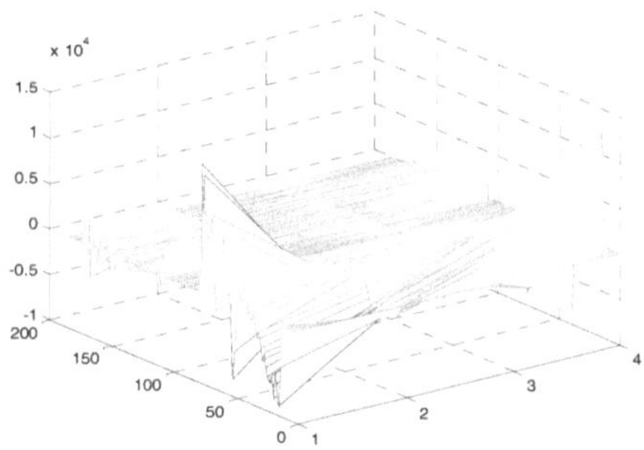

Fig. 7. Ridgelet transform of previous frame

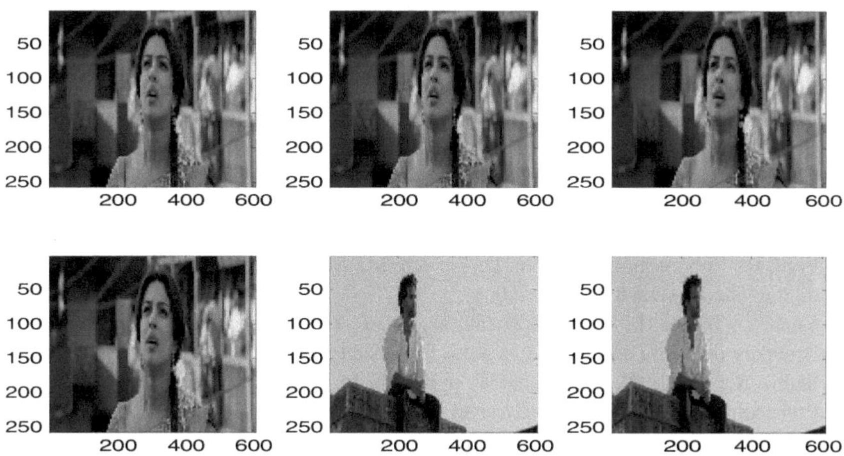

Fig. 8. Figure showing change in shot boundary

Fig. 8 depicts shot boundary detection between the frame numbers 20 to 30. Frame difference is identified exactly between 23 and 24.We can see in the figure hands appear in frame number 24 as compared to frame 23. Previously we have seen that the finite ridgelet transform of the different frames is different pattern, it is more suitable for describing the signals with line or super-plane singularities. Finite ridgelet transform is a discrete orthonormal version of ridgelet transform

4 Conclusion

In this paper we have presented novel approach based on finite ridgelet transform to the detection of shot boundaries. We have first removed the effect of illumination change, as often illumination disturbance is mistaken as shot boundaries. The illumination disturbance is removed using DCT and DWT. In the second part finite ridgelet transform technique is employed for shot boundary detection. Ridgelet transform is a new directional resolution transform and it is more suitable for describing the signals with line or super-plane singularities. Finite ridgelet transform is a discrete orthonormal version of ridgelet transform. It can be used as building block in obtaining new schemes which can deal with natural frames. Experimental results show that the proposed method effectively detects shot boundaries Our future work will be focused on eliminating disturbances caused due to large object and camera motion, one of the major challenge in shot boundary detection.

References

[1] Zhang, H.J., Kankanhalli, A., Smoliar, S.W.: Automatic partitioning of full-motion video. Multimedia Systems 1, 10–28 (1993)
[2] Boreczky, J.S., Rowe, L.: Comparison of video shot boundary detection techniques. In: Proceedings IS&T/SPIE Storage and Retrieval for Still Image and Video Databases IV, vol. 2670, pp. 170–179 (February 1996)
[3] Lienhart, R.: Reliable transition detection in videos: A survey and practitioners guide. International Journal of Image and Graphics 1(3), 469–486 (2001)
[4] Hanjalic, A.: Shot-boundary detection: Unraveled and resolved? IEEE Transaction Circuits System Video Technology 12(2), 90–105 (2002)
[5] Gargi, U., Kasturi, R., Strayer, S.H.: Performance characterization of video shot-change detection methods. IEEE Transaction Circuits Systems Video Technology 10(1), 1–13 (2000)
[6] Ford, R., Robonson, C., Temple, D., Gelach, M.: Metrics for short boundary detection in digital video system 8, 37–46 (2000)
[7] Yuan, J., Wang, H., Xiao, L., Zheng, W., Li, J., Lin, F., et al.: A formal study of shot boundary detection on circuit & systems for video technology 17(2), 168–186 (2007)
[8] Sethi, K., Patel, N.: A statistical approach to scene change detection. In: SPIE, Proceedings on Storage and Retrieval for Image & Video Database III, vol. 2420, pp. 329–338 (1995)
[9] Yuan, J., Wang, H., Xiao, L., Zheng, W., Li, J., Lin, F., Zhang, B.: A formal study of shot boundary detection. IEEE Transaction on Circuits & Systems for Video Technology 17(2), 234–239 (2007)

[10] Peter, T.: The Radon Transform-Theory and Implementation. PhD thesis, Dept. of Mathematical Modelling Section for Digital Signal Processing of Technical University of Denmark (1996)
[11] Do, M.N., Vetterli, M.: Finite Ridgelet Transform for image Representation. IEEE Transaction on Image Processing 2(1) (2003)
[12] Yan- Li, Shengqian-Wang, Chengchi-Deng: Redundant Ridgelet Transform and its Applications to Image Processing. In: Fifth International Conference on Information Assurance and Security (2009)
[13] Candues, E.J., Donoho, D.: Ridgelet a Key to higher to Higher Dimensional intermittency. Phil. Trans. R. Soc. Lond. A. Great Britain (1999)
[14] E. J. Candµes, Ridgelets: Theory and Applications. Ph.D. thesis, Department of Statistics, Stanford University (1998)
[15] Hoilund, C.: The Radon Transform. Aalborg University, VGIS (2007)

Enhanced Algorithms for VMTL, EMTL and TML on Cycles and Wheels

Nissankara Lakshmi Prasanna[1] and Nagalla Sudhakar[2]

[1] ANU and CSE Department,
Vignan's LARA Institute of Technology and Science,
Vadlamudi, Guntur, Andhra Pradesh, India
prasanna.manu@gmail.com
[2] Bapatla Engineering College, Bapatla, Guntur, Andhra Pradesh, India
Suds.nagalla@gmail.com

Abstract. This paper deals with the labeling of vertices and edges of a graph. Let G be a graph with vertex set V and edge set E, where |V| be the number of vertices and |E| edges of G. The two bijection methods for which we have designed algorithms are as follows. Initially A bijection λ_1:V∪E → {1, 2… |V| + |E|} is called a Vertex-Magic Total Labeling (VMTL) if there is a vertex magic constant vk such that the weight of vertex m is, $\lambda_1(m) + \Sigma_{n\in A(m)}\lambda_1(mn) = vk, \forall m \in V$ Where A(m) is the set of vertices adjacent to x. In the similar fashion the bijection λ_2:V∪E → {1, 2… |V| + |E|} is called Edge-Magic Total Labeling (EMTL) if there is a edge magic constant ek such that the weight of an edge es(mn), $\lambda_2(m) + \lambda_2(n) + \lambda_2 e(mn) = ek, \forall e \in E$. A resultant Graph which consists of both VMTL and EMTL are said to be Total Magic Labeling (TML) for different vertex magic constant and edge magic constant values. Baker and Sawada proposed algorithms to find VMTLs on cycles and wheels. In this paper we enhanced these algorithms and also we proposed new algorithms to generate EMTLs and TMLs of cycles and wheels. We used the concept variations and sum set sequences to produce VMTLs and EMTLs on cycles and wheels. Also we designed modules to identify TML's.

Keywords: Magic labeling, Magic labeling algorithms, Magic constant, Cycles, Wheels, Vertex Magic Total Labeling, Edge Magic Total Labeling, Total Magic Labeling etc.

1 Introduction

General definitions of cycles and wheels, magic labeling, vertex magic total labeling, edge magic total labeling, total magic labeling are as follows. Cycle is a graph where there is an edge between the adjacent vertices only and the vertex is adjacent to last one. Wheel is a Cycle with central hub, where all vertices of cycle are adjacent to it. Labeling is the process of assigning integers to graph elements under some constraint. If the constraint is only vertex set V then it is called vertex magic, if the constraint is the edge set then it is called edge magic, if the constraint is on both vertices and edges

S.C. Satapathy et al. (eds.), *ICT and Critical Infrastructure: Proceedings of the 48th Annual Convention of CSI - Volume II*, Advances in Intelligent Systems and Computing 249,
DOI: 10.1007/978-3-319-03095-1_19, © Springer International Publishing Switzerland 2014

it leads to the total magic labeling i.e. Let G be a graph with vertex set V and edge set E, where |V| be the number of vertices and |E| edges of G. A general reference for graph theoretic notations is [3]. Let G (V, E) be a finite, simple and undirected graph. The graph G has a vertex set V ∈ V(G) and edge set E ∈ E(G). We denote e = |E| and v = |V| the standard graph theoretic notation is followed. In this paper we deal only with cycles and wheels. The labeling of a graph is a map that takes graph elements such as vertices and edges to numbers (usually non-negative integers). Here the domain is a set of all vertices and edges giving rise to total labeling. The most complete recent survey of graph labeling is by [2]. Sedlacek introduced the magic labeling concept in 1963.

Here we are giving algorithmic implementations to the two bijection methods. If A bijection λ_1:V∪E → {1, 2… |V| + |E|} is called a Vertex-Magic Total Labeling (VMTL) if there is a vertex magic constant vk such that the weight of vertex m, $\lambda_1(m) + \Sigma_{n \in A(m)} \lambda_1(mn) = vk, \forall m \in V$ Where A(m) is the set of vertices adjacent to x. This labeling was introduced by McDougall et al. [5] in 2002. Another bijection λ_2:V∪E → {1, 2… |V| + |E|} is called Edge-Magic Total Labeling (EMTL) if there is a edge magic constant ek such that the weight of an edge emn, $\lambda_2(m) + \lambda_2(n) + \lambda_2 e(mn) = ek, \forall e \in E$ This is described in [4]. By assigning different combinations of labels to vertices and edges, it is possible to construct magic labeling with different magic constants on the same graph. A lower bound for a VMTL is obtained by applying the largest |V| labels to the vertices, while an upper bound is found by applying the smallest |V| labels to the vertices. The following formula gives lower and upper bound for vertex magic constant without taking into account the structure of the graph [6]. $13n^2+11n+2 \leq 2(n+1)k \leq 17n^2+15n+2$

This is a vertex magic constant vk limit equation. Once the structure of the graph is taken into account, additional limits may be found. The set of integers which are delimited by these upper and lower bounds is the feasible range. The values which are the magic constant for some VMTL of a graph form the graph's spectrum. Therefore the spectrum is a subset of the feasible range. For a Cycle, these limits are given by H.R. Andersen et al. [7] in 2002 as given as $5n+3 \leq 2k \leq 7n+3$.

In order to develop this theory strong Baker and Sawada [1] gave algorithms that generate all non-isomorphic VMTLs for cycles and wheels.

2 Background

There are two types of connected graphs. A cycle graph C_n is a connected graph where from n vertices every vertex is adjacent to exactly two other distinct vertices. A wheel graph W_n is a cycle graph Cn with central vertex called hub where all vertices if cycle are connected to hub.i.e. Wn=Cn+hub.

Detailed description for Vertex Total Magic Labeling (VTML) is the assignment of labels (integers) in the range {1,2...... +ve} to components of graph such that each vertex weight (λ1) is same and a constant generally referred as vertex magic constant (vk). Here the weight of vertex refers to the label applied to that vertex and all edges connected to it. Here A() consist all adjacent vertices. Example of VMTL on C4 is given in fig.1a. The expression for vertex m as follows. $\lambda_1(m) + \Sigma_{n \in A(m)} \lambda_1(mn) = vk, \forall m \in V$ Edge Total Magic Labeling (ETML) is an assignment of labels in the range {1,2... v+e} to components of graph such that each edge weight (λ2) is same and a edge magic constant (ek). The weight of a vertex refers the label applied to that vertex and all edges connected to it. Here is the expression for an edge (m,n) where m and n are end vertices. Example of EMTL on W4 is given in fig.1b. $\lambda_2(m)+\lambda_2(n)+\lambda_2(emn)=ek, \forall e \in E$

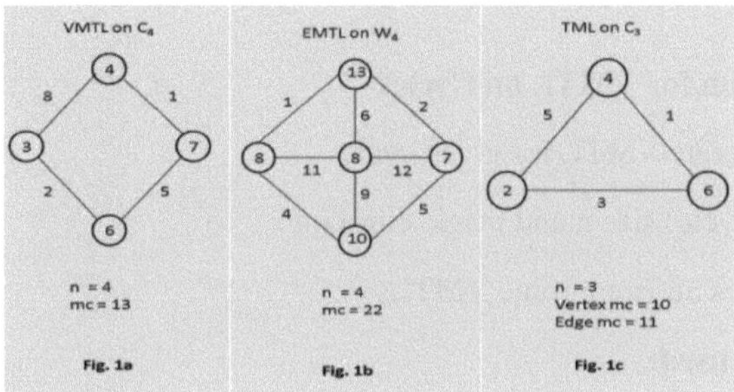

Fig. 1. (a) VMTL on C_4 (b) EMTL on W_4 (c) TML on C_3

A graph which consists of both VMTL and EMTL is said to be Total Magic Labeling (TML). Example of TML on C_3 is given in fig.1c. Now the process to develop this model is to design a sumset concept. We use the concept of variations to design a sumset. A variation which is permuted combination without repetition. The variations of size r chosen from a set of n different objects are the permutations of combinations of r. the number of variations of size r chosen from n objects equals the number of combinations of size r multiplied by the r! Permutations. The equation for variation is as follows $np_r = n!/(n-r)!$

Permutation without repetition generally referred as variations .Variations are used for order of sumset procedure so we should not go for combinations. One thing to remind here is already assigned label is never considered so permutations with repetitions are not allowed. The best solution for this problem is permutation without repetition i.e, variations. In the previous work, algorithms are developed by using trail and error method which checks all possible solutions. We are using the same concept but we are assembling some filters to get better method. These are explained in next session.

3 Magic Labeling Algorithms on Cycles

Here in session 3.1 we deal with cycles. 3.1.1 gives details of algorithm description to generate VMTL/EMTL on cycle. Next in 3.1.2 we have a module which uses above results to identify TMLs. 3.2 deals with wheels. Here we need separate algorithms for

VMTL and EMTL. 3.2.1 gives algorithm for VMTL on wheels. 3.2.2 describes algorithm for EMTL on wheels. With the results from 3.2.1 and 3.2.2, we got TML's on wheels. Each algorithm is followed with results for given cycle size and magic constant.

3.1 Cycle

Cycle is a closed path/circle. We find the correct label that is either VMTL or EMTL. This always depends on starting point. If the starting point is a vertex, the assignment of numbers will be VMTL or if the starting point is an edge, it results EMTL.

3.1.1 Algorithm for VMTL on Cycles

Algorithm to generate VMTL for given cycle

Input: Graph (Cycle) size n and magic constant k

Output: Generates all isomorphic VMTLs

Other Variables used:

Availability[x]: An array which decides whether a label is available or already used.

Initial assumptions: All labels are available.

Algorithm:

1. set labels range as {1, 2……, 2*n}.
2. for i:1 to 2*n
   if there is a variation(2*n)p_3,with available labels, whose sum is k.
   set them as labels of last edge, first vertex and first edge.
   previous vertex=1 current vertex=2
   for j: 1 to 2*n
   if there is a variation(2*n)p2, with available labels, whose sum is k.
   set them as λ1(current vertex) and λ1(current edge).
   previous vertex=current vertex   current vertex++
   if (current vertex=n)
   if(only one available label + label assigned current vertex+ label assigned to last edge =k)
   then assign label to λ1(currentvertex) and display solution.
   otherwise make the labels assigned to previous vertex as available.
   previous vertex--   and current vertex--
   display "work finished".
3. stop.

This algorithm results same as Baker and Sawada algorithm with less computation burden. At the beginning stage itself the total number of branched items are to be examined and reduced to half. So the total computation burden is reduced to half.

3.1.2 Algorithm for TML on Cycle

Suppose we got the solution from 3.1 is VMTL, let us check whether this is an EMTL for a different magic constant ek or vice versa. If this condition satisfies we can print the solution as TML.

Input: G(V,E) with VMTL (from 3.1)

Output: prints TML if exists.

Algorithm:

1. ek= λ1(vn)+ λl(e1)+ λ1(v1)
2. count=1;
3. for i:2 to n if λ1(vi-1)+ λ1(ei) + λ1(vi) = ek count=count+1
4. if count=n them display "TML exist"
   else display "TML not exist".

From the above module we checked the cycles, which are of different sizes (n=3to10) and we found that there is only TML with n=3 vertex magic constant as 10 and edge magic constant as 11.

3.2 Wheels

Wheel is a cyclic in structure with a central hub. The edges from hub to all vertices are termed as spokes. We designed different algorithms for these two. TML's always depends on result of this both VMTL and EMTL.

3.2.1 Algorithm for VMTL on Wheels

Input: Graph (Wheel) size n and vertex magic constant vk

Output: Generates all VMTLs for given graph size and magic constant.

Other Variables used: *Availability[x]:* An array which decides whether a label is available or already used.

Initial assumptions: All labels are available.

Algorithm:

1. set labels range as {1, 2……, 3*n+1}.
2. for i: 1to3*n+1
   if there is a variation (3*n+1)p_n, with available labels whose sum is vk
   (it should not be an isomorphic set)
   for j:1 to 3*n+1
   if there is a variation(3*n+1)p_n, with available labels, whose sum is vk-
   λ1(spoke j). set them as λ1(last edge), λ1(first vertex) and λ1(first edge).
   previous vertex=1 current vertex=2
   for k:2 to n

if there is a variation(3*n+1)p_2, with available unused labels, whose sum is vk-λ1(spoke-k) then set them as labels of currentvertex and current edge.
previous vertex=current vertex and current vertex= current_vertex+1
if (current vertex=n)
if(only available label+ label assigned current vertex+ label assigned current spoke + to last edge=vk) display "got solution"
otherwise make the labels assigned to previous vertex as available.
previous vertex-- and current vertex--.
display "finished work".

3. stop.

By implementing this algorithm also we got the results of Baker and Sawada with less computation burden. The first step reduces it by n(no.of spokes). Later the procedure for this is same as the cycle which has given 50% reduction in computation effort. So this algorithm reduces total work by 2n. So this is proposed as the best reduction approach.

3.2.1 Algorithm for EMTL on Wheels

Input : Graph (Wheel) size n and vertex magic constant vk

Output : Generates EMTLs for given graph size and magic constant.

Other Variables used:

Availability[x]: An array which decides whether a label is available or already used.

Initial assumptions: All labels are available.

Algorithm:

1. set labels range as {1, 2……, 3*n+1}.
2. for i: 1 to 3*n+1
   generate variation set (3*n+1)p3 with unused and available labels whose sum is ek and all those sets must start with common number. count the number of sets generated as setcount.
   if setcount>=n for each variation set setcount p_n generate matrix of size n*3 call method chkforsoluiton (matrix) otherwise report as "work finished".
3. stop.

chkforsolution (matrix)

1. fix the common first number as λ2(hub).
2. assign second column λ2(spokes) and third column to λ2(vertices).
3. if for all i=1 to n λ2(vi)- λ2(v(i+1)%n) is available label
   display "got solution".
4. rearrange matrix elements(2, 3 columns only) again call chkforsolution(matrix) until there is no possible re-arrangement.
5. Stop.

Table 1. The part of results obtained by implementing the above module

Wheel size=4		Wheel size=5		Wheel size=6	
Magic constant	#E MTLs	Magic constant	#E MTLs	Magic constant	#E MTLs
16	2	19	2	22	3
18	2	20	2	23	2
19	1	21	4	24	6
20	4	22	1	25	6
21	6	23	3	26	4
22	3	24	6	27	11
23	1	25	4	28	5
24	1	26	4	29	8
26	1	27	6	30	0
		28	3	31	8
		29	1	32	5
		30	4	33	11
		31	1	34	4
		32	2	35	6
				36	6
				37	2
				38	4
Total #EMTL's	21		43		91

3.2.2 Algorithm for TML on Wheel

For all VMTL/ EMTLs from 3.2.1and 3.2.2 we checked whether this is EMTL/VMTL for a different magic constant ek/vk. If this condition satisfies we can print the solution as TMLs for given VMTL. The same work continues for EMTL also by interchanging vertices and edges.

Input: G(V,E) with VMTL

Output: prints TML if exists.

Algorithm:

1. Initialize ek, ek= λ2(vn-1)+ λ2(en-1)+ λ2(v1)
2. count=1;
3. for i:2 to n if λ2(vi-1)+ λ2 (ei) + λ2(vi) = ek count=count+1
4. for i:1 to n if λ2(vi)+ λ2 (si) + λ2 (hub) = ek count=count+1

5. if count=2*n them display "Given is TML"
   else display "Given is not TML".
   With this module we can get number of TML's for given graph size and any of vertex magic constant or edge magic constant.

4 Conclusion and Future Work

These algorithms give the accurate results for this approach. The vertex magic total labeling and Edge magic labeling modules works on the cycles and wheels. Here we reduced the computation effort by pruning many branches at initial stages itself. Pruning at this stage impacts a lot on computation burden. The pruning is continued at each stage of generating new label set which yields great impact on total computation burden. We propose to continue this approach by applying these techniques on other structures of graphs as further work. We hope it gives better results when compared to existing approach for VMTLs. Also EMTLs and TMLs of other graph structures can be computed.

References

1. Baker, A., Sawada, J.: Magic labelings on cycles and wheels. In: Yang, B., Du, D.-Z., Wang, C.A. (eds.) COCOA 2008. LNCS, vol. 5165, pp. 361–373. Springer, Heidelberg (2008)
2. Gallion, J.A.: A dynamic survey of graph labeling. Electronic J. Combinations 14, DS6 (2007)
3. West, D.B.: An introduction to graph theory. Prentice-Hall (2004)
4. Gray, I., McDougall, J., Wallis, W.D.: On Vertex labeling of complete Graphs. Bull., Inst., Combin., Appl. 38, 42–44 (2003)
5. MacDougall, J.A., Miller, M., Slamin, Wallis, W.D.: Vertex Magic Total Labelings of Graphs. Utilitas Math. 61, 3–21 (2002)
6. MacDougall, J.A., Miller, M., Wallis, W.D.: Vertex-magic total labelings of wheels and related graphs. Utilitas Mathematica 62, 175–183 (2002)
7. Andersen, H.R., Hulgaard, H.: Boolean Expression Diagrams. Information and Computation 179, 194–212 (2002)

Improved Iris Recognition Using Eigen Values for Feature Extraction for Off Gaze Images

Asim Sayed[1], M. Sardeshmukh[1], and Suresh Limkar[2]

[1] Department of Electronics & Telecommunication, SAOE, Kondhwa, Pune, India
[2] Department of Computer Engineering, AISSMS IOIT, Pune, India
{asim27902,sureshlimkar}@gmail.com,
manojsar@rediffmail.com

Abstract. There are various Iris recognition and identification schemes known to produce exceptional results with very less errors and at times no errors at all but are patented. Many prominent researchers have given their schemes for either recognition of an Iris from an image and then identifying it from a set of available database so as to know who it belongs to. The Gabor filter is a preferred algorithm for feature extraction of Iris image but it has certain limitations, hence Principal Component Analysis (PCA) is used to overcome the limitations of the Gabor filter and provide a solution which achieves better results which are encouraging and provide a better solution to Gabor filters for Off Gaze images.

Keywords: Gabor Filter, Principal Component Analysis, Iris Recognition, Iris Identification, False Acceptance Rate, False Rejection Rate, Eigen Values, Eigen Vectors.

1 Introduction

The Iris is a very integral organ of our body, though it is very delicate it is highly protected by every individual. There are many reflexes of our body which help protect the eyes due to the delicate nature of the organ as compared to other biometric features of the human body. To add to the advantages the Iris have a high bearing capacity to carry information with an in information density of 3.2 measurable bits per mm2 [1]. Though our vision system processes up to 5 billion bits per second with great accuracy and precision this speed is unmatched by any computer data processing unit [2]. Hence Iris is used as a biometric measure to authenticate people since the Irides are unique to every individual and the fact that no two Irides of the same person match which means there is a vast diversity and distinction in every iris, which is an added advantage to use the Iris as a biometric measure [3]. The following figure gives an idea about the general structure of the Iris which contains the Pupil (in the center), the Sclera (The white part) and the Iris which is between them, the Iris pattern is clearly shown in the image shown in fig. 1. Among the various physical biometric applications used for personal authentication in highly secured environment, Iris patterns have attracted a lot of attention for the last few decades in

S.C. Satapathy et al. (eds.), *ICT and Critical Infrastructure: Proceedings of the 48th Annual Convention of CSI - Volume II*, Advances in Intelligent Systems and Computing 249,
DOI: 10.1007/978-3-319-03095-1_20,

biometric technology because they have stable and distinctive features for personal identification [4]. Iris recognition systems, in particular, are gaining interest because the iris's rich texture offers a strong biometric cue for recognizing individuals.

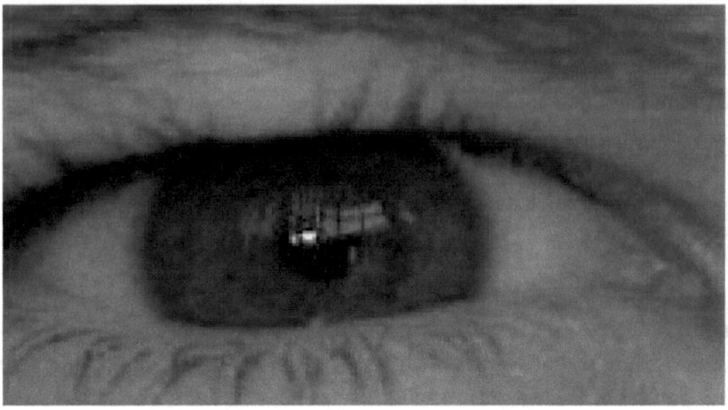

Fig. 1. Image of the Human Iris

2 CASIA Iris Database

With the pronounced need for reliable personal identification iris recognition has become an important enabling technology in our society [5]. Although an iris pattern is naturally an ideal identifier, the development of a high-performance iris recognition algorithm and transferring it from research lab to practical applications is still a challenging task. Automatic iris recognition has to face unpredictable variations of iris images in real-world applications. For example, recognition of iris images of poor quality, nonlinearly deformed iris images, iris images at a distance, iris images on the move, and faked iris images all are open problems in iris recognition [6]. A basic work to solve the problems is to design and develop a high quality iris image database including all these variations. Moreover, a novel iris image database may help identify some frontier problems in iris recognition and leads to a new generation of iris recognition technology [7]. Hence the Database used here is from Chinese Academy of Sciences Institute of Automation also dubbed as (CASIA – Iris).

3 Existing Potent Algorithms

Current iris recognition systems claim to perform with very high accuracy. However, these iris images are captured in a controlled environment to ensure high quality. Daugman proposed an iris recognition system representing an iris as a mathematical function [7]. Mayank Vatsa proposed a support-vector-machine-based learning algorithm selects locally enhanced regions from each globally enhanced image and combines these good-quality regions to create a single high-quality iris image [8]. Daugman also proposed algorithms for iris segmentation, quality enhancement, match score fusion, and indexing to improve both the accuracy and the speed of iris recognition [9]. Leila Fallah Araghi used Iris Recognition based on covariance of discrete wavelet using Competitive Neural Network (LVQ) [9]. A set of Edge of Iris profiles are used to build a covariance matrix by discrete wavelet transform using Neural Network. It is found that this method for Iris Recognition design offers good class discriminacy. In order to get good results and more accuracy for iris Recognition for human identification the Back Propagation Algorithm is used. Dr. J. Daugman makes use of an integro-differential operator for locating the circular iris and pupil regions, and also the arcs of the upper and lower eyelids [7][24]. It is defined as-

$$\mathrm{Max}(r,xp,y0)=|G_{\sigma}(r)\ \partial/\partial r\ (\oint_{r},x_0,y_0(I(x,y)/2\pi r)ds)| \quad (1)$$

Where I(x,y) is the eye image, r is the radius to search for, $G_{\sigma}(r)$ is a Gaussian smoothing function, and s is the contour of the circle given by r, x_0, y_0. The operator searches for the circular path where there is maximum change in pixel values, by varying the radius and centre x and y position of the circular contour. The operator is applied iteratively with the amount of smoothing progressively reduced in order to attain precise localization. Eyelids are localized in a similar manner, with the path of contour integration changed from circular to an arc. The integro-differential can be seen as a variation of the Hough transform, since it too makes use of first derivatives of the image and performs a search to find geometric parameters [7]. Since it works with raw derivative information, it does not suffer from the thresholding problems of the Hough transform. However, the algorithm can fail where there is noise in the eye image, such as from reflections, since it works only on a local scale [9].

4 Segmentation

The segmentation module detects the pupillary and limbus boundaries and identifies the regions where the eyelids and eyelashes interrupt the limbus boundary's contour. A good segmentation algorithm involves two procedures, iris localization and noise reduction. In performing the preceding edge detection step, Wildes et al. bias the derivatives in the horizontal direction for detecting the eyelids, and in the vertical direction for detecting the outer circular boundary of the iris. The fact that the eyelids are usually horizontally aligned, and also the eyelid edge map will corrupt the circular iris boundary edge map if using all gradient data can be useful for segmenting the Iris [9]. Taking only the vertical gradients for locating the iris boundary will reduce influence of the eyelids when performing circular Hough transform, and not all of the edge pixels defining the circle are required for successful localization. Not only does this make circle localization more accurate, it also makes it more efficient, since there are less edge points to cast votes in the Hough space [9].

The region of interest can be found out very easily due to this technique this technique can be used to find the edges the filter. The technique used to find the edges in this case is the Sobel Edge filter. Fig. 2 shows the Sobel edge filter of the original Image, it has clearly sorted the pupil and the Iris boundaries. In this figure the eye lashes are also visible along with the light used to photograph the image these are the noise contents in an image and hinder the recognition and identification process as the

data is significantly lost due to these characteristics contained in an image. An algorithm for noise cancellation must be used or the pupil must be masked with a black colored sphere, CASIA database also has provisions to use such masked pupil images [6].

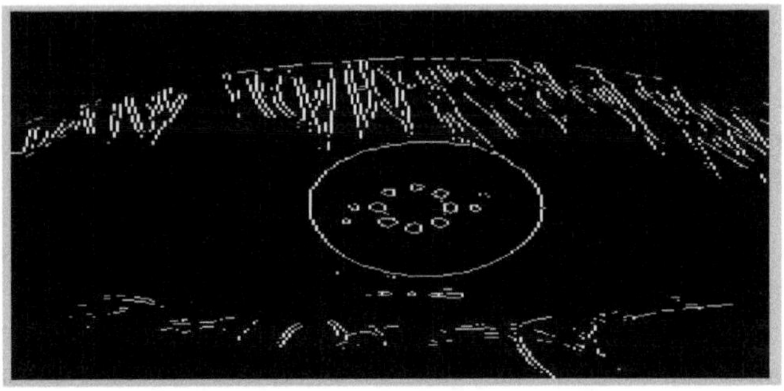

Fig. 2. Sobel Edge Filter of Original Image [22]

5 Gabor Filters

Gabor has shown the existence of a 'Quantum Principle' for information: such that for 1D signals the conjoint time frequency domain must be quantized in such a manner that no filter or signal should acquire a certain minimal area in it [10]. Here the minimal area refers to the tradeoff between the time and frequency resolution. He also discovered that the best tradeoff between these functions was provided by Gaussian modulated complex exponentials. Hence Gabor proposed elementary functions which are generated by varying the modulating wave frequency and keeping the Gaussian constant.

Prof J. Daugman has described the Gabor functions can be modeled to function the way the mammalian brains perceive vision in the visual cortex [11]. Hence the analysis by the Gabor functions is similar to the perception in the visual system applied by the human brain [12]. The modulation of sine wave with Gaussian will result in localization in space to a good extent whereas a certain amount of loss will occur with localization in frequency [13]. Hence use of quadrature pair of Gabor Filters can be used to decompose the signals where real part will be specified by Gaussian modulated cosine function and the imaginary part will be specified by the Gaussian modulated sine function [14]. The real and imaginary components are known as even symmetric and odd symmetric components respectively [15][16]. The frequency of the sine/cosine wave signifies the frequency of the filter whereas the width of the Gaussian signifies the bandwidth of the filter [16]. A 2-D Gabor filter used for spatial coordinates (x, y) in image processing is represented as-

$$G(x, y) = e^{-\pi[(x-x_0)^2 / \alpha^2+(y-y_0)^2/\beta^2]} * e^{-2\pi i[u_0(x-x_0) +v_0 (y-y_0)]} \qquad (2)$$

Where (x_o,y_o) specify position in the image, (α,β) specify the effective width and length, and (u_o, v_o) specify modulation, which has spatial frequency ω_0. It is evident that Gabor filters have a significant advantage in terms of spatial locality and orientational selectivity as they are optimally localized in the frequency and the spatial domains [16]. Even though these parameters are ideal for their selection for feature extraction there are a few parameters which need to be considered and may not be the best bet for the application pertaining to all images, hence the draw backs for using Gabor filter is expensive like the dimensionality of the Image filtered using the Gabor filter is much bigger than the initial size of the original image here the dimension are bigger and it can be noted that the that this image size is greatly increased if it is not down sampled [17]. Another shortcoming of the Gabor Filter is that it faces the problem of orthogonality as different filters from different banks are not orthogonal to each another. Hence the information encoded by the Gabor Filter may be redundant and may negatively affect the accuracy of the classifier. The accuracy may vary depending on the filter being used from different filter banks

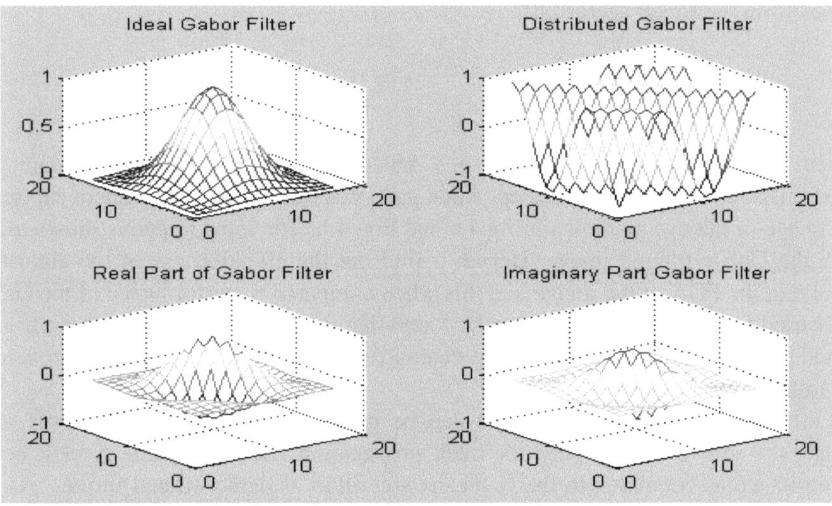

Fig. 3. a)Ideal Gabor Filter components b) Distributed Gabor filter components c) Real part of Gabor Filter components of the input image d)Imaginary part of Gabor filter Components. [22]

6 Principal Component Analysis

PCA is used to reduce the dimensionality of the data set containing large number of inter related variables, although retaining the variations in the data sets as much as possible [18]. It can be thought of a mathematical procedure that uses an orthogonal transform to convert a set of observations of possibly correlated variables into a set of values of linearly uncorrelated variables called principal components. The number of principal components is less than or equal to the number of original variables [20]. This transformation is defined in such a way that the first principal component has the largest possible variance (that is, accounts for as much of the variability in the data as possible), and each succeeding component in turn has the highest variance possible under the constraint that it be orthogonal to (i.e., uncorrelated with) the preceding components. Principal components are guaranteed to be independent only if the data set is jointly normally distributed [19]. PCA is sensitive to the relative scaling of the original variables.

PCA is the simplest of the true eigenvector-based multivariate analyses. Often, its operation can be thought of as revealing the internal structure of the data in a way that best explains the variance in the data. If a multivariate dataset is visualized as a set of coordinates in a high-dimensional data space (1 axis per variable), PCA can supply the user with a lower-dimensional picture, a "shadow" of this object when viewed from its most informative viewpoint. This is done by using only the first few principal components so that the dimensionality of the transformed data is reduced. PCA is closely related to factor analysis. Factor analysis typically incorporates more domain specific assumptions about the underlying structure and solves eigenvectors of a slightly different matrix. These values are also known as Eigen values and set of these values form the Eigen vector [20].

7 Proposed Algorithm

When applying the Gabor filter to CASIA Iris database the appropriate Iris segmentation was found out, except when gaze was not towards the camera capturing the image of the eye as shown in fig.4 a and Fig. 4 b. The input image is shown along with the Gabor filtered image. Hence to find out the effectiveness of the algorithm involving the PCA of the image and this when compared to performance of the Gabor filter used for the same image it can be found that the proposed algorithm for off gaze would perform in a better manner and guarantee better results which can be proven by the following results.

Further a classification algorithm can be used to properly classify the off gaze image, the use of Neural network Back propagation (BP) algorithm gives a better performance as compared to the K means algorithm as demonstrated further. As the BP algorithm is an effective learning tool and the database which is used is comparatively small, therefore it hardly affects the time requirements.

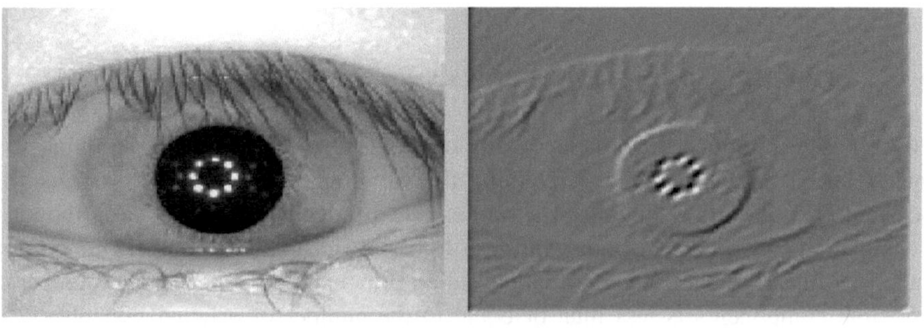

Fig. 4. a) Input image used for as mask b) Gabor filtered Image [22]

The filtered image using Gabor filter is then applied with Hough's transform such that the area of interest which is the Iris and the pupil stand out and the features can be marked with Hough circles. Here the result of Gabor filter deteriorates as compared with that of Principal Component Analysis. The image used is S1072L02.jpg which is the input image shown in fig. 4a hence there are 7 such images taken at an interval of 2 to 4 seconds the point of using this database is to verify whether the algorithm correctly identifies the irides of these images which are identical. The result of Hough transform on Gabor filtered image is given by fig. 5a When PCA algorithm was implemented on the same image it gave a better segmentation of the Iris and the Pupil. It is evident here that the PCA algorithm yields results in a better manner hence allowing for better segmentation than in the case of Gabor filter as shown in fig. 5b.

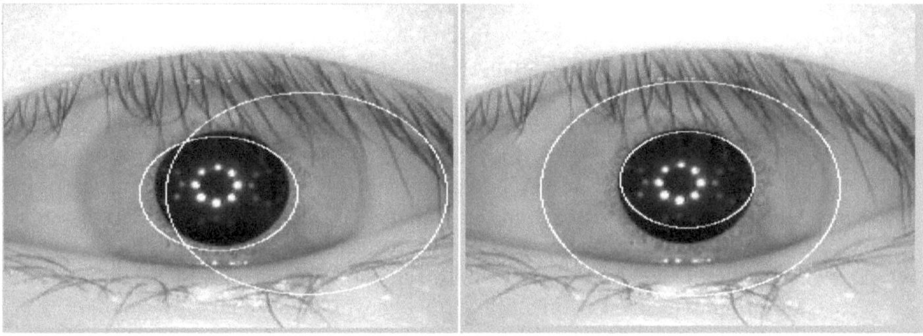

Fig. 5. a) Hough circle on Gabor filtered Image b) Hough's transform applied to image for segmentation after PCA [22]

8 Conclusion

The use of PCA over Gabor filter is recommended as in the case of Gabor filter the output file created with the filter is of more size but as generically speaking the PCA was designed to eliminate these and preserve the variation serves as a better alternative to the Gabor filter, but as researches in gesture and expression domain have found that Gabor performs as good as PCA in gesture and expression detection [25]. Such is not the case with Iris recognition and one can say that due to the biometric of the eye demanding more features to be extracted as compared to frame by frame analysis in the expression detection and identification. PCA outperforms Gabor filters even though there are few researches which rightly point out using Gabor filters along with PCA to enhance the functionality of segmentation of Iris images [25][26]. It can rightly concluded that using Eigen vectors of the PCA algorithm the segmentation process became more accurate and noise was eliminated compared to Gabor filters for Off Gaze Images.

References

[1] Burghardt, T.: Bakk Medien-Inf. Inside iris recognition. Diss. Master's thesis, University of Bristol (2002)

[2] Daugman, J.: How iris recognition works. IEEE Transactions on Circuits and Systems for Video Technology 14(1), 21–30 (2004)

[3] Daugman, J.: Recognising persons by their iris patterns. In: Li, S.Z., Lai, J.-H., Tan, T., Feng, G.-C., Wang, Y. (eds.) SINOBIOMETRICS 2004. LNCS, vol. 3338, pp. 5–25. Springer, Heidelberg (2004)

[4] Wildes, R.P.: Iris recognition: an emerging biometric technology. Proceedings of the IEEE 85(9), 1348–1363 (1997)

[5] Specification of CASIA Iris Image Database(ver 1.0), Chinese Academy of Sciences (March 2007), http://www.nlpr.ia.ac.cn/english/irds/irisdatabase.htm

[6] Phillips, P.J., Bowyer, K.W., Flynn, P.J.: Comments on the CASIA version 1.0 iris data set. IEEE Transactions on Pattern Analysis and Machine Intelligence 29(10), 1869–1870 (2007)

[7] Daugman, J.: New methods in iris recognition. IEEE Transactions on Systems, Man, and Cybernetics, Part B: Cybernetics 37(5), 1167–1175 (2007)

[8] Vatsa, M., Singh, R., Noore, A.: Improving iris recognition performance using segmentation, quality enhancement, match score fusion, and indexing. IEEE Transactions on Systems, Man, and Cybernetics, Part B: Cybernetics 38(4), 1021–1035 (2008)

[9] Ahmad AL-Allaf, O.N., AbdAlKader, S.A., Tamimi, A.A.: Pattern Recognition Neural Network for Improving the Performance of Iris Recognition System

[10] Masek, L.: Recognition of human iris patterns for biometric identification. Diss. Master's thesis, University of Western Australia (2003)

[11] Gabor, D.: Theory of Communication. J. IEE 93, 429–459 (1946)

[12] Daugman, J.: Complete Discrete 2-D Gabor Transforms by Neural Networks for Image Analysis and Compression. IEEE Trans. on Acoustics, Speech, and Signal Processing 36(7), 1169–1179 (1988)

[13] Zhu, Y., Tan, T., Wang, Y.: Biometric personal identification based on iris patterns. In: Proceedings of the 15th International Conference on Pattern Recognition, vol. 2. IEEE (2000)

[14] Ma, L., Wang, Y., Tan, T.: Iris recognition based on multichannel Gabor filtering. In: Proc. Fifth Asian Conf. Computer Vision., vol. 1 (2002)

[15] Daugman, J., Downing, C.: Gabor wavelets for statistical pattern recognition. In: The Handbook of Brain Theory and Neural Networks. MIT Press (1998)

[16] Daugman, J.G.: Six formal properties of two-dimensional anisotropie visual filters: Structural principles and frequency/orientation selectivity. IEEE Transactions on Systems, Man and Cybernetics 5, 882–887 (1983)

[17] Grigorescu, S.E., Petkov, N., Kruizinga, P.: Comparison of texture features based on Gabor filters. IEEE Transactions on Image Processing 11(10), 1160–1167 (2002)

[18] Dunn, D., Higgins, W.E.: Optimal Gabor filters for texture segmentation. IEEE Transactions on Image Processing 4(7), 947–964 (1995)

[19] (2013), The Wikipedia website https://en.wikipedia.org/wiki/Principal_component_analysis

[20] Dorairaj, V., Schmid, N.A., Fahmy, G.: Performance evaluation of iris-based recognition system implementing pca and ica encoding techniques. In: Defense and Security. International Society for Optics and Photonics (2005)

[21] Jolliffe, I.: Principal component analysis. John Wiley & Sons, Ltd. (2005)
[22] MATLAB version R2011b. The Mathworks Inc., Pune (2011)
[23] Khalil, M.R., et al.: Personal Identification with Iris Patterns.
[24] Daugman, J.G.: Biometric personal identification system based on iris analysis. U.S. Patent No. 5,291,560 (March 1, 1994)
[25] Chung, K.-C., Kee, S.C., Kim, S.R.: Face recognition using principal component analysis of Gabor filter responses. In: Proceedings of the International Workshop on Recognition, Analysis, and Tracking of Faces and Gestures in Real-Time Systems. IEEE (1999)
[26] Ahonen, T., Hadid, A., Pietikainen, M.: Face description with local binary patterns: Application to face recognition. IEEE Transactions on Pattern Analysis and Machine Intelligence 28(12), 2037–2041 (2006)

Database Model for Step-Geometric Data – An Object Oriented Approach

A. Balakrishna[1], Chinta Someswararao[2], and M.S.V.S. Bhadri Raju[2]

[1] Dept. of Mechanical Engineering, SRKR Engg. College, Bhimavaram, AP, India
[2] Department of CSE, SRKR Engg. College, Bhimavaram, AP, India
prof.adavi@gmail.com

Abstract. Information systems in today's manufacturing enterprises are distributed. Data exchange and share can be performed by computer network systems. Enterprises are performing operations globally and e-manufacturing enterprises not only obtain online information but also organize production activities. The present manufacturing scenario demands the communication among different CAD/CAM/CAE systems usually involves a huge amount of data, different formats, and proprietary platforms. To deal with such difficulties, the tool we need must be equipped with the some feature like platform independency. To full fill this, we developed an object oriented database tool to retrieve and store the GEOMETRY model data from a STEP file using JAVA. This tool operates around a model that uses a well-defined configuration management strategy to manage and storage of GEOMETRY data.

Keywords: Object Oriented, Database Model, JAVA, STEP, CAD/CAM, Geometric Data.

1 Introduction

Information systems and computer technologies are playing key role in modern industries and business. Every day the companies make strategic business decisions to improve their position in the market. They examine the business value chain to improve the product innovation, customer intimacy, and operational efficiency. The product development is one of the key weapons in the war for a competitive advantage. The policy in the product development is in the form of five 'rights', viz. the right information, in the right format, for the right people, in the right location, and at the right time. The design and development of the product in small-scale and large-scale industries are managed with CAD/CAM/CAE systems, these systems are heterogeneous nature. For many years manufacturing has been seeking to exchange product model data by defining an extended entity relationship model covering the life cycle of geometrically defined products. The life cycle was defined to be from initial conceptual design to detailed design, to manufacturing, to maintenance and final disposal. The systems to be supported life cycle phases were to include all kinds of engineering design systems including Computer Aided Design (CAD), Computer Aided Engineering (CAE), Computer Aided Manufacturing (CAM), Computerized

S.C. Satapathy et al. (eds.), *ICT and Critical Infrastructure: Proceedings of the 48th Annual Convention of CSI - Volume II*, Advances in Intelligent Systems and Computing 249,
DOI: 10.1007/978-3-319-03095-1_21,

Numerical Control (CNC) and Product Data Management (PDM)[1-5]. In this paper, a database model for step-geometry has been developed for geometrical data and communicated this data in Client/Server environment. The main objectives are

- To share CAD data files by different users by using a Standard for the Exchange of Product model data (STEP) as the standard to represent product information.
- To avoid format mismatch by the development of a Translator to store STEP Geometry Data into Oracle Database

2 Introduction to STEP

This standard, ISO 10303 [6-8], is informally known as STEP (STandard for the Exchange of Product model data). Its scope is much broader than that of other existing CAD data exchange formats, notably the Initial Graphics Exchange Specification (IGES), a US standard that has been in use for nearly twenty years. Whereas IGES was developed primarily for the exchange of pure geometric data between computer aided design (CAD) systems, STEP is intended to handle a much wider range of product-related data covering the entire life-cycle of a product.

The development of STEP has been one of the largest efforts ever undertaken by ISO. Several hundred people from many different countries have been involved in the work for the past sixteen years, and development is as active now as it ever has been in the past. STEP is increasingly recognized by industry as an effective means of exchanging product-related data between different CAD systems or between CAD and downstream application systems [9].

3 Literature Survey

In a CAD/CAM International context, there are several existing standards for data exchange, such as Initial Graphics Exchange Specification (IGES), SET, VDA-FS, EDIF, etc. The most popular exchange standard in use is the IGES. Although IGES is best supported as an interchange format for geometric information, it cannot fulfill the completeness requirement in representing product data.

Bhandar. M.P., et al.[10] proposed an infrastructure for sharing manufacturing information for the virtual enterprise. They use the STEP model data as the standard to represent complete information of a product throughout its life cycle. It integrates geometric representation and adds additional information, such as the process models, for different stages of the product development. And they used the Common Object Request Broker Architecture (CORBA) as the communication tool, and the World Wide Web (WWW) as their infrastructure. They categorize transmission protocols for translating data on the network system into two groups: the high level programming languages and the low level function calls that are embedded in the operating systems. CORBA is the high level programming language used as an interface that handles objects in a network environment and generates communication programs for both client and server, so both sides can manage their objects directly. With CORBA, local clients can transmit, manipulate objects, and send messages.

Regli[11] discusses the feature of Internet-enabled CAD systems. He brings out two features that Internet tools should have: access to information, access to tools and collaborators. Smith and Muller[12] discussed the database used by CAD/CAM systems. They focus on obtaining a multi-view database system for information sharing for establishing a Concurrent Engineering environment. E.Ly[13] builds a distributed editing system on the network so that the editing processes can be carried out on it. He uses a WWW browser as the working platform and the JAVA as the programming language. Evans and Daniel[14] discusses JAVA ability comparing the communication abilities between the traditional COI and the JAVA'S CORBA in a WWW browser.

Brown and Versprille[15] discuss issues on information sharing through the network among different CAx systems. They focus on feature extraction methods of traditional CAD/CAM databases. They use CORBA as the tool to transmit features, and store them in an object-oriented database system and suggested using other tools other than CORBA for transmission, such as Microsoft's ActiveX components. Kimuro et.al[16] discuss a Continuous Acquisition of Logistic Support (CALS) environment with CAD databases using agents. They use such technology to query distributed CAD data-bases as a centralized database system.

4 Optimization Model for STEP File

As depicted in Figure 1, optimization models are constructed for database, which in turn is constructed from the EXPRESS[17] entity database using descriptive models and aggregation methods. The class of optimization models selected to analyze of models implicitly defines the EXPRESS entity database. This model helps in.

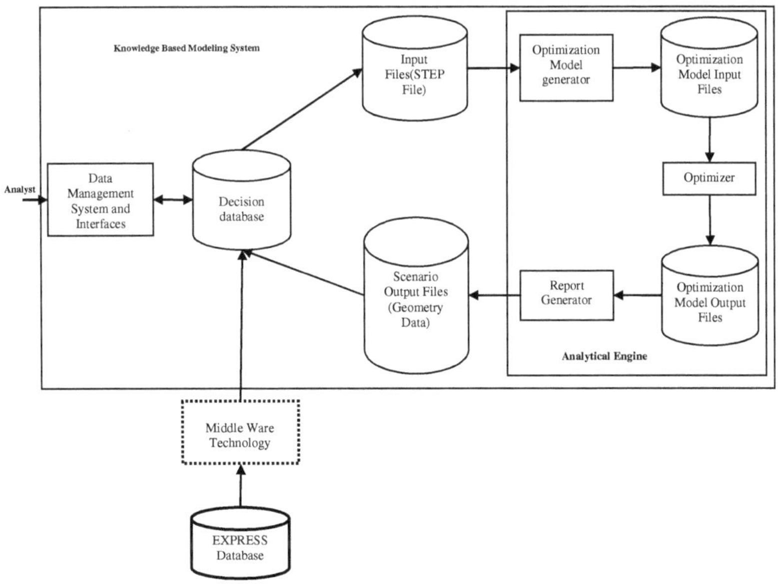

Fig. 1. Optimization model

- Organizational behaviorists have studied the relationship between exploration of new possibilities and the exploitation of old certainties in studying how organizations learn in adapting to a changing world.
- Exploration includes activities described as search, risk taking, experimentation, discovery, and innovation.
- Exploitation includes activities described as refinement, production, efficiency, implementation, and execution. Organizations that over-emphasize exploration are liable to suffer the costs of experimentation without reaping many benefits.
- Their exploratory activities produce an excess of underdeveloped new ideas and a dearth of distinctive competencies.

5 Data Management

Generally Conceptual data models are generally used for engineering information modeling at a high level of abstraction. However, engineering information systems are constructed based on logical database models. So at the level of data manipulation, that is, a low level of abstraction, the logical database model is used for engineering information modeling. Here, logical database models are often created through mapping conceptual data models into logical database models. In this data management, we used conceptual design for stores the PDM, geometric data from STEP file.

i. Database Systems

In this paper we use the Oracle 10g as the database. Oracle 10g as like the other relational database systems such as Ms Access and DB2 are used by engineering organizations to store and manage configuration control data. The strength of relational systems is in their ability to store large amounts of data in a highly normalized, tabular form, and to perform efficient queries across large data sets. Relational systems use SQL for both data definition and data manipulation.

ii. STEP Data Mapping To ORACLE 10G

The Oracle 10g implementation uses the mapping for STEP data from EXPRESS entities to the relational model. Each entity is mapped to a table with columns for attributes. Each table has a column with a unique identifier for each instance. Attributes with primitive values are stored in place, and composite values like entity instances, selects, and aggregates are stored as foreign keys containing the unique instance identifier. The corresponding EXPRESS Entity database designs are shown in Tables 1,2.

iii. Design of Geometric Entity Database

Geometric database having fields id, entity, entity_data0, entity_data1,….,entity_data m as shown in Table 2. Purpose of this database is to store Geometric data after separating the data from STEP file. The first field id have the information of the entity, entity have the information of entity, entity_data0 have information of the entity data, and so on.,

Table 1. Express Entity Database Design

Field Name	Data Type
entity	Text
Data1	Text
Data_type1	Text
Data2	Text
Data_type2	Text
....	
Data m	Text
Data_type m	Text

Table 2. Geometric Entity Database Design

Field Name	Data Type
id	Text
entity	Text
entity_data0	Text
entity_data1	Text
entity_data2	Text
.....	
entity_data m	Text

6 Implementation

In this paper an interactive user interface program is developed to extract STEP – Geometric data from neutral format STEP file using JAVA language as shown in In this work an interface program is developed to extract Product data(PDM), Geometric data from neutral format STEP file using Java. EXPRESS Schema entity definitions for Product and Geometry data are stored in Oracle 10g and these are used in backend for validation[17]. The Geometry data as Cartesian-point, oriented-edge, edge-curve, vertex, axis-placement information etc, information extracted from STEP file as per back end Express Schema entities database. The extracted data is inserted into Geometry database. Template is designed using Java 1.7 for the execution of interface program shown Figure 2.

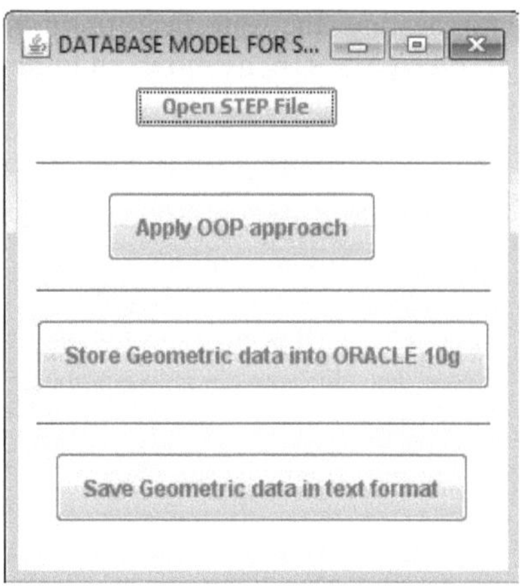

Fig. 2. Object Oriented Database Construction model

Process for Design the Geometric DATABASE MODEL

Step1: Select the STEP file and save it in variable STEPVARIABLE

Step2: Open the file STEPVARIABLE in read mode

Step3: Read a line from STEPVARIABLE and assigned to a variable STEPLINE

Step4: If STEPLINE equals to null GOTO Step 8

Step5: Separate the entity, entitydata1,entitydata2,…., entitydata m from STPLN in following manner.

- i Read character by character from STEPLINE
- ii Extract the entity, assigned to a variable ENTITY
- iii Extract the entity data , assigned to a variable ENTITYDATA1
- iv Extract the next entity data, assigned to variable ENTITYDATA2, …ENTITYDATAM until end of the STEPLINE

Step6: Search the back end EXPRESS entity database for ENTITY in column "entity".

Step7: If found store the ENTITY along with the ENTITYDATA1, ENTITYDATA2, . ., ENTITYDATAM in the Database "Geometry"

Step8: Otherwise increment the counter in STEPVARIABLE and GOTO Step 3

Step9: STOP

7 Case Study

In this work, following model is designed using the solid modeling techniques. A Gas-Regulator consider for testing the interface program. The model of Gas-Regulator is created using CATIA V5 R16, as shown in Figure 3.

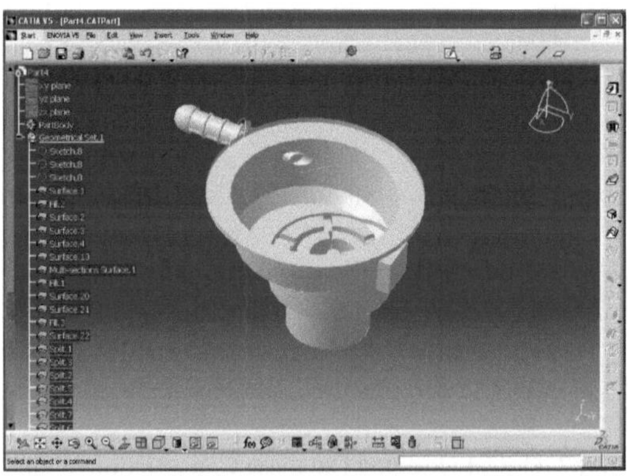

Fig. 3. Assembly of GAS-Regulator

Here we presents STEP file for GAS-Regulator model.

```
ISO-10303-21;
HEADER;
FILE_DESCRIPTION(('CATIA V5 STEP Exchange'),'2;1');
FILE_NAME('C:\\Documents and Settings\\Administrator\\Desktop\\STEP
FILES\\GASREGULATORBODY.stp','2013-07-20T05:35:15+00:00',('none'),('none'),'CATIA Version 5 Release 16
(IN-10)','CATIA V5 STEP AP203','none');
FILE_SCHEMA(('CONFIG_CONTROL_DESIGN'));
ENDSEC;
/* file written by CATIA V5R16 */
DATA;
#5=PRODUCT('Part4',",",(#2)) ;
#1=APPLICATION_CONTEXT('configuration controlled 3D design of mechanical parts and assemblies') ;
#14=PRODUCT_DEFINITION(' ',' ',#6,#3) ;
#16=SECURITY_CLASSIFICATION(' ',' ',#15) ;
………..
#88=CARTESIAN_POINT('Axis2P3D Location',(-21.6430454254,-41.,16.7157007856)) ;
#92=CARTESIAN_POINT('Limit',(-28.3930454254,-41.,16.7157007856)) ;
……….
#6104=DIRECTION('Axis2P3D XDirection',(0.988395467625,0.,-0.151902598982)) ;
#6108=DIRECTION('Axis2P3D Direction',(0.,-1.,0.)) ;
……………..
#7062=ORIENTED_EDGE(",*,*,#7059,.F.) ;
#7129=ORIENTED_EDGE(",*,*,#7101,.T.) ;
…….
#7065=OPEN_SHELL('Surface.18',(#7192)) ;
#7193=OPEN_SHELL('Surface.19',(#7320)) ;
#98=SHELL_BASED_SURFACE_MODEL('NONE',(#97)) ;
#4101=SHELL_BASED_SURFACE_MODEL('NONE',(#4100)) ;
………….
#821=EDGE_CURVE(",#820,#813,#818,.T.) ;
#826=EDGE_CURVE(",#820,#406,#825,.T.) ;
……………
#7228=VERTEX_POINT(",#7227) ;
#7286=VERTEX_POINT(",#7285) ;
#7288=VERTEX_POINT(",#7287) ;
………
ENDSEC;
END-ISO-10303-21;
```

Data Base has no information before executing above program as shown in Table 3.

Table 3. Geometric Entity Database

id	entity	entity_data0	entity_data1	…..	entity_datam

Table 4. Data in the Geometric Entity Database

id	entity	entity_data0	entity_data1	…	entity_data m
88	CARTESIAN_POINT	'Axis2P3D Location'	21.6430454254,-41.,16.7157007856		
6104	DIRECTION	'Axis2P3D XDirection'	0.988395467625		
7191	FACE_BOUND	‘ ‘	,#7188		
7288	VERTEX_POINT	‘ ‘	#7287		

After execution of this program, above empty database shown in table 3 is filled with geometric data as shown in table 4. This information is useful for further processing.

8 Conclusions

Now-a-days knowledge demanding production is arisen in engineering industry especially in design, analysis and manufacturing. In which new information requirements are proposed in data knowledge sharing and reusing in richer information types like CAD/CAM. To overcome the some problem arisen in the above said industries especially in STEP, in this paper, we proposed a new method to maintain object oriented STEP-Geometric database using object oriented approach with JAVA. This approach maintains STEP-GEOMETRY Database as well as propose easily understandable user interface. In future we will apply this method for STEP-GEOMETRY XML Ontology and downstream applications.

Acknowledgments. This work is partially supported by AICTE, Government of INDIA, Technology Bhavan, New Delhi-110016, Ref.No.20/AICTE/RFID/RPS (POLICY-1)14/2013-14.

References

1. Balakrishna, A., Someswararao, C.: Implementation of Interactive Database System for STEP-Geometric Data from Express Entities. In: IEEE Conference on Computer Science and Technology, pp. 285–289 (2010)
2. Balakrishna, A., Someswararao, C.: Implementation Of Interactive Data Knowledge Management And Semantic Web For Step-Data From Express Entities. In: Third IEEE Conference on Emerging Trends in Engineering and Technology, pp. 537–542 (2010)
3. Balakrishna, A., Someswararao, C.: PDM data classification from STEP- an object oriented String matching approach. In: IEEE Conference on Application of Information and Communication Technologies, pp. 1–9 (2011)
4. Balakrishna, A., Someswararao, C.: Development of a Manufacturing database System for STEP-NC data from express Entities. International Journal of Engineering Science and Technology 2, 6819–6828 (2010)
5. Balakrishna, A., Someswararao, C.: Fuzzy Approach to the Selection of Material Data in Concurrent Engineering Environment. International Journal of Engineering, 921–927 (2011)
6. ISO, ISO 10303:1994 - Industrial Automation Systems and Integration - Product Data Representation and Exchange (The ISO web site is at `http://www.iso.ch/cate/cat.html` - search on 10303 to find a list of parts of the standard)
7. Owen, J.: STEP: An Introduction, Information Geometers, 2nd edn., Winchester, UK (1997) ISBN 1-874728-04-6
8. Kemmerer, S.J. (ed.): STEP: The Grand Experience. NIST Special Publication SP 939, US Government Printing Office, Washington, DC 20402, USA (July 1999)
9. PDES Inc., Success stories, `http://pdesinc.aticorp.org`

10. Bhandarkar, M.P., Downie, B., Hardwick, M., Nagi, R.: Migrating from IGES to STEP: one to one translation of IGES drawing to STEP drafting data. Computers in Industry, 261–277 (2000)
11. Regli, W.C.: Internet-enabled computer aided design. IEEE Internet Computing 1(1), 39–50 (1997)
12. Smith, G.L., Muller, J.C.: PreAmp - a pre-competitive in intelligent manufacturing technology: an architecture to demonstrate concurrent engineering and information sharing. Concurrent Engineering: Research and Application, 107–115 (1994)
13. Ly, E.: Distributed JAVA applet for project management on the web. IEEE Internet Computing, 21–26 (1997)
14. Evans, E., Daniel, R.: Using JAVA applets and CORBA for multi-user distributed application. IEEE Internet Computing, 43–55 (1997)
15. Brown, D., Versprille, K.: The OCAI initiative (open cax architecture and interoperability). In: Proceedings of the CALS Expo. International 1997, Tokyo, pp. 37–41 (1997)
16. Kimuro, T., Akasaka, H., Nishino, Y.: New approach for searching distributed databases by agent technology. In: Proceedings of the CALS Expo. International 1997, Tokyo, pp. 13–22 (1997)
17. `http://www.nist.gov`

[illegible]

A Proposal for Color Segmentation in PET/CT-Guided Liver Images

Neha Bangar and Akashdeep Sharma

University Institute of Engineering and Technology, Panjab University, Chandigarh, India
neha.bangar1@gmail.com, akashdeep@pu.ac.in

Abstract. Automatic methods for detection and segmentation of tumor have become essential for the computer-oriented diagnosis of liver tumors in images. Tumor Segmentation in grayscale medical images can be difficult since the intensity values between tumor and healthy tissue is very close. Positron emission tomography combined with computed tomography (PET/CT) provides more accurate measurements of tumor size than is possible with visual assessment alone. In this paper, a new method for the detection of liver tumor in PET/CT scans is proposed. The images are denoised using median filter and binary tree quantization clustering algorithm is used for segmentation. Finally image dilation and erosion, boundary detection, ROI selection and shape feature extraction are applied on the selected cluster to identify the shape of the tumor.

Keywords: Denoising, Liver segmentation, Tumor Detection, Binary Tree Quantization Algorithm, Shape Feature Extraction.

1 Introduction

One of the major causes of death in humans is considered to be liver tumor [31]. It is very important to detect tumors at early stages for the survival of the patients. As the number of images in medical databases is large, analyzing all images manually is very difficult and so, computer oriented surgery has become one of the major research subjects [1] and this has led to medical imaging modalities such as X-ray, CT [2], Magnetic Resonance Imaging (MRI) [3, 4], SPECT [5], PET [6] and ultrasound [7, 8].

Till-date, so many methods had been proposed for detection and segmentation of tumor in liver CT and MRI images but not much of the work has been done on PET/CT liver images. *Amir H. Foruzan et.al* [9] proposed a knowledge-based technique for liver segmentation in CT data for liver initial border estimation in which they started with image simplification, then searched rib bones, connected them together to find ROI of liver. They then used split thresholding technique to segment the images. Different colors were assigned to objects present in ROI, the split-threshold step and the objects that were found in 75% of right part of the abdomen. After this a colored image was obtained in which liver had a specific color from where liver boundary was extracted. In the method of *Xing Zhang et.al* [10],

S.C. Satapathy et al. (eds.), *ICT and Critical Infrastructure: Proceedings of the 48th Annual Convention of CSI - Volume II*, Advances in Intelligent Systems and Computing 249,
DOI: 10.1007/978-3-319-03095-1_22, © Springer International Publishing Switzerland 2014

automatic liver segmentation included average liver shape model localization in CT via 3D generalized Hough Transform, subspace initialization of Statistical Shape Model through intensity and gradient profile and then deforming the model to liver contour through optimal surface detection method based on graph theory. *Laszlo Rusko et.al* [11] method automatically segments the liver using region-growing facilitated by pre- and post-processing functions, which considers anatomical and multi-phase information to eliminate over and under-segmentation. *Hassan Masoumi et.al* [12] presented Automatic liver segmentation in MRI images using an iterative watershed algorithm and artificial neural network to extract features of the liver region. But all these methods resulted only in liver boundary and did not detect liver tumor. *O. Lezoray et.*al [13] proposed an unsupervised clustering technique in which watershed operates on distance function to centers of class for determining the number of classes. In this method, segmentation of colored image considered pairwise color projections where each of these projections is analyzed to look for the dominant colors of 2-D histogram and to fully automate the segmentation, energy function was used to quantify the quality of the segmentation. But the difficulty with the histogram method is to identify peaks and valleys in the image. *Marisol Martinez-Escobar et.al* [14] first colorized the pixels representing tumor and healthy tissues and then used threshold method for segmentation to detect tumor to overcome the problem as faced in histogram. But these methods are either performed on CT or MRI images or the images are first colored and then segmentation is applied. Since morphological changes always proceed metabolic changes and are detected through imaging modalities like CT or MRI, PET is expected in enabling an early assessment of response to treatment. 18F-FDG PET has been reported to give earlier response for tumor detection than CT [15].

Positron emission tomography (PET) with 18FFluorodeoxyglucose (18F-FDG) is widely suggested method medical imaging as numerous tumors are diagnosed very accurately which has improved the decision for therapy consideration and assessing patients having cancer at different stages in the last two decades [16, 17]. It is based on the tumor specific high intracellular accumulation of the glucose analog fluorodeoxyglucose (18F-FDG) [18]. It gives tumor's physiological information and its metabolic activities [19]. PET/CT provides functional and anatomical imaging within a single scanner in a single scanning session [20]. Though PET has been replaced by PET/CT, most of the segmentation work to detect tumor has been done on PET only [21-24].

Baardwijk et. al [25] provided the advantage of combining the information from PET and CT images for the segmentation purpose. *Potesil et. al* [26] proposed a method using initial hot spot detection and segmentation in PET to provide tumor structure appearance and shape model to classify voxels in CT. *Xia et. al* [27] proposed an automated segmentation method for brain PET/CT images with MAP-MRF model achieved by solving a maximum a-posteriori (MAP) problem using expectation-maximization (EM) algorithm with simulated annealing. But this method suffered from long execution times due to simulated annealing. *Yu et. al* [28] proposed the co-registered multimodality pattern analysis segmentation system (COMPASS) to extract texture features from PET/CT and then used decision-tree based K-nearest-neighbor classifier to label each voxel as either "normal" or "abnormal" and the performance was compared with threshold methods: SUV value and signal/background ratio. *Yu et. al* [29] further provided the effectiveness of using the co-registered segmentation in order to identify textural features to distinguish tumor from healthy tissue in head and neck regions. Most of the approaches use standard uptake value (SUV), a semi-quantitative normalization of FDG uptake in PET images and measures the FDG concentration normalized by decay corrected injection dose/gram body mass, to detect tumor. In these approaches, hot spots are detected as local regions with high SUV values, and segmentation methods apply threshold relating to the maximum SUV values. But these approaches are basically used to investigate the value and application of FDG PET/CT in clinical practices to detect tumors.

In this paper, we have used PET/CT images and propose a segmentation method which is based on binary tree quantization clustering rather than primitive methods like thresholding and then boundary detection and shape feature extraction is done for liver tumor detection in PET/CT images.

2 Proposed Approach

Our proposed method consists of 6 steps as shown in figure1:

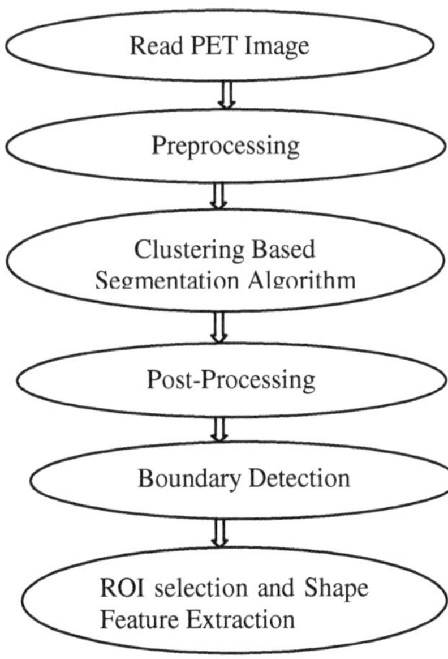

Fig. 1. Flowchart of the Proposed Method

2.1 Preprocessing

PET/CT images may contain noise in them so to remove noise we denoised the image using 2D-median filter in which depending upon the intensity, the pixels in the neighborhood window are ranked and the median i.e. the middle value becomes the output value for the center pixel. As the output value is from the neighboring values, new unknown values are not created near the edges, so median filtering is more effective when our main aim is to simultaneously reduce noise and preserve the edges. To apply median filter on colored images, number of colors is firstly determined in the image and then median filter is applied on each color separately. All the denoised components are combined to get the final denoised image.

2.2 Segmentation

The proposed segmentation algorithm is based on binary tree quantization algorithm performed on PET/CT images. To segment the image, we cluster the pixels determining various statistics. In Binary tree quantization method, partition with the largest eigenvalue is chosen to be split. Eigenvalue is the scalar which is associated with eigenvector where eigenvector is a vector that changes its magnitude but not its direction. The splitting axis is the one that passes through the centroid of all the colors in that region and must be perpendicular to the direction of the maximum total squared variation which is derived from the eigenvalue and eigenvector of the covariance matrix of the data in that particular region. To conclude:

All the pixels are placed in the same cluster in the beginning. The cluster is then split around its mean value projection on the first principle component. The parameters to the clustering algorithm are the number of clusters, K, to use which is set/input by the user, and the first clusterC_1.

1. Initialize the first cluster, C_1
2. Calculate the mean, μ_1, and the covariance matrix $\sum_1$ of the cluster C_1
3. For i=2 to K do
4. Find the cluster,C_n with the largest eigenvalue and its associated eigenvector e_n.
5. Split C_n in two sets along the mean value projections on the eigenvector, $C_i = \{x \in C_n : e_n^T z_n \leq e_n^T \mu_n\}$ and update the original cluster with other half $C_n^* = C_n - C_i$.
6. Compute mean and covariance matrix of the two halves obtained in step2 as $\mu_n^*, \sum_n^*, \mu_i \; and \sum_i$.

2.3 Post-processing Steps

The post-processing step involves dilation and erosion morphological operations on the selected cluster. Morphological operations apply a structuring element which is basically a matrix of 0's and 1's only and can have any shape and size. Pixels having value as 1 defines the neighborhood to the input image and so create an output image of the same size. In this operation, each pixel value in the output image is compared to corresponding pixel in the input image with its neighbors. The erosion step removes stray disconnected regions (removes pixels on object boundaries) and dilation fills in holes within the segmented region or it adds pixels to both inner and outer boundaries of regions. These are used for boundary detection of the liver tumor described in the next section.

2.4 Boundary Detection

To detect the boundaries of the object of interest we use erosion and dilation as post processing steps. For this following steps are performed:

a) subtracting eroded image from original image
b) subtracting original image from dilated image and
c) subtracting eroded image from dilated image

These steps output 3 boundary images and we select one out of them that best suits our need.

2.5 ROI Selection and Shape Feature Extraction

After boundary detection, Region of interest is selected using GUI based polygon method which selects a polygonal region of interest within an image [30]. Then shape feature extraction is performed on the selected ROI to calculate area, equiv-diameter, pixel-list and diameter of the tumor.

3 Results

The binary tree quantization method was evaluated on PET/CT liver images obtained from PGI, Chandigarh. Figure2 shows the segmentation results of binary tree quantization method. Fig.2-(a) shows the original PET/CT image, Fig.2-(b) is the first cluster obtained which shows all organs except liver, Fig.2-(c) is the second cluster showing brightened area of liver and tumor than the original image, Fig.2-(d) shows the third cluster which outputs only the tumor inside the liver, Fig.2-(e) is the fifth cluster showing the liver boundary, Fig.2-(f-g) clusters show the rough boundary of the detected tumor.

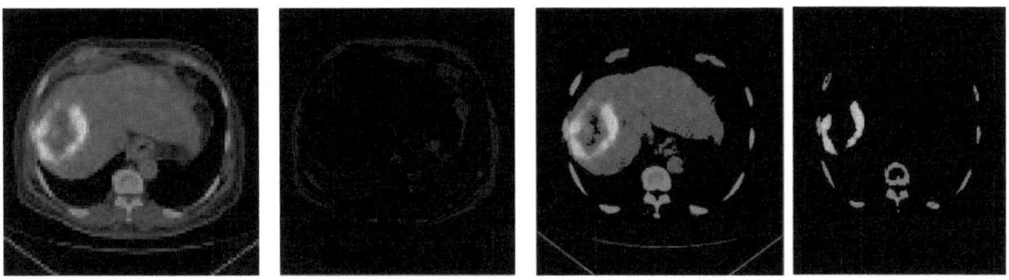

(a) (b) (c) (d)

Fig. 2. Binary Tree Quantization outputs: (a) Original Image; (b) Cluster1; (c) Cluster2; (d) Cluster3; (e) Cluster4, (f) Cluster5; (g) Cluster6

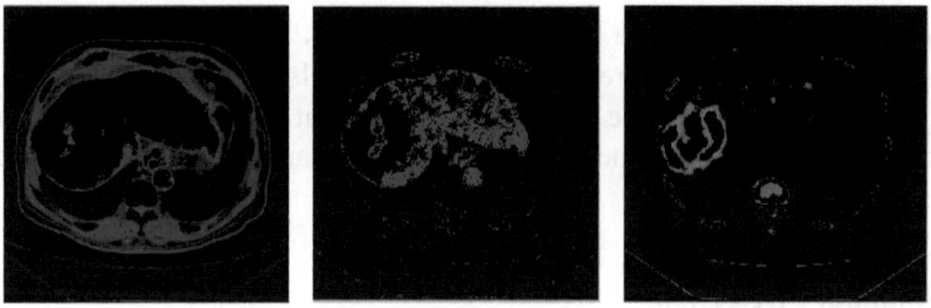

(e) (f) (g)

Fig. 2. *(continued)*

4 Conclusion

In this paper, we have proposed a binary tree quantization clustering algorithm for segmentation. Results have been provided for the segmentation where as operations of image dilation and erosion, boundary detection, ROI selection and shape feature extraction have been listed. The method is giving good results as it is giving bright tumor regions in output image.

References

1. Masuda, Y., Tateyama, T., Xiong, W., Zhou, J., Wakamiya, M., Kanasaki, S., Furukawa, A., Chen, Y.W.: Liver Tumor Detection in CT images by Adaptive Contrast Enhancement and the EM/MPM Algorithm. In: 18th IEEE Conference on Image Processing, pp. 1421–1424 (September 2011)
2. Hounsfield, G.N.: Computerized Transverse Axial scanning Tomography: Part 1, Description of the System. British Journal of Radiology 46, 1016–1022 (1973)
3. Lipinski, B., Herzog, H., Kops, E.R., Oberschelp, W., Muleer-Gartner, H.W.: Expectation Maximization Reconstruction of Positron Emission Tomography Images using Anatomical magnetic Resonance Information. IEEE Transaction on Medical Imaging 16, 129–136 (1997)
4. Bazille, A., Guttman, M.A., Mcveigh, E.R., Zerhouni, E.A.: Impact of Semiautomated versus Manual Image Segmentation Errors on Myocardial Strain Calculation by Magnetic Resonance Tagging. Investigative Radiology 29, 427–433 (1994)
5. Anger, H.: Use of Gamma-Ray Pinhole Camera for viva studies. In: A Nature Conference on Nuclear Reprogramming and the Cancer Genome, vol. 170, pp. 200–204 (1952)
6. Ouyang, X., Wang, W.H., Johnson, V.E., Hu, X., Chen, C.T.: Incorporation of Correlated Structural Images in PET Image Reconstruction. IEEE Transactions on Medical Imaging 13, 627–640 (2002)
7. Akgul, Y.S., Kambhamettu, C., Stone, M.: Extraction and Tracking of the Tongue Surface from Ultrasound Image Sequences. In: 1998 IEEE Computer Society Conference on Computer Vision Pattern Recognition, pp. 298–303 (June 1998)
8. Abeyratne, U.R., Petropulu, A.P., Reid, J.M.: On modeling the Tissue Response from Ultrasonic B-scan. IEEE Transactions on Medical Imaging 2, 479–490 (1996)

9. Foruzan, A.H., Zoroofi, R.A., Hori, M., Sato, Y.: A Knowledge-based Technique for Liver Segmentation in CT Data. Computerized Medical Imaging and Graphics 33, 567–587 (2009)
10. Zhang, X., Tian, J., Deng, K., Yongfang, W., Xiuli, I.: Automatic Liver Segmentation Using a Statistical Shape Model With Optimal Surface Detection. IEEE Transactions on Biomedical Engineering 57, 2622–2626 (2010)
11. Rusko, L., Bekes, G., Fidrich, M.A.: Automatic Segmentation of the Liver from Multi- and Single-Phase Contrast-Enhanced CT. Medical Image Analysis 13, 871–882 (2009)
12. Masoumi, H., Behrad, A., Pourmina, M.A., Roosta, A.: Automatic liver segmentation in MRI Images using an Iterative Watershed Algorithm and Artificial Neural Network. Biomedical Signal Processing and Control 7, 429–437 (2012)
13. Lezoray, O., Charrier, C.: Color Image Segmentation using Morphological Clustering and Fusion with Automatic Scale Selection. Pattern Recognition Letters 30, 397–406 (2009)
14. Escobar, M.M., Foo, J.L., Winer, E.: Colorization of CT images to Improve Tissue Contrast for Tumor Segmentation. Computers in Biology and Medicine 42, 1170–1178 (2012)
15. Necib, H., Garcia, C., Wagner, A., Vanderleinden, B., Emonts, P., Hendlisz, A., Flamen, P., Buvat, I.: Detection and Characterization of Tumor Changes in 18FFDG Patient Monitoring using Parametric Imaging. J. of Nucl. Med. 52, 354–361 (2011)
16. Lartizien, C., Francisco, S.M., Prost, R.: Automatic Detection of Lung and Liver Lesions in 3-D Positron Emission Tomography Images: A Pilot Study. IEEE Transactions on Nuclear Science 59, 102–112 (2012)
17. Changyang, L., Wanga, X., Xiaa, Y., Eberlb, S., Yinc, Y., Feng, D.D.: Automated PET-guided Liver Segmentation from Low-Contrast CT Volumes using Probabilistic Atlas. Computer Methods and Programs in Biomedicine 107, 164–174 (2011)
18. Blechacz, B., Gores, G.J.: PET scan for Hepatic Mass. Hepatology 52, 2186–2191 (2010)
19. Ming, X., Feng, Y., Guo, Y., Yang, C.: A New Automatic Segmentation Method for Lung Tumor Based on SUV threshold on 18F-FDG PET images. In: 2012 IEEE International Conference on Virtual Environments Human-Computer Interfaces and Measurement Systems (VECIMS), pp. 5–8 (July 2012)
20. Yong, X., Stefan, E., Lingfeng, W., Michael, F., David, D.D.: Dual-Modality brain PET-CT image segmentation based on adaptive use of functional and anatomical information. Computerized Medical Imaging and Graphics 36, 47–53 (2011)
21. Belhassen, S., Zaidi, H.: A Novel Fuzzy C-means Algorithm for Unsupervised Heterogeneous Tumor Quantification in PET. Medical Physics 37, 1309–1324 (2010)
22. Geets, X., Lee, J.A., Bol, A., Lonneux, M., Gregoire, V.: A gradient-based method for segmenting FDG-PET images: methodology and validation. Eur. J. of Nucl. Med. Mol. Imaging 34, 1427–1438 (2007)
23. Hatt, M., Rest, C.L., Turzo, A., Roux, C., Visvikis, D.: Fuzzy Logically Adaptive Bayesian Segmentation Approach for Volume Determination in PET. IEEE Transactions on Medical Imaging 28, 881–893 (2009)
24. Li, H., Thorstad, W.L., Biehl, K.J., Laforest, R., Su, Y., Shoghi, K.I., Donnelly, E.D., Low, D.A., Lu, W.: A Novel PET Tumor Delineation Method based on Adaptive region-Growing and Dual-Front Active Contours. Medical Physics 35, 3711–3721 (2008)
25. Baardwijk, A., Bosmans, G., Boersma, L.: PET-CT based Auto-contouring in Non- Small-Cell Lung Cancer correlates with Pathology and reduces Interobserver Variability in the Delineation of the Primary Tumor and involved Nodal Volumes. International Journal of Radiation and Oncology, Biology and Physics 68, 771–778 (2007)

26. Potesil, V., Huang, X., Zhou, X.: Automated Tumor Delineation using Joint PET/CT information. In: Proc. SPIE International Symposium on Medical Imaging: Computer-Aided Diagnosis, vol. 65142 (March 2007)
27. Xia, Y., Wen, L., Eberl, S., Fulham, M., Fend, D.: Segmentation of Dual Modality Brain PET/CT images using MAP-MRF model. In: 2008 IEEE 10th Workshop on Multimedia Signal Processing, pp. 107–110 (October 2008)
28. Yu, H., Caldwell, C., Mah, K.: Automated Radiation targeting in head-and-neck cancer using Region-based Texture Analysis of PET and CT images. International Journal of Radiation and Oncology, Biology and Physics 75, 618–625 (2009)
29. Yu, H., Caldwell, C., Mah, K., Mozeg, D.: Coregistered FDG PET/CT-based Textural Characterization of Head and Neck Cancer for Radiation Treatment Planning. IEEE Transactions on Medical Imaging 28, 374–383 (2009)
30. Gunjal, B.L., Mali, S.N.: ROI Based Embedded Watermarking of Medical Images for Secured Communication in Telemedicine. International J. Comp. and Commun. Eng. 68, 815–820 (2012)
31. Centre for Control and Information Services, National Centre, Japan, `http://ganjoho.jp/public/statistics/pub/statistics01.html`

Audio Segmentation for Speech Recognition Using Segment Features

Gayatri M. Bhandari[1], Rameshwar S. Kawitkar[2], and Madhuri P. Borawake[3]

[1] J.J.T. University and
JSPM's Bhivarabai Sawant Institute of Tech. & Research(W),
Pune, India
gayatri.bhandari1980@gmail.com
[2] Sinhgad Institute of Technology, Pune
rskawitkar@rediffmail.com
[3] J.J.T. University and
PDEA, College of Engg., Pune
madhuri.borawake@gmail.com

Abstract. The amount of audio available in different databases on the Internet today is immense. Even systems that do allow searches for multimedia content, like AltaVista and Lycos, only allow queries based on the multimedia filename, nearby text on the web page containing the file, and metadata embedded in the file such as title and author. This might yield some useful results if the metadata provided by the distributor is extensive. Producing this data is a tedious manual task, and therefore automatic means for creating this information is needed. In this paper an algorithm to segment the given audio and extract the features such as MFCC, SF, SNR, ZCR is proposed and the experimental results shown for the given algorithm.

Keywords: Audio segmentation, Feature extraction, MFCC, LPC, SNR, ZCR.

1 Introduction

Audio exists at everywhere, but is often out-of-order. It is necessary to arrange them into regularized classes in order to use them more easily. It is also useful, especially in video content analysis, to segment an audio stream according to audio types.

In many applications we are interested in segmenting the audio stream into homogeneous regions. Thus audio segmentation is the task of segmenting a continuous audio stream in terms of acoustically homogenous regions [4]. The goal of audio segmentation is to detect acoustic changes in an audio stream. This segmentation can provide useful information such as division into speaker turns and speaker identities, allowing for automatic indexing and retrieval of all occurrences of a particular speaker. If we group together all segments produced by the same speaker we can perform an automatic online adaption of the speech recognition acoustic models to improve overall system performance.

S.C. Satapathy et al. (eds.), *ICT and Critical Infrastructure: Proceedings of the 48th Annual Convention of CSI - Volume II*, Advances in Intelligent Systems and Computing 249,
DOI: 10.1007/978-3-319-03095-1_23, © Springer International Publishing Switzerland 2014

1.1 Segmentation Approaches

In typical segmentation methods are categorized into three groups, namely energy-based, metric-based, and model-based. The energy-based algorithm only makes use of the running power in time domain. On the other hand, both the metric-based and the model-based method are based on statistical models, say, multivariate Gaussian distributions. That means, rather than using the feature values directly, the running means and variances of them are modeled by a multidimensional Gaussian distribution.

1.1.1 Energy-Based Algorithm

The energy-based algorithm can be very easily implemented. Silence periods, that measured by the energy value and a predefined threshold, are assumed to be the segment boundaries. However, since there is no direct connection between the segment boundaries and the acoustic changes, this method can be problematic for many applications, such as gender detection, and speaker identification, etc.

1.1.2 Model-Based Algorithm

In the model-based algorithm, statistical distribution models are used for each acoustic class (e.g., speech, music background, noise background, etc.) The boundaries between classes are used as the segment boundaries. Typically, Bayesian Information Criterion (BIC) is used to make the decision if the changing point turns out, which is essentially a hypothesis testing problem.

1.1.3 Metric-Based Algorithm

In the metric-based algorithm, statistical distribution models are also used for modeling the feature space. Gaussian model is a typical choice, but some other distributions can also be used. For example, Chi-squared distribution are found to be appropriate and with less computational cost in. The sound transition is measured by the distance between the distributions of two adjacent windows. The local maximum of distance value suggests a changing point.

According to this here six different classes of audio are defined.

1. Speech: This is pure speech recorded in the studio without background such as music.

2. Speech over Music: This category includes all studio speech with music in the background.

3. Telephone Speech: Some sections of the program have telephonic interventions from the viewers. These interventions are mixed in the program's main audio stream as a wide band stream.

4. Telephone Speech over Music: The same as previous class but additionally there is music in the background.

5. Music: Pure music recorded in the studio without any speech on top of it.

6. Silence
Real-time speaker segmentation is required in many applications, such as speaker tracking in real-time news-video segmentation and classification, or real-time speaker adapted speech recognition. Here real-time, yet effective and robust speaker segmentation algorithm based on LSP correlation analysis can be done. Both the speaker identities and speaker number are assumed unknown. The proposed incremental speaker updating and segmental clustering schemes ensure this method can be processed in real-time with limited delay.

1.2 Related Work

Several methods have been developed for audio segmentation. Chen identifies two types of segmentation approaches namely, classification-dependent segmentation (CDS) and classification-independent segmentation (CIS) [1]. CDS methods are problematical because it is difficult to control the performance [1].

CIS approaches can be further separated into time-domain and frequency-domain depending upon which audio features they use, or supervised and unsupervised approaches depending on whether the approach requires a training set to learn from prior audio segmentation results. CIS may also be defined as model-based or non-model based methods.

In model-based approaches, Gaussian mixture models (GMM) [4], [5], Hidden Markov Models (HMM) [6], Bayesian [7], and Artificial Neural Networks (ANN) [8] have all been applied to the task of segmentation. Examples of an unsupervised audio segmentation approach can be found in [7] and [9]. These unsupervised approaches test the likelihood ratio between two hypotheses of change and no change for a given observation sequence. On the other hand, the systems developed by Ramabhadran et al. [6] and Spina, and Zue [4] must be trained before segmentation. These existing methods are limited because they deal with limited and narrow classes such as speech/music/noise/ silence.

Audio segmentation methods based on a similarity matrix, have been employed for broadcasting news, which is relatively acoustic dissimilar, and for music to extract structures or music summarizations. The accuracy evaluation of these methods was undertaken with specific input audios and has not been previously reported for use

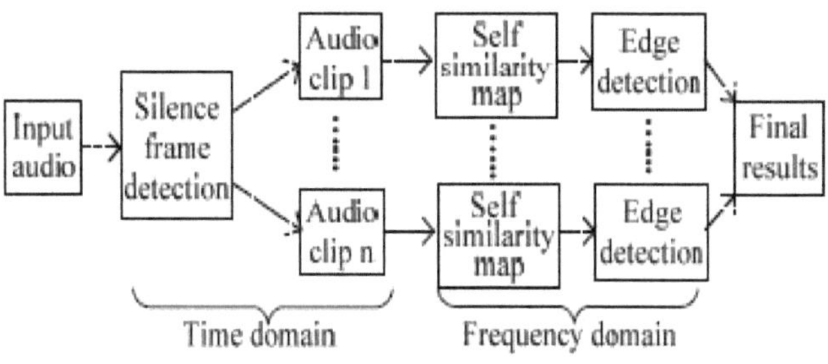

Fig. 1. Frame work for audio segmentation method

with audio files in a non-music/non-speech database. This paper introduces a two phase unsupervised model-free segmentation method that works for general audio files. In this paper, we discuss the process by which we developed and evaluated an efficient CIS method that can determine segment boundaries without being supplied with any information other than the audio file itself.

2 Feature Analysis

Fig. 2 shows the basic processing flow of the proposed approach that integrates audio segmentation and speaker segmentation. After feature extraction, the input digital audio stream is classified into speech and nonspeech. Nonspeech segments are further classified into music, environmental sound, and silence, while speech segments are further segmented by speaker identity [2].

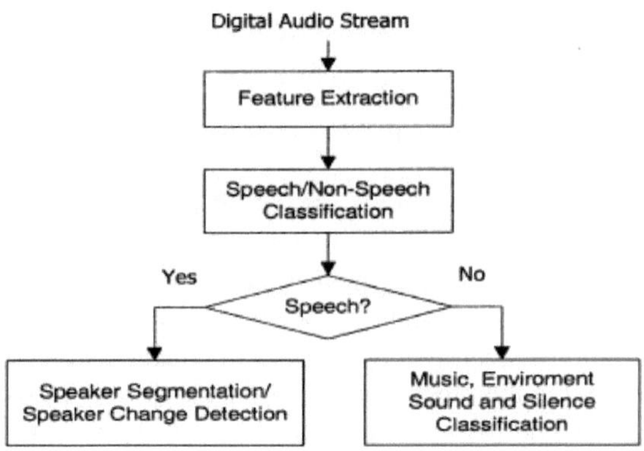

Fig. 2. Basic Processing flow of audio content analysis

3 Objectives of the Proposed Research

1. The objective of audio segmentation for classifying the audio components into Speech Music , Non-speech , noise , silence along with the other major features such as MFCC , SF , LPC , SNR , HZCRR etc and can be transferred over the network and by analyzing these audio features the reconstruction of audio signal should be more accurate.
2. We proposed to use thirteen features in time, frequency, and cepstrum domains and model-based (MAP, GMM, KNN, etc.) classifier, which achieved an accuracy rate over 90% on real-time discrimination between speech and music. As in general, speech and music have quite different spectral distribution and temporal changing patterns, it is not very difficult to reach a relatively high level of discrimination accuracy.
3. Further classification of audio data may take other sounds into consideration besides speech and music.

4. We also proposed an approach to detect and classify audio that consists of mixed classes such as combinations of speech and music together with environment sounds. The accuracy of classification is more than 80%.
5. An acoustic segmentation approach was also proposed where audio recordings to be segmented into speech, silence, laughter and non-speech sounds.

We have to use cepstral coefficients as features and the Hidden Markov model (HMM) as the classifier. We propose a MGM-based (Modified Gaussian Modelling) hierarchical classifier for audio stream classification. Compared to traditional classifiers, MGM can automatically optimize the weights of different kinds of features based on training data. It can raise the discriminative capability of audio classes with lower computing cost.

4 System Flow

Fig. 3 shows the flowchart of proposed audio segmentation and classification algorithm. It is a hierarchical structure. In the first level, a long audio stream can be segmented into some audio clips according to the change of background sound by MBCR based histogram modeling. Then a two level MGM (Modified Gaussian modeling) classifier is adopted to hierarchically put the segmented audio clips into six pre-defined categories in terms of discriminative background sounds, which is pure speech (PS), pure music (PM), song (S),speech with music (SWM), speech with noise (SWN)and silence (SIL).

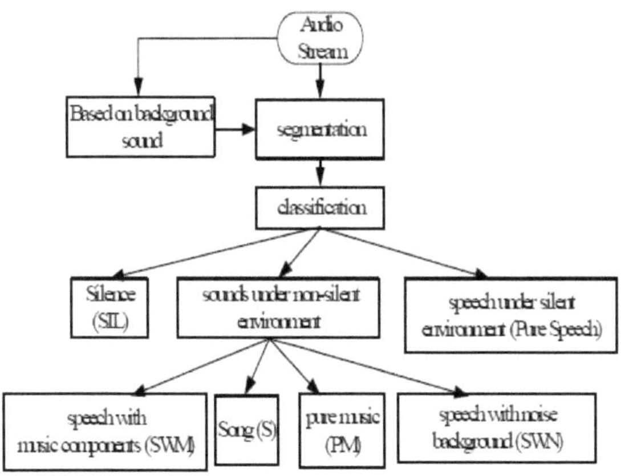

Fig. 3. The flowchart of segmentation and classification algorithm

5 Segmentation Algorithm

Since background sounds always change with the change of scenes, the acoustic skip point of an audio stream may be checked by background sounds. As shown in Fig. 2, the MBCR feature vectors are firstly extracted from the audio stream. We set a sliding window which consists of two sub-windows with equal time length. The window on input signal is shifted with a range of overlapping. Then two histograms are created from each sliding sub-windows. The similarity between two sub-windows can be measured by histogram matching. The skip point can thus be detected by searching the local lowest similarity below a threshold.

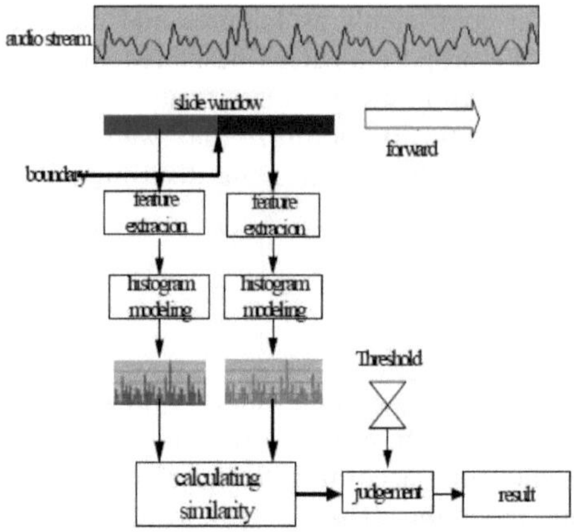

Fig. 4. Block diagram of Segmentation algorithm

The proposed algorithm for audio segmentation segment the audio into different parameters as described before also feature extraction algorithm separates out the different audio features such as MFCC, LPC, SF, SNR, and HZCRR

6 Feature Extraction

Considering the lower frequency spectrum is too sensitive to even a bit of changes of the scenes and speakers, it could cause segmented clips too small. It will have effects on succeeding audio classification. We, thus, use Multiple sub-Bands spectrum Centroid relative Ratio (MBCR) [5] over 800Hz as basic feature. This feature may depict centroid movement trend in a time frequency-intensity space. Its mathematical description can be described as follows.

$$SCR(i,j) = \frac{SC(i,j)}{\underset{J=1:N}{\mathrm{Max}}\,(SC(i,j))} \tag{1}$$

$$SC(i,j) = \frac{f(j) * FrmEn(i,j)}{\sum_{k=1}^{N} FrmEn(i,k)} \tag{2}$$

where $SCR(i, j)$ is MBCR of the ***ith*** frame and the ***jth*sub**-band, $SC(i, j)$ is the frequency Centroid of the ***ith*** frame and the ***jth*** sub-band, and N denotes the number of frequency sub-bands. The element of $f(j)$ is the normalized central frequency.

$$FrmEn(i,j) = \log(\int_{\omega_L(j)}^{\omega_H(j)} |F(i,\omega)| d\omega) \tag{3}$$

where $\omega L(j)$ and $\omega H(j)$ are lower and upper bound of sub-band j respectively, $F(i,\omega)$ represent denotes the Fast Fourier Transform (FFT) at the frequency ω and frame i, and $| F(i,\omega) |$ is square root of the power at the frequency ω and frame i.

7 Results

We conducted a series of experiments based on proposed audio segmentation and classification approach. The performance was evaluated on the recordings of real TV program. The segmentation and classification results were evaluated by the recall rateδ, accuracy rateξ, and average precisionη. These are defined as

$$\delta = \frac{\textit{the number of correctly objects}}{\textit{the number of objects that should be correct}}$$

$$\xi = \frac{\textit{the number of correctly objects}}{\textit{the number of all get objects}}$$

$$\eta = \frac{\delta * \xi}{0.5*(\delta + \xi)}$$

We pre-defined six categories as audio classes, which is pure speech (PS), pure music (PM), song (S), speech with music (SWM), speech with noise (SWN) and silence (SIL).

Table 1. The results of first level classification

Algorithm	Audio type	Accuracy	Recall	Precision
Equal time	Pure Speech (PS)	**85.15%**	**85.63%**	**87.62%**
	Silence (SIL)	**97.10%**	**86.14%**	**91.29%**
	Others	**77.95%**	**95.08%**	**85.67%**
MBCR	Pure Speech (PS)	**91.33%**	**93.65%**	**92.47%**
	Silence (SIL)	**98.22%**	**92.97%**	**95.52%**
	Others	**85.68%**	**95.45%**	**90.3%**

8 Conclusions

In above method, we have presented comparative analysis of on feature extraction using segmentation techniques. Different parameters such as audio type, accuracy, recall factor and precision has been evaluated for pure speech, silence etc.

The above classification can be extended for other feature such as Spectrum , Spectral Centroid , MFCC , LPC , ZCR , SNR , Moments, Beat Histogram , Beat Sum , RMS etc in order to precisely segment all these features in order to reduce the storage capacity which is under process. The above algorithms can be modified to extract other than above features which are not mentioned here.

References

[1] Peiszer, E., Lidy, T., Rauber, A.: Automatic Audio Segmentation: Segment Boundary and Structure Detection in Popular Music (2008)
[2] Cook, G.T.P.: Multifeature Audio Segmentation for Browsing and Annotation. In: Proc.1999 IEEE Workshop on Applications of Signal Processing to Audio and Acoustics, New Paltz, New York, pp. W99-1–W99-4 (1999)
[3] Lu, G.: Indexing and Retrieval of Audio: A Survey, pp. 269–290 (2001)
[4] Zhang, J.X., Whalley, J., Brooks, S.: A Two Phase Method for general audio segmentation (2004)
[5] Foote, J.: Automatic Audio Segmentation Using A Measure of Audio Novelty
[6] Julien, P., José, A., Régine, A.: Audio classi_cation by search of primary components, pp. 1–12
[7] Lu, L., Zhang, H.-J., Jiang, H.: Content Analysis for Audio Classification and Segmentation. IEEE Transaction on Speech and Audio Processing, 504–516 (2002)
[8] Lu, L., Li, S.Z., Zhang, H.-J.: Content based audio segmentation using Support Vector Machines (2008)
[9] Aguilo, M., Butko, T., Temko, A., Nadeu, C.: A Hierarchical Architecture for Audio Segmentation in a Broadcast News Task, pp. 17–20 (2009)
[10] Cettolo, M., Vescovi, M., Rizzi, R.: Evaluation of BIC-based algorithms for audio segmentation, pp. 147–170. Elsevier (2005)
[11] Goodwin, M.M., Laroche, J.: Audio Segmentation by feature space clustering using linear discriminant analysis and dynamic programming (2003)
[12] Haque, M.A., Kim, J.-M.: An analysis of content-based classification of audio signals using a fuzzy c-means algorithm (2012)
[13] Mesgarani, N., Slaney, M., Shamma, S.A.: Discrimination of Speech From Nonspeech Based on Multiscale Spectro-Temporal Modulations, pp. 920–930 (2006)
[14] Krishnamoorthy, P., Kumar, S.: Hierarchical audio content classification system using an optimal feature selection algorithm, pp. 415–444 (2010)
[15] Panagiotis, S., Vasileios, M., Ioannis, K., Hugo, M., Miguel, B., Isabel, T.: On the use of audio events for improving video scene segmentation
[16] Abdallah, S., Sandler, M., Rhodes, C., Casey, M.: Using duration Models to reduce fragmentation in audio segmentation 65, 485–515 (2006)
[17] Cheng, S.-S., Wang, H.-M., Fu, H.-C.: BIC-BASED Audio Segmentation by divide and conquer
[18] Yong, S.: Audio Segmentation, pp. 1–4 (2007)

[19] Matsunaga, S., Mizuno, O., Ohtsuki, K., Hayashi, Y.: Audio source segmentation using spectral correlation features for automatic indexing of broadcast news, pp. 2103–2106
[20] Sainath, T.N., Kanevsky, D., Iyengar, G.: Uusupervised audio segmentation using extended Baum-Welch Transformations, I 209-I 212 (2007)
[21] Giannakopoulos, T., Pikrakis, A., Theodoridis, S.: A Novel Efficient Approach for Audio Segmentation (2008)
[22] Zhang, Y., Zhou, J.: Audio Segmentation based on Multiscale audio classification, pp. IV-349–IV-352 (2004)
[23] Peng, Y., Ngo, C.-W., Fang, C., Chen, X., Xiao, J.: Audio Similarity Measure by Graph Modeling and Matching, pp. 603–606
[24] Harchaoui, Z., Vallet, F., Lung-Yut-Fong, A., Cap, O.: Regularized Kernel-Based ApproachToUnsupervised Audio Segmentation

[illegible]

A Modified Speckle Suppression Algorithm for Breast Ultrasound Images Using Directional Filters

Vikrant Bhateja, Atul Srivastava, Gopal Singh, and Jay Singh

Deptt. of Electronics & Communication Engg., SRMGPC, Lucknow-227105 (U.P.), India
{bhateja.vikrant,atul.srivastava216,
gopal.singh13492,jaysinghy6k}@gmail.com

Abstract. Speckle noise in ultrasound images (US) is a serious constraint leading to false therapeutic decision making in computer aided diagnosis. This highlights the utility of speckle suppression along with due preservation of edges as well as textural features while processing breast ultrasound images (for computer aided diagnosis of breast cancer). This paper presents a modified speckle suppression algorithm employing directional average filters for breast ultrasound images in homogeneity domain. The threshold mechanism during the process is adjusted using the entropies of foreground and background regions to ensure appropriate extraction of textural information. Simulation results demonstrate significantly improved performance in comparison to recently proposed methods in terms of speckle removal as well as edge preservation.

Keywords: Speckle removal, Law's textural energy measure, Maximum entropy, Canny, Edge preservation factor.

1 Introduction

Breast cancer is one of the most threatening diseases being the second leading cause of cancer related deaths in women. The discovery of a lump in the breast is typically the first sign of breast cancer. There are two existing screening modalities to detect breast cancer in women namely ultrasonography and mammography. Ultrasound is non-invasive, harmless and is coming forward as an important adjunct to mammography. However, the problem with ultrasound images is that they suffer from speckle noise which makes the edges unclear and deteriorates the image quality. Speckle noise generally occurs when details in the tissues are small and cannot be resolved by high wavelength ultrasound waves hence masking the details of the image [1]. Therefore, it is very necessary to reduce speckle in order to cut down the frequency of false positives that lead to unnecessary and painful biopsies. A lot of methods are proposed in recent years to deal with speckle noise. Linear filtering techniques like Gaussian and Gabor filters [2] are not effective because they blur the image, fading the edges as well as textural patterns. Non linear median filter [3] fails to detect speckle when speckles' size exceeds that of median filtering window. Homomorphic filter makes use of a Butterworth filter to remove speckle [4] .Wiener filter [5] works very well for filtering additive noises; as speckle is a multiplicative noise, it yields limited

S.C. Satapathy et al. (eds.), *ICT and Critical Infrastructure: Proceedings of the 48th Annual Convention of CSI - Volume II*, Advances in Intelligent Systems and Computing 249,
DOI: 10.1007/978-3-319-03095-1_24, © Springer International Publishing Switzerland 2014

performance to achieve its suppression. Wavelet based approaches [6]-[8] for speckle reduction employ wavelet shrinkage along with thresholding techniques. However, these approaches are limited, as the pre-selected threshold may not work suitably for all scales. Tong *et al.* proposed a speckle reduction method based on optical design by utilizing angle diversity [9]. A. Jain *et al.* [10]-[12] worked upon denoising mixture of speckle as well as impulse noises. V. Bhateja *et al.* [13]-[14] used non linear and polynomial filtering to enhance masses in different imaging modality namely, breast mammography. In work of Rahman *et al.* speckle filtering was achieved using concept of Extra Energy Reduction (EER) [15] to limit extra energy from an image and provides a new gray level to each reconstructed pixel. A. Gupta *et al.* presented speckle reducing approach for SAR (Synthetic Aperture Radar) images [16]-[17] using anisotropic diffusion filters. In another work [18]-[19] devised speckle filter based on local statistics, contourlet transform and non-linear adaptive thresholding. This paper proposes a speckle reduction algorithm for breast US images using modified directional average filters. A threshold mechanism based on maximum entropy is utilized to segment images into homogenous and non-homogenous regions. The proposed algorithm attempts to suppress speckle along with edge preservation as depicted by the obtained results. This paper is organized as follows: Section 2 explains the proposed denoising algorithm. Section 3 presents the results and discussion and finally conclusions are drawn in Section 4.

2 Proposed Speckle Suppression Algorithm

The procedural methodology for proposed speckle suppression algorithm can be discussed with the help of various modules. The breast US images used at the input in the present work are obtained from Prince of Wales Hospital Ultrasound Database [20]. These images are pre-processed which requires normalization to gray scale. The proposed algorithm is then applied to the pre-processed US images which consists of three modules: Extraction of Features, 2-D Homogeneity Histogram & Thresholding and proposed Directional Average Filter. A description of these modules is given in subsections to follow:

2.1 Features Extraction

It is a well known fact that different tissues, muscles and masses have separate textural properties. These textural characteristics can help to detect any abnormalities, if present and thereby facilitating in diagnosis of various diseases. In proposed method, Law's Textural Energy Measures (TEM) [21] are utilized to obtain the textural information of an US image. In this method of textural extraction, generally five features are used to completely describe the textural characteristics. These are edges, mean gray level, ripples, spots and waves. All these features have separate row matrices which are used to detect the presence of them utilizing combinations of different 1 x 5 kernels (mentioned in (1)).

$$E = [-2\ -4\ 0\ 4\ 2],\ M = [2\ 8\ 12\ 8\ 2],\ R = [2\ -8\ 12\ -8\ 2],\ S = [-2\ 0\ 4\ 0\ 2],\ W = [-2\ 4\ 0\ -4\ 2] \quad (1)$$

The four 5 x 5 masks can be then obtained as: M^T X E, M^T X S, E^T X M, S^T X M where: M^T indicates transpose of row matrix M. All four masks are applied on an US image to obtain four different textural specific matrices. Finally, textural value is calculated pixel wise in a Pythagorean summation manner [22]. If four textural matrices obtained are *a(i,j), b(i,j), c(i,j) and d(i,j)* respectively; then their Pythagorean sum (texture value) is given as below :

$$t(i,j) = \sqrt{a(i,j)^2 + b(i,j)^2 + c(i,j)^2 + d(i,j)^2} \tag{2}$$

2.2 2-D Homogeneity Histogram and Thresholding

The textural domain matrix obtained in sub-section 2.1 can then be further used to convert the gray scale image into homogeneity domain using Eq. (3).

$$H(i,j) = K(1 - T(i,j)) \tag{3}$$

where: K is a constant to normalize homogeneity values in the range of [0, K] and $T(i,j)$ is the normalized version of textural information matrix $t(i,j)$. $H(i,j)$ denotes the homogeneity domain matrix used to generate 2D histogram [23] also called homogram; it is a critical part of homogeneity based discrimination in an US image. Further, a threshold is needed which will act as the basis for differentiation of homogenous and non-homogenous portions of the US image. Many thresholding techniques based on clustering, histogram shape and attribute similarities are in existence [24] but entropic thresholding is a better technique than those listed previously. Threshold will be determined utilizing the principle of maximum entropy [25]. Now Eq. (4) determines the threshold using maximal entropy.

$$T = Arg\max(H) \tag{4}$$

where: H is the summation of entropy images of foreground and background. Once the threshold is obtained, the pixels having homogeneity and local mean homogeneity values greater than the threshold T are termed as 'Homogenous' and if it is less than T then they are named as 'Non-Homogenous'.

2.3 Modified Directional Average Filtering

Once the image is divided into homogeneous and non-homogeneous regions; the homogenous pixels are left unprocessed where as non-homogenous regions are processed using the proposed directional average filtering templates given in (5)-(7). This modified directional filtering template is iteratively applied on non-homogenous regions as shown in (8). Number of iterations of directional average filter depends upon the Homogeneity Ratio (*HR*). *HR* is defined as the ratio of number of homogenous pixels to total number of pixels in an image. Larger value of *HR* indicates greater homogeneity.

All directional mask: $$A=\frac{1}{25a}\begin{pmatrix} a & a & a & a & a \\ a & a & a & a & a \\ a & a & a & a & a \\ a & a & a & a & a \\ a & a & a & a & a \end{pmatrix}. \tag{5}$$

Horizontal mask: $$B=\frac{1}{5a}\begin{pmatrix} 0 & 0 & 0 & 0 & 0 \\ 0 & 0 & 0 & 0 & 0 \\ a & a & a & a & a \\ 0 & 0 & 0 & 0 & 0 \\ 0 & 0 & 0 & 0 & 0 \end{pmatrix} \tag{6}$$

Vertical mask: $$C=\frac{1}{5a}\begin{pmatrix} 0 & 0 & a & 0 & 0 \\ 0 & 0 & a & 0 & 0 \\ 0 & 0 & a & 0 & 0 \\ 0 & 0 & a & 0 & 0 \\ 0 & 0 & a & 0 & 0 \end{pmatrix} \tag{7}$$

$$D(NHs)=\begin{Bmatrix} conv(NHs,A) \text{ for } E_x = E_y \\ conv(NHs,B)\, forE_x > E_y \\ conv(NHs,C)\, forE_x < E_y \end{Bmatrix} \tag{8}$$

Where: *NHs* represents the non homogenous region of the US image; *conv* (.) denotes the convolution operator; E_x and E_y be the horizontal and vertical edge values. The selection of the directional filtering template from (5)-(7) for convolution is made based on the obtained edge values (E_x and E_y) using Canny operator [26]. Hence, in this manner the non homogenous regions sets are dealt with making the US image more homogenous and noise free with each iteration of the modified directional filter. It is worth noting that the values of the parameter *a* in (5)-(7) can be determined using the values of quality evaluation metrics.

2.4 Quality Evaluation Metrics

In order to evaluate the degree of speckle suppression performed by the proposed algorithm the quality evaluation metrics should be chosen to account for both noise filtering as well as edge preservation. Hence, in the present work three quality metrics are used for quality evaluation of reconstructed US images. They are Peak Signal-to-Noise Ratio *(PSNR* in dB*)* [27], Coefficient of Correlation *(CoC)* and Edge Preservation Factor *(EPF)* [28]. Higher value of *PSNR* indicates better quality of reconstructed image and lower mean squared error. The coefficient of correlation *(CoC)* is a measure to find the degree of correlation among reconstructed and images. Finally, Edge Preservation Factor *(EPF)* is the measure of how effectively the edges are being preserved during the process of speckle reduction. Higher the value of *EPF*, greater is the edge preservation. Higher values of all the three parameters: *PSNR, CoC* and *EPF* will justify the overall robustness of the proposed algorithm. Image quality evaluation metrics used above and some proposed recently [29]-[35] can be used for speckle removal algorithms.

3 Results and Discussions

The proposed algorithm is simulated using the Eq. in (1) - (8) on Breast US images obtained from Prince of Wales Hospital Ultrasound Database [20]. Eq. (1)-(4) are used to discriminate between homogenous and non-homogenous regions of the image whereas Eq. (5)-(7) deal with proposed directional average filter. Value of parameter a used in our experiments is 1. Value of K in (3) is taken as 100 so that homogeneity values lie between 0 and 100. Also, value of HR is selected as 0.95. The values of these tuning parameters are adaptively selected based on the maximal values of quality parameters: *PSNR, CoC* and *EPF* respectively.

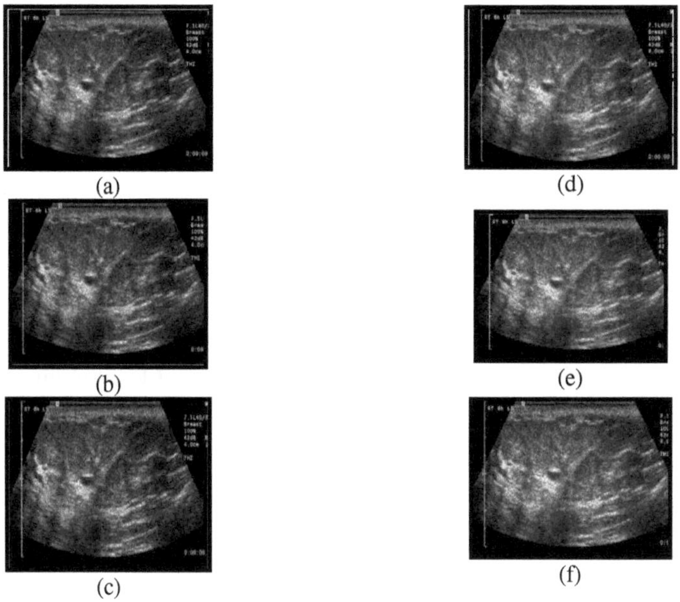

Fig. 1. (a) ,(b) and (c) are noisy US images with noise variance: 0.01, 0.02 and 0.03 respectively and (d), (e) and (f) are the corresponding denoised images

Table 1. Quality Evaluation for proposed speckle removal algorithm for different speckle variance levels

Noise Variance	*PSNR (in dB)*	*CoC*	*EPF*
0	35.3643	0.9967	0.9916
0.01	29.3702	0.9870	0.9407
0.02	26.8745	0.9771	0.8958
0.03	25.2535	0.9669	0.8537
0.04	24.1029	0.9571	0.8169
0.1	20.3556	0.9031	0.6569

It is clear from the table that the proposed algorithm maintains a high degree of correlation even at noise variance levels as high as 0.1. In Fig. 2(a), (b) and (c) edges are unclear and hazy due to the speckle noise. The image quality is improved in Fig. 2(d), (e) and (f) as reasonable amount of speckle filtering is achieved along with preservation of edges and textural features. This can be also quantitatively verified using high *EPF* values in Table 1. The proposed algorithm yields *CoC* as high as 0.9967; also the *PSNR* decays slowly as the noise variances are increased. In other speckle reduction approaches like Wavelet based methods [6]-[8] *CoC* values are generated generally in the range of 0.5-0.6. Frost, Kuan and Lee filters [11] have *PSNR* nearly 24dB for speckle noise variances of 0.02-0.03. In the works of Y. Guo *et al.* [36] obtained values of *EPF* are in range 0.8-0.9 for low variances but the proposed denoising algorithm increases the value of *EPF* even further to the ranges 0.96-0.99. This demonstrates the effectiveness of the proposed algorithm in comparison to recent works both in terms of speckle suppression as well as edge preservation.

4 Conclusion

The proposed algorithm mainly focuses on two problems i.e. removal of speckle noise and preservation of edges as well as textural information. In the present work, non-homogenous pixels are recursively filtered with a modified directional average filter which does not affect the homogenous regions thus preventing edge blurring. The threshold selection mechanism has been automated using the entropies of foreground and background regions to ensure appropriate extraction of textural information. This has been validated by higher values of *CoC* and *EPF* obtained during simulation. The values of performance metrics indicates the effectiveness of speckle suppression algorithm in comparison to other methods. It can prove to be helpful for radiologist and doctors in diagnostic decision making [37]and reading medical images with better precision.

References

1. Bhateja, V., Devi, S.: An Improved Non-Linear Transformation Function for Enhancement of Mammographic Breast Masses. In: Proc. of (IEEE) 3rd International Conference on Electronics & Computer Technology (ICECT 2011), Kanyakumari (India), vol. 5, pp. 341–346 (2011)
2. Chen, L., Lu, G., Zhang, D.: Effects of Different Gabor Filter Parameters on Image Retrieval by Texture. In: Proc. of IEEE 10th International Conference on Multi-Media Modelling, Brisbane, Australia, pp. 273–278 (January 2004)
3. Perreault, S., Hébert, P.: Median Filtering in Constant Time. IEEE Trans. on Image Processing 16(6), 2389–2394 (2007)
4. Solbo, S., Eltoft, T.: Homomorphic Wavelet Based Statistical Despeckling of SAR Images. IEEE Trans. on Geoscience and Remote Sensing 42(4), 711–721 (2004)
5. Strela, V.: Denoising via Block Wiener Filtering in Wavelet Domain. In: 3rd European Congress of Mathematics, Barcelona, Spain (July 2000)
6. Kaur, A., Singh, K.: Speckle Noise Reduction by Using Wavelets. In: National Conference on Computational Instrumentation, Chandigarh, India, pp. 198–203 (March 2010)

7. Mohideen, S.K., Perumal, S.A., Sathik, M.M.: Image Denoising Using Discrete Wavelet Transform. International Journal of Computer Science and Network Security 8(1), 213–216 (2008)
8. Solbo, S., Eltoft, T.: A Stationary Wavelet Domain Wiener Filter for Correlated Speckle. IEEE Trans. on Geoscience and Remote Sensing 46(4), 1219–1230 (2008)
9. Tong, Z., Chen, X., Akram, M.N., Aksnes, A.: Compound Speckle Characterization Method and Reduction by Optical Design. Journal of Display Technology 8(3), 132–137 (2012)
10. Jain, A., Bhateja, V.: A Novel Image Denoising Algorithm for Suppressing Mixture of Speckle and Impulse Noise in Spatial Domain. In: Proc. of IEEE 3rd International Conference on Electronics and Computer Technology, Kanyakumari, India, pp. 207–211 (April 2013)
11. Singh, S., Jain, A., Bhateja, V.: A Comparative Evaluation of Various Despeckling Algorithms for Medical Images. In: Proc. of (ACMICPS) CUBE International Information Technology Conference & Exhibition, Pune, India, pp. 32–37 (September 2012)
12. Jain, A., Singh, S., Bhateja, V.: A Robust Approach for Denoising and Enhancement of Mammographic Breast Masses. International Journal on Convergence Computing 1(1), 38–49 (2013)
13. Bhateja, V., Urooj, S., Pandey, A., Misra, M., Lay-Ekuakille, A.: A Polynomial Filtering Model for Enhancement of Mammogram Lesions. In: Proc. of IEEE International Symposium on Medical Measurements and Applications, Gatineau(Quebec), Canada, pp. 97–100 (May 2013)
14. Bhateja, V., Urooj, S., Pandey, A., Misra, M., Lay-Ekuakille, A.: Improvement of Masses Detection in Digital Mammograms employing Non-Linear Filtering. In: Proc. of IEEE International Multi-Conference on Automation, Computing, Control, Communication and Compressed Sensing, Palai-Kottayam, Kerala, India, pp. 406–408 (2013)
15. Rahman, M., Motiur Kumar, P.K., Borucki, B., Nowinski, K.S.: Speckle noise reduction of ultrasound images using Extra-Energy Reduction function. In: International Conference on Informatics, Electronics and Vision, Dhaka, Bangladesh, pp. 1–6 (May 2013)
16. Gupta, A., Tripathi, A., Bhateja, V.: Despeckling of SAR Images via an Improved Anisotropic Diffusion Algorithm. In: Satapathy, S.C., Udgata, S.K., Biswal, B.N. (eds.) Proceedings of Int. Conf. on Front. of Intell. Comput. AISC, vol. 199, pp. 747–754. Springer, Heidelberg (2013)
17. Gupta, A., Ganguly, A., Bhateja, V.: A Noise Robust Edge Detector for Color Images using Hilbert Transform. In: Proc. of (IEEE) 3rd International Advance Computing Conference (IACC 2013), Ghaziabad (U.P.), India, pp. 1207–1212 (2013)
18. Gupta, A., Tripathi, A., Bhateja, V.: De-Speckling of SAR Images in Contourlet Domain Using a New Adaptive Thresholding. In: Proc. of IEEE 3rd International Advance Computing Conference (IACC), Ghaziabad (U.P.), India, pp. 1257–1261 (2013)
19. Bhateja, V., Tripathi, A., Gupta, A.: An Improved Local Statistics Filter for Denoising of SAR Images. In: Thampi, S.M., Abraham, A., Pal, S.K., Rodriguez, J.M.C. (eds.) Recent Advances in Intelligent Informatics. AISC, vol. 235, pp. 23–29. Springer, Heidelberg (2014)
20. http://www.droid.cuhk.edu.hk/service/ultrasound/usdiagnostic.htm
21. Gonzalez, R.C., Woods, R.E.: Digital Image Processing, 2nd edn. Prentice Hall (2002)
22. http://g.lemaitre58.free.fr/pdf/vibot/scene_segmentationinterpretation/cooccurencelaw.pdf

23. Zhang, J., Hu, J.: Image Segmentation Based on 2D Otsu Method with Histogram Analysis. In: International Conference on Computer Science and Software Engineering, Wuhan, Hubei, pp. 105–108 (December 2008)
24. Sezgin, M., Sankur, B.: Survey Over Image Thresholding Techniques and Quantitative Performance Evaluation. Journal of Electronic Imaging 13(1), 146–165 (2004)
25. Azarbad, M., Ebrahimzade, M.A., Izadian, V.: Segmentation of Infrared Images and Objectives Detection Using Maximum Entropy Method Based on the Bee Algorithm. International Journal of Computer Information Systems and Industrial Management Applications 3, 26–33 (2011)
26. Gupta, A., Ganguly, A., Bhateja, V.: An Edge Detection Approach for Images Contaminated with Gaussian and Impulse Noises. In: Mohan, S., Suresh Kumar, S. (eds.) ICSIP 2012. LNEE, vol. 222, pp. 523–533. Springer, Heidelberg (2012)
27. Jain, A., Bhateja, V.: A Versatile Denoising Method for Images Contaminated with Gaussian Noise. In: Proc. of CUBE Int. Information Technology Conf. & Exhibition, Pune, India, pp. 65–68 (2012)
28. Chumning, H., Huadong., C.W.: Edge preservation evaluation of digital speckle filters. In: IEEE International Geoscience and Remote Sensing Symposium, June 24-28, vol. 4, pp. 2471,2473 (2002)
29. Gupta, P., Srivastava, P., Bharadwaj, S., Bhateja, V.: A HVS based Perceptual Quality Estimation Measure for Color Images. ACEEE International Journal on Signal & Image Processing (IJSIP) 3(1), 63–68 (2012)
30. Gupta, P., Srivastava, P., Bhardwaj, S., Bhateja, V.: A Novel Full Reference Image Quality Index for Color Images. In: Satapathy, S.C., Avadhani, P.S., Abraham, A. (eds.) Proceedings of the InConINDIA 2012. AISC, vol. 132, pp. 245–253. Springer, Heidelberg (2012)
31. Gupta, P., Tripathi, N., Bhateja, V.: Multiple Distortion Pooling Image Quality Assessment. Inderscience Publishers International Journal on Convergence Computing 1(1), 60–72 (2013)
32. Gupta, P., Srivastava, P., Bharadwaj, S., Bhateja, V.: A Modified PSNR Metric based on HVS for Quality Assessment of Color Images. In: Proc. of IEEE International Conference on Communication and Industrial Application (ICCIA), Kolkata (W.B.), pp. 96–99 (2011)
33. Jain, A., Bhateja, V.: A Full-Reference Image Quality Metric for Objective Evaluation in Spatial Domain. In: Proc. of IEEE International Conference on Communication and Industrial Application (ICCIA), Kolkata (W. B.), India, pp. 91–95 (2011)
34. Bhateja, V., Srivastava, A., Kalsi, A.: Fast SSIM Index for Color Images Employing Reduced-Reference Evaluation. In: Satapathy, S.C., Udgata, S.K., Biswal, B.N. (eds.) FICTA 2013. AISC, vol. 247, pp. 451–458. Springer, Heidelberg (2014)
35. Bhateja, V., Singh, G., Srivastava, A.: A Novel Weighted Diffusion Filtering Approach for Speckle Suppression in Ultrasound Images. In: Satapathy, S.C., Udgata, S.K., Biswal, B.N. (eds.) FICTA 2013. AISC, vol. 247, pp. 459–466. Springer, Heidelberg (2014)
36. Guo, Y., Cheng, H.D., Tian, J., Zhang, Y.: A Novel Approach to Speckle Reduction to Ultrasound Image. In: Proc. of 11th Joint Conference on Information Sciences, China, pp. 1–6 (December 2008)
37. Bhateja, V., Misra, M., Urooj, S., Lay-Ekuakille, A.: A Robust Polynomial Filtering Framework for Mammographic Image Enhancement from Biomedical Sensors. IEEE Sensors Journal, 1–10 (2013)

An Efficient Secret Image Sharing Scheme Using an Effectual Position Exchange Technique

Amit Dutta[1] and Dipak Kumar Kole[2]

[1] Department of Information Technology,
St. Thomas' College of Engineering & Technology, Kolkata
to.dutta@gmail.com
[2] Department of Computer Science and Engineering,
St. Thomas' College of Engineering & Technology, Kolkata
dipak.kole@gmail.com

Abstract. In image secret sharing scheme the specific image is distributed among a group of n participants. Each gets a shared image that has no visual relationship with the original one. The entire set or their subsets are used to reconstruct the secret image. The sharing schemes are useful for group authentication. In this paper, we proposed a novel approach to image secret sharing scheme where a secret image is transformed into an entirely new image by repositioning intensity values of each pixel using an effective position exchange technique. The transformed image undergoes a relatively less overhead negative transformation to make it more unpredictable. Finally n random numbers are generated for creating n shares in a way that their summation yields '1' and they lay in some predefined interval. Each stage provides an additional level of security. The proposed method is simple and efficient and also ensures complete secrecy and reconstruction.

Keywords: Shared image, negative transformation and position exchange technique.

1 Introduction

The primitive work on secret sharing was done by G. R. Blakley [1] and Adi Shamir [2]. Blakley applied finite geometries to formulate and solve the problem where Shamir's solution is based on the property of polynomial interpolation in finite fields. In 1994, Moni Naor and Adi Shamir [3] proposed a new type of secret sharing scheme called visual cryptography scheme. In this scheme an image was encrypted into n shares in such a way that no computational devices were required to decode, where a human visual system is used to recognize the secret image by superimposing all the shares directly.

In a (k, n) threshold technique of visual cryptographic scheme, a secret is encoded into n shares, and any k (less or equal to n) share is used to reconstruct the secrete image but same would not be possible for any k-1 or less shares. Until the year 1997, although the transparencies could be stacked to recover the secret image without any

DOI: 10.1007/978-3-319-03095-1_25,

computation, the revealed secret images (as in [4] [8] [6] [3]) were all black and white. In [7], Hwang proposed a new visual cryptography scheme which improved the visual effect of the shares. Hwang's scheme is very useful when we need to manage a lot of transparencies; nevertheless, it can only be used in black and white images. Adhikari and Bose [10] constructed a new visual cryptographic scheme using Latin squares.

Zhi Zhou, Gonzalo R.Arce and Giovanni Di Crescenzo [11] have proposed the concept of halftone visual cryptography based on blue-noise dithering principles. Feng Liu, Chuankan Wu and Xijun Lin [12], proposed the step construction visual cryptography scheme based on optimal pixel expansion. Liu and Wu [13], have proposed a visual cryptographic scheme in which they embedded random shares into meaningful covering shares. A very interesting scheme was again proposed by Ran-Zan Wang, Shuo-Fang Hsu [14]. In this method, an additional tag is used in the shares. Sian-Jheng Lin and Wei-Ho Chung [15], suggested a scheme where the number of share can be changed dynamically to include new shares without disturbing the original shares. A way to prevent cheating was suggested by Hu and Tzeng [16], where one visual cryptography scheme is converted into another scheme with minimum overhead. Visual Cryptography is not limited to binary and gray-level images. It can further be used to color images as well. Color VCS can also be performed by the use of symmetric key [17], proposed by B. SaiChandana and S.Anuradha. In [9], Verheul and Van Tilborg used the concept of arcs to construct a colored visual cryptography scheme, where users could share colored secret images. Daoshun Wang, Lei Zhang, Ning Ma and Xiaobo Li [18] have constructed a method of secret sharing scheme using simple Boolean operations. In this paper we are introducing a pixel position exchange technique to create a new image from the secret image and negative of that image is used to create shared image.

This paper is organized as follows. In Section 2, preliminaries about the input and output obtained are discussed. The proposed method is explained in Section 3. Experimental results and conclusion are listed in Section 4 and Section 5 respectively.

2 Preliminaries

The input used to describe the proposed method as well as the output obtained are discuss here.

Input: We consider grayscale image as original secret image. The following matrix A of order MxN is consider as input image.

$$A = [\ a_{ij}\ ]_{MxN}\text{, where } i = 1,2,3,\ldots,M \text{ and } j=1,2,3,\ldots,N$$

The value of each a_{ij} represents the intensity value of different pixels of the image and each a_{ij} lies between 0 and 255.

Output: Outputs are different secret shares s_1, s_2, $s_3 \ldots s_{n-1}$ and s_n. Each secret share image is also a matrix of order MxN.

Note: There will be two intermediate output image produced by the method before creating *n* shares of secret images, both are having same size as above.

3 Proposed Method

Most of the methods in this area have directly used the original secret image as input for creating shared images. However, this paper has proposed a totally unique approach. First, the original image ‘A’ is transformed into an entirely new image ‘B’ by altering the intensity value of each pixel. In the second stage a negative transformation is applied to the transformed image ‘B’ to get another image ‘C’. Henceforth a random method is used to generate *n* share.

3.1 Sharing Phase

We create two subsequent images from original secret image in following section 3.1.1, 3.1.2 and section 3.1.3 for creating n shared image; as discussed in the proposed method.

3.1.1 Pixel Position Exchange Method. Here we exchange the intensity value of the pixel staying in odd position in each row an even position in each column, in following way,

i) By interchanging the intensity value in the *odd position* of first row from beginning of the row with the corresponding position from the end of the row till it reaches the middle of the row. This is applied over all other rows in the image.
ii) Now interchanging the intensity value in the *even position* of first column from beginning of the column with the corresponding position from the end of the column until it reaches the middle of the column.

Applying this technique on original secret image ‘A’ to get image ‘B’.

3.1.2 Negative Transformation Used on Image ‘B’ to Get Image ‘C’. To make the original image ‘A’ more unpredictable one, we use a negative transformation,

$$s = (255 - r) \quad (1)$$

on image ‘B’, where r and s are the corresponding intensity value of image ‘B’ and transformed negative image ‘C’.

3.1.3 Using Random Method to Generate n Share from Image ‘C’. In this section we generate n random numbers in the range (0, 1) for each pixel of the last transformed image ’C’ such that summation of them is ‘1’, for creating n shared image.

Let, ‘m’ be the intensity value of any pixel of the image ‘C’ such that

$$r_1 + r_2 + \cdots + r_{n-1} + r_n = 1, \quad \forall\, r_i \; in \; (0,1) \quad (2)$$

Therefore, $r_1 m + r_2 m + \cdots + r_{n-1} m + r_n m = m$ (3)

So, we get *n* number of partition of the number *m*. The method use to choose each random numbers is refined in the following way,

I. It follow the equation (2) and
II. Two boundary values are omitted in each of the following cases.

r_1: $in\ (0,1)$
r_2: $in\ (0, 1 - r_1)$
.....
.....
r_{n-1}: $in\ (0, 1 - r_1 - r_2 - \ ... - r_{n-3} - r_{n-2})$
$r_n = 1 - \ (r_1 + r_2 + \ ... + r_{n-2} + r_{n-1})$

The value of r_n depends on the values of previously choosen $(n - 1)$ random numbers. This way we split all the pixel of the transformed image 'C'. So first partition of each pixel will used to create first share, second partition is used to construct the second share and so on. This way we create all n shares s_1, s_2, $s_3 ... s_{n-1}$ and s_n .

3.2 Reconstruction Phase

To reconstruct the original image from n share s_1, s_2, $s_3 ... s_{n-1}$ and s_n . At first summing the corresponding pixel position of all n share to get the transformed image 'C'. Now using the negative image transformation we get the image 'B'. Applying the same transformation defined earlier in section 3.1.1, we gets back the original image 'A'.

3.3 Proposed Algorithm

In this section the overall algorithm of the proposed method is discussed where an input is a grayscale secret image and output are *n* shared images.

Step1: Read the input grayscale image, $A=[\ a_{ij}\ ]_{MxN}$ where *i=1,2,3,....,M* and *j=1,2,3,...,N*

Step2: Interchanging the intensity value of each pixels in the odd position from start of each row with corresponding pixel from end of the row.

Step2.1: let row number *j*=1
Step2.2: check row number *j* less or equal to *M*
Step2.3: let column number *i*=1
Step2.4: check column no *i* less or equal to *N/2*
Step2.5: interchange the intensity value of ith position with $N-(i-1)^{th}$ position
Step2.6: increase *i* by 2
Step2.7: repeat step 2.4 to 2.6
Step2.8: increase row number *j* by 1
Step2.9: repeat step 2.2 to 2.8

Step3: Interchanging the intensity value of each pixels in the *even* position from start of each *column* with corresponding pixel from end of the *column*. It is similar to step 2.

Step4: Input image in this step is 'B'. Using the transformation *s=255-r* to get the image 'C', where *r* and *s* are the corresponding intensity value of image 'B' and 'C'.

Step5: Now for each pixel of image 'C' we are generating $(n-1)$ random numbers in the range (0, 1), so that it satisfies the condition stated in proposed method section 3.1.3.

The nth number r_n is generated as, $r_n = 1 - (r_1 + r_2 + \cdots + r_{n-1} + r_{n-1})$

Step6: Each *n*-partition of each pixel will from the respective *n* share image.

Step7: Recombining all shares by means of adding the respective pixel positions, the negative image is obtain and then again applying the pixel intensity exchange method will return the original image.

We are illustrating the said algorithm with an example given below,

Example: Suppose the original grayscale secret image is represented by the matrix M of order 4x4, shown Fig. 1(a), applying pixel position exchange technique we obtain the matrix M' in Fig.1(b) and using negative transformation discussed in equation (1) on the matrix M' in Fig. 1(b) produces M" in Fig. 1(c).

$$\begin{pmatrix} 140 & 143 & 142 & 139 \\ 138 & 144 & 155 & 141 \\ 123 & 234 & 134 & 135 \\ 122 & 255 & 240 & 243 \end{pmatrix} \quad \begin{pmatrix} 139 & 143 & 142 & 140 \\ 135 & 243 & 134 & 123 \\ 141 & 144 & 155 & 138 \\ 243 & 255 & 246 & 122 \end{pmatrix} \quad \begin{pmatrix} 116 & 112 & 113 & 115 \\ 120 & 12 & 121 & 132 \\ 114 & 111 & 100 & 117 \\ 12 & 0 & 9 & 133 \end{pmatrix}$$

a b c

Fig. 1. (a) Matrix M of order 4x4. (b) The matrix M' is obtain by applying pixel position exchange methodology on the matrix M in Fig. 1(a). (c) The matrix M" is obtain by applying negative transformation on the matrix M' in Fig. 1(b).

Let we have 3-share to generate. For the first pixel value(116) of M" , the two random numbers r1 and r2 are say, 0.3 in the range (0, 1) and 0.1 in the range(0, 0.7). Therefore, r3 =0.6 as $\mathrm{r}_1 + \mathrm{r}_2 + \mathrm{r}_3 = 1$.

We can write, 116=0.3*116 + 0.1*116 + 0.6*116=(35+12+69). So, intensity value of first pixel of first share is 35 that of second share is 12 and that of third share is 69. We are considering ceiling function for this purpose. Same approach is used to construct all the pixels of all shares.

In reconstruction phase, adding the corresponding pixel position of each share then applies negative transform on it; finally applying position exchange technique, gives the original image.

4 Experimental Results

We have implemented the above algorithm in MATLAB version 7.8.0.347(R2009a), 32-bit(win32) in windows XP professional with service pack 2. The processor used is Core 2 duo , 3.06GHz and 2.00GB RAM. The following result of creation of shares from secret image or reconstructing the secret is done immediately. We shows one such result using the grayscale gif image of "Lena" having 512x512 pixels in the Fig. 2(a). After applying pixel intensity exchange technique we get the Fig. 2(b). Making the negative transformation of image in Fig. 2(b) we get Fig. 2(c). Finally from Fig. 2(c) we generate randomly the three shares in Fig. 2(d), Fig. 2(e) and Fig. 2(f). In reconstruction phase we get the image in Fig. 2(g). The quality of the reconstructed image is same as that of original as the method in completely reversible.

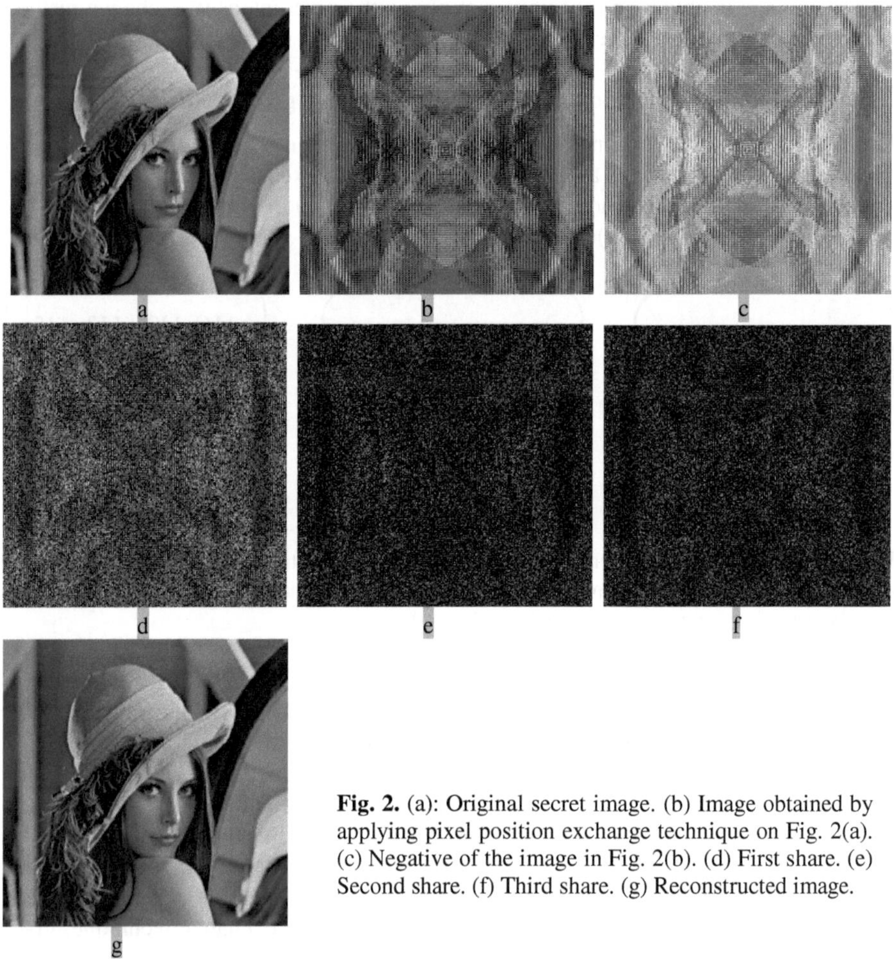

Fig. 2. (a): Original secret image. (b) Image obtained by applying pixel position exchange technique on Fig. 2(a). (c) Negative of the image in Fig. 2(b). (d) First share. (e) Second share. (f) Third share. (g) Reconstructed image.

5 Conclusion

The output obtained using position exchange methodology is quite unpredictable as the experimental result shows. The time complexity of the linear negative transformation that we have used for intermediate phase 'is very low' as compare to other types in this category. An interval based random numbers are used here to reduce the chance of unwanted data in the computation. The method uses the concept of (n,n)-schema, i.e, n share is generate from the original secret image and all n share is mandatorily required to reconstruct the original, n-1 share is unable to reconstruct the original. For future work we could consider the (n,k) schema, where only k $(k \leq n)$ share is used to reconstruct the original. Size of all the shared images is same as of the original secret image. There is no loss of intensity in the reconstructed image as the entire procedure is reversible. The paper can be useful in many application domains where group authentication is required.

References

1. Blakley, G.R.: Safeguarding Cryptographic keys. In: Proceeding of AFIPS 1979 National Computer Conference, New York, NY, vol. 48, pp. 313–317 (June 1979)
2. Shamir, A.: How to Share a Secret. Communications of the ACM 22(11), 612–613 (1979)
3. Naor, M., Shamir, A.: Visual Cryptography. In: De Santis, A. (ed.) EUROCRYPT 1994. LNCS, vol. 950, pp. 1–12. Springer, Heidelberg (1995)
4. Ateniese, G., Blundo, C., Santis, A.D., Atinson, D.: Atinson, Constructions and Bounds for Visual Cryptography. In: Meyer auf der Heide, F., Monien, B. (eds.) ICALP 1996. LNCS, vol. 1099, pp. 416–428. Springer, Heidelberg (1996)
5. Chang, C., Tsait, C., Chen, T.: A New Scheme for Sharing Secret Colour Images in Computer Network. In: Proceeding of International Conference on Parallel and Distributed Systems, pp. 21–27 (July 2000)
6. Droste, S.: New Results on Visual Cryptography. In: Koblitz, N. (ed.) CRYPTO 1996. LNCS, vol. 1109, pp. 401–415. Springer, Heidelberg (1996)
7. Hwang, R., Chang, C.: Some Secret Sharing Schemes and their Applications, Ph. D. dissertation of the Department of Computer Science and Information Engineering, National Chung Cheng University, Chiayi, Taiwan (1998)
8. Ateniese, G., Blundo, C., Santis, A.D., Atinson, D.: Visual Cryptography for General Access Structures. Information and Computation 129(2), 86–106 (1996)
9. Verheul, E., Tilborg, H.V.: Constructions and Propertics of k out of n Visual Secret Sharing Schemes. Designs, Codes and Cryptography 11(2), 179–196 (1997)
10. Adhikari, A., Bose, M.: A new visual cryptographic scheme using Latin squares. IEICE Transactions on Fundamentals E87-A, 1998–2002 (2004)
11. Zhou, Z., Arce, G.R., Crescenzo, G.D.: Halftone Visual Cryptography. IEEE Transactions on Image Processing 15(8) (August 2006)
12. Liu, F., Wu, C., Lin, X.: Step Construction of Visual Cryptography Schemes. IEEE Transactions on Information Forensics and Security 5(1) (March 2010)
13. Liu, F., Wu, C.: Embedded Extended Visual Cryptography Schemes. IEEE Transactions on Information Forensics and Security 6(2) (June 2011)
14. Wang, R.-Z., Hsu, S.-F.: Tagged Visual Cryptography. IEEE Signal Processing Letters 18(11) (November 2011)

15. Lin, S.-J., Wei-Ho: A Probabilistic Model of (t,n) Visual Cryptography Scheme With Dynamic Group. IEEE Transactions on Information Forensics and Security 7(1) (February 2012)
16. Hu, C.-M., Tzeng, W.-G.: Cheating Prevention in Visual Cryptography. IEEE Transactions on Image Processing 16(1) (January 2007)
17. SaiChandana, B., Anuradha, S.: A New Visual Cryptography Scheme for Color Images. Journal of Engineering Science and Technology 2(6), 1997–(2000)
18. Wang, D., Zhang, L., Ma, N., Li, X.: Two secret sharing schemes based on Boolean operations (received July 15, 2005); (received in revised form October 4, 2006); (accepted November 10, 2006)

Image Fusion Technique for Remote Sensing Image Enhancement

B. Saichandana[1], S. Ramesh[2], K. Srinivas[3], and R. Kirankumar[4]

[1] GITAM Institute of Technology, GITAM University, Visakhapatnam
[2] Sri Vahini Institute of Science and Technology, Tiruvuru
[3] Siddhartha Engineering College, Vijayawada
[4] Krishna University, Machilipatnam

Abstract. Remote sensing image enhancement algorithm based on image fusion is presented in this paper, in-order to resolve problems of poor contrast and sharpness in degraded images. The proposed method consists of two essentials: the first, the techniques of image sharpening, dynamic histogram equalization and fuzzy enhancement technique were respectively applied to the same degraded image. Second, these obtained three images, which individually preserved the enhancement effect of either of these techniques, are fused into a single image via different fusion rules. This method enhances the contrast well in the remote sensing image without introducing severe side effects, such as washed out appearance, checkerboard effects etc., or undesirable artifacts. The experiment results indicate that the proposed algorithm integrates the merits of dynamic histogram equalization, fuzzy enhancement and sharpening effectively and achieves a considerable efficiency in the enhancement of degraded images exhibiting both blurred details and low contrast. The qualitative and quantitative performances of the proposed method are compared with other methods producing better quality enhanced image

Keywords: Image Fusion, Image Enhancement, Histogram Equalization, Fuzzy Image Enhancement, Image Processing.

1 Introduction

Image enhancement is an important technique in the image preprocessing field. Image enhancement is a fundamental task applied in image processing to improve the interpretability and appearance of the image providing better input image for further input image processing task. In general, raw remote sensing images have a relatively narrow range of brightness values; hence, contrast enhancement is frequently used to enhance the image for better interpretation and visualization. Many image enhancement algorithms have been developed to improve the appearance of images. Image enhancement can be clustered into two groups namely frequency domain and spatial domain methods. In the frequency domain method [1], the enhancement is conducted by modifying the frequency transform of the image, taking more computation time even with fast transformation technique, which is unsuitable for real

S.C. Satapathy et al. (eds.), *ICT and Critical Infrastructure: Proceedings of the 48th Annual Convention of CSI - Volume II*, Advances in Intelligent Systems and Computing 249,
DOI: 10.1007/978-3-319-03095-1_26, © Springer International Publishing Switzerland 2014

time application. In the spatial domain method, image pixels are directly modified to enhance the image.

Image fusion is the combination of two or more different images to form a new image by using a certain algorithm [2]. The objective of image fusion exists in combining multiple source images into a fused image that exhibits more useful information than the individual source image. For about two decades, image fusion has emerged as a promising image processing technique in many fields, like remote sensing and medicine [3].

In the proposed approach, image fusion is utilized to enhance degraded remote sensing images by combining the performance of image sharpening, Dynamic Histogram Equalization and Fuzzy Enhancement Technique. First, image sharpening, Dynamic Histogram Equalization and Fuzzy Enhancement Technique are respectively applied to the same degraded image in order to obtain three processed images. Then these three images are fused through special rules to get the single enhanced image. The experiment results show that the proposed algorithm greatly improves both contrast and sharpness of degraded images and concurrently provides adequate details and appropriate contrast in the enhanced images. The block diagram of proposed enhancement algorithm is shown in figure 1. This paper is organized as follows: section 2 presents Image Sharpening technique, section 3 presents Dynamic Histogram Equalization technique, section 4 presents Fuzzy Enhancement technique, section 5 presents Experimental results and finally section 6 report conclusions.

2 Image Sharpening

Sharpening can elevate an image's sharpness via compensating contours and emphasizing edges. The sharpening process adopts the following 3 X 3 Laplacian template:

$$W=\begin{bmatrix} 0 & -1 & 0 \\ -1 & 4 & -1 \\ 0 & -1 & 0 \end{bmatrix}$$

This widely used sharpening template comes from the superimposition of an image f(i,j) and its Laplacian ∇^2 f(i,j), which has the discrete form of the following :

$$\nabla^2 f(i,j) = 4f(i,j) - f(i-1,j) - f(i+1,j) - f(i,j-1) - f(i,j+1). \tag{1}$$

3 Dynamic Histogram Equalization

The Dynamic Histogram Equalization [4] is a smart contrast enhancement technique that takes control over the effect of traditional Histogram Equalization so that it performs the enhancement of an image without making any loss of details in it. The Dynamic Histogram Equalization (DHE) process is divided into three steps. In the first step, it divides the whole histogram of the input image into a number of sub-histograms until it ensures that no domination of higher histogram components on lower histogram components. The partitions are done based on local minima. It makes partitions by taking the portion of histogram that falls between two local minima.

Mathematically if $h_0, h_1, h_2 \ldots\ldots h_m$ are (m+1) gray levels that correspond to (m+1) local minima, then the first sub-histogram will take the gray level range $[m_0, m_1]$, the second will take $[m_1+1, m_2]$ and so on. If there is some dominating portion in the sub-histogram, the DHE splits the sub-histogram in to smaller sub-histogram based on the mean ('μ') and standard deviation ('σ') of the gray level frequencies of each sub-histogram region. If the number of consecutive gray levels having frequency within the range from $(\mu - \sigma)$ to $(\mu + \sigma)$ becomes more than 68.3 of the total frequency of all gray levels in a sub-histogram, then there is no dominating portion in the sub-histogram. If the percent is less than the 68.3 then there is a dominating portion, DHE splits the sub-histogram into three smaller sub-histograms at gray levels $(\mu - \sigma)$ and $(\mu + \sigma)$.

In the second step, the gray level allocation [5] is done based on the dynamic range of the gray levels in the input image. If the dynamic range is low, then based on the ratio of the span of gray levels that the input image sub-histogram occupy, the DHE allocates a particular range of gray levels over which it may span in the output image histogram.

$$span_k = m_k - m_{k-1}$$
$$range_k = \frac{span_k}{\sum span_k} * (L-1) \quad (2)$$

Where $span_k$ = dynamic gray level range used by sub-histogram k in the input image.

m_k = kth local minima in the input image histogram.

$range_k$ = dynamic gray level range for sub-histogram k in output image.

If the dynamic range of gray levels in the input image is high, then a scaled value of cumulative frequencies (c_f) of the gray levels in the sub-histogram is used to perform actively in the allocation process of grayscale ranges among sub-histograms.

$$factor_k = span_k * (\log cf_k)^x$$
$$range_k = \frac{factor_k}{\sum factor_k} * (L-1) \quad (3)$$

Where cf_k = summation of all histogram values of kth sub-histogram.

x= amount of emphasis given on frequency.

In the third step, conventional histogram equalization is applied to each sub-histogram, but its span in the output image histogram is allowed to confine within the allocated gray level range that is designated to it. Therefore any portion of the input image histogram is not allowed to dominate in histogram equalization.

4 Fuzzy Enhancement Technique

The fuzzy image enhancement technique [9] is done by maximizing fuzzy measures contained in the image. The fuzzy measures used are entropy (E) and index of fuzziness (IOF). The original image of size M x N has intensity levels xi in the range of [0 L-1] can be considered as a collection of fuzzy singletons in the fuzzy set notation . The Entropy of a fuzzy set is calculated as

$$E(A) = \frac{1}{n \ln 2} \sum_i S(\mu_A(x_i)), \quad i = 1, 2, \ldots, n,$$

with the Shannon's function

$$S(\mu_A(x_i)) = -\mu_A(x_i) \ln [\mu_A(x_i)] - [1 - \mu_A(x_i)] \ln [1 - \mu_A(x_i)]. \tag{4}$$

Where $\mu_A(x_i)$ represents the membership or grade of belonging μ_A of x_i and x_i being the element (gray scale intensity at the pixel) of the set. The index of fuzziness is calculated as

$$IOF = \frac{2}{MN} \sum_{i=0}^{m-1} \sum_{j=0}^{n-1} \min\{p_{i,}(1 - p_{i,})\} \tag{5}$$

where

$$p_i = sim[\frac{\pi}{2} \times (1 - \frac{\mu(m)}{\mu_{\max}})]$$

and $\mu(m)$ is the S-shaped member ship function with parameters a, b and c are specified to ensure the member-ship function maximizes the information contained in the image.

$$\mu(m) = \begin{cases} 0 & \text{for } m \le a \\ \frac{m-a}{(b-a)(c-a)} & \text{for } a < m \le b \\ 1 - \frac{(m-c)^2}{(c-b)(c-a)} & \text{for } b < m \le c \\ 1 & \text{for } m \ge c \end{cases} \tag{6}$$

The parameters a, b and c are given by equations

$$\begin{aligned} a &= \alpha E_{\max} \\ b &= \beta |IOF_{\max} - E_{\max}| \\ c &= \gamma IOF_{\max} \end{aligned} \tag{7}$$

where α , β and γ are the membership factors that are chosen to obtain optimum S-membership function if fuzzified image. IOFmax and Emax are maximum index of fuzziness and maximum entropy respectively.

The calculated membership function transformed the image intensity levels from the spatial domain to fuzzy domain. The original image has been transformed and most of the regions in the image contained mixed region of overexposed and underexposed regions. Therefore, a parameter called 'exposure' [1] is introduced to denote percentage of the image gray levels is underexposed and overexposed. The exposure is normalized in the range of [0 1]. If the value of exposure is less than 0.5, it denotes that the image contains more underexposed region than overexposed region. The threshold is determined to divide the image into two parts. The threshold is given by equation

$$T = \theta L(1\text{-Exposure}) \tag{8}$$

where T and θ are threshold and exposure operator respectively. The exposure operator, θ is defined to obtain optimum threshold for enhancement. The threshold, T which is in the range [0 L-1] divides the gray levels into two regions which are [0, T-1] for underexposed region and [T, L-1] for overexposed region. The membership function (i.e fuzzified image) is then modified to further enhance the fuzzified image.

$$\mu_{enh} = \begin{cases} \sqrt{\tau\mu(m)} & \text{for } \mu(m) < T \\ [\mu(m)]^2 & \text{for } \mu(m) \geq T \end{cases} \tag{9}$$

The gray levels of the image are heap near the maximum gray level and minimum gray level for overexposed and underexposed regions. A power-law transformation operator is defined for the improvement of the overexposed region of the image. The intensities of the membership function in overexposed region are improved by modifying their membership function in this region. Meanwhile the underexposed regions have the exposure values less than 0.5 and thus only need a gradual amount of saturation enhancement. The membership function is modified using saturation operator of square root.

5 Fusion Rules

Fusion rules play a crucial role in image fusion applications. Researchers have developed some fusion rules for various applications [6] [7]. Let P, Q and R respectively represent the images- sharpened image, dynamic histogram equalized image and fuzzy enhanced image. Also F denotes the fused result of P, Q and R.

The following rule is applied for the fused image F:

$$F(i,j) = p_1 * P(i,j) + p_2 * Q(i,j) + p_3 * R(i,j)$$
$$p_1 + p_2 + p_3 = 1. \tag{10}$$

Where p_1, p_2 and p_3 are weight coefficients that can adjust the portion of P, Q and R to control the brightness of fused image. Experiments suggest that $p_1=0.2$, $p_2=0.4$ and $p_3=0.4$ usually provide satisfactory results for the fusion of three images.

6 Experimental Results

In this section, we present the experimental results of the proposed algorithm on two remote sensing images. The qualitative results of the proposed algorithm are shown in figure 2and figure 3. First, sharpening, Dynamic Histogram Equalization and Fuzzy Enhancement Technique are respectively applied to the same degraded image in order to obtain three individually enhanced images. Then these three images are fused through special rules to get the single enhanced image with suitable contrast and abundant details. The fused image combines the merits of three images, both improving the contrast and enhancing the edges at the same time.

Quantitative analysis is a numerically oriented procedure to figure out the performance of algorithms without any human error. Statistical parameters [8] such as Entropy, Standard deviation and Average Gradient are used for quantitative assessment of enhancement performance of our fusion-based algorithm. The larger entropy indicates the enhanced images provide more image information. Also the larger standard Deviation and average gradient demonstrate that the enhanced images reveal more gray levels and exhibit higher definition. The quantitative values are shown in table 1 and table 2.

Table 1. Quantitative Assessment of Proposed Algorithm on Fig 1

Figure	Entropy	Standard Deviation	Average Gradient
Fig 2-(b)	3.0812	63.467	7.327
Fig 2-(c)	4.081	72.346	10.234
Fig 2-(d)	4.872	74.891	14.234
Fig 2-(e)	5.876	76.891	25.669

Table 2. Quantitative Assessment of Proposed Algorithm on Fig 2

Figure	Entropy	Standard Deviation	Average Gradient
Fig 3-(b)	2.812	53.231	6.467
Fig 3-(c)	3.481	62.246	9.224
Fig 3-(d)	4.872	68.811	13.134
Fig 3-(e)	7.876	74.671	24.139

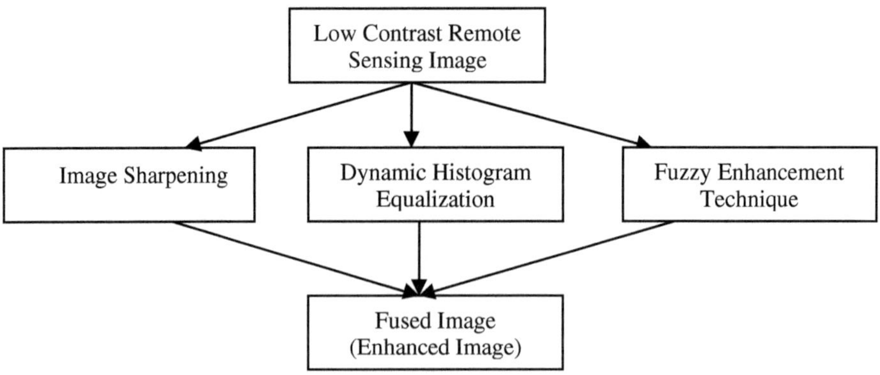

Fig. 1. Flow diagram of the proposed Method

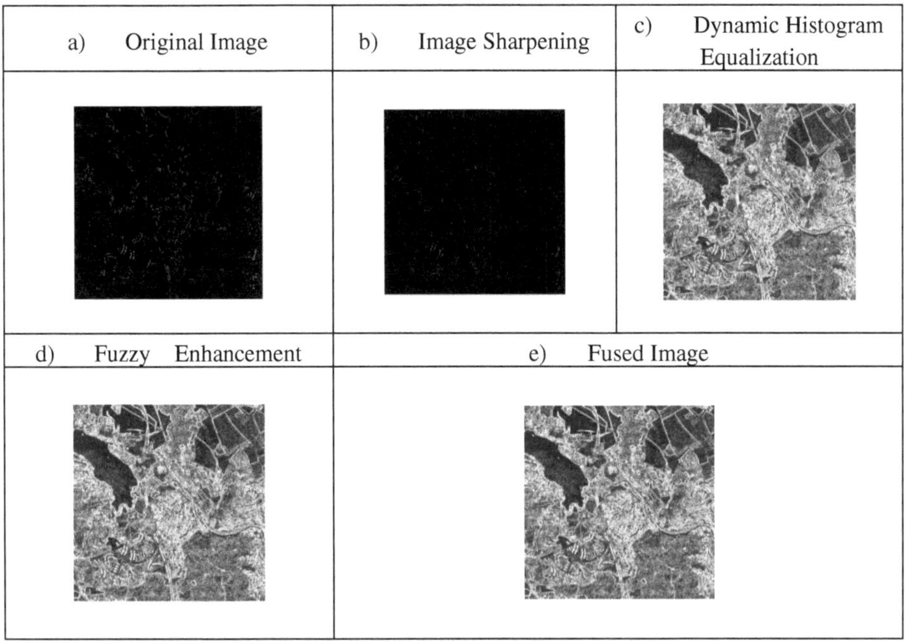

Fig. 2. Qualitative Analysis of Proposed Algorithm

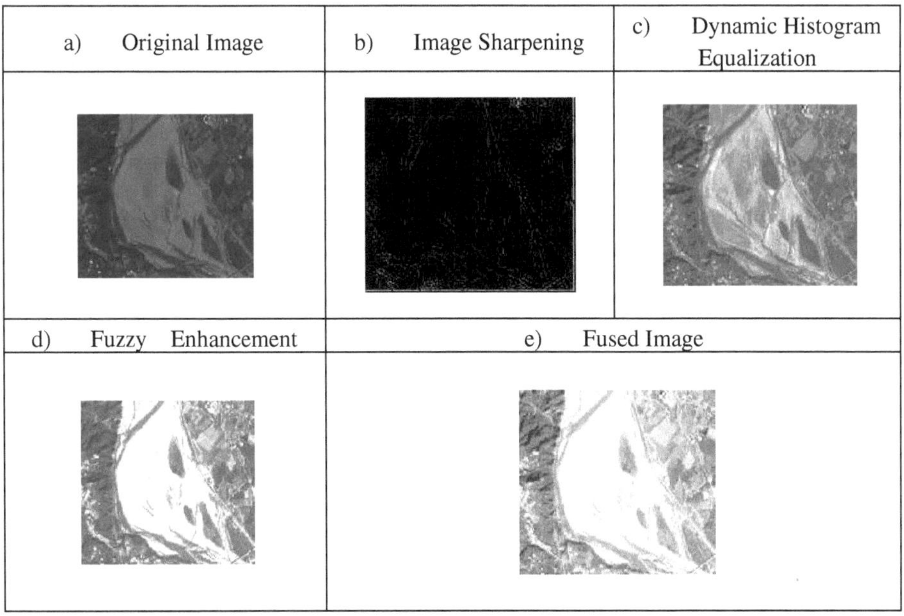

Fig. 3. Qualitative Analysis of Proposed Algorithm

7 Conclusions

Aiming at problems of poor contrast and blurred edges in degraded images, a novel enhancement algorithm is proposed in present research. First, image sharpening, Dynamic Histogram Equalization and Fuzzy enhancement technique are respectively applied to the same degraded image in order to obtain three processed individual images. Then these three images are fused through special rules to get the single enhanced image. Experiment results prove that the proposed enhancement algorithm can efficiently combine the merits of dynamic histogram equalization, fuzzy enhancement and sharpening, improving both the contrast and sharpness of the degraded image at the same time.

References

1. Hanmandlu, M., Jha, D.: An Optimal Fuzzy System for Color Image Enhancement. IEEE Transactions on Image Processing 15, 2956–2966 (2006)
2. Pohl, C., Van Genderen, J.L.: Multisensor image fusion in remote sensing: Concepts, methods and applications. International Journal of Remote Sensing 19(5), 823–854 (1998)
3. Pei, L., Zhao, Y., Luo, H.: Application of Wavelet-based Image Fusion in Image Enhancement. In: 2010 3rd International Congress on Image and Signal Processing (2010)
4. Abdullah-Al-Wadud, M., Kabir, H., AliAkber Dewan, M., Chae, O.: A Dynamic Histogram Equalization for Image Contrast Enhancement. IEEE (2007)
5. Wang, Y., Chen, Q., Zhang, B.: Image enhancement based on equal area dualistic subimage histogram equalization method. IEEE Tran. Consumer Electron. 45(1), 68–75 (1999)
6. Tao, G.Q., Li, D.P., Lu, G.H.: Study on Image Fusion Based on Different Fusion Rules of Wavelet Transform. Infrared and Laser Engineering 32(2), 173–176 (2003)
7. Qiang, Z.X., Peng, J.X., Wang, H.Q.: Remote Sensing Image Fusion Based on Local Deviation of Wavelet Transform. Huazhong Univ. of Sci. & Tech. (Nature Science Edition) 31(6), 89–91 (2003)
8. Liu, C.C., Hu, S.B., Yang, J.H., Guo, X.: A Method of Histogram Incomplete Equalization. Journal of Shandong University (Engineering Science) 33(6), 661–664 (2003)
9. Hasikin, K., Isa, N.A.M.: Enhancement of low contrast image using fuzzy set theory. In: 2012 14th International Conference on Modeling and Simulation. IEEE (2012)

Variant Nearest Neighbor Classification Algorithm for Text Document

M.S.V.S. Bhadri Raju[1], B. Vishnu Vardhan[2], and V. Sowmya[3]

[1] Department of CSE, SRKR Engg. College, Bhimavaram, AP, India
[2] Department of IT, JNTUCE, Jagityala, AP, India
[3] Department of CSE, GRIET, Hyderabad, AP, India
sowmyaakiran@gmail.com

Abstract. Categorizing the text documents into predefined number of categories is called text classification. This paper analyzes various ways of applying nearest neighbor classification for text documents. Text document classification categorizes the documents into predefined classes. In this paper, cosine similarity measure is used to find the similarity between the documents. This similarity measure is applied on term frequency-Inverse document frequency vector space model representation of preprocessed Classic data set. The documents that are most similar to a document are said to be nearest neighbors of that document. In this work, nearest neighbors and k nearest neighbor classification algorithms are used to classify the documents into predefined classes and classifier accuracy is measured.

Keywords: Text document classification, Stemming Algorithm, Cosine Similarity Measure, Classifier Accuracy.

1 Introduction

Classification is one of the significant techniques of data mining. Classification is a supervised learning as the class labels are known. Classification categorizes the data into number of groups.

Due to the availability of increasing number of electronic documents and the growth of World Wide Web document classification is gaining importance. The government electronic repositories news, news articles, biological databases, chat rooms, digital libraries, online forums, electronic mail and blogs repositories are the resources of unstructured and semi-structured information.[1]

Text classification is the primary requirement of text retrieval systems, which retrieve text in response to a user query, and text understanding systems, which transform text in some way such as producing summaries, answering questions or extracting data [11].

Text classification plays a crucial role in information management tasks, because there is an exponential growth of electronic documents in the web. Algorithms that improve the accuracy and efficiency of the classifier are required.[2]

S.C. Satapathy et al. (eds.), *ICT and Critical Infrastructure: Proceedings of the 48th Annual Convention of CSI - Volume II*, Advances in Intelligent Systems and Computing 249,
DOI: 10.1007/978-3-319-03095-1_27, © Springer International Publishing Switzerland 2014

The task of data mining is to automatically classify documents into predefined classes based on their content. Many algorithms have been developed to deal with automatic text classification [10].

Classification algorithms are mainly categorized into eager and lazy learners [3]. Decision tree induction, Rule based classification are eager learners whereas k-nearest neighbor, rule based classification are lazy learners. Classification is a two-step process. Eager learners in the first step, that is in the training step the classifier model is built by applying classification algorithm on training set. In the later step this model uses this classifier model to generate class labels of the test set. Lazy learners will not build any classifier model in the training step but in the testing step the classification algorithm is used to generate the class labels of the test set by using the training set.

For measuring the classifier accuracy the dataset is divided into training and testing set. That means the class labels of training set and test set are known. In the testing step, eager learner uses classifier model to generate the new labels for the test set where as lazy learner the classification algorithm is applied on the training set to generate the new class labels of the test set. The new class labels generated are compared to already existing class labels to find the accuracy of the classifier. If the new class label of an object is equal to already existing class label of that object in the testing set then the object is correctly classified. If not the object is incorrectly classified. The accuracy of the classifier depends on the number of objects correctly classified.

The dataset is divided into two sets training and testing by using uniform distribution and non-uniform distribution. In uniform distribution, form each category the same number of objects will be chosen randomly for both training set and testing set. In non-uniform distribution the objects are chosen randomly regardless of category for both training and testing set. The objects in both the sets are different.

2 Document Representation

In the vector space model the documents are represented as vectors in a common vector space. [4]

The document encoding techniques are Boolean, Term Frequency and Term Frequency with Inverse Document Frequency.

2.1 Term Frequency-Inverse Document Frequency

Term Frequency-Inverse Document Frequency(TF-IDF) vector space model is used for the purpose of analysis. The inverse document frequency term weight is one way of assigning higher weights to these more discriminative words. IDF is defined via the fraction N/n_i, where, N is the total number of documents in the collection and n_i is the number of documents in which $term_i$ occurs.

Thus, the TF-IDF representation of the document d is:

$$d_{tf-idf} = [\ tf_1 \log(n / df_1),\ tf_2 \log(n / df_2), \ldots\ldots, tf_D \log(n / df_D)] \qquad (1)$$

To account for the documents of different lengths, each document vector is normalized to a unit vector (i.e., $||d_{tf\text{-}idf}||=1$).

3 Similarity Measures

Similar documents are found to classify the documents into predefined number of groups. A variety of similarity or distance measures, such as cosine similarity, Manhattan distance and Euclidean distance are available for finding the similarity between the documents [5].

3.1 Cosine Similarity Measure

The cosine function is used to find the similarity for two documents d_i and $d_j$as follows

$$\cos\left(d_i, d_j\right) = \frac{d_i . d_j}{|| d_i |||| d_j ||} \quad (2)$$

Cosine dissiilarity = 1-cos(d_i , d_j)

When the cosine dissimilarity value is 0 the two documents are identical, and 1 if there is nothing in common between them (i.e., their document vectors are orthogonal to each other).

4 Related Work

Words in the documents have morphological variants in bag of words representation. Therefore the stemming algorithms are applied on these morphological words to reduce the number of words in the vector space model. Instead of using the morphological variants of words the stemmed words are used as key terms in the vector space model [6].

Porter stemming algorithm is most extensively used stemmer which was introduced by Porter (1980). The suffixes are removed from a word by applying a set of rules iteratively.

There are many classification algorithms for classifying the objects into predefined number of categories. Classification algorithms are mainly categorized into eager and lazy learners. The prerequisite for classification algorithms is the data should be divided into training and testing sets. Classification is a two-step process: one is training step and the other is testing step. Lazy learner does not perform anything in the training step except in the testing step where as eager learners will perform more work in the training step rather than testing step. [7]

5 Classification Algorithm

K nearest neighbor algorithm is used to classify the objects into predefined groups based on nearest neighbors. Classification is a two-step process: training step and test step. Therefore the dataset is divided into training set and test set based on two different approaches 1.non-uniform distribution and 2.uniform distribution. In the classification dataset one of the attribute is declared as class label attribute. 20% of the data from the dataset is taken as test set and remaining 80% as training set. In the non-uniform distribution, from the whole dataset 20% of the data is taken as testing set and the remaining 80% of the data as training set irrespective of category. But in the uniform distribution from each category 20% of the data is considered for training and remaining for testing.

The nearest neighbor classification algorithm is applied in different ways: One is by finding the nearest neighbors and other is by finding k-nearest neighbors [8].Nearest neighbor classification algorithm is a lazy learner. As it is a lazy learner it does not perform anything in the training step except collecting the training data set. In the testing step the test set is taken and dissimilarity between the objects of the training set and test set is calculated for finding the nearest neighbors and k-nearest neighbors. If the dissimilarity between the objects is less that means both the objects are more similar. After finding the nearest neighbor objects, based on the class label of these objects the new class labels for the testing objects is determined. Now the new class labels of these testing objects are compared with the previous class labels of that objects respectively to measure the accuracy of the classifier. If the classifier accuracy is more we can adopt this algorithm for classifying the new incoming objects.

In the k-nearest neighbors, k is the predefined number of nearest neighbors required for finding the class label of the objects where as in the nearest neighbors it will not depend on the predefined number of nearest neighbors. Both the algorithms are explained below. These algorithms work based on the dissimilarity between the objects mentioned in section 3.

Nearest neighbor algorithm:

Step 1: Generate the dissimilarity matrix between testing objects (as rows) and training objects (as columns).

Step 2: Repeat

- Step 2.1: Find the minimum value in a row (i.e., for a testing object) of step 1.
- Step 2.2: Store the corresponding training objects where ever the minimum value found in step 2.1 is existed.(these training objects are nearest neighbors to the testing object)
- Step 2.3:By considering the class labels of the nearest training objects found in step 2.2 the new class label for the testing object is determined.(i.e., the new class label of the testing object = maximum number of times a particular class label is repeated among nearest neighbor training objects)

Step 3: Until all the testing objects (rows) are completed

Step 4: Now the new labels and the old labels of the testing objects are compared to find the accuracy of the classifier.

K- Nearest neighbor algorithm:

Step 1: Generate the dissimilarity matrix between testing objects (as rows) and training objects (as columns).

Step 2: Give the value of k (i.e., predefined number of nearest neighbors)

Step 3: Find the k similar training objects for each testing object by using the dissimilarity matrix generated in step 1.(i.e., these k similar training objects are k nearest neighbors)

Step 4: By considering the class labels of these k-nearest neighbors found in step 3 the new class label for the testing objects are determined. (i.e., the new class label of the testing object = maximum number of times a particular class label is repeated among k-nearest neighbors)

Step 5: Now the new labels and the old labels of the testing objects are compared to find the accuracy of the classifier.

We need to preprocess the documents before applying the algorithm. Preprocessing consists of steps like removal of stop words, perform stemming, prune the words that appear with very low frequency etc. and build vector space model.

6 Classification

In this section, by using bench mark classic dataset a series of experiments are carried out to categorize the documents into predefined number of categories by using the algorithms explained in section 5 and to inspect the accuracy of those classification algorithms. This section first describes the dataset, preprocessing of the dataset and evaluation of the classifier.

6.1 Dataset

In this work the experiments are carried out on with one of the bench mark dataset i.e., Classic dataset collected from uci.kdd repositories. There are four different collections CACM, CISI, CRAN and MED in Classic dataset. For experimenting 800 documents are considered after preprocessing the total 7095 documents. [9]

The documents consisting of single words are meaningless to consider as document dataset. So, the documents with the words less than the mean length of the documents belonging to that category are deleted. That is file reduction on each category is applied. After applying the file reduction strategy we got the valid documents. 200 documents from each category are collected from these valid documents summing to 800 documents.

6.2 Pre-processing

In the preprocessing step the plain text documents are taken as input and output a stream of tokens. These tokens are used in the construction of vector space model. The stop words are removed and stemming algorithm is applied for reducing the number of words in vector space model. The words with low frequency and high frequency are removed from the vector space model. In this work, the accuracy of the nearest neighbors and k nearest neighbors are evaluated.

6.3 Evaluation

The accuracy of a classifier is measured on the test set that is the percentage of the test set objects that are correctly classified [6]. Out of 800 documents, as said in the section 5 20% of the documents are chosen as testing set and remaining 20% of the documents are chosen as training set. That is 160 documents are chosen as testing set and 640 documents are chosen as testing set from 800 documents. A dissimilarity matrix is generated for the testing documents versus training objects by using cosine dissimilarity measure. The matrix generated is 160 X 640. The training set and test set are separated by performing the non-uniform distribution and uniform distribution as mentioned in section 5. In the uniform distribution, from 200 documents of each category 40 documents are chosen randomly as testing set summing to 160(4*40) and the remaining 160 from each category are taken as testing set summing to 640(4*160). In non-uniform distribution from the 800 documents 160 documents are chosen randomly for the testing set and the remaining as training set regardless of category.

The code uses 10 fold cross validation for checking the accuracy of the classifier. The accuracy of nearest neighbors with non uniform distribution, nearest neighbors with uniform distribution, k-nearest neighbors with non-uniform distribution and k-nearest neighbors with uniform distribution are given in the table 1, table 2, table3 and table 4 respectively. For the k-nearest neighbors, the accuracy of the classifier are shown for different values of k. The accuracy of the classifier is above 90% for the nearest neighbors and k nearest neighbor algorithms and for both uniform and non-uniform distributions of data set discussed in this paper. By experimenting the accuracy of the classifier is more for the value k=1 in the k-nearest neighbor classification algorithm for both uniform sampling and non-uniform distribution.

Accuracy = documents correctly classified / total number of documents in the test set (5)

Table 1. Nearest neighbors with non-uniform distribution

Categories of documents	Documents chosen by non-uniform distribution for test set	Documents that are correctly classified
CACM	41	33
CISI	32	29
CRAN	48	45
MED	39	37
Total	160	144
Accuracy		90.0%

Table 2. Nearest neighbors with uniform distribution

Categories of documents	Documents chosen by uniform distribution for test set	Documents that are correctly classified
CACM	40	30
CISI	40	38
CRAN	40	39
MED	40	38
Total	160	145
Accuracy		90.625%

Table 3. k- Nearest neighbor with non-uniform distribution

k-values	k=1		k=2		k=3	
Catego-ries of docume nts	Documents chosen for test set	Documents that are correctly classified	Documents chosen for test set	Documents that are correctly classified	Documents chosen for test set	Documents that are correctly classified
CACM	46	45	37	35	36	33
CISI	38	35	41	36	42	37
CRAN	33	33	43	42	47	47
MED	43	41	39	37	35	33
Total	160	154	160	150	160	150
Accuracy		96.25%		93.75%		93.75%

Table 4. k- Nearest neighbor with uniform distribution

k-values	k=1		k=2		k=3	
Categor-ies of docum-ents	Documents chosen for test set	Documents that are correctly classified	Documents chosen for test set	Documents that are correctly classified	Documents chosen for test set	Documents that are correctly classified
CACM	40	37	40	38	40	36
CISI	40	39	40	34	40	35
CRAN	40	38	40	39	40	40
MED	40	39	40	37	40	39
Total	160	153	160	148	160	150
Accuracy		95.625%		92.5%		93.75%

7 Conclusions and Future Work

In this study we found that nearest neighbor and k-nearest neighbor classification algorithms work wells for text document classification. The results showed that finding one most nearest neighbor is better than finding the many nearest neighbors in classifying or finding the class labels of the data. Thus leading to the best accuracy of the classifier when k=1. Though the experimental results proved good we would like to extend this work for larger datasets with more classes. In future work we intend to incorporate ontology based text classification. We also intend to investigate all classification algorithms both eager learners, and lazy learners and discover the best algorithm which is more efficient and accurate for text document classification.

References

1. Khan, A., Bahurdin, B.B., Khan, K.: An Overview of E-Documents Classification. In: 2009 International Conference on Machine Learning and Computing IPCSIT, vol. 3. IACSIT Press, Singapore (2011)
2. Shang, W., Huang, H., Zhu, H., Lin, Y., Qu, Y., Wang, Z.: A Noval Feature Selection Algorithm for text catogorization. Elsevier, Science Direct Expert System with Application 33(1), 1–5 (2006)
3. Aha, D. (ed.): Lazy learning. Kluwer Academic Publishers (1997)
4. Salton, G., Wong, A., Yang, C.S.: A vector space model for automatic indexing. Communications of the ACM 18(11), 613–620 (1975)

5. Huang, A.: Similarity Measures for Text Document Clustering. Published in the Proceedings of New Zealand Computer Science Research Student Conference (2008)
6. Porter, M.F.: An algorithm for suffix stripping. Program 14(3), 130–137 (1980)
7. Han, J., Kamber, M.: Dat Mining concepts and techniques. Elsevier Publishers
8. Jarvis, R.A., Patrick, E.A.: Clustering Using a Similarity Measure Based on Shared Nearest Neighbors. IEEE Transactions on Computers C-22(11) (November 1973)
9. Sandhya, N., Sri Lalitha, Y.: Analysis of Stemming Algorithm for Text Clustering. IJCSI International Journal of Computer Science 8(5(1)) (September 2011)
10. Kruengkrai, C., Jaruskulchai, C.: A Parallel Learning Algorithm for Text Classification. In: The Eighth ACM SIGKDD International Conference on Knowledge Discovery and Data Mining (KDD 2002), Canada (July 2002)
11. Kamruzzaman, S.M., Haider, F., Hasan, A.R.: Text Classification Using Data Mining

Naive Bayes for URL Classification Using Kid's Computer Data

Anand Neetu

Lingayas University and
Dept. of Computer Sc.
Maharaja Surajmal Institute, Delhi
neetuanand77@rediffmail.com

Abstract. The vast size of the World Wide Web (WWW) nowadays makes it the largest database ever existed. One of the most important functions of the Internet is information retrieval. This research explores a new data source called personal (kids') browsing data(PBD).The purpose of this study is to assist information retrieval on the Internet by applying data mining techniques. The use of data mining in this domain can be seen as the application of a new technology to an acknowledged problem. Several techniques exist in data mining: association rule, classification, cluster, sequential, and time series. However, the ultimate purpose of WUM is to discover useful knowledge from Web users' interactive data. In this paper we intend to focus on the classification task. Using only the URL of a web page its category can be identified. The advantage of doing classification using only URLs is its high speed.

Keywords: Classification, WUM, Naïve Bayes classification, URL fragmentation.

1 Introduction

The log file is like a box of treasure waiting to be exploited containing valuable information for the web usage mining system. The "next generation" of monitoring comprehends the intrinsic drawbacks of server-side inferences, data mining, data filling, and data manipulation. As an alternative, contemporary research has turned to software augmentation on the client to give a more granulated view of the user activities. We designed and developed a new system for client side to find the interest and usage pattern of a particular user. Data mining and World Wide Web are two important and active areas of current researches. Data mining has recently emerged as a growing field of multidisciplinary research. It combines disciplines such as databases, machine learning, artificial intelligence, statistics, automated scientific discovery, data visualization, decision science, and high performance computing. Web Mining, has been the focus of several recent research projects and papers. Web data mining can be defined as applying data mining techniques to automatically discover and extract useful information from the World Wide Web. Web data mining

S.C. Satapathy et al. (eds.), *ICT and Critical Infrastructure: Proceedings of the 48th Annual Convention of CSI - Volume II*, Advances in Intelligent Systems and Computing 249,
DOI: 10.1007/978-3-319-03095-1_28, © Springer International Publishing Switzerland 2014

can be divided in three general categories: web content mining, web structure mining and lastly web usage mining. Here the focus is on the later area of web data mining that tries to exploit the navigational traces of the users in order to extract knowledge about their preferences and their behaviour. Web Usage Mining (WUM) is an active area of research and commercialization. The task of modelling and predicting a user navigational behaviour on Internet can be useful in quite many web applications such as web caching, web page recommendation, web search engines and personalization. An important topic in Web Usage Mining is classification of URL of users to find their surfing behaviour. By analyzing the characteristics of the classification, web users can be understood better. The structure of the paper is as follows. In Section 2, we present related background information. Section 3 presents the methods proposed by this paper. In Section 4, we describe the mining technique followed to evaluate our methods and we present experimental results and conclusion in Section 5.

2 Related Work

The idea of using URL for classification is not new but our concept of first generating the user data and then applying classification is new. The approach used by the different researchers is as follows:[1] Their goal is to investigate whether visual features of HTML web pages can improve the classification of fine-grained genres. Innately it seems that it is helpful and the challenge is to extract visual features that capture the layout characteristics of the genres. A corpus of web pages from disparate e-commerce sites was generated and manually classified into several genres. Experiments confirm that using HTML features and particularly URL address features can improve classification beyond using textual features alone.[2] They showed that the automated classification of Web pages can be much enhanced if, instead of looking at their textual content, they considered URL link's and the visual placement of those links on a referring page(tree like structure). They developed a model and algorithm for machine learning using tree-structured features. This method is applied in automated tools for recognizing and blocking Web advertisements and for recommending "interesting" news stories to a reader. Their algorithms are more faster and more accurate than those based on the text content of Web documents.[3]Proposed a lightweight system to detect malicious websites online based on URL lexical and host features and call it MALURLs. The system relies on Naïve Bayes classifier as a probabilistic model to detect if the target website is a malicious or benign. It introduced new features and employs self-learning using Genetic Algorithm to improve the classification speed and precision.[4]They identified a potential deficiency of MNB in the context of unbalanced datasets and shown that per-class word vector normalization presents a way to address the problem. Their results show that normalization can indeed significantly improve performance. It has been shown that MNB with class vector normalization is very closely related to the standard centroid classifier for text classification if the class vectors are normalized to unit length, and verified the relationship empirically. [5] Their research explores a new data source called intentional browsing data (IBD) for potentially improving the efficiency of WUM applications. IBD is a category of online browsing actions, such as "copy", "scroll", or "save as," and is not recorded in Web log files. Consequently, their research aims to build a basic understanding of IBD which will lead to its easy adoption in WUM research and practice. Specifically, this paper formally defines IBD and clarifies its relationships with other browsing data via a proposed taxonomy. In order to make IBD available like Web log files, an online data collection mechanism for capturing IBD is also proposed and discussed.

3 Research Methodology

An ideal monitoring framework would address several goals: non-interference with the user activity, flexible data collection and the ability to integrate monitoring with other client-side applications. The steps for monitoring and classification activity are given as:

Monitoring Module: The module was constructed and installed on the client (Kid's)

computer to capture and store the log data. Every time the client opened a web page its URL is get stored into the database.

Log data collection: Created database structure for storing usage record of the client.

Data preparation: It is a two steps procedure. In the first step data pre-processing occurs, it is the process of cleaning and transforming raw data sets into a form suitable for web mining. In the second step data transformation take place, it plays the most important role in the process of data collection and pre-processing, since, all subsequent data mining tasks are based on the results of this process. Through the processes of data cleaning, selection, integration and transformation, the raw web log is transformed into well-structured data file which are ready for subsequent data mining tasks.

Classification Discovery: Classification technique is applied to classify various URLs into different categories.

Classification Analysis: Analysis of the result.

4 Mining Techniques

4.1 Classification

In data mining, classification is one of the most focused sub-areas which can be directly used for prediction purpose. Broadly described, classification is a two-step method. In the first step, a representation is built illustrating a predetermined set of data classes or concepts. The prototype is constructed by examining databank tuples described by attributes. Each tuple is assumed to belong to a predefined class, as determined by one of the attributes, called the class label attribute. In the context of classification, data tuples are also referred to as samples, examples, or objects. The data tuples are then analyzed to build the model collectively form the training data set. The distinct tuples making up the training set are referred to as training samples and are randomly selected from the sample population. In the second step, the model is used for classification. First, the predictive accuracy of the model (or classifier) is estimated. The accuracy of a model on a given test set is the percentage of test set samples correctly classified by the model. For each test sample, the known class label is compared with the learned model's class prediction for that sample. If the accuracy of the model is considered acceptable, the model can be used to classify future data tuples or objects for which the class label is not known.

4.2 Classification Techniques

Decision Tree. This technique recursively separates observations in branches to construct a tree for the purpose of improving the prediction accuracy. Most decision tree classifiers perform classification in two phases: tree-growing (or building) and tree-pruning. Popular decision tree algorithms include ID3, C4.5, C5, and CART.

Nearest Neighbors' Algorithm. It is considered as statistical learning algorithms and it is extremely simple to implement and leaves itself open to a wide variety of variations. K-Nearest Neighbor is one of the best known distance based algorithms, it has different version such as closest point, single link, complete link, K-Most Similar Neighbor etc.

Genetic Algorithms. It is a powerful technique for solution of various combinatorial or optimization problems. They are more an instrument for scientific research rather than a tool for generic practical data analysis.

Nonlinear Regression Methods (NR). It is based on searching for a dependence of the target variable in the form of function. This method has better chances of providing reliable solutions in medical diagnostics applications.

Support Vector Machines. It is the most robust and successful classification algorithms. They are based upon the idea of maximizing the margin i.e. maximizing the minimum distance from the separating hyper plane to the nearest example.

Maximum Likelihood Estimation (MLE). It deals with finding the set of models and parameters that maximises this probability. With statistical classification the usual procedure is to select the class, that most probably generates the partial observation and then use that class' distributions for the unknowns. This is generally satisfactory only if the probability of the chosen class is representative.

Naïve Bayesian. It is a successful classifier based upon the principle of Maximum a Posteriori (MAP).

4.3 URL Fragmentation

URL stands for uniform resource locator or formerly the universal resource locator. URLs are extremely good features for learning. The characteristic's which make it suitable for classification are as follows:

- It is easy to extract them and they are fairly steady. Each URL maps uniquely to web page or document, and any fetch document must have a URL. In contrast, other web features like anchor text, alt tags, and image sizes, are optional and not unique to a document.
- URLs can be read without downloading the target document, this helps us to implement classification more quickly.
- URLs have an instinctive and simple mapping to certain classification problems.

In the classification with URL, the specific keyword in URL is important factor. Each URL is split into a sequence of strings of letters at any punctuation mark, number or other non-letter character. The URL has three main parts: protocol, hostname and path. For example, http://engg.entrancecorner.com/exams /5833-ipucet-2013-application-form.html. The protocol is: http://", the hostname is: "engg.entrancecorner.com" and the path is: "exams/5833-ipucet-2013-application-form. html", would be split into the tokens http, engg, entrancecorner, com, exams, 5833, ipucet, 2013, application, form. This methodology is quicker than classic web page classification, as the pages themselves do not have to be fetched and analysed. There is a lot of learning algorithms which has been applied to text classification including Naive Bayes (NB), Decision Tree (DT), k-nearest neighbour (k-NN), Support Vector Machines (SVM), Neural Networks (NN), and Maximum Likelihood Estimation (MLE).In this paper we discuss Naïve Bayes algorithm for performing classification.

4.4 Naïve Bayesian for URL Classification

Naïve Bayes classifier is built on the Bayesian theory which is known as simple and effective probability classification method and has become one of the important contents in the text categorization. The fundamental idea is the use of feature items and categories of conditional probability to estimate a given document category probability. The naïve Bayes classifier is the simplest of these models; in that it assumes that all attributes of the examples are independent of each other given the context of the class. The method of calculation of probability can be divided into maximum Likelihood model (MLM), multivariate Bernoulli model (MBM), Multinomial mode (MM), Poisson model (PM) and so on. Multinomial naive Bayes (MNB) is a popular method for document classification due to its computational efficiency and relatively good predictive performance.

Bayesian classifiers use **Bayes theorem**, which says

$$p(cj \mid d) = p(d \mid cj)\, p(cj)\, p(d)$$

•$p(cj \mid d)$ = probability of instance d being in class cj,
•$p(d \mid cj)$ = probability of generating instance d given class cj.
•$p(cj)$= probability of occurrence of class cj.
•$p(d)$= probability of instance d occurring.
To simplify the task, naïve Bayesian classifiers assume attributes have independent distributions, and thereby estimate $p(d|cj) = p(d1|cj) * p(d2|cj) * \ldots.* p(dn|cj)$.
The Naive Bayes classifier is often represented as this type of graph…

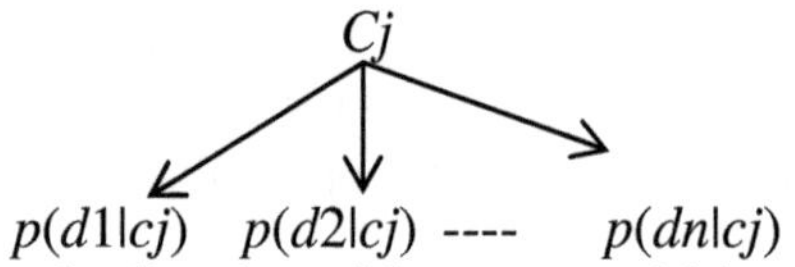

Note the direction of the arrows, which state that each class causes certain features, with a certain probability.

Pros and Cons of Using Naïve Bayes

***Pros*:**

- Fast to train (single scan). Fast to classify
- Very easy to program and intuitive
- Very easy to deal with missing attributes
- Not sensitive to irrelevant features
- Handles real and discrete data and streaming data as well

Cons:

- Assumes independence of features

5 Experimental Results

This experiment is carried on the open source data mining tool of Weka. Weka contains tools for data pre-processing, classification, regression, clustering, association rules, and visualization. It is also well-suited for developing new machine learning schemes. In this experiment, for Kid's Internet usage monitoring, software developed in VB is installed on the target computer which allow parents to observe and bound their children's usage of many applications, websites visited and online searches, social networking behaviour and other programs. Log data of client computer is collected and stored in a database. After pre-processing a classification approach can be applied. We demonstrate that our classification model based on Kid's internet usage is one of the first experiment with this type of data and performs better than other classification methods significantly. Specially, our model can solve the problems which other methods for text classification that uses contents of web page to classify the web page face as we are using URL only, not the contents. The database recorded by our software system is shown below:

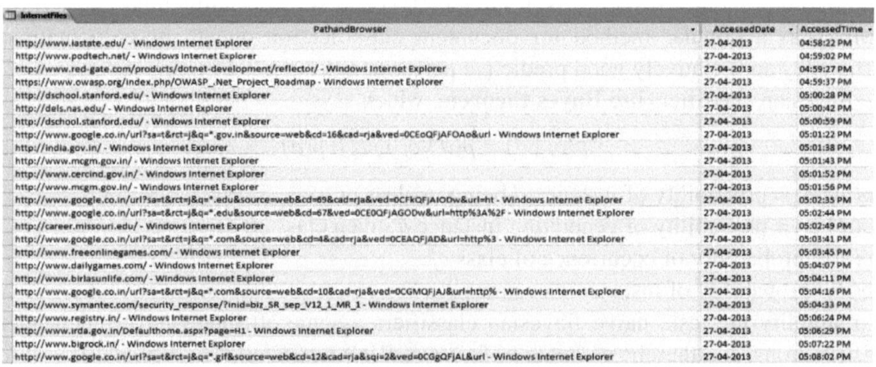

InternetFiles

PathandBrowser	AccessedDate	AccessedTime
http://www.iastate.edu/ - Windows Internet Explorer	27-04-2013	04:58:22 PM
http://www.podtech.net/ - Windows Internet Explorer	27-04-2013	04:59:02 PM
http://www.red-gate.com/products/dotnet-development/reflector/ - Windows Internet Explorer	27-04-2013	04:59:27 PM
https://www.owasp.org/index.php/OWASP_.Net_Project_Roadmap - Windows Internet Explorer	27-04-2013	04:59:37 PM
http://dschool.stanford.edu/ - Windows Internet Explorer	27-04-2013	05:00:28 PM
http://dels.nas.edu/ - Windows Internet Explorer	27-04-2013	05:00:42 PM
http://dschool.stanford.edu/ - Windows Internet Explorer	27-04-2013	05:00:59 PM
http://www.google.co.in/url?sa=t&rct=j&q=*.gov.in&source=web&cd=16&cad=rja&ved=0CEoQFjAFOAo&url - Windows Internet Explorer	27-04-2013	05:01:22 PM
http://india.gov.in/ - Windows Internet Explorer	27-04-2013	05:01:38 PM
http://www.mcgm.gov.in/ - Windows Internet Explorer	27-04-2013	05:01:43 PM
http://www.cercind.gov.in/ - Windows Internet Explorer	27-04-2013	05:01:52 PM
http://www.mcgm.gov.in/ - Windows Internet Explorer	27-04-2013	05:01:56 PM
http://www.google.co.in/url?sa=t&rct=j&q=*.edu&source=web&cd=69&cad=rja&ved=0CFkQFjAIODw&url=ht - Windows Internet Explorer	27-04-2013	05:02:35 PM
http://www.google.co.in/url?sa=t&rct=j&q=*.edu&source=web&cd=67&ved=0CE0QFjAGODw&url=http%3A%2F - Windows Internet Explorer	27-04-2013	05:02:44 PM
http://career.missouri.edu/ - Windows Internet Explorer	27-04-2013	05:02:49 PM
http://www.google.co.in/url?sa=t&rct=j&q=*.com&source=web&cd=4&cad=rja&ved=0CEAQFjAD&url=http%3 - Windows Internet Explorer	27-04-2013	05:03:41 PM
http://www.freeonlinegames.com/ - Windows Internet Explorer	27-04-2013	05:03:45 PM
http://www.dailygames.com/ - Windows Internet Explorer	27-04-2013	05:04:07 PM
http://www.birlasunlife.com/ - Windows Internet Explorer	27-04-2013	05:04:11 PM
http://www.google.co.in/url?sa=t&rct=j&q=*.com&source=web&cd=10&cad=rja&ved=0CGMQFjAJ&url=http% - Windows Internet Explorer	27-04-2013	05:04:16 PM
http://www.symantec.com/security_response/?inid=biz_SR_sep_V12_1_MR_1 - Windows Internet Explorer	27-04-2013	05:04:33 PM
http://www.registry.in/ - Windows Internet Explorer	27-04-2013	05:06:24 PM
http://www.irda.gov.in/Defaulthome.aspx?page=H1 - Windows Internet Explorer	27-04-2013	05:06:29 PM
http://www.bigrock.in/ - Windows Internet Explorer	27-04-2013	05:07:22 PM
http://www.google.co.in/url?sa=t&rct=j&q=*.gif&source=web&cd=12&cad=rja&sqi=2&ved=0CGgQFjAL&url - Windows Internet Explorer	27-04-2013	05:08:02 PM

Fig. 1. Internet usage database of kid's computer

5.1 Creating the Dataset

The total URL collected from the kids' computer by using monitoring module for 5 days are 150. The collected URLs are converted into, list of tokens that are used to train the classifier. Since we are going to classify the record for kids Internet usage, the data is classified into 5 classes mainly Education, Games, Social, Sports and News. After preprocessing the sample of the recorded URL's with their assigned classes are shown below:

URL	Sports	Games	Education	Social	News	Class
url107		1				games
url108			1			education
url109			1			education
url110		1	1			games
url111		1	4			education
url112			3			education
url113			1			education
url114			4			education
url115				1		social
url116				2		social

Fig. 2. Data after preprocessing step

The above data is then used as an input to Weka classifier. After using Naïve Bayesian classifier the result is as shown below

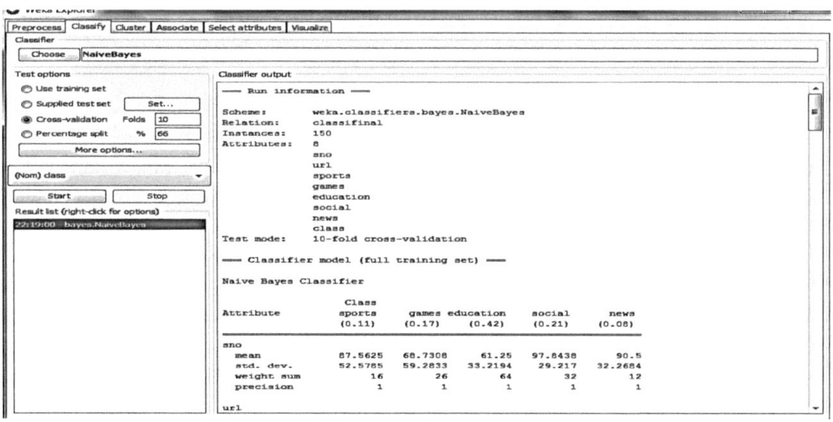

Fig. 3. Result after applying Naïve Bayesian Classification

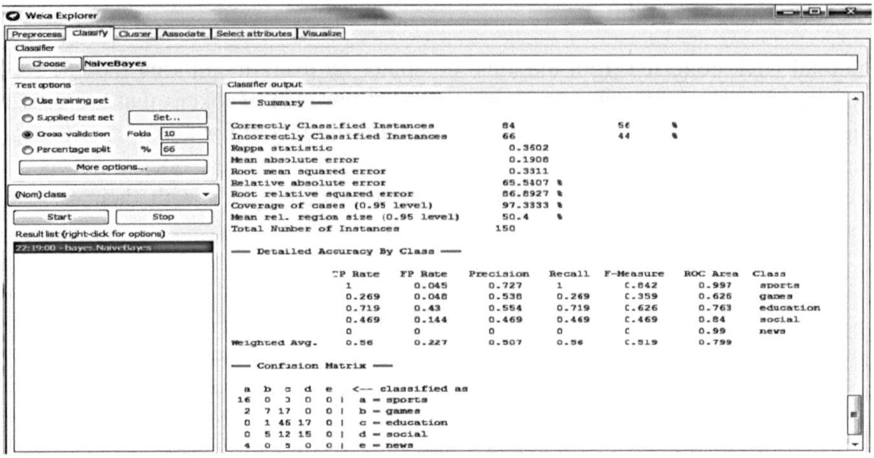

Fig. 4. Result of applying Naïve Bayes Classification

5.2 Final Words and Future Scope

Our experimental results indicate that the techniques discussed here are promising, and bear further investigation and development. Our future work in this area involves conducting experiments with various types of transactions derived from user sessions. Even though this was a small scale and early experiment, many useful conclusions can be drawn. URL based categorization is extremely efficient both in time and space, as the data examined is small in comparison to other approaches. The proposed classification model opens up some interesting directions for future research. Furthermore various other classification techniques can be applied and the comparisons will be done to show the best one in this area.

References

1. Levering, R., Cutler, M., Yu, L.: Using Visual Features for Fine-Grained Genre Classification of Web Pages. In: Proceedings of the 41st Hawaii International Conference on System Sciences (2008)
2. Shih, L.K., Karger, R.D.: Using URLs and Table Layout for Web Classification Tasks. In: WWW 2004 (May 17-22, 2004)
3. Aldwairi, M., Alsalman, R.: MALURLS: A Lightweight Malicious Website. Emerging Technologies in Web Intelligence 4(2) (2012)
4. Frank, E., Bouckaert, R.R.: Naive Bayes for Text Classification with Unbalanced Classes. In: Fürnkranz, J., Scheffer, T., Spiliopoulou, M. (eds.) PKDD 2006. LNCS (LNAI), vol. 4213, pp. 503–510. Springer, Heidelberg (2006)
5. Tao, Y.H., Hong, T.P., Su, Y.M.: Web usage mining with intentional browsing data. Journal of Expert Systems with Applications 34, 1893–1904 (2008)

Semantic Framework to Text Clustering with Neighbors

Y. Sri Lalitha[1] and A. Govardhan[2]

[1] Department of CSE, Hyderabad
srilalitham.y@gmail.com
[2] Department of CSE, JNTUH College of Engineering, Hyderabad

Abstract. Conventional document clustering techniques use bag-of-words to represent documents, an often unsatisfactory representation, as it ignores the relationships between words that do not co-occur literally. Including semantic knowledge in text representation we can establish the relations between words and thus result in better clusters. Here we apply neighbors and link concept with semantic framework to cluster documents. The neighbors and link provides the global information to compute the closeness of two documents than simple pair wise similarity. We have given a framework to represent text documents with semantic knowledge and proposed Shared Neighbor Based Semantic Text Clustering algorithm. *Our experiments on Reuters, Classic and real-time datasets shows significant improvement in forming coherent clusters.*

Keywords: Similarity Measures, Coherent Clustering, Neighbor Clustering, WordNet.

1 Introduction

Tremendous growth in the volume of text documents available on the internet, digital libraries, news sources and company-wide intranets has lead an increased interest in developing methods that can help users to effectively navigate, summarize and organize this information with the ultimate goal of helping them to find the relevant data. Fast and high quality document clustering algorithms play an important role. This ever increasing importance of document clustering and the expanded range of its applications lead to the development of number of new and novel algorithms.

Section 2 deals with related work on text clustering with WordNet. Section 3 describes the similarity measure with neighbors. Section 4 presents proposed SNBSTC algorithm. Experimental results analysis is dealt in Section 5 and Section 6 concludes and discusses the future work.

2 Related Work

In this section we mainly focus on WordNet based text clustering works. Conventional text clustering algorithms with neighbors in [9, 5, 1] use bag of words representation and thus ignores semantic relationships among terms. As a result, if

S.C. Satapathy et al. (eds.), *ICT and Critical Infrastructure: Proceedings of the 48th Annual Convention of CSI - Volume II*, Advances in Intelligent Systems and Computing 249,
DOI: 10.1007/978-3-319-03095-1_29, © Springer International Publishing Switzerland 2014

two documents use different collections of terms to represent the same topic, they can be assigned to different clusters, even though the terms they use are meaningfully same associated in other forms. To incorporate semantic relationship among terms we use WordNet[7] an on-line lexical reference system. WordNet covers semantic and lexical relations between terms and their meanings, such as synonymy, polysemy, and hyponymy/hypernymy. With the incorporation of such background knowledge by converting terms into term meanings we can establish relationships between terms in document datasets. In Information Retrieval techniques WordNet is applied in finding the semantic similarity of retrieved terms [8]. In [10] they combine the WordNet knowledge with fuzzy association rules, in [15], they extend the bisecting k-means using WordNet. The techniques proposed in [14,15] add all available information to the representation of the text documents, so that the dimension of the text database is increased, and additional noise is added by the incorrect senses retrieved from WordNet. In [11] For a given set of terms(commonly noun) all the hypernyms of each term were extracted, then weight them appropriately, and finally chose representative hypernyms that seem to extend the overall meaning of the set of given terms. This approach depends entirely on the weighting formula that will be used during the process which depends on the depth and frequency of appearance. In [12] NSTC the semantic relation between terms using WordNet is included and reweighted VSM with a measure of relatedness and term frequency. They used this new VSM with neighbors and link similarity measure presented in [14] to cluster text documents using kmeans algorithm. In this paper we propose a document clustering algorithm with a framework to represent Semantic Information and apply neighbor and link concept.

3 Document Representation

The documents are represented as vectors in a vector space model VSM(bag-of-words) where each document, d, is a vector (set of document "terms"). Each document is represented by the (TF) vector, $dt_f = (tf_1, tf_2, \ldots, tf_n)$, where tf_i is the frequency of the ith term in the document. Here we are using *Inverse document frequency* (IDF) which gives higher weight to terms that only occur in a few documents. IDF is defined as the fraction N/df_i, where, N is the total number of documents in the collection and df_i is the number of documents in which term i occurs.

3.1 Similarity Measures

For clustering to happen, a clear measure of closeness of the pair of objects is essential. If clusters formed are to be meaningful, the similarity measure should be invariant to transformations natural to the problem domain. A variety of similarity or distance measures like Cosine, Jaccard, Pearson Correlation and Euclidean distance measures have been proposed and widely applied to text documents [2, 3].

Neighbors and Link : Let sim(di, dj) be a similarity function capturing the pairwise similarity between two documents, di, and dj, and have values between 0 and 1, with a larger value indicating higher similarity. For a given threshold Θ, di and dj are defined as neighbors of each other if $Sim(di, dj) \geq \Theta$ with $0 \leq \Theta \leq 1$, where Θ is a user-defined threshold to control how similar a pair of documents should be in order to be considered as neighbors of each other. Neighbor matrix of n document dataset is an $n \times n$ adjacency matrix M, in which an entry M[i, j] is 1 or 0 depending on whether documents d_i and d_j are neighbors or not [4]. The number of neighbors of a document d_i in the data set is denoted by $N[d_i]$ and it is the number of entries whose values are 1 in the ith row of the neighbor matrix M.

The value of the link function $link(d_i, d_j)$ is defined as the number of common neighbors between d_i and d_j [4] and it derived by multiplying the ith row of the neighbor matrix M with its jth column

$$link(di, dj) = \sum_{i=1}^{n} M[i, m] * M[m, j] \quad (1)$$

Thus, if $link(d_i, d_j)$ is large, then probability of d_i and d_j being in the same cluster is more. Since the measures [Cosine/Jaccard/Pearson] measure pair wise similarity between two documents, hence using it alone is considered as a local clustering while involving the link function can be considered as a global clustering approach [4].

3.2 Similarity Measure with Link

When a document d_i shares a group of terms with its neighbors and a document d_j shares another group of terms with many neighbors of d_i, even if d_i and d_j are not considered similar by the similarity measures, their neighbors show how close they are. Based on these, in [5] a similarity measure making use of cosine and link functions is proposed. Here we extended it to jaccard and pearson measures. The similarity measures for jaccard is as follows

$$f(\mathrm{di}, \mathrm{cj}) = \alpha \times \frac{link(di, cj)}{Lmax} + (1 - \alpha) \times \mathrm{jac}(di, cj), \quad where\ 0 \leq \alpha \leq 1 \quad (2)$$

where, *Lmax* is the largest possible value of link(di, cj), and α is the coefficient set by the user. In case of k-means algorithm, where all the documents in the dataset are involved in the clustering process, the largest possible value of link(di, cj) is n, total number of documents in the dataset, which means entire dataset are neighbors of both di and cj and in case of bisecting K-means, the largest possible value of link(di, cj) is the number of documents in the selected cluster cj and the smallest possible value of link(di, cj) is 0, where di and cj do not have any common neighbors. The Lmax to normalize the link values so that the value of link(di, cj)/ Lmax always falls in the range of [0, 1]. With $0 \leq \alpha \leq 1$, the value of f(di, cj) is between [0,1] for all the cases. The above equation shows that the sum of weighted values of the jaccard and link functions are used to evaluate the closeness of two documents, and a larger value of f(di, cj) indicates, that they are closer. When α is set to 0, the similarity measure becomes the jaccard function, and becomes the link function when α is 1. Since the jaccard function and the link function together evaluate the closeness of two documents in different aspects, our new similarity measures are more comprehensive. During the clustering process, iteratively each document is assigned to the cluster whose centroid is most similar to the document, so that the global criterion function is maximized. We use new matrix M' which is an n x (n + k) matrix, in which an entry M'[i, n+j] is 1 or 0 depending on whether a document di and a centroid cj are neighbors or not. The expanded neighbor matrix is shown Fig. 2. Link(di, cj) is given by

$$link(di, cj) = \sum_{m=1}^{n} \mathrm{M}'[\mathrm{i}, \mathrm{m}] \times \mathrm{M}'[\mathrm{m}, \mathrm{n}+\mathrm{j}] \tag{3}$$

Table 1. Expanded Neighbor matrix (M') of dataset S with ϴ=0.3 & k=3

	d_1	d_2	d_3	d_4	d_5	d_6	c_1	c_2	c_3
d_1	1	0	0	0	0	1	1	0	0
d_2	0	1	0	0	1	0	0	1	0
d_3	0	0	1	1	1	0	0	0	1
d_4	0	0	1	1	1	1	1	0	1
d_5	0	1	1	1	1	1	0	1	1
d_6	1	0	0	1	1	1	1	0	0

4 Semantic Text Clustering

K-Means and its variants require initial cluster centers (Seed points) to be specified. If a document is a neighbor of two documents then we say it is a shared neighbor of those two documents. We use a procedure to find seeds with link function.

Find shared neighbors of d_i, d_j represented as SHN(d_i, d_j) as follows:

NB_List(d_i,) where $1 \leq k \leq n$ is a collection of documents with $M[i,k] \geq 1$.
SHN(d_i, d_j) = NB_List(d_i) ∩ NB_List(d_j,) where $i \neq k$, and $1 \leq k \leq n$.
Count the total no. of SHN(d_i, d_j)

Arrange the Shared Neighbor combination in descending order. Find the disjoint or minimum intersection combinations. Consider the combinations with minimum intersect into the list. Let SHN(d3, d4) ∩ SHN(d1, d4) = 0, then consider (d3, d4) and (d1, d4) as possible candidates for seeds with k=2. Find the highest similarity document from the combination of shared neighbor candidates and consider that document as the initial centroid of one cluster. Highest Similarity of candidates (di, dj) is calculated as follows

$$HS(di, dj) = max(\sum_{p=1}^{n} sim(di, dp), max\left(\sum_{p=1}^{n} sim(dj, dp)\right) \quad (4)$$

4.1 Semantic Framework for Document Representation

A word refers to a term in text documents and a word meaning refers to the lexicalized concept that a word can be used to express [7]. Enriching the term vectors with lexicalized concepts from WordNet resolves synonyms and introduces more general concepts in identifying related topics. For instance, a document about 'chair' may not be related to a document about 'dressing table' by the clustering algorithm if there are only 'chair' and 'dressing table' in the term vector. But, if more general concept 'furniture' is added to both documents, their semantic relationship is revealed. We believe that using lexicalized concepts in text clustering not only improves clustering but also reduces the high dimensionality problem of text documents.

Terminology

We use different terms to express the same meaning called synonyms. A term meaning in WordNet is represented by a synonym set (SynSet) a set of terms which are synonyms. In this paper, a SynSet is denoted by SyS, e.g. SyS_1 = {chair, seat}, ''seat'' is a synonym of ''Chair'', so they are interchangeable in documents. When the term matching is performed to find the frequency of a term in a text database, the frequency of terms ''seat'' and ''chair'' can be summed up if we treat these two terms as same in a document. If it appears in different documents, then these documents may be placed in same cluster. Synsets are interconnected with semantic relations forming a large semantic network such as hyponym/hypernym which defines *isA* relationship between concepts also called as subset/superset relationship. For ex., the hypernym of a synset {armchair, folding chair} is {Chair}, and the hypernym of {Chair} is {Furniture}. Documents containing ''arm chair'' may share the same topic with other documents containing ''folding chair'' or ''Chair''. With simple term matching, we may lose the relatedness in these terms. If we use ''→'' to represent this relationship, then these three synsets could be related as {arm chair, folding chair}''→''{Chair} ''→''{Furniture}. In this case, {chair} is a direct hypernym of {arm chair, folding chair} and {furniture} is an inherited hypernym of {arm chair, folding chair}. Such binding of a synset with its hypernym in this paper is called as *Synset Hypernym Component* denoted by SHC. The SHC and SyS relation is denoted as $SyS_i \in SHC_j$. In this work we propose text clustering approach with SyS and SHC relation, as this is the main relation group of WordNet hierarchy[13].

Since more than one synonym exists for terms, identifying the right synonym that a term expresses in a certain context is a difficult task. Here, we use a Synonym Group (SG), which is a collection of SHC, to find the correct meaning of a term. The relationship between SHC and SG is denoted as $SHC_j \in SG_h$. For a SyS_i, if $SyS_i \in SHC_j$ and $SHC_j \in SG_h$, then it is denoted as $SyS_i \in SG_h$. For example there is a SG, $SG_1 = \{SHC_1, SHC_2\}$, where $SHC_1 = \{SyS_1 \rightarrow SyS_2\}$, $SHC_2 = \{SyS_3 \rightarrow SyS_4\}$,

SyS_1 = {arm chair}, SyS_2 = {chair}, SyS_3 = {folding chair, lounge }, and SyS_4 = {furniture}. We expect that one of the SyS belonging to SG_1 is the real term meaning of the term ''chair'' in this document. In our approach unrecognized terms in WordNet are taken as it is into the unique terms list as these terms may capture important information and is helpful in clustering.

In WordNet multiple synonyms of a given term are ordered from the most to the least frequently used. For each SyS selected, atleast two SyS with one direct hypernym synset is retrieved. In this way, each term has its SG containing at least one SHC. For example, a document d_1 = <t_1,t_2,t_3 > can be converted to a sequence of SG as d_1' = <SG_1,SG_2,SG_3>, where SG_1 = {$SyS_1 \rightarrow SyS_2$, $SyS_3 \rightarrow SyS_4$}, SG_2 = {$SyS_5 \rightarrow SyS_8$, $SyS_6 \rightarrow SyS_7$}, SG_3 = {$SyS_9 \rightarrow SyS_5$}.

After the above step, many SG have more than one SHC since WordNet assigns more than one term meaning to a term, thus adds noise to the database, and affect the clustering accuracy. To estimate the real meaning of the text for each term, the SHC containing the SyS with high frequencies are considered. If a SyS appears in more than one SG we merge the SHC into a new SG and discard the SHC from existing SGs. We Prune the SG that have the document frequency of the SyS less than user defined threshold. For example consider the SGs SG_1, SG_2 with SG1 as {mining->defense, mining->production} and SG2 as {growth->organic process, growing->production} after merge and replace the synsets are SG1={mining->defense}, SG2={growth->organic process} and new synset SG3={mining, growing->production}. Since all the SG sharing the same synset are replaced by one SG which contains only one SHC, the number of unique SG in the document can be reduced.

4.2 SNBSTC Algorithm

In [16,6] various cluster split strategies proposed for Bisecting K-means, cluster to be split is based on the compactness of the cluster which is measured using shared neighbors. We split the least compact cluster ie., the cluster with least shared neighbors. Calculate strength of a cluster and split that cluster that has less strength. Let ‖cj‖ be the length of jth cluster and TSHj be the shared neighbors of cluster j that is we are considering only local neighbors.

$$Strength_j = TSH_j / \| c_j \|$$
$$\text{Min}(Strength_j) \text{ where } 1 \leq j \leq k$$

Build a document - SG matrix where each row represents a document and each column represents the tfidf value of the corresponding SG. This semantic matrix is used to find shared neighbors and in Clustering document. The proposed algorithm named as "*Shared Neighbors Based Semantic Text Clustering (SNBSTC)*".

SNBSTC Algorithm

```
{
  Let D be the dataset
  For each doc di ∈ D
```

```
{
  Identify Unique_Terms(di)
        For each UWk ∈ Unique_Terms(di)
              {
                 SG_Coll ← Prepare_SG(UWk)
                 Add SG_Coll to SG_List
              }
              Merge_SG_Coll ← Merge_SG(SG_List)
              Add Merge_SG_Coll to Merged_SG_List
}
  Generate_Unique_SG(Dt)
  Build Document - SG matrix
  Build Neighbor matrix
  Let SM be sim. measure with neighbor inf.
  Let K be the no. of clusters
    Dt  = D
    p= 2   // Bisecting K-means, initially splits D into 2
    While (p<=k) {
        split the cluster Dt into 2
              call SH_INI_CEN(2, Dt)
               // finds 2 initial centers from Dt
    Step (i)
        for each di ∈ Dt
     {
                      Cj←di  iff  SM(di, cj) is maximum
     where 1≤j≤2 , cj is the jth centroid and Cj is jth Cluster
     }
        recompute the centroids
   goto step(i) till convergence.

    // Find the cluster to be split with neighbor
              cid = call SHNSplit(p, clustersp)
   // cid is cluster that should be split
              p++    // p is the no. of clusters
                     // clustersp  is p clusters formed
}
Prepare_SG(uw)
{
  Step(ii)
    If (SyS(uw))
    If( HyS (uw) != null)
                    SHC = (SyS→HyS)
              else
                    SHC = (SyS→derived(HyS))
              SG = concat(SG, SHC)
                    Endif
  Repeat Step (ii)
        //till the required no.of SHC obtained
        Return SG
```

```
}
Merge_SG(SG_list)
{
 For each SyS ϵ d(i)
 {
  If (frequency of SyS > 1 in SG_list)
   If(SHC(X→Y)ϵSG_List[i] && SHC(V→X)ϵSG_List[j])
  {
  SG_List[new] = Combine (SG_List[i] , SG_List [j])
                  //ie {{V → X}→Y} into a new SG_List
        discard SHC from SG_Lists [i] and SG_List[j]
        add SG_List[new] to SG_List
  }
 Return SG_List
}
```

5 Experiment Results

This section first describes the characteristics of the datasets, then presents and analyzes the experiment results. We use "Reuters 21578, Classic from uci.kdd and DT a dataset of 200 documents of research papers containing four different categories, collected from web. Table 2 summarizes all the considered datasets.

Table 2. Datasets

Sno	Data Sets	No. of Docs.	No. of Classes	Avg. Class Size	Num. Of Unique Terms	Num. of Unique SG
1	Dt	197	4	148	5412	4046
2	Reu1	180	6	108	5532	3953
3	Reu2	300	1	99	6517	4001
4	Reu3	140	1	158	4617	3165
5	CL_1	800	4	203	17980	12514
6	CL_2	200	1	163	5966	4208
7	CL_3	400	2	221	10059	7201

Table 3. Classic_set Entropy

	Cosine	Jaccard	Pearson
CL_1	0.1492	0.1231	0.1908
CL_2	0.1976	0.1793	0.2086
CL_3	0.2173	0.2057	0.2271

We have applied Shared Neighbor based clustering to Semantic Neighbors proposed in [12] NSTC and Rank Based Clustering proposed in NSTC with the Semantic Framework called Semantic Groups proposed in this work. We denote *R_SN, S_SN* for the semantic approach proposed in [12], where *R* is Rank Based and *S* is Shared Neighbor and *R_SG, S_SG* for the Semantic Framework approach, Semantic Groups proposed in this work. Table 3 shows entropy on classic_sets with all three similarity measures and neighbors on proposed Semantic framework (SG) and noticed that Jaccard and Cosine form well clusters than Pearson function. Table 4 shows the time taken by all methods considered and concludes that proposed (SG) method takes less time to complete the clustering process.

Table 4. Time taken by R_SN, S_SN, R_SG and S_SG methods

	R_SN	S_SN	R_SG	S_SG
Dt	296.772	311.205	117.266	113.818
CL_1	10011.5	7538.927	897.204	819.723
CL_2	1138.795	1104.19	694.503	687.418
CL_3	282.488	306.047	149.476	145.693
Reu_1	418.269	344.154	177.825	176.763
Reu_2	276.079	240.958	193.259	135.192
Reu_3	410.959	408.162	342.739	321.922

Figure 1 Depicts the dimension reduction with proposed semantic frame work (SG). With (SN) approach the term matrix is reweighted with semantic knowledge, hence dimensions will not reduce.

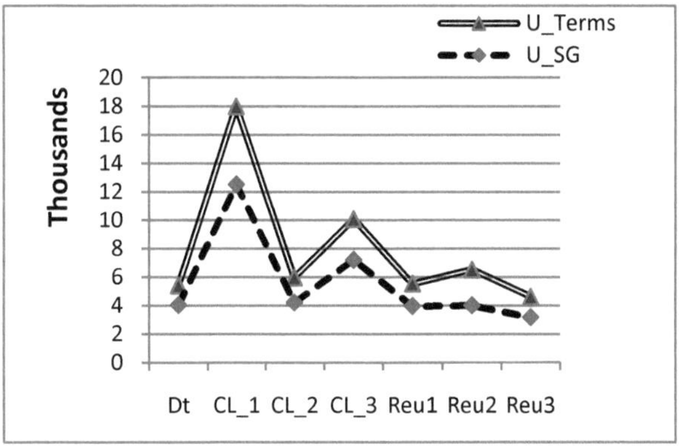

Fig. 1. Dimension Chart

Figure 2 shows the time chart on all datasets with Shared_SG and Rank_SN approaches and can concluded that SG approach takes less time than Semantic approach proposed in [12]

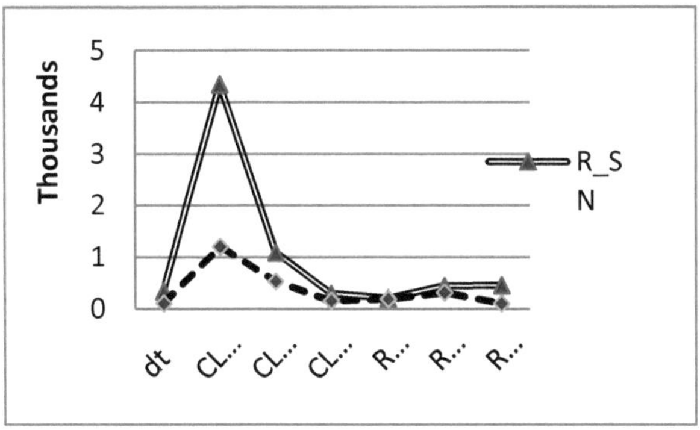

Fig. 2. R_SN vs S_SG Time Chart

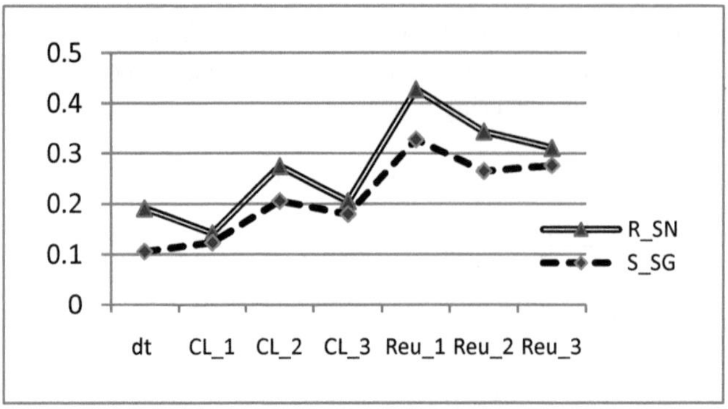

Fig. 3. Entropy of R_SN Vs S_SG

Figure 3 indicates entropy graph of R_SN and S_SG semantic approaches and noticed our approach (SG) shows better results.

6 Conclusions and Future Work

This work presents a semantic framework and a shared neighbors approach to text clustering, applied to Bisecting K-means clustering. The proposed method is compared with a method that reweights the VSM with semantic knowledge and observed that proposed is efficient in terms of performance and cluster quality. We can concluded that neighbor information with back ground knowledge will improve text clustering. In our future work we study various semantic representations with neighbors and notice their effect on clustering and classification also explore neighbor concept to present a method that produces more accurate.

References

1. Gowda, K.C., Krishna, G.: Agglomerative clustering using the concept of mutual nearest neighborhood. Pattern Recognition 10(2), 105–112 (1978)
2. Huang, A.: Similarity Measures for Text Document Clustering. Published in the Proceedings of New Zealand Computer Science Research Student Conference (2008)
3. Strehl, A., Ghosh, J., Mooney, R.: Impact of similarity measures on web-page clustering. In: AAAI 2000: Workshop on Artificial Intelligence for Web Search (July 2000)
4. Guha, S., Rastogi, R., Shim, K.: ROCK: a robust clustering algorithm for categorical attributes. Information Systems 25(5), 345–366 (2000)
5. Luo, C., Li, Y., Chung, S.M.: Text Document Clustering based on neighbors. Data and Knowledge Engineering 68, 1271–1288 (2009)
6. Steinbach, M., Karypis, G., Kumar, V.: A comparison of document clustering techniques. In: KDD Workshop on Text Mining (2000)
7. Fellbaum, C. (ed.): WordNet: An Electronic Lexical Database. MIT Press (1998)
8. Varelas, G., Voutsakis, E., Raftopoulou, P., Petrakis, E., Milios, E.: Semantic similarity methods in WordNet and their application to information retrieval on the web. In: Proceedings of the 7th Annual ACM International Workshop on Web Information and Data Management, pp. 10–16 (2005)
9. Jarvis, R.A., Patrick, E.A.: Clustering using a similarity measure based on shared near neighbors. IEEE Transactions on Computers C-22(11), 1025–1034 (1973)

10. Chen, C.-L., Tseng, F.S.C., Liang, T.: An integration of fuzzy association rules and wordNet for document clustering. In: Theeramunkong, T., Kijsirikul, B., Cercone, N., Ho, T.-B. (eds.) PAKDD 2009. LNCS, vol. 5476, pp. 147–159. Springer, Heidelberg (2009)
11. Bouras, C., Tsogkas, V.: W-kmeans clustering news articles using WordNet. In: Setchi, R., Jordanov, I., Howlett, R.J., Jain, L.C. (eds.) KES 2010, Part III. LNCS (LNAI), vol. 6278, pp. 379–388. Springer, Heidelberg (2010)
12. Danish, M., Shirgahi, H.: Text document clustering using semantic neighbors. Journal of Software Engineering (2011)
13. Li, Y., Chung, S.M., Holt, J.D.: Text document clustering based on frequent word meaning sequences. Journal of Data and Knowledge Engineering 64, 381–404 (2008)
14. Hotho, A., Staab, S., Stumme, G.: Ontologies improve text document clustering. In: Proceedings of the 3rd IEEE International Conference on Data Mining, pp. 541–544 (2003)
15. Sedding, J., Kazakov, D.: WordNet-based text document clustering. In: Proc. of COLING-Workshop on Robust Methods in Analysis of Natural Language Data (2004)
16. Jain, A.K., Dubes, R.C.: Algorithms for Clustering Data. Prentice Hall, Englewood Cliffs (1988)

Multi-Agent System for Spatio Temporal Data Mining

I.L. Narasimha Rao[1], A. Govardhan[2], and K. Venkateswara Rao[3]

[1] Department of CSE, Aurora's Technological and Research Institute, Uppal, Hyderabad-98
[2] School of IT, JNTU, Hyderabad
[3] Department of CSE, CVR College of Engineering, Ibrahimpatnam, RR Dt., A.P, India
ilnrao@yahoo.com, govardhan_cse@jntuh.ac.in, kvenkat.cse@gmail.com

Abstract. Spatiotemporal data comprises of states or position for an object, an event in space over time. Huge amount of the data is available in various application areas such as environment monitoring, traffic management, and weather forecast. This data might be collected and stored at various locations at different points of time. Many challenges are posed in analytical processing and mining of such data due to complexity of the spatiotemporal objects and their relationships with each other in both temporal and spatial dimensions. More scalable and practical approach in this context is distributed analysis and mining of the spatiotemporal data.

Multi-Agent System deals with applications which need distributed problem solving. The behavior of the agents is based on the data observed from various distributed sources. Since the agents in multi-agent system are generally distributed and have reactive and proactive characteristic, It is appealing to combine distributed spatiotemporal data mining with multi-agent system.

The core issues and problems in multi-agent distributed spatiotemporal data analysis and mining do not concern specific data mining techniques. Its core issues and problems are related to achieving collaboration in the multi-agent system indented for distributed spatiotemporal data analysis and mining. This paper is intended to describe architecture of multi-agent system for spatiotemporal data mining and identify issues involved in realization of such system and also to review technologies available for developing such system.

Keywords: Spatiotemporal, Multi Agent, Data mining.

1 Introduction

Curiosity to examine and understand the nature of data has been increasing significantly in all sectors of business because benefits that are brought by having useful information in hand for decision making are recognized. In addition to this, enormous historical data owned and managed by various organizations with the help of good quality software played significant role in raising this curiosity too. Many kinds of spatiotemporal applications require spatial and temporal data for modeling various entities. Availability of location based services and mobile computing enabled

S.C. Satapathy et al. (eds.), *ICT and Critical Infrastructure: Proceedings of the 48th Annual Convention of CSI - Volume II*, Advances in Intelligent Systems and Computing 249,
DOI: 10.1007/978-3-319-03095-1_30, © Springer International Publishing Switzerland 2014

the collection of spatial and temporal data. This data managed in business information systems is a key and dependable resource for decision makers. But analysis of the data in spatiotemporal context and its use within the decision making processes need research. Moreover, this data might be collected and stored at several locations at different points of time in various formats. For example, the NASA Earth Observatory System (EOS) stores, manages and distributes many datasets at EOS Data and Information System (EOSDIS) sites. A pair of Landset 7 and Terra spacecraft alone generates approximately 350 GB of EOSDIS data per day [1]. This huge volume of spatiotemporal data available may often hide potentially useful and interesting patterns and trends. Manual analysis and examination of this voluminous data is quite difficult and often impossible. The tools, concepts and techniques provided by spatiotemporal data mining are quite useful in this context. Spatiotemporal data mining is an upcoming research field devoted to the design and development of computational techniques and their application to spatiotemporal analysis.. It involves integration of techniques from various disciplines such as machine learning, database and data warehouse technology, statistics, neural networks, data visualization, pattern recognition, spatial and temporal analysis.

A multi-agent system [2] contains many intelligent agents that interact with each other. The agents are autonomous entities with cooperative interaction to accomplish a common goal. If multiple agents exist in an environment on several machines and the tasks required to be completed cannot be executed by a single agent, then multi-agent system is required for collaboration, cooperation, and control and communication among those agents to complete the tasks. Multi-agent architectures are for multiple agents which have a common goal.

A communication language such as Knowledge Query Manipulation Language (KQML) of agents defines the semantics for communication between agents. It also defines protocols to constrain messages that agents can send to one another.

2 Related Work

BODHI [1,3,5], PADMA [3,4,5], JAM [3,5], PAPYRUS [1,3,5], KDEC [4], JBAT are the agent based distributed data mining systems which are more prominent and representative. BODHI and JAM are agent based meta-learning systems designed for data classification. Both these systems are developed using Java. PADMA (Parallel Data Mining Agents) system demonstrated that agent based data mining tools are suitable for exploiting benefits of parallel computing. The objective of PADMA and PAPYRUS is to integrate knowledge discovered from various sites, minimize network communication and maximize local computation. PADMA is used to find hierarchical clusters in document categorization. PAPYRUS is a clustering multi-agent system in which both data and results can be exchanged among the agents as per the given MAS strategies. H. Baazaoui Zghal et.al, [6] proposed a framework for data mining based multi-agent and applied it to spatial data.

Mattias Klusch et al proposed KDEC [5] clustering scheme based on MAS approach. It uses a distributed density-based clustering algorithm. JBAT is the most recent clustering system based on MAS approach. It can be used for both non-distributed and distributed data clustering. It uses K-means clustering algorithm.

K. Venkateswara Rao et al [7] surveyed various kinds of spatiotemporal data mining tasks for discovering different types of knowledge from spatiotemporal datasets, identified issues [8] in representation, processing, analysis and mining of spatiotemporal data and elaborated an architecture framework for spatiotemporal data mining system [8].

3 Multi-Agent System For Spatiotemporal Data Mining

3.1 Architecture

The Architecture of a Multi-agent system for spatiotemporal data mining is shown in Fig. 1. Each agent in multi-agent system generally contains interface module, process module and knowledge module. The interface module is responsible for communicating with other agent or with the environment. The knowledge module provides necessary knowledge to be proactive or reactive in various scenarios. The process module does necessary processing to make decisions. Different types of agents in multi-agent based spatiotemporal data mining system are Interface agent, Facilitator agent, Broker agent, Data agent, Pre-processing agent, Data mining agent, Result validation agent. These agents are briefly described below.

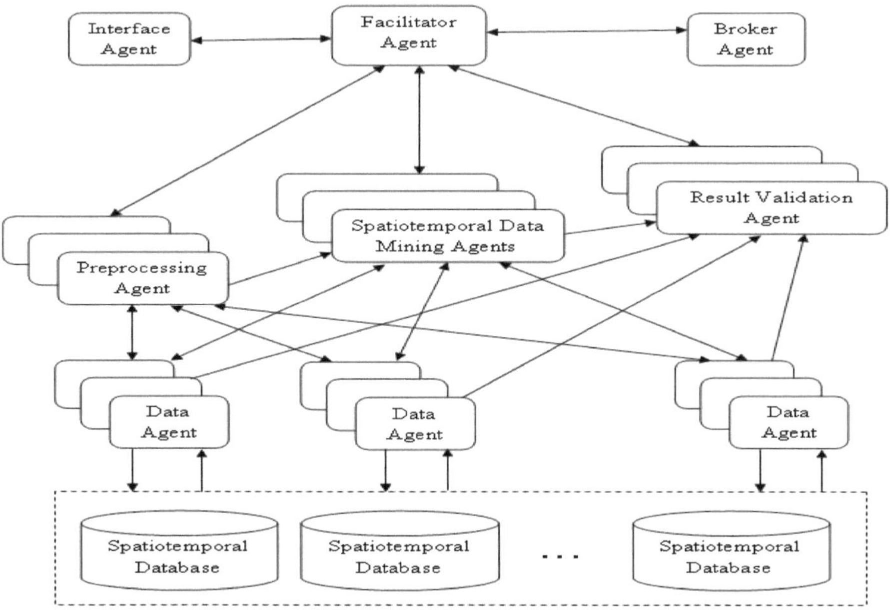

Fig. 1. Architecture of a Multi-agent System for Spatiotemporal Data Mining

Interface Agent: This agent communicates to user. It accepts user requirements and provides data mining results to him. The interface module is responsible for getting input from the user as well as inter-agent communication. Methods in the process module capture user input and communicate it to the facilitator agent. The knowledge module manages history of user's interaction and their profiles.

Facilitator Agent: This agent takes the responsibility of activation and synchronization of various agents. It receives a request from the interface agent and produce a work plan. It elaborates various tasks to be completed in the work plan and ensures the work plan is completed. It communicates the results to the interface agent.

Broker Agent: This agent maintains names, ontology and capabilities of all the agents which are registered with it to become a part of the multi-agent based spatiotemporal data mining system. The broker agent receives a request from the facilitator agent and responds with names of appropriate agents which have the capabilities requested.

Data Agent: This agent manages meta-data about each spatiotemporal data source. It is responsible for resolving data definition and data representation conflicts. It retrieves required data needed by pre-processing agents, data mining agents or result validation agents. It facilitate data retrieval for pre-defined and ad-hoc queries that are generated from the user requests.

Pre-processing Agent: This agent does required data cleansing prior to the data usage by spatiotemporal data mining agents. It embodies data cleansing methods and data preparation techniques required for each of the data mining algorithms.

Data Mining Agent: Spatiotemporal data mining methods and techniques are implemented by data mining agents. Knowledge module of this agent manages Meta data such as "which method is suitable for a particular kind of problem" and "format of input and input requirements for each spatiotemporal data mining method". The knowledge is used by the processing module while initiating and carrying out the data mining activity. The process module also captures data mining results and communicates them to result validation agent or facilitator agent.

Result Validation Agent: This agent gets the results from the spatiotemporal data mining agents. It performs the validation operations on the data mining results. The agent is able to process the results to fulfill various presentation and visual representation software. It maintains details about visualization primitives and report templates that are used to present the results.

3.2 Spatiotemporal Data Mining Process in the Multi-Agent System

A spatiotemporal data mining task is submitted to the system through interface agent. The interface agent communicates the same to the facilitator agent. When facilitator agent receives the request from the interface agent, it then negotiates with the broker agent to find out which agents to be launched for the task. The spatiotemporal data mining tasks which are launched are responsible for completing the task while the facilitator agent continues to process other request from the interface agents. When the data mining agents complete the task, then the results are passed to results validation agent which in turn validates the results and communicates to the facilitator agent. The facilitator agent passes the results to the interface agent which presents them to the use.

3.3 Design Issues

Following are the design issues to be considered while developing agent-based systems.

- How data mining tasks are scheduled and how synchronization among those tasks is to be achieved.
- How priority of the data mining tasks is to be decided by the agents.
- How the internal structure of the agents to be made persistent.
- How to make the agents be collaborated.
- How a change in data causes behavioral changes to the agents.
- How communication and messaging can be done among agents. What protocols can be used.
- What are the useful hierarchies of the agents.

4 Distributed Spatiotemporal Data Mining System Architecture

An Architecture specified for distributed data mining system [1] is extended for distributed spatiotemporal data mining and described in Fig 2.

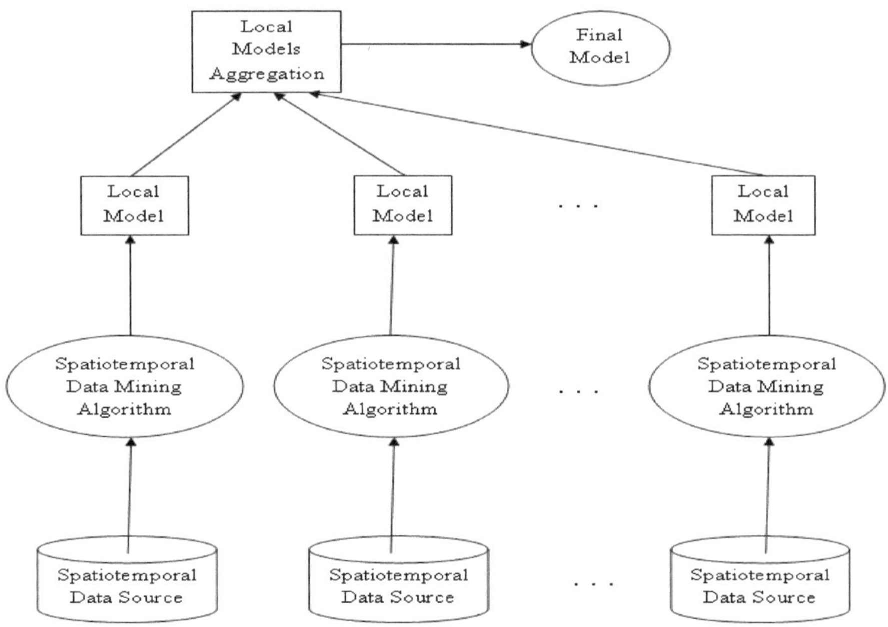

Fig. 2. Generic Architecture of Distributed Spatiotemporal Data Mining System

Distributed spatiotemporal data at different sites is retrieved, processed, analyzed and mined by spatiotemporal data mining algorithms to generate local models or patterns. These local models are aggregated by coordinator site to produce the final global model. Multi-agent systems can be used for distributed computing [9], communication and data integration services [10]. Vladimir Gorodetsky et.al, [11] used multi-agent technology for distributed data mining and classification. The architecture in Fig. 2 need to be refined to incorporate multiple agents to enhance the system performance. This requires integration of multi-agent system for spatiotemporal data mining specified in Fig. 1 with the architecture specified in Fig. 2.

5 Agent Technologies

Technologies that are available for developing multi-agent systems are reviewed briefly in this section. Agent technologies are the basic building blocks required to implement agent applications that interact with database applications and other agents in distributed environment. The .Net framework has built-in capabilities such as remoting, serialization and the enterprise services which support the construction of agent-based systems. Java-based agent technology involves necessary tools and support for realizing large-scale multi-agent systems using the Java platform. There are many tools and frameworks such as JADE (Java Agent Development Framework) [12] , Aglets, Bee-gent (Bonding and Encapsulation Enhancement agent), Zeus, IBM Agent Building Environment (ABE) developed by different organizations and vendors for building agent-based systems using Java technology.

JADE is a open source available under Gnu license. This is completely coded in the Java language. It provides Application Programming Interface (API) to access its features to implement multi-agent systems. The built-in tools in JADE supports the debugging and deployment of agent-based systems. JADE is compliant with the FIPA specifications. FIPA (Foundation for Intelligent Physical Agents) is an international organization established in 1996. It was accepted as the IEEE computer society's eleventh standard committee to develop a collection of standards relating to agent technology. FIPA ACL [12] is an Agent Communication Language (ACL) defined by FIPA. AUML [13,14], Agent UML - an extension of Unified Modeling Language, can be used to model Agent-based Systems.

6 Conclusions

This paper described nature of spatiotemporal data and significance of multi-agent system for discovering patterns and trends from such data. Generic architecture of multi-agent system for spatiotemporal data mining is discussed and the process involved in mining spatiotemporal mining spatiotemporal knowledge using such system is also described. The issues involved in developing the multi-agent system are identified. An architecture for distributed spatiotemporal data mining and necessity of integrating the multi-agent system with that architecture is explained.

Future work involves performing following two steps for each of the spatiotemporal data mining tasks.

1. Refining the architecture specified for multi-agent system for spatiotemporal data mining, modeling the system using AUML and implement it for the given spatiotemporal data mining task using JADE.
2. Test the system using suitable spatiotemporal data sets from various domains.

Currently the authors are modeling and developing the specified system for spatiotemporal data classification. The system that is being developed needs to be tested with various spatiotemporal datasets from different areas of applications in future.

References

1. Park, B.-H., Kargupta, H.: Distributed Data Mining: Algorithms, Systems, and Applications. In: Data Mining Handbook, pp. 341–358 (2002)
2. Maalal, S., Addou, M.: A new approach of designing Multi-agent Systems. International Journal of Advanced Computer Science and Applications 2(1), 148–157 (2011)
3. Klusch, M., Lodi, S., Moro, G.: Issues of Agent-Based Distributed Data Mining. In: Proceeding of Second International Joint Conference on Autonomous Agents & Multiagent Systems, AAMAS, Melboune, pp. 1034–1035 (2003)
4. Kargupta, H., Hamzaoglu, I., Stafford, B.: Scalable, Distributed Data Mining – An Agent Architecture. In: Proceedings of KDD 1997, pp. 211–214 (1997)
5. Klush, M., Lodi, S., Moro, G.: The role of Agents in Distributed Data Mining: Issues and Benefits. In: IEEE/WIC International Conference on Intelligent Agent Technology, pp. 211–217 (October 2003)
6. Baazaoui Zghal, H., Faiz, S., Ben Ghezala, H.: A framework for Data Mining Based Multi-agent: An Application to Spatial Data. World Academy of Science, Engineering and Technology, 22–26 (2005)
7. Venkateswara Rao, K., Govardhan, A., Chalapati Rao, K.V.: Spatiotemporal Data Mining: Issues, Tasks and Applications. International Journal of Computer Science and Engineering Survey (IJCSES) 3(1), 39–52 (2012) ISSN: 0976-2760 (online), 0976-3252 (print)
8. Venkateswara Rao, K., Govardhan, A., Chalapati Rao, K.V.: An Architecture Framework for Spatiotemporal Data Mining System. International Journal of Software Engineering & Applications (IJSEA) 3(5), 125–146 (2012) ISSN: 0975 - 9018 (online), 0976-2221 (print)
9. Zhang, Z., et al.: Multiagent System for Distributed Computing, Communication and Data Integration Needs in the Power Industry. IEEE Power Engineering Society Meeting 1, 45–49 (2004)
10. Corchado, J.M., Tapia, D.I., Bajo, J.: A Multi-agent Architecture for Distributed Services and Applications. International Journal of Innovative Computing, Information and Control 8(4), 2453–2476 (2012)
11. Gorodetsky, V., Karsaeyv, O., Samoilov, V.: Multi-agent Technology for Distributed Data Mining and Classification. In: Proceedings of the IEEE/WIC International Conference on Intelligent Technology (2003)
12. Bellifemine, F., Poggi, A., Rimassa, G.: JADE: A FIPA-Compliant Agent Framework. In: Proceedings of the Fourth Conference on the Practical Application of Intelligent Agents and Multi-Agent Technology, pp. 97–108 (April 1999)

13. Dinsoreanu, M., Salomie, I., Pusztai, K.: On the Design of Agent-based Systems using UML and Extensions. In: Proceedings of the 24th International Conference on Information Technology Interfaces, pp. 205–210 (2002)
14. Bauer, B., Odell, J.: UML 2.0 and agents: how to build agent-based systems with the new UML standard. Engineering Applications of Artificial Intelligence 18(2), 141–157 (2005)

Cropping and Rotation Invariant Watermarking Scheme in the Spatial Domain

Tauheed Ahmed, Ratnakirti Roy, and Suvamoy Changder

Department of Computer Applications, National Institute of Technology, Durgapur, India
{tauheed.ahmd,rroy.nitdgp,suvamoy.nitdgp}@gmail.com

Abstract. Digital Watermarking is an information hiding technique in which an identifying piece of information is embedded into a cover work so as to identify the ownership of some important information or document. It is widely used for enforcing copyright infringement protection for various types of digital data including images. In images, various watermarking techniques have been proposed over time. The watermarking techniques are often susceptible to geometric attacks such as rotation and cropping. This paper proposes an image watermarking scheme which uses the geometric properties of an image to ensure invariance of the watermark to rotation and cropping. It also incorporates a checksum based mechanism for tracking any distortion effect in the cover work. Efforts have been given to ensure that the proposed algorithm conforms to high Robustness and Fidelity which are the primary quality requirements for any Digital Watermarking system.

Keywords: Digital Image Watermarking, Cropping Invariance, Rotation Invariance, Fidelity, Robustness, Checksum.

1 Introduction

Today, internet is the easiest and the most popular medium of communication. Its easy accessibility and availability has made it the most convenient way of data transmission. However, this flexibility of communication also adheres with it the challenge of preserving authenticity of the transmitted data over the internet [1, 2]. Intellectual property rights infringement is one of the major risks that digital data over the internet incurs. Watermarking is one of the best ways to defend against such risks.

In the context of digital watermarking [3] the object which needs to be preserved is called the *work* and the hidden message is the *watermark*. Digital Watermarking [4] Algorithm is composed of three parts: watermark embedding algorithm, the watermark detection algorithm and the watermark extraction algorithm [5].Text files, audio, videos, images, etc. are some of the well-known objects which require copyright protection such that its illegal use and distribution can be prohibited [6]. We can incorporate such security measures by embedding watermark in these objects.

Digital Watermark techniques can be categorized into spatial and transform domain [7]. Spatial domain watermarking techniques are simple and have low computing complexity, since no frequency transform is done. Frequency domain

S.C. Satapathy et al. (eds.), *ICT and Critical Infrastructure: Proceedings of the 48th Annual Convention of CSI - Volume II*, Advances in Intelligent Systems and Computing 249,
DOI: 10.1007/978-3-319-03095-1_31, © Springer International Publishing Switzerland 2014

watermarking embeds the watermark into the transformed image and has comparatively high embedding complexity [8].

Watermark preservation has always been an important focus for the owner of the copyrighted object. The aim of digital watermarking is to keep the watermark objects in such a way that very existence of the hidden data is *imperceptible* to an adversary [9]. In addition to imperceptibility, there are some desirable characteristics that a watermark should possess, related to robustness. *First*, the watermark should be resilient to standard manipulations (cropping, rotation) [10] of unintentional as well as intentional nature. *Second*, it should be statistically irremovable, that is, a statistical analysis should not produce any advantage from the attacking point of view [9]. These issues can be handled by hiding the watermark in the *most emphasized portion* of an image and making the watermarking system invariant to image manipulations.

This paper aims to present a watermarking technique in the spatial domain that utilizes the geometric properties of an image to render it invariant to rotation thereby resisting geometric attacks. It also incorporates a checksum [11] based mechanism to track distortions on the cover work which ensures that distortive attacks on the image will not pass undetected.

2 Related Researches

Robustness and fidelity are the key issues in the development of image watermarking algorithm. In view of the importance of digital images copyright protection [12], the digital watermarking technology is appropriate for preventing digital intellectual property rights infringement. Some of the most relevant research works in the digital image watermarking domain are presented next.

Most of the watermark based algorithms are invariant against certain types of geometric distortions. This scenario can be held in the rotation invariant security image watermarking algorithm [13] based on steerable pyramid transform. In this the rotation invariance and robust watermarking are achieved concurrently on the same transform domain. The watermarks are embedded into an oriented sub band at angle θ, which can be interpolated with base filter kernels and used as a key in watermark detection to increase the security of watermark; the watermark detector is designed based on the steerable vector HMM model.

2D-FRFT [14] is a rotation invariant technique which combines 2D signal with the addition and rotation invariant properties of Two-Dimensional Fractional Fourier Transform in order to improve robustness and security of digital image watermark. However, the Two-Dimensional Fractional Fourier Transform computation is complicated and thus increases the overall computational complexity.

All in one rotation scale and translation invariant spread spectrum digital image watermarking [15] is a Fourier-Mellin transform-based algorithm, used for digital image watermarking. To implement this algorithm it requires performing phases of transformation ranging from Fourier transform, Log-Polar transform and Mellin Transform at the end which increases its computational complexity.

3 Proposed Method

The proposed method focuses on the extraction of the watermark invariant of geometric attacks such as rotation and cropping with a provision to detect cropping and other distortions like pixel manipulations, selective color modification etc. Such a feature is important because even if the image is cropped leaving the watermarked position intact, still the original piece of work suffers distortion. Such distortions reduce the originality of the work in spite of the watermark being intact. To ensure the integrity of the image a checksum number is generated using different pixels of the original work serving as feature points. The checksum is then watermarked into the most prominent feature of the cover work called the *subject* of the image. Embedding in such an area renders the system invariant to cropping as an adversary would preferably not distort the main subject of the image and the watermark can be extracted as long the subject remains intact. The proposed technique can detect almost most of the types of distortions that might be subjected to a cover work.

However, in case the cover work is rotated, the checksum will detect the distortion but will not be able to detect the degree of rotation. As a solution to this, we propose a second technique to detect and reverse rotation geometric attack using the geometric properties of an image. Detection is done using the method based on the angle of major axis with the x-axis. The orientation of major axis is extracted to detect if there is any rotation as the major axis of image can either be along x-axis or along y-axis. Fidelity also being a primary requirement of any watermarking system, efforts has been given to embed the watermark with an efficient embedding scheme known as *Matrix Encoding* [16].The overall proposed algorithm is pictorially shown in Fig.1.

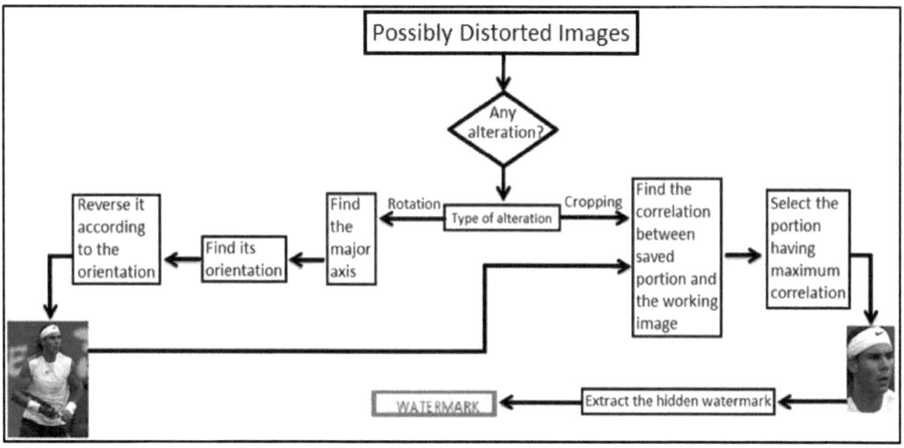

Fig. 1. Block Diagram of the proposed mechanism

3.1 Checksum Generation

Checksum is a simple error-detection scheme in which each transmitted message is accompanied by a value based on the number of set bits in the message. The receiver then applies the same formula to check if the received message is garbled or not. Thus using this checksum mechanism cropping detection can be done easily.

Let *C* be the cover work and *I* be the image block which contains the subject of the image ($I \subseteq C$). Let R be the remaining portion of the cover work such that,

$$R = C - I \quad (1)$$

Let R_i denote any plane of the image segment *R*. Then $R_i \in \{R_r, R_g, R_b\}$ where R_r, R_g, R_b denote the red, green and blue planes respectively. For checksum generation at first we select random pixel values from the three planes of *R* covering its entire portion. These pixel values are then stored separately. The overall checksum generation technique is mathematically describes as follows:

Let *a* be an array of numbers and a_i, a_{i+1} be two consecutive elements of this array. Now the checksum is generated in two steps.

i) Perform bitwise XOR[17] operation on a_i, $a_{i+1}$to give z_i

$$z_i = a_i \oplus a_{i+1} \quad (2)$$

ii) Evaluate the checksum *Z*

$$Z = z_i + z_{i+1} \quad (3)$$

$$checksum = Z \quad (4)$$

where, *checksum* is the error detecting value.

Therefore, whenever any cropping is done in the cover work it is detectable using the checksum value available with us. The mechanism is pictorially shown in Fig 2.

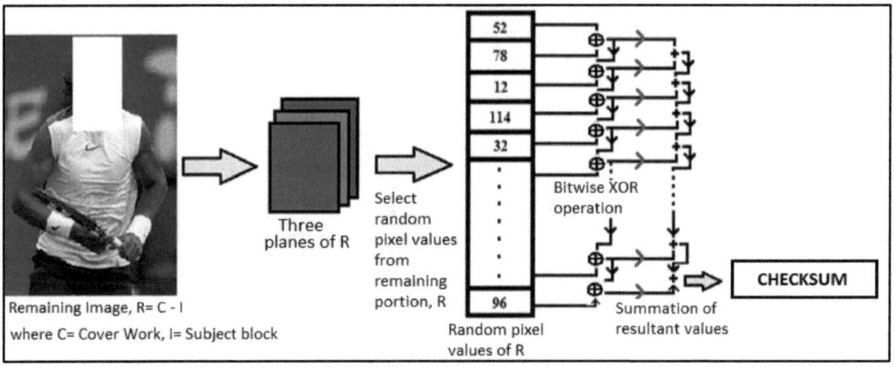

Fig. 2. Checksum generation to check distortion

3.2 Embedding Algorithm

The *subject* of the image is chosen for embedding the watermark. It can be face of person, building, or any other inanimate object on which the entire image remains focussed. Let the image block denoting the subject is *I*. Once the subject of the image is chosen the embedding process begins. The embedding algorithm is as follows:

Input: An image *C* that needs to be watermarked, Watermark image *W*.
Output: Watermarked image *W*.
Algorithm:
Step 1: Extract each plane of *I*. Let I_r, I_g, I_b be the red, blue and green plane of *I*.
Step 2: Find *R* using (1). From each plane of *R* generate random pixel values.
Step 3: Using these pixel values generate *checksum*. Append the size dimensions of the original cover with it. Next, extract each plane of watermark image, *W*.
Step 4: Generate the random locations to hide the watermark and the *checksum*.
Step 5: Calculate the number of bits to be embedded, say *length* and use a counter to track how many bits are already embedded, say *counter*. Set *counter*: =1.
Step 6: For each image plane I_r, I_g, I_b perform the following:
While (*length-counter>2)*
1. Take three values from target plane and get their LSBs as a_1, a_2, a_3 and two bits from the respective watermark plane w_1, w_2.
2. Determine the bit to be changed using matrix encoding mechanism.
3. Change the bit as indicated by the result of previous step.
4. Set *counter*: =*counter* + 2.

End While
End For
Step 7: Save the modified image block *I* as *M* (say*)*. It will be used for watermark extraction in the extraction phase. Next save the final watermarked image.

3.3 Extraction Algorithm

Once the watermarked image is received in the receiver site the extraction phase begins. Extraction algorithm works in three stages. In the first stage check for rotation and cropping is done. While in the second and third stage detection of watermarked position and the actual extraction is being performed respectively. The three stages are described as follows:

Input: Watermarked Image *W*, Watermark Length *S*, Pseudo Random number generator key.
Output: Embedded Watermark.

3.3.1 Check for Distortion

Any alteration is detected using geometric properties of the image which lowers its computational complexity. Detection is done using the method based on the angle of major axis with the x-axis as it can either be along x-axis or y-axis. To detect rotation, firstly the four corner points of the image are found. The angle of rotation is the angle of major axis with the *x*-axis. Detection of the four corner co-ordinates of image helps us to decide whether the image's major axis is along abscissa or ordinate. If there is no rotation the major axis makes an angle of 0^0or 90^0 with abscissa. If the image's major axis makes an angle other than 0^0or 90^0 then it signifies that image has been subjected to rotation. Checking for rotation can be done in the following manner:

Step 1: Detect all four corner co-ordinates of actual image *W*.
Step 2: Measure the length and width of the image using the corner co-ordinates of the image and decide the major axis as shown below.

if (length>width)
Major axis =length;
else
Major axis=width;

Step 3: Calculate the slope *m*, angle of rotation, *theta* of the major axis using following formulae:

$$m = \frac{Y_2 - Y_1}{X_2 - X_1}$$

$$theta = tan^{-1} m$$

where (X_1, Y_1) *and* $(X_2, Y_2)$are the co-ordinate locations of the major axis of image.

Step 4: Rotate the image in the reverse direction by the detected angle.

3.3.2 Cropping Detection

Watermarking techniques are also prone to attacks such as cropping. Cropping attacks not only removes visible watermarks but also distorts the image. Such attacks put the survival of the watermark at a risk. In order to detect any sort of cropping done in the cover work, we use a checksum based mechanism. Let *W'* be the watermarked, but cropped image.

Step 1: Extract the size of the cover work from the watermarked image *W'*.
Step 2: Following the procedure described in 3.1, calculate the checksum for *W'*.
Step 3: Next this *checksum* is compared with the original checksum value.
Step 4: The comparison result determines if there is any cropping or not.

3.3.3 Locating the Watermark

Extraction method requires detection of the actual watermarked portion of the image. For this, the altered image *T* is first divided into blocks of size of *M (*defined in 3.1*)* and the modified portion, *M* saved during embedding is evaluated. Steps of detecting watermark position are described below:

Step 1: Get the altered image *T* to be tested from the previous stage.
Step 2: Get the modified portion *M*.
Step 3: Find the size of both the images. Let X_T, Y_T be the length and width of *T* and X_M, Y_M be the length and width of *I*.
Step 4: For $i=1$ to $X_T - X_M$
For $j=1$ to $Y_T - Y_M$
Take the image portion of size X_M, Y_M and modified portion *M*. The watermarked portion of image can be found by correlating these two images.
End For
End For
Step5: The portion which gives a correlation ≈ 1 or *1*has the chances of hidden watermark. Let the portion be *H*.

3.3.4 Watermark Extraction

The extraction algorithm is the reverse of the embedding process. The extraction algorithm is applied on *H*, detected in above stage. Extraction steps are as follows:

Step 1: Calculate the number of bits needs to be extracted using message length, say *l*.

Step 2: Keep track of number of bits already extracted using a counter, say *counter.*
Set *counter:* =1;
Step3: For each $H_i \in \{H_r, H_g, H_b\}$ perform the following:
While (*l-counter>2)*
1. Take three values from target plane and get their LSBs as a_1, a_2, a_3.
2. Extract the hidden bits using the equations given below:
$$x_1 = a_1 \oplus a_3$$
$$x_2 = a_2 \oplus a_3$$
3. Store the extracted bit in an array *A*.
4. Set *counter:* = *counter* + 2.

End while
End For
Step 4: Combine the extracted bit of each plane to get the actual watermark.

4 Experimental Results and Analysis

To illustrate the working of the algorithm, the standard test image of *Nadal* (613X833) has been taken as the cover work and a test watermark image (120X22) is embedded in it. The algorithm has been tested for *Robustness* (Invariance to rotation and cropping) and *Fidelity* of the watermarked image.

4.1 Robustness

Experiments are done against rotated and cropped images. To implement this, the image is rotated at different angles and cropped at different portions. Extraction of the watermark is successful until the embedding portion remains unaltered.

The results in Table 1 shows that the proposed method is invariant to cropping as long the subject of the image remains intact. The checksum values signify that the distortion in the image can be readily detected using the proposed technique. Similarly to detect whether the image has been subjected to rotation or not, the major axis of the image is ascertained and its orientation is used to determine the degree of rotation.

Table 2 shows that the proposed method can detect the watermark successfully when the cover work has been subjected to rotation attacks. The results also signify that the time required to detect rotation and extract the watermark from the rotated image is also quite low. The experimental results are given in Table 1&2.

Table 1. Results of extracting watermark from cropped images

Original Watermarked Image	Cropped Image	Cropped Portion	Extraction of watermark	Cover Check-sum	Altered Image Checksum	Extraction Time (sec)
13x833	409x833	Left	Yes	27113	20732	6.1666
613x833	420x833	Right	Yes	27113	21334	5.5383
613x833	371x833	Left and Right	Yes	27113	23221	6.9464
613x833	339x545	All directions	Yes	27113	25705	3.7131

Table 2. Results of extracting watermark from rotated images

Original Watermarked image	Slope of major axis	Rotated angle (degrees)	Watermark Extracted	Extraction Time(sec)
613x833	1.7266	59.92	Yes	0.875903
613x833	1	45	Yes	0.922463
613x833	0.5792	30.07	Yes	0.884155
613x833	1.4194	-30.01	Yes	0.917858

4.2 Fidelity

The cover image distortion produced due to embedding can be measured in terms of *Peak Signal to Noise Ratio (PSNR)* and *Mean Squared Error (MSE)*. These two error measuring metrics can be defined as follows:

$$MSE = \frac{1}{MXN}\sum_{i=1}^{M}\sum_{j=1}^{N}(x_{ij} - y_{ij})^2 \tag{5}$$

$$PSNR = 10\, log_{10}\left(\frac{255^2}{MSE}\right) dB \tag{6}$$

Where x_{ij} and y_{ij} in the *MSE* equation are the pixel values in the cover and stego image respectively and *M, N* are the dimensions of the cover image along the horizontal and vertical axes.

Table 3. Fidelity and Embedding Time for Various Sizes of the Secret Information

Size	MSE	PSNR (dB)	Time (sec.)
60x60	0.0317	63.1147	0.241428
64x64	0.0360	62.5712	0.266114
70x70	0.0429	61.8020	0.319990
80x80	0.0561	60.6380	0.442880
85x85	0.0630	60.1389	0.469763
120x22	0.0234	64.4462	0.175098
Average	**0.04218**	**62.1185**	**0.319212**

The above mentioned mechanism for the embedding of watermark first takes the image subject and then uses that portion to hide the watermark. The actual process of embedding has been pictorially represented in the Figure 3.

Fig. 3. Identifying subject and embed watermark

The algorithm for extraction can extract the watermark even though the image has suffered from geometric distortions such as rotation or cropping. Our experiment tests both the cases for extracting the watermark. Extraction of the watermark from a cropped work and a rotated work is pictorially shown in Figure 4 and 5 respectively.

The algorithm for extraction can extract the watermark even though the image has suffered from geometric distortions such as rotation or cropping. Our experiment tests both the cases for extracting the watermark. Extraction of the watermark from a cropped work and a rotated work is pictorially shown in Figure 4 and 5 respectively.

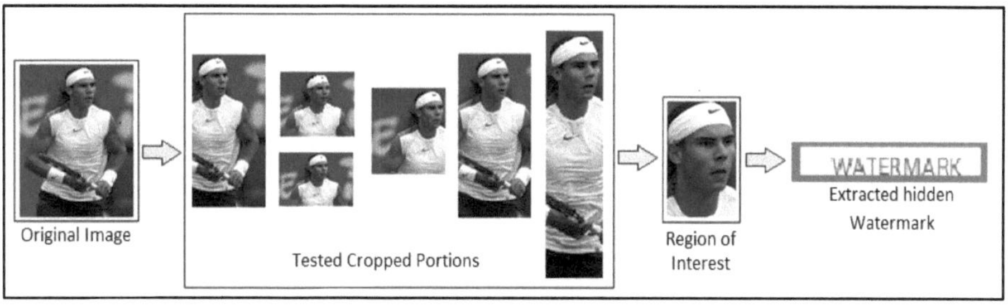

Fig. 4. Watermark extraction from cropped image

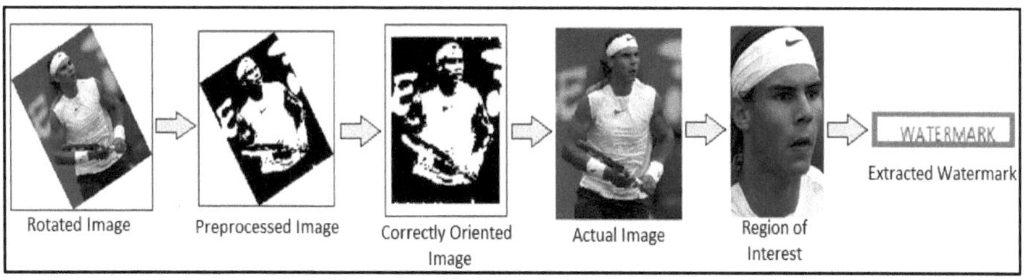

Fig. 5. Extracting watermark from the rotated image

5 Comparison between Algorithms

To evaluate the performance of the proposed algorithm we need to compare it with some of the already known related algorithms. Evaluation is done on the ability to resist geometrical attacks, rotation and cropping and the domain in which the algorithm is implemented.

5.1 Comparison Based on Resisting Geometrical Attacks and Time Complexity

The primary aim of watermarking is that in any condition the watermarked portion should not be compromised, but sometimes even though the watermark is present we are unable to locate it because of some geometrical transformation performed over the image. There are several algorithm which claims to resist such geometrical transformations. The comparison between these methods and the proposed one is given in Table 4.

Similarly, comparison based on Time complexity of an algorithm quantifies the amount of time taken by an algorithm to run as a function of the length of the string representing the input. Here the complexity of the algorithm is compared based on the operation each algorithm is using. Time complexity of spatial domain algorithm is found to be lower than that of transform domain algorithm mainly because no frequency transformation is done in the spatial domain.

Table 4. Comparison between watermarking schemes based on resisting geometrical attacks

Algorithm	Domain	Technique	Invariance	
			Rotation	Cropping
RIA[6]	Transform	Steerable Pyramid[18]	Yes	No
2D-FRFT[7]	Transform	Fractional Fourier	Yes	No
RST IA[8]	Transform	Fourier-Mellin[19]	Yes	No
Proposed Algorithm	Spatial	Manipulation of Geometric Properties	Yes	Yes

Table 5. Comparison between different watermarking schemes based on time complexity

Algorithm	Mechanism Used	Average Complexity
RIA[6]	Steerable pyramid[18]	$O(MN)$
2D-FRFT[7]	Fourier Transform	$O(N \log N)$
RST Invariant Algorithm[8]	Fourier-Mellin[19]	$O(N^2)$
Proposed Algorithm	Image Geometric Properties	$O(N)$

6 Conclusion

In this paper a watermarking technique is proposed which can extract the watermark even though the image has suffered from geometrical distortions like cropping or rotation. Tests have been done with the modified versions of the image by rotating at different angles and cropping to various sizes.

The major advantage of this technique is that it can extract a watermark from both cropped and rotated image with a lower time complexity than many other methods. Further a novel approach of checksum has been identified to detect distortion.

Future work will focus on automatic detection of the subject of the image to hide watermark. It will also emphasize on optimizing the capacity and complexity of the algorithm.

References

1. Dekker, M.: Security of the Internet. The Froehlich/Kent Encyclopedia of Telecommunications 15, 231–255 (1997), Source: `http://www.cert.org/encyc_article/` (last accessed July 12, 2013)
2. Artz, D.: Digital Steganography: Hiding Data within Data. IEEE Internet Computing Journal (June 2001)
3. Digital Watermarking, Source: `http://www.alpvision.com/watermarking.html/` (last accessed July 12, 2013)
4. Petitcolas, F.A.P.: Digital Watermarking. In: Becker, E., et al. (eds.) Digital Rights Management. LNCS, vol. 2770, pp. 81–92. Springer, Heidelberg (2003)
5. Zhang, Y.: Digital Watermarking Technology: A Review. In: ETP International Conference on Future Computer and Communication. IEEE (2009)
6. Stokes, S.: Digital copyright: law and practice, Source: `http://www2.law.ed.ac.uk/ahrc/script-ed/vol6-3/gs_review.asp` (last accessed July 18, 2013)
7. Silman, J.: Steganography and Steganalysis: An Overview. SANS Institute (2001)
8. Serdean, C.V., Tomlinson, M., Wade, J., Ambroze, A.M.: Protecting Intellectual Rights: Digital Watermarking in the wavelet domain. In: IEEE Int. Workshop Trends and Recent Achievements in IT, pp. 16–18 (2002)
9. Cox, I., Miller, M., Bloom, J., Miller, M.: Watermarking applications and their properties. Information Technology: Coding and Computing, 6–10 (2000)
10. Attacks on Digital Watermarks: Classification. Estimation-based Attacks and Benchmarks, Source: `http://citeseerx.ist.psu.edu/viewdoc/download?doi=10.1.1.84.5061&rep=rep1&type=pdf` (last accessed July 23, 2013)
11. Checksum, Source: `http://www.accuhash.com/what-is-checksum.html` (last accessed July 17, 2013)
12. Xuehua, J.: Digital Watermarking and its Application in Image Copyright Protection. Intelligent Computation Technology and Automation (ICICTA), 114–117 (2010)
13. Ni, J., Zhang, R., Huang, J., Wang, C., Li, Q.: Rotation-Invariant Secure Image Watermarking Algorithm Incorporating Steerable Pyramid Transform. In: 5th International Workshop IWDW, Jeju island Korea, pp. 246–260 (2006)
14. Gao, L., Qi, L., Yang, S., Wang, Y., Yun, T., Guan, L.: 2D-FRFT based rotation invariant digital image watermarking. In: IEEE International Symposium on Multimedia, pp. 286–289 (2012)
15. Joseph, J.K., Ruanaidh, O., Thierry, P.: Rotation, Scale and Translation Invariant Digital Image Watermarking. In: Proc. International Conference on Image Processing, pp. 536–539 (1997)

16. Crandall, R.: Some Notes on Steganography. Posted on Steganography Mailing List (1998), Source:
`http://www.dia.unisa.it/~ads/corsosecurity/www/CORSO-0203/steganografia/LINKS%20LOCALI/matrix-encoding.pdf`
(last accessed July 17, 2013)
17. The Bitwise Operators Source:
`http://www.cs.cf.ac.uk/Dave/PERL/node36.html`
(last accessed July 10, 2013)
18. Simoncelli, E.P., Freeman, W.T.: The steerable pyramid: a flexible rchitecture for multi-scale derivative computation. In: 2nd Annual IEEE Intl. Conference on Image Processing, Washington, DC (October 1995)
19. Moller, R., Salguero, H., Salguero, E.: Image recognition using the Fourier-Mellin Transform. LIPSE-SEPI-ESIME-IPN, Mexico

Satellite Image Fusion Using Window Based PCA

Amit Kumar Sen[1], Subhadip Mukherjee[2], and Amlan Chakrabarti[3]

[1] Dept. of Information Technology, IMPS College of Engg. and Tech., India
amitsen41@gmail.com
[2] CMC Ltd., Kolkata, India
subhadip.mukherjee@cmcltd.com
[3] A.K. Choudhury School of Information Technology, University of Calcutta, India
acakcs@caluniv.ac.in

Abstract. Building suitable image fusion techniques for remote sensing application is an emerging field of research. Though there exist quite a few algorithms in this domain, but still there is a scope of improvement in terms of quality of the fused image and reduction in the complexity of the fusion algorithms. In this paper, we have proposed a new adaptive fusion methodology, which is a modified form of the principle component analysis (PCA) technique based on a window technique. Our proposed method gives higher fusion quality compared to some of the existing standard methods, in terms of image quality and promises to be less complex. For our experiment, we have used the high spatial resolution panchromatic (PAN) image and the multispectral (MS) image, as available from remote sensing satellites such as SPOT5.

Keywords: Image Fusion, principal component analysis, remote sensing, panchromatic image, multispectral image.

1 Introduction

Satellite image consists of photograph of earth and other planets captured by artificial satellite. Satellite images have several applications in meteorology, agriculture, forestry, land space, biodiversity, conservation, regional planning, education, intelligence and warfare. Generally remote sensing satellites like IKONOS-2, Quick Bird, LANDSAT ETM, SPOT-5 [1] are normally provided with sensors with one high spatial resolution panchromatic (PAN) and several multispectral (MS) bands. There are several reasons for not capturing the images merely in high resolution: the most important of them being the incoming radiation energy and the data volume collected by the sensor (Zhang, 2004).

Image fusion is the process of combining relevant information from two or more images into a single image. The resulting image happens to be more informative than each of the individual images. In the domain satellite image based remote sensing; the goal of the fusion process is to obtain a high-resolution multispectral image, which combines the spectral characteristic of the low-resolution data (MS) with the spatial

S.C. Satapathy et al. (eds.), *ICT and Critical Infrastructure: Proceedings of the 48th Annual Convention of CSI - Volume II*, Advances in Intelligent Systems and Computing 249,
DOI: 10.1007/978-3-319-03095-1_32,

resolution of the panchromatic image. Commonly, image fusion techniques can be categorized into two groups, namely discrete wavelet transform based and statistical based.

In statistical based image fusion, the common techniques are principal component analysis (PCA) [4] based, intensity-hue-saturation (IHS) [6], CS (Component Substitution) based [2,7], PCA-NSCT (Principle Component Analysis-Non Subsampled Contourlet Transform) method [5] and Gram-Scmidt method [11]. Wavelet Transform method for fusion can be found in [2] and Additive Wavelet Transform (AWL) can be found in [12].

In this work, our proposed fusion method is based on a modified form of PCA, key contributions in this work can be briefed as follows:

- Proposal of a new image fusion technique based on window based PCA.
- Our algorithm overcomes the drawback of the traditional PCA based technique.
- Our fusion technique proves to be better than the existing techniques proposed in recent literatures, in terms of the information content of the resultant fused image.

2 Window Based PCA For Image Fusion

PCA is a statistical based approach [8], which transforms or projects the features from the original domain to a new domain (known as PCA domain), where the features are arranged in the order of their variance. From the greatest variance of the data we can find the first principal component, the second greatest variance leads to second principal component, and so on. Fusion process is achieved in the PCA domain by retaining only those features that contain a significant amount of information.

Our proposed fusion technique is a window based approach, which is a modification over the existing PCA. Here, we first divide the images into some static window blocks. Then we find the principal eigenvector for each of the window blocks and perform fusion on the two corresponding window blocks of the two images to be fused. This assures the selection of the principal component in each of the window blocks, i.e. from all the spatial regions of the image.

2.1 Key Steps of the Technique

The key steps of our proposed methodology for image fusion can be briefed as follows:

Step 1: Let the PAN and the MS images are represented as ‘A’ and ‘B’ respectively. Both the PAN and MS images are split into n window blocks. The information, which is acquired from the fused image and quality matrix of the fused image is mostly depended on n.

Step 2: The row and column of every window block is arranged to create a data vector, i.e. for a n window block for PAN image, we create a data vector $x_1, x_2,\ldots, x_n$ and similarly $y_1, y_2,\ldots, y_n$ for the MS image.

Step 3: Evaluation of the covariance between every window block of both MS and PAN images. For the ith window block of both the images the covariance is calculated as shown in Eqn. 1:

$$C = \begin{bmatrix} cov(x_i\ x_i) & cov(x_i\ y_i) \\ cov(y_i\ x_i) & cov(y_i\ y_i) \end{bmatrix} \quad (1)$$

Step 4: The Eigen value of all the Eigen vectors are calculated from the covariance matrix and the Eigen vector that has the maximum value for each of the window blocks is called the principle Eigen Vector i.e. $(x_1, y_1)^T$, $(x_2, y_2)^T$, ... , $(x_n, y_n)^T$.
Step 5: The approximate weight of every window block is calculated using the following formula in Eqn. 2 and 3:

$$W(A_i) = \frac{x_i}{x_i+y_i} \quad (2); \qquad W(B_i) = \frac{y_i}{x_i+y_i} \quad (3);$$

Step 6: Adding the approximation weight of the two corresponding window blocks of PAN and MS images, as shown in Eqn.4.

$$F = A * W(A_i) + B * W(B_i) \quad (4)$$

Step 7: Arranging all the fused window blocks and obtaining the final fused image.

The generation of the weighting function W is a key step in this technique. The value of W depends on the number of window blocks as well as the information content of each window. If the number of window block is m then we can write $W \alpha 1/m$.

2.2 Key Features of the Technique

There is a basic difference between traditional PCA based image fusion and our proposed window based PCA methodology. In PCA based image fusion PCA is evaluated between the two source MS and PAN image, so there can be a chance that the values of the principal components occur in the same region of the image. But, in our proposed window based PCA method first we split both MS and PAN image into several window blocks. Then we find principal components from two corresponding window blocks of MS and PAN images, and hence the principal component is taken over all the window blocks. The advantages of our algorithm over all other algorithms are the following:

1. The algorithm works on each of the blocks so less number of principal components is left for fusion than in traditional PCA technique.
2. If algorithm is performed over smaller block size then it gives better result but the trade-off is the processing time.
3. The de-merit of the traditional PCA technique lies in the fact that the PCA components evaluated from the MS images depend on the correlation between the spectral bands of the whole image. So, the performance of PCA can vary with images having different correlation between the MS bands. Also there are chances that all the principle component values may occur from the same region of the MS image, so the total spectral cover may be unachieved. The process of windowing

reduces all these issues as it inspects over smaller regions of the image and hence disparity between the different regions of an image is better handled.

4. In respect to the CS method where the histogram is only considered not the principal components, there is chance of loss the information quality.
5. In respect to wavelet transform method where it takes selected bands for the fusion, our method can perform better as here the entire image information is considered.

3 Quality Index

The evaluation procedures are based on verification of the preservation of spectral characteristics and the improvement of the spatial resolution. In order to assess the spatial and spectral effect of the image several statistical evaluation methods have been proposed for assessing image quality [9, 10]. We make use of the following parameters:

Correlation Coefficient: It measures the degree of correlation between the fused and the reference images.

$$cc = \frac{\sum_{i=1}^{M}\sum_{j=1}^{N}\left(f(x,y)-\mu_f\right)\left(r(x,y)-\mu_r\right)}{\sqrt{\sum_{i=1}^{M}\sum_{j=1}^{N}\left(\left(f(x,y)-\mu_f\right)^2\left(r(x,y)-\mu_r\right)^2\right)}} \quad (5)$$

Entropy: It measures the richness of information in the fused image.

$$E = -\sum_{i=0}^{L} h(i)\log_2 h(i) \quad (6)$$

Mutual Information: It measures the information shared between the fused image and the reference image using histograms.

$$MI = \sum_{i=o}^{L}\sum_{j=0}^{L} h_f r(i,j)\log_2 \frac{h_f,r(i,j)}{h_f(i)h_r(j)} \quad (7)$$

Mean Square Error (MSE): It measures the spectral distortion introduced by the fusion process.

$$MSE = \frac{1}{M \times N}\sum_{x=1}^{M}\sum_{y=1}^{N}\left(f(x,y)-r(x,y)\right)^2 \quad (8)$$

Relative Shift Mean: It measures the amount of information added or lost during fusion process.

$$\frac{1}{M \times N}\sum_{x=1}^{N}\sum_{y=1}^{N}\frac{r(x,y)-f(x,y)}{f(x,y)} \quad (9)$$

Standard Deviation (SD): It measures the contrast of image.

$$SD = \sqrt{\frac{1}{M \times N}\sum_{x=1}^{M}\sum_{y=1}^{N}\left(f(x,y)-\mu\right)^2} \quad (10)$$

Warping Degree: It measures the level of optical spectral distortion.

$$WD = \frac{1}{M \times N}\sum_{x=1}^{M}\sum_{y=1}^{N}|f(x,y) - r(x,y)| \tag{11}$$

Peak Signal Noise Ratio (PSNR): It measures the ratio of maximum information and noise of fused image.

$$PSNR = 10\log_{10}\frac{D^2MN}{\sum_{x=1}^{M}\sum_{y=1}^{N}(f(x,y) - r(x,y))^2} \tag{12}$$

4 Results and Analysis

For evaluating the performance of our proposed technique and for comparison with the other existing techniques, we use some the quality metrics as discussed in the previous section. To estimate the performance of our proposed algorithm, we have used the images from satellite sensors like IKONOS and SPOT5.

IKONOS: The IKONOS imagery used for fusion performance estimation was acquired on November 19, 2001.

SPOT-5: The sensor has a 5-m-resolution PAN image and a 10-m-resolution MS image. A high-spatial-resolution image of 2.5 m is obtained by using two 5-m-resolutions PAN images acquired simultaneously the data set was acquired on June 28, 2002.

Fig. 1 and Fig. 2 shows source PAN and MS images from IKONOS and SPOT5 respectively and the resultant fused images obtained from our proposed technique as well as others. MATLAB 7.0 have been used for the simulation. From Table 2 and Table 4 it can be observed that the correlation coefficient, entropy and mutual information measure of the resultant fused IKONOS and SPOT5 images generated by our technique, are higher than that of the traditional CS, PCA, wavelet transform method and Gram Schmidt methods. This proves that higher number of spectral bands as well as higher information content exists in the fused images, produced by our technique. The measure of the SD for our technique is very close to the other techniques, which means that the noise profile of our resultant image is very similar to that of the others. Warping degree of the fused images generated by our proposed technique is lower than that of the others, which proves the better sharpness of the images. MSE is also least in our method, which suggest that our technique generates higher quality images. Our results are not best in terms of PSNR, which suggests that some pre-processing in terms of noise filtering is needed. The best values in Table 2 and Table 4 are bold-faced. Table 1 and Table 3 shows that result for Window Based PCA method for 64 window blocks produce better result than same for 16 window blocks and 16 window blocks produce better result than 4 window blocks.

In our methodology we assume that number of window blocks is n, i.e. we calculate covariance matrix and Eigen value for n number of window blocks, create n number of Eigen vectors and calculate approximate weight for every window blocks from these Eigen vectors, so in general complexity of our algorithm is ~$O(n^2)$.

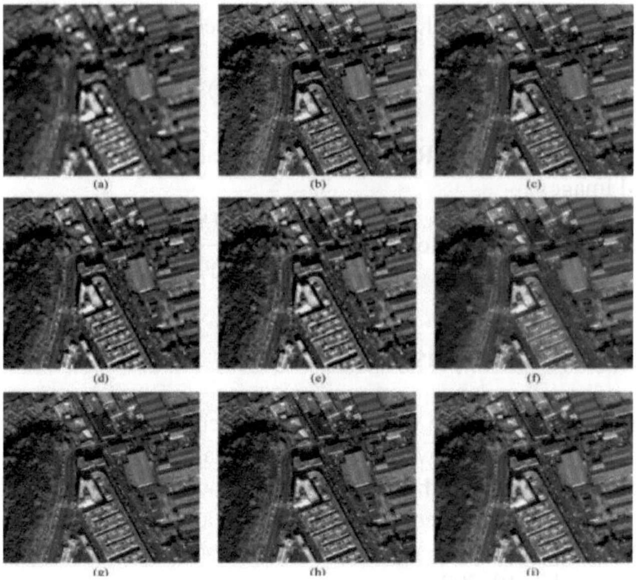

Fig. 1. IKONOS image (a) MS (b) PAN (c) Proposed Method (Window based PCA with 64 window blocks) (d) Normal PCA (e) NSCT-PCA (f) Wavelet Transfofm (g) CS method (h) Gram Schimdt Method (i) AWL Method

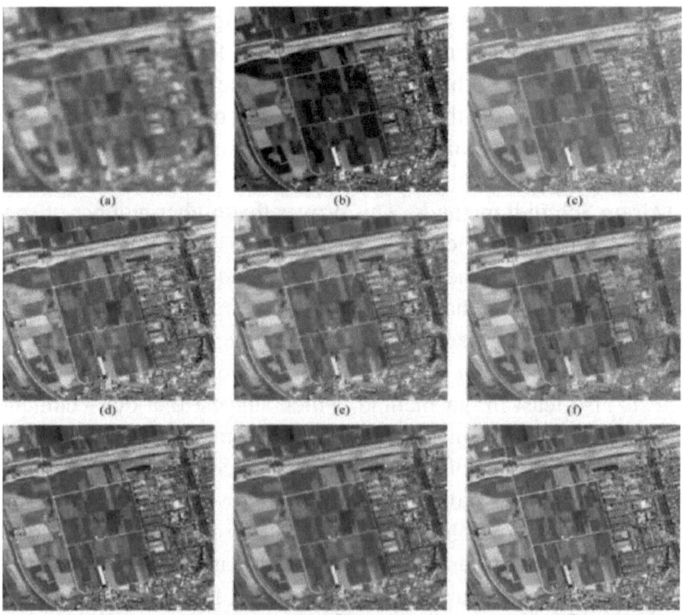

Fig. 2. Spot5 image (a) MS (b) PAN (c) Proposed Method (Window based PCA with 64 window blocks) (d)Normal PCA (e) NSCT-PCA (f) Wavelet Transfofm (g) CS method (h) Gram Schimdt Method (i) AWL Method

Table 1. Quality measure of fused IKONOS image with different window blocks

Image Quality	Window Based PCA (64 Window Blocks)	Window Based PCA (16 Window Blocks)	Window Based PCA (4 Window Blocks)
Correlation Coefficient	0.85	0.895	0.89
Entropy	0.52	0.515	0.51
Mutual Information	0.45	0.445	0.44
Mean Square Error	0.47	0.475	0.48
Standard Deviation	0.004	0.004	0.004
Wrapping Degree	0.21	0.22	0.23
Peak Signal to Noise Ratio	0.48	0.475	0.47

Table 2. Quality comparison of fused IKONOS image using window based PCA and other method

Correlation Coefficient	Proposed Window Based PCA	Traditional PCA	NSCT-PCA	Wavelet Transform	CS Method	Gram-Schmidt Method
Correlation Coefficient	**0.895**	0.89	0.89	0.88	0.87	0.86
Entropy	**0.52**	0.48	0.5	0.46	0.43	0.4
Mutual Information	**0.45**	0.4	0.42	0.37	0.35	0.33
Mean Square Error	**0.47**	0.5	0.49	0.52	0.54	0.55
Standard Deviation	0.004	0.004	0.004	**0.003**	**0.003**	**0.003**
Wrapping Degree	**0.21**	0.25	0.23	0.27	0.3	0.32
Peak Signal to Noise Ratio	0.48	0.47	0.48	**0.53**	0.5	0.49

Table 3. Quality measure of fused SPOT5 image with different window blocks

Image Quality	Window Based PCA (64 Window Blocks)	Window Based PCA (16 Window Blocks)	Window Based PCA (4 Window Blocks)
Correlation Coefficient	0.92	0.92	0.92
Entropy	0.85	0.845	0.84
Mutual Information	0.74	0.735	0.73
Mean Square Error	0.77	0.775	0.78
Standard Daviation	0.005	0.005	0.005
Warpping Degree	0.33	0.34	0.35
Peak Signal to Noise Ratio	0.77	0.765	0.76

Table 4. Quality comparison of fused SPOT5 image using window based PCA and other method

Correlation Coefficient	Proposed Window Based PCA	Traditional PCA	NSCT-PCA	Wavelet Transform	CS Method	Gram-Schmidt Method
Correlation Coefficient	**0.92**	**0.92**	**0.92**	0.91	0.90	0.89
Entropy	**0.85**	0.8	0.82	0.74	0.71	0.68
Mutual Information	**0.74**	0.7	0.71	0.67	0.65	0.61
Mean Square Error	**0.77**	0.8	0.79	0.82	0.84	0.88
Standard Daviation	0.005	0.005	0.005	0.004	0.004	**0.003**
Warpping Degree	**0.33**	0.39	0.38	0.41	0.44	0.40
Peak Signal to Noise Ratio	0.77	0.76	0.77	**0.82**	0.8	0.79

5 Conclusions

In this paper a new fusion algorithm based on window based PCA has been proposed for satellite image fusion. We examined our algorithm on satellite images from IKONOS and Quick Bird sources and compared with the traditional PCA, CS, NSCT-PCA, Wavelet and Gram-Schmidt methods. The results justify that our proposed technique produces better quality of fused image. In future, we wish to work with other varieties of satellite images like LANDSAT ETM and QUIK BIRD etc. We would also like to work on feature based techniques for image fusion in future.

References

1. Wang, J., et al.: Review of Satellite Remote Sensing Use in Forest Health Studies. The Open Geography Journal 3, 28–42 (2010)
2. Li, H., Manjunath, S., Mitra, S.: Multisensor image fusion using the wavelet transform. Graphical Models and Image Processing 57(3), 235–245 (1995)
3. Gonzalez-Audicana, M., Saleta, J.L., Catalan, R.G., Garcia, R.: Fusion of Multispectral and Panchromatic Images Using Improved IHS and PCA Mergers Based on Wavelet Decomposition. IEEE Trans. on Geosci. and Remote Sens. 42(6), 1291–1299 (2004)
4. Zheng, Y., Hou, X., Bian, T., Qin, Z.: Effective Image Fusion Rules Of Multi-scale Image Decomposition. In: Procedings of the 5th International Symposium on Image and signal Processing and Analysis, pp. 362–366 (2007)
5. Shi, H., Tian, B., Wang, Y.: Fusion of Multispectral and Panchromatic Satellite Images using Principal Component Analysis and Nonsubsampled Contourlet Transform. In: Processings of 10th International Conference on FSKD, pp. 2312–2315 (2010)
6. Tu, et al.: A fast intensity-hue-saturation fusion technique with spectral adjustment for IKONOS imagery. IEEE Geoscience and Remote Sensing Letters 1(4), 309–312 (2004)
7. Choi, J., Yu, K., Kim, Y.: A New Adaptive Component-Substitution-Based Satellite Image Fusion by Using Partial Replacement. IEEE Transactions on Geoscience and Remote Sensing 49(1), 295–309 (2011)

8. Smit, L.I.: A tutorial on Principal Component Analysis, pp. 1–27 (2002)
9. Li, S., Li, Z., Gong, J.: Multivariate statistical analysis of measures for assessing the quality of image fusion. International Journal of Image and Data Fusion 1(1), 47–66 (2010)
10. Yakhdani, M.F., Azizi, A.: Quality assessment of image fusion techniques for multisensory High resolution satellite images (case study: irs-p5 and irs-p6 Satellite images). In: Wagner, W., Székely, B. (eds.) ISPRS TC VII Symposium – 100 Years ISPRS, Vienna, Austria, July 5-7, vol. XXXVIII, Part 7B, pp. 204–208. IAPRS (2010)
11. Aiazzi, B., Baronti, S., Selva, M., Alparone, L.: MS + Pan image fusion by an enhanced Gram-Schmidt spectral sharpening. In: Bochenek, Z. (ed.) New Developments and Challenges in Remote Sensing, pp. 113–120. Mill Press, Rotterdamp (2007)
12. Nunez, E., Otazu, X., Fors, O., Prades, A., Palà, V., Arbiol, R.: Multiresolution-based image fusion with adaptive wavelet decomposition. IEEE Transactions on Geoscience and Remote Sensing 37(3), 1204–1211 (1999)

Rough Set Approach for Novel Decision Making in Medical Data for Rule Generation and Cost Sensitiveness

P.K. Srimani[1] and Manjula Sanjay Koti[2]

[1] Dept. of Computer Science & Maths,
Bangalore University, Bangalore., India
profsrimanipk@gmail.com
[2] Dept. of MCA, Dayananda Sagar
College of Engineering, Bangalore & Research Scholar, Bharathiar University,
Coimbatore India
man2san@rediffmail.com

Abstract. Data mining techniques can be applied in the area of Software Engineering for getting improved results. Medical data mining has great potential for exploring the hidden patterns in medical data and these patterns can be utilized for clinical diagnosis. Analysis of medical data is often concerned with the treatment of incomplete knowledge, with management of inconsistent pieces of information. In the present study, the theory of Rough set is applied to find dependence relationship among data, evaluate the importance of attributes, discover the patterns of data, learn common decision-making rules, reduce the redundancies and seek the minimum subset of attributes so as to attain satisfactory classification. It is concluded that the decision rules with and without reducts) generated by the rough set induction algorithms (Exhaustive, Covering, LEM2 and GA) not only provide new medical insight but also are useful for medical experts to analyze the problem effectively and find optimal cost .

Keywords: Genetic algorithm, Induction algorithms, Medical data mining, Rough set, Rule generation, Reducts.

1 Introduction

Software Engineering (SE) is the computing field concerned with the designing, developing, implementing, maintaining and modifying software. Software Engineering data consists of sequences, graphs, and text. SE concerns computer-based system development; which includes system specification, architectural design, integration, and deployment. With the increasing importance of SE, the difficulty of maintaining, creating and developing software has also risen. Challenges in SE include requirement gathering, systems integration and evolution, maintainability, pattern discovery, fault detection, reliability and complexity of software development [1], [2]. There are seven steps in the process: data integration, data cleaning, data selection, data transformation, data mining, pattern evaluation and knowledge

S.C. Satapathy et al. (eds.), *ICT and Critical Infrastructure: Proceedings of the 48th Annual Convention of CSI - Volume II*, Advances in Intelligent Systems and Computing 249,
DOI: 10.1007/978-3-319-03095-1_33, © Springer International Publishing Switzerland 2014

presentation. Data mining techniques that can be applied in improving SE include generalization, characterization, classification, clustering, associative tree, decision tree or rule induction, frequent pattern mining [3]. Improvement of software productivity and quality are an important goal of software engineering. The process of finding useful patterns or meaning in raw data has been called knowledge discovery in databases [4], [5].

Medical data mining has great potential for exploring the hidden patterns in the data sets of the medical domain and these patterns can be utilized for clinical diagnosis. However, the available raw medical data are widely distributed, heterogeneous in nature and voluminous. These data need to be collected in an organized form and made available to a Hospital Information System (HIS). Actually, medical databases have accumulated large quantities of information about patients and their medical conditions. Relationships and patterns within this data could provide new medical knowledge. The genetic algorithms offers an attractive approach for solving the feature subset selection problem. A genetic algorithm (GA) is a search technique used in computing to find true or approximate solutions to optimization and search problems and are categorized as global search heuristics. This finds applications in cost-sensitive design of classifiers for tasks such as medical diagnosis, computer vision, among others. The GA-based approach to feature subset selection does not rely on monotonicity assumptions that are used in traditional approaches to feature selection which often limits their applicability to real-world classification and knowledge acquisition tasks.

The success of machine learning associated with medical data sets is strongly affected by many factors and one such factor is the quality of the data which depends on irrelevant, redundant and noisy data. Thus, when the data quality is not excellent, the prediction of knowledge discovery during the training process becomes an arduous task. The existing intelligent techniques [6], [7] of medical data analysis are concerned with (i) Treatment of incomplete knowledge (ii) Management of inconsistent pieces of information and (iii) Manipulation of various levels of representation of data. This difficulty is minimized by feature selection which identifies and removes the irrelevant and redundant features in the data to a great extent.

Further, Intelligent methods such as neural networks, fuzzy sets, decision trees and expert systems are applied to the medical fields [8],[9] but cannot derive conclusions from incomplete knowledge or can manage inconsistent information.

In recent years Classical rough set theory developed by [10] has made a great success in the field of knowledge acquisition. We have used Pima data set for our study, which has been widely used in machine learning experiments and is currently available through the UCI repository of standard data sets. The present investigation on Rough sets is organized as follows: Related work, Methodology, Experiments and Results. Finally the conclusions are presented.

1.1 Applications

Rough sets have been proposed for a very wide variety of applications. In particular, the rough set approach seems to be important for Artificial Intelligence and cognitive sciences, especially in machine learning, knowledge discovery, data mining, expert systems, approximate reasoning and pattern recognition. Rough set rule induction algorithms generate decision rules [11],[12] which not only provide new medical insight but also are useful for medical experts to analyze the problem effectively. These decision rules are more useful for medical experts to analyze and gain understanding into the problem at hand.

2 Literature Survey

In [13], the author proposes a rough set algorithm to generate diagnostic rules based on the hierarchical structure of differential medical diagnosis. Here, the author discusses the characterization of decision attributes which is extracted from databases and the classes are classified into several generalized groups with respect to the characterization. There are two kinds of sub-rules: classification rules for each generalized group and the induced rules for each class within each group. Finally, those two parts are integrated into one rule for each decision attribute. The induced rules can correctly represent experts' decision processes. The authors [14] use a rough set approach for identifying a patient group in need of a scintigraphic scan for subsequent modeling. This approach depends on the distance function between existing values and these values can be calculated by distance function between the conditions attributes values for the complete information system and incomplete information system. Here, the author [15] compares rough set-based methods, in particular dynamic reducts, with statistical methods, neural networks, decision trees and decision rules. He analyzes medical data, i.e. lymphography, breast cancer and primary tumors, and finds that error rates for rough sets are fully comparable as well as often significantly lower than that for other techniques. In [16], the application of rough set classification algorithm exhibits higher classification accuracy than decision tree algorithms. The generated rules are more understandable than those produced by decision tree methods. Some of the other works include [17], [18].

3 Dataset Description

We have used Pima data set for our study, which has been widely used in machine learning experiments and is currently available through the UCI repository of standard data sets. To study the positive as well as the negative aspects of the diabetes disease, Pima data set can be utilized, which contains 768 data samples. Each sample contains 8 attributes which are considered as high risk factors for the occurrence of diabetes, like Plasma glucose concentration, Diastolic blood pressure (mm Hg), Triceps skin fold thickness (mm), 2-hour serum insulin (mu U/ms), Body mass index (weight in kg/(height in m)) Diabetes pedigrees function and Age (years). All the 768 examples were randomly separated into a training set of 576 cases (378, non-diabetic and 198, diabetic) and a test set of 192 cases (122 non-diabetic and 70 diabetic cases) [19].

4 Methodology

The present study illustrates how rough set theory could be used for the analysis of medical data especially for generating classification rules from a set of observed samples of the Pima data set. In rough set theory, knowledge is represented in information systems. An information system is a data set represented in a tabular form called decision table in which each row represents an object (for eg., a case or an event) and each column represents an attribute. In order to determine all the reducts of the data that contains the minimal subset of attributes that are associated with a class label for classification. The Rough Set reduction technique is employed. In a knowledge system reducts are often used at the data preprocessing stage during the attribute selection process. It is important to note that reduct is not unique and in a decision table multiple reducts may exists. The core of a decision table which consists of essential information is certainly contained in every reduct. In other words, a reduct generated from the original data set should contain the core attributes. During the attribute selection process reduct and core are the commonly used since the main purpose of rough set theory is to select the most relevant attributes with regard to the classification task and to remove the irrelevant attributes. The set of attributes which is common to all reducts is called the core which is possessed by every legitimate reduct, and hence consists of essential attributes which cannot be removed from the information system without causing collapse of the equivalence-class structure. In other words, a core is absolutely necessary for the representation of the categorical structure.

4.1 Rule Induction

It is emphasized that the number of all minimal consistent decision rules for a given decision table can be exponential with respect to the size of decision table. Three heuristics have been implemented in RSES:

4.1.1 Genetic Algorithm

One can compute a predefined number of minimal consistent rules with genetic algorithm that comprises permutation encoding and special crossover operator [21].

Genetic algorithm – Pseudo code

```
Choose initial population
 Evaluate the fitness of each individual in the population
 Repeat
  Select best-ranking individuals to reproduce
  Breed new generation through crossover and mutation
  (genetic operations) and give birth to offspring
  Evaluate the individual fitnesses of the offspring
  Replace worst ranked part of population with offspring
 Until <terminating condition>
```

4.1.2 Exhaustive Algorithm

This algorithm realizes the computation of object oriented reducts (or local reducts). It has been shown that some minimal consistent decision rules for a given decision table S can be obtained from objects by reduction of redundant descriptors. The method is based on Boolean reasoning approach.

```
exhaustive(intsol,intdepth)
{ if(issolution(sol))
printsolution(sol)
else
{ solgenerated=generatesolution()
  exhaustive(solgenerated,depth+1) }
```

4.1.3 Covering Algorithm

This algorithm searches for minimal (or very close to minimal) set of rules which cover the whole set of objects.

```
            Inputs: labeled training dataset D
  Outputs: ruleset R that covers all instances in D
  Procedure:
    Initialize R as the empty set
    for each class C {
    while D is nonempty {
    Construct one rule r that correctly classifies some
    instances in D that  belong to class C and does not
    incorrectly classify any non-C instances
      Add rule r to ruleset R
      Remove from D all instances correctly classified by
r}}
     return R
```

4.1.4 LEM2 Algorithm

It is to be noted that LEM2 algorithm is another kind of covering algorithm which has been used in the present study [20].

5 Experiments and Results

The results of the experiments conducted on the Pima data set by using rough set approach are presented and discussed in this section.

Table 1. Rules through reduct

Algorithm	No. of reducts	Length of Reduct Min	Max	Mean
Exhaustive	32	3	5	3.8
Genetic	10	3	4	3.4

Table 2. Rule generation

Algorithm	No. of rules	Length of Rules Min Max Mean			Accuracy (%)	Coverage	Filtered rules	Length of Rules Min Max Mean		
Exhaustive	16364	3	5	3.8	71.8	0.152	20	3	5	3.6
Genetic	5110	3	4	3.4	78.16	0.1	10	3	3.6	3.1

Table 1 represents reduct generation through Exhaustive and Genetic algorithms. It is found that the number of reducts is more in the case of exhaustive. Here the length of the reduct is the number of descriptors in the premise of reducts. Table 2 represents rule generation through reducts by using these algorithms and also the classification results. The length of the rule is the number of descriptors in the premise of rules. Here, the accuracy happens to be more in the case of genetic algorithm although the coverage is less when compared to the exhaustive algorithm. Table 3 represents direct rule generation (i.e., without reducts) by using Exhaustive, Genetic, Covering and LEM2 algorithms and accuracy is presented in Table 3. Here LEM2 algorithm has the highest accuracy of 76% even though its coverage is less. Table 4 presents the length of the rules.

Table 3. Rule generation-Direct

Algorithms	Rules	Filtered rules	Accuracy (%)	Coverage	Std.dev.
Exhaustive	5861	855	67.2	1	-
Covering	357	150	64.4	0.734	-
Lem2	300	114	76	0.293	-
Genetic	3574	749	64.26	0.990	3.14

In the case of reducts generation based on genetic algorithms implemented through RSES, it is found that the results will be non-deterministic and this is certainly the consequence of a genetic algorithm paradigm. Accordingly different accuracies for different executions will be obtained for the same data sets. Therefore, it is absolutely necessary to experiment many times and compute the average accuracy and the standard deviation. Further, it is suggested that only those results for which small values of standard deviation are obtained would be acceptable. The same procedure is followed in Tables 1-3. Nowhere in the literature, results corresponding to the genetic algorithm implemented through RSES is available.

Table 4. Rule generation-Direct

Algorithms	Length of rules Min	Max	Mean
Exhaustive	1	4	2.1
Covering	1	1	1
Lem2	2	6	3.5
Genetic	1	4	2

In medical world, for any disease to be diagnosed there are some tests to be performed and each and every test can be considered as a feature. By the process of feature selection, the performance of tests that are highly expensive and irrelevant could be avoided, which in turn reduces the cost associated with the diagnosis and helps the patients and the doctors to a great extent. In processing the medical data, choosing the optimal subset of features is important, not only to reduce the processing cost but also to improve the classification performance of the model built from the selected data.

Table 5 predicts the following: with GA (i) the number of features reduced is 4 (ii) accuracy is 74.8% (iii) Time is 0.02 ms (iv) ROC value being 79.1% and (v) optimal cost is 25%.

Table 5. Optimal Cost prediction

Sl. No	Algorithm	Original features	Features reduced	Accuracy (%)	ROC (%)	Cost (units)	Average Cost(%)	Time (ms)
1	With GA	8	4	74.8	79.1	193	25%	0.02
2	Without GA	8	-	73.8	75.1	201	26%	0.06

6 Conclusion

In the context of many practical problems like medical diagnosis, feature subset selection introduces a multi-criteria (viz., accuracy of classification, cost and risk associated with classification) optimization problem which facilitates the meaningful pattern recognition. These criteria strongly depend on the selection of attributes that describe the patterns. In fact, the main idea of feature selection is to select a subset of input variables by eliminating the feature/attributes with little or no predictive information. In the performance of data mining and knowledge discovery activities, rough set theory has been regarded as a powerful, feasible and effective methodology since its very inception in 1982. Generally, medical data contains irrelevant features, uncertainties and missing values. Accordingly, the analysis of such medical data deals with incomplete and inconsistent information with the tremendous manipulation at different levels. In this context, it is emphasized that rough set rule induction algorithms are capable of generating decision rules which can potentially provide new medical insight and profound medical knowledge. By taking into consideration all the above aspects, the present investigation is carried out. The results clearly show that rough set approach is certainly a useful tool for medical applications.

The results of the present investigation strongly reveal the following: (i) the number of reducts is more in the case of exhaustive algorithm (ii) the accuracy happens to be more in the case of genetic algorithm although the coverage is less when compared to the exhaustive algorithm (iii) LEM2 algorithm has the highest accuracy of 76% even though it's coverage is less (iv) the number of features reduced is 4, accuracy is 74.8% , Time is 0.02 ms , ROC value being 79.1% and optimal cost is 25% with and without GA. Finally, it is concluded that the results obtained from GA implemented through RSES is of non-deterministic nature and hence accurate results can be obtained only by computing the average of several runs for the same data set. This is first of its kind and not found in the literature.

Acknowledgement. Mrs. Manjula Sanjay Koti is grateful to Dayananda Sagar College of Engineering, Bangalore and Bharathiar University, Coimbatore for providing the facilities to carry out the research work.

References

1. Clarke, J.: Reformulating software as s search problem. IEEE Proc.: Software 150(3), 161–175 (2003)
2. Fern, X.L., Komireddy, C., Gregoreanu, V., Burnett, M.: Mining problem-solving strategies from HCL data. ACM Trans. Computer- Human Interaction 17(1), Article 3, 1–22 (2010)
3. DePree, R.W.: Pattern recognition in software engineering. IEEE Computer, 48–53 (1983)
4. Han, J., Kamber, M.: Data Mining Concepts and Techniques. Elsevier Inc., San Francisco (2007)
5. Piate, U.M., Tsky-Shapiro, G., Smyth, P., Uthurusamy, F.R.: From Data Mining to Knowledge Discovery: An Overview. In: Advances in Knowledge Discovery and Data Mining, pp. 1–36 (1996)
6. Lavrajc, N., Keravnou, E., Zupan, B.: Intelligent Data Analysis in Medicine and Pharmacology. Kluwer Academic Publishers (1997)
7. Wolf, S., Oliver, H., Herbert, S., Michael, M.: Intelligent data mining for medical quality management. In: Proceedings of the Fifth International Workshop on Intelligent Data Analysis in Medicine and Pharmacology (IDAMAP 2000), Berlin, Germany (2000)
8. Lin, J.-C., Wu, K.-C.: Using Rough Set and Fuzzy Method to Discover the Effects of Acid Rain on the Plant Growth. JCIT 2(1), 48 (2007)
9. Leung, Y., Fischer, M.M., Wu, W.-Z., Mi, J.-S.: A rough set approach for the discovery of classification rules in interval-valued information systems. International Journal of Approximate Reasoning 47(2), 233–246 (2008)
10. Skowron, A., Pawlak, Z., Komorowski, J., Polkowski, L.: A rough set perspective on data and knowledge. In: Kloesgen, W., Żytkow, J. (eds.) Handbook of KDD, pp. 134–149. Oxford University Press (2002)
11. Lin, T.Y.: From rough sets and neighborhood systems to information granulation and computing in words. In: Proceedings of European Congress on Intelligent Techniques and Soft Computing, pp. 1602–1607 (1997)

12. Lin, T.Y., Yao, Y.Y., Zadeh, L.A. (eds.): Rough Sets, Granular Computing and Data Mining. Physica-Verlag, Heidelberg (2002)
13. Tsumoto, S.: Mining diagnostic rules from clinical databases using rough sets and medical diagnostic model. Inform. Sci. 162, 65–80 (2004)
14. Komorowski, J., Ohrn, A.: Modelling prognostic power of cardiac tests using rough sets. Artif. Intell. Med. 15, 167–191 (1999)
15. Hu, K.Y., Lu, Y.C., Shi, C.Y.: Feature ranking in rough sets. AI Commun. 16(1), 41–50 (2003)
16. Ding, S., Zhang, Y., Xu, L., Qian, J.: A Feature Selection Algorithm Based on Tolerant Granule. JCIT 6(1), 191–195 (2011)
17. Sriman, P.K., Koti, M.S.: A Comparision of different learning models used in data mining for medical data, The Smithsonian Astrophysics Data System. In: AIP Conf. Proceedings, vol. 1414, pp. 51–55 (2011), doi:10.1063/1.3669930
18. Srimani, P.K., Koti, M.S.: Cost Sensitivity analysis and prediction of optimal rules for medical data. IJMIE, 1641–1647 (2012)
19. UCI repository, `http://archive.ics.uci.edu/ml/`
20. Grzymała-Busse, J.W., Hu, M.: A comparison of several approaches to missing attribute values in data mining. In: Ziarko, W.P., Yao, Y. (eds.) RSCTC 2000. LNCS (LNAI), vol. 2005, pp. 378–385. Springer, Heidelberg (2001), doi:10.1007/3-540-45554-X_46
21. Goldberg, D.: Genetic Algorithms in Search, Optimization, and Machine Learning. Addison-Wesley, New York (1989)

Position Paper: Defect Prediction Approaches for Software Projects Using Genetic Fuzzy Data Mining

V. Ramaswamy[1], T.P. Pushphavathi[2], and V. Suma[3]

[1] Dept. of CSE, BIET, Davanagere, India
researchwork04@yahoo.com
[2] Jain University, Bangalore, RIIC, Dayanada Sagar Institute, Bangalore, India
acepushpa@yahoo.co.in
[3] Dept. of CSE, RIIC, Dayanada Sagar Institute, Bangalore, India
sumavdsce@gmail.com

Abstract. Despite significant advances in software engineering research, the ability to produce reliable software products for a variety of critical applications remains an open problem. The key challenge has been the fact that each software product is unique, and existing methods are predominantly not capable of adapting to the observations made during project development. This paper makes the following claim: Genetic fuzzy data mining methods provide an ideal research paradigm for achieving reliable and efficient software defect pattern analysis. A brief outline of some fuzzy data mining methods is provided, along with a justification of why they are applicable to software defect analysis. Furthermore, some practical challenges to the extensive use of fuzzy data mining methods are discussed, along with possible solutions to these challenges.

Keywords: Data Mining, Fuzzy Clustering, Software Engineering, Random forest, Metrics, Software Quality, Project Management.

1 Introduction

The development of a software product is an inherently complex venture, in that defect prediction is a multifaceted area of research. The software industry is a multi-billion dollar contributor to the economic growth and technological advancement of modern society [1].Bohem and Sullivan attributed the importance of software economics to three reasons: the alteration of the dynamics of technology innovation; the increased impact of software-enabled change in organizations; and the identification of value creation as the key to success [2]. The quality and reliability of software products are hence of paramount importance, and thus software testing is a key component of the software development life-cycle. Despite significant advances in testing methods, the increased use of software has only increased the cost of debugging defective software. Existing software testing methods are still unable to provide high-quality software products.

S.C. Satapathy et al. (eds.), *ICT and Critical Infrastructure: Proceedings of the 48th Annual Convention of CSI - Volume II*, Advances in Intelligent Systems and Computing 249,
DOI: 10.1007/978-3-319-03095-1_34, © Springer International Publishing Switzerland 2014

According to a survey carried out by the Standish Group, an average software project exceeded its budget by 90 percent and its schedule by 222 percent (Chaos Chronicles, 1995). Software Projects are not risk-free ventures. According to the Standish Group's study in 2004, only 29% of projects succeeded, 53% were challenged (delivered late, over budget and/or with less than the required features and functions) and 18% failed (cancelled prior to completion or delivered and never used) [4],[5].This survey took place in mid 90s and contained data from about 8000 projects. These statistics shows the importance of measuring the software early in its life cycle and taking the necessary precautions before these results come out. For the software projects carried out in the industry, an extensive metrics program is usually seen unnecessary and the practitioners start to stress on a metrics program when things are bad or when there is a need to satisfy some external assessment body. A systematic literature review of defect prediction studies published from January 2000 to December 2012 aims to identify and analyse the models used to predict faults in source code. The outcome of this review reveals evaluation of how studies have developed used and validated models for fault prediction in software engineering [15].With the continuous increase of software size and complexity, software quality assurance is becoming increasingly important. One way to improve software quality is software defect prediction, which is also an efficient means to relieve the effort in software code inspection or testing. Under this situation only need to inspect or test a part of software artefacts and ignore the remainders. By doing so, an organization's limited resources could be reasonably allocated with the goal to detect and correct the greatest number of defects in software. Therefore, software defect prediction has been widely studied, and a number of methods have been proposed to address this problem ([1]-[8]). Of these methods, classification is a popular approach for software defect prediction, including analogy-based approaches [9], [10], tree-based methods [11],[12], statistical procedures [13],[14] etc. These classification methods are relatively straightforward transplants from the fields of data mining and machine learning, and they follow the same path of using software metrics, which are extracted from a software source code, as candidate factors to categorize the software components as defective or no defective. This is achieved by building a binary classifier with historical defect data to predict the defect proneness of new software components.

It has been stated that software defect pattern analysis is among the most challenging domains for structured machine learning over the next few years [11]. Data mining algorithms, especially the Random Forest ensemble methods, are appropriate for adapting to the behavior of the observed (debugging) data. This paper advocates the use of fuzzy data mining methods for achieving software defects, primarily because these methods can model the inherent stochasticity of software testing and other real-world systems.

2 Background

A software defect is an error, flaw, mistake, failure, or fault in a computer program or system that produces an incorrect or unexpected result, or causes it to behave in unintended ways [24][25].Software defects are expensive in terms of quality and cost.

Moreover, the cost of capturing and correcting defects is one of the most expensive software development activities. It will not be possible to eliminate all defects but it is possible to minimize the number of defects and their severe impact on the projects. To do this a defect management process needs to be implemented that focuses on improving software quality via decreasing the defect density. A little investment in defect management process can yield significant profits[23].Track and detect potential software defects early stage is critical in many high assurance systems. However, building accurate quality estimation models is challenging because noisy data usually degrades trained model's performance [26]. The two general types of noise in software metrics and quality data exist. One relates to mislabeled software modules, caused by software engineers failing to detect, forgetting to report, or simply ignoring existing software faults. The other pertains to deficiencies in some collected software metrics, which can lead to two similar (in terms of given metrics) software modules for different fault proneness labels. Removing such noisy instances can significantly improve the performance of calibrated software quality estimation models. Therefore, investigative the problematic software module before calibrating any software quality estimation models is desirable [28].

2.1 Software Defect Prediction

Software defect prediction is the process of locating defective modules in software. To produce high quality software, the final product should have as few defects as possible. Early detection of software defects could lead to reduced development costs and rework effort and more reliable software. So, the study of the defect prediction is important to achieve software quality. The most discussed problem is software defect prediction in the field of software quality and software reliability. As Boehm observed finding and fixing a problem after delivery is 100 times more expensive than fixing it during requirement and design phase. Additionally software projects spend 40 to 50 percent of their efforts in avoidable rework [25].

2.2 Data Mining and Machine Learning Techniques

Data mining techniques and machine learning algorithms are useful in prediction of software bug estimation. Machine learning models and Data mining techniques can be applied on the software repositories to extract the defects of a software product. Common techniques include random forest, decision tree learning, Naïve Bayesian classification and neural networks, j48 and ONER etc.

2.3 A Fuzzy-Logic Approach to Software Metrics and Models

A significant motivation for using fuzzy logic is the ability to estimate required defect much earlier in the development process. Since many of the independent variables in software metric models are either difficult to quantify (for example complexity), or are only known to a rough degree (such as system size), the use of fuzzy variables seems intuitively appealing. It is our conjecture here that project managers are in fact able to classify systems using fuzzy variables with reasonable levels of both accuracy and consistency. However, the above-mentioned description does establish the applicability of fuzzy c means clustering (FCM), random forest methods to software defect testing challenges. Recent papers in the software defects literature have shown that such machine learning formulations lead to significantly better performance than the existing methods. However, the utilization of these methods as a research paradigm for formulating software defect prediction challenges needs further investigation. The goal of this paper is to stimulate interest in the software quality testing community, and to promote the use of FCM, random forest machine learning methods for formulating challenges in the field of software project success prediction.

3 Practical Challenges

Data mining methods using genetic algorithm and fuzzy clustering techniques have the capability to transform the field of software defect prediction analysis, by offering probabilistic solutions to major challenges in the field. However, some practical challenges need to be overcome in order to enable the widespread use genetic fuzzy data mining methods. Most software engineering data mining studies rely on well-known, publicly available tools, further many such tools are general purpose and should be adapted to assist the particular task at hand. However, Software engineering researchers may lack the expertise to adapt or develop mining algorithms or tools, while data mining researchers may lack the background to understand mining requirements in the software engineering domain. On promise way to reduce this gap is to foster close collaborations between the software engineering community (requirement providers) and data mining community (solution providers). This research effort represents one such instance.

Another main challenge is that, in real-world software projects, software fault measurements (such as fault proneness labels) might not be available for training a software quality-estimation model. This happens when an organization is dealing with a software project type it's never dealt with before. In addition, it might not have recorded or collected software fault data from a previous system release. So, how does the quality assurance team predict a software project's quality without the collected software metrics? The team can't take a supervised learning approach without software quality metrics such as the risk-based class or number of faults. The estimation task then falls on the analyst (expert), who must decide the labels for each software module. Cluster analysis, an exploratory data analysis tool, naturally addresses these two challenges.

Software Projects success criteria based on the details of past literatures [16-22], the software projects as categorized into two groups as successful and unsuccessful from the project managers or experts opinion point of view. The project is completed on time and on budget, with all features and functions originally specified is a successful project and the project is completed and functioning but exceeds budget, time estimate, and with fewer features and functions than originally specified or the project is cancelled before completion or never implemented is a unsuccessful. In this paper it cover along three dimensions (software engineering, data mining, and future directions)[24-28]:Issues: What types of software engineering data are available for mining? and how they should be customized to fit the requirements and characteristics of software engineering data.Which software engineering tasks can benefit from mining software engineering data?How are data mining techniques used in software engineering?What are the challenges in applying data mining techniques to software engineering data?Which data mining techniques are most suitable for specific types of software engineering data?What are the freely available data sources and data mining and analysis tools (e.g.WEKA [28])?What software repositories and datasets should be mined for defect prediction?How can we resolve the problem of ceiling effects as well as imbalanced and highly skewed datasets?How can we get better results in identifying defects from large features and high level software modules?

Research Questions (RQ)

RQ1 What is the context of the defect prediction model?

To understand the environment or which the prediction model was developed. Examine context primarily in terms of the origin of systems and the programming language that the model has been developed and tested on. This contextual information allows us to discuss the applicability and generalisability of models.NASA's publicly available software metrics data have proved very popular in developing fault prediction models. And has the advantage that researchers are able to replicate and compare results using different approaches based on the same data set. However, although the repository holds many metrics and is publicly available, it is not possible to explore the source code or trace back how the metrics were extracted. It is also not always possible to identify if any changes have been made to the extraction and computation mechanisms over time.

RQ2 What variables have been used in defect prediction models?

This question identifies the independent and dependent variables of the model. It shows the range of predictors of fault proneness used in models as well as the form that these tasks. It can also report on how thoroughly some variables are investigated compared to others, results shows that researchers are still struggling to find reliable and useful metrics as an input to their fault prediction models. Researchers continue to search for predictor variables that correlate to faults independent of project context the metric that performs best will always depend on the context. However it remains some distance from determining which model is most general, i.e., that works in most situations.

RQ3 What modeling methods have been used in the development of defect prediction models?

This question identifies whether models are based on regression, machine learning or other approaches. It allows us to discuss the popularity and effective use of particular modeling methods. However, despite the substantial level of research activity and the many different models developed in the area of fault prediction, there is still no consensus in the research community as to which approach is most applicable to specific environmental circumstances.

RQ4 How do studies measure the performance of their models?

Understanding how prediction studies measure and validate model performance gives an indication of how confident we can be in these results. Analyse several aspects of model performance (e.g. significance levels, data balance, accuracy) to discuss how well models are able to predict faults in code. However, despite the substantial level of research activity and the many different models developed in the area of fault prediction, there is still no consensus in the research community as to which approach is most applicable to specific environmental circumstances.

4 Methodology

The software metric data gives us the values for specific variables to measure a specific module/function or the whole software. When combined with the weighted error/defect data, this data set becomes the input for a machine learning system. A learning system is defined as a system that is said to learn from experience with respect to some class of tasks and performance measure, such that its performance at these tasks improves with experience (Mitchell, 1997). To design a learning system, the data set in this work is divided into two parts: the training data set and the testing data set. Some predictor functions are defined and trained with respect to Multi-Layer Perceptions and Decision Tree algorithms and the results are evaluated with the testing data set.

Problem Statement

Two types of research can be studied on the code based metrics in terms of defect prediction. The first one is predicting whether a given code segment is defected or not. The second one is predicting the magnitude of the possible defect, if any, with respect to various viewpoints such as density, severity or priority. Estimating the defect causing potential of a given software project has a very critical value for the reliability of the project. This research is primarily focused on the second type of predictions. But it also includes some major experiments involving the first type of predictions. Given a training data set, a learning system can be set up. This system would come out with a score point that indicates how much a test data and code segment is defected. After predicting this score point, the results can be evaluated with respect to popular performance functions. The two most common options here are the Mean Absolute Error (mae), the Mean Squared Error (mse), Net reliability. The mae is generally used for classification, while the mse is most commonly seen in function approximation. In this research we use mse since the performance function for the results of the experiments aims second type of prediction. Although mae could be a good measure for classification experiments, in our case, due to the fact that our output values are zeros and ones to use some custom error measures. Establish a method for identifying software defects using classifier methods in order to define its success. In this work NASA's Metrics Data Program (MDP) as software metrics data are used at the initial level. Classify the dataset with fuzzy c means clustering, then optimize the defect factor using genetic algorithm by setting up crucial parameters as fitness function. Finally, an improved random forest ensemble classifier for this research and will use it along with earlier developed modules to make a hybrid approach and best amongst all.

5 Conclusion

Faults in software systems continue to be a major problem. Knowing the causes of possible defects as well as identifying general software process areas that may need attention from the initialization of a project could save money, time and work. The possibility of early estimating the potential faultiness of software could help on planning, controlling and executing software development activities. There are many metrics and technique available for investigate the accuracy of fault prone classes which may help software organizations for planning and performing defect prediction analysis activities. As the complexity and the constraints under which the software is developed are increasing, it is difficult to produce software without faults. Such faulty software classes may increase development & maintenance cost, due to software failures and decrease customer's satisfaction.When a software system is developed, the majority of faults are found in a few of its modules. In most of the cases, 55% of faults exist within 20% of source code. It is, therefore, much of interest is to find out fault-prone software modules at early stage of a project [28].

To summarize, this paper advocates the use of genetic fuzzy data mining methods as a widespread research paradigm, in order to address the representation, inference and learning challenges in the field of software projects defect prediction. The ability of these methods to explicitly account for the unique characteristics of each software product will enable researchers to make significant inroads on the hard challenges in the domain. As a result, robust software products will be created, which will have a major impact on the economic growth and technological advancement of modern society.

References

1. Gallaher, M., Kropp, B.: Economic impacts of inadequate infrastructure for software testing. Technical report, National Institute of Standards and Technology (May 2002)
2. Bohem, B.W., Sullivan, K.: Software economics: A roadmap. In: International Conference on Software Engineering, Limerick, Ireland, pp. 319–343 (2000)
3. Bertolino, A., Strigini, L.: On the Use of Testability Measures for Dependability Assessment. IEEE Trans. Software Engineering 22(2), 97–108 (1996)
4. Bishop, M.: Neural Networks for Pattern Recognition. Oxford University Press (1995)
5. Boetticher, G.D., Srinivas, K., Eichmann, D.: A Neural Net-Based Approach to Software Metrics. In: Proceedings of the Fifth International Conference on Software Engineering and Knowledge Engineering, San Francisco, pp. 271–274 (1993)

6. CHAOS Chronicles, The Standish Group - Standish Group Internal Report (1995)
7. Cusumano, M.A.: Japan's Software Factories. Oxford University Press (1991)
8. Diaz, M., Sligo, J.: How Software Process Improvement Helped Motorola. IEEE Software 14(5), 75–81 (1997)
9. Dickinson, W., Leon, D., Podgurski, A.: Finding failures by cluster analysis of execution profiles. In: ICSE, pp. 339–348 (2001)
10. Fenton, N., Neil, M.: A critique of software defect prediction models. IEEE Transactions on Software Engineering 25(5), 675–689 (1999)
11. Groce, A., Visser, W.: What went wrong: Explaining counterexamples. In: Ball, T., Rajamani, S.K. (eds.) SPIN 2003. LNCS, vol. 2648, pp. 121–135. Springer, Heidelberg (2003)
12. Jensen, F.V.: An Introduction to Bayesian Networks. Springer (1996); Henry, S., Kafura, D.: The Evaluation of Software System's Structure Using Quantitative Software Metrics. Software Practice and Experience 14(6), 561–573 (1984)
13. Hudepohl, P., Khoshgoftaar, M., Mayrand, J.: Integrating Metrics and Models for Software Risk Assessment. In: The Seventh International Symposium on Software Reliability Engineering (ISSRE 1996) (1996)
14. Mitchell, T.M.: Machine Learning. McGrawHill (1997); Neumann, D.E.: An Enhanced Neural Network Technique for Software Risk Analysis. IEEE Transactions on Software Engineering, 904–912 (2002)
15. Yuriy, B., Ernst, M.D.: Finding latent code errors via machine learning over program executions. In: Proceedings of the 26th International Conference on Software Engineering, Edinburgh, Scotland (2004)
16. Beecham, S., Hall, T., Bowes, D., Gray, D., Counsel, S., Black, S.: A Systematic Review of Fault Prediction Approaches used in Software Engineering
17. Shan, X., Jiang, G., Huang, T.: A framework of estimating software project success potential based on association rule mining. IEEE, doi:978-1-4244-4639-1/09/$25.00 ©2009
18. Pinto, J.K., Slevin, D.P.: Project success: definitions and measurement techniques. Project Management Journal 19, 67–72 (1988)
19. Jones, C.: Patterns of large software systems: failure and success. IEEE Computer 28, 86–87 (1995)
20. Baccarini, D.: The logical framework method for defining project success. Project Management Journal 30, 25–32 (1999)
21. Linberg, K.R.: Software developer perceptions about software project failure: a case study. The Journal of Systems and Software 49, 177–192 (1999)
22. Weka 3: Data Mining Software in Java, `http://www.cs.waikato.ac.nz/ml/weka/`
23. Zhang, H., Kitchenham, B., Jeffery, R.: Achieving Software Project Success: A Semi-quantitative Approach. In: Wang, Q., Pfahl, D., Raffo, D.M. (eds.) ICSP 2007. LNCS, vol. 4470, pp. 332–343. Springer, Heidelberg (2007)
24. Martin, N.L., Pearson, J.M., Furumo, K.A.: IS Project Management: Size, Complexity, Practices and the Project Management Office. In: Proceedings of the 38th Annual Hawaii International Conference on System Sciences (HICSS 2005) - Track 8, p. 234b (2005)
25. Weber, R., Waller, M., Verner, J.M., Evanco, W.M.: Predicting Software Development Project Outcomes. In: Ashley, K.D., Bridge, D.G. (eds.) ICCBR 2003. LNCS, vol. 2689, pp. 595–609. Springer, Heidelberg (2003)
26. King, M.A., Elder IV, J.F.: Evaluation of fourteen desktop data mining tools. In: IEEE International Conference on Systems, Man, and Cybernetics, vol. 3, pp. 27–29 (1998)
27. Diehl, S., Gall, H., Hassan, A.E.: Guest editors introduction: special issue on mining software repositories. Empirical Software Engineering 14(3), 257–261 (2009)
28. Mendonca, M., Sunderhaft, N.L.: Mining software engineering data: A survey. A DACS state-of-the-art report, Data & Analysis Center for Software, Rome, NY (1999)

A Fast Algorithm for Salt-and-Pepper Noise Removal with Edge Preservation Using Cardinal Spline Interpolation for Intrinsic Finger Print Forensic Images

P. Syamala Jaya Sree[1] and Pradeep Kumar[2]

[1] Department of Computer Science & IT
Jaypee University of Information Technology, Waknaghat, 173215, India
jayasree.syamala@gmail.com
[2] Department of Electronics and Communication Engineering
Jaypee University of Information Technology, Waknaghat, 173215, India
erpradeep_tiet@yahoo.co.in

Abstract. The accuracy of a proper Biometric Identification and Authentication Systems in Image Forensics depends on the image quality to arrive at a reliable and accuracy result. To get a noise-free fingerprint image, they are applied under the pre-processing and filtering tasks. The Fingerprint Recognition system is often demanded by the accuracy factor. In this paper an attempt is made to evaluate the filtering techniques in the removal of Salt & Pepper Noise. This work proposes a faster and an efficient way to remove salt-and-pepper impulse noise and also the edge-preserving regularization of the henceforth obtained finger print noise free image. In this paper, we propose a two phase mechanism where the noisy pixels are identified and removed in the first phase and only these noisy pixels are involved in cardinal spline edge regularization process in the second phase. Promising results were found even for Noise levels as high as 90% with the proposed algorithm. The results were found to be much better than the previously proposed nonlinear filters or regularization methods both in terms of noise removal as well as edge regularization for image forensics.

Keywords: Image Forensics, Digital Finger Print Images, Edge-preserving regularization, salt-and-pepper impulse noise, cardinal splines.

1 Introduction

The use of multimedia processing and network technologies has facilitated the increase in security demands of multimedia contents. Traditional image content protection schemes use extrinsic approaches, such as watermarking or fingerprinting. However, under many circumstances, extrinsic content protection is not possible. Therefore, there is great interest in developing forensic tools via intrinsic fingerprints to solve these problems [1]. The widespread availability of photo editing software has made it easy to create visually convincing digital image forgeries. To address this problem, there has been much recent work in the field of digital image forensics.

S.C. Satapathy et al. (eds.), *ICT and Critical Infrastructure: Proceedings of the 48th Annual Convention of CSI - Volume II*, Advances in Intelligent Systems and Computing 249,
DOI: 10.1007/978-3-319-03095-1_35, © Springer International Publishing Switzerland 2014

There has been little work, however, in the field of anti-forensics, which seeks to develop a set of techniques designed to fool current forensic methodologies. The addition of salt and pepper noise / additive noise has been used by many researchers to solve the problem of intrinsic finger print systems for digital forensics and different technical methods have been proposed to remove impulse noise.

For the removal of noise, error minimizing methods [2]–[10] have been used successfully to preserve the edges and the details in the images. These methods failed in the case of impulse noise. Moreover the restoration will alter basically all pixels in the image, including those that are not corrupted by the impulse noise.

In this paper, a "decision-based" cum "detail-preserving" technique is put forward and applied to finger print images that is efficient and faster than the previously proposed methods [5] for salt-and-pepper noise removal and subsequent edge-preservation. The noisy pixels were first found out in the initial stage and then they were replaced by interconnecting the surrounding uncorrupted pixels using Cardinal splines with the help of a kernel. Later the indices of these noisy candidates were noted. To these indices in the output generated in the initial stage, the nearest neighbors were selected and they were interconnected with cubic Cardinal splines. B-splines have been widely used in signal and image processing [4]. Many researchers like Unser [4-6] have applied them into various field of image processing like compression, transformation of images [9]. Cardinal splines are used for the interpolation since they have the property to pass through the control points. The Cardinal splines also have local propagation property that they are not affected by the change of single control point. The main reason for using Cardinal splines is that they have compact support with good continuity [8]. We have applied this algorithm to finger print images since it is a very efficient method when compared to the existing methods of noise removal in finger print imagery.

The outline of this paper is as follows:. The proposed denoising algorithm and edge preserving algorithm is elaborated in section 2. The significant measures are discussed in Section 3. The results have been discussed in section 4 and conclusions are discussed in Section 5.

2 Proposed Methodology and Algorithm

The Cardinal Splines [8] are interpolating piecewise cubic polynomials with specified end point tangents at the boundary of each cross section. The slope at the control point is calculated from the co-ordinates of the two adjacent control points.

2.1 Algorithm for Noise Removal

The Cardinal Splines[8] have been used for removing the impulse noise from the given image. For every pixel of the image, we first extract the pixels which are not corrupted by the noise (the pixels which are not 0 or 255). Initially, as described in Table 1, we start with the initial window size according to the percentage of noise. We require four pixel values for the interpolation of the corrupted pixel. If in any kernel, we have noise free pixels less than four then we increase the size of the kernel as per the requirement. When we increase the size of the kernel then we should keep the values which were there in the initial kernel. This is mainly because naturally we the best interpolation when the interpolating pixels are nearest to the corrupted pixel which is being replaced. This is in turn because of the less local variation in pixel intensity.

If X(i.j) be the original image and Y(i,j) be the image corrupted by the salt and pepper noise. For every pixel of Y(i,j) ,we start with the initial kernel S(x,y)∈Y(i,j). If S(x,y) contains the four pixels unaffected by the noise then use those pixel values to find the value of the centre pixel on which window is centered . Let L be the number of noise free pixels in the S(x,y) . Let Z be a vector containing all the pixels unaffected by the noise.

If L<4 then increase the size of the S(x,y) accordingly.

Case 1 : If L=3, we do not increase the size of the initial window but replicate the third value of the Z to get the fourth value . Then interpolate the new value using Cardinal Splines.

Case 2 : If L=2 then increase the size of the S(x,y) by adding two to its initial size and extract the noise free pixels in Z. Since we need at least four noise free pixels, we need two more of them . Then we repeat the interpolation on these pixels. If we do not get enough noise pixels even after we increase the window size, we save the noise free pixels which are obtained when we increased the window size for the first time and then we add further noise free pixels by further increasing the window size. We continue this process untill we get the required number of noise free pixels. But the basic idea is that we select those noise free pixels for interpolation which are closer to the noisy pixel which is going to be replaced.

Case 3: If L=1 then we increase the size of window by adding four to its initial size. If we don't obtain the required number of noise free pixels when we increase the window size for the first time, we continue the above said process in case 3 where the required noise free pixels were less when the window size was increased for the first time.

Case 4 : If L=0 then we are increasing the window by adding six to its initial size. Again we follow the same procedure that is followed in case 2 and case 3 if the required noise free pixels are less when we increase the window size for the first time in this case.

Suppose S(x,y) was 3x3 and if L=2 then new size of the S(x,y) become 5x5. While increasing the size of the window and extracting the new noise free pixels , all pixels those present in the initial window must be used in the interpolation. If S(x,y) is the initial window and S'(x,y) be the new window after the increment in the size then we update Z vector in such a way that the values from the S(x,y) should be considered first and then the remaining values must be taken from S'(x,y).

The initial window size should be considered according to the given table.

Table 1. Initial Window Size

Noise %	Initial Window Size (S_{xy})
n< 55	3 X 3
55≤ n<80	5X5
n>=90	7X7

After calculating the value of every pixel we get the output image O(i,j). But initially Y(i,j) contains noise free pixels have to be replaced into O(i,j) so that we get a good improvement in the PSNR.

2.2 Algorithm for Edge Preservation

Then the output image is now considered. In this image we consider only those pixels in which noise was added which can be known by collecting the indices of the noisy pixels immediately after the original was corrupted with salt-and-pepper impulse noise, and replace them with the value obtained by interpolating the nearest elements using Cardinal Splines. Thus the final restored image is obtained.

3 Significant Measures

3.1 Peak Signal to Noise Ratio (PSNR)

We have tested our proposed algorithm for different levels of noise ranging from as low as 5% to as high as 95%. The experimental results have been gauged using the mean absolute error (MAE) and peak signal-to-noise ratio (PSNR) measures that have been given below.

$$MAE = \frac{1}{mn}\sum_{x=0}^{m-1}\sum_{y=0}^{n-1} | A(x,y) - R(x,y) | \tag{1}$$

where *A* and *R* are the original and the restored images having a resolution of m*n.

$$PSNR = 10 \log_{10}\left(\frac{\max^2}{MSE}\right) \tag{2}$$

Where max is the maximum possible pixel value of the image and its value is 255 in the case of a grayscale image included in the meta data of the online version.

4 Results and Discussion

Test Image 1 : Finger Print Image 1

Original Finger Print Image 1

70% Noisy Finger Print Image1

Fig. 1. Finger Print Image1. The maximum values of α and β were found at 70% noise level.

Phase 1

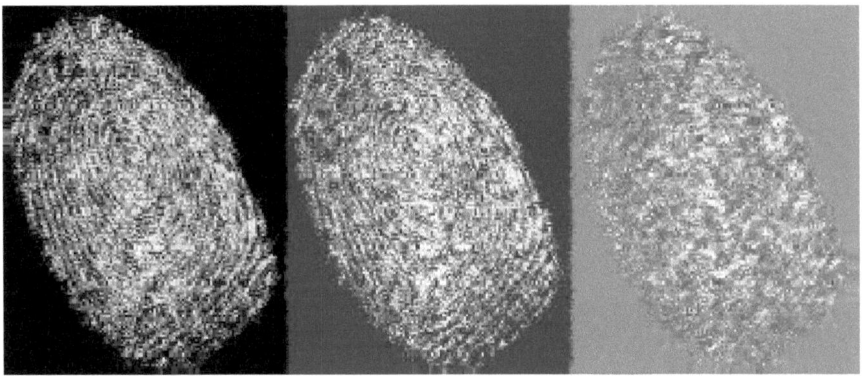

Fig. 2. Restored FP image1 at 50% noise (15.12 dB) 70% noise (10.39 dB)90 %(7.45 dB)

Phase 2

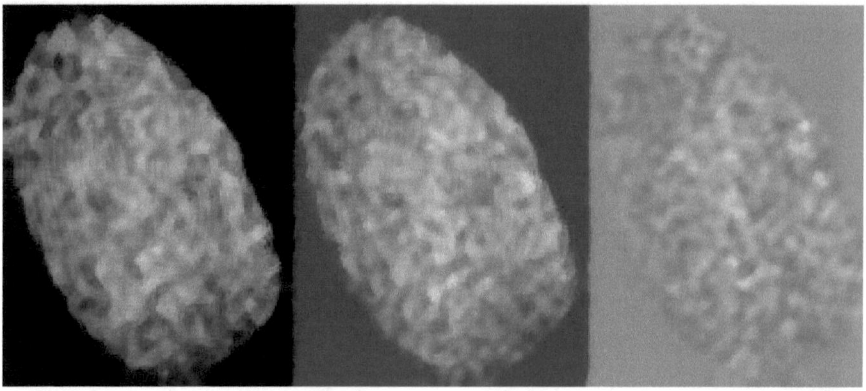

Fig. 3. Restored FP image1 at 50% noise (20.12 dB) 70% noise (15.39 dB)90 %(12.45 dB)

Test Image 2 : Finger Print Image 2

Original Finger Print Image 2

70% Noisy Finger Print Image2

Phase1

Fig. 4. Restored FP image2 at 50% noise (16.21 dB) 70% noise (11.39 dB)90 %(8.55 dB)

Phase 2

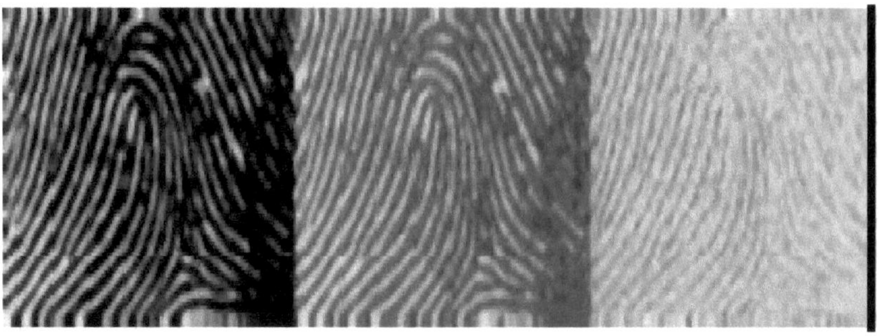

Fig. 5. Restored FP image2 at 50% noise (21.12 dB) 70% noise (16.89 dB)90 %(13.45 dB)

5 Discussion

The simulations of the algorithm were run for the finger print images 1 and 2. All the images were 8-bit grayscale images. The dynamic ranges for all the images were 0-255. Salt-and-pepper impulse noise ranging from 10% to 90% was added and the images were tested with the proposed algorithm. For each and every image the maximum value of PSNR was found out by varying the parameters α and β. $\beta$was varied from the values -3.6 to +1.8 in intervals of 0.2 while α was varied from 0 to 1 in intervals of 0.1.

6 Conclusion

In this paper, we put forward a decision-based, edge-detail preserving algorithm which ticks both the characters of computational efficiency and image restoration capacity. It maintains a great balance between both of them. Our results outperform the other existing filters visually, quantitatively and temporally. Thus it is a very good dependable algorithm for Salt-and-pepper impulse noise removal. This work has been the extension of our previous work [5] where the same has been applied to intrinsic forensic images. The Algorithm has been implemented in MATLAB version 7,1.5 GHz Dual Core Intel Processor and we found that it is very fast since the time taken for running the Algorithm is 8 sec.

References

1. Sabrina Lin, W., Tjoa, S.K., Vicky Zhao, H., Ray Liu, K.J.: Digital Image Source Coder Forensics Via Intrinsic Fingerprints. IEEE Transactions On Information Forensics And Security 4(3), 460–475 (2009)
2. Chan, R.H., Ho, C.-W., Nikolova, M.: Salt-and-Pepper Noise Removal by Median-Type Noise Detectors and Detail-Preserving Regularization. IEEE Trans. on Image Processing 14(10) (October 2005)

3. Wang, Z., Zhang, D.: Progressive switching median filter for the removal of impulse noise from highly corrupted images. IEEE Trans. Circuits Syst. II., Analog Digit. Signal Process 46(1), 78–80 (1999)
4. Unser, M.: Splines: A Perfect Fit for Signal and Image Processing. IEEE Signal Processing Magazine 16(6), 24–38 (1999)
5. Syamala Jayasree, P., Bodduna, K., Kumar, P., Siddavatam, R.: An Expeditious cum Efficient Algorithm for Salt-and-Pepper Noise Removal and Edge-Detail Preservation using Cardinal Spline Interpolation. Elsevier Journal of Visual Communication and Image Representation (Under review, 2013)
6. Unser, M., Aldroubi, A., Eden, M.: Fast B-Spline Transforms for Continuous Image Representation and Interpolation. IEEE Transactions on Pattern Analysis and Machine Intelligence 13(3), 277–285 (1991)
7. Meijering, E., Unser, M.: A Note on Cubic Convolution Interpolation. IEEE Transactions on Image Processing 12(4), 477–479 (2003)
8. Jaiswal, T., Siddavatam, R.: Image noise cancellation by lifting filter using second generation wavelets. In: Proceedings of IEEE International Conference on Advances in Recent Technologies in Communication and Computing, Kerala, India, October 27-28, pp. 667–671 (2009)
9. Hearn, D., Pauline Baker, M.: Computer Graphics with OpenGL, 3rd edn. Pearson Publishers (2009)
10. Gonzalez, R.C., Woods, R.E.: Digital Image Processing. Prentice-Hall, Englewood Cliffs (2002)

An Approach to Predict Software Project Success Based on Random Forest Classifier

V. Suma[1], T.P. Pushphavathi[2], and V. Ramaswamy[3]

[1] Dept. of CSE, RIIC, Dayanada Sagar Institute, Bangalore, India
sumavdsce@gmail.com
[2] Jain University, Bangalore, RIIC, Dayanada Sagar Institute, Bangalore, DBIT, Bangalore, India
acepushpa@yahoo.co.in
[3] Dept. of CSE, BIET, Davanagere, India
researchwork04@yahoo.com

Abstract. The success or failure of a software project depends on the product's quality and reliability. The predictions of defects are important since it helps direct test effort, reduce costs and improve the quality of software. Software defects are expensive in terms of quality and cost. Data mining techniques and machine learning algorithms can be applied on these repositories to extract the useful information. This paper presents a software defect prediction model based on Random Forest (RF) ensemble classifier, which is more robust and beneficial for large-scale software system. The difference in the performance of the proposed methodology over other methods is statistically significant. Two fold information, one is RF is efficient irrespective of the domain of applications that is from the point of project, complexity of project, domain of project. Second is this inference enabled to predict the success level of projects. RF is travels light to project managers to predict the success of the projects based on the mining carried out using RF from empirical investigations.

Keywords: Data Mining, Clustering, Software Engineering, Random forest, Metrics, Software Quality, Project Management.

1 Introduction

The success or failure of a software project depends on the product's quality and reliability. Software practitioners collect software metrics and defect data during the software development process, and then analyse the data for defect prediction modelling.The use of data mining in software engineering represents a swing from verification-driven data analysis approaches to discovery-driven data analysis approaches. In the former approach, a decision maker must hypothesize the existence of information of interest, collect this information, and test the posed hypothesis against the information collected. Due to the size and complexity of data repositories nowadays, this approach is not sufficient to efficiently explore the data available in an organization. Discovery-driven approaches sieve through large amounts of data and automatically

S.C. Satapathy et al. (eds.), *ICT and Critical Infrastructure: Proceedings of the 48th Annual Convention of CSI - Volume II*, Advances in Intelligent Systems and Computing 249,
DOI: 10.1007/978-3-319-03095-1_36, © Springer International Publishing Switzerland 2014

(or semi-automatically) discover important information hidden in the data. The choice of which data mining technique one should use at a given point in time depends on the domain expert's goals and the tasks one wants to perform to achieve those goals. The growing dependency on software-based systems increases the need to deliver high quality and reliable software products. Detecting software defects in a timely manner prior to system deployment can reduce software maintenance costs and avoid tarnishing the reputation of an organization [2].There is lot of work done in prediction of the fault proneness of the software systems. But it is the severity of the faults that is more important than number of faults existing in the developed system as the major faults matters most for a developer and those major faults needs immediate attention [1].Ensemble selection has recently appeared as a popular ensemble learning method, not only because its implementation is fairly straightforward, but also due to its excellent predictive performance on practical problems [3].The selection of metrics type is dependent on the programming hypothesis used in the project and research targets.The key research problem is which data mining techniques are most suitable for specific types of software engineering data?

Organization of the paper is as follows, Section 2 specifies the related work in the domain of data mining and software engineering. Section 3 provides research methodology followed during this work. Section 4 provides overview of mining software engineering projects data. Section 5 indicates methods for effective project management. Section 6 specifies the performance measurements. Experimental results and summary of this part of research is briefed in Section 7.

2 Related Work

The development of technology has always forced several researchers to progress towards improvement for achieving quality of project.Authors in [4] for software quality estimation, software development practitioners typically construct quality-classification or fault prediction models using software metrics and fault data from a previous system release or a similar software project. Engineers then use these models to predict the fault proneness of software modules in development.Authors in [5] main contribution is the development of an automated way of assigning fault-proneness labels to the modules and the removing the subjective expert opinion. There is no heuristic step in this model as needed in k-means clustering based fault prediction approaches.Authors in [6] express data mining for secure software engineering improves software productivity and quality; software engineers are increasingly applying data mining algorithms to various software engineering tasks. However mining software engineering data poses several challenges, requiring various algorithms to effectively mine sequences, graphs and text from such data. However mining software project data poses several challenges, requiring various algorithms to effectively mine sequences, graphs and text from such data Using well established data mining techniques, practitioners and researchers can explore the potential of this valuable data in order to better manage their projects and do produce higher-quality software systems that are delivered on time and within budget.

Authors in [7] presents the various challenges in software engineering(SE) data mining such as requirements unique to SE data, complex data and patterns, large scale data and just in time mining. Much work needs to be done to adapt general-purpose mining algorithms or develop specific algorithms to satisfy the unique requirements of SE data and tasks. Further, the scope of SE tasks that can benefit from data mining must be expanded as well as the range of SE data that can be mined.Authors in [8] explains the Random Forest(RF) algorithm for constructing the tree, RF was proposed by Breiman [9] and constructs a forest of multiple trees and each tree depends on the value of a random vector. For each of the tree in the forest, this random vector is sampled with the same distribution and independently. The model predicted using LB technique yields the best AUC with value 0.806. Hence, software researchers may use the machine learning techniques specially boosting algorithms for predicting faulty classes in early phases of software development.A Random Forest is a classifier consisting of a collection of tree-structured classifiers [3]. The random forest classifies a new object from an input vector by examining the input vector on each tree in the forest. Each tree casts a unit vote at the input vector by giving a classification. The forest selects the classification having the most votes over all the trees in the forest [3]. Authors in [11] compare the methodology with many existing approaches using the same data sets. The proposed approach to predicting fault prone modules runs efficiently on large data sets and it is more robust to outliers and noise compared to other classifiers. Hence, authors express it is especially valuable for the large systems.

3 Research Methodology

An empirical investigation related to this research is carried on several product based software industries of varying production capabilities. The empirical data of various projects developed at the companies under study is obtained at Document Management Repository of respective companies. Additionally, modes of data collection include interactions with project developing team, quality assurance departments, defect management centers etc. The projects that are collected for the investigation purpose are a non critical application which includes projects of ERP (Enterprise Resource Planning), Business, Retail, Medical and web applications.

Hypothesis 1: The data collected from the companies are at CMM level 5.

Hypothesis 2: The projects that are collected for the investigation purpose are a non critical application in various domains, which includes projects such as Enterprise Resource Planning (ERP), web, medical, retail and business applications. These projects are further developed using Oracle database and Java based tools in Linux Operating system environment.

4 Research Work

The empirical data which is collected from the industries are pre-processed and formatted using Waikato Environment for Knowledge Analysis (WEKA) data mining tool where the selected data is converted into Attribute Relation File Format (ARFF) which is readable from WEKA. The transformed data thus obtained is subjected to data mining process.In this case study, the data sets were collected from non critical projects such as ERP (Enterprise Resource Planning), business, retail, medical and web applications.Equation 1 infers that prior to model building for classification it has labelled each attribute in the dataset either as success (no defect) or failure (defect) as follows:

$$\text{Project Success Rate} = \{\text{Success: \#defects} = 0;\text{Failure: \#defects} >=1\} \quad (1)$$

These two classes represent the binary target classes for training and validating the prediction models in order to analyze the software project tasks such as defect detection, debugging, testing etc. Here the defect prediction is performed using the ensemble random forest to the prediction of defect prone modules. Table I lists the attributes and its data types for this case study. The case study includes the five data sets were collected from non critical projects as shown in table II.The research work includes for collecting the sampled relevant metric data, the metrics needed for prediction of software defect severity, also analyze and refine metrics data using data mining tool. The fault-prone modules constitute only a small portion in the data sets. Random forests, trying to minimize overall error rate, will keep the error rate low on the large class while letting the smaller classes have a larger error rate. This will obviously impose problems for software quality prediction, because many defect-prone modules will be misclassified as non-defect prone ones and hence might be released into the later phase of the software life cycle.

Table 1. Attributes and its data types in each data set

Project No	Name of attribute	Data Type
1	Project development time(Person Hours)	N
2	Function point	N
3	Defect captured	N
4	Defect Estimation	N
5	Project Success	C

Table 2. Data Sets used in the Case Studies

Project Type	# projects	Language
ERP	498	Java
Web	10,885	Oracle
Medical	2,209	C
Retail	618	C++
Business	1,209	Java

5 Random Forests

Random Forests is the unique learning machine that has no need of an explicit test sample because of its use of bootstrap sampling for every tree. This ensures that every tree in the forest is built on about 63% of the available data, leaving the remaining approximately 37% for testing the OOB (out-of-bag) data. Regression models will be evaluated by mean squared error or mean absolute error. Binary classification models will normally be evaluated by AUC/ROC (area under the ROC curve), and multinomial classification models will be evaluated by some averaging of the misclassification rates across the classes.

The reason for RF classifier is their accuracy. RF is much better than individual decision trees and are on par with or better than neural networks and support vector machines. The SPM software suite incorporates many automation features which make tree ensembles easy to build, require little data preparation, and are fast and reliable. By combining outputs from multiple models into a single decision to boost our prediction power and create a model for conviction.Procedure used as follows: MDP datasets used are ERP, Medical, Retail and Business and Web applications. Preprocess the data set attribute values according to equation 1.For each dataset apply RF algorithm using 10-fold cross validation.The accuracy measurements used on datasets are f-measure, Mean Absolute Error (MAE) and Mean Square Error(MSE) for classification accuracy and function optimization on datasets.

6 Performance Measures

The accuracy indicators are adapted in these experiments to evaluate and compare the aforementioned ensemble algorithms, which are classification accuracy (f-measure) and with a binary classification confusion matrix, in which true positive (TP) which is be predicted to positive and is actual positive, false positive (FP) which is be predicted to positive but is actual not, true negative (TN) which is be predicted to negative and is actual negative, false negative (FP) which is be predicted to negative but is actual not. So we had AC=(TP+TN)/(TP+FP+TN+FN). AUC is area under ROC curve. That is also the integral of ROC curve with false positive rate as x axis and true positive rate as y axis. If ROC curve is more close to top-left of coordinate, the corresponding classifier must have better generalization ability so that the corresponding AUC will be lager. Therefore, AUC can quantitatively indicate the generalization ability of corresponding classifier.In cross-validation, it needs to decide on a fixed number of folds or partitions of the data. Then the data is split into three approximately equal partitions. That is, use two-thirds of the data for training and one-third for testing, and repeat the procedure three times so that in the end, every instance has been used exactly once for testing. This is called threefold cross-validation, and if stratification is adopted as well. 10-fold cross validation was used in all the experiments.

7 Experimetal Results

Random Forest is a top-down tree-structured classifier built using recursive partitioning technique in which the measurement space is successively split resembling a terminal node in the tree ([8], [10]). However, metrics are the sources of measurement to predict the quality of the project which is vital in software engineering. Therefore, this research aimed at applying Random Forest technique upon non critical projects applications to evaluate the quality of these projects. Table 3 infers the performance analysis of Random Forest ensemble classifier. According to accuracy measurements mentioned in section 6 are f-measure, True Positive Rate (TPR) and Receiver Operator Curve (ROC), also MAE and MSE results are showed in Table 3.

Table 3. Performance analysis of Random Forest

Projects	Performance			MAE	MSE
	f-measure	TPR	ROC		
Business	0.79	0.8	0.82	0.31	0.40
ERP	0.79	0.8	0.82	0.31	0.40
Medical	0.80	0.7	0.85	0.29	0.25
Retail	0.81	0.8	0.84	0.27	0.30
Web	0.85	0.5	0.86	0.20	0.28

TPR: True Positive Rate;
ROC- Receiver Operator Curve;
MAE: Mean Absolute Error;
MSE: Mean Square Error;

Figure 2 infers that RF classifier has better accuracy with projects applications. In web application project dataset TPR is given 50% accuracy but ROC is 86% , therefore by experimental observation of three parameters, performance of RF is better classifier for large data.

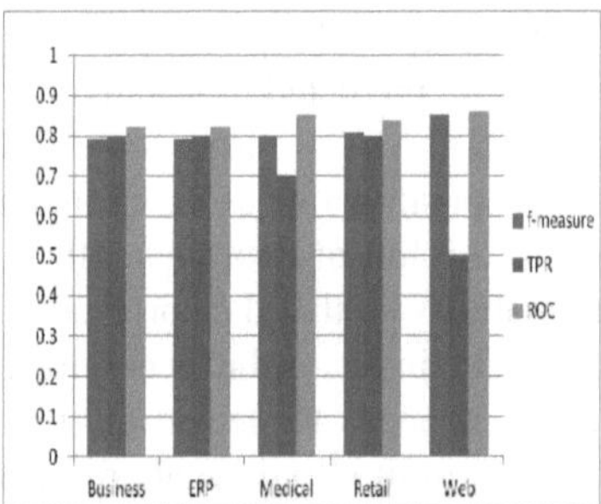

Fig. 1. Accuracy f-measure,TPR and ROC

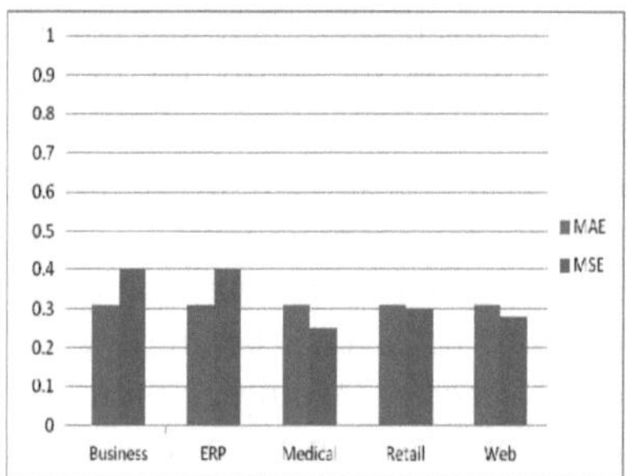

Fig. 2. Performance of MAE and MSE

Figure 3 infers low values of MAE and MSE means more classification and function optimization accuracy.Table IV illustrates comparative analysis of RF with different domains of projects. However RF classifier has good performance. Figure 4 infers its percentage of success levels.

Table 4. Comparative performance analysis of Random Forest

Projects	f-measure
Business	0.79
ERP	0.79
Medical	0.80
Retail	0.81
Web	0.85

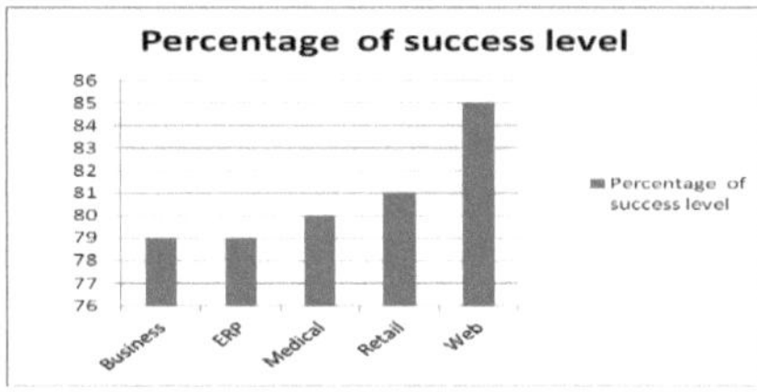

Fig. 3. Percentage of success level of projects

From experimental observations random forest is a good candidate for software quality prediction, especially for large-scale systems, because of it is reported to be consistently accurate, when compared with current classification algorithms, runs efficiently on large data sets, it has an efficient method for estimating missing data and retains accuracy when a large portion of the data is missing. It gives estimates of which attributes are important in the classification and it is more robust.

8 Conclusion

In this paper, we propose an efficient software project mining model for the prediction of performance of the project, depending upon dataset of the software project data collected from the various companies. Approach like this helps in the decision making capability of the project managers/developers of the company prior to starting of a project. The experimental observation of random forest classifier shows the efficient defect prediction of a project becomes simple. It believes that if the proposed procedure followed, thus the investigational observations of random forest is a good entrant for software quality prediction, especially for large-scale systems, because of it is reported to be consistently accurate, when compared with current classification algorithms [8], runs efficiently on large data sets. It gives estimates of which attributes are important in the classification and it is more robust. It has a potential to enhance the statistical validity of future experiments.

Acknowledgments. The authors would like to acknowledge the software companies involved in this study and the project managers for their invaluable help in providing necessary information for our work under the framework of the Non-Disclosure Agreement.

References

1. Sandhu, P.S., Malhotra, U., Ardil, E.: Predicting the Impact of the Defect on the Overall Environment in Function Based Systems. World Academy Of Science, Engineering And Technology (56), 140–143 (2009)

2. Seliya, N., Khoshgoftaar, T.M.: Software quality estimation with limited fault data:a semi-supervised learning perspective. Software Qual. J. 15, 327–344 (2007), doi:10.1007/s11219-007-9013-8
3. Sun, Q., Pfahringer, B.: Bagging Ensemble Selection. Department of Computer Science. The University of Waikato Hamilton, New Zealand
4. Zhong, S., Khoshgoftaar, T.M., Seliya, N.: Analyzing Software Measurement Data with Clustering Techniques. Florida Atlantic University (March/April 2004), http://www.computer.org/intelligent
5. Catal, C., Sevim, U., Diri, B., Member, AENG: Software Fault Prediction of Unlabeled Program Modules. In: Proceedings of the World Congress on Engineering, WCE 2009, London, U.K, July 1-3, vol. I (2009)
6. Krishna Prasad, A.V., Rama Krishna, S.: Data Mining for Secure Software Engineering – Source Code Management Tool Case Study. International Journal of Engineering Science and Technology 2(7), 2667–2677 (2010)
7. Xie, T., Thummalapenta, S., Lo, D., Liu, C.: Data Mining for Software Engineering. IEEE Published by the IEEE Computer Society (August 2009), 0018-9162/09/$26.00 © 2009
8. Malhotra, R., Singh, Y.: On the Applicability of Machine Learning Techniques for Object Oriented Software Fault Prediction
9. Fenton, N., Ohlsson, N.: Quantitative analysis of faults and failures in a complex software system. IEEE Transactions on Software Engineering 26(8), 797–814 (2000)
10. Seliya, N.: Software quality analysis with limited prior knowledge of faults. Graduate Seminar, Wayne State University, Department of Computer Science (2006), http://www.cs.wayne.edu/graduateseminars/gradsem_f06/Slides/seliya_wsu_talk.ppt
11. Ma, Y., Guo, L., Cukic, B.: A Statistical Framework for the Prediction of Fault-Proneness
12. Boehm, B., Basili, V.R.: Software Defect Reduction Top 10 List. Software Manangement (January 2001)
13. Canul-Reich, J., Shoemaker, L., Hall, L.O.: Ensembles of Fuzzy Classifiers. In: The Proceedings of IEEE International Conference on Fuzzy Systems (FUZZ-IEEE 2007), July 23-26. Imperial College, London (2007)
14. Provost, F., Fawcett, T.: Robust Classification for Imprecise Environments. Machine Learning 42(3), 203–231 (2001)

An Efficient Approach to Improve Retrieval Rate in Content Based Image Retrieval Using MPEG-7 Features

K. Srujan Raju, K. Sreelatha, and Shriya Kumari

CMR Technical Campus, Medak (dist.), Hyderabad, India
{ksrujanraju,shriya.thakur}@gmail.com,
sreelathak2000@yahoo.com

Abstract. Content Based Image Retrieval is a technique of automatic indexing and retrieving of images .In order to improve the retrieval performance of images, this paper proposes an efficient approach for extracting and retrieving color images. The block diagram of our proposed approach to content-based image retrieval (CBIR) is given first, and then we introduce Histogram Euclidean distance (L2 distance) ,Cosine Distance and Histogram Intersection which are used to measure the image level similarity . The combined features of finding a set of weights of different distances represented by mini-max algorithm is used to declare the best match to the query image. Query images are used to retrieve images similar to the query image .Comparison and analysis of performance evaluation for features and similarity measures for the retrieved images with ground truth are shown. This proposed retrieval approach demonstrates a promising performance. Experiment shows that combined features are superior to any one of the three.

Keywords: Content Based Image Retrieval, Edge Histogram Descriptor(EHD), L2 distance, Cosine Distance, Histogram Intersection mini-max algorithm, MPEG-7.

1 Introduction

Content-based image retrieval has become a prominent research topic because of the proliferation of video and image data in digital form. Fast retrieval of images from large databases is an important problem that needs to be addressed. Image retrieval systems attempt to search through a database to find images that are perceptually similar to a query image.

CBIR aims to develop an efficient Visual Content Based technique to search, browse and retrieve relevant images from large scale digital image collections. Low-level visual features of the images such as color and texture are especially useful to represent and to compare images automatically [1]. A reference is made to MPEG-7 standard, by using color histogram descriptor and edge histogram descriptor. Most proposed CBIR techniques automatically extract low-level features (e.g. color, texture, shapes and layout of objects) to measure the similarities among the images differences.

S.C. Satapathy et al. (eds.), *ICT and Critical Infrastructure: Proceedings of the 48th Annual Convention of CSI - Volume II*, Advances in Intelligent Systems and Computing 249,
DOI: 10.1007/978-3-319-03095-1_37, © Springer International Publishing Switzerland 2014

In this paper, a technique for evaluating the performance measures for colored images with ground truths is calculated and the best performance measure is displayed.

2 Content Based Image Retrieval Framework

The block diagram of our proposed approach to CBIR is shown in Fig. 1. In the CBIR system, the relevance between a query and any target image is ranked according to a similarity measure .This is computed from the visual features [1] [3]. The similarity comparison is performed based on visual content descriptors including color histogram and edge histogram. The key steps in the figure 1 are features extraction and similarity measure on content.

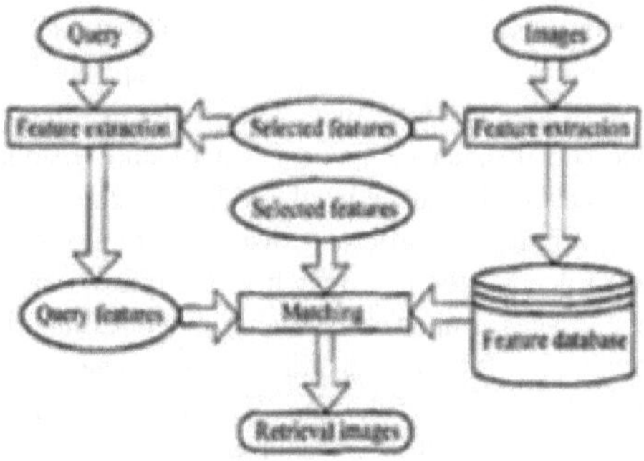

Fig. 1. Diagram for Content Based Image Retrieval System

3 Feature Extraction

3.1 Color Histogram

Color is an important visual attribute for both human perception and computer vision and one of the most widely used visual features in image or video retrieval. An appropriate color space and color quantization must be specified along with a histogram representation of an image for retrieval purpose. Histogram describes the global distribution of pixels of an image. The main advantage of a color histogram is its small sensitivity to variations in scale, rotation and translation of an image. Images can be in RGB, HSV or HSB .

For our approach we use an effective way to finding the similarity between the images by implementing the various distance measures as given in part 4 and comparing the distance measures and we propose an algorithm in part 4 which declares a fast solution to find the best match to the query image. This algorithm can be implemented to images having ground truths and can help in giving faster retrieval solutions.

3.2 Color Histogram Calculation

We define color histograms for images with ground truth as a set of bins where each bin denotes the probability of pixels in the image being of a particular color. A color histogram H for a given image is defined as a vector as

$$H = \{H[0],H[1],H[2],\ldots\ldots\ldots,H[i],\ldots H[n]\}$$

Where 'i' represents color in the color histogram i which is the number of pixels in color 'i' in the image and 'n' is the number of bins in the color histogram.

Typically, each pixel in an image will be assigned to a bin of a color histogram of that image, so for the color histogram of an image, the value of each bin is the number of pixels that has the same corresponding color. In order to compare images of different sizes, color histograms should be normalized. The normalized color histogram 'H' is defined as:

$$H' = \{H'[0],H'[1],H'[2],\ldots\ldots\ldots,H'[i],\ldots H'[n]\}$$

Where $H'[i]=H'[i]\,/p$, where p is the total number of pixels in an image. An ideal color space quantization presumes that distinct colors should not be located in the same Sub-cube and similar colors should be assigned to the same Sub-cube. Using few colors will decrease the possibility that similar colors are assigned to different bins, but it increases the possibility that distinct colors are assigned to the same bins, and that the information content of the images will decrease by a greater degree as well. On the other hand, color histograms with a large number of bins will contain more information about the content of images, thus decreasing the possibility of distinct colors which will be assigned to the same bins. However, they increase the possibility that similar colors will be assigned to different bins. Therefore, there is a trade-off in determining how many bins should be used in color histograms.

Converting RGB to HSV

The value is given by

$$V = \frac{R+G+B}{3}\ . \tag{1}$$

where the quantities R, G and B are the amounts of Red, Green and Blue components normalized to the range [0, 1]. The value is therefore just the average of the three color components. The saturation is given by

$$S = 1 - \frac{\min(R,G,B)}{I} = 1 - \frac{3}{R+G+B}\min(R,G,B)\ . \tag{2}$$

where the min(R; G; B) term is really just indicating the amount of white present. If any of R, G or B are zero, then there is no white and we have a pure color. The hue is given by

$$H = cos^{-1}\left(\frac{\frac{1}{2}\{(R-G)+(R-B)\}}{\{(R-G)^2 + (R-B)(G-B)\}\, 1/2}\right) \qquad (3)$$

3.3 Edge Histogram Calculation

One way of representing an important edge feature is to use a histogram. An edge histogram in the image space represents the frequency and the directionality of the brightness changes in the image. This Edge Histogram Descriptor (EHD) proposed for MPEG-7 expresses only the local edge distribution in the image. The MPEG-7 edge histogram is designed to contain only 80 bins describing the local edge distribution. These 80 histogram bins are the only standardized semantics for the MPEG-7 EHD. To improve the retrieval performance, global edge distribution is used. A given image is first Sub-divided into 4x4 Sub-images, and local edge histograms for each of these Sub-images are computed.

Edges are broadly grouped into five categories: Vertical, Horizontal, 45 degree diagonal, 135 degree diagonal, and Isotropic (non directional) as seen in Fig. 2. Thus, each local histogram has five bins corresponding to the above five categories. The image partitioned into 16 sub images results in 80 bins.

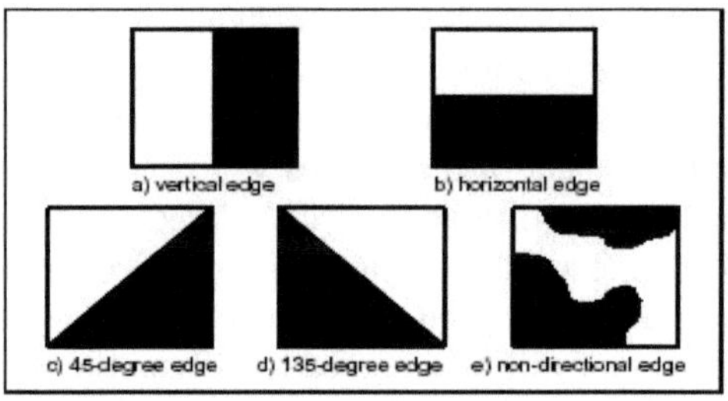

Fig. 2. Five Image Types

The semantics of the histogram bins form the normative part of the MPEG-7 standard descriptor. Specifically, starting from the Sub-image at (0,0) and ending at (3,3) , 16 Sub-images are visited in the raster scan order and corresponding local histogram bins are arranged accordingly.

For each Image-block, we determine whether there is at least an edge and which edge is predominant. When an edge exists, the predominant edge type among the five edge categories is also determined [2] . Then, the histogram value of the corresponding edge bin increases by one. Otherwise, for the monotone region in the image, the Image-block contains no edge. Each Image-block is classified into one of the five types of edge blocks or a non edge block .Each histogram bin value is normalized by the total number of Image-blocks including the non edge blocks. This in turn, implies that the information regarding non edge distribution in the Sub-image (smoothness) is also indirectly considered in the EHD. Now, the normalized bin values are quantized for binary representation. To extract directional edge features, we need to define small square Image-blocks in each Sub-image as shown in Fig. 3.

Specifically, we divide the image space into Non-overlapping square Image-blocks and then extract the edge information from them. The purpose of fixing the number of Image-blocks is to cope with the different sizes (resolutions) of the images. The size of the Image-block is assumed to be a multiple of 2.

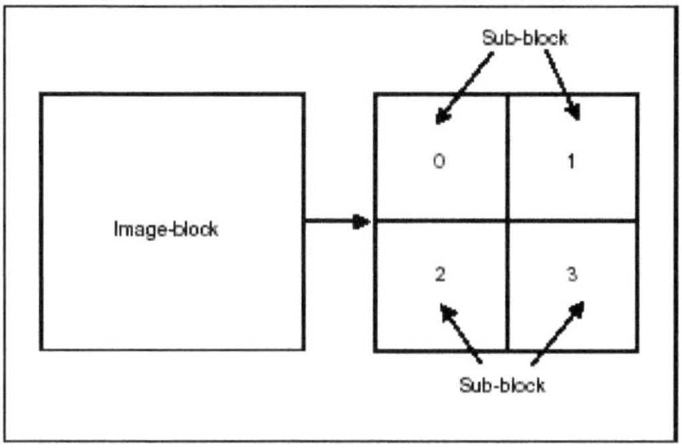

Fig. 3. Sub-blocks and their labeling

Simple method to extract an edge feature in the Image-block is to apply Digital filters in the spatial domain. First divide the Image-block into four Sub-blocks. Then, by assigning labels for four Sub-blocks from 0 to 3, it can be represented as the average gray levels for four Sub-blocks at $(i,j)^{th}$ Image-block as $a_0(i,j)$, $a_1(i,j)$, $a_2(i,j)$, and $a_3(i.j)$, respectively. The Filter coefficients for vertical, horizontal, 45-degree diagonal, 135-degree diagonal, and non-directional edges as $f_v(k)$, $f_h(k)$, $f_{d-45}(k)$, $f_{d-135}(k)$, and $f_{nd}(k)$ respectively, where k = 0,1,2,3 represents the location of the Sub-blocks. Now, the respective edge magnitudes $m_v(i,j)$, $m_{d-45}(i,j)$, $m_{d-135}(i,j)$, and $m_{nd}(i,j)$for $(i,j)^{th}$ image-block can be obtained as follows from eq. (4) to eq. (8). Within each Sub-image, the edge types are arranged in order as shown in (Fig 4).

$$m_v(i,j) = \left| \sum_{k=0}^{3} a_k(i,j) \times f_v(k) \right|. \quad (4)$$

$$m_h(i,j) = \left| \sum_{k=0}^{3} a_k(i,j) \times f_h(k) \right|. \quad (5)$$

$$m_{d-45}(i,j) = \left| \sum_{k=0}^{3} a_k(i,j) \times f_{d-45}(k) \right|. \quad (6)$$

$$m_{d-135}(i,j) = \left| \sum_{k=0}^{3} a_k(i,j) \times f_{d-135}(k) \right|. \quad (7)$$

$$m_{nd}(i,j) = \left| \sum_{k=0}^{3} a_k(i,j) \times f_{nd}(i,j) \right|. \quad (8)$$

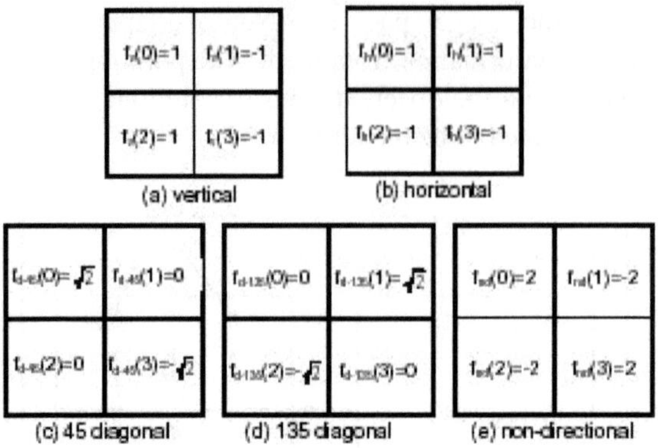

Fig. 4. Filter Coefficients for edge detection

The maximum value among five edge strengths obtained from (4) to (8) in the Image-block is considered to have the corresponding edge in it. Otherwise, the Image-block contains no edge.

$$max\{m_v(i,j), m_h(i,j), m_{d-45}(i,j), m_{d-135}(i,j), m_{nd}(i,j)\} \ . \quad (9)$$

4 Distance Measures

We implemented three kinds of histogram distance measures for a histogram H(i), i = 1,2,.., N.

(i) **L-2 Distance**

Defined as:

$$d(q,t) = \left[\sum_{m=1}^{N}(h_q(m) - h_t(m))^2\right]^{\frac{1}{2}} . \quad (10)$$

This Metric is uniform in terms of the Euclidean distance between vectors in feature space, but the vectors are not normalized to unit length.

(ii) **Cosine Distance:** If we normalize all vectors to unit length, and look at the angle between them, we have cosine distance,

Defined as:

$$d(q,t) = \frac{2}{\pi}\cos^{-1}\left[\frac{\sum_{m=1}^{N} h_q(m)h_t(m)}{\min(||h_q||, ||h_t||)}\right] \quad (11)$$

(iii) **Histogram Intersection :** The denominator term is needed for Non-normalized histogram features.

Defined as:

$$d'_{q,t} = 1 - \frac{\sum_{m=0}^{M-1} \min\,(h_q[m], h_t[m])}{\min\,(|h_q|, |h_t|)} \quad . \tag{12}$$

4.1 Combining Features and Making Decisions

This is a proposed algorithm to search for the maximum distance over the weight space which gives a faster solution to declare the best match to the query image . Here, it formulates the problem of combining different features as a problem of finding a set of weights of different feature distances, and then presents a Mini-Max algorithm in finding the Best-matching image.

Mini-Max Combination

If we take a query image q, with images in the database as i, and K features and thus K kinds of distances then,

$$d_k(q,i), k = 1, \ldots K, i = 1, \ldots N \; . \tag{13}$$

Assume we are going to combine them as a weighted sum of all the distances, i.e. the distance for an image in the database is written as:

$$D(q,i) = \sum_{k=1}^{K} w_k d_k\,(q,i) \quad . \tag{14}$$

$$\sum_{k=1}^{K} w_k = 1, w_k \geq 0, \forall\, k = 1,2, \ldots\ldots K \quad . \tag{15}$$

Now we want to search for a vector w that satisfies eq.14 and the resulting distance measure is "most close" to our subjective criteria. There are two candidate approaches.

1)Assign a set of weights based on the perceptual judgment of the designer on some image set (training) though in some cases such set of weights may perform poorly on certain new datasets.

2) The second approach is having no assumption about the subjective judgment of a user. For this , choose the image that minimizes the maximum distance over all valid set of weights as the best match. For every image i, searching for the maximum distance over the weight space turns out to be a linear program thus having a fast solution ie.: Maximize eq. 13 subject to eq.14 where all d's are the constants and $w_{k,k=1,\ldots.K}$ are unknown. The image with the minimum "Max-Distance" is declared as the best match to the query image. The Max -distance of every image i, is a linear function of w over [0, 1] . Thus the maximum either lies at w=0 or w=1, and comparing $d_c(q,i)$ and $d_e(q,i)$ is sufficient.

$$D(q,i) = wd_c(q,i) + (1-w)d_e(q,i), 0 \le w \le 1 \ . \tag{16}$$

Then we rank the maximum of $d_c(q,i)$and $d_e(q,i)$ for all i, and take n images with the least distance as our return result.

4.2 The Distance Measure of Combined Features

When we use combined features for image retrieval, the corresponding distance measure becomes different as a result of different feature extraction. The retrieval rate increases drastically.

5 Performance Evaluations in Content Based Image Retrieval

Testing the effectiveness of the Content Based Image Retrieval about testing how well it can retrieve similar images to the query image and how well the system prevents the return results that are not relevant to the source at all in the user point of view can be done as follows.

1) A sample query image is selected from one of the image category in the database.
2) The result images are returned, and a count of how many images are returned and how many of the returned images are similar to the query image are taken.
3) It is determining whether or not two images are similar and this is purely up to the user's perception. (Human perceptions can easily recognize the similarity between two images although in some cases, different users can give different opinions).
4) After images are retrieved, the system's effectiveness is determined. To achieve this, two evaluation measures are used here. The first measure is called recall. It is a measure of the ability of a system to present all relevant items. The recall is calculated as given below

$$Recall = \frac{Number\ of\ relevant\ items\ retrieved}{Number\ of\ relevant\ items\ in\ collection} \ .$$

The second measure is called precision. It is a measure of the ability of a system to present only relevant items. The equation for calculating precision is given below.

$$\text{Precision} = \frac{Number\ of\ relevant\ items\ retrieved}{Total\ number\ of\ items\ retrieved} \ .$$

The number of relevant items retrieved is the number of the returned images that are similar to the query image in this case. The number of relevant items in collection is the number of images that are in the same particular category with the query image. The total number of items retrieved is the number of images that are returned by the system.

6 Experimental Evaluation

A database consisting of different types of images ie.20 images with ground truths have been given and implemented in the system. A query image is selected to match the data set The relevant images are retrieved in the screen based on the given query image using different features of the color images. The result using different features are given below in fig(a),fig(b) and fig(c). The results are tabulated into table 2 . The precision and recall of all queries are obtained for the feature sets and the comparison between the distance measures to show the improved precision rate are tabulated for different data sets. The results show that the cosine distance measure gives better results than the L2 distance and histogram intersection.

Query image (theme: Buildings and cities)

Euclidean Distance (L2 distance)

Fig. a) Result set using Euclidean Distance

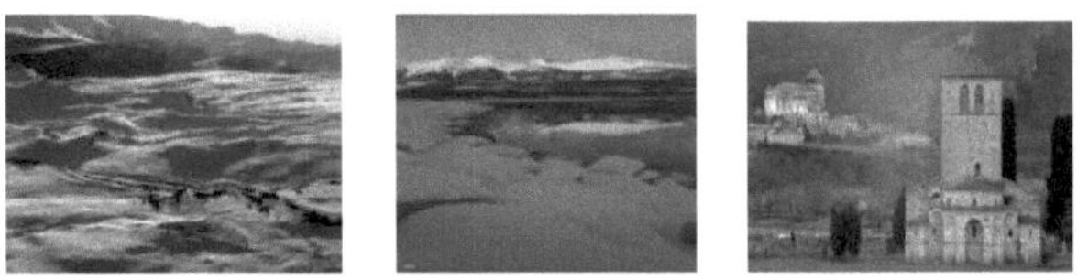

Using Histogram Intersection

Fig. b) Result set using Histogram Intersection

Using cosine distance

Fig. c) Result set using Cosine distance

Table 1. Comparision of Precision Rate

Different Parameters	Glass	Apple	Building and cities
Euclidean Distance	66%	86%	89%
Histogram Intersection	82%	81%	33%
Cosine Distance	63%	86%	53%

7 Conclusion

Efficient retrieval of images using different MPEG-7 Features is proposed in this paper. We have used a database consists of different image data sets with ground truths . The performance of the system is measured by using different features .The Cosine distance measure has given better results than the Histogram Intersection measure and Euclidean Distance Measure. The Mini-Max algorithm is proposed which directs us towards achieving faster and improved retrieval rate for obtaining a better precision value.

Acknowledgment. The Successful Completion of any task would be incomplete without expression of simple gratitude to the people who encouraged our work. Words are not enough to express the sense of gratitude towards everyone who directly or indirectly helped in completing this task.

I am thankful to my organization CMR Technical Campus, which provided good facilities to accomplish my work and would like to sincerely thank our Chairman , Director, Dean, HOD and my faculty members for giving great support, valuable suggestions and guidance in every aspect of my work.

References

[1] Shi, D.C., Xu, L., Han, U.: Image retrieval using both color and texture features. The Journal Of China Universities Of Posts and Telecommunications 14 Supplement, Article ID 1005-8885 (October 2007), S1-0094-06

[2] Won, C.S., Park, D.K., Park, S.-J.: Efficient use of MPEG-7 Edge Histogram Descriptor. ETRI Journal 24(1) (February 2002)
[3] Tahoun, M.A., Nagaty, K.A., El-Arief, T.I., Mohammed, M.A.-M.: A Robust Content-Based Image Retrieval System Using Multiple Features Representations. IEEE (2005)
[4] Bormane, D.S., Madugunki, M., Bhadoria, S., Dethe, C.G.: Comparison of Different CBIR Techniques. In: 2011 IEEE Conference, RSCOE, Pune, India (2011)

Performance Evaluation of Multiple Image Binarization Algorithms Using Multiple Metrics on Standard Image Databases

Sudipta Roy[1], Sangeet Saha[2], Ayan Dey[2], Soharab Hossain Shaikh[2], and Nabendu Chaki[1]

[1] Department of Computer Science & Engineering
[2] A.K. Choudhury School of Information Technology
University of Calcutta, Kolkata, West Bengal, India
sudiptaroy01@yahoo.com,
{sangeet.saha87,deyayan9,soharab.hossain}@gmail.com,
nabendu@ieee.org

Abstract. The area of image binarization has matured to a significant extent in last few years. There has been multiple, well-defined metrics for quantitative performance estimation of the existing techniques for binarization. However, it stills remains a problem to benchmark one binarization technique with another as different metrics are used to establish the comparative edges of different binarization approaches. In this paper, an experimental work is reported that uses three different metrics for quantitative performance evaluation of seven binarization techniques applied on four different types of images: Arial, Texture, Degraded text and MRI. Based on visually and experimentally the most appropriate methods for binarization of images have been identified for each of the four classes under consideration. We have used standard image databases along with the archived reference images, as available, for experimental purpose.

Keywords: Iterative Partitioning method, Image Thresholding, Reference Image, Misclassification Error, Relative Foreground Area Error.

1 Introduction

Gray scale image and binary image are the two key variations among digital images. In a gray scale image a particular pixel can take an intensity value between 0 to 255 where in a binary image it could take only two values, either 0 or 1. The procedure to convert a gray scale image into a binary image is known as image binarization. It has diverse applications in many research areas especially for document image analysis, medical image processing and other types of captured image processing. Binarization is often the first stage in multi-phase image processing and image analysis applications. Binarization is crucial for medical image analysis. Almost all the medical image processing techniques need to produce a binary form of original image. The benefit of having a binary image is that it reduces the complexity of the data and simplifies the process of image segmentation. Another important research area is restoration of

S.C. Satapathy et al. (eds.), *ICT and Critical Infrastructure: Proceedings of the 48th Annual Convention of CSI - Volume II*, Advances in Intelligent Systems and Computing 249,
DOI: 10.1007/978-3-319-03095-1_38,

ancient degraded document by using a proper binarization method. Degradations in document images result from poor quality of paper, printing process, ink blot and fading, document aging, redundant marks etc. The goal of document restoration is to remove some of these artifacts and recover an image i.e. close to one that would obtain under ideal printing and imaging condition.

Different techniques have been proposed for image binarization. Most of these are good for a particular image category. In this paper, the performances of existing binarization techniques are elaborately discussed and analyzed considering four different kinds of image databases. The primary goal has been to find the most suitable binarization method for specific image type(s). The evaluation is done on standard image databases [9,10,11] by both visual inspection as well as measuring the performance metrics [1, 2] with respect to reference images created following the method presented in [1].

Seven existing binarization algorithms are compared in this paper. Those are Otsu[3], Niblack[4], Kapur[5], Bernsen[6], Sauvola[7], Th-mean[8], and the Iterative Partitioning method proposed in [1]. The four different kinds of image databases taken for experimental evaluation they are *Arial images, Texture images, MRI images and Degraded Document images* from standard image databases. Except for the MRI database, the reference images are formed by using a majority voting scheme [1] computed for the seven binarization techniques mentioned above. However, majority voting produces visibly very poor quality of binary image for the MRI images. Thus, the reference images for MRI have been taken by using photo editing software (Photoshop CS2, version 1.0). The main goal of the study is to categorize the best binarization algorithm for a particular image type by quantitative analysis based on experimental results.

The rest of this paper is organized as follow: in section 2, the existing works on different image binarization techniques have been discussed. Section 3 presents test image-databases, metrics used for quantitative performance analysis and the experimental results obtained on different images from the selected databases. The paper ends with concluding remarks in section 4.

2 Background

Image binarization is typically treated as a thresholding operation on a grayscale image. It can be classified as global and local thresholding operations. In the global methods (global thresholding), a single calculated threshold value is used to classify image pixels into object or background classes, while for the local methods (adaptive thresholding), information contained in the neighbourhood of a particular pixel guides the threshold value for that pixel.

Otsu's method, as proposed in [3], is a global thresholding method and based on discriminate analysis. Also Kapur's method [5] is an extension of Otsu's method. Niblack [4] proposed a method that calculates a pixel wise thresholding by shifting a rectangular window across the image. This method varies the threshold over the image, based on the local mean and local standard deviation. Bernsen's method [6] correctly classifies poor quality images with inhomogeneous background and is therefore suitable for text shadow boundaries removal. This method calculates the local threshold value based on the mean value of the minimum and maximum intensities of pixels within a window. Sauvola's method [7] overcomes the shortcomings of

Niblack's method. A new Iterative partitioning binarization method is proposed in [1] which uses Otsu's method for calculating threshold.

In present paper, authors use the process of iterative partitioning [1] as a framework (as in Fig.1) for evaluating the performances of a number of other binarization methods.

3 Performance Metrics and Experimental Results

Here we compare different binarization methods visually as well as using multiple performance evaluation metrics for each of four category of images like Degraded text, Arial, Texture, Magnetic Resonance imaging (MRI). We want to classify which method will give us proper binarization for different types of images.

The binarization methods have been tested with a variety of images and to evaluate the performance of a new methodology [1]. We have taken images from standard image databases for experimental purpose. The Arial, Texture image sets are taken from USC_SIPI [9] and degraded document test images are taken from DIBCO database[10]. The MRI image is collected from internet and a database [11]. We used performance evaluation metric for example Misclassification error (ME), Relative Foreground Area Error (RAE) and F-measure [1].

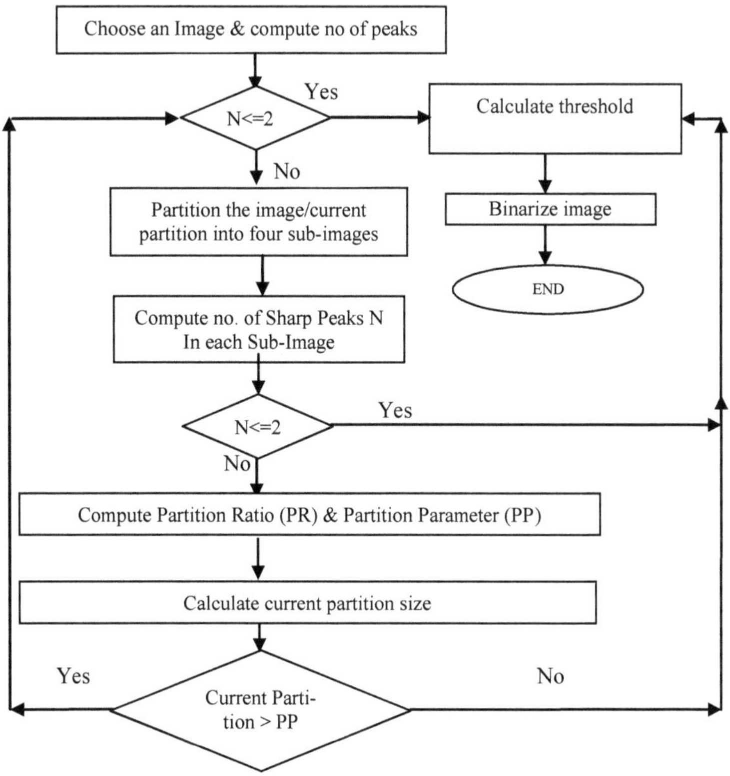

Fig. 1. Iterative Partitioning as a framework

3.1 Performance Evaluation Methodology

Building Reference Images

Initially reference image is formed by the majority voting [1] scheme using five methods e.g. Otsu [3], Niblack [4], Bernsen [6], Sauvola [7] and Iterative partitioning [1]. In order to generate the reference image, all the binary images have been consulted pixel-by-pixel. Each pixel in the reference image is set to 1 if the respective pixel of most of the methods is 1 in their respective binary image. Otherwise the pixel is set to 0.

A measure is proposed in [12], in which total seven binarization method has been used i.e. Otsu, Niblack, Bernsen, Sauvola, Th-Mean, Kapur and iterative as a framework, the threshold of each algorithm is being calculated and only those algorithms are take to majority voting for reference image creation in which its threshold doesn't deviate more than deviational parameter from the average value of all threshold values pre-calculated but this method also not free from failure of forming proper reference images mainly for Arial, MRI and texture. Finally, the reference image has been created by majority voting scheme [2] with two more methods i.e. Kapur and Th-mean. It has been observed that this new method is giving satisfactory results for creation of proper reference images in most of cases like degraded, Arial, and texture images but failed in case of MRI images, thus we manually create the reference image with the help of photo editing tool.

Results and Analysis

Initially the evaluation mechanism started with existing five binarization algorithm Otsu[3], Niblack[4], Bernsen[6], Sauvola[7] individually on each images of four different image category. The Iterative partition method [1] is taken as a framework and applied on different algorithms instead of Otsu as a basic binarization technique as proposed in [1]. The performance is evaluated for each category as shown in Fig1.

The performance evaluation is done in two ways, at first observation is made by visual inspection and later to verify the observation by studying different metrics ME, RAE, F- measure as stated in the later. A reference image is created using the majority voting scheme. Output images for Arial image type applying existing binarization technique, using iterative as a framework and there corresponding reference image individually are shown in fig 2[a-h].

Here we have seen visually that Otsu, Bernsen produce good results; Niblack and Sauvola are not applicable and iterative portioning does not produce good result. For overall Ariel image databases by visual inspection Bernsen, Otsu, produce better results but this visual inspection does not match with our performance evaluation based on metric basis classification; this is because of improper reference image.

Due to improper reference image we cannot match visually with quantifiably for degraded, texture and MRI images. For overall degraded image databases by visual inspection Sauvola, Bernsen, Otsu, and iterative portioning produce better results. For overall texture image databases by visual inspection Bernsen, Otsu, and iterative portioning produce better results.

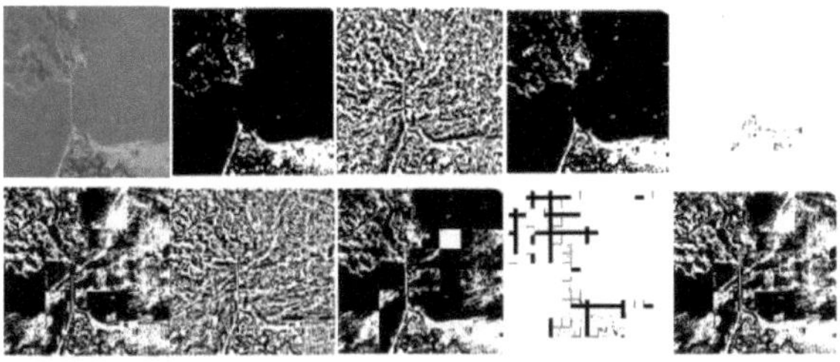

Fig. 2 [a-h]. Output of different binarization methods on Arial Images; a) Original image, binary images by b) Otsu, c) Niblack, d) Bernsen, e) Sauvola, and iterative partitioning using f) Otsu, g) Niblack, h) Bernsen, i) Sauvola, j) Reference imageby majority voting scheme

For overall MRI image databases by visual inspection Bernsen, Otsu, produce better results. But all those visual inspection does not match with our performance evaluation metric basis classification; thus we failed to reach in a concrete conclusion to specify which algorithm is suitable for which particular types of image this is because of improper reference image.

As an alternate measure, the resizing and cropping in the order $2^k \times 2^k$ of test images did not affect much for creation of a reference image. The choice of certain size lies on the fact that iterative portioning method [1] sub divides each image if sharp peak>2; but this process did not support for improving our reference image creation.

After thorough sequence of experiments, two experiments are proven to be giving good results in case of constructing ground truth image. The first measure is 30% deviation methodology described above [12] in which total seven binarization method has been used but this method also not free from failure of creating proper reference images mainly for Ariel, MRI and texture.

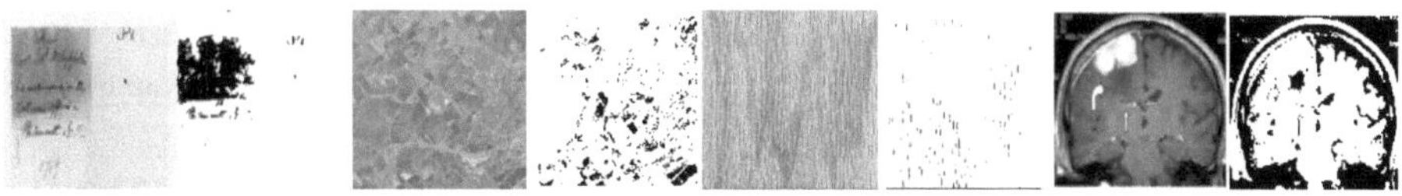

Fig. 3[a-h]. Improper reference image corresponding is different type of original image; a) degraded image and b) reference image by 30% deviation, c) Arial image and d) its equivalent reference image by 30% deviation, e) Texture image and f) its equivalent reference image by 30% deviation, g) MRI of brain image, h) its equivalent reference image by 30% deviation

The second one and present experiment, the reference image has been created by majority voting scheme as described in [1] but with total 7 binarization algorithm i.e.in additions with two new methods i.e. Kapur and Th-mean.

i) On Arial Images: According to the results listed in the tables 1,2 and 3 and by visual inspection it can be said that Kapur, Otsu, Bernsen's methods are preferred methods. Iterative partitioning using all existing algorithm is not suitable for Arial images at all. The reason behind this can be concluded as follows as Arial image is the satellite view or view taken from top of an area at a particular instant, so the intensity distribution in different portions of a real image are not uniform. As a result iterative partitioning using existing methods is not suitable here. Otsu, Bernsen, Kapur's methods are giving better results to both visually and metrics wise.

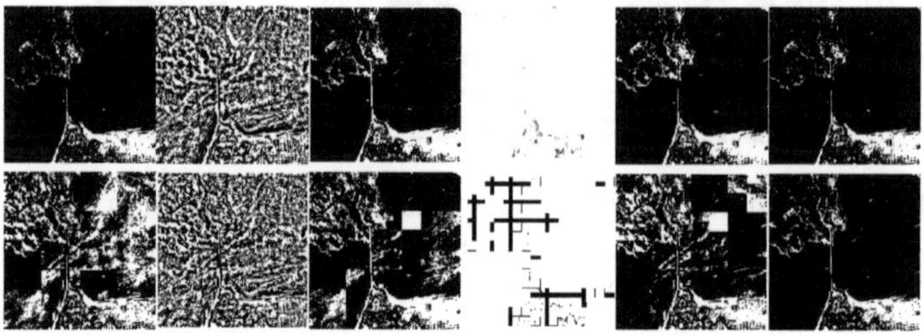

Fig. 4[a-l]. Output of different binarization methods on Arial Images by; a) Otsu, b) Niblack, c) Bernsen, d) Sauvola,, e) Kapur, f) Th-mean and iterative partitioning using g) Otsu, h) Niblack, i) Bernsen, j) Sauvola, k) Kapur l) Reference image by majority voting scheme

Table 1. For ME Measurement of Arial Images

Image	Otsu	Niblack	Bernsen	Sauvola	Kapur	Th-mean	Iter_ Otsu	Iter_ Nib	Iter_ Bern	Iter_ Sauv	Iter_ Kapu
A1	0.02	0.45	0.01	0.87	0.01	0.02	0.29	0.45	0.14	0.81	0.15
A2	0.03	0.22	0.06	0.26	0.05	0.03	0.19	0.23	0.04	0.39	0.14
A3	0.12	0.13	0.14	0.40	0.27	0.10	0.08	0.15	0.14	0.44	0.16
A4	0.06	0.14	0.13	0.32	0.27	0.05	0.11	0.15	0.12	0.41	0.12
A5	0.01	0.25	0.01	0.56	0.19	0.01	0.20	0.28	0.11	0.55	0.19

Table 2. For RE Measurement of Arial Images

image	Otsu	Niblack	Bernsen	Sauvola	Kapur	Th-mean	Iter_ Otsu	Iter_ Nib	Iter_ Bern	Iter_ Sauv	Iter_ Kapur
A1	16.06	76.92	6.61	87.27	3.97	16.06	68.19	77.09	52.03	85.94	52.08
A2	2.62	15.81	8.07	26.86	6.10	3.90	18.22	16.22	0.03	7.58	7.88
A3	20.24	0.89	15.12	39.99	30.93	1.39	9.30	0.65	17.38	24.61	6.09
A4	6.95	7.703	13.63	33.71	29.83	17.1	14.78	7.07	14.62	15.59	11.10
A5	3.17	29.53	1.81	59.24	48.31	8.39	15.59	29.99	14.68	49.98	17.64

Table 3. For F-measure of Arial Images

image	Otsu	Niblack	Bernsen	Sauvola	Kapur	Th-mean	Iter_ Otsu	Iter_ Nib	Iter_ Bern	Iter_ _Sauv	Iter_ Kapur
A1	95.67	34.81	93.15	21.25	93.57	95.67	45.16	34.47	61.84	22.80	60.97
A2	85.80	82.67	86.51	70.91	82.71	86.20	84.13	82.33	84.73	72.90	83.76
A3	92.49	88.18	84.36	69.00	75.49	92.99	92.56	87.52	83.65	72.07	88.14
A4	90.40	87.43	82.40	70.76	73.39	90.88	89.95	86.52	82.33	73.02	88.88
A5	91.45	71.02	90.33	50.96	78.41	91.06	76.60	69.73	81.49	54.82	76.70

ii) On Degraded Documenut Images: Sauvola, Kapur and Otsu's methods are preferred. Bernsen also produces good results. Iterative partition method using Otsu, Kapur and Bernsen existing algorithm is suitable for degraded document images except a few images, However, Niblack and iterative Niblack are not suitable. So it has been seen that all local thresholding algorithms are working better than global thresholding techniques. If the histogram of the text overlaps with that of the background, they result in improper binary images. Various local binarization methods like Sauvola and Kapur's methods are working well as these methods use local information around a pixel to classify it as either text or background. Results of some degraded document images on different methods and their corresponding reference images using majority voting scheme are shown below.

Fig. 5[a-l]. Output of different binarization methods on Degraded Documents by; a)Otsu, b)Niblack, c)Bernsen, d)Sauvola, e)Kapur, f)Th-mean methods and iterative partitioning using g)Otsu , h)Niblack, i)Bernsen, j)Sauvola, k)Kapur method, l)Reference image by majority voting

Table 4. For ME Measurement of Degraded Images

image	Otsu	Niblack	Bernsen	Sauvola	Kapur	Th-mean	Iter_ _Otsu	Iter_ Nib	Iter_ Bern	Iter_ Sauv	Iter_ Kapur
D1	0.13	0.28	0.08	0.08	0.05	0.13	0.20	0.28	0.06	0.10	0.07
D2	0.01	0.17	0.05	0.09	0.02	0.015	0.03	0.17	0.05	0.10	0.04
D3	0.01	0.18	0.03	0.07	0.01	0.01	0.05	0.19	0.03	0.07	0.04
D4	0.00	0.23	0.01	0.01	0.00	0.00	0.16	0.24	0.01	0.04	0.07
D5	0.09	0.19	0.10	0.14	0.14	0.09	0.08	0.20	0.08	0.18	0.09

Table 5. For RAE Measurement of Degraded Images

image	Otsu	Niblack	Bernsen	Sauvola	Kapur	Th-mean	Iter_ _Otsu	Iter_ Nib	Iter_ Bern	Iter_ Sauv	Iter_ Kapur
D1	14.23	26.74	4.13	8.63	5.02	14.23	20.47	27.13	6.29	4.74	2.82
D2	1.34	18.30	5.20	9.40	2.25	1.76	2.96	18.63	6.19	7.15	2.41
D3	0.61	18.24	3.33	6.94	1.66	0.8	4.73	19.74	3.86	6.26	1.70
D4	0.21	24.05	1.29	1.27	0.30	0.29	16.62	24.75	1.36	1.09	7.73
D5	11.72	15.17	7.27	15.15	15.28	11.72	7.69	16.25	8.52	8.37	4.11

Table 6. For F-Measure Measurement of Degraded Images

image	Otsu	Niblack	Bernsen	Sauvola	Kapur	Th-mean	Iter_Otsu	Iter_Nib	Iter_Bern	Iter_Sauv	Iter_Kapur
D1	86.70	83.92	83.59	81.44	83.32	86.70	85.74	83.87	82.66	82.39	85.27
D2	98.21	95.83	95.61	93.35	97.10	98.29	98.25	95.67	95.05	93.48	97.59
D3	96.96	94.99	95.29	93.39	97.01	96.99	96.83	94.50	95.01	93.37	96.50
D4	89.57	88.77	88.93	88.94	89.42	89.60	89.71	88.62	88.89	89.11	90.04
D5	94.64	91.67	89.93	86.63	86.55	94.64	94.27	91.30	90.24	87.63	93.22

iii) On Texture Images: Similarly, it has been seen form the experimental results that Kapur, Otsu, Bernsen methods are preferred methods. Iterative partitioning using those algorithm in the framework is suitable for Texture images except Sauvola's method because Texture image generally contains uniform distribution of intensity i.e. within a texture image different portion are almost same type and hence application of iterative partitioning method using Kapur, Otsu, Bernsen methods is giving satisfactory results than the Kapur, Otsu, Bernsen methods itself.

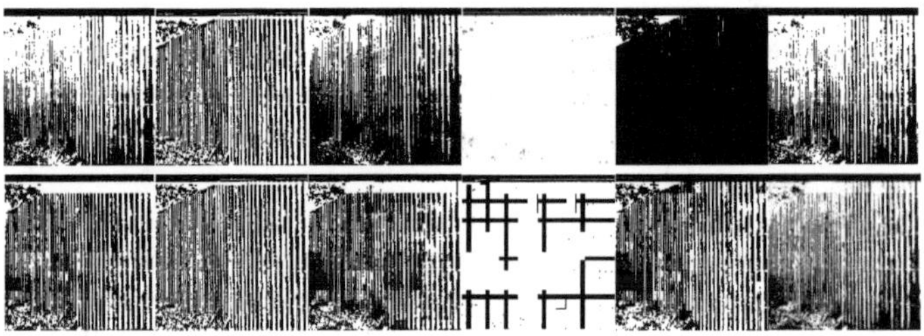

Fig. 6[a-l]. Output of different binarization methods of Texture image by; a)Otsu, c)Niblack, d)Bernsen, e)Sauvola, f)Kapur, g)Th-mean methods and iterative partitioning usingh)Otsu , i)Niblack, j)Bernsen, k)Sauvola, l)Kapur methods, m) Reference image by majority voting

Table 7. For ME Measurement of Texture Images

Image	Otsu	Niblack	Bernsen	Sauvola	Kapur	Th-mean	Iter_Otsu	Iter_Nib	Iter_Bern	Iter_Sauv	Iter_Kapur
T1	0.02	0.19	0.04	0.07	0.03	0.02	0.07	0.20	0.03	0.13	0.06
T2	0.01	0.33	0.04	0.39	0.03	0.01	0.22	0.34	0.15	0.41	0.33
T3	0.00	0.10	0.00	0.08	0.08	0.01	0.06	0.12	0.05	0.09	0.08
T4	0.03	0.06	0.04	0.21	0.10	0.03	0.04	0.07	0.03	0.32	0.06
T5	0.03	0.10	0.07	0.13	0.10	0.03	0.04	0.11	0.05	0.24	0.06

Table 8. For RAE Measurement of Texture Images

images	Otsu	Niblack	Bernsen	Sauvola	Kapur	Th-mean	Iter_Otsu	Iter_Nib	Iter_Bern	Iter_Sauv	Iter_Kapur
T1	2.23	20.88	4.69	7.61	3.41	2.82	6.97	22.94	2.13	1.13	5.36
T2	2.79	3.24	7.44	42.46	5.61	2.78	12.30	4.76	10.83	37.54	19.70
T3	0.29	4.08	0.33	9.39	9.86	1.21	6.20	5.60	4.81	10.52	9.48
T4	3.94	0.57	5.26	25.92	14.26	5.25	4.45	0.88	1.15	2.26	5.16
T5	3.01	11.33	8.25	15.30	12.41	3.8	5.49	12.08	6.28	0.22	6.08

Table 9. For F-Measure Measurement of Texture Images

images	Otsu	Niblack	Bernsen	Sauvola	Kapur	Th-mean	Iter_Otsu	Iter_Nib	Iter_Bern	Iter_Sauv	Iter_Kapur
T1	95.55	93.89	93.21	91.68	93.84	95.62	95.40	93.38	94.42	91.95	95.46
T2	92.42	76.07	92.53	66.01	89.39	92.42	80.18	75.43	83.53	67.40	75.645
T3	99.61	96.96	99.49	94.74	99.16	99.66	96.17	96.64	96.95	94.10	99.101
T4	96.12	94.29	96.45	81.60	88.75	96.45	95.75	93.66	94.77	82.17	95.33
T5	96.28	95.489	92.03	88.16	89.81	96.46	96.33	94.99	93.15	88.24	95.72

iv) On MRI: After evaluating the performance of existing binarization methods on each image of the MRI database the conclusion can be drawn that Kapur, Bernsen, and Otsu's methods are preferred over others. Iterative partition method using all existing algorithm and other methods are not suitable for MRI images. In *MRI images* there must be a dark background with the actual portion of MRI image. When different binarization methods are applied on that, then as a result the algorithms binarize whole image without accurately detecting the region of interest giving a black background, so some method fails to give proper segmentation. Iterative partitioning method is not applicable for this image category. Kapur, Bernsen produce good results and for some images the performance of Otsu's method is satisfactory.

Figure 8 shows sample test images.

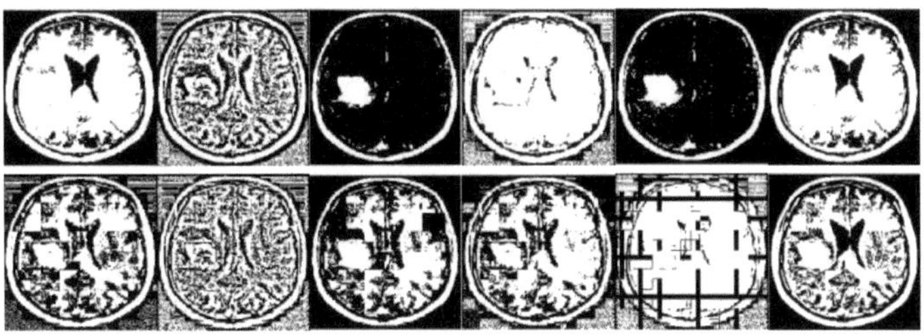

Fig. 7[a-l]. Output of different binarization methods of MRI image by; a)Otsu, b)Niblack, c)Bernsen, d)Sauvola, e)Kapur, f)Th-mean and iterative partitioning using g)Otsu , h)Niblack, i)Bernsen, j)Sauvola, k)Kapur method, l)Reference image by majority voting scheme

Table 10. For ME Measurement of MRI Images

Image	Otsu	Niblack	Bernsen	Sauvola	Kapur	Th-mean	Iter_Otsu	Iter_Nib	Iter_Bern	Iter_Sauv	Iter_Kapur
M1	0.24	0.23	0.17	0.25	0.16	0.24	0.22	0.23	0.21	0.23	0.22
M2	0.23	0.21	0.16	0.25	0.12	0.23	0.20	0.22	0.20	0.23	0.20
M3	0.16	0.22	0.12	0.28	0.12	0.16	0.16	0.21	0.15	0.25	0.19
M4	0.17	0.18	0.14	0.18	0.14	0.17	0.16	0.18	0.15	0.18	0.18
M5	0.21	0.22	0.17	0.22	0.17	0.21	0.21	0.21	0.20	0.21	0.23

Table 11. For RAE Measurement of MRI Images

Ima-ge	Otsu	Niblack	Bernsen	Sauvola	Kapur	Th-mean	Iter_Otsu	Iter_Nib	Iter_Bern	Iter_Sauv	Iter_Kapur
M1	13.36	3.16	53.62	20.66	63.37	13.36	13.74	3.13	17.19	0.25	10.46
M2	40.95	33.01	15.40	51.04	72.16	40.95	26.51	33.90	21.92	40.07	24.06
M3	17.49	33.33	84.44	54.36	94.29	17.49	7.900	33.96	33.51	46.16	14.32
M4	13.44	27.06	85.40	8.2	94.11	13.44	47.14	25.44	60.88	24.01	26.28
M5	10.53	13.18	69.79	1.64	92.86	10.53	15.50	13.68	30.49	13.27	2.30

Table 12. For *F-Measure Measurement of MRI Images*

Image	Otsu	Niblack	Bernsen	Sauvola	Kapur	Th-mean	Iter_Otsu	Iter_Nib	Iter_Bern	Iter_Sauv	Iter_Kapur
M1	3.06	3.44	2.07	2.98	1.54	3.06	3.32	3.51	3.23	3.55	3.50
M2	6.67	7.04	8.31	5.86	3.73	6.66	7.51	6.98	7.69	6.71	7.62
M3	9.16	7.70	0.58	5.87	0.07	9.16	9.80	7.73	9.03	6.65	8.97
M4	8.16	7.62	0.32	7.96	0.88	8.16	7.72	7.64	6.53	8.43	7.95
M5	9.44	9.03	3.03	9.18	0.05	9.44	9.46	9.21	9.15	9.45	9.23

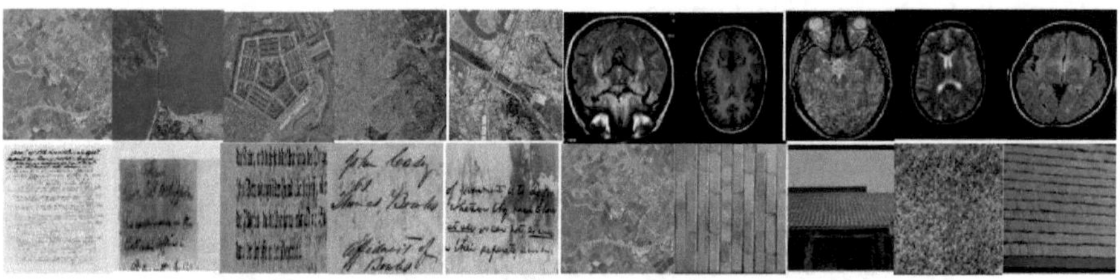

Fig. 8. Sample Test Images

4 Conclusions

The selection of proper binarization algorithm for digital images is a challenging task. It is tough to reach any conclusion by quantitative study using diverse metrics as this kind of objective measure does not always corroborate with the visual qualitative finding for different metrics applied on diverse types of images. On the other hand, qualitative estimation alone lacks standardization and hence often falls short of a trusted opinion. The size of the image database plays an important role for experimental study. With more images, we can exhaustively apply algorithms and the experimental model gains robustness. The work in this paper may be extended by using larger image databases with diverse images and relevant reference images. Further, the presented experimental framework may be used to benchmark newer image binarization algorithms against the existing methods that we have implemented here.

References

1. Shaikh, S.H., Maity, A.K., Chaki, N.: A New Image Binarization Method using Iterative Partitioning. Journal on Machine Vision and Applications 24(2), 337–350 (2013)
2. Sezgin, M., Sankur, B.: Survey over image thresholding techniques and quantitative performance evaluation. Journal of Electronic Imaging 13(1), 146–165 (2004)
3. Otsu, N.: A Threshold Selection Method from Gray Level Histograms. IEEE Transactions on Systems, Man, and Cybernetics, SMC-9, 62–66 (1979)
4. Niblack, W.: An Introduction to Digital Image Processing, pp. 115–116. Prentice Hall, Eaglewood Cliffs (1986)
5. Kapur, J.N., Sahoo, P.K., Wong, A.K.C.: A New Method for Gray-Level Picture Thresholding Using the Entropy of the Histogram. Computer Vision, Graphics, and Image Processing 29, 273–285 (1985)
6. Bernsen, J.: Dynamic thresholding of gray level images. In: ICPR 1986: Proceedings of the International Conference on Pattern Recognition, pp. 1251–1255 (1986)
7. Sauvola, J., Pietikainen, M.: Adaptive document image binarization. Pattern Recognition 33(2), 225–236 (2000)
8. Gonzalez, R.C., Woods, R.E.: Digital Image Processing, 2nd edn. Prentice Hall, New Jersey (2002)
9. USC-SIPI Image Database, University of Southern California, Signal and Image Processing Institute, `http://sipi.usc.edu/database/`
10. Library of Congress website, `http://www.loc.gov/` & DIBCO database
11. BrainWeb: Simulated Brain Database, `http://www.bic.mni.mcgill.ca/brainweb`
12. Dey, A., Shaikh, S.H., Saeed, K., Chaki, N.: Modified Majority Voting Algorithm towards Creating Reference Image for Binarization. In: International Conference on Computer Science, Engineering and Applications (ICCSEA 2013) (2013)
13. Kefali, A., Sari, T., Sellami, M.: Evaluation of several binarization techniques for old Arabic documents Images. In: The First International Symposium on Modeling and Implementing Complex Systems (MISC 2010), Constantine, Algeria, pp. 88–99 (2010)
14. Gatos, B., Pratikakis, I., Perantonis, S.J.: Adaptive degraded document image binarization. Pattern Recogn., 317–327 (2006)
15. Sontasundaram, K., Kalavathi, I.: Medical Image Binarization Using Square Wave Representation. In: Balasubramaniam, P. (ed.) ICLICC 2011. CCIS, vol. 140, pp. 152–158. Springer, Heidelberg (2011)
16. Banerjee, J., Namboodiri, A.M., Jawahar, C.V.: Contextual Restoration of Severely Degraded Document Images. In: Proceedings of IEEE Computer Society Conference on Computer Vision and Pattern Recognition (CVPR 2009), Miami Beach, Florida, USA, pp. 20–25 (June 2009)

17. Leedham, G., Varma, S., Patankar, A., Govindaraju, V.: Separating text and background in degraded document images – a comparison of global thresholding techniques for multistage thresholding. IEEE Computer Society
18. N.V.: A binarization algorithm for historical manuscripts. In: 12th WSEAS International Conference on Communications, Heraklion, Greece, pp. 23–25 (July 2008)
19. Smith, E.H.B.: An analysis of binarization ground truthing. In: 9th IAPR International Workshop on Document Analysis Systems (2010)
20. Stathis, P., Kavallieratou, E., Papamarkos, N.: An evaluation technique for binarization algorithms. J. Univ. Comput. Sci. 14(18), 3011–3030 (2008)

A Signal Processing Approach for Eucaryotic Gene Identification

Mihir Narayan Mohanty

ITER, Siksha 'O' Anusandhan University, Bhubaneswar, Odisha, 751030, India
mihirmohanty@soauniversity.ac.in

Abstract. Signal processing has a great role in innovative developments in technology. Its techniques have been applied mostly in every field of science and engineering. In the field of bioinformatics it has to play an important role in the study of biomedical applications. Accurate prediction of protein coding regions (Exons) from genomic sequences is an increasing demand for bioinformatics research. Many progresses made in the identification of protein coding regions during the last few decades. But the performances of the identification methods still required to be improved. This paper deals with the identification of protein-coding regions of the DNA sequence mainly focus on analysis of the gene introns. Applications of signal processing tools like spectral analysis, digital filtering of DNA sequences are explored. It has been tried to develop a new method to predict protein coding regions based on the fact that most of exon sequences have a 3-base periodicity. The period-3 property found in exons helps signal processing based time-domain and frequency domain methods to predict these regions efficiently. Also, an efficient technique has been developed for the identification of protein coding region based on the period-3 behavior of codon sequences. It is based on time domain periodogram approach. Here it has been identified the protein coding regions, wherein we reduced the background noise significantly and improve the identification efficiency. In addition to this also comparison is done between time domain periodogram and the existing frequency based techniques. Simulation results obtained are shown the effectiveness of the proposed methods. This proves that the DSP techniques have important applications in obtaining useful information from these gene sequences.

Keywords: Signal Processing Method, DNA, Exon, Frequency domain methods, GCF, TDP.

1 Introduction

Various methods of signal processing applied to molecular biology is popular in recent years. For the past two decades, the major area of application of digital signal processing is in the field of bio-informatics, especially for the analysis of genomic sequences. Genomic signal processing deals with applying signal processing techniques to sequences occurring in life, particularly the DNA sequences. Such techniques become extremely useful in obtaining information from the large sets of data, in the form of the human genome. Gene analysis is becoming a major topic because

S.C. Satapathy et al. (eds.), *ICT and Critical Infrastructure: Proceedings of the 48th Annual Convention of CSI - Volume II*, Advances in Intelligent Systems and Computing 249,
DOI: 10.1007/978-3-319-03095-1_39, © Springer International Publishing Switzerland 2014

of their implication in a large number of diseases, mainly those involved in the mechanisms of heredity. If the gene location, composition and structure is known then the genetic component will be responsible for the considered disease. The problem of gene prediction is an important problem in the field of bioinformatics [1]. It presents many difficulties especially in eukaryotes where exons (coding regions) are interrupted by introns (non-coding regions).

The main goal of most of the signal processing researchers is to predict the gene locations in a genomic sequence maximizing the prediction accuracy. Identifications of protein-coding regions (exons) in eukaryotic gene structure are one of the future challenges for analysis of the genomic sequence. Many digital signal processing methods are proposed for identification of genes till date, each having some advantages and disadvantages. In addition, it is indispensable to develop different identification methods since combining different methods may greatly improve the identification accuracy.

2 DNA Representation

Recently, a number of numerical DNA sequence representations have evolved in order to transform the DNA sequence analysis problems from the traditional string processing domain to the discrete signal processing domain. On the other hand, the coding regions (exons) detection problem has received a special attention due to the 3-base periodicity property of exons which can be easily detected using simple discrete signal processing techniques. The 3-base periodicity in the nucleotide arrangement is evidenced as a sharp peak at frequency $f = 1/3$ in the frequency domain power spectrum. Many digital signal processing techniques have been used to automatically distinguish the protein coding regions (exons) from non-coding regions (introns) in a DNA sequence. DNA sequence is expressed as a string of characters. A DNA sequence is comprised of gene and intergenic regions. In Procaryotes, genes mostly occur as uninterrupted stretches of DNA. In Eucaryotes, genes consists of exons (Protein-coding regions) separated by introns (Protein non-coding regions). Only exons participate in protein synthesis.

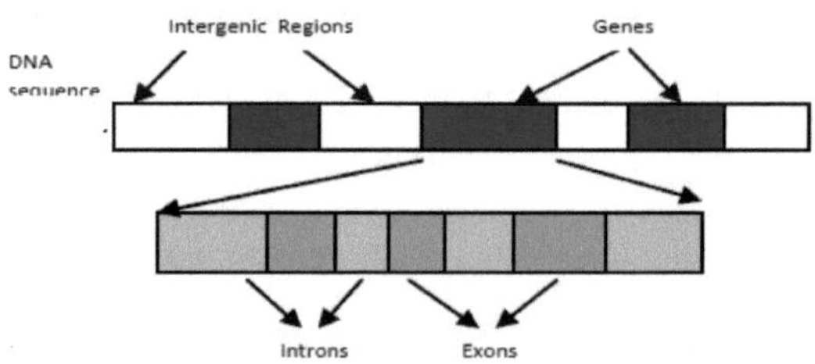

Fig. 1. A DNA sequence showing genes.exons

DNA sequence is symbolically represented by a character string consisting of four alphabets, A, T, C, and G, representing four distinct nucleotide bases, Adenine, Thymine, Cytosine and Guanine, respectively.

But to apply digital signal processing methods for the analysis of gene sequence, these characters are needed to be assigned some numerical values. In recent years, a number of schemes have been introduced to map DNA characters (i.e., A, C, G, and T) into numeric values. These each offer different properties and map the DNA sequences into between one and four numerical sequences. However an ideal mapping should preserve the period-3 property of the DNA sequence. Some possible desirable properties of a DNA representation include:

- Each nucleotide has equal 'weight' (e.g., magnitude), since there is no biological evidence to suggest that one is more 'important' than other.
- Distances between all pairs of nucleotides should be equal, since there is no biological evidence to suggest that any pair is 'closer' than another.
- Representations should be compact; in particular redundancy should be minimized.
- Representations should allow access to a range of mathematical analysis tools.

3 Signal Processing Based Exon Prediction Methods

Genomic information is discrete in nature. This enables the application of digital signal processing techniques for the analysis of genome sequence. It has been studied that protein-coding regions (exons) of DNA sequences exhibit a period-3 property due to the codon structure. This basic property of DNA sequence is used as the foundation for all DSP methods for exon prediction. There has numerous time-domain and frequency-domain algorithms been evolved for the analysis of genome sequence. Here first focus on frequency-domain methods, and then discuss their short-comings and then go for time-domain methods.

3.1 Frequency-Domain Algorithms

The periodicity of 3 suggests that —the discrete-Fourier transform of a gene sequence gives dominant peaks in protein-coding regions [2, 3]. Thus above the peaks, predefined threshold value are identified as exons. This section gives a brief review on some of the frequency-domain based gene prediction methods.

Sliding Window DFT

The discrete Fourier transform is the earliest method for the analysis of DNA sequence. In this approach, the DNA sequence is expressed as four binary indicator sequences corresponding to base A, T, C, and G.

For a sequence 'GTAATCGGTACAT......' the indicator sequence $X_A[n]$ = 0011000001010..... Similarly, other base indicator sequences are obtained.

The DFT of a length-N block of $X_A[n]$ is defined as

$$X_A[k] = \sum_{n=0}^{N-1} X_A(n) e^{-j2\pi kn/N}, 0 \leq k \leq N-1 \quad (1)$$

Where X[k] is a frequency-domain sequence of length N. Similarly, DFTs $X_T[k]$, $X_C[k]$, and $X_G[k]$ are obtained using equation(1).Due to the period-3 property, the DFT coefficients corresponding to k=N/3 (where N is chosen to be a multiple of 3 e.g. 351) are large [3]. Thus the plot of spectral content (SC) measure

$$SC[k] = |X_A[k]|^2 + |X_T[k]|^2 + |X_C[k]|^2 + |X_G[k]|^2 \quad (2)$$

would show a peak at the sample value k=N/3. But the plot of

$$SC[k] = |a.X_A[k] + t.X_T[k] + c.X_C[k] + g.X_G[k]|^2 \quad (3)$$

would provide better result [4] than the one given in equation (2).The later measure is known as "optimized spectral content measure".

In this paper, it has been assumed that, a=t =c=g=1. Also some may assign any optimal complex values to a, t, c, g. Apart from these measures, various modifications of the SC measures are available like Spectral Rotation (SR) measure [5], Paired Spectral Content (PSC) measure [6] , Paired and Weighted Spectral Rotation (PWSR) measure [7].The calculation of the DFT at a single frequency k=N/3 is sufficient. The strength of the peak varies depending on the gene. The window sequence is then slide by one or more bases and S[N/3] is recalculated. The window length can range from hundred to few thousand. The window length N should be large enough so that the periodicity effect dominates the background 1/f noise. However, a large window takes longer computation time and poorer base-domain resolution.

Representations such as Voss, Z-curve, Tetrahedron introduces redundancy in the DNA numerical representation. To avoid these problems, two methods were proposed, i.e., Paired-Numeric, and Frequency of Nucleotide-Occurrence [8]. The DNA character string can also be mapped into a numerical sequence using genetic algorithm [9].

4 Proposed Methods

4.1 Coding Statistics of DNA Sequence

For the identification of protein coding regions in the DNA based on period-3 property, DNA character string is foremost translated in four different numerical sequences called indicator sequence, one for each base. This representation is called Voss Representation [10]. A comparative study of various mapping techniques using Anti-notch filter has been done in [11] for better identification of protein coding regions. Among various mapping techniques Voss Representation, Z-curve and tetrahedron mapping techniques give almost similar results and provides maximum identification accuracy. For detailed analysis Voss mapping technique is preferred due to its simplicity. DNA character string $D(n)$ is converted into binary signals $U_A(n)$, $U_C(n)$, $U_G(n)$, $U_T(n)$, which tell us about presence or absence of corresponding nucleotides at any location n.

Let us consider a DNA character string as: $d(n)$ = [ACTGGCTACGTTAAGC]

Then the different binary signals will be as follows:

$u_A(n)$ = [1000000100001100]
$u_C(n)$ = [0100010010000001]
$u_G(n)$ = [0001100001000010]
$u_T(n)$ = [0010001000110000]

Presence of exons in the four binary indicator sequences is determined by the Discrete Fourier Transform method [12-16]. Let the Discrete Fourier transform of the indicator sequence of length N be defined by

$$U_x(k) = \sum_{n=0}^{N-1} u_x(n).e^{-\frac{j2\pi kn}{N}}, \qquad 0 \leq k \leq N\text{-}1 \tag{4}$$

for x = A,T,C and G. Then the spectral content is given by the absolute value of power of DFT coefficients

$$S(k) = \sum_{k=0}^{N-1} |U_x(k)|^2 \tag{5}$$

The presence of coding regions is indicated by a peak at $k = N/3$ in the plot of S (k) versus k. This method can be effectively used to predict the exon regions of the DNA sequences. In this method coding regions are identified by evaluating S [$N/3$] over a window of N samples, then sliding the window by one or more samples and recalculating S [$N/3$]. It is carried out over the entire DNA sequence. The peaks in the spectra obtained by the sliding window DFT correspond to the protein-coding regions. This approach involves large computations which can pose difficulty for online evaluation of protein coding regions.

4.2 Time Domain Periodogram (TDP)

Time domain periodogram provides as a byproduct a well behaved estimate of signal intensity. TDP algorithm can be implemented in integer arithmetic using only additions and comparisons. It is derived and shown in Figure 2.

The time domain periodogram algorithm (TDPA) is just like the Average Magnitude Difference Function (AMDF) operates on the signal waveform in the time domain. The principles of the TDPA can be explained in the following way [2, 12]. In order to test whether DNA samples $s(1)$, $s(2)$, . .., $s(n)$ contain a period of length N, we write the DNA samples in matrix form with rows containing sub-sequences of length equal to period being tested 'k'.

$$\begin{bmatrix} S(1) & \cdots & S(N) \\ \vdots & \ddots & \vdots \\ S((M-1)(N+1)) & \cdots & S(MN) \end{bmatrix}$$

The columns are then summed, and the maximum and minimum of the resulting vector are then used to derive the final estimate of the degree of periodicity at period k, as follows:

$$TDP_{vector}[k] = \sum_{n=1}^{N/k} b[kn - (k-1), \ldots, \sum_{n=1}^{N/k} b[kn] \tag{6}$$

$$TDP[k] = \max(TDP_{vector}[k] - \min(TDP_{vector}[k]) \tag{7}$$

Practically, TDP[k] will produce a peak if correlation exists at period $k = 3$. It can be shown [17] that for large N, TDP[k] has a very sharp peak, enabling accurate detection of periodicity. For efficient implementation, equation (6) can be simplified to:

$$TDP[k] = max\,(TDP_{vector}[k]). \tag{8}$$

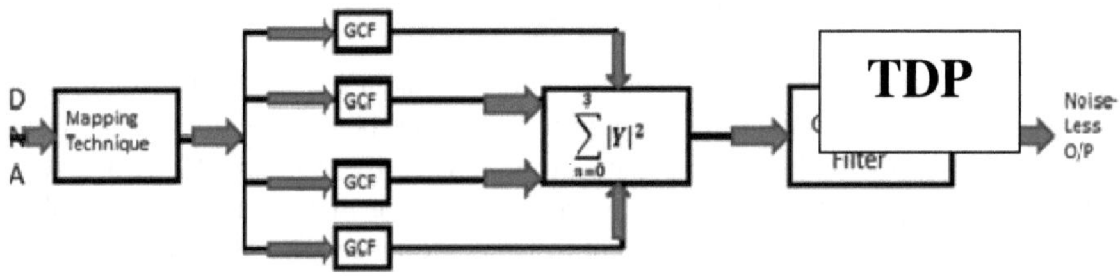

Fig. 2. Block diagram of proposed method (TDP method)

5 Results and Discussion

The genomic sequences of several genes of different organisms were taken for the identification of exons using various methods. The performance of proposed algorithm is compared with the DFT spectral content method. Comparative performance is illustrated in Figure 3 and 4.These figures illustrate the suppression of noise present in non-coding regions. For evaluation of gene structure prediction programs different measures of prediction have been discussed. A very simple algorithm based on direct capturing of noise and removing it from resultant power spectrum has been developed for improving accuracy in the detection of protein coding regions by DFT spectral content method.

The single indicator sequence using paired numeric properties of nucleotides is used as numerical representation [11].

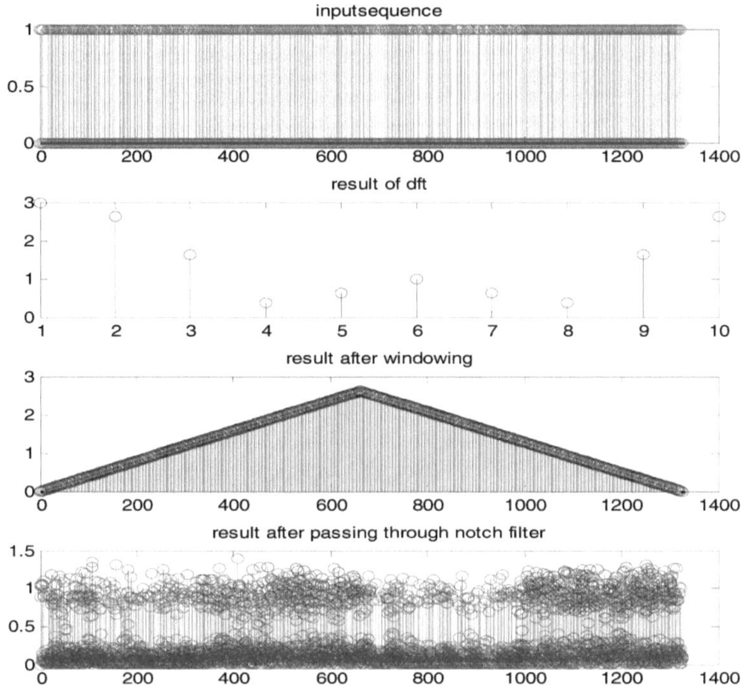

Fig. 3. The input gene sequence and its DFT

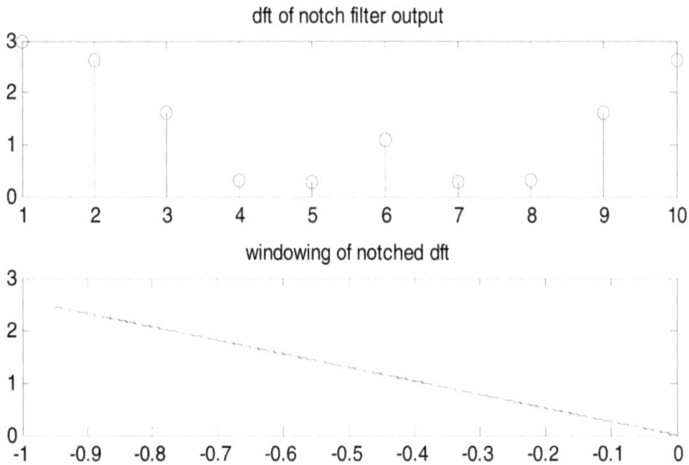

Fig. 4. Windowed DFT and output with Notch filter

The GCF and TDP algorithms show high peak at exon location as compared to existing methods as shown in figures. Figure 5 shows the exon prediction results for gene F55F11.4a with the accession no: AF099922 in the C.elegans chromosome –III using GCF algorithm. Figure 6 shows the result when the TDP algorithm is used wherein we can observe that background noise is reduced to a significant level. The five peaks corresponding to the exons can be seen at the respective locations (1....111, 1600....1929, 3186....3449, 4537....4716, 6329....6677).

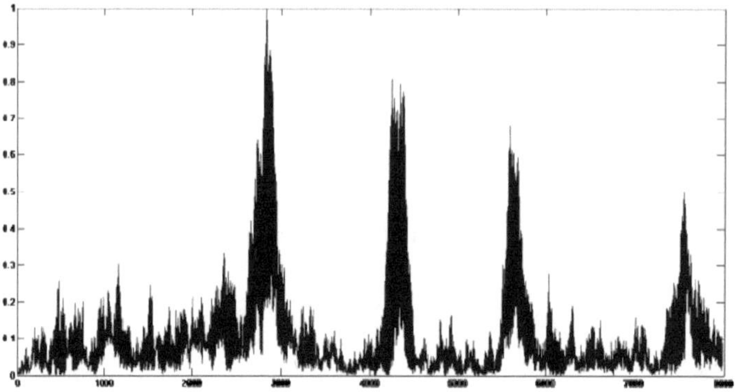

Fig. 5. Result of existing GCF (C. elegans)

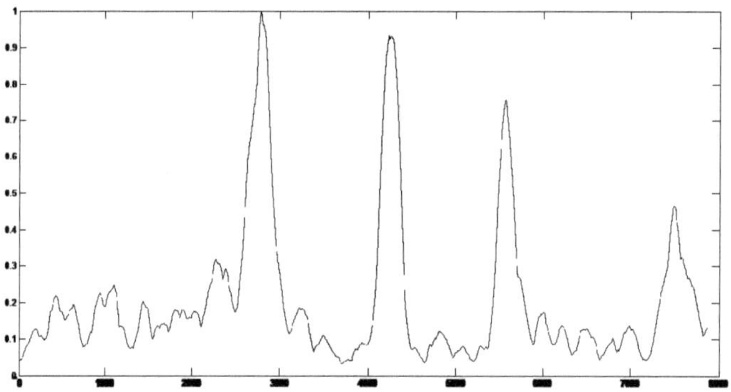

Fig. 6. Result of GCF followed by TDP (C. elegans)

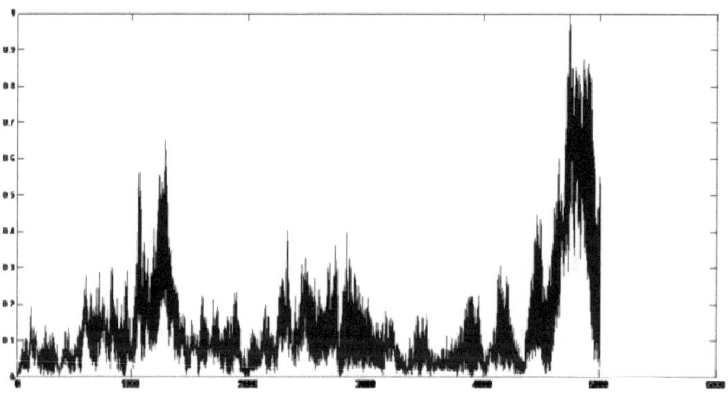

Fig. 7. Result of existing GCF (AB003306)

The output of the generalized comb filter (GCF) as shown in Figure 5 is not of much advantage as background noise is very prominent in the output. To achieve a more efficient identification of exons in the DNA sequence these background noise is to be reduced remarkably. Figure 6 shows its result with TDP method. Similarly, Figure 7 and 8 show the exon identification result for gene AB00306 using GCF, and TDP methods respectively. This indicates a sharp peak at its exon location (1020….1217, 2207….2513, 4543….4832).

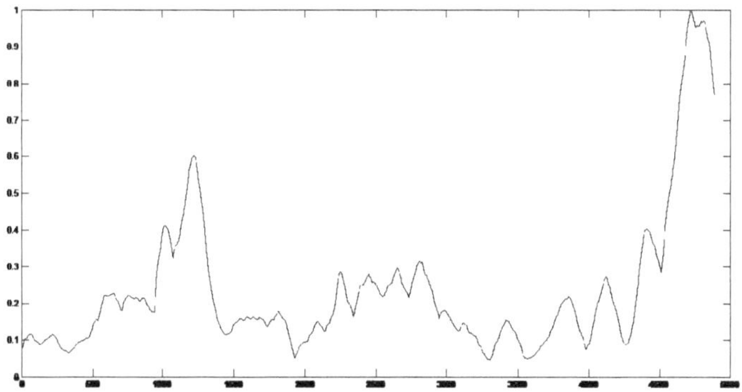

Fig. 8. Result of proposed GCF followed by TDP (AB003306)

The TDP algorithm senses the exons effectively by showing high peak at gene locations. Again these methods detect all the exons at their respective locations. TDP algorithm is found to be more efficient than other two approaches for gene identification as it reduces the background noise considerably.

6 Conclusion

The inability of classical DSP techniques like Fourier Transforms and time-frequency analysis to eliminate the noise and identify properly and detecting smaller length exons is tackled. The accuracy was achieved by using proposed method. In the existing method GCF all the five exons are depicted by the sharp peaks; however the background noise is the factor of concern in this approach. In order to improve the identification accuracy, GCF followed by TDP approach has proposed wherein all the exons at different protein coding regions are identified effectively. This proposed method reduces the background noise significantly with improved identification accuracy. Hence it can be used as an alternative to other DSP approach for better identification of exons. There are some obvious peaks at the locations of biological function segments with period property and no peaks at other regions. Adapting some other coding methods into DSP techniques can also be experimented with as future work.

References

1. Mount, D.W.: Bioinformatics: Genome and Sequence Analysis, 2nd edn. Cold Spring Harbor Laboratory Press, New York (2004)
2. Anastassiou, D.: Genomic signal processing. IEEE Signal Processing Magazine 18(4), 8–20 (2001)
3. Anastassiou, D.: DSP in Genomics. In: Proc. IEEE Int. Conf. Acoustics, Speech, Signal Processing, Salt lake City, Utah, USA, pp. 1053–1056 (May 2001)
4. Voss, R.F.: Evolution of long-range fractal correlations and 1/f noise in DNA base sequences. Phy. Rev. Lett. 68(25), 3805–3808 (1992)
5. Nair, A.S., Sreenathan, S.P.: A Coding measure scheme employing electron-ion interaction pseudo potential (EIIP). Bioinformation by Bioinformatics Publishing Group
6. Ahmad, M., Abdullah, A., Buragga, K.: A Novel Optimized Approach for Gene Identification in DNA Sequences. Journal of Applied Sciences, 806–814 (2011)
7. Deng, S., Chen, Z., Ding, G., Li, Y.: Prediction of Protein Coding Regions by Combining Fourier and Wavelet transform. In: Proc. of the 3rd IEEE Int. Congress on Image and Signal Processing, pp. 4113–4117 (October 2010)
8. Akhtar, M., Epps, J., Ambikairajah, E.: Signal Processing in sequence analysis: Advances in Eukaryotic Gene Prediction. IEEE Journal of Selected Topics in Signal Processing 2(3), 310–321 (2008)
9. Biju, V.G., Mydhili, P.: Genetic Algorithm Based indicator sequence method for exon prediction. In: Proc. of the IEEE int. Conf. on Advances in Computing, Control, & Telecommunication Technologies, pp. 856–858 (December 2009)
10. Voss, R.: Evolution of Long-Range Fractal Correlations and 1/f Noise in DNA Base Sequences. Physical Review Letters 68(25), 3805–3808 (1992)
11. Hota, M.K., Srivastava, V.K.: Identification of Protein coding regions using Antinotch filters. Digital Signal Processing 22(6), 869–877 (2012)
12. Silverman, B.D., Linsker, R.: A Measure of DNA Periodicity. Journal of Theoretical Biology 118(3), 295–300 (1986)
13. Tiwari, S., Ramachandran, S., Bhattacharya, A., Bhatta-charya, S., Ramaswamy, R.: Identification of Probable Genes by Fourier Analysis of Genomic Sequences. Bioinformatics 13(3), 263–270 (1997)
14. Anastassiou, D.: Digital Signal Processing of Bio-molecular Sequences. Technical Report, Columbia University, 2000-20-041 (April 2000)
15. Anastassiou, D.: Frequency-Domain Analysis of Biomolecular Sequences. Bioinformatics 16(12), 1073–1082 (2000)
16. Fickett, J.W.: The gene identification problem: an overview for developers. Comput. Chem. 20, 103–118 (1996)
17. Vaidyanathan, P.P., Yoon, B.J.: The role of signal processing concepts in genomics and proteomics. J. Franklin Inst. 341, 111–135 (2004)

Addressing Analyzability in Terms of Object Oriented Design Complexity

Suhel Ahmad Khan and Raees Ahmad Khan

Department of Information Technology
Babasaheb Bhimrao Ambedkar University, Lucknow, UP, India
ahmadsuhel28@gmail.com, khanraees@yahoo.com

Abstract. This paper defines analyzability in terms of design complexity. A model has been developed to evaluate analyzability in terms of object oriented design complexity. The evolved model has also been implemented and validated with realistic data. An experimental data shows the impact of design complexity on analyzability of software in accordance with its anticipated influence and importance.

Keywords: Analyzability, Complexity, Object oriented design.

1 Introduction

Every sort of security is a decisive concern for each functional or nonfunctional activity of an organization that means *'Business Processes Security is a key to success of any venture'*. Many of the hardware and software security tools are designed to provide security trial during online activities of an organization but still there is a need of a robust security system to afford threats of to and fro data transactions. In current scenario organizations are involved in developing pioneer software tools or resources to face customary happening security pressure and to protect their business process against the nasty approach. Security defects are part of software development [1].

Secure development is generally associated with the process of designing reliable, stable, bug and vulnerability free software. Nowadays, software is becoming more vulnerable to attacks because of its increasing complexity, connectivity and extensibilities. Increased network connectivity, complexity of code has led to an increase number of security breaches-something that can damage secure software development programs [2].

Design complexity is not the only factor that makes things hard to understand but with enough complexity anything can become harder to understand. Complexity should not be exceeded to a certain limit. For any complex software application or design the effort estimation or rectification of cause of failure is complicated. Due to enough complexity the things are harder to understand and there is a possibility of increased defects. As complexity increases, the analyzability of software decreases till critical point is achieved.

S.C. Satapathy et al. (eds.), *ICT and Critical Infrastructure: Proceedings of the 48th Annual Convention of CSI - Volume II*, Advances in Intelligent Systems and Computing 249,
DOI: 10.1007/978-3-319-03095-1_40, © Springer International Publishing Switzerland 2014

2 Analyzability and Complexity

According to ISO/9126 the analyzability is defined as *'the capability of the software product to be diagnosed for deficiencies or causes of failures in the software, or for the parts to be modified to be identified'* [3]. The rule of analyzability permits the use of unambiguous design. It provides an ability to diagnose deficiencies or reason of a breakdown and essential requisite for modification to implement required changes in design in a realistic period of time. According to the definition of Analyzability, it's a process in which we find out the root cause of failure or deficiencies [4, 5]. Analyzability discusses the effort estimation and reason of a breakdown. This can be revealed as follows:

1. Effort to diagnose & rectifying the defects
2. Cause of failure

One of the most valuable means of addressing the basic root cause of malicious software is to improve the current flawed development process. During the development process, one of the most influencing factors of the system's security is the complexity of its design. The effects of complexity and analyzability with quality and its attributes are discussed in figure 1.

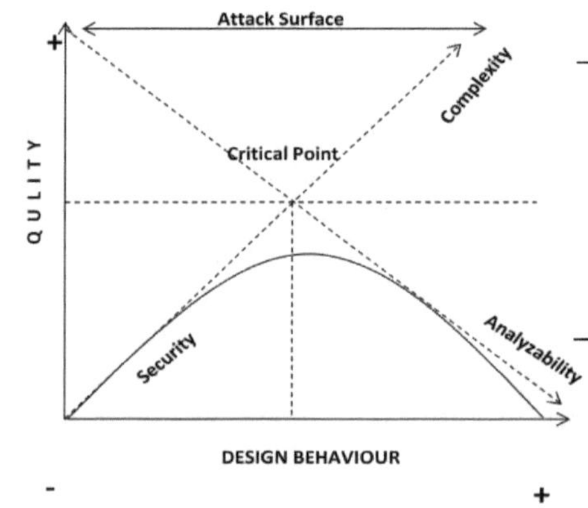

$$A \; \alpha \; \frac{K}{C} \quad ¥ \quad A > 0, C > 0,$$

$$\text{Analyzability } A = K * e^{-\alpha * \text{Complexity}} \text{------(A)}$$

Where C= Complexity
A= Analyzability

Fig. 1. Impact diagram of complexity and analyzability

3 Related Review Work

A paper entitled 'Introducing Temporal Analyzability Late in the Lifecycle of Complex Real Time Systems', written by Anders Wall, Johan Andersson, Jonas Neander, Christer Norstrom, Martin Lembke, of the Department of Computer Engineering, Malardalen University, Sweden, regarding an approach to rediscover analyzability with temporal behavior into complex real time control system at ABB robotics. This works advocates that early stage handling is the most appropriate way to minimize system complexity. It also proposed analytical models and expresses execution times as statistical distributions for existing systems. The proposed work contributes with prototype ART-ML modeling language as well as tools for measuring execution time and length of message queues in the existing system. A case study of robot controller at ABB Robotics is engaged to simulate the temporal behavior. The proposed model was abstract in terms of both functional dependencies and temporal behavior. The impact of altering the behavior of a software component can be visible to all products that use it. Maintaining such a complex system requires careful analyses to be carried prior to adding new functions or redesigning parts of the system to not introduce unnecessary complexity and thereby increasing both the development and maintenance cost [6].

A research paper regarding complexity and analyzability which explore the views regarding complexity is 'if the time shapes of the relation between the system are of higher order (e.g. A system connected by relationships involving third order delays in more complex than one involving simple continuous integration). According to the researchers point of view the complexity can be defined through a number of elements of state space, the range of the values for state space, the degree of connectivity of the state space and nonlinearity in functional and time shapes of relationship between state space elements. The theories that express the realization concept to understand the particular stage of complexity by understanding implication of theories through equilibrium, sensitivity and transit response is being highlighted in this particular work [7].

Quantifying the Analyzability of Software Architectures, a research paper contributed by Eric Bouwers, Jos´e Pedro Correia, Arie van Deursen, Joost Visser, by Software Improvement Group, Amsterdam, in WICSA, 2011, exploring the component decomposition having strong influence on system's analyzability, but it raises the question how much the system can be decomposed. The results of this contribution conclude an empirical exploration of how systems are decomposed into top-level components. The discussion concludes the requirement of a metric to measure the balance of components which is usable across all life-cycle phases of a project. In which manner the metric is correlated with the opinion of experts about the analyzability of a software system. The whole works argue only for analyzability and use of decomposition methodology for better analyzability through reduced complexity. No direct discussion is being used to control analyzability or address analyzability in terms of design complexity or real complex software applications [8].

On the basis of Factor-Criteria-Metric quality model, the key attributes like analyzability, changeability and stability are identified to correlate with structural complexity metrics for the object-oriented software maintainability. Kiewkanya combined eight types of relationship to define metrics for two sub-characteristics of the object-oriented design maintainability, that is, understandability and modifiability. These relationships are generalization, aggregation, composition, common association, association class, dependency, realization, and class complexity. The proposed work also discusses the effects of relationships on stability and the effects of complexity on analyzability to complete three maintainability sub-characteristics. Complexity is classified as structural complexity and cognitive complexity. Cognitive complexity is defined in terms of readability and understandability of the software by the human. Understandability and modifiability are comparable to analyzability and changeability respectively [9].

4 Model Formulation

Analyzability describes the capability to discover the primary cause of failure in the software design or model. Basic engineering concept for designing software are that, they should be simple, follow strong model with hygiene interfaces, reduced amount of bugs and easy to use to maintain. In the software design model, analyzability describes the capability to discover the primary cause of failure. Analyzability discussion is an early indicator of software security. The minimum time taken to identify failures is an excellent sign to use the optimization programs for improved security. Concerning these issues on analyzability establish a direct relationship with design complexity. The values of analyzability and software design complexity are calculated by proposed method [5, 10] on the given dataset [11, 12].

$$\textbf{Analyzability A= 0.328 * e}^{-0.34\ \text{Complexity}} \text{ ------------ } \textbf{(B)}$$

5 Model Implementation

This part of the paper reviews, how well the model effectively quantifies the analyzability & Design Complexity of object oriented design using the class diagram. An object oriented class diagram of different versions with the complete data set is collected and calculated values are plotted to show the model effectiveness. The calculated values of equation are displayed in table 1.

Table 1. Values of complexity & analyzability

	Complexity	Analyzability
CD6	2.10	0.157
CD2	2.20	0.154
CD1	2.33	0.149
CD3	2.60	0.134
CD5	5.66	0.045
CD4	8.14	0.019

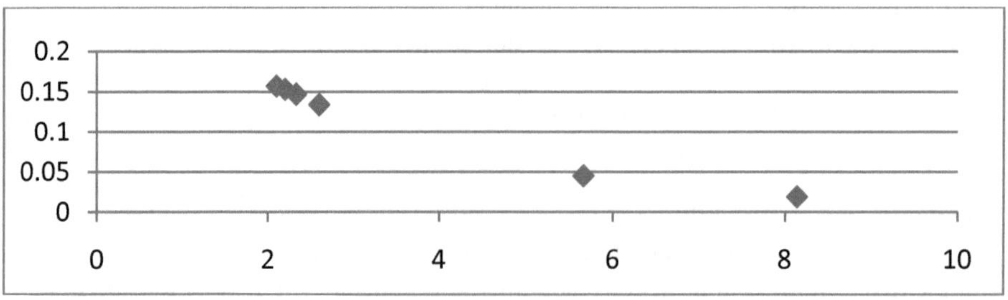

Fig. 2. Plotted values of complexity & analyzability

The result of analyzability calculation with respect to object oriented design complexity is displayed in table 1. The plotted values between analyzability and complexity are depicted in figure 2. The interpretation of plotted values is that as design complexity increases, the analyzability of software decreases. Complexity has a negative impact on analyzability of object oriented software. The complexity is closely related to the analyzability. More complex software makes systems unreliable, because design complications reduce the software analyzability. High analyzability predicts that software is easier to understand and maintainable. The Pearson correlation in table 2 provides a strong justification of our assumption that equation of analyzability in terms of design complexity is highly acceptable.

Table 2. Pearson correlation table

Correlations			
		A	B
A	Pearson Correlation	**1**	**-.962****
	Sig. (2-tailed)		**.000**
	N	**13**	**13**
B	Pearson Correlation	**-.962****	**1**
	Sig. (2-tailed)	**.000**	
	N	**13**	**13**

**. Correlation is significant at the 0.01 level (2-tailed).

6 Empirical Validation

The empirical studies are an essential part of software engineering practices to evaluate the proposed techniques for appropriate execution. It reviews the need of improvement for effectiveness and efficiency in used practice. Empirical validation is the best practice to claim the model acceptance. In view of this fact, an experimental validation of the proposed model for analyzability evaluation model has been carried out using sample tryouts. In order to validate analyzability, the data is collected from different class hierarchies of online shopping management system for ten projects. The known analyzability rating for the given projects (P_1-P_{10}) is shown in table 3[13].

Table 3. Known analyzability rating

Project No.	P_1	P_2	P_3	P_4	P_5	P_6	P_7	P_8	P_9	P_{10}
Analyzability Rating	10	1	7	4	2	5	6	9	8	3

Using the same set of data for the given projects (P_1-P_{10}), analyzability was computed using the proposed Model and ranked. The details are shown in table 4.

Table 4. Calculated analyzability rating using model

Project No.	P_1	P_2	P_3	P_4	P_5	P_6	P_7	P_8	P_9	P_{10}
Analyzability Rating	8	1	6	5	3	7	4	7	9	2

Table 5. Computed ranking, actual ranking and their correlations

Projects → / **Analyzability Ranking** ↓	P_1	P_2	P_3	P_4	P_5	P_6	P_7	P_8	P_9	P_{10}
Computed Ranking	8	1	6	5	3	7	4	7	9	2
Known Ranking	10	1	7	4	2	5	6	9	8	3
$\sum d^2$	4	0	1	1	1	4	4	4	1	1
r_s	0.976	1	0.994	0.994	0.994	0.976	0.976	0.976	0.994	0.994
r_s **>0.781**	√	√	√	√	√	√	√	√	√	√

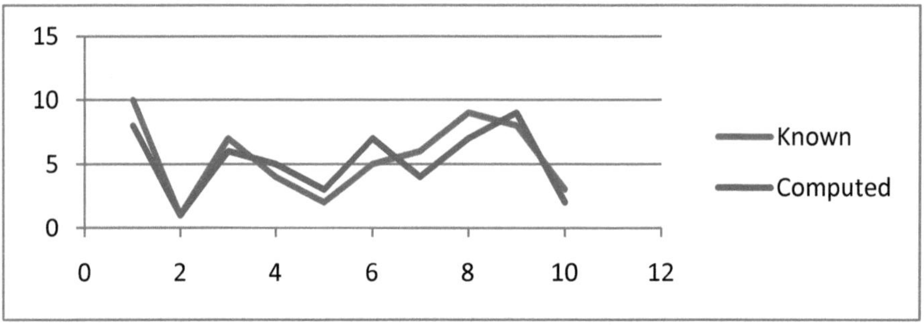

Fig. 3. Actual Rating Vs. computed Rating

Speraman's Rank Correlation coefficient r_s was used to test the significance of correlation between calculated values of analyzability using model and it's 'Known Values'. The r_s was computed using the formula given as under:

$$r_s = 1 - \frac{6\sum d^2}{n(n^2-1)} \qquad -1.0 \le r_s \le +1.0$$

Where'd' is the difference between calculated values and 'Known Values' of analyzability and n is the number of Software Projects (n=10) used in the experiment. Pairs of these values with correlation values r_s above [±.781] are checked in the table 5. The figure 3 presents the graphical representation of actual and computed ranks of used model. The correlation is acceptable with high degree of confidence, i.e. at the

99%. Therefore, it concludes proposed that Model is reliable and valid in the context. However, the study needs to be standardized with a larger experimental tryout for better acceptability and utility.

7 Conclusion and Future Work

This paper has presented a prescriptive approach for analyzability quantification using object oriented design complexity. It is an effort to measure complexity of objects oriented design using available method by examining the parameters which controls the design complexity. A summarized discussion on the basis of plotted values of analyzability and complexity describes the fact that complexity increases, when analyzability diereses. The applied validation analysis on this study concludes that proposed approach is acceptable. The contribution of work is summarized as follows:

- A relation between design complexity and analyzability has been taken into consideration.
- A detailed discussion is being presented on the impact of analyzability and complexity.
- The proposed approach is implemented through a realistic data taken from [10, 11, 12].
- The applied approach is empirically validated through experimental tryouts with its probable weight and significance.

In future, analyzability will be addressed in terms of security of object oriented designs with the help of class diagrams. Security of software can be judged on the basis of software analyzability.

References

1. Wysopal, C.: Building Security into Your software Development Lifecycle (January 30, 2008), http://www.scmagazineus.com/building-security-into-your-software-development-lifecycle/article/104705/
2. Software Technologies for Embedded and Ubiquitous Systems, 1st edn. Springer (November 14, 2007) ISBN-10: 3540756639
3. International Organization for Standardization. ISO/IEC 9126-1: Software Engineering-product Quality-Part-1: Quality Model (2001)
4. Perepletechikov, M., Ryan, C., Tari, Z.: The Impact of SERVICE Cohesion on the Analyzability of Service Oriented Software. IEEE Transactions on Services Computing 3(2) (April-June 2010)
5. Khan, S.A., Khan, R.A.: Analyzability Quantification Model of Object Oriented Design. In: International Conference on Computer, Communication, Control and Information Technology, C3IT 2012, vol. 4, pp. 536–542. Published in Science Direct by Elsevier, Procedia Technology (2012)

6. Wall, A., Andersson, J., Neander, J., Norström, C., Lembke, M.: Introducing Temporal Analyzability Late in the Lifecycle of Complex Real Time Systems. In: Chen, J., Hong, S. (eds.) RTCSA 2003. LNCS, vol. 2968, pp. 513–528. Springer, Heidelberg (2004)
7. Webref: http://www.faculty.ucr.edu/~hanneman/dynamics/chapter%205.pdf
8. Bouwers, E., Pedro Correia, J., van Deursen, A., Visser, J.: Quantifying the Analyzability of Software Architectures. In: WICSA 2011. Technical Report Series Report TUD-SERG-2011-005 (2011) ISSN 1872-5392
9. Kiewkanya, M.: Measuring Object-Oriented Software Maintainability in Design Phase Using Structural Complexity and Aesthetic Metrics. PhD thesis, Chulalongkorn University, Thailand (2006)
10. Mustaf, K., Khan, R.A.: Quality Metric Development Framework. Journal of Computer Science 1(3), 437–444 (2005) ISSN: 1549-3636
11. Olivas, G.M.J., Piattini, M., Romero, F.: A Controlled Experiment for Corroborating the Usefulness of Class Diagram Metrics at the early phases of Object Oriented Developments. In: Proceedings of ADIS 2001, Workshop on Decision Support in Software Engineering (2001)
12. Genero, M., Piattini, M., Calero, C.: An Empirical Study to Validate Metrics for Class Diagrams, web reference: http://citeseerx.ist.psu.edu/viewdoc/summary?doi=10.1.1.11.4858
13. Khan, S.A., Khan, R.A.: Integrity Quantification Model for Object Oriented Design. ACM SIGSOFT Software Engineering Notes 37(2), 1–3 (2012), http://doi.acm.org/10.1145/2108144.2108154, doi:10.1145/2108144.2108154

An Approach for Automated Detection and Classification of Thyroid Cancer Cells

R. Jagdeeshkannan, G. Aarthi, L. Akshaya, Kavya Ravy, and Subramanian

R.M.K. Engineering College, Kavaraipettai – 601206, Chennai, India
dr_rjk@hotmail.com,
{aarthig.27,akudazz121,subramaniankalyan}@gmail.com

Abstract. Cancer that forms in the thyroid gland (an organ at the base of the throat that makes hormones that help control heart rate, blood pressure, body temperature, and weight) is known as the Thyroid Cancer. The prognosis of thyroid cancer is related to the type of cancer. This approach is based on a developed computer analyzer of images that is aimed at automated processing, detection and classification of thyroid cancer cells. It also classifies the type of thyroid cancer present in the digital image of the cell. Classification, a data mining function which accurately predicts the target class for each case in the data. Template matching technique is used in order to match on pixel by pixel basis. It works by sliding the template across the original image. As it slides, it compares or matches the template to the portion of the image directly under it. By this technique the defected cells are well enhanced and the types of thyroid cancer are distinguished. Thus by depending upon the decision rule, all pixels are classified in a single class and hence the types of thyroid cancer cells are well categorized. This method enhances the cell in an efficient manner and effectively separates the cancer cells from the background. This demonstrates the potential effectiveness of the system that enables the medical practitioners to perform the diagnostic task and proper medication.

1 Introduction

Cancer is a class of diseases characterized by out-of-control cell growth. Thyroid cancer is a type of cancer that originates from follicular or Para follicular thyroid cells. Thyroid cancers can be classified according to their histopathological characteristics [4].

Papillary is the most common type of thyroid cancer. It arises from the follicle cells of the thyroid. Medullary thyroid cancer produces too much calcitonin, a hormone that can be detected in the blood at a very early stage. Anaplastic thyroid cancer is a rare and rapidly growing cancer. It is typically very hard to treat due to its aggressive nature, but there are treatment options. The earlier the detection the more are the chances of survival [10].

Data mining [2] (the analysis step of the "Knowledge Discovery in Databases" process, or KDD) an interdisciplinary subfield of computer science is the computational process of discovering patterns in large data sets. Data mining is a

S.C. Satapathy et al. (eds.), *ICT and Critical Infrastructure: Proceedings of the 48th Annual Convention of CSI - Volume II*, Advances in Intelligent Systems and Computing 249,
DOI: 10.1007/978-3-319-03095-1_41, © Springer International Publishing Switzerland 2014

process by which previously unknown information and patterns are extracted from large quantities of data. In mining the inputs are predominantly cleaned, transformed and searches the data using algorithms.

This application that is developed mainly focuses on increasing the speed of detection of thyroid cancer cells using Lab VIEW and MATLAB together [1]. With Lab VIEW it is very easy to program different tasks that are performed in parallel by means of multithreading. MATLAB is a numerical computing environment. This approach helps in differentiating the normal cells from the cancer affected cells [3]. The error in recognition of the cancer cell is also minimized due to this. The application also identifies the type of thyroid cancer by classifying them based on template matching using correlation. Due to the increasing number of cancer cases, there is a growing need for this application which will benefit the society by faster detection and diagnosis and early treatment thereby improving survival rates.

2 Literature Survey

Thyroid cancer forms in the thyroid gland in throat. The pioneering work of Leung.C.C, et al[4] describes the assessment of tumor cells in the prediction of behavior of thyroid cancer. According to the authors a fuzzy edge detection method is used and effectively separates the cells from the background. However the preprocessing techniques were not discussed by them. The preprocessing steps include the cleaning of data from external noise. The work proposed by Pei-Yung Hsiao, et al [5] is able to provide various levels of noise smoothing and reduction, which are highly desired in early stages of the image processing flow. Their work results in high-resolution processing. Since they have implemented a 2D Gaussian filter using hardware shifters and adders, edges are not preserved in the digital image. The gray scale pixels after noise removal is transformed into two tone pixels. An apparatus designed by Yoshiaki Hanyu[12] processes two-tone image data so as to magnify a relevant image and smooth a boundary line between a zone consisting of first-tone pixels of two-tone pixels constituting the relevant image and a zone consisting of second-tone pixels of the two-tone pixels. The spatial domain so far is being changed to frequency domain. Fast Fourier Transform is applied to convert an image from the image (spatial) domain to the frequency domain. According to Raghu Muthyalam[7], FFT turns the complicated convolution operations into simple multiplications and then an inverse transform is applied in the frequency domain to get the result of the convolution. As a next step of these "fill hole" of Lab VIEW function is done. The "Reject Border", another Lab VIEW function rejects the border of the images and the cells along the boundary.

Next is the particle analysis which analyzes the particle and reduces the pixels content in each particle [13]. By minimizing the pixels in a particle it helps to classify the image efficiently. We can also use the operation to create masks for any particle or for particle boundaries. Further the proposed approach of Singhal, A,et al[9] implements the Principle Component Analysis (PCA) which reduce the input data in order to make classification easy and allowing large databases to be processed in a relatively small amount of time. This data mining technique reduces the dimensions of the image. In the classification process, the image after PCA with minimum pixel values is compared with a template image corresponding to the characteristic of each cancer. The paper penned by Sarvaiya, J.N [8] describes medical image registration by template matching based on Cross Correlation. Their objective was to establish the correspondence between the reference and template images. But the matched area is not highlighted. Here Template Matching is done by pixel by pixel basis. It gives the output highlighting the match mentioning the type of thyroid cancer if present.

3 Description

The motivation behind this approach has been mentioned extensively in the literature survey. This work focuses on the detection of the defected cells and shows the proposed methodology for classification of the thyroid cancer types.

The digital image of the cell is fed into the Lab VIEW software. We use image processing tools for pre-processing which is explained in chapter 4. The pre-processing techniques that are applied to the images can be summarized as given below.

1. In digitization noise could be introduced. Data cleaning is the process of removing inconsistencies or noise in order to improve the quality of image for examination [5].
2. In Data Transformation, data are transformed or consolidated into forms appropriate for mining. It transforms the single intensity valued pixels into plurality [12]. It also includes transforming an image from one domain to another domain [7].
3. Two Image enhancement techniques are employed here. The first one fills the holes or the unwanted space in the particle. The dataset we have considered almost comprises of images in which the particles are sticking along with the border. The second technique rejects the border in order to give a clear picture for further process.
4. Complex data analysis and mining on huge amounts of data may take a very long time, making such analysis impractical or infeasible. Data reduction techniques have been helpful in analyzing reduced representation of the dataset without compromising the integrity of the original data and yet producing the quality knowledge [9].

The most important process of this system is classifying the types of thyroid cancer. The intent of the classification process is to categorize all pixels into a single class. This categorized data of the image is then used to compare with a template image which then performs correlation on every match [8]. The objective of classification is to uniquely identify the defected cells and effectively highlight the types of thyroid cancer. The algorithm of this approach is depicted in fig.1.

4 Overview of the System

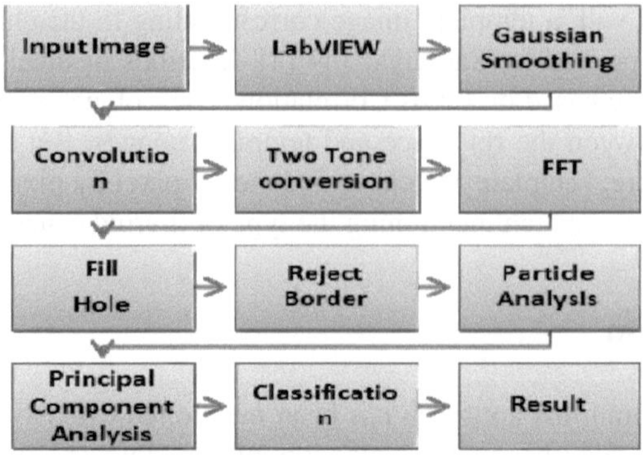

Fig. 1. System Architecture

4.1 Data Cleaning

4.1.1 Gaussian Smoothing

The effect of Gaussian smoothing is to blur an image. The Gaussian blur is for calculating the transformation to apply to each pixel in the image. The Gaussian outputs a weighted average of each pixel's neighborhood, with the average weighted more towards the value of the central pixels. Because of this, a Gaussian provides gentler smoothing and preserves edges better.

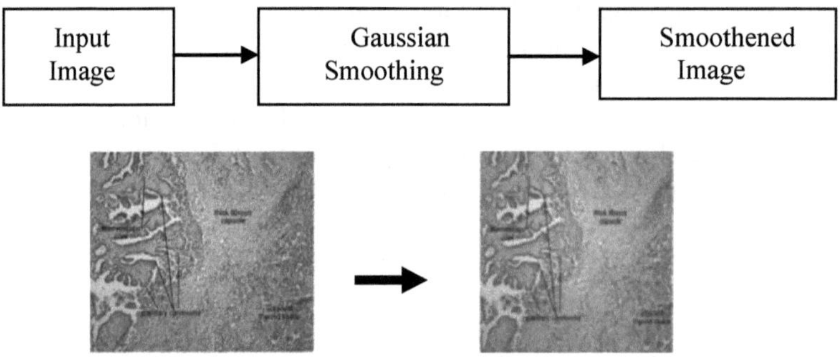

Fig. 2. Gaussian Smoothing

4.1.2 Convolution

It is a neighborhood operation in which each pixel is the weighted sum of neighboring input pixels. The IMAQ Convolute tool is used as a convolution filter here. Features are defined by a n x m matrix that is applied to the gray scale image where n=3 and m=3. There are a few rules about the filter:

- Its size has to be uneven, so that it has a center, for example 3x3, 5x5 are ok.
- It doesn't have to, but the sum of all elements of the filter should be 1 if you want the resulting image to have the same brightness as the original.
- If the sum of the elements is larger than 1, the result will be a brighter image, and if it's smaller than 1, a darker image.

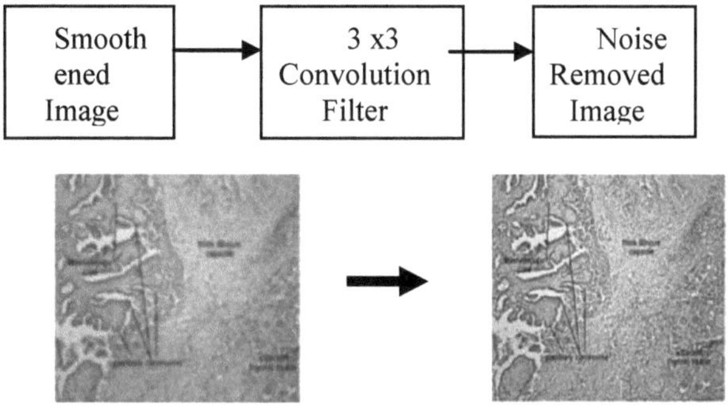

Fig. 3. Convolution

4.2 Data Transformation

4.2.1 Two-Tone Conversion

Two tone conversions is a process of transforming the gray scale image to binary image. It smoothens the boundary line between a zone consisting of first-tone pixels of two-tone pixels constituting the relevant image and a zone consisting of second-tone pixels of the two-tone pixels. The IMAQ AutoBThreshold2 and IMAQ Morphology are the tools used for transforming the image.

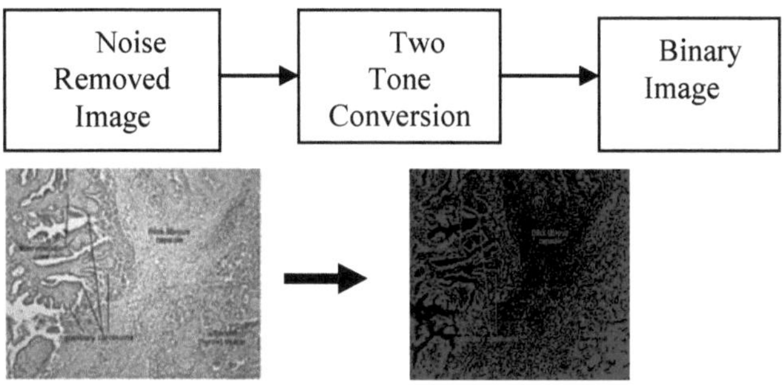

Fig. 4. Two-Tone Conversion

4.2.2 Fast Fourier Transform

Fast Fourier Transform (FFT) is an efficient implementation of image processing techniques. It is applied to convert an image from the image (spatial) domain to the frequency domain. When input image is given, here the number of frequencies in the frequency domain is equal to the number of pixels in the image or spatial domain. An inverse transform using IMAQ Inverse FFT is then applied in the frequency domain to get the result of the convolution. The inverse transform re-transforms the frequencies to the image in the spatial domain. Fourier Transform decomposes an image into its real and imaginary components which is a representation of the image in the frequency domain.

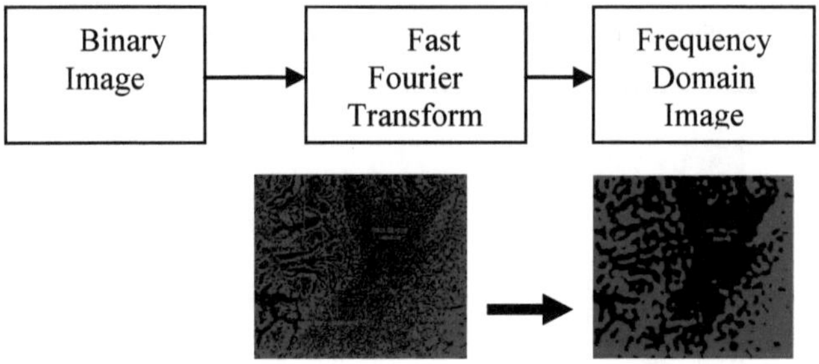

Fig. 5. Fast Fourier Transform

4.3 Image Enhancement

4.3.1 Fill Hole

It fills the holes found in a particle using IMAQ Fill Hole function. The holes are filled with a pixel value of 1. The source image must be an 8-bit binary image. The holes found in contact with the image border are never filled because it is impossible to determine whether these holes are part of a particle.

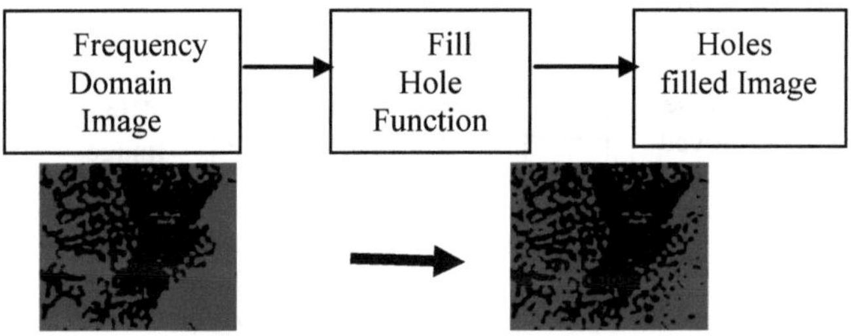

Fig. 6. Fill Hole

4.3.2 Reject Border

The IMAQ Reject Border function removes the border and the particle touching the border of the image. Particles that touch the border may have been truncated by the choice of the image size. In particle analysis this will help to give desired results.

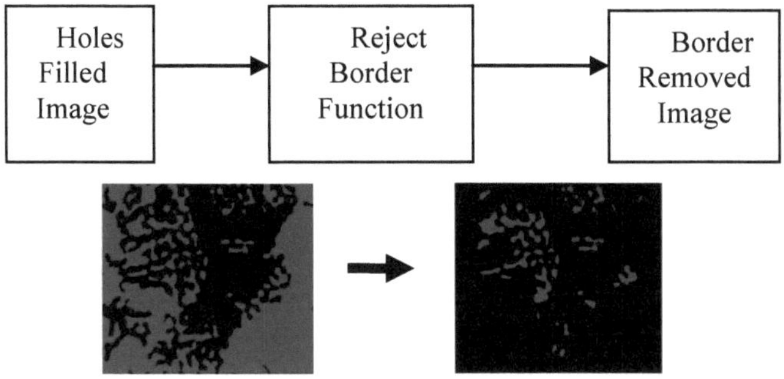

Fig. 7. Reject Border

4.4 Data Reduction

4.4.1 Particle Analysis

The particle size and shape are characterized by using image technique IMAQ Particle Filter of Lab VIEW. This analysis is performed on a binary image. A particle in the binary image may consist of one or more pixels whose level is zero. A particle is defined to consist minimum number of pixels. After executing this operation using IMAQ Particle Analysis the output will show the total number of particles.

4.4.2 Principle Component Analysis

Principle Component analysis is a useful technique for image compression .Here we encounter a situation where there are a large number of pixels for an image in the database. The accuracy and reliability of a classification or prediction model will suffer if we include highly correlated variables. This procedure analyzes the principal components of the input variables.

5 Classifications

The classification task begins with a data set in which the class assignments are known. The input data, also called the training set, consists of multiple records each having multiple attributes. This is then compared with the input image to detect the presence of thyroid cancer. To enhance the classification process correlation is combined with template matching. The two variables are the corresponding pixel values in the template and source image. Correlation helps in finding the exact pixels that match with the template image and hence it gives accurate results. Template matching using correlation is given in fig.8.

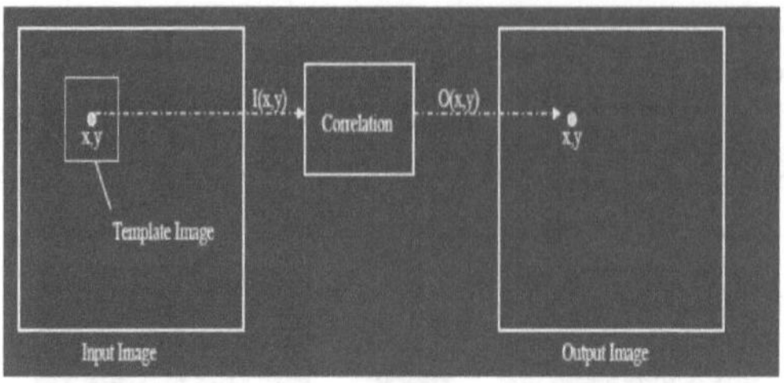

Fig. 8. Template Matching with Correlation

We employ PCA for the reduction of pixel values of the images. The database contains records of the normal cell and the four types of cancer cells. MATLAB code is implemented for the classification and matching technique. The code corresponds to the following

- The input image, template image and the pixel value are given as input.
- Each and every pixel value in the input image are scanned and compared to the given input pixel value.
- Then the classification is done according to the type of thyroid cancer.
- Template matching technique used for classifying.
- Finally the correlation is performed with the template image.

The input values are given to the artificial neural network for training. A neural network consists of an interconnected group of artificial neuron. The dataset training used for this approach is done by MATLAB. The Neural Pattern Recognition (npr) tool of MATLAB is used for this system. The input to the network consists of the training tuples and the output is their associated target values. The architecture of the neural network consists of three layers an input layer, a hidden one and an output layer. The number of nodes in the input layer is equal to the number of the tuples. The network is shown in the fig.9. Here we consider 1000 values for input with 5 target nodes corresponding to the normal cell and the four types of thyroid cancers.

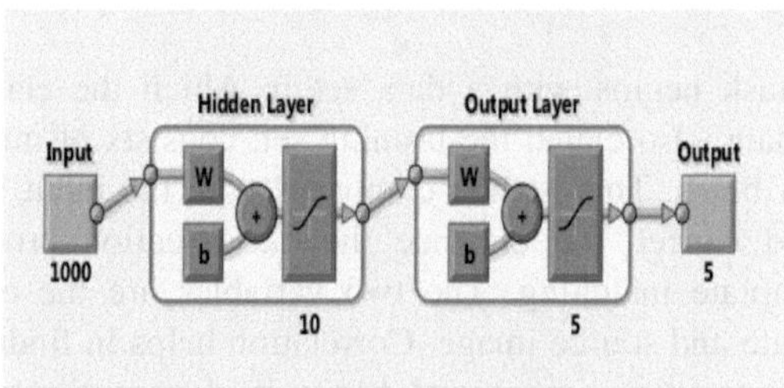

Fig. 9. Knowledge Discovery Process

6 Results and Discussion

The challenges in the thyroid cancer are evaluated in the test bed. It involves the keen examination of the digitized cell image [3]. The input image papillary thyroid for instance is taken. It is sensitive to noise and other inequalities which is shown in fig.5. Hence the preprocessing is done. As a result of applying Gaussian filter [5] the image gets blurred and edges are smoothened as shown in fig.6. The Convolution takes the sum of the product value for every pixel and hence the random noise is removed which will result in an image as shown in fig.7. This image has finite dimension. The binarizationprocess [12] is used to convert the pixels of binary values and the output is shown in fig.8. The output of this will contain an image as in fig.9. The enhancement techniques employed results in the effective detection of the defected cell as shown in fig.10. By using particle analysis [13] in this approach it will reduce the number of pixels in each particle. It is shown in the fig.11.The accuracy (AC) is the proportion of the total number of predictions that were correct. The classification accuracy here is 81.8% which is proved by training the tuples from database using artificial neural network. The performance plot of this approach is shown in fig12. Performance of the system is commonly evaluated using the confusion matrix. The strength of a confusion matrix is that it identifies the nature of the classification errors well. The training and validation of the confusion matrix is as shown below.

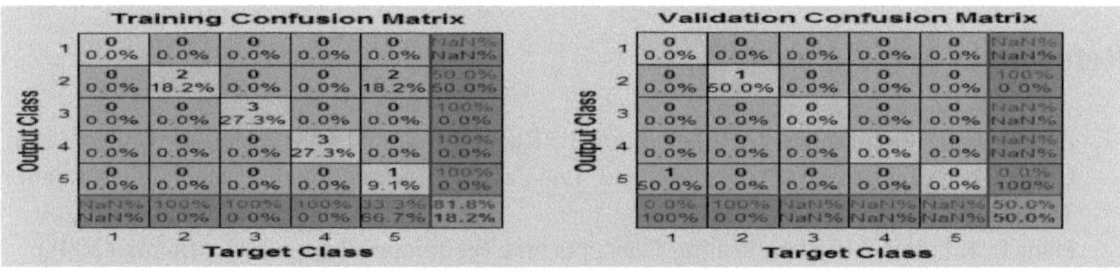

Fig. 10. Template Reading and Correlation

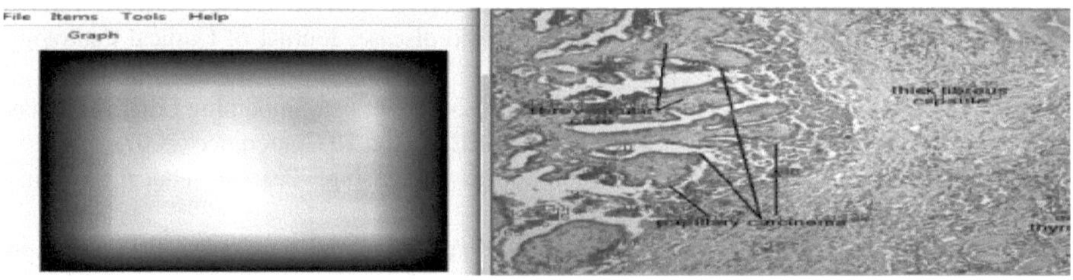

Fig. 11. Correlation Output and Template Matched Image

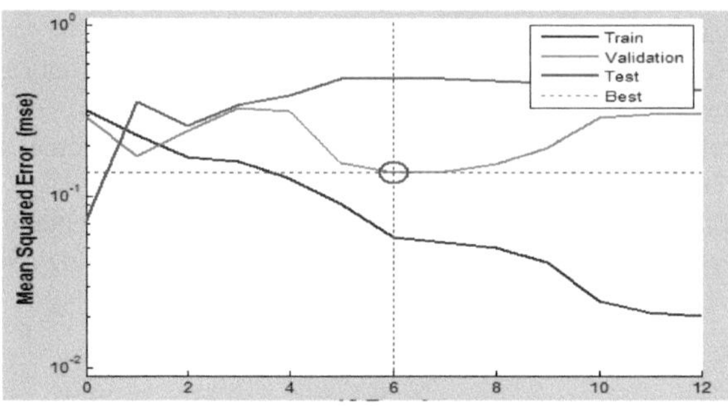

Fig. 12. Performance Slot

7 Conclusion

Remarkable advances have occurred in recent years in understanding the thyroid cancer. As it can be seen here the defected cells are extracted among the other cells, very well. Thus by depending upon the decision rule, all pixels are classified in a single class and hence the types of thyroid cancer cells are well categorized. This method enhances the cell in an efficient manner and effectively separates the cancer cells from the background. This technique can be used in large scale cells examination and can improve the accuracy and speed by means of MATLAB & LABVIEW. It can be used as a tool for follow-up diagnosis. This approach will help to aid the diagnosis process by automatically detecting the cancer cells in thyroid images. Future work also includes the use of this analysis method to design a neural network for other cancer cell recognition in medical images and the evolvement of the proposed time efficient scheme for application in an integrated real time system for the assessment of the thyroid gland.

References

1. Zadeh, H.G., Janianpour, S., Haddadnia, J.: Recognition and Classification of the Cancer Cells by Using Image Processing and Lab VIEW. International Journal of Computer Theory and Engineering (2009)
2. Han, J., Kamber, M.: Data Mining,Concepts and Techniques. Morgan Kaufmann (2001)
3. Leung, C.C., Chan, F.H.Y., Lam, K.Y., Kwok, P.C.K., Chen, W.F.: Thyroid cancer cells boundary location by a fuzzy edge detection method. In: Proceedings of the 15th International Conference on Pattern Recognition (2000)
4. LiVolsi, V.A., Baloch, Z.W.: Pathology of thyroid disease. Journal of Clinical Pathology (2007)
5. Hsiao, P.-Y., Chou, S.-S., Huang, F.-C.: Generic 2-D Gaussian Smoothing Filter for Noisy Image Processing. In: TENCON 2007 - 2007 IEEE Region 10 Conference (2007)
6. Gonzalez, R.C., Woods, R.E.: Digital Image processing, Pearson education. First Impression (2009)
7. Muthyalam, R.: Implementation of Fast Fourier Transform for Image Processing in DirectX 10, `http://Article-software.intel.com`
8. Sarvaiya, J.N.: Image Registration by Template Matching Using Normalized Cross-Correlation. In: International Conference on Advances in Computing, Control, & Telecommunication Technologies, ACT 2009 (2009)
9. Singhal, A., Seborg, D.E.: Pattern matching in historical batch data using PCA. IEEE Control Systems (2002)
10. Falk, S.: Thyroid Disease: Endocrinology, Surgery, Nuclear Medicine, and Radiotherapy, 2nd edn. (1997)
11. Wang, Y.Y., Sun, Y.N., Lin, C.C.K., Ju, M.S.: Nerve Cell Segmentation via Multi-Scale Gradient Watershed Hierarchies. Engineering in Medicine and Biology Society (2006)

12. Hanyu, Y.: Apparatus and method for processing Two-Tone Image data so as to smooth image and convert each image pixel into plurality of pixel. U.S Patent NO:5,812,742 (1998)
13. http://www.wavemetrics.com/products/igorpro/imageprocessing/imageanalysis/particleanalysis.html
14. http://www2.cs.uregina.ca/~dbd/cs831/notes/confusion_matrix/confusion_matrix.html

Quality Validation of Software Design before Change Implementation

Aprna Tripathi, Dharmender Singh Kushwaha, and Arun Kumar Misra

Department of Computer Science and Engineering
MNNIT Allahabad
Allahabad, India
{rcs1051,dsk,akm}@mnnit.ac.in

Abstract. Change in any software application seems to be very simple when someone demands it, but the complexity of the task appears during implementation phase. Change implementation is acceptable when it includes the requested functionality as well as improves and preserves the quality of software. Complexity is an important issue for software development as it affects the cost, effort and quality of the product. Change complexity could be helpful in analyzing the impact of change. This paper focuses on measuring the software complexity that analyzes the quality of software design. To measure the complexity, UML class diagram is considered. Additionally, the goal of this paper is to identify the best possible design among various possible solutions to implement requested change in the existing system that does not adversely affects the software design of the existing software. Here, the change type considered is only new requests made by user or client.

Keywords: Software change management, cohesion, coupling, complexity and software quality.

1 Introduction

Change is a continuous process. Change may be requested by different sources and may have different types. It seems to be very simple when someone demands it, but the complexity of the task appears when it moves towards implementation phase. Software maintenance phase adopts the change activity.

Software maintenance is basically a post development activity but most of the times it consumes 40-70% of the overall development costs. This variation of cost from 40 to 70 depends on many factors. One of the factors is the design of the change that is carried out to implement the change. The design structure depends how the change is understood by the designer. There may be more than one way to implement a single change in the existing system. Thus, it could be a good practice to analyze all the possible designs (if possible) before actual implementation. When a change is requested, it is not only the issue where to make the change. However, how the change will be processed, is also an important consideration in order to maintain the quality parameter of the software in terms of reliability, understandability, reusability and maintainability.

S.C. Satapathy et al. (eds.), *ICT and Critical Infrastructure: Proceedings of the 48th Annual Convention of CSI - Volume II*, Advances in Intelligent Systems and Computing 249,
DOI: 10.1007/978-3-319-03095-1_42, © Springer International Publishing Switzerland 2014

Complexity is an important issue for software development as it affects the cost, effort and quality of the product. Change complexity could be helpful in analyzing the impact of change. In addition, effort required in implementing change, testing of change and effort required for complete system testing after the change has been implemented can be estimated. Cohesion and coupling are very useful property for estimating the quality of any software. S. Pfleeger [5] defines coupling as the degree of dependence among components and cohesion as the degree to which all elements of components, directed towards a single task and all elements directed towards that task are contained in a single component. High coupling could makes modification process difficult while high cohesion is preferable for software. In the proposed work, coupling and cohesion are considered as the basis to estimate the quality of design for requested change.

This paper proposes an approach to analyze the complexity of change that is based on the combined concept of cohesion and coupling. Section 2 discusses relevant related work. Section 3 details proposed approach. Section 4, 5 and 6 discusses the case studies, results and validity respectively. Section 7, the last section of the paper, outlines conclusions and future work.

2 State-of-The-Art

Buckley et al. [1] propose taxonomy of software change. As a change is requested, it is analyzed and then after the approval of the decision committee it is further processed. An impact is the effect of one object on another. Software Change Impact analysis (SCIA) is used to determine the scope of change requests as a basis for resource planning, effort estimation and scheduling. Various software change models are discussed by Ghosh [12]. A huge literature is reviewed for finding the change design validation phase in the software change process model [13], but none of the model incorporates the change design quality validation before change implementation. The change impact analysis identifies the effect of requested change on the existing system. The affects may be in term of software functionality or quality of the software. Angelis et al. [2] discusses the importance of the change impact analysis issues during software change implementation process. Arnold and Bohner [3] define a three-part conceptual framework to compare different impact analysis approaches and assess the strengths and weaknesses of individual approach. Gethers et al. [4] propose a framework for impact analysis based on the degree of automation and developer augmentation information, which is obtained from system maintenance scenario. The Pfleeger and Atlee [5] focus on the risks associated with the change and state, "Impact Analysis (IA) is the evaluation of many risks associated with the change, including estimates of the effects on resources, effort, and schedule". However, effect on non-functional properties of the system such as maintainability, readability, reusability etc are also important to analyze before implementing the change. Complexity of the system has the ability to measure the non- functional parameter. Banker et al. [6] examine the relationships between software complexity and software maintainability in commercial software environments. Author proposes a framework to enable researchers (and managers) to assess such products and techniques more quickly by introducing software complexity as a factor linking software development tools and techniques and software maintenance costs. There are various metrics proposed by various authors for measuring the software complexity. Hassan et al. [7] proposes complexity metrics that are based on the code change process instead of on the code. They conjecture that a complex code change process negatively affects it's product, i.e., the software system. Mc Cabe [8] uses a fundamental assumption that the software complexity is related to the number of control paths generated by the code. Reddy and A. Ananda Rao [9] proposes three metrics: dependency oriented complexity metric for structure (DOCMS(R)), dependency oriented complexity metric for an artifact causing ripples DOCMA (CR)), and dependency oriented complexity metric for an artifact affected by ripples (DOCMA(AR)).

The most favorable parameters for measuring the software complexity found in the literature are the coupling and cohesion. Chidamber et al. [10, 11] say that classes are coupled if methods or instance variables in one class are used by the other. Coupling between Object-classes (CBO) for a class is the number of other classes coupled with it. For measuring the cohesiveness author also proposes the metric Lack of Cohesion in Methods (LCOM) Number of non-similar method pairs in a class of pairs. Li et al. [11] proposes Data Abstraction Coupling (DAC) for a class is the number of attributes having other classes as their types. To compute the software change complexity is still challenging. The main objective of this paper is to propose and implement a novel approach for analyzing the change and design a formula for computing the software change complexity.

3 Proposed Approach

To compute the complexity of requested change, there are following steps.

Step 1. Identification of coupling and cohesion level for each class.
Step 2. Weight assignment to identified levels of cohesion and coupling for each class of existing system
Step 3. Compute coupling and cohesion weight of existing system.
Step 4. Repeat step 1, 2 and 3 for the modified design of the system that includes the requested change.
Step 5. Find the deviations of coupling and cohesion between existing system and modified system.
Step 6. Analyzing the change complexity.
Step 7. Decision about acceptance or rejection of modified design.

Step 1. To identify the coupling and cohesion type for each class of the existing system, the coupling and cohesion classification given by Myers is considered.
Step 2. After acknowledging the type of cohesion and coupling of each class inside the software, a weight is assigned to each relation that is in between two classes inside class diagram of the system corresponding to its type of cohesion or coupling. In Table 1 and Table 2 the type of cohesion as well as coupling are represented by level of coupling and cohesion with its corresponding weight.

Table 1. Coupling Levels with their Weight

Level of Coupling	Weight *(Wcoupling)*	
Content	6	High ↑
Common	5	
Control	4	
Stamp	3	
Data	2	Low
Uncoupled	1	

Table 2. Cohesion Levels with their Weight

Level of Cohesion	Weight (Wcohesion)	
Coincidental	8	Low ↑
Logical	7	
Temporal	6	
Procedural	5	
Communicational	4	
Sequential	3	
Informational	2	
Functional	1	High

Step 3. Let N1 be the number of external links, C1 be the number of classes in between classes in the existing. A class may have NK1 internal links in existing classes. Where, N1, C1, and NK1 are positive integer. Each link carries a weight according to the table 1 and 2.Thus, Coupling and cohesion weight of existing system are :

$$\text{CupWexisting} = \sum_{j=1}^{N1} W_{Ecoupling(j)} \quad (1)$$

$$\text{CohWexisting} = \sum_{i=1}^{C1} \sum_{m=1}^{NK1} \mathrm{W}_{\mathrm{Ecohesion(i,m)}} \quad (2)$$

Step 4. Step 1, 2 and 3 are repeated to find the cohesion and coupling level existed in the modified design. Then, to every identified cohesion and coupling a weight is assigned. It is assumed that N2 be the number of external links in between classes C2 be the number of classes in the changed system respectively. A class may have NK2 number of internal links in changed classes, where, N2, C2, and NK2 are positive integer. Each link carries a weight according to the table 1 and 2. Thus, coupling and cohesion weight of changed system are expressed as:

$$\text{CupWchanged} = \sum_{k=1}^{N2} \mathrm{W}_{\mathrm{Ccoupling(k)}} \quad (3)$$

$$\text{CohWchanged} = \sum_{i=1}^{C2} \sum_{m=1}^{NK2} \mathrm{W}_{\mathrm{Ccohesion(i,m)}} \quad (4)$$

Step 5. DiffCupW is the difference in coupling and cohesion weight of the existing system and changed system are expressed as:

$$\text{DiffCupW} = \text{CupWchanged} - \text{CupWexisting} \tag{5}$$

$$\text{DiffCohW} = \text{CohWchanged} - \text{CohWexisting} \tag{6}$$

Step 6. The qualitative impact of coupling and cohesion on the system is summarize in the following tables 3:

Table 3. Coupling- Cohesion preference levels

	Increase	Decrease
Coupling	Non-Preferable	Preferable
Cohesion	Preferable	Non-Preferable

Effect of coupling and cohesion on system complexity is summarized in Table 4.

Step 7. After step 6, if the value of qualitative impact on existing system is preferable, the change is implementation according to the modified design. In case it is non-preferable, designer explored towards alternate design for implementing change in the existing system.

Table 4. Combined effect of coupling and cohesion on system complexity

Case Id	Case Description	Qualitative Impact on Existing System
CA1	If DiffCupW and DiffCohW both zero	Preferable
CA2	If DiffCupW and DiffCohW both positive $\lvert$ DiffCupW $\rvert < \lvert$ DiffCohW $\rvert$	Preferable
CA3	If DiffCupW and DiffCohW both negative DiffCupW $\rvert > \lvert$ DiffCohW $\rvert$	Preferable
CA4	If DiffCupW is positive, DiffCohW is negative	Non-Preferable
CA5	If DiffCupW is negative, DiffCohW is positive	Preferable
CA6	If DiffCupW is zero and DiffCohW is positive	Preferable
CA7	If DiffCupW is zero and DiffCohW is negative	Non-Preferable
CA8	If DiffCupW is positive and DiffCohW is Zero	Non-Preferable
CA9	If DiffCupW is negative and DiffCohW is Zero	Preferable

4 Case Studies

To validate the proposed approach, three case studies are considered. One among three is discussed here in details. For other two, summarized results are shown at the end of this section.

Case Study 1: The E*xisting System:* There is a Vehicle information system (VIS) that automates the vehicle information. to implement the requirements class diagram shown id fig. 1 is designed. Proposed *Change:* It is requested that also show the details of manufacturing company name and date of manufacturing of the product when details of the vehicle are printed

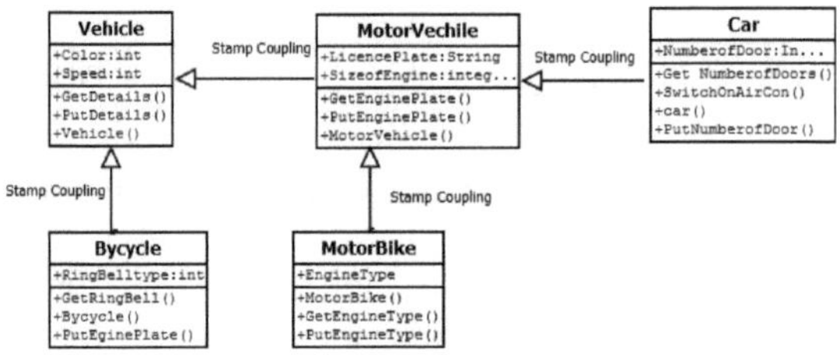

Fig. 1. Class Diagram of Existing VIS

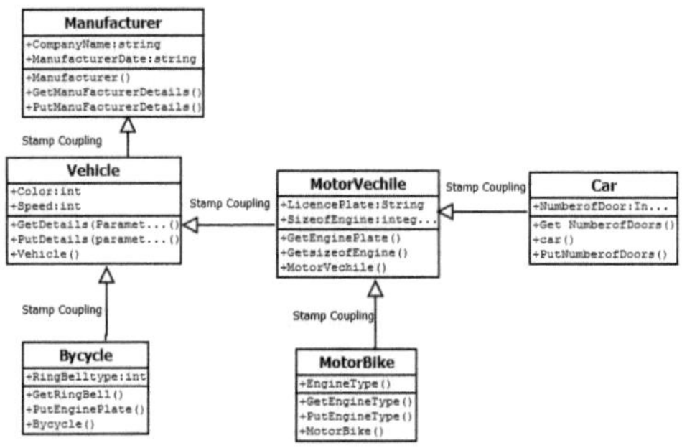

Fig. 2. Class Diagram of Changed VIS (Method 1)

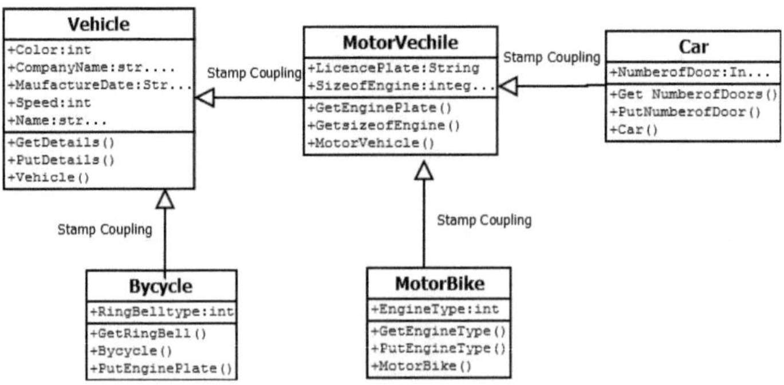

Fig. 3. Class Diagram of Changed VIS (Method 2)

Two possible ways to implement the requested change are discussed in this section. Fig. 2 and 3 shows the possible class diagram after incorporating the change in existing class diagram.

In case 1, there is a change in coupling since a new class is added in the existing system. From Eq. 1, 2 and 5 $C_{up}W_{existing} = 12$, $C_{up}W_{changed} = 15$ and $Diff_{CupW} = 3$. For computing the value of DiffCohW, we consider Eq. 3, 4 and 6. For this we referrer to the Table 2, for Existing system $C_{oh}W_{existing} = 45$, $C_{oh}W_{changed} = 45$ and $Diff_{CohW} = 0$

Finally, we have $Diff_{CupW} = 3$ and $Diff_{CohW}$ is 0. It falls under case 8 from Table 4 i.e. the CA8. The above results show that there is a change in the coupling, the cohesiveness of the system is not affected. Thus, it concluded that this design for implementing requested change is non-preferable, and it is requested that designer try to explore for some alternate solution for incorporating requested change. Fig.3 shows class diagram that is the alternative method for implementing the change. During analysis, it is found that none of the new class is added in the existing system while few changes like addition of attributes in vehicle class, deletion of method from car class in the existing class diagram. Thus, Eq. 1, 2 and 5 CupWexisting= 12, $C_{up}W_{changed}$=12 and $Diff_{CupW}$= 0. For computing the value of $Diff_{CohW}$, we consider Eq. 3, 4 and 6. For this, we refer the Table 2, for Existing system $C_{oh}W_{existing} = 45$, $C_{oh}W_{changed} = 45$ and $Diff_{CohW} = 0$.Finally, we have $Diff_{CupW} = 0$ and $Diff_{CohW}$ is 0. It falls under case 1 from Table 3 i.e. the CA1.

The above results show that neither the coupling nor the cohesiveness is changed. Thus, it concluded that this design for implementing requested change is preferable. And the implementation phase of the requested can be proceeded.

5 Validation

To validate the proposed approach, a metrics plug-in is considered. Coupling and cohesion metrics are analyzed for the same case study through metrics plug-in [14]. Lack of Cohesion in Methods (Total Correlation) of Existing System, Changed system- Method 1 (With manufacture class) and Changed system- Method 2 Without manufacture class are 117%, 117% and 301% respectively.

Here, the high value of Lack of Cohesion in Methods (LCOM) shows the strong depends of the variables used in the method.

6 Results

From section 5, it is clear that the changed system method 2 is more acceptable to implement requested change in the existing system than changed system method 1. Since the changed system method 2 enhances the cohesion of the existing system after incorporating change while the method 1 increases the coupling level. Thus, it validates that the proposed method gives the more efficient results for making the decision about acceptance/ rejection of the design to implement a change.

7 Conclusion and Future Work

The software quality and success of the software are correlated with each other. Thus, it becomes necessary that before change implementation in the existing system, a thorough analysis to analyze quality to the software at design phase is necessary. This paper proposes a fundamentally new approach that seeks a systematic solution to analyze complexity of the software at design level. In this work, the software coupling and cohesion is computed to analyze the software complexity before change implementation. It gives a new dimension to the analysts for analyzing the requested change.

References

1. Buckley, J., Mens, T., Zenger, M., Rashid, A., Kniesel, G.: Towards a taxonomy of software change: Research Articles. J. Softw. Maint. Evol. 17(5), 309–332 (2005)
2. Angelis, L., Wohlin, C.: An Empirical Study on Views of Importance of Change Impact Analysis Issues. IEEE Trans. Softw. Eng. 34(4), 516–530 (2008)
3. Arnold, R.S., Bohner, S.A.: Impact Analysis - Towards A Framework for Comparison. In: Proceedings of the Conference on Software Maintenance, Los Alamitos, CA, pp. 292–301 (September 1993)
4. Gethers, M., Kagdi, H., Dit, B., Poshyvanyk, D.: An adaptive approach to impact analysis from change requests to source code. In: Proceedings of the 26th IEEE/ACM International Conference on Automated Software Engineering (ASE 2011), pp. 540–543. IEEE Computer Society, Washington, DC (2011)
5. Pfleeger, S.L., Atlee, J.M.: Software Engineering Theory and Practice Upper Saddle River. Prentice-Hall, New Jersey (2006)
6. Banker, R.D., Datar, S.M., Zweig, D.: Software complexity and maintainability. In: DeGross, J.I., Henderson, J.C., Konsynski, B.R. (eds.) Proceedings of the Tenth International Conference on Information Systems (ICIS 1989), pp. 247–255. ACM, New York (1989)
7. Hassan, A.E.: Predicting faults using the complexity of code changes. In: Proceedings of the 31st International Conference on Software Engineering (ICSE 2009), pp. 78–88. IEEE Computer Society, Washington, DC (2009)
8. McCabe, T.J.: A Complexity Measure. IEEE Transactions on Software Engineering SE-2(4) (December 1976)
9. Reddy Reddy, K., Ananda Rao, A.: Dependency oriented complexity metrics to detect rippling related design defects. SIGSOFT Softw. Eng. Notes 34(4), 1–7 (2009)
10. Chidamber, S.R., Kemerer, C.K.: Towards a Metrics Suite for Object Oriented Design. In: Proceedings of 6th ACM Conference on Object Oriented Programming, Systems, Languages and Applications (OOPSLA 1991), Phoenix, Arizona, pp. 197–211 (1991)
11. Li, W., Henry, S.: Object-Oriented metrics that predict maintainability. Journal of Systems and Software 23(2), 111–122 (1993)
12. Ghosh, S.M., Sharma, H.R., Mohabay, V.: Analysis and Modeling of Change Management Process Model. International Journal of Software Engineering and Its Applications 5(2) (April 2011)
13. Olsen, N.: The software rush hour. IEEE Software Magazine, 29–37 (September 1993)
14. `http://eclipse-metrics.sourceforge.net/`

Testing Approach for Dynamic Web Applications Based on Automated Test Strategies

Chittineni Aruna[1] and R. Siva Ram Prasad[2]

[1] Acharya Nagarjuna University and Dept. of CSE,
KKR & KSR Institute of Technology and Sciences, Guntur Dist,
Andhra Pradesh, India
chittineni.aruna@gmail.com.
[2] Dept. of CSE and Head, Dept. of IBS, Acharya Nagarjuna University,
Guntur Dist, Andhra Pradesh, India

Abstract. Presently there is a problem with testing of web applications. Fault tolerant is the main aspect for the people with research-orientation. They are searching for better techniques by testing the fault tolerant applications. Previously Different fault localization algorithms such as Ochiai were implemented for automated test strategies. Auto test generation strategy, is a boon to validate different quality applications in time. However, their working scenario was restricted to stand-alone applications only. Later, Auto test generation strategy is combined with source mapping and using an extended domain for conditional and function-call statements to generate automated test suits. Recently an enhanced Ochiai i.e., fault localization algorithms was proposed which has the ability to handle web applications as well, but Ochiai driven oracles offer rigid support by offering static analysis services to only PHP applications. We propose a new approach to extend the Ochiai algorithm with Metamorphic testing strategies to develop an integrated framework that can offer support beyond PHP and such as Java/HTML/JavaScript. Metamorphic testing observes that even if the executions do not result in failures, they still bear useful data. Exploitation higher approaches, we tend to develop unique test-generation strategies that are geared towards manufacturing test suites which have supreme or maximal fault-localization effectiveness in many internet technologies and a sensible implementation validates our claim

Keywords: Dynamic Testing, Metamorphic Relations, Test cases, Metamorphic Testing.

1 Introduction

Testing is an objective and independent view of the software, allow the business for understanding the risks of that software implementation [1]. Testing strategies include process of executing an application with finding software bugs. Testing is the process of verifying a computer program or application to meet the requirements for its designing and development. It can be implemented with same characteristics [2, 3] for satisfying customer requirements. Testing can be implemented at any time in the

S.C. Satapathy et al. (eds.), *ICT and Critical Infrastructure: Proceedings of the 48th Annual Convention of CSI - Volume II*, Advances in Intelligent Systems and Computing 249,
DOI: 10.1007/978-3-319-03095-1_43, © Springer International Publishing Switzerland 2014

software development process. Traditionally most of the test strategies effort occurs after the requirements have been defined and the coding process has been completed. Web applications are developed with the combination of several programming languages, such as JavaScript and PHP with embedded SQL commands. Java Script is used for describing client side applications and PHP is used for server applications [4, 5]. After development, an application, gives required output in the form of dynamically html pages, which require additional scripts to be executed. If applicants makes mistakes and introduce bugs, it results crashes in dynamically generated HTML pages [4]. The output of a Web application can be displayed in a browser [5].

The goal of testing is to find the bugs that are the reason for web application crashes as in HTML. The causes of some faults may terminate the application, when a Web application calls an undefined function. The HTML output presents an error message and the application execution is halted. For this different fault localization techniques are needed.

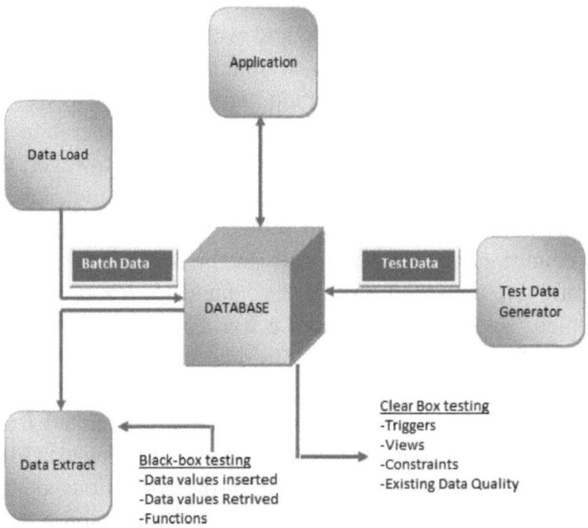

Fig. 1. Testing Architecture in Databases

Above diagram shows data administration testing with server side web applications. By physical exertion the management flows within every execution, faults that are discovered throughout the execution are recorded. Until to attaining ample coverage of the statements within the application, the method is recurrent.

However when a test fails, developers need to find the location of the fault in the source code before they can fix the problem. In recent years, a number of automated techniques have been proposed to assist programmers with this task, which is usually called *fault localization*. Many *fault-localization techniques* attempt to predict the location of a fault by applying statistical analyses to data obtained from the execution of multiple tests. The effectiveness of existing fault-localization technique can be measured by determining how many statements need to be inspected, on average, until the fault is found. In later techniques, test cases are produced by combining concrete and symbolic execution to generate passing and failing runs instead of assuming the existence of a test suite with passing and failing test cases. However, existing approaches are mainly offering their support to only PHP applications.

More recently, a *metamorphic testing* method was proposed by Chen et al. [19,20]. It has been proposed as a property-based test case selection strategy. It is based on the intuition that even if no failure is revealed by a test case selected according to some strategies, it still has useful information. Thus, follow-up test cases should be further constructed from the original test cases with reference to some necessary conditions of the problem to be implemented. Such necessary properties guiding the construction of follow-up test cases are known as *metamorphic relations*.

This paper proposes a novel approach which is based on metamorphic testing along with ochiai algorithm. Proposed system produces an integrated framework that offers rigid support beyond PHP such as java, HTML, JSP, JavaScript.

2 Related Work

We review related work that uses machine learning approaches as pseudo-oracles, as well as related work on metamorphic testing and other approaches to alleviating the test oracle problem [21]. Different tools with automated test suite generation techniques are studied.

Kie˙zun et al .present a dynamic tool, Ardilla [6], to create SQL and XSS attacks. Their tool uses dynamic tainting, concolic execution, and attack-candidate generation and validation. Their tool reports only real faults. However, Kie˙zun *et al.* focuses on finding security faults. Mean while this paper concentrate on functional correctness.

McAllister et al. [7] also tackles the problem of testing interactive web application. Mainly their approach attempts to follow user interactions. However, their approach to handling persistent state relies on instrumenting one particular web application framework.

Wassermann et al. [8] present a concolic testing tool for PHP. The goal of their work is to automatically identify security vulnerabilities caused by injecting malicious strings. Their tool uses a framework of finite state transducers and a specialized constraint solver.

Halfond and Orso [9] use static analysis of the server-side implementation logic to extract a web application's interface, which means the set of input parameters and their potential values. However, they implemented their technique for JavaScript.

Park et al. [10] recently described an approach for fault localization in concurrent Java programs in which occurrences of non-serializable access patterns are correlated with failures using the Jaccard formula.

Jiang et al. [11] study the impact of test-suite reduction strategies on fault-localization effectiveness. Baudry et al. [12] study how the fault localization effectiveness of a test suite can be improved by adding tests. They propose the notion of a dynamic basic block, which is a set of statements that is covered by the same tests. In addition, a related testing criterion that aims to maximize the number of dynamic basic blocks.

Yu et al. [13] also study the impact of several test-suite reduction strategies on fault localization. They conclude that statement-based reduction approaches negatively affect fault localization effectiveness, nevertheless that vector-based reduction approaches, which aim to preserve the set of statement vectors exercised by a test suite, have negligible effects on effectiveness.

3 Existing Approach

In an auto test generation strategy, the approach is quite a boon to validate quality applications in time. However, their working scenario was restricted to stand-alone applications only. Later, Auto test generation strategy is combined with source mapping and using an extended domain for conditional and function-call statements to generate automated test suits. Recently an enhanced Ochiai i.e., fault localization algorithms was proposed which has the ability to handle web applications as well, but Ochiai driven oracles offer rigid support by offering static analysis services to only PHP applications. Working procedure of the ochiai algorithm is shown in figure-2.

1: Invoke to system.
2: Consider all the statements in PHP for test generation.
3: Generate similarity functions for each statement with
    Suspicious Rating S_{com}= S_{map}=1>S_{alg}>0.5.
    S_{map}=Source Mapping with each Statement (S_{com})
4: Calculate Similarity Coefficient for each statement using Suspicious faults in each statement (S_{com}).
5: Each Suspicious fault was calculate rating incremented in every statement as follows
    for(int Sr;S_{com}>1;Sr++)
    Sr = Suspicious Rate
6: The algorithm predicts the location of fault by computing each statement.
7: Perform numerous experiments with different systems by repeat Steps 3&4 for each system.
8: Obtain Numerical results.

Fig. 2. Ochiai Implementation Procedure for test generation in Fault localization applications

Different existing techniques are implemented SHADOW interpreter version for generating efficient results. These techniques simultaneously perform concrete program execution using concrete values and symbolic execution process with associated variables. They implemented the following fault localization techniques as extensions for shadow interpreter.

Statement Coverage: All fault localization techniques use the percentage of falling and passing tests executing a given statement to calculate the statements suspiciousness.

Source Mapping: Creating the mapping by recording the line number of the originating PHP statement.

The figure-3 shows the performance of existing techniques, based upon the faults results obtained. Kindly, when observation done on each system

- SM (Source Mapping) results with low suspicious rating (nearly it lies between 10 to 30 percentage out of 70),
- Ochiai performs medium faults generation compared to source mapping.
- Apply source mapping with Ochiai, will generate efficient generation of faults (it is equivalent to maximum efficiency).

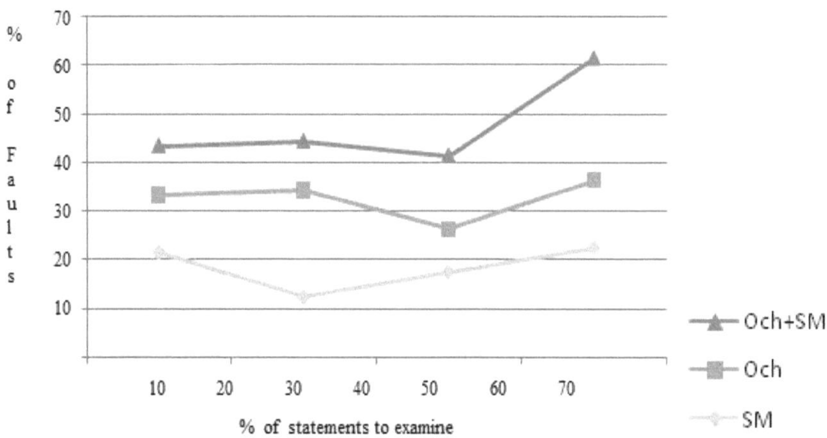

Fig. 3. Effectiveness comparison of different fault-localization techniques

Evaluation on some open source applications validates the rise in generation of test cases from 6.5 compared to 46.8, which is a significant increase. In that cases Ochiai algorithm does not support due to insufficient Suspicious Rating of Similarity Co-efficient, Due to these considerations a novel frame work is proposed for doing excellent test efficiency.

This paper likes to enhance any of the fault localization algorithms like Tarantulas[14], Ochiai [15,17], and Jaccard[16] which may has the ability to handle web applications. Using above approaches i.e., Ochiai, we develop a novel test-generation strategies which are geared toward producing test suits that have maximal fault-localization effectiveness. By exertion the control flows within every execution, faults are determined and recorded. Until the comfortable coverage of the statements within the application attained, the method is continual.

4 Proposed Approach

Currently the variant Ochiai driven oracles offers rigid support by offering static analysis services to only PHP applications. This paper proposes to extend the Ochiai algorithm with Metamorphic testing strategies to develop an integrated framework that can offer support beyond PHP such as Java/HTML/JavaScript [4] [5]. Instead of employing any oracles for initiating testing, we propose to implement metamorphic testing.

4.1 Metamorphic Testing

Metamorphic testing is a technique for the verification of software output without a complete testing. Procedure for proposed technique is presented in figure-4. Metamorphic testing observes that although the executions do not end in failures, they still bear helpful information. It has been proposed as a property-based test case selection strategy. It is based on the intuition that even if no failure is revealed by a test case selected according to some strategies, it still has useful information. Follow-up test cases ought to be created from the original set of test cases with relation to designated necessary properties of the desired functions. Such necessary properties of the functions are known as *metamorphic relations*. The subject program is verified through metamorphic relations (MR).

Metamorphic set contains the program logics, which is the implementation of metamorphic relations, to compute follow-up test cases based on an incoming (or outgoing) message, and evaluates test results consistent with the enforced metamorphic relations. In metamorphic testing, the properties are not limited to identity relations. It has been applied, for instance, to verify the convergence of solutions of partial differential equations with respect to the refinement of grid points [21]. When compared with data diversity, a further difference is that other test cases used in data diversity are basically *expressed* forms of the original test cases. This constraint is necessary because the technique is applied in fault tolerance, with the objective of applying alternate ways to process the original test case but using the same program. In metamorphic testing, although other test cases are also derived from the original test cases, they are not limited by this constraint.

4.2 Metamorphic Relations

In this section, the MRs is outlined with a tendency to anticipate classification algorithms to exhibit, and outline them additionally formal as follows.

MR-0: Consistence with Affine Transformation. The result should be the same if we apply the same arbitrary affine transformation function, f(x) = kx + b, (kx= 0) to every value x to any subset of features in the training data set S and the test case ts.

MR-1.1: Permutation of Sophisticated Class Labels. Assume that we have a class-label permutation function Perm () to perform one-to-one mapping between a class label within the set of labels L to different label in L. If the source case result is li, apply the permutation function to the set of corresponding class labels C for the follow-up case, the results of the follow-up case ought to be Perm (li).

MR-1.2: Permutation of the Attribute. If we have a tendency to permute the m attributes of all the samples and therefore the test data, the result ought to stay unchanged.

MR-2.1: Addition of Uninformative Attributes. An uninformative attribute is one that is equally related to every class label. For the source input, suppose we tend to get the result ct = li for the test case ts. In the follow-up input, we tend to add an uninformative attribute to S and respectively a replacement attribute in st. The selection of the actual value to be added here is not necessary as this attribute is equally related with the class labels. The output of the follow-up test suits should still be li.

MR-2.2: Addition of Informative Attributes. For the source input, suppose we get the result ct = li for the test case ts. In the follow-up input, we tend to add an informative attribute to S and ts specify that attribute is powerfully related to class li and equally related with all different classes. The output of the follow-up test case ought to still be li.

These metamorphic relations are integrated with Ochiai algorithm to propose the best suited fault tolerant technique for web applications. Following is the procedure for proposed approach.

1: Consider a program under test P; collect the set of programs descriptions Dp that represents the programs interacting with P.
2: Design a metamorphic relations MR_i applicable to test P.
3: Implement MRi in the metamorphic set MS of the P.
4: Repeat Steps -2 to Step-3, until no more metamorphic relation is needed for testing.
5: For each available successful test case t_o, do
   i. MS uses applicable MR_i to construct the following-up test case t_f of t_o.
   ii. MS invokes P to execute t_f.
   iii. MS obtains the final results t_f
   iv. If MS detect a failure by using MR_i, then report the failure and go to Step (step-7).
   v. Repeat Steps-5(i) to step-5(iv), until no more applicable MR_i.
6: Report that no failure is found.
7: Exit

Fig. 4. Procedure for Proposed approach

In Step-1, collect the program description of the program under test. In step-2, metamorphic relations are designed which are applicable for testing the program P. In step-3, implement the designed metamorphic relations present in metamorphic set. The above two steps i.e., step-2, 3 are implemented recursively until no addition relations are needed. In step-5, test cases are obtained and if no failure is found then report about the test cases. If failure found then exit, and re-apply the metamorphic relations.

It is unlikely for a single MR to detect all possible faults. Therefore, four MRs that are quite different from one another with a view to detecting various faults were used here that are discussed in section-4.2 Finding smart and sensible MRs requires knowledge of the problem domain, understanding of user requirements or necessities, in addition some creative thinking [3]. These MRs are identified according to equivalence and inequality relations among regular expressions. So this kind of testing facilitates in an automated addressing of all possible forms of failures in most web technologies. This paper work differs from most previous research on fault localization in that it does not assume the existence of a test suite with passing and failing test cases. Previous work focused exclusively on finding failures by identifying inputs that cause an application to crash or produce malformed HTML.

This paper addresses the problem of determining where in the source code changes neeto be made in order to fix the detected failures. Program dicing was introduced, a method for combining the information of different program slices. The idea behind the scheme is once a program computes an accurate value for variable x and an incorrect value for variable y, the fault is probably going to be found in statements that are within the slice w.r.t. y, however not within the slice w.r.t. x. Variations. Use of set-union, set-intersection, and nearest neighbor strategies for fault localization; these all work by scrutiny execution traces of passing and failing program runs.

5 Analysis

According to evaluations of some open source applications, they validates that there is a rise in generation of test cases from 6.5 compared to 46.8, which is a significant increase. Existing techniques i.e., fault localization techniques, doesn't support due to

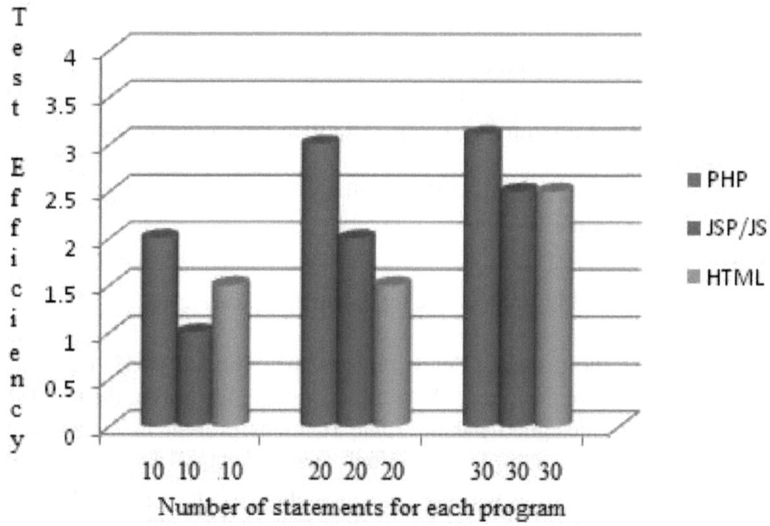

Fig. 5. HTML, PHP & JSP Unit test cases generation Using Existing Ochiai

insufficient Suspicious Rating of Similarity Co-efficient. In particular, existing techniques are mainly applicable to PHP code. If there is any code related to HTML in PHP applications then additional interpreters are used. Even though, it will generate only limited set of test cases which leads to inefficient testing. The below figure shows the testing efficiency of existing system with respect to no. of statements to examine.

In above graph, clearly observe when ever test generation for each statement in HTML, PHP, and JSP examples is not suitable for current fault locations in web applications. In the figure 5, invariant test cases are formed when existing approach Ochiai was considered which was discussed in section-3.

In order to overcome the problem of existing system, a novel integrated framework was proposed. Proposed system supports not only PHP but also HTML, Java Scripts, JSP, which generate test cases or test suites efficiently within mean time with respect to no. of statements examined. Metamorphic testing uses metamorphic relation, which generate test cases for different kind of applications. The below figure shows the testing efficiency of existing system with respect to no. of statements to examine.

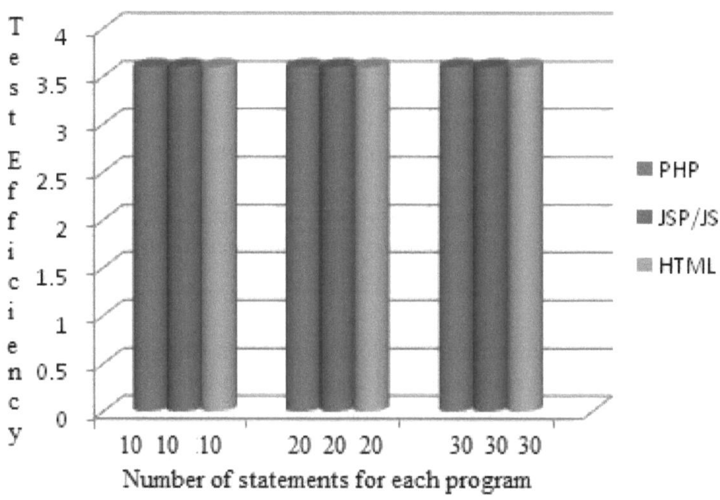

Fig. 6. HTML, PHP&JSP Unit test cases generation Using Proposed Ochiai with Metamorphic relations

Compared to existing graph results of Ochiai there is a difference in test case generation. In proposed system, for each and every program maximum test cases are drawn from metamorphic relations. So, maximum efficiency is generated for each and every program as shown in figure-6.

6 Implementation and Results

Proposed tool is developed with the help of net beans IDE. To develop the tool, Ochiai algorithm is used with metamorphic relations to increase the efficiency of the test cases generation. In this testing process, metamorphic relations are given as input in DLL format for testing each method with equivalent attributes and parameters. Using the proposed system, the test cases are generated effectively for both HTML and PHP applications by giving program as input .

6.1 Dealing With Sample java Program

In this section, testing results for accessing simple java program are described. In this way, the overall processing is calculated for every method present in the java program with compiler execution time. After applying the proposed system to sample java program, the following are some of test cases generated, which are discussed in table 1.

Table 1. Test cases

Test Method	Expected Output	Result
Invoke Application without libraries	**Application with disabled features**	**Pass**
Invoke Application with libraries	**Application with complete features**	**Pass**
Over all process Testing	**Should Generate Varying Test Cases based upon the operations and volume of the Code**	**Pass**

These are useful for generating complete description of every method present in simple java program. By using proposed testing methodology present in the testing process, the testing results are calculated for every method and every statement in the program.

6.2 Dealing with HTML and PHP

In priori approaches, Apollo[17,18] was used and shadow Interpreter based on Zend PHP interpreter are implemented that consequently presents concrete testing process with concrete values, and a symbolic testing execution with symbolic values. This paper proposes a new system, for developing the HTML format sequences in the testing methodologies for getting results in the invariant process.

When dealing with either HTML or PHP, related code is given as input to proposed system. After giving the input, the efficient test cases generated and obtained from the code. Evaluation of code is done by verifying every method with related attributes and parameters. After evaluation, Test cases are generated.

Consider a HTML application code with a function called myfunction() which contains 3 parameters. Operation performed in myfunction() are adding, subtracting and multiplying the 3 parameters. When this application code is given as input to tool, then verification process starts on every method; however, given code contain only single method. Now test cases are generated using metamorphic relations; while generating the test cases all possible conditions are verified i.e., Whether given parameters are integers or characters or alphanumeric, and soon. After generating all test cases to given code, those test cases are used to run application without any fault or improper termination.

7 Conclusion

In recent years, a number of automated techniques have been proposed to assist programmers for finding and fixing the faults in a program, which is usually called fault localization. Existing analyses require and uses fault localization algorithms such as Ochiai. In an auto test generation strategy, this approach is kind of boon to validate quality applications in time. It has a tendency to develop a unique test-generation strategy that is double geared toward manufacturing test suites that have greatest fault-localization effectiveness. This paper proposes to extend the Ochiai algorithm with Metamorphic testing strategies to develop an integrated framework that may offer support beyond PHP such as Java/HTML/JavaScript. Instead of employing any oracles for initiating testing, this paper proposes and implements metamorphic testing. Metamorphic testing is a technique for the verification of software output without a complete testing oracle.

References

1. Hiett, E., Mee, R.: Going Faster: Testing the Web Application. IEEE Software 19(2), 60–65 (2002)
2. Ye, L.: Model-Based Testing Approach for Web Applications (2007)
3. Di Lucca, G.A., Fasolino, A.R.: Testing Web-based applications: The state of the art and future trends. Information and Software Technology 48, 1172–1186 (2006)
4. Web Application Developer's Guide, by Borland Software Corporation
5. Artzi, S., Møller, A., Dolby, J., Jensen, S., Tip, F.: A Framework for Automated Testing of Javascript Web Applications. Proceedings in Int'l Conf. Software Engineering (2011)
6. Kieżun, A., Guo, P., Jayaraman, K., Ernst, M.: Automatic creation of SQL injection and cross-site scripting attacks. In: Proceedings of International Conference of Software Engineering (ICSE) (2009)
7. McAllister, S., Kirda, E., Kruegel, C.: Leveraging user interactions for in-depth testing of web applications. In: Lippmann, R., Kirda, E., Trachtenberg, A. (eds.) RAID 2008. LNCS, vol. 5230, pp. 191–210. Springer, Heidelberg (2008)
8. Wassermann, G., Yu, D., Chander, A., Dhurjati, D., Inamura, H., Su, Z.: Dynamic test input generation for web applications. In: Proceedings of the ACM/SIGSOFT International Symposium on Software Testing and Analysis (ISSTA 2008), pp. 249–260 (2008)

9. Halfond, W.G.J., Orso, A.: Improving test case generation for Web applications using automated interface discovery. In: ESEC-FSE (2007)
10. Park, S., Vuduc, R.W., Harrold, M.J.: Falcon: Fault Localization in Concurrent Programs. In: Proc. 32nd ACM/IEEE Int'l Conf. Software Eng., pp. 245–254 (2010)
11. Jiang, B., Zhang, Z., Tse, T., Chen, T.Y.: How Well Do Test Case Prioritization Techniques Support Statistical Fault Localization. In: Proc. 33rd Ann. IEEE Int'l Computer Software and Applications Conf. (July 2009)
12. Baudry, B., Fleurey, F., Le Traon, Y.: Improving Test Suites for Efficient Fault Localization. In: Osterweil, L.J., Rombach, H.D., Soffa, M.L. (eds.) Proc. 28th Int'l Conf. Software Eng., pp. 82–91 (2006)
13. Yu, Y., Jones, J.A., Harrold, M.J.: An Empirical Study of the Effects of Test-Suite Reduction on Fault Localization. In: Proc. Int'l Conf. Software Eng., pp. 201–210 (2008)
14. Jones, J.A., Harrold, M.J., Stasko, J.: Visualization of test information to assist fault localization. In: ICSE, pp. 467–477 (2002)
15. Abreu, R., Zoeteweij, P., van Gemund, A.J.C.: An evaluation of similarity coefficients for software fault localization. In: PRDC 2006, pp. 39–46 (2006)
16. Chen, M.Y., Kiciman, E., Fratkin, E., Fox, A., Brewer, E.: Pinpoint: Problem Determination in Large, Dynamic Internet Services. In: Proc. Int'l Conf. Dependable Systems and Networks, pp. 595–604 (2002)
17. Artzi, S., Kieżun, A., Dolby, J., Tip, F., Dig, D., Paradkar, A., Ernst, M.D.: Finding bugs in dynamic web applications. In: ISSTA, pp. 261–272 (2008)
18. Artzi, S., Kieżun, A., Dolby, J., Tip, F., Dig, D., Paradkar, A., Ernst, M.D.: Finding bugs in web applications using dynamic test generation and explicit state model checking. IEEE Transactions on Software Engineering (2010)
19. Chen, H.Y., Tse, T.H., Chan, F.T., Chen, T.Y.: In black and white: an integrated approach to class-level testing of object oriented programs. ACM Transactions on Software Engineeringand Methodology 7(3), 250–295 (1998)
20. Chen, H.Y., Tse, T.H., Chen, T.Y.: TACCLE: a methodology for object-oriented software testing at the class and cluster levels. ACM Transactions on Software Engineering and Methodology 10(1), 56–109 (2001)
21. Chen, T.Y., Cheung, S.C., Yiu, S.M.: Metamorphic testing: a new approach for generating next test cases. Technical Report HKUST-CS98-01. Department of Computer Science, Hong Kong University of Science and Technology, Hong Kong (1998)

Study on Agile Process Methodology and Emergence of Unsupervised Learning to Identify Patterns from Object Oriented System

Mulugu Narendhar[1] and K. Anuradha[2]

[1] Department of CSE, BSIT,
Hyderabad, A.P, India
mnarender53@gmail.com
[2] Department of CSE, GREIT,
Hyderabad, A.P, India
kodalianuradha@yahoo.com

Abstract. Data mining is knowledge extraction for secure software engineering, improves the quality and productivity, poses several challenges, requiring various algorithms to effectively mine text, graph from such database. Fact that building models in the context of the framework one of the task data miners, almost important though all other tasks associated with data mining. Data mining techniques are tackling the right business problem, must understand the data this is available and turn noisy data into data from which we can build robust models. It is important to be aware data mining is really what we might call an agile model. The concept of agility comes from the agile software engineering principles includes increment development of the business requirements and need to check whether the requirement satisfies with the client inputs our testing and rebuilding models improves the performance. For software engineering code execution, code changes list of bugs and requirement engineering our system uses mining techniques to explore valuable data patterns in order to meet better projects inputs and higher quality software systems that delivered on time. Our research uses frequent mining, pattern matching and machine learning applied to agile software architecture model in gathering and also extracting security requirements best effort business rules for novel research.

Keywords: Agile Model, Data Mining, Software Engineering, Architecture & Design Pattern.

1 Introduction

Features of data mining such as database tasks is the fact that knowledge acquired from mining, classification association rules and retrieval of relevant information produce the most accurate results. Similar in software engineering we can see two distinct ways data mining information retrieval and supervised & unsupervised methods are applied. Software engineering is artifact of software development that are

S.C. Satapathy et al. (eds.), *ICT and Critical Infrastructure: Proceedings of the 48th Annual Convention of CSI - Volume II*, Advances in Intelligent Systems and Computing 249,
DOI: 10.1007/978-3-319-03095-1_44, © Springer International Publishing Switzerland 2014

document hierarchies, code repositories, bug report data bases for the purpose of learning new interesting information about the underlying patterns.

Application data mining use the term certain related activities and techniques from machine learning, natural language processing and information retrieval in different processes in software lifecycle with the automating and improving performance on the tasks involved. Mining techniques are typically applicable to intensive tasks and are capable of speeding up the performance the order of magnitude. Some many benefits of applying data mining to specific tasks a human analyst must always access correct results of the automated methods. As the goal data mining methods is improvement of the process observe that only result is seen by others is generated by the analyst. Today's mainstream software processes do not effectively address two major issues on cost and time there is emerging producing new processes known as "Agile Software Process" [1] . The new processes focus more on people interactions development of code than on documentation, this work presents how mining techniques impact on agile software process to extract text mining, graph mining.

Frequent patterns are item sets that subsequently appear in data set with frequency for example a set of items such as brad and jam that appear together in a transaction data set is a frequent item set. Frequent pattern mining was introduction by Agrawal for market basket analysis in the form of association rule mining for instance customer buying bread how likely they are going buy jam, sugar on the same trip to the supermarket. Frequent Patterns: Many commercial applications need pattern that are more complicated than a frequent item sets such patterns go beyond sets and sequences toward trees, graphs. Among the various kinds of graph patterns frequent substructure are the basic patterns, we have two types of frequent pattern mining one is [2] Apriority-based approach and other is pattern-growth.

Apriori is frequent substructure mining algorithm suppose we are searching for small size of graph and will proceed in bottom up manner here at each iteration the size redirects frequent substructures is increased by one. These are first generated by joining two similar but slightly different sub graphs that were discovered already.Apriori frequent substructure mining algorithm uses a vertex based candidate generation method that increases the substructure size by one vertex at each iteration. Two size k frequent graphs are joined only when two graphs have the same size (k-1) sub graphs.

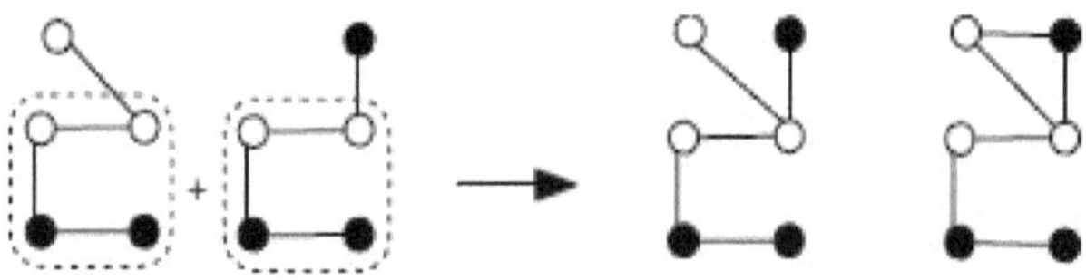

Fig. 1. Two sub graphs joined by two chains

The graph size means the number of vertices in a graph and new candidate includes the common size (k-1) sub graph and the additional vertices from the two size k patterns. FSG is an edge-based candidate generation strategy that increases the sub structure size by one edge in each iteration. Two size patterns are merged if and only if they share the same sub graph with k-1 edges.

2 Software Architecture

Architecture is a process that defines the solutions final meets the goals of end product technical and functionally while optimizing common quality attributes such as performance and security. Architecture encompasses the set of tasks and decisions about the system including the selection of structural elements and their interfaces by which the system is composed behavior as specified in collaboration among those elements composition of these structural and behavioral elements into larger subsystems and an architectural style [3]. It also involves functionality usability resilience performance reuse the tradeoffs and aesthetic concerns.

Like common problems, key scenarios, complex structure, software must be built on a solid foundation, modern tools and platform help to simple task of building applications but they do not replace the need to design application carefully based on specific scenario. Systems should be designed with consideration for the user, system and business goals. For each of these should outline key scenario and identify important quality attributes and satisfaction and dissatisfaction where possible.

Software architecture use viewpoints such as functional, information, deployment to guide the process of capturing the architecture as a set of views with the development of each view being by the use of a specific viewpoint. Architecture focuses on how the major elements and components within an application are used or interact with major elements and components within the application. The selection of data structures and algorithms implementation details of components are design concern.

Application architecture to build a bridge between requirements and technical requirements by understanding use cases and then finding way to implement those use cases in the system software. The architecture goal is to identify the requirements that affect the structure of the application which reduces the business risks associated with building technical solutions.

To develop architecture we need to follow below steps:

Expose the structure of the system but hide the implementation details.
Realize all of the use cases and scenarios.
Try to address the requirements of various stakeholders.
Handle the functional and quality requirements.

Software architecture is classifies as shown here. Artifact driven defines the relation between subsystems, groups the related artifact in subsystems these are the architecture components. Use case driven identifies fundamental classes from use cases and groups these classes in packages are the architecture components. Pattern starts with requirement specification which select appropriate patterns from a pattern base are architecture components. Pattern is a generic and reusable design solution for recurring problems in a context.

Software pattern is design rationale constraints facilitate reuse and sharing of models and design artifacts to fix a particular problem. Developers do not invent software patterns they discover patterns from building systems. Each pattern states problem to which the pattern is applicable constraints of using the pattern in practice and often some other information about the pattern. Developers need to understand software pattern does not require knowledge of programming skills only require a extra effort in order to recurring solutions to specific problem.

3 Problem Definition

In this research, our descriptive approach primarily used to generate structural and behavioral perspective explanation of the study. The project may include the mining techniques in agile method software architecture design. In previous work we identify the problem in software engineering that reliability of code, complexity of structure, more effort and cost we analyze the problem mining techniques are efficient to solution

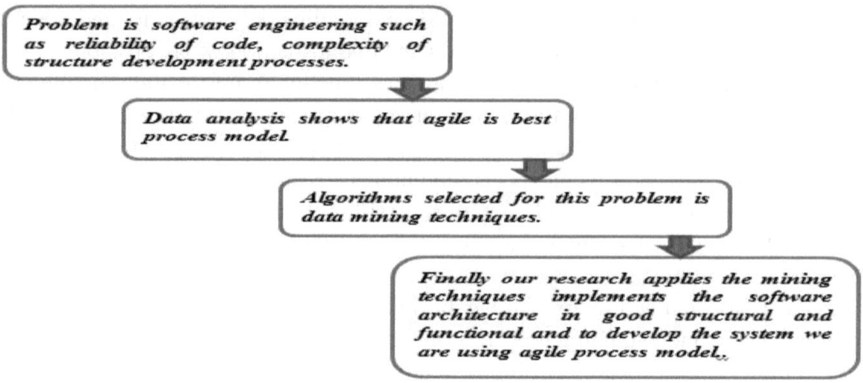

Fig. 2. Flow diagrams for proposed Work

Software architecture & design pattern is common problem in design phase of software engineering. Discovery of pattern engineering in design or coding phase represents a program understanding process, identifying the patterns using unsupervised learning would be useful to find objects in architecture & design improves quality attributes of system.

4 Agile Process Technique

Computer programmers, computer scientists, software engineers, and management scientists have been trying to solve computer programming problems such as productivity, quality, reliability, customer satisfaction, cost effectiveness, and time-to-market for more than five decades. The concept has been introduced in the 1960's and now, the transition to agile processes is a growing trend that will have lasting effects on the industry and the people involved.

4.1 Managing Constraints, Scope and Quality

Agile Methodology manages changes and constraints. The way the project is managed will allow the customer to minimize cost of change while taking into account constraints.

In Agile environment we can identify two types of constraints:

External constraints: The project manager is to handle the external constraints that are imposed by the project environment and the customer through contractual terms. The project manager is still guarantying those constraints imposed on time, budget quality and scope.

Internal constraints: The project manager rely on the team to handle the internal constraints which are often more technical. "Agile projects engage people with profound knowledge of the system; team is typically diverse of generalizing specialist.

In Agile Methodology, Typically, when the project manager is defining the project scope with the customer and other stakeholders, the project charter, very often, consists of several white paper boards with color-coded markings and post-its. In traditional project management, before proceeding to the initiation phase, the scope has to be well defined and understood by the project manager, whereas agile methodology is demanding more than just understanding the scope.

4.2 Agile Estimation Algorithm

Agile Software Methodology estimation process followed by the cost, size and duration CSD algorithm for the same Agile software process. Agile estimation process is classified into two phases: Early Estimation (EE) and Iterative Estimation (IE) as shown in fig.3. An estimator can update the estimates whenever the uncertainty in requirements reduces. The purpose of EE is to identify the scope of the project by envisaging the upfront with just enough requirements. This provides just enough understanding to put together an initial budget and schedule without paying the price of excessive documentation. EE also provides the platform for fix budget agile project and generates a satisfaction amongst the stakeholders by providing range of estimate to reflect temporal risk.IE is an iterative activity that starts at the beginning of iteration to incorporate new requirements/ changes. Agile estimation process considers only clear specified requirements and other factors that affect the estimation to derive CSD of the project as shown in Fig. 3. Here, estimation process starts after the prioritization of requirements and EE is just to understand the CSD involved in project. After finalization of project, iteration planning starts and continues till all the requirements/ changes required by customers are exhausted. EE and IE use story points for estimating CSD. Existing AEMs use expert opinion and historical data for evaluating story points. In this section, we propose *CAEA* to evaluate story points after the inclusion of various CSD affecting factors in AEM. Computation of variables in various projects to establish the fact that inclusion of vital factors in agile estimation generates realistic and more precise CSD estimation. An algorithm *CAEA* is based on above constructive agile estimation and computes the story points for CSD estimation. It incorporates the vital factors such as project domain, performance, configuration etc. as discussed in Section 3. We have graded the intensity of these factors on the scale of low, medium and high based upon the complexity of the project. It is preferred to map the intensity levels with mathematical series such as square series (1, 4, 9) or Fibonacci series (2, 3, 5). Square series has been proved to be the most preferred series in agile estimation since it provides realistic level of accuracy for complex and ill-defined project.

Formal description of proposed AEA is described as follows:

Algorithm: AEA

This algorithm computes the of story points on the basis of the input as the grades of vital factors of a project

Step 1: Vital factors of project are identified on the grade of low, medium and high

Using square series or Fibonacci series.

Step 2: Compute sum of all grades of various factors for a project denoted as Unadjusted Value (UV).

Step 3: Decompose the project in small tasks or stories.

Step 4: Assign Story Point (SP) to each story based upon the size.

Step 5: Compute New Story Point (NSP) by using equation (1).

Step 6: Compute SOP by using equation (2).

Step 7: Compute DOP through equation (3).

$$NSP = SP + 0.1 * UV \quad \text{------(1)}$$

$$\text{Size of Project(SOP)} = \sum_{i=1..n} NSP_i \quad \text{------(2)}$$

$$\text{Duration of project (DOP)} = SOP / Velocity \quad \text{------(3)}$$

Where SP story point of a story, UV is unadjusted value,
NSPi is NSP of ith the story and n is total number of stories of project,
SOP is size of project,
Velocity is number of NSP developed in iteration.

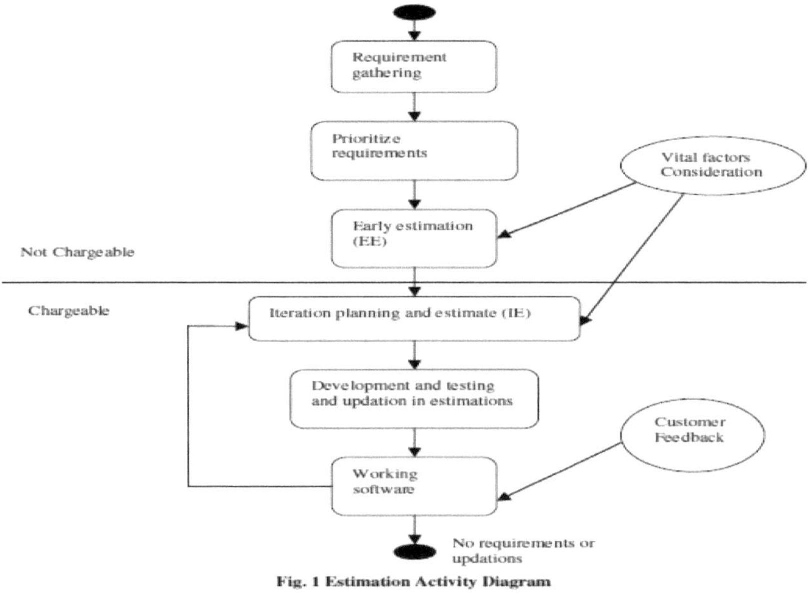

Fig. 3. Flow diagram of Agile Process

4.3 Agile Software Process Methodologies

XP and Scrum iterative approach process models are widely used in the Agile Philosophy

4.3.1 Extreme Programming

Extreme programming(XP), concentrates on the development rather than managerial aspects of software projects. XP was designed so that organizations would be free to adopt all or part of the methodology. XP projects start with a release planning, followed by several iterations, each of which concludes with user acceptance testing. User representative part of the XP team can add detail requirements as the software is being built. This allows requirements to evolve as both users and developers define what the product will look like. In release plan, team breaks up the development tasks into iterations. Release plan defines each iteration plan, which drives the development for that iteration. At the end of iteration, conduct the acceptance tests against the user stories. If they find bugs, fixing the bugs becomes a step in the iteration. If users decide that enough user stories have been delivered, the team can choose to terminate the project before all of the originally planned user stories have been implemented.

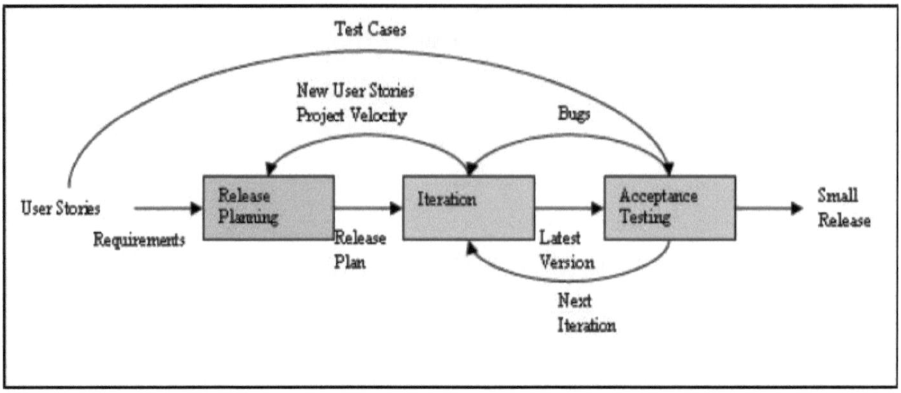

Fig. 4. Extreme Programming Process Model shows a simplified version of XP. Full XP includes many steps in release planning, iteration, and acceptance.

Integrate Often: Development teams must integrate changes into the development baseline at least once a day.

Project Velocity: Velocity is a measure of how much work is getting done on the project. This important metric drives release planning and schedule updates.

Pair Programming: All code for a production release is created by two people working together at a single computer. XP proposes that two coders working together will satisfy user stories at the same rate as two coders working alone, but with much higher quality.

User Story: A user story describes problems to be solved by the system being built. These stories must be written by the user and should be about three sentences long. User stories do not describe a solution, use technical language, or contain traditional requirements-speak, such as "shall" statements. Instead, a sample user story might go like this: Search for customers. The user tells the application to search for customers. The application asks the user to specify which customers.

4.3.2 Scrum

Scrum is the term for a huddled mass of players engaged with each other to get a job done. In software development, the job is to put out a release. Scrum for software development came out of the rapid prototyping community because prototypes wanted a methodology that would support an environment in which the requirements were not only incomplete at the start, but also could change rapidly during development.

Scrum project is a backlog of work to be done. This backlog is populated during the planning phase of a release and defines the scope of the release. After the team completes the project scope and high-level designs, it divides the development process into a series of short iterations called sprints. Each sprint aims to implement a fixed number of backlog items. Before each sprint, the team members identify the backlog items for the sprint. At the end of a sprint, the team reviews the sprint to articulate lessons learned and check progress. During a sprint, the team has a daily meeting called a scrum. Each team member describes the work to be done that day, progress from the day before, and any blocks that must be cleared. To keep the meetings short, the scrum is supposed to be conducted with everyone in the same room—standing up for the whole meeting. When enough of the backlog has been implemented so that the end users believe the release is worth putting into production, management closes development. The team then performs integration testing, training, and documentation as necessary for product release.

Scrum development process concentrates on managing sprints. Before each sprint begins, the team plans the sprint, identifying the backlog items and assigning teams to these items. Teams develop, wrap, review, and adjust each of the backlog items. During development, the team determines the changes necessary to implement a backlog item. The team then writes the code, tests it, and documents the changes. During wrap, the team creates the executable necessary to demonstrate the changes. In review, the team demonstrates the new features, adds new backlog items, and assesses risk. Finally, the team consolidates data from the review to update the changes as necessary.

5 Clustering

Cluster based on the information found in the data describing the instances, combining similar objects in one cluster and dissimilar objects in other cluster. In many applications clusters are not well different from another most cluster seeks a result a crisp classification of the data into non-overlapping groups.

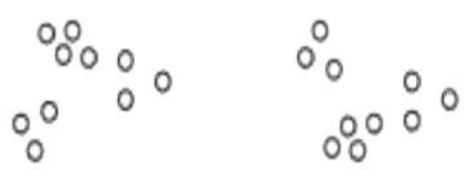

Fig. 5. (a) Instance

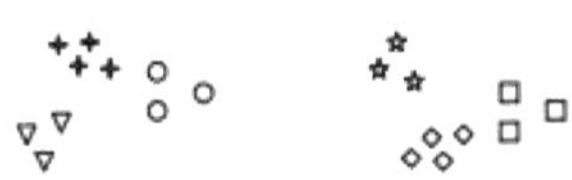

Fig. 5b. Many clusters

Figure a consider instances of in different ways that they can be divided into clusters. If we allow clusters to be nested then the interpretation of structure of these points is that each of which has many sub clusters. Cluster is a classification of instances from the data does not allow previously assigned class labels except perhaps for verification. Cluster is different from pattern recognition and decision analysis seeks to find rules for in a given set of pre-defined objects.

Hierarchical construct the clusters by recursively partitioning the instance in top-down or bottom-up approach classified.

Agglomerative clustering, each object initially represents a cluster of its own clusters then clusters are successively merged until the desired cluster structure is obtained. Divisive cluster is all objects initially belong to one cluster then cluster divided into sub clusters which are successively divided into their own sub-clusters.

Hierarchical cluster results in dendrogram representing the nested grouping of objects and similarity levels at which groupings change.

Complete link clustering also called the diameter maximum method or the further neighbor methods that consider the distance between two clusters to be equal to the longest distance from any member of one cluster to any member of the other cluster.

Average link clustering method that consider distance between two clusters to be equal to the average distance from any number of clusters to any number of the cluster.

Partitioning methods relocates instances by moving them from one cluster to another starting from an initial.

For the partition can be of K-means [15] & K-mediod. The purpose solution is based on K-means (Unsupervised) clustering combine with Id3 Decision Tree type of Classification (Supervised) under mentioned section describes in details of K-means & Decision Tree. K-means [3], [14] is a centroid based technique each cluster is represented by the center of gravity of the cluster so that the intra cluster similarity is high and inters cluster similarity is low. This technique is scalable and efficient in processing large data sets because the computational complexity is O(nkt) where n-total number of objects, k is number of clusters, t is number of iterations and k<<n and t<<n.

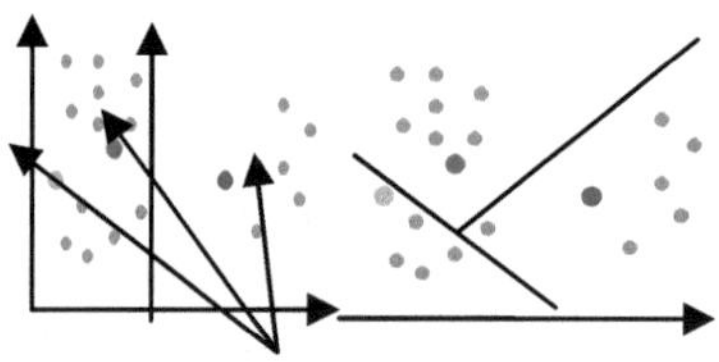

Seeds

(a) Un clustered Data Instances (b) Resultant Clusters

Fig. 6. Formation of clusters using seed points

C. K-mean algorithm

1. Select k centroids arbitrarily (called as seed as shown in the figure) for each cluster C_i,$i\varepsilon$ [1, k]
2. Assign each data point to the cluster whose centroid is closest to the data point.
3. Calculate the centroid $C_i$of cluster $C_{i,}$$i\varepsilon$ [1, k] In short
4. Repeat steps 2 and 3 until no points change between clusters. A major disadvantage of K means is that one must specify the clusters in advance and further the algorithm is very sensitive of noise, mixed pixels and outliers. The definition of means limit the application to only numerical variables.

6 Experiment Setup

Software design patterns offer elegant solutions to common problems in software engineering. From a programming standards and a reverse engineering perspective the discovery of patterns in a software artifact represents a step in the program understanding process.

Consider the Object-oriented system as input.

Apply data mining unsupervised learning to identify the patterns in the system.

Clustering is grouping the similar patterns in one cluster and dissimilar pattern in other cluster.

In clustering we choose any one of cluster either Hierarchal or Partitioned which is best to identify the pattern in object oriented system.

Then will show the end product of software engineering is reliable, cost effective with high positive rates.

7 Conclusion

The proposed pattern recognition system provides a solution to emphasizing design patterns and applied to the architecture. The model also has extensive contributions to the fields of Agile process Methodology, web Intelligence, Recommendation Systems, and Information Systems and reviews the software engineering. Our work extends to investigate the methods that generate and implement user instance from object oriented system. The present work is framework need to develop implementation of software architecture & design patterns from object oriented system with high positive rates and best quality attributes.

References

[1] Dekhtyar, A., Hayes, J.H., Menzies, T.: Text is Software Too. In: Proceedings of the International Workshop on Mining of Software Repositories (MSR), Edinborough, Scotland, pp. 22–27 (May 2004)

[2] Hayes, J.H., Dekhtyar, A., Osbourne, J.: Improving Requirements Tracing via Information Retrieval. In: Proceedings of the International Conference on Requirements Engineering (RE), Monterey, California, pp. 151–161 (September 2003)

[3] Hayes, J.H., Dekhtyar, A., Sundaram, K.S., Howard, S.: Helping Analysts Trace Requirements: An Objective Look. In: Proceedings of the International Conference on Requirements Engineering (RE), Kyoto, Japan (September 2004)

[4] Schwaber, K., Beedle, M.: Agile Software Development with Scrum. Prentice Hall (2001)

[5] Beck, K., et al.: The Agile Manifesto (2001), `http://www.agilemanifesto.org` (downloaded March 6, 2009)

[6] Babchuk, N., Goode, W.J.: Work incentives in a self-determined group. American Sociological Review 16(5), 679–687 (1951)

[7] Badani, M.: Mapping Agile development practices to Traditional PMBOK (June 4, 2001), `http://www.pmiissig.Org/pds/DOCS/BadaniBioandAbstractMappingAgileDevelopmentPractices.doc` (accessed February 29, 2007)

[8] Sliger, M.: Survival guide to going Agile, Rally Software Development Corporation, p. 8 (2006)

Data Collection, Statistical Analysis and Clustering Studies of Cancer Dataset from Viziayanagaram District, AP, India

T. Panduranga Vital[1], G.S.V. Prasada Raju[2], D.S.V.G.K. Kaladhar[3], Tarigoppula V.S. Sriram[4], Krishna Apparao Rayavarapu[3], P.V. Nageswara Rao[5], S.T.P.R.C. Pavan Kumar[3], and S. Appala Raju[1]

[1] CSE, Raghu Engineering College, Visakhapatnam, India
[2] Dept. of Computer Science, School of Distance Education, Andhra University, Visakhapatnam, India
[3] Dept. of Bioinformatics, GITAM University, Visakhapatnam, India
[4] Raghu Engineering College, Visakhapatnam, India
[5] Department of Computer Science and Engineering, GIT, GITAM University, Visakhapatnam, India

{vital2927,dr.dowluru,ramjeesis,ks.rayavarapu, srimath.pavan,s.a.raju.cse}@gmail.com, gsvpraju2011@yahoo.com, nagesh@gitam.edu

Abstract. Cancer detection is one of major research that can be processed through datasets and data mining techniques. The data has been collected from Vizianagaram district (Village) during 2013 with 328 instances and 28 attributes (Gender, Age, Cancer Type, Family_members, Drinking, Smoking, Tea, Coffee, perfumes, Morning_eat, Travelling, Wake_up, Sleep, Tensions, Cool_drinks, Icecream, Height, weight, hair_loss, Marital, milk, bath, Oil, Fast_food, other diseases, Mobile, Sports, Mosquito_replents). The dataset has been analyzed using weka version 3.6.3and Orange softwares v2.7. The histogram shows higher instances for Lung cancer (56), Mouth (40), Bone (40), Skin (32), and Colon (24). There are more number of instances observed in Males (53.7%) compared with females (46.3%). The disease in married people are more (61%) compared to unmarried (39%) with average age groups observed at 33.78±10.12, Height as 159.02±9.79 cms and weight as 61.55±11.69 Kgs. Nearly 90.2% patients has no other diseases, 136 patients (41.5%) prefer drinking alcohol, 72 patients (22%) prefer smoking, 208(63.4%) prefer drinking tea, 96 (29.3%) prefer drinking coffee, 216(65%) prefer taking rice, 80(24.4%) prefer taking cool drinks, no person like ice creams, 88 (26.8%) prefer taking milk, 238(63.4%) prefer taking sunflower oil in cooking and 68(26.8%) prefer taking fast food. The data shows hair loss, use of mobile phones and mosquito repellents as major factors in cancer. It concludes that Age, Gender, Height, weight, marital status, tea, walking, hairloss, mobile and mosquito repellents are major factors/attributes in cancer occurrence.

Keywords: Cancer, Statistical analysis, Clustering, Vizianagaram.

S.C. Satapathy et al. (eds.), *ICT and Critical Infrastructure: Proceedings of the 48th Annual Convention of CSI - Volume II*, Advances in Intelligent Systems and Computing 249, DOI: 10.1007/978-3-319-03095-1_45, © Springer International Publishing Switzerland 2014

1 Introduction

The innovative biomedical technologies through prediction of cancer survivability have been a challenging research problem for many researchers. The advancement of hardware and software technologies from the present decades of the connected research, much progress has been recorded in several related fields. For instance, better explanatory prognostic factors can be measured and recorded with low cost computer with high volumes of better quality data. The data being collected and stored automatically and finally processed to better analytical methods. These huge amounts of data are being processed effectively and efficiently using various algorithms and techniques. There is a need for critical review the statistical analysis that focuses on cancer-related clinical outcomes [2].

The EUROCARE-3 database contains 6.5 million of data from cancer patients diagnosed from 1978 to 1994 with populations covered by 67 cancer registries in 22 European countries [3]. EUROCARE-4 database has provided information regarding the data on 151,400,000 cancer patients diagnosed from 1995-1999 populations covered by 93 cancer registries in 23 countries [4]. The information of cancer datasets through statistical approaches may improve patients' and physicians' understanding of the potential benefits of adjuvant therapy [5].

A multivariate analysis of the data like age, gender and performance status, can evaluate the effects on symptom profile. The ten most prevalent symptoms were anorexia, constipation, dry mouth, dyspnea, early satiety, easy fatigue, lack of energy, pain, weakness, and weight loss. The other symptoms like anxiety, blackout, bloating, constipation, depression, headache, hoarseness, nausea, pain, vomiting, sedation, sleep problems, satiety, and sleep problems may also be associated with the disease [6]. Gender has been reported as 59% men and 41% women where more men smoked and were heavier smokers than women [7].

Attributes like body mass index, cigarette use, alcohol intake, and other possible risk factors can crack unknown reasons for predicting incidence in increasing cancer [8, 9]. Alcohol and cigarettes were significant risk factors whereas body mass index was inversely associated with risk for adenocarcinomas of the esophagus and gastric cardia. The present studies like to conduct on Gender, Age, Cancer Type, Family members suffered with cancer, Drinking Alcohol, Smoking Cigarettes, taking Tea, taking Coffee, using perfumes, food take in Morning, Travelling mode, Wakeup time, Sleeping time, facing any tensions, take Cool drinks, take Ice cream, height, weight, have hair loss, Marital status, take milk, bathing water type, use oil, take Fastfood, have any other diseases, use Mobile phone, like Sports, use Mosquito repellents in cancer patients.

2 Methodology

The data has been collected from Vizianagaram district during 2013 with 328 instances and 28 attributes (Gender, Age, Cancer Type, Family_members, Drinking, Smoking, Tea, Coffee, perfumes, Morning_eat, Travelling, Wake_up, Sleep,

Tensions, Cool_drinks, Icecream, Height, weight, hair_loss, Marital, milk, bath, Oil, Fast_food, other diseases, Mobile, Sports, Mosquito_replents).

The dataset has been analyzed using weka and Orange software's. The frequency of patients has been analyzed from the collected data from Vizianagaram district. The questionnaire has been framed based on the present personal profile, living and food habits, travelling mode and usage of previous history, and internal and external factors.

Fig. 1 shows k-means clustering model and Fig. 2 shows MDS Model.

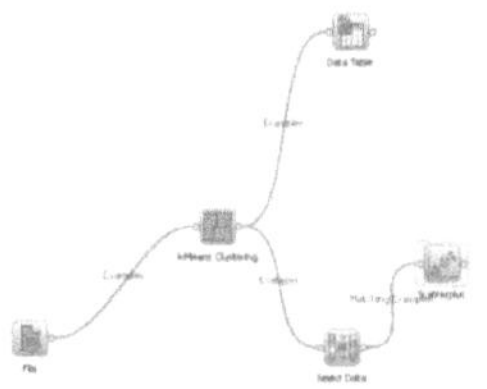

Fig. 1. k-Means Clustering connection

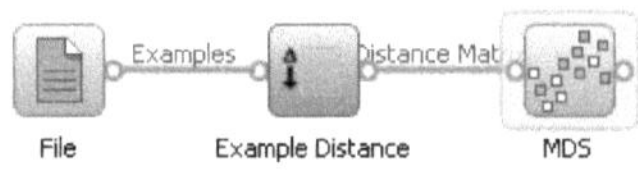

Fig. 2. MDS Model

The data for Gender has been seen from EUROCARE-3 and EUROCARE-4 datasets for finding the Gender as major attribute.

3 Results and Discussions

The data has been provided the statistical relationship of the collected datasets. Fig. 3 shows the number of instances with different types of cancers. The histogram shows higher instances for Lung cancer (56), Mouth cancer (40), Bone cancer (40), Skin cancer (32), and Colon cancer (24). Most of the other cancers like Cervical, Thyroid, Parathyroid, Childhood, Breast, Malignance, Liver, Gall bladder, Heart, throat and appendix are also prevalent in village areas like Vizianagaram.

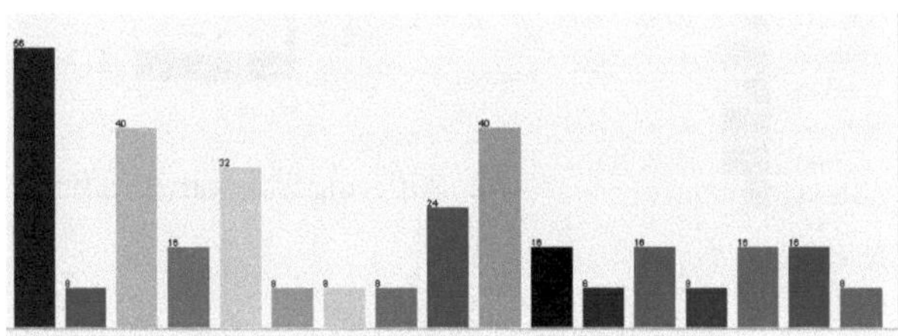

Fig. 3. Cancer types and frequency

Fig. 4 shows personnel profile of the collected data from cancer patients. There are more number of instances observed in Males (53.7%) compared with females (46.3%). The disease in married people are more (61%) compared to unmarried (39%). The average age groups observed at 33.78±10.12, Height as 159.02±9.79 cms and weight as 61.55±11.69 Kgs.

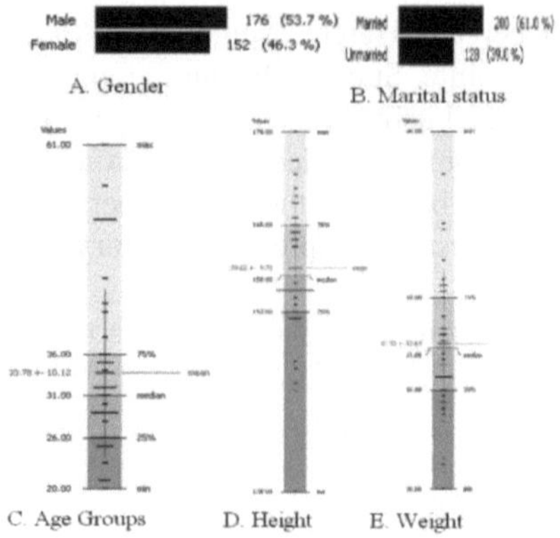

Fig. 4. Personnel Profile

Fig. 5 has shown the cancer relationship like number of instances, family history and present suffering with other diseases. There is more number of lung cancer (17%) patients observed based on survey. Nearly 14.6% of family members have previous history of suffering with this disease. Presently 2.4% of patients suffering with diabetes, 4.9% suffering with low BP, 2.4% with other diseases and 90.2% patients has no other diseases.

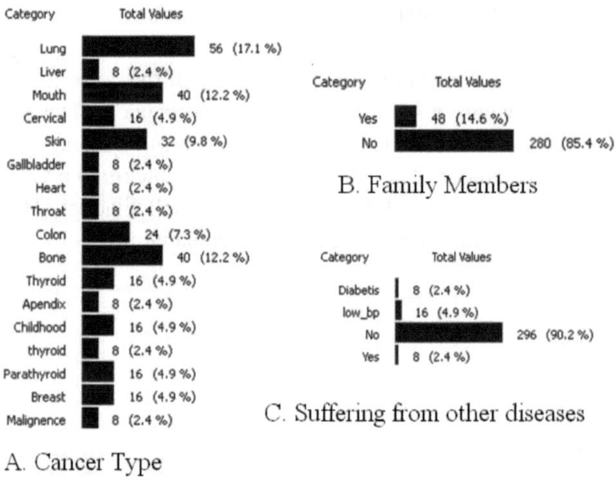

Fig. 5. Cancer relationship

Fig. 6 has shown the food habits of cancer patients. 136 patients (41.5%) prefer drinking alcohol, 72 patients (22%) prefer smoking, 208(63.4%) prefer drinking tea, 96(29.3%) prefer drinking coffee, 216(65%) prefer taking rice, 80(24.4%) prefer taking cool drinks, no person like ice creams, 88 (26.8%) prefer taking milk, 238(63.4%) prefer taking sunflower oil in cooking and 68(26.8%) prefer taking fast food.

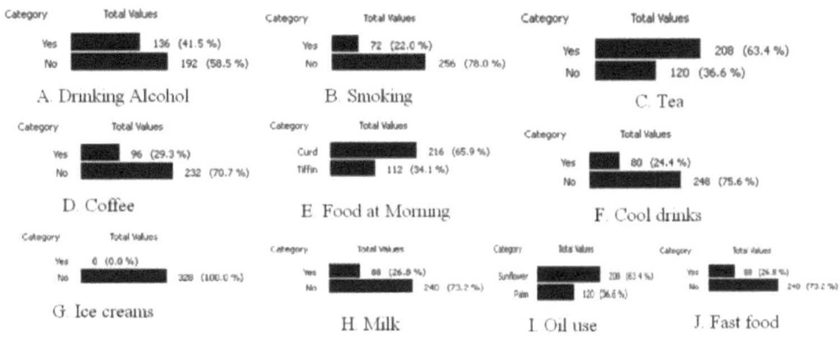

Fig. 6. Food Habits

Fig. 7 showing living conditions of cancer patients. Walking, bathing with hot water may be major factors observed based on survey.

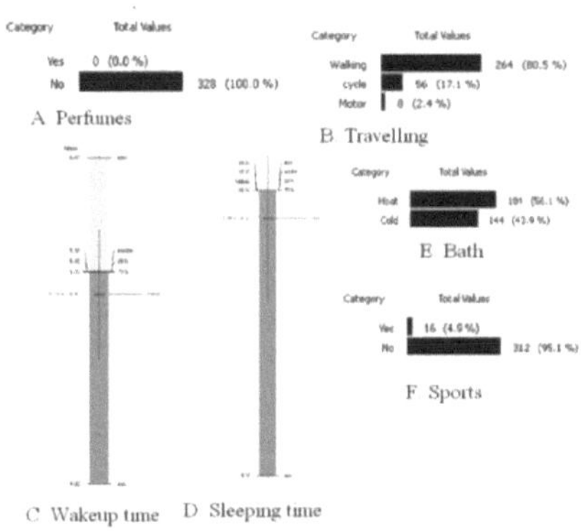

Fig. 7. Living Habits

Fig. 8 provides the internal factors like tensions and hairloss, external factors like Mobile phones and mosquito repellents. The data shows hair loss, use of mobile phones and mosquito repellents as major factors in cancer.

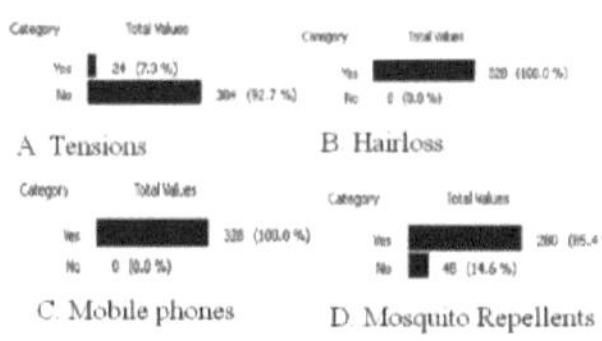

Fig. 8. Internal and External factors

Fig. 9 shows the scatter plot for k-Means cluster for Gender and Type of cancer. Most of the cancer occurrences are present at Lung cancer for males, and mouth and breast for females.

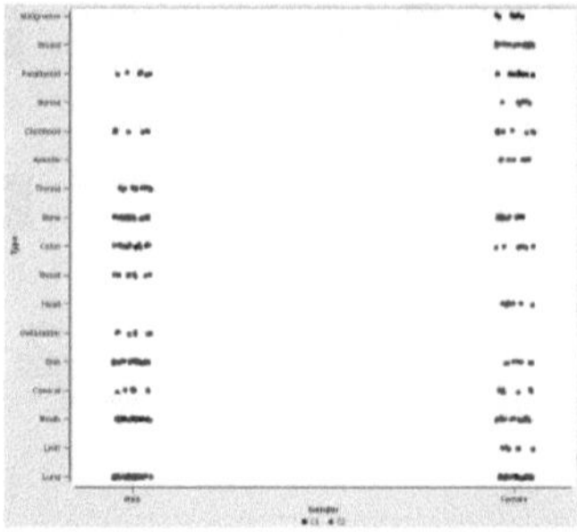

Fig. 9. k-Means cluster plot

Fig. 10 has shown MDS plot for cancer patient dataset. Multidimensional scaling (MDS) provides a protrusion into a plane fitted to the known distances between the points provided in the dataset

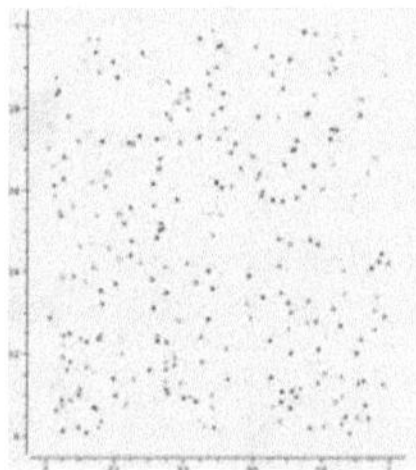

Fig. 10. MDS Plot

Fig. 11 provides Cluster analysis using SimpleKMeans. The data “0” denotes for absence of cancer and “1” denotes for presence of the cancer.

Cluster centroids:

Attribute	Full Data (328)	Cluster# 0 (152)	1 (176)
Gender	Male	Female	Male
Age	32.7805	32.9474	32.6364
Type	Lung	Lung	Lung
Family_members	No	No	No
Drinking	No	No	Yes
Smoking	No	No	No
Tea	Yes	Yes	Yes
Coffe	No	No	No
perfumes	No	No	No
Morning_eat	Curd	Curd	Curd
Travelling	Walking	Walking	Walking
Wake_up	5.1634	5.2316	5.1045
Sleep	10.2024	10.1947	10.2091
Tensions	No	No	No
Cool_drinks	No	No	No
Icecream	No	No	No
Height	159.0244	159.3684	158.7273
weight	61.561	62.9474	60.3636
hair_loss	Yes	Yes	Yes
Marital	Married	Married	Unmarried
milk	No	No	No
bath	Heat	Cold	Heat
Oil	Sunflower	Palm	Sunflower
Fast_food	No	No	No
otherdiseases	No	No	No
Mobile	Yes	Yes	Yes
Sports	No	No	No
Mosquito_replents	Yes	Yes	Yes

Clustered Instances

0 152 (46%)
1 176 (54%)

Fig. 11. Cluster analysis using SimpleKMeans

Based on FarthestFirst for Cluster centroids, the Clusters with Cluster 0 observed for Male, 52.0, Skin, No, No, No, Yes, No, No, Tiffin, Walking, 5.3, 10.3, No, No, No, 169.0, 63.0, Yes, Married, No, Cold, Sunflower, Yes, No, Yes, No and Yes for above attributes. Cluster 1 instances are Female, 35.0, Appendix, Yes, No, No, Yes, Yes, No, Curd, Walking, 5.3, 10.3, No, Yes, No, 155.0, 62.0, Yes, Unmarried, No, Heat, Palm, No, No, Yes, No and Yes. There are 216(66%) as unclassified instances and 112 (34%) as truly classified instances.

Based on CfsSubsetEval, the selected attributes are Age (2), Morning_eat (10), Height (17), weight (18): with cancer data (4).

Table 1 shows the number of cases from Europe with gender and Age classes as attributes. There is more number of males compared with females with cancer. The dataset from vizianagaram has also shown similar information but other details are missing in EUROCARE. Hence the present study has shown good information on cancer.

Table 1. Number of cancer cases from Europe (EUROCARE-3 and EUROCARE-4)

Period	Site	Country	Registry	Sex	Age class	Duration	Cases
1983-1985	All Cancers	EUROPE	Europe	Males	All ages 15-99	1	395114
1983-1985	All Cancers	EUROPE	Europe	Females	All ages 15-99	1	386374
1986-1988	All Cancers	EUROPE	Europe	Males	All ages 15-99	1	421819
1986-1988	All Cancers	EUROPE	Europe	Females	All ages 15-99	1	417113
1989-1991	All Cancers	EUROPE	Europe	Males	All ages 15-99	1	435439
1989-1991	All Cancers	EUROPE	Europe	Females	All ages 15-99	1	435391
1992-1994	All Cancers	EUROPE	Europe	Males	All ages 15-99	1	458422
1992-1994	All Cancers	EUROPE	Europe	Females	All ages 15-99	1	449864
1990-1994	All Cancers	EUROPE	Europe	Males	All ages 15-99	1	911574
1990-1994	All Cancers	EUROPE	Europe	Females	All ages 15-99	1	878319
1995-1999	All Cancers	EUROPE	Europe	Males	All ages 15-99	1	1316229
1995-1999	All Cancers	EUROPE	Europe	Females	All ages 15-99	1	1256199

Analysis of cancer datasets is an important research in the present decades in techniques like data mining and biomedical applications [10, 11]. Prevalence symptoms differ with attributes like age, gender, and cancer site [12]. Colorectal cancer varies considerably by age, gender, and race by using the largest aggregation of cancer incidence data [13]. The present studies on collection, statistical analysis and clustering studies provided good results with Age, Morning_eat, Height and weight as selective attributes.

4 Conclusion

The survey and analysis with attributes like Age, Gender, Height, weight, Marital status, Tea, Walking, Hair loss, Mobile and Mosquito repellents are major factors/attributes in cancer occurrences and diagnosis. Further analysis has to be conducted in villages for the other cancer types.

Acknowledgements. Author would like to thank management and staff of Raghu Engineering College and GITAM University Visakhapatnam, India for their kind support in bringing out the above literature and providing lab facilities.

References

1. Delen, D., Walker, G., Kadam, A.: Predicting breast cancer survivability: a comparison of three data mining methods. Artificial Intelligence in Medicine 34(2), 113–127 (2005)
2. Dupuy, A., Simon, R.M.: Critical review of published microarray studies for cancer outcome and guidelines on statistical analysis and reporting. Journal of the National Cancer Institute 99(2), 147–157 (2007)
3. Capocaccia, R., Gatta, G., Roazzi, P., Carrani, E., Santaquilani, M., De Angelis, R., Tavilla, A.: The EUROCARE-3 database: methodology of data collection, standardisation, quality control and statistical analysis. Annals of Oncology: Official Journal of the European Society for Medical Oncology/ESMO 14, v14 (2003)
4. De Angelis, R., Francisci, S., Baili, P., Marchesi, F., Roazzi, P., Belot, A., Crocettih, E., Puryi, P., Knijnc, A., Colemanj, M., Capocacciaa, R.: The EUROCARE-4 database on cancer survival in Europe: data standardisation, quality control and methods of statistical analysis. European Journal of Cancer 45(6), 909–930 (2009)
5. Gill, S., Loprinzi, C.L., Sargent, D.J., Thomé, S.D., Alberts, S.R., Haller, D.G., Benedetti, J., Francini, G., Shepherd, L.E., Seitz, J.F., Labianca, R., Chen, W., Cha, S.S., Heldebrant, M.P., Heldebrant, R.M.: Pooled analysis of fluorouracil-based adjuvant therapy for stage II and III colon cancer: who benefits and by how much? Journal of Clinical Oncology 22(10), 1797–1806 (2004)
6. Walsh, D., Donnelly, S., Rybicki, L.: The symptoms of advanced cancer: relationship to age, gender, and performance status in 1,000 patients. Supportive Care in Cancer 8(3), 175–179 (2000)
7. Visbal, A.L., Williams, B.A., Nichols III, F.C., Marks, R.S., Jett, J.R., Aubry, M.C., Edell, E.S., Wampfler, J.A., Molina, J.R., Yang, P.: Gender differences in non–small-cell lung cancer survival: an analysis of 4,618 patients diagnosed between 1997 and 2002. The Annals of thoracic surgery 78(1), 209–215 (2004)
8. Valero de Bernabé, J., Soriano, T., Albaladejo, R., Juarranz, M., Calle, M.E., Martínez, D., Domínguez-Rojas, V.: Risk factors for low birth weight: a review. European Journal of Obstetrics & Gynecology and Reproductive Biology 116(1), 3–15 (2004)
9. Courtenay, W.H.: Behavioral factors associated with disease, injury, and death among men: Evidence and implications for prevention. The Journal of Men's Studies 9(1), 81–142 (2000)
10. Kaladhar, D.S.V.G.K., Chandana, B., Kumar, P.B.: Predicting cancer survivability using Classification algorithms. LMT 34(65.7), 96–106 (2011)
11. Kaladhar, D.S.V.G.K., Pottumuthu, B.K., Rao, P.V.N., Vadlamudi, V., Chaitanya, A.K., Reddy, R.H.: The Elements of Statistical Learning in Colon Cancer Datasets: Data Mining, Inference and Prediction. Algorithms Research 2(1), 8–17 (2013)
12. Donnelly, S., Walsh, D.: The symptoms of advanced cancer. Seminars in Oncology 22(2), 67 (1995)
13. Wu, X.C., Chen, V.W., Steele, B., Ruiz, B., Fulton, J., Liu, L., Carozza, S.E., Greenlee, R.: Subsite-specific incidence rate and stage of disease in colorectal cancer by race, gender, and age group in the United States, 1992–1997. Cancer 92(10), 2547–2554 (2001)

Classification on DNA Sequences of Hepatitis B Virus

H. Swapna Rekha and P. Vijaya Lakshmi

Dept. of Computer Science and Engineering,
SSCE, Chilakapalem
{swapnarekha23,vijji032003}@gmail.com

Abstract. Bioinformatics is the application of computer technology to the management of biological information. Computers are used to gather, store, analyze and integrate biological and genetic information which can then be applied to gene-based drug discovery and development. One of the challenges is to identify genomic markers in Hepatitis B Virus (HBV) that are associated with HCC (liver cancer) development by comparing the complete genomic sequences of HBV among patients with HCC and those without HCC. In this study, a data mining framework, which includes molecular evolution analysis and classification, is introduced. Our research group has collected HBV DNA sequences, either genotype B or C, from over 200 patients specifically for this project. In the molecular evolution analysis and clustering, three subgroups have been identified in genotype C and a clustering method has been developed to separate the subgroups. A new classification method by Nonlinear Integral has been developed. Good performance of this method comes from the use of the fuzzy measure and the relevant nonlinear integral. The non additivity of the fuzzy measure reflects the importance of the feature attributes as well as their interactions. A thorough comparison study of these two methods with existing methods is detailed. For genotype B, genotype C subgroups C1, C2, and C3, important mutation markers (sites) have been found, respectively. These two classification methods have been applied to classify never-seen-before examples for validation. The results show that the classification methods have more than 70 percent accuracy and 80 percent sensitivity for most data sets, which are considered high as an initial scanning method for liver cancer diagnosis.

1 Introduction

In Asia, infection of Hepatitis B virus (HBV) is a major health problem. At least 10 percent of the Chinese population (120 million people) are HBV carriers, and up to 25 percent of HBV carriers will die as a result of HBV- related complications including liver cirrhosis and hepato-cellular carcinoma (HCC), i.e., liver cancer. Chronic infection by the HBV causes an increased risk of hepato- cellular carcinoma (HCC) by morethan100-fold The PCR technology was used to differentiate the nucleotide variant [6]. Because the focus of this paper is on the study of data mining techniques, the selection process and criteria of patients and the research experiments run by our Biochemistry Department will not be discussed in detail.

S.C. Satapathy et al. (eds.), *ICT and Critical Infrastructure: Proceedings of the 48th Annual Convention of CSI - Volume II*, Advances in Intelligent Systems and Computing 249,
DOI: 10.1007/978-3-319-03095-1_46, © Springer International Publishing Switzerland 2014

HBV DNA sequences were taken from 13 patients. Keum et al. amplified a conserved core region and a surface antigen region of HBV DNA by PCR from sera of 27 Korean chronic hepatitis B patients for detecting hepatitis B virus mutants. Our project is one of the biggest HBV DNA full-sequence collection and analysis studies of its kind. We have collected DNA sequences from 98 Control (normal) and 100 HCC (cancer) patients specifically for this project. The DNA sequences of HBV are not exactly the same for each group, and they possess some individual nucleotide mutations that may or may not be related to HCC. From previous studies, HBV can be divided into seven genotypes where each of them has more than 8 percent difference of nucleotides from the others. In Hong Kong, genotypes B and C are the predominant types, and all the examples we have are of these two genotypes.

For genotype B, genotype C subgroups C1, C2 and C3, mutation markers (sites) have been found respectively. The classification methods of these three groups have been applied to classify 122 never-seen-before samples. The results show that the classification methods have more than 80% accuracy, which is considered high as an initial scanning method for liver cancer diagnosis. To reduce the noise of genotypic difference among the sequences collected, we propose to analyze these DNA examples in each genotype separately. Classification is one of the most studied data mining tasks. The goal attribute might be the prediction of whether or not a patient has cancer, while the predictive feature attributes might be the mutation sites of the patient's virus DNA. The focus of this study is to identify genetic marker(s) for liver cancer (HCC) from HBV DNA sequences.

The aim of this study is to develop a data mining framework which contains an appropriate classifier for liver cancer based on HBV DNA and clinical data. We develop two new algorithms based on rule learning (RL) and nonlinear Integral (NI). We then carry out a thorough comparative study on these two new models with existing classifiers. The basic units of virus DNA are nucleotides. There are four different types of nucleotides found in DNA. The four nucleotides are given one-letter abbreviations as shorthand for the four bases, which are A, G, C and T respectively.

An example of DNA strand is: ATGGGCTAATCCTCTAATCTGG

There are 3215 units in each HBV virus DNA strand. Each unit is named as a particular site which is ordered from sites 1 to 3215.

Classification is the most studied data mining task.. In this paper, we identified the important mutation sites (mar- kers) in the HBV sequences that could have caused or been related to liver cancer. We use information entropy for finding genetic markers of HCC in the HBV genome data and propose a new classification model based on nonlinear integrals. This paper is organized as follows: Section 2 describes the data mining framework which includes the new rule learning and the nonlinear integral classification models in detail. All the methods and data sets used in this project are detailed in Section 3. The experimental results and the comparative studies are presented in Section 4. Section 5 concludes with the summary and the discussion of some directions for future work.

2 Data Mining Framework

The data mining framework developed is shown in Fig. 1. There are nine modules. After the molecular evolutionary analysis, the data are passed to the Clustering Module to check whether clusters exist based on the phylogenetic tree analysis. If clusters are found, each cluster will be analyzed separately for potential genetic marker sites because it will minimize the noise produced by the genotype differences and give much better classification accuracy. For each cluster (or genotype), the data are divided into training and test sets. The training examples are then passed to the Feature Selection Module to find the useful features (genetic marker sites) for classification. The features selected are also sent to the preprocessing module to extract the values of these features in the testing data set for testing in the Classification module. Finally, the prediction results of the classifier are verified and evaluated based on the testing examples. If the evaluation results are unsatisfactory, i.e., stopping criteria are not satisfied, the learning process is repeated starting from the feature selection; otherwise, the classifier will be validated by never-seen-before examples. The following sections will explain how the features are selected and also the basic principles of the classifier.

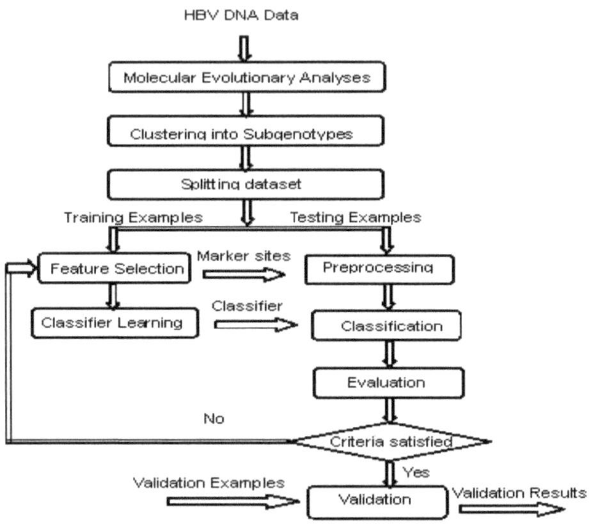

Fig. 1. Data Mining Framework

2.1 Molecular Evolutionary Analysis

Serum examples from 49 patients infected with HBV genotype C, as determined by previous genotype-specific restriction fragment length polymorphism analysis, were studied. All serum examples were kept in an 80 C freezer for storage. All patients were ethnic Chinese and were followed up in the Hepatitis Clinic of the Prince of Wales Hospital (Hong Kong). All patients were positive for hepatitis B surface antigen for at least six months and had no evidence of hepatocellular carcinoma. Sixty-nine full- genome nucleotide sequences of HBV genotype C and 12 full-genome nucleotide sequences of nongenotype C HBV were also retrieved from the GenBank database for comparison. All reference sequences from GenBank were derived from patients with chronic hepatitis B; HBV nucleotide sequences from patients with acute hepatitis B hepatocellular carcinoma or patients treated with antiviral agents were excluded. The geographical origins of patients harboring different HBV genotype C genomes in GenBank were retrieved from the respective original publications and the descriptions in the GenBank database. The full-genome nucleotide sequences of the isolates of HBV genotype C from our center were compared with those of the isolates of HBV genotype C and nongenotype C HBV retrieved from the GenBank database. Nucleotide se- quences are multiple-aligned using ClustalW version 1.83 and corrected manually by visual inspection. Genetic distances are estimated by Kimura's two-parameter method and the phylogenetic trees are constructed by the neighbor- joining method [3]. The reliability of the pairwise comparison and phylogenetic tree analysis is assessed and assured by bootstrap resampling with 1,000 replicates. Phylogenetic and molecular evolutionary analyses are done using MEGA version 3.0.

2.2 Current Classification Algorithms

There are several common classification models such as Na¨ıve Bayesian Network [5], Decision Tree, Neural Networks, and Rule Learning using Evolutionary Algorithm [6]. The learning processes of Na¨ıve Bayesian Networks and Decision Tree are faster. However, they cannot cope well with feature interactions. Neural Networks are treated as black box learning and it is difficult for a human to understand or interpret the classification explicitly. However, Rule Learning using Evolutionary Algorithm performs a global search and can cope with feature interactions better than the previous classification models [7]. Also, the classification rules generated are simple and easily interpretable by human experts who frequently use the same reasoning approach very much similar to the rules. Therefore, the Rule Learning Using Evolutionary Algorithm approach is clearly a better choice in terms of interpretability of the knowledge acquired through the classifier learned.

Rule learning tries to learn rules from a set of training data (examples). It can be modeled as a search problem of finding the best rules that classify the training examples with minimum error. However, the search space can be very large; a robust search algorithm is required. Here, Generic Genetic Programming (GGP) [8], which is a type of the Evolutionary Algorithms (EA), is adopted as our search and optimization algorithm. First, a population is initialized by generating individuals (a set of rules) randomly. A fitness function is used to evaluate how good an individual is, that is, how many cases it can classify correctly. Then, some individuals are selected to evolve (generate) new individuals with the genetic operators. Individuals become better and better through the evolution process until the termination criterion is met.

The input is the training data set, and the output is a rule set, which can classify the training data with higher accuracy. We assume that there are n features (attributes),

X=x1, x2,…,xn and K classes C=c1,c2…..,ck for each attribute xj,one of its mj values can be taken each rule includes two components :the anticident

(IF part) and the consequence (THEN part), as follows

If (x1=v1) Λ (x2=v2) Λ …… Λ(xl=vl) THEN class is C=ck, where the anticident includes l(l=[1,n]) unique attributes x1,x2,….xl €{x1,x2,….,xn},v1,v2…,vl€ {A;G;T:C} and ck,k€{1,2,…k}, is a certain class to which the object to be classified .In our case we have only two classes namely HCC and CONTROL .there are l unique attributes present in and n-1 attributes absent from each rule. Each attribute present in the anticident can only take one of its possible values, {A; G; T; C}. All the rules in the output rule set are connected by ELSE IF, meaning that the order of application of the rules must be followed.

We use a simple example to illustrate the rules deduced by the Rule Learning. For HBV data set B, which will be introduced in the following section, we have learned the rules for diagnosing liver cancer (HCC) and nonliver cancer (CONTROL) cases. The rules are given as follows:

IF A1762 and G1764 and C53, then HCC,
ELSE IF T1762 and A1764 and CG2712, then HCC,
ELSE IF T1762 and A1764 and T2712 and C2525, then HCC,
ELSE CONTROL.

Although Rule learning based on EA can interpret the interaction of features, the degree of the interaction cannot be analyzed exactly by a measure. Hence, we introduce the Fuzzy Measure to describe the interaction with respect to the classification. A new classification model is proposed based on Nonlinear Integrals with respect to signed Fuzzy Measure in the following section.

2.3 Classification Based on Nonlinear Integrals

In classification, we are given a data set consisting of N example records, called the training set, where each record contains the value of a classifying attribute Y and the value of feature attributes x1 ; x2 ; ... ; xm . Positive integer N is the data size. The classifying attribute indicates the class to which each example belongs, and it is a categorical attribute with values coming from an unordered finite domain. The set of all possible values of the decisive attribute is denoted by C = c1 ; c2 ; ... ; cK , where each ck ; k = 1; 2; ... ; K, refers to a specified class. The feature attributes are numerical, and their values are described by an m-dimensional vector, (f(x1),f(x2),…f(xm)). The range of the vector, a subset of n-dimensional euclidean space, is called the feature space. Thus, the j example record consists of the jth observation for all feature attributes and the classifying attribute, and is denoted by (fj(x1),fj(x2)…fj(xm),Yj);j = 1; 2; ... ;N [9].

In this section, a method of classification based on nonlinear integrals will be presented. It can be viewed as an idea of projecting the points in the feature space onto a real axis through a nonlinear integral, and then using a one- dimensional classifier to classify these points according to a certain criterion optimally. Our classifying attributes holding the discrete value of A, C, G, or T is numericalized to be a virtual variable. All of these are realized under the guidance of an adaptive genetic algorithm [10]. Good performance of this method comes from the use of the fuzzy measure and the relevant nonlinear integral, since the non-additivity of the fuzzy measure reflects the importance of the feature attributes, as well as their inherent interactions, toward the discrimination of the points. In fact, each feature attribute has its, respective, important index reflecting its amount of contribution toward the decision. Furthermore, the global contribution of several feature attributes to classification is not just the simple sum of the contribution of each feature to the decision, but may vary nonlinearly. A combination of the feature attributes may have a mutually restraining or a complementary synergy effect on their contributions toward the classification decision. Hence, the fuzzy measure defined on the power set of all feature attributes is a proper representation of the respective importance of the feature attributes and the interactions among them, and a relevant nonlinear integral is a good fusion tool to aggregate the information coming from the individual and the combinations of the feature attributes for the classification. The following are the details of these basic concepts and the mathematical model for the classification problem.

2.4 Fuzzy Measures and Nonlinear Integrals

The use of the Choquet integrals with respect to a signed fuzzy measure has been shown as an efficient approach to aggregate information from attributes via a nonadditive set function [5]. Let X = {x1, ..., xn} represent the attributes of the sample space and Π(X) denote the power set of X.The signed fuzzy measure μis defined as a set function μ: Π(X) → (−∞, ∞), where μ(0/) = 0.Let μi, i = 1, ..., 2n– 1, denote the values of the set function μ on the nonempty sets in Π(A), and f denote a given function, where f (x1), ..., f(xn) represent the values of each attribute for one observation. The procedure of calculating the generalized Choquet integral is given in [14], summarized as follows. Let {x1′ x2′ , ..., xn′ } be a permutation of (x1, x2, ..., xn) such that f (x1), f (x2′), ... f (xn′) is in nondecreasing order. That is,f (x1′) ≤f (x2′) ≤ ... ≤f(xn′)The Choquet integral with respect to fuzzy measure μ is defined as (c)◊f dμ= ∑nj = 1[f (xj′) − f(xj−1′)] μ ({xj′, xj + 1′, ..., xn′}),where f (x0′) = 0 and (c) indicates Choquet integral. Let ω: X → [0, 1] be a nonnegative weight function on the attributes such that ∑ni ω(xi) =1.

2.5 GA-Based Adaptive Classifier

The next step is to find an appropriate formula that projects the n-dimensional feature space onto a real axis L such that each point f =(f1 ; f2 ; ... ; fm) becomes a value of the virtual variable that is optimal with respect to the classification. In such a way, each classification boundary is just a point on real axis L.

The classification process can be divided into two parts for implementation:

Step 1. The nonlinear integral classifier depends on the fuzzy measure μ, so the first step is to determine the optimal values of μ by using the GA tool. In fact, the fitness function comes from the linear classifier used in the second procedure. It is an iterative process. The optimal Fuzzy measure will be the out put to the next step

Step 2. When the fuzzy measure is determined, the virtual value can be obtained using the Nonlinear Integral. Then, we can classify these virtual values on real axis using a linear classifier.

The following section focuses on the above problems.

GA-based learning fuzzy measure. Here, we discuss the optimization of the fuzzy measure under the criterion of minimizing the corresponding global misclassification rate, which is obtained in the second part above.

In our GA model, we use a variant of the original function f, f ' = a + bf , where a is a vector to shift the coordinates of the data and b is a vector to scale the values of predictive attributes. Each chromosome represents fuzzy measure vector, shifting vector a, and scaling vector b. A signed fuzzy measure is 0 at empty set. If there are m attributes in training data, a chromosome has 2m -1 + 2m genes which are set to random real values at initialization. Genetic operations used are traditional ones. At each generation, for each chromosome, all variables are fixed and the virtual values of all training data are calculated using a nonlinear integral. The fitness function can be defined as follows:

$$\text{Fitness} = w1 * \text{accuracy} + w2 * \text{sensitivity};$$

Where w1 and w2 are the adjustment parameters given by users. Accuracy and sensitivity are determined in the second part of the model.

Linear classifier for the virtual values. After determining the fuzzy measure, shifting vector a, scaling vector b, and the respective classification function from the training data in GA, points in the m-dimensional feature space are projected onto a real axis using a nonlinear integral. We use Fisher's linear discriminate function to perform classification in this one-dimensional space [10]. Positive and negative centroids for projected data are determined by the following formulas

$$.m_+=\frac{\sum_{i:yi=1} xi}{\sum_{i:yi=1} 1}, \quad m_-=\frac{\sum_{i:yi=-1} xi}{\sum_{i:yi=-1} 1},$$

Ronald Fisher defined the scatter matrices as

$$S_{+\_} = \sum_{xi,yi=\pm 1}(xi - m \pm)(xi - m\pm)'$$

$S_w=S_+ + S_-$ is called the Within-Class Scatter Matrix. Similarly the Between – Scatter Matrix can be defined as

$$S_B \equiv (m_+ - m_-)(m_+ - m_-)'$$

Hence, this result in an equivalent expression for Fisher's discriminate criterion is a ratio between two quadratic forms as

$$J(W)=\frac{w'S_B W}{w'S_W W}$$

in which w represents the direction of the projection space, i.e., the one-dimensional space. We can solve the programming problem by maximizing J (w). The optimal w can be represented as $w = S_w^1 * (m_+ - m_-)$. So the fisher's discriminate function is formulated as

$$y = w * (x - n_+ * m_+ - n_- * m_-),$$

in which $n\pm$ is the sum of observations in each class, respectively. Finally, a threshold needs to be fixed in order to define a complete classifier.

3 Methods

We applied EA-based Rule Learning [10] and Nonlinear Integral classifiers to classify the HBV DNA data into liver cancer (HCC) and normal (CON, control) classes, As mentioned before, we conduct a detailed study on the Rule Learning and Nonlinear Integral classifier separately. In this section, we will give brief descriptions about these classical methods of classification and the data sets used. Then, the implementation details of the Nonlinear Integral (NI) classifier and the evaluation methodology will be introduced.

3.1 Methods Description

The following paragraphs are the brief descriptions of the four classical methods used to compare with our new methods.

3.1.1 Decision Tree [11]

A decision tree is a tree-structured classifier. The Decision Tree method learns decision tree using a recursive tree growing process. Each test corresponding to an attribute is evaluated on the training data using a test criteria function. The test criteria function assigns each test a score based on how well it partitions the data set. The test with the highest score is selected and placed at the root of the tree. The subtrees of each node are then grown recursively by applying the same algorithm to the examples in each leaf. The algorithm terminates when the current node contains either all positive or all negative examples. We used the widely available package—See5.0, which is the state-of-the- art of the Decision Tree classifier.

3.1.2 Neural Network

An Artificial Neural Network (ANN), or commonly just called neural network (NN), is an interconnected group of artificial neurons that uses a mathematical or computational model for information processing based on a connectionistic approach to computation. In most cases, an NN is an adaptive system that changes its structure or weights of the interconnections based on external and internal information (stimuli) that flows through the network.

In more practical terms, NNs are nonlinear statistical data modeling for decision-making and classification tools. They can be used to model complex relationships between inputs and outputs or to find patterns in data. However, it is essentially a black box approach, and it is not easy to interpret how they function.

3.1.3 Naı¨ve Bayes [13]

A Na¨ıveBayes classifier is a simple probabilistic classifier based on applying Bayes' theorem with strong (na¨ıve) independence assumptions. A more descriptive term for the underlying probability model would be "independent feature model."

Depending on the precise nature of the probability model, na¨ıveBayes classifiers can be trained very efficiently in a supervised learning setting. In many practical applica- tions, parameter estimation for na¨ıve Bayes models uses the method of maximum likelihood; in other words, one can work with the naïve Bayes model without believing in Bayesian probability or using any Bayesian method.

3.2 Implementation Details of Nonlinear Integral

To implement the learning algorithm of our new classifier based on nonlinear integrals, we use the GA tool in Matlab v7.2 Programming combined with Fisher's discriminate function programming [14]. All the parameters of our GA in our experiments are shown in Table 1. We set the generation limit to be 100 as the stopping criteria.

3.3 Data Description

The data set contains 98 control patients and 100 HCC patients. The HBV DNA sequences are obtained specifically for this study from these patients carefully selected by our medical experts to minimize the demographic bias. There are four data sets corresponding to the different clusters, namely B, C1, C2, and C3. The numbers of patients for each data set are shown in Table 3 in which the last column represents the proportion of each data set. For each data set, an independent validation set is prepared to evaluate the performance of the classifiers. Table 2 shows the number of patients of the validation data sets.

Table 1. Summary of HBV Data set

Datasets	CON	HCC	Total	%
B	49	37	86	43.878
C1	10	16	26	13.265
C2	18	22	45	20.408
C3	19	25	44	22.449
Total	96	100	196	

Table 2. Summary of Validation data set

Datasets	CON	HCC	Total	%
B	40	18	58	47.934
C1	8	8	16	13.223
C2	15	5	20	16.529
C3	19	8	27	22.314
Total	82	39	121	

Evaluation Methodology

In classifying an unknown case, depending on the class predicted by the classifier and the true class of the patient (Control or HCC), four possible types of results can be observed for the prediction as follows:

1. True positive—the result of the patient has been predicted as positive (Cancer) and the patient has cancer.
2. False positive—the result of the patient has been predicted as positive (Cancer) but the patient does not have cancer.
3. True negative—the result of the patient has been predicted as negative (Control), and indeed, the patient does not have cancer.
4. False negative—the result of the patient has been predicted as negative (Control) but the patient has cancer.

Let TP, FP, TN, and FN, respectively, denote the number of true positives, false positives, true negatives, and false negatives. For each learning and evaluation experiment, Accuracy, Sensitivity, and Specificity defined below are used as the fitness or performance indicators of the classification:

$$Accuracy = (TP + TN)/(TP + TN + FP + FN);$$
$$Sensitivity = TP/(TP + FN);$$
$$Specificity = TN/(TN + FP):$$

4 Experimental Results

In this section, we first present the results of EA-based Rule Learning [39] and Nonlinear Integral classifiers to classify the HBV DNA data into liver cancer (HCC) and normal (CON, control) classes, and then compare them with several traditional classification methods which include See5.0 (Decision Tree) [11], Neural Network [11], SVM [12], and Na¨ıve Bayes [13]. As mentioned before, we do a detailed study on the Rule Learning and Nonlinear Integral classifier separately because of the importance of their high inter- pretability of the models representing the knowledge acquired through the learning processes. The biochemists and doctors can see explicitly and clearly the influences of the mutated sites or markers and their potential interactions toward the formation of liver cancer.

For each data set, we will use the five attributes which include those selected by the Rule Learning method, and in some cases.

Datasets	Performance	With FS Train	With FS Test	Without FS Train	Without FS Test
B	Accuracy	0.771	0.683	0.687	0.617
	Sensitivity	0.858	0.738	0.337	0.233
	Specificity	0.708	0.646	0.942	0.893
C1	Accuracy	0.922	0.790	0.826	0.515
	Sensitivity	1.000	0.920	0.937	0.715
	Specificity	0.798	0.600	0.650	0.190
C2	Accuracy	0.871	0.750	0.793	0.626
	Sensitivity	0.888	0.700	0.813	0.698
	Specificity	0.854	0.800	0.769	0.530
C3	Accuracy	0.828	0.732	0.813	0.750
	Sensitivity	0.843	0.705	0.718	0.643
	Specificity	0.808	0.765	0.937	0.900

For reducing computational complexity, we reduce the number of attributes by including the feature selection method. We compared the results of Nonlinear Integral with and without feature selection in Table 6. It shows that feature selection is very useful.

4.1 Comparison between NIC and RL

Table 7 shows the comparison results of Rule Learning and the NIC, and Table 8 shows the comparison results of our methods with several classic methods on data sets B, C1, C2, and C3. The results of RL and NIC for each data set and a validation set, which contains the never-seen-before cases, are shown in Table 4.

In Table 4, sensitivity results of NIC are higher than those of RL in most cases and other values are comparable. Since sensitivity is more important for doctors to diagnose, the performance of NIC is considered to be better than that of RL. Furthermore, NIC can not only determine the important sites (markers) with regard to the diagnosis but also give their degrees of contribution in real values, which are relatively meaningful in biomedical research. This will be described in the following section.

Datasets	Performance	RL Train	RL Test	RL Valid-ation	NIC Train	NIC Test	NIC Valid-ation
B	Accuracy	0.716	**0.716**	0.769	0.771	0.683	0.721
	Sensitivity	0.730	0.731	0.800	0.858	**0.738**	0.742
	Specificity	0.706	0.707	0.750	0.708	0.646	0.697
C1	Accuracy	0.808	**0.800**	0.917	0.922	0.790	0.712
	Sensitivity	0.812	0.790	1.000	1.000	**0.920**	0.854
	Specificity	0.800	0.800	0.857	0.798	0.600	0.570
C2	Accuracy	0.775	**0.775**	0.917	0.871	0.750	0.712
	Sensitivity	0.700	**0.700**	1.000	0.888	**0.700**	0.854
	Specificity	0.850	0.850	0.857	0.854	0.800	0.570
C3	Accuracy	0.773	**0.770**	0.647	0.828	0.732	0.639
	Sensitivity	0.720	**0.717**	0.700	0.843	0.705	0.721
	Specificity	0.842	0.835	0.571	0.808	0.765	0.523

Note: RL=Rule Learning; NIC=Nonlinear Integral Classifier

5 Discussions and Future Work

In this paper, Classification for DNA sequence biological data sets has been presented. It has been applied to the Hepatitis B Virus DNA data sets which are among the largest in the world and have been collected by our medical school specifically for this project. We have developed a framework for markers discovery. This frame- work has incorporated two algorithms, NIC and RL. Both classifiers can explicitly give the importance of the markers and their interactions and have shown good performance in cancer prediction.

Moreover, the details of the new classification method based on nonlinear .Besides the high interpretability of the Nonlinear Integrals Classifier, the experimental results have shown that it is one of the best classifiers especially in terms of sensitivity. It is very useful for preliminary diagnosis and screening test of liver cancer caused by HBV. In our model, we use GA for optimization which provides multimodal solutions containing sets of best solutions. Finally, we have used a regularization method to get a solution with the fewest nonzero fuzzy measure values. It can provide some important individual and combinations of key markers of the HBV DNA sequences.

Our findings have been validated by independent data sets in the validation process. To confirm the biological role of these mutations, further experimental work using in situ mutagenesis of replicative HBV clones on their carcinogenicity in animal and cell line models will be required.

However, even though we have generated one of the largest data sets, the example sizes of the data sets are still small (less than 100) for each case. It is a challenge for the classifier based on nonlinear integral to avoid overtraining.

References

[1] Chan, H.L.Y., Tse, C.H., Ng, E.Y.T., Leung, K.S., Lee, K.H., Tsui, K.W., Sung, J.J.Y.: Phylogenetic, Virological and Clinical Characteristics of Genotype C Hepatitis B Virus with Tcc at Codon 15 of the Precore Region. J. Clinical Microbiology 44(3), 681–687 (2006)

[2] Ciancio, A., Smedile, A., Rizzetto, M.: Identification of HBV DNA Sequences that Are Predictive of Response to Lamivudine Therapy. Hepatology 39, 64–73 (2004)

[3] Kimura, M.: A Simple Method for Estimating Evolutionary Rates of Base Substitutions through Comparative Studies of Nucleotide Sequences. J. Molecular Evolution 16, 111–120 (1980)

[4] Orito, E., et al.: Geographic Distribution of Hepatitis B Virus (HBV) Genotype in Patients with Chronic HBV Infection in Japan. Hepatology 34, 590–594 (2001)

[5] Eugene, C.: Bayesian Network without Tears. AI Magazine 12(4), 50–63 (1991)

[6] Freitas, A.A.: A Survey of Evolutionary Algorithms for Data Mining and Knowledge Discovery. In: Ghosh, A., Tsutsui, S. (eds.) Advances in Evolutionary Computation. Springer (2002)

[7] Wong, M.L., Leung, K.S.: Data Mining Using Grammar Based Genetic Programming and Applications. Kluwer Academic Publishers (January 2000)

[8] Xu, K.B., Wang, Z.Y., Heng, P.A., Leung, K.S.: Classification by Nonlinear Integral Projections. IEEE Trans. Fuzzy Systems 11(2), 187–201 (2003)

[9] Wong, M.L., Leung, K.S.: Genetic Logic Programming and Applications. IEEE Expert 10(5), 68–76 (1995)
[10] Data Mining Tools See5 and C5.0, Software (May 2006), http://www.rulequest.com/see5info.html
[11] SAS1EnterpriseMiner (EM), http://www.sas.com/technologies/analytics/datamining/miner/
[12] Chang, C.C., Lin, C.J.: LIBSVM: A Library for Support Vector Machines, Software, http://www.csie.ntu.edu.tw/~cjlin/libsvm
[13] Borgelt, C.: Bayes Classifier Induction, Software (2009), http://fuzzy.cs.uni-magdeburg.de/~borgelt/bayes.html
[14] Zhang, H.: The Optimality of Naive Bayes. In: Proc. 17th Int'l Florida Alliance of Information and Referral Services (FLAIRS) Conf. (2004)
[15] Van Der Walt, C.M., Barnard, E.: Data Characteristics That Determine Classifier Performance. In: Proc. 16th Ann. Symp. Pattern Recognition Assoc. of South Africa, pp. 160–165 (2006), http://www.patternrecognition.co.za
[16] Leung, K.S., Ng, Y.T., Lee, K.H., Chan, L.Y., Tsui, K.W., Mok, T., Tse, C.H., Sung, J.: Data Mining on DNA Sequences of Hepatitis B Virus by Nonlinear Integrals. Proc.

An Effective Automated Method for the Detection of Grids in DNA Microarray

P.K. Srimani[1] and Shanthi Mahesh[2,*]

[1] Dept. of Computer Science & Maths,
Bangalore University, Director R&D, BU. Bangalore-560078, Karnataka, India
[2] Dept. of Information Science & Engineering,
Atria Institute of Technology, Bangalore-560024, Karnataka, India
shanthi_md@yahoo.co.in

Abstract. Microarray is a technology which allows biologists to potentially monitor the activity of all the genes of an organism. Microarrays, widely recognized as the next revolution in molecular biology, enables scientists to analyze genes, proteins and other biological molecules on a genomic scale. Image processing is the first step in knowledge discovery from the microarray. The process of extracting features consists of three stages: gridding, segmentation and quantification. Gridding is to assign each spot with individual coordinates. This paper presents a fully automatic grid configuration algorithm for detecting the microarray image spots as input, and makes no assumptions about the size of the spots, and number of rows and columns in the grid. The approach is based on the detection of an optimum sub image. This method is capable of processing the image automatically and does not demand any input parameters. Experimental result shows that this method is highly efficient method of gridding that uses intensity projection profile.

Keywords: Microarray, Gridding, Spot Intensity, Expression Level, Segmentation.

1 Introduction

Image processing is the first step in knowledge discovery from the microarray. The process of extracting features consists of three stages: Gridding, segmentation and quantification. Gridding is to assign each spot with individual coordinates. As the technologies for the production of high quality microarray advances swiftly, quantification of microarray data becomes a major task. Gridding is the first step in the analysis of microarray images for loading the sub-arrays and individual spots within each sub-array. Segmentation is to classify the pixels into foreground, background and others(eg., noise)[1].Microarrays, widely recognized as the next revolution in molecular biology, enables scientists to analyze genes, proteins and other biological molecules on a genomic scale [2, 3]. Microarrays are relatively new

* Corresponding author.

S.C. Satapathy et al. (eds.), *ICT and Critical Infrastructure: Proceedings of the 48th Annual Convention of CSI - Volume II*, Advances in Intelligent Systems and Computing 249,
DOI: 10.1007/978-3-319-03095-1_47,

[4, 5] but they are already extremely popular. Biologists and physicians have enthusiastically embraced this technology, and they are currently producing an unprecedented quantity of microarray data. Image processing and analysis is an important aspect of microarray experiments, one which have a potentially large impact on the identification of differentially expressed genes. Image processing for microarray images includes three tasks: spot gridding, segmentation and information extraction. In the analysis of microarray experiments gridding techniques based on distribution of pixel intensities play an important role, since automizing this process leads to high throughput analysis, including segmentation [6], normalization, and clustering. Roughly speaking, gridding consists of determining the spot locations in a microarray image. The method is based on the Orientation Matching Transform (OMT) presented in [7] and until now it has been evaluated solely on synthetic images generated for the purpose. The contribution of this paper is the adaptation of the technique to make it suitable and robust for the treatment of real images of microarrays that present much more difficulty to the gridding.

2 Related Work

In [8] the authors have presented a comparative analysis of measurement of intensity of microarray spot using the intensity transformation methods such as gray level, logarithmic, gamma and contrast stretching. In [9] presented an automatic approach based on texture analysis characterization techniques is proposed to localize spots in microarray images. The method estimates the displacement vectors which characterize the texture, which is achieved by means of applying the generalized Hough transform on the 2D autocorrelation function. In [10], have presented a fully automatic grid alignment algorithm for detecting the microarray image spots. The approach is based on the detection of an optimum sub-image. In [11] have presented an accurate fully automatic gridding method for locating sub-array and individual spots using intensity projection profile of the most suitable subimage. The method is capable of processing the image without any user intervention. The [author 12] have proposed a parameterless and fully automatic approach that first detects the sub-grids given the entire microarray image, and then detects the locations of the spots in each sub-grid. In [author 13], have proposed an efficient and simple automatic gridding method for microarray image analysis. This method gives high accuracy and a promising technique for an efficient and automatic gridding the noisy microarray images.

3 Preprocessing

First step in using the DNA microarray array is to extract ribonucleic acid (RNA) from the cells; RNA indicates which genes are currently active. The RNA is processed to form fluorescently labeled cDNAs known as probes that will hybridize to their corresponding targets in the microarray. Typically, control and test RNA samples are processed on the same array using two different dye tagged probes [14,

15, 16]. The final step of the laboratory process is to produce an image of the surface of the hybridized array. The microarray is then scanned by activation with laser at appropriate wavelength to excite each dye. The relative fluorescence between each dye on each spot is then recorded and a composite image may produce. By comparing gene expression in normal and disease cells, microarrays can be used to identify disease genes for the development of therapeutic drugs. The main goal of array image processing is to measure the intensity of the spots and quantify the gene expression values based on these intensities. Segmentation is to classify the pixels into foreground, background and others. Quantification is to compute unique intensity values for each spot, which are related to the quantity of mRNA present in the solution that hybridize at the particular location of a microarray substrate.

4 Methodology

The microarray images are downloaded from Stanford Microarray Database (SMD) that stores raw and normalized data from microarray experiments, and provides the interface to retrieve, analyze and visualize their data. Experiments are conducted using MATLAB.

The following steps are required for the processed system:
(a). Read the Microarray Image, (b). Crop region of interest (get Xmin,Ymin,Width and Height of the cropped image), (c). Display Red & Green layer the image, (d).Spot finding (Display Red & Green layers), (e).Create Mean Horizontal profile, (f). Autocorrelation, (g).Remove background , (h). Segment the peaks, (i).Find the centers, (j).Estimate division between the spots, (k).Transpose and repeat, (l).Segmentation using thresholding, (m).Calculate spot intensity & expression level, (n).Display red and green intensity values, (0).Display expression level for the image.

Most of the microarray images consist of low intensity features that are not well distinguishable from the back ground, these problems lead to errors that propagate to all stages of statistical analysis. So, we suggested a pre-processing step to the microarray image to overcome those problems. Finally, we presented the effect of the proposed gridding method results. Our method starts by cropping a chosen microarray sub-image then converting it to grayscale. Then, the pre-processing step is applied using histogram equalization, to obtain high contrast between the foreground (spots) and the background. Next we computed the mean horizontal profile MH(y) of the image f (x, y) (dimensions X and Y, pixel x = (x, y). Then followed by autocorrelation that profile in order to enhance it. From the peak values of the auto correlated profile, we obtain the spot to spot estimated interval. The next step in the enhancement of the mean horizontal profile was to use a top-hat filter with a morphological flat, linear structuring element of length equal to the obtained estimated interval. The top-hat filter is defined as the difference between an image and its opened version. It enhances the-details that would otherwise be hidden in low contrast regions.

5 Experiment and Results

Gridding is the first step in the analysis of microarray images for locating the subarrays and individual spots within each subarray This paper presents a fully automatic gridding method for locating subarrays and individual spots using the intensity projection profile of the most suitable subimage. Experimental results show that the method is capable of gridding microarray images with irregular spots and varying surface intensity distribution. Gridding refers to the process of locating each sub arrays within a microarray image. The global parameters required for accurately locating subarrays are width and height of each subarray as well as spacing between them. These parameters have been estimated using the intensity projection profiles of the binary reference image generated in the preprocessing step. Horizontal and vertical intensity projection profiles of binary reference image are the sum of pixel intensities along each row and column respectively.

a. Read the Microarray Image

   MATLAB can read many standard image formats including TIFF, GIF and BMP using the *imread* command.

b. Crop region

   Next *imcrop* function has to use to extract a region of interest. Figure 1 shows the cropped image from the original image. This image was stored in RGB format. We are only interested in the red and green planes. To extract the red plane, index layer 1, for the green plane, layer 2.

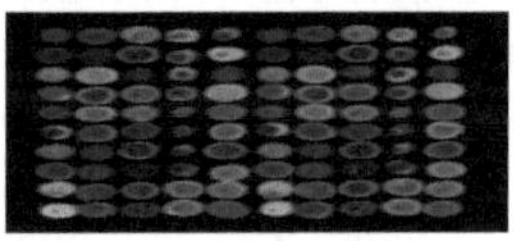

Fig. 1. Croped Image

c. Display red & green layers

   This image was stored in RGB format. We are only interested in the red and green planes. Figure 2 shows the red and green layer of the cropped image. To extract the red plane, simply index layer 1, for the green plane, layer 2. Figure 3 shows the RGB image converted to grayscale.

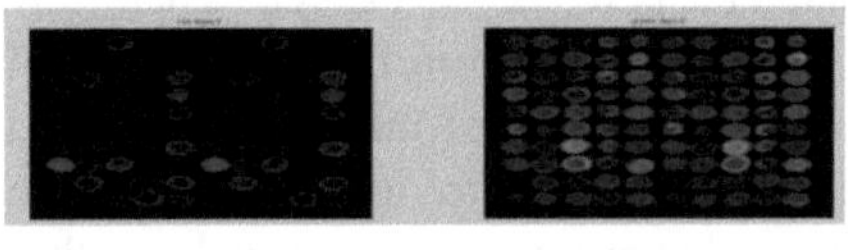

Fig. 2. Red and Green layer

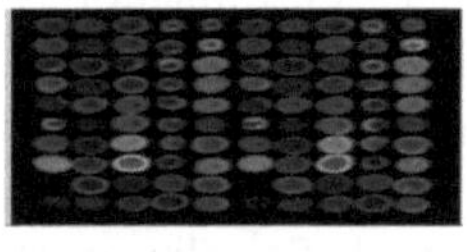

Fig. 3. Gray scale image

d. Create horizontal profile and Estimate spot spacing by autocorrelation.

   Figure 4 shows the horizontal profile. This will help us identify where the centers of the spots are and where the gaps between the spots can be found. Ideally the spots would be periodically spaced consistently printed, but in practice they tend to have different sizes and intensities, so the horizontal profile is irregular. Autocorrelation to enhance the self similarity of the profile. The smooth result promotes peak finding and estimation of spot spacing. Figure 5 shows the autocorrelation.

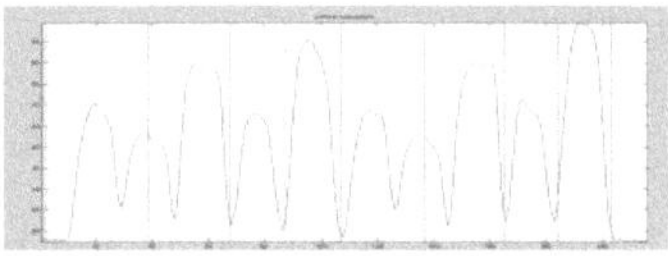

Fig. 4. Horizontal Profile

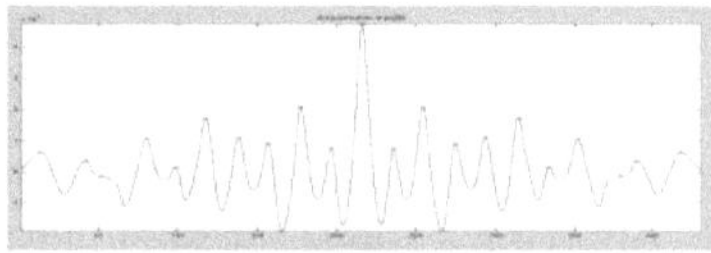

Fig. 5. Autocorrelation

e. Segment peaks

Now that we have clean and anchored gaps between the peaks, we can number each peak region with the bwlabel command. Figure 6 shows the segment peaks. These regions were segmented by thresholding with im2bw. The threshold value was automatically determined by statistical properties of the data using graythresh. Figure 7, Centroids of the peaks can be exacted. These correspond to the horizontal centers of the spots.

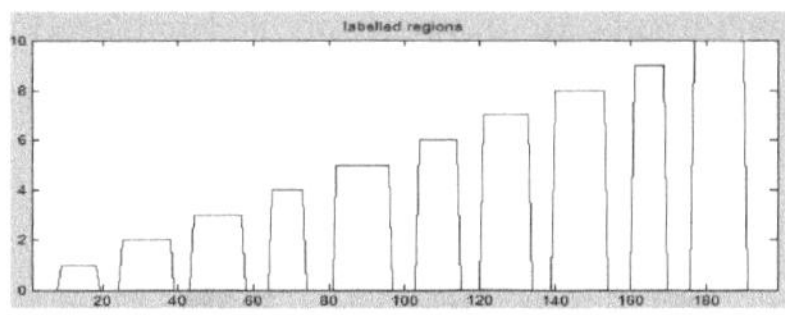

Fig. 6. Segment peaks

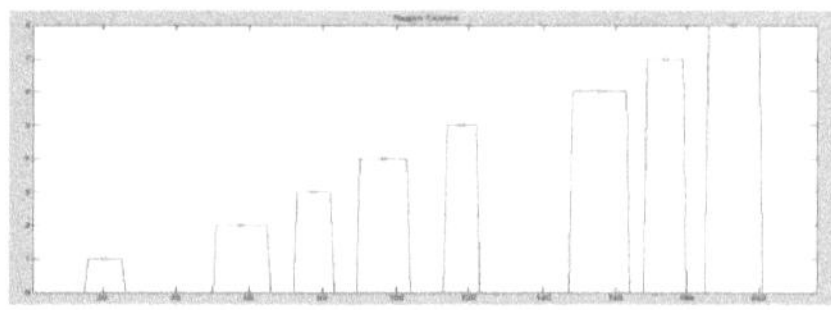

Fig. 7. Centroid

f. Determine divisions between spots

The midpoint between adjacent peaks provides grid point locations.

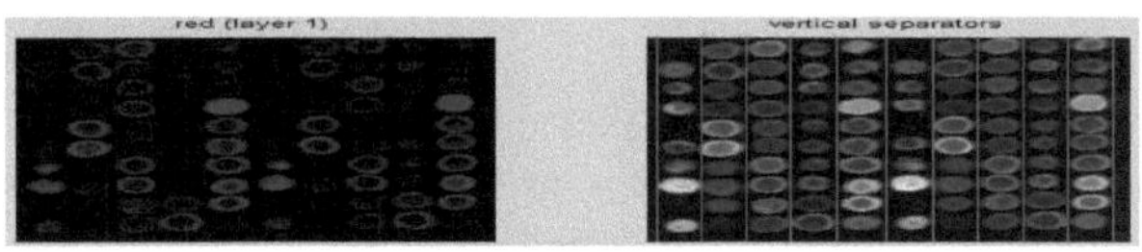

Fig. 8. Vertical separators for red layer

g. Transpose and repeat

Similar to the vertical grid, we can do the same for the horizontal spacing. This is done by simply transpose the image and repeat all the steps used above.

h. Put bounding boxes around each spot

Once rectangular grid is obtained, using pairs of neighboring grid points we can form bounding box regions to address each spot individually is shown in figure 9. The position and size coordinates each of the bounding box, were tabulated for convenience into a 4-column matrix called ROI, which stands for regions of interest.

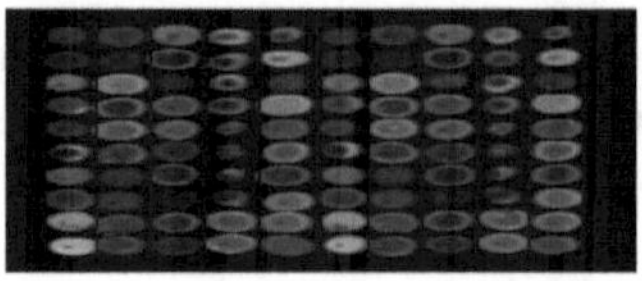

Fig. 9. Gridding

i. Segment spots from background by thresholding
   Applying a single threshold level to the whole image so all spots are detected equally is generally a good idea. However, in this case is doesn't work so well due to large differences in spot brightness.

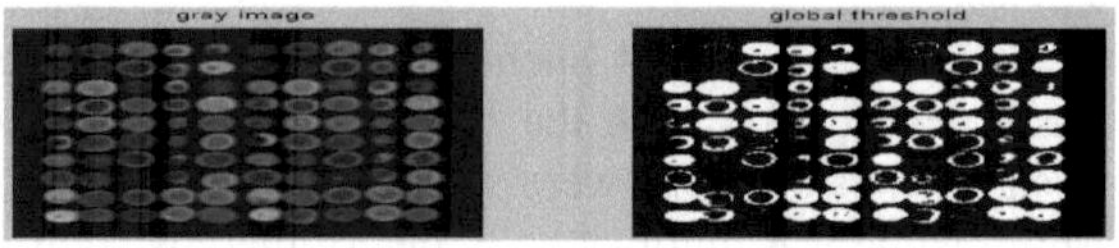

Fig. 10. Segmentation from background

j. Apply logarithmic transformation then threshold intensities
   One approach to equalize large variations in magnitude is by transforming intensity values to logarithmic space, which works better but some weak spots are still missed.

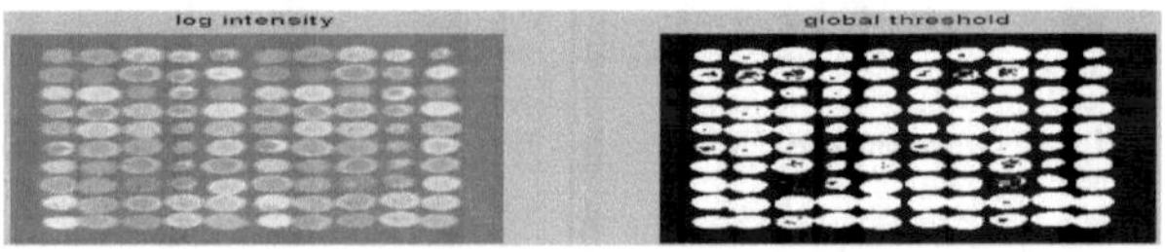

Fig. 11. Logarithmic Transformation and global thresholding

k. Try local thresholding instead
   Alternatively, the bounding boxes can be used to determine local threshold values for each spot is shown in figure 12. The code is a little more sophisticated, requiring looping and indexing. Unfortunately, the results are mixed. Weak spots showed up well but spots with bright perimeters were as bad as the original global threshold before log space transformation.

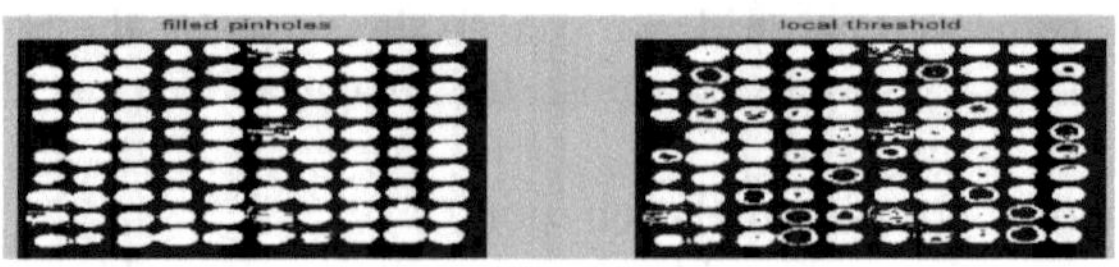

Fig. 12. Local thresholding

l. Label spot masks by bounding box
If the gridding went well, all spots should be a single color. The results here are pretty good. There is still room for improvement.

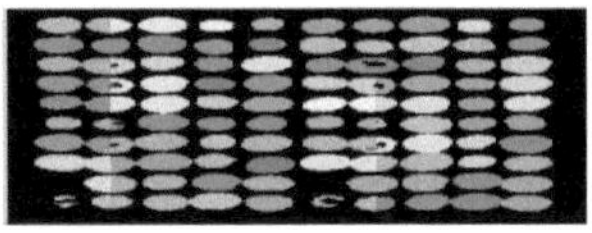

Fig. 13. Label spots

m. Extract first spot for measurement
Examine the first spot closely to see how we can measure its red and green intensities, and ultimately quantify its gene expression value. Figure 14 shows the first spot. The measurement technique can then be repeated for all spots.

Fig. 14. Single spot

n. Measure spot intensity & relative expression level
Calculate the nominal intensity over the spot for both the red and green layers. A measure of gene expression level can then be calculated from the two color intensities. Here a simple log-ratio measurement is shown.

intensity = 32 81; expression Level = -0.9287

u. Remove background, calculate again and compare measurements
It is noticed that, the background intensity around the spot was not zero. This could bias results. To see how much difference it makes, we can perform background subtraction around all spots, again using imtophat but this time in 2D on the image using a disk shaped structuring element. Then we can calculate color intensities and relative expression level again to see what effect background bias had on the measurement.

intensity = 14 70; expressionLevel = -1.6094

V. Set up graphical display for results
It is helpful to see red and green intensity values overlaid onto the respective color images to gain confidence that measured intensities make sense. It is also be helpful to overlay quantitative expression levels onto the original image to provide additional visual assurance of measurement results. The rectangular grid also helps correlate measured values between images.

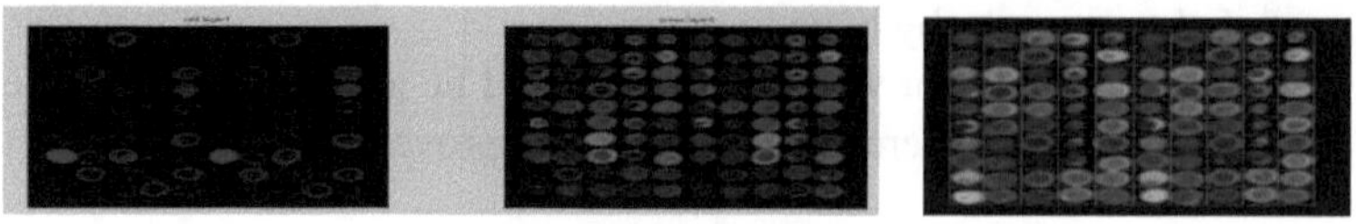

Fig. 15. Gridded image

W. Repeat measurement for all spots
Spot extraction and intensity calculation for all the spots in the grid is repeated. Figure 16 shows expression level of each spots. Here the measured values were tabulated as additional columns beside the ROI positions for each spot into a new matrix called spotData.

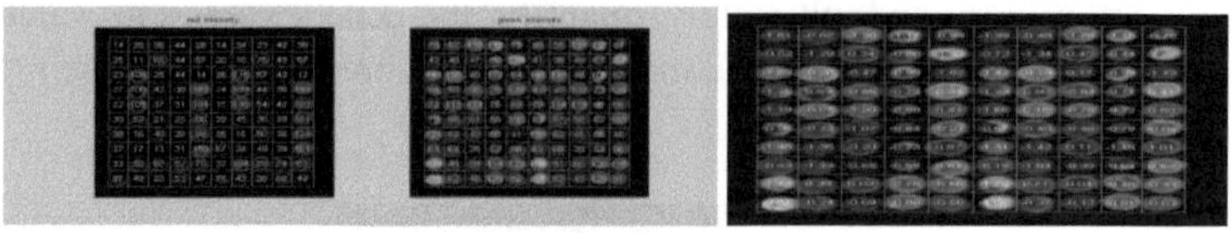

Fig. 16. Expression level of each spot

6 Conclusion

It is found that, the background intensity around the spot was not zero. This could bias results. To see how much difference it makes, we can perform background subtraction around all spots, again using imtophat but this time in 2D on the image using a disk shaped structuring element. Then we can calculate color intensities and relative expression level again to see what effect background bias had on the measurement. In this case the measurement shows more down regulation with background removed.

It is helpful to see red and green intensity values overlaid onto the respective color images to gain confidence that measured intensities make sense. It is also be helpful to overlay quantitative expression levels onto the original image to provide additional visual assurance of measurement results. The rectangular grid also helps correlate measured values between images.

References

1. Wu, S., Yan, H.: Microarray image processing based on clustering and morphological analysis. In: Proceedings of the First Asia-Pacific Bioinformatics Conference on sBioinformatics 2003, vol. 19, pp. 111–118 (2003)
2. Draghici, S.: Data Analysis Tools for DNA Microarrays. Chapman and Hall/CRC (2003)
3. M. Schena Microarray Analysis. Wiley-Liss (2002)
4. DeRisi, J.L., Iyer, V.R., Brown, P.O.: Exploring the metaboblic and genetic control of gene expression on a genomic scale. Science 278, 680–686 (1997)

5. Schena, M., Shalom, D., Davis, R., Brown, P.: Quantitative monitoring of gene expression patterns with a complementary DNA microarray. Science 270, 467–470 (1995), Stekel, D.: Microarray bioinformatics. Cambridge University Press, Cambridge (2003), Bowtell, D., Sambrook, J.: DNA microarrays: A molecular cloning manual. Cold Spring Harbor Laboratory Press (2003),
6. Rueda, L., Qin, L.: An Improved Clustering-Based Approach for DNA Microarray Image Segmentation. In: Campilho, A.C., Kamel, M.S. (eds.) ICIAR 2004. LNCS, vol. 3212, pp. 17–24. Springer, Heidelberg (2004)
7. Ceccarelli, M., Petrosino, A.: The Orientation Matching Transform Approach to Circular Object Detection. In: Proceedings of IEEE International Conference on Image Processing, pp. 712–715 (2001)
8. Anandhavalli, M., Mishra, C., Ghose, M.K.: Analysis of Microarray Image Spot Intensity: A Comparative Study. International Journal of Computer Theory and Engineering 1(5), 1793–8201 (2009)
9. Larese, M.G., Gomez, J.C.: Automatic Spot Addressing in cDNA Microarray Images. JCS&T 8(2) (July 2008)
10. Deepa, J., Thomas, T.: Automatic Gridding of DNA Microarray Images Using Optimum Sub-image. International Journal of Recent Trends in Engineering 1(4) (May 2009)
11. Deepa, J., Thomas, T.: A New Gridding Technique for High Density Microarray Images Using Intensity Projection Profile of Best Sub Image. Computer Engineering Intelligent Systems 4(1) (2013) ISSN 2222-1719
12. Rueda, L., Rezaeian, I.: A Fully automatic gridding method for cDNA microarray images. BMC Bioinformatics (April 21, 2011)
13. Labib, F.E.-Z., Fouad, I., Mabrouk, M., Sharawy, A.: An Efficient Fully Automated Method for Gridding Microarray Images. American Journal of Biomedical Engineering 2(3), 115–119 (2012)
14. Sorin, D.: Data analysis tool for DNA Microarrays. Mathematical biology and medicine series. Chapman&Hall/CRC, London (2003)
15. Stekel, D.: MicroarrayBioinformatics. Cambridge University Press, NewYork (2003)
16. Lonardi, S., Luo, Y.: Gridding of microarray images. In: Proceedings of IEEE Computational Systems Bioinformatics Conferences, CSB 2004 (2004), doi:0-7695-2194-0/04

Software Safety Analysis to Identify Critical Software Faults in Software-Controlled Safety-Critical Systems

Ben Swarup Medikonda[1] and P. Seetha Ramaiah[2]

[1] Department of Computer Science and Engineering
Vignan's Institute of Information Technology (VIIT),
Visakhapatnam, India
[2] Department of Computer Science and Systems Engineering
Andhra University,
Visakhapatnam, India
{bforben,psrama}@gmail.com

Abstract. Modern systems are increasingly software intensive because of the progress of technology and the proliferation of computers in everyday life. Computers control everything possible from microwave ovens to complex weapon systems. However, software can have a severe impact on the safety of systems, as some high profile accidents like *Therac-25* and *Ariane5* have shown. Despite the risks, software increasingly is making its way into safety-critical systems. A general purpose software engineering process is insufficient by itself to produce safe and reliable software. While traditional testing and other dynamic analysis techniques are best for uncovering functional errors they are inadequate whenever a computer-based system can cause injury or death. Therefore, software for safety-critical systems must deal with the hazards identified by safety analysis in order to make the system safe, risk-free and fail-safe. Certain critical software faults in critical systems can result in catastrophic consequences such as death, injury or environmental harm. The focus of this paper is a new approach to software safety analysis based on a combination of two existing fault removal techniques. A comprehensive software safety analysis involving a combination of Software Failure Modes and Effects Analysis (SFMEA) and Software Fault Tree Analysis (SFTA) is conducted on the software functions of the critical system to identify potentially hazardous software faults. A prototype safety-critical system - Railroad Crossing Control System (RCCS), incorporating a microcontroller and software to operate the train on a track circuit is described

Keywords: software safety, safety-critical systems, software faults, software safety analysis.

1 Introduction

A safety-critical system is one that has the potential to cause accidents. Software is hazardous if it can cause a hazard i.e. cause other components to become hazardous or

S.C. Satapathy et al. (eds.), *ICT and Critical Infrastructure: Proceedings of the 48th Annual Convention of CSI - Volume II*, Advances in Intelligent Systems and Computing 249,
DOI: 10.1007/978-3-319-03095-1_48, © Springer International Publishing Switzerland 2014

if it is used to control a hazard. Software is deemed safe if it is impossible or at least highly unlikely that the software could ever produce an output that would cause a catastrophic event for the system that the software controls. Examples of catastrophic events include loss of physical property, physical harm, and loss-of-life. Software engineering of a safety-critical system requires a clear understanding of the software's role in, and interactions with, the system [1,2].

1.1 Software-Induced Failures in Real-Life

Computers are increasingly being introduced into safety-critical systems and, as a consequence, have been involved in accidents. Some well known incidents are the massive overdoses given by the computer-controlled radiation therapy machine Therac-25 [3] with resultant death and serious injuries, during the mid-eighties; European Space Agency's Ariane 5 rocket explosion [4] during lift-off in June 1996, and SeaLaunch rocket failure [5] during lift off in March 2000. Recent examples include the following: on 7 October 2008, Qantas Flight 72 from Singapore to Perth made an emergency landing following an inflight accident featuring a pair of sudden uncommanded pitch-down manoeuvres that resulted in serious injuries to many of the occupants. The Australian

Transport Safety Bureau (ATSB) said that incorrect information from the faulty computer triggered a series of alarms and then prompted the Airbus A330's flight control computers to put the jet into a 197-metre nosedive [6].

All these examples indicate that accidents still take place despite all the measures taken to prevent them. Since complete elimination of unforeseen hazards is not always possible, what we need is a fail-safe design which, in the event of a failure, allows the system to fail in a safe way, causing no harm or at least the minimum level of danger. To meet the fail-safe requirements, rigorous safety analysis is required to identify potential hazards and take corrective measures during the entire system development life cycle.

There are many software fault removal techniques in literature. The most frequent classification is by differentiating between static and dynamic techniques [8]. Different authors focus on probabilistic based approaches (like the Markov modeling method), or statistical, approaches like statistical testing, software reliability models [9]. However most of the fault removal techniques are non-probabilistic. In some standards, static techniques require formal methods and proofs based on mathematical demonstrations. Other standards and literature classify these techniques in functional and logical terms [10] or by just mentioning functional testing like in [11] or structural testing, like in [12].

None of the fault removal techniques like algorithm analysis, control flow analysis, Petri-Net analysis, reliability block diagrams, sneak circuit analysis, event tree analysis, FMEA and FTA can be considered apt and complete in all respects, when used in isolation. A way out of this is to analyse how to combine individual techniques so that the fault removal process is significantly improved. One of the most effective combinations is FMEA+FTA. The literature [9,10] already mentions that FTA technique can be associated effectively with other practices like FMEA. Their greatest advantage is in combination with each other. FMEA concentrates in identifying the severity and criticality of failures and FTA in identifying the causes of faults. FMEA technique is a fully bottom-up approach and FTA has a fully complementary top-down approach. Moreover, these two techniques are directly compatible with system level techniques.

In this paper, we propose a system-level approach to software safety analysis for critical systems that combines two existing fault removal techniques – FMEA and FTA to identify and eventually remove software faults at successive software development phases. We have applied our safety approach to a model railroad crossing control system to validate its effectiveness. We also compare how the safety-specific software development of a critical system is distinct from the traditional non-safety-specific software development.

The rest of this paper is organized as follows: section 2 describes the Railroad Crossing Control System (RCCS). Section 3 applies the safety analysis using SFMEA and SFTA techniques to RCCS. Section 4 addresses the hardware and software development issues of RCCS. Section 5 presents an analysis of the experimental results and section 6 concludes the discussion.

2 Railroad Crossing Control System (RCCS)

Crossing gates on a full-size railroads are controlled by a complex control system that causes the gates to be lowered to prevent access to the crossing shortly before a train arrives and to be raised to allow access to resume after the train has departed. RCCS is a prototype, real-time, safety-critical railroad-crossing control system composed of several software-controlled hardware components.

2.1 RCCS Interfaces

The main interfaces of the microcontroller, which hosts and runs the embedded software, are shown below in Figure 1. The main inputs to the microcontroller are signals from the 7 sensors on the track, the 2 gates at the railroad intersection, the track- change lever, and the 3 signal lights. The main outputs of the micro-controller are control signals for the train, Gate1 Gate 2, track change lever, signal lights, LCD display. The values of these output signals are determined using different algorithms combining the input signals that are constantly updated and read by the software.

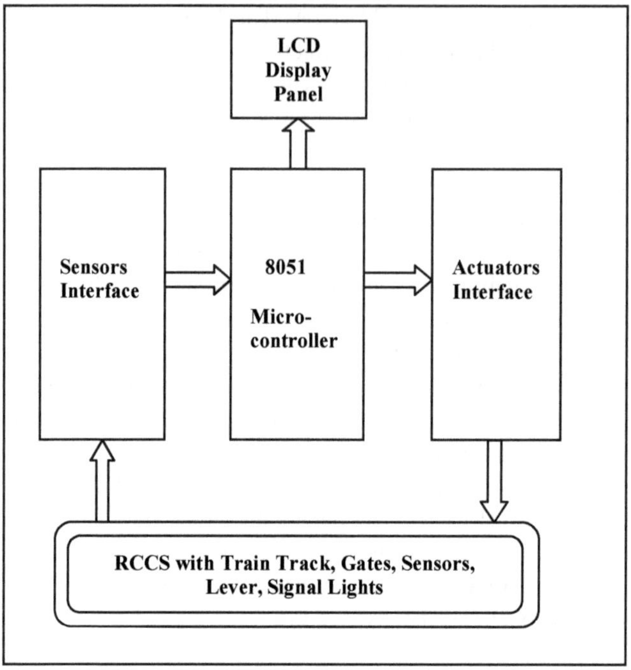

Fig. 1. External interfaces of RCCS microcontroller

The main functionality of RCCS is listed in Table 1.

Table 1. RCCS System Functions – Key Areas

RCCS System Functions
• Control the overall operation of train on the track circuit
• Control the opening and closing of Gate 1 and 2 at the railroad intersections
• Control the track lever to change the track route from the outer to the inner loop
• Check the internal health of all the subsystems
• Control the train operation at the Signal Lights
• Monitor all the sensors on the track circuit

3 Software Safety Analysis of RCCS

The safety analysis of RCCS software functions takes place in three sequential steps.

- **Software Failure Mode and Effects Analysis (SFMEA)**

This analysis is performed in order to determine the top events for lower level analysis. SFMEA analysis will be performed following the list of failure types. SFMEA will be used to identify critical functions based on the applicable software specification. The severity consequences of a failure , as well as the observability requirements and the effects of the failure will be used to define the criticality level of the function and thus whether this function will be considered in further deeper criticality analysis. The formulation of recommendations of fault related techniques that may help reduce failure criticality is included as part of this analysis step.

- **Software Fault Tree Analysis (SFTA)**

After determining the top-level failure events, a complete Software Fault Tree Analysis shall be performed to analyse the faults that can cause those failures. This is a top down technique that determines the origin of the critical failure. The top-down technique is applied following the information provided at the design level, descending to the code modules . SFTA will be used to confirm the criticality of the functions (as output from SFMEA) when analyzing the design and code (from the software requirements phase, through the design and implementation phases) and to help:

- Reduce the criticality level of the functions due to software design and / or coding fault-related techniques used (or recommended to be used)

- Detail the test-case definition for the set of validation test cases to be executed.

- **Evaluation of Results**

The evaluation of the results will be performed after the above two steps in order to highlight the potential discrepancies and prepare the recommended corrective measures.

3.1 SFMEA Analysis of RCCS

The SFMEA, a sample of which is shown in the Table 2 below presents some software failure modes defined for RCCS. The origin and effects of each failure mode are analyzed identifying the top level events for further refinement, when the consequence of this failure could be catastrophic for this system. Three top events were singled out for further analysis of failure mode Gate not closed as train is passing through railroad intersection.

Table 2. Example of SFMEA table for RCCS

Failure Mode	Possible Causes	Effect	Severity of risk	Prevention And Compensation
Gate not closed as train is passing through	a) sensor not detected by s/w b) gate motor mechanism is defective c)s/w gives wrong command d)s/w gives right command at wrong time	Train collision with passing road traffic leading to accidents	Critical	Software first checks the working status of gates each time the train is about to cross the gates
Track change lever is not activated to change train route	a) sensor is not detected by s/w b)track lever motor mechanism is defective c)s/w gives wrong command to lever d)s/w gives right command at wrong time e)s/w fails to give a command to activate lever	Train fails to change its path from the outer track circuit to the inner track circuit leading to accident	Critical	Software first checks the working status of the track lever each time the train is about to enter the inner track loop
Control program software is corrupted	a) logic fault b) interface fault c) data fault d)calculation fault e) memory fault	Unpredictable sequence of operations leading to accident	Critical or Catastrophic	Algorithm logic is verified for accuracy. Data Structures and Memory overflow is checked.

3.2 SFTA Analysis of RCCS

The fault tree is a graphical representation of the conditions or other factors causing or contributing to the occurrence of the so-called top event, which normally is identified as an undesirable event. A systematic construction of the fault tree consists in defining the immediate cause of the top event. These immediate cause events are the immediate cause or immediate mechanism for the top event to occur. From here, the immediate events should be considered as sub-top events and the same process should be applied to them. All applicable fault types should be considered for applicability as the cause of a higher level fault. This process proceeds down the tree until the limit of resolution of tree is reached, thereby reaching the basic events, which are the terminal nodes of the tree. Figure 2 shows the sample fault tree for the top event *Gate Not Closed* at the railroad intersection.

3.3 Recommendations to Design and Coding

From the software safety analysis we have conducted, some of the major critical events that might occur and the corresponding safety properties the RCCS software has to implement, and which are controlled by the embedded software in the microcontroller are listed below.

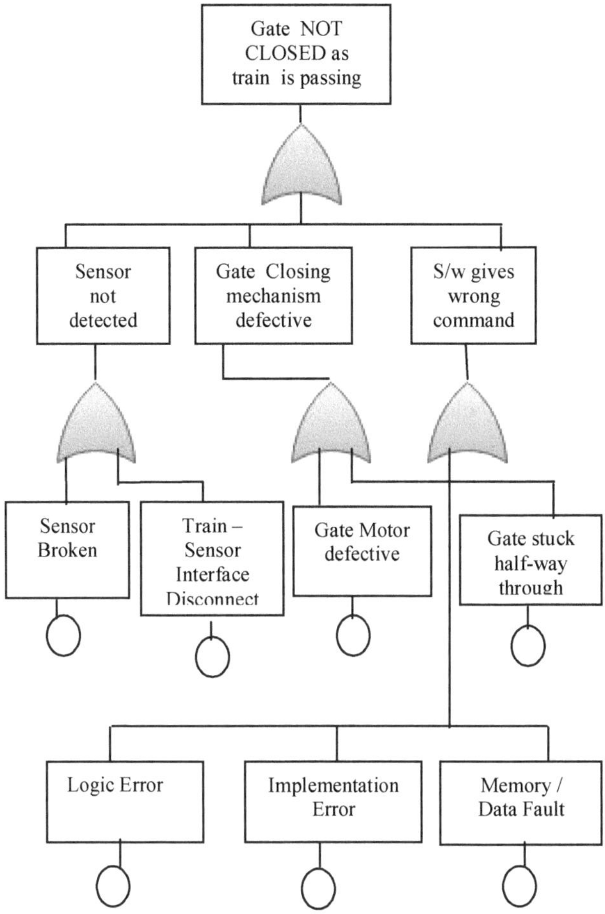

Fig. 2. Software Fault Tree sample for top event *Gate Not Closed* at the railroad intersection

- The software shall make sure that the 2 gates on either side of the railroad intersection operate correctly – ie. opening and closing the gates, at the proper time. The consequences of failure to do so are very severe, since it can result in the train and road traffic collision, leading to death.
- The software shall make sure that the train changes its path from the outer track circuit to the inner track circuit by correctly operating the track change lever at the right time. Failure to do so can have severe consequences leading to collision with another train that may be stationary on the outer track.
- The software shall prevent the running operation of the train if it detects that the gates at the intersection have not been fully closed.
- The software shall prevent the running operation of the train, if the train engine detects any physical obstacle just ahead of it, either at the mid-section of the railroad intersection or at any point on the track path, just ahead of the engine. Failure to do so can lead to collisions.

- The software shall prevent the running operation of the train if a Red signal is displayed in the Signal Light alongside the track. Failure to do so can lead to accidents.

4 RCCS Prototype Development

RCCS hardware and software development is described in this section.

4.1 RCCS Hardware Components

RCCS model consists of the following main components: train, railway track, sensors, gates, microcontroller, signal lights, and a track-change lever. A brief description of each component is given below.

Train: The train is powered by a power supply relay. When the power is initially switched on, the train begins movement along the track when the metallic wheels of the train receive power. The train comes to a halt at the position where the power to the tracks is switched off.

Sensors: These are used to detect the location of the train on the tracks. Altogether RCCS employs seven sensors. Two pairs of sensors detect the train position before and after the gates. A set of two sensors relate to track change where the track splits into two directions. One sensor gives the train position with reference to the platform, which is the starting point of the train movement. Information from each of the sensors is passed to controller.

***Controller*:** An 8051 is used as a controller for RCCS. RCCS software that controls the overall operation of the system is stored in the memory of the controller. The controller continuously monitors the sensors and controls the gate actuators, track change lever, and the signal lights.

Gates: RCCS has two sets of gates on either side of the track layout. The gate receives signals from the controller. When it receives *lower* command, arms of the gate moves down and close the gate, preventing the road traffic at the intersection. When the gate receives *raise*, it moves up allowing the traffic to pass through. The gates are operated by means of a motor-based mechanism.

Signal Lights*:* RCCS contains three train signals, erected beside the track. One signal is at the platform to signal a halt at the platform. The other two signals are placed just before the point of convergence of the inner track and outer track, which lead to the platform.

4.2 RCCS Software Development

The safety-specific version of RCCS controller program used the same techniques as the non-safety version with the addition of the following safety-specific analysis: preliminary hazard analysis, and design-level hazard analysis, FMEA and FTA analyses. These techniques target the specification and designs. The goal here is to determine if the inclusion of these methods reduces the number of latent safety-critical faults relative to non-safety specific methods.

The software safety-based development involves preliminary software hazard analysis, which among other things identifies software hazards, ie. the states in the software that can lead to an accident. Without identifying the hazards, we have little assurance that the hazards will not occur. Therefore, preliminary software hazard analysis is an important first step in verifying safety-critical software systems. Once the hazard list exists, the verification process can continue by applying several static and dynamic verification techniques. Static techniques include failure modes and effects analysis (FMEA), and fault-tree analysis (FTA).

After static verification, software engineers must dynamically verify the software's safety (ie. safety testing). Safety-critical testing of RCCS can be done by separating the code into two risk groups. Group one includes hazards that are catastrophic or critical. Group two includes hazards that are marginal or negligible. More testing effort should be spent on those code sections dealing with hazards related to group one.

5 Experimental Results and Analysis

In view of the comprehensive safety analysis, and specification and implementation the safety properties during RCCS design and development, the expected result was that safety-specific RCCS development would produce a software system with fewer latent safety-critical faults than traditional non-safety specific techniques alone. This is due to the belief that the safety-specific techniques will prevent safety-critical faults in the specifications and designs that the traditional techniques have a tendency to miss. Figure 3 and Figure 4 show the RCCS laboratory prototype developed in the lab.

During the operation of RCCS, the safety-specific development version of RCCS clearly demonstrated the fulfillment of the safety properties. For example, if the gate at the railroad intersection is not closed at all, or partially closed, as the train is about to pass through the intersection, the controller software makes the train come to a halt. Only after confirming that the gate is fully closed does the software allow the train to pass through the railroad intersection. On the other hand, in the non-safety version of RCCS, the controller software allows the train to pass through the intersection without confirming whether the gate is actually closed or not, assuming that the gate function will operate without failure, leading to a major accident.

Likewise, in the safety-version of RCCS, when the train is changing its track route from the outer loop to the inner loop, the software first confirms whether the track change lever is fully activated and operational. If the track lever is stuck halfway through and the rails connection to the inner loop is incomplete, the software makes the train come to a halt. In the case of the non-safety version, the software allows the train to change route without confirming the health status of the track lever, leading to an accident. The safety version also demonstrated a preliminary check of the internal health of all the RCCS subsystems – the gates mechanism, track lever operation, sensors, signal light LEDs, displaying the health status on the LCD display panel.

Fig. 3. Train Crossing Gate1 (background) as Gate 2 is Open (foreground)

Fig. 4. Train Crossing Gate2 (foreground) as Gate 1 is Open (background)

6 Conclusion

This paper discussed a new approach to software safety analysis to identify critical software faults for a prototype software-controlled safety-critical system. A comprehensive software safety analysis involving a combination of SFMEA and SFTA techniques was conducted on the software functions of the critical system to identify potentially hazardous software faults. The safety properties of the prototype railroad crossing control system were identified as part of the safety-critical requirements. These safety requirements were incorporated in the design and development of a railroad crossing control system (RCCS). We also briefly compared safety-specific and non-safety specific techniques at developing RCCS. The non-safety version of RCCS broadly focused on achieving the functional behavior of the system. The safety-specific version clearly demonstrated that the software safety properties identified in RCCS specification were fully met in the working system.

References

1. Lutz, R.R.: Software Engineering for Safety: a Roadmap. In: Proceedings of the Conference on The Future of Software Engineering, Limerick, Ireland, June 04-11, pp. 213–226 (2000)

2. Knight, J.C.: Safety Critical Systems: Challenges and Directions. In: Proceedings of the 24th International Conference on Software Engineering (ICSE), Orlando, Florida (2002)
3. Leveson, N.G., Turner, C.S.: An investigation of the Therac-25 accidents. IEEE Computer 26(7), 18–41 (1987)
4. Gleick, J.: The New York Times Magazine (December 1, 1996)
5. Gray, D.M.: Frontier Status Report #203 (May 19, 2000), http://www.asi.org
6. http://en.wikipedia.org/wiki/Qantas_Flight_72
7. http://news.bbc.co.uk/2/hi/science/nature/4381840.stm
8. IEEE STD 1012, IEEE Standard for Software Verification and Validation Plans, The Institute of Electrical and Electronics Engineering, Inc. USA (1986)
9. Leveson, N.G.: Safeware: System Safety and Computers. Addison-Wesley (1995)
10. Herman, D.S.: Software Safety and Reliability Basics: Software Safety and Reliability: Techniques, Approaches, and Standards of Key Industrial Sectors, ch. 2. Wiley-IEEE Computer Society Press (2000)
11. EN50128 Railway Applications: Software for Railway Protection and Control Systems. CENELEC
12. DO-178B/ED-12B Software Considerations in Airborne Systems and Equipment Certification, RTCA, EUROCAE (December 1992)
13. IEEE Std. 610.12-1990, Standard Glossary of Software Engineering Terminology
14. Tribble, A.C., et al.: Software Safety Analysis of a Flight Guidance System. In: Proceedings of the 21st Digital Avionics Systems Conference (DASC 2002), Irvine, California, October 27-31 (2002)

Mutual Dependency of Function Points and Scope Creep towards the Success of Software Projects: An Investigation

K. Lakshmi Madhuri and V. Suma

Research and Industry Incubation Centre,
Dayananda Sagar College of Engineering,
Bangalore
{madhuri.vethamoorthy,sumavdsce}@gmail.com

Abstract. The Project Management strategies in the current corporate scenario are not helping towards successful project management. Project scope is one of the major project management knowledge areas in software industries. However, with existing techniques of project management, it is yet a challenge to effectively manage the scope creep. However, function point analysis is deemed to be one of the popular factors that modulate the project management. This paper therefore aims at analysing the impact of scope creep in terms of function points towards realization of project success. Projects from two widely developed domains namely ERP and Financial are studied to comprehend above said aim. The investigation results indicate existence of transitivity relation between scope creep, function points and project success. This knowledge further enables the project manager to accordingly plan for the generation of expected success level of projects during software development process.

Keywords: Project Management, Scope Creep, Function Points, Software Engineering, Software Quality.

1 Introduction

Quality Software is a success factor for every software organization. Since the competition in the IT world is high, it is essential for all IT industries to insist on the quality of their products. Software Engineering is a vital domain for production of high quality software. Factors such as cost, time, technology, resources, company policies and standards are the key constrains for the development of high quality software. Further, quality also depends upon the project domain, its complexity and mode of developing the project. It is well proven that customer satisfaction is completely based on the high quality retained; hence the effective project management is making an impact. The skills and experience of the project manager in handling the resources is significant in a project success. According to the authors in [9] project managers provides reasons for failure to be due to scope creep which

S.C. Satapathy et al. (eds.), *ICT and Critical Infrastructure: Proceedings of the 48th Annual Convention of CSI - Volume II*, Advances in Intelligent Systems and Computing 249,
DOI: 10.1007/978-3-319-03095-1_49, © Springer International Publishing Switzerland 2014

influences various factors such as schedule slip, improper planning, changing or new requests, quality failing, etc. Therefore, Project Manager focuses on the scope of the project which is one of the three factors of the project success iron triangle. Function Points is considered as one of the major tool for measuring the scope of the project [10]. Function points play a vital role in estimating the factors of the project management such as time, cost, managing scope creep etc. Function Point Analysis is one of the proven, consistent methods for estimating the software project complexity. It is obvious that all changes to the scope will effect on the number of function points [10]. In this research paper we aim at investigating the dependency of function points and scope creep towards the success of project investigation.

2 Literature Survey

To endure the competition and produce quality software it is apparent to insist on continual research and development process, the Software Engineering Book of Knowledge (SWEBOK) has introduced seven knowledge areas in project management techniques such as project integration management, project scope management, project time management, project cost management, project quality management, project human resource management and project communications management as the most important project management techniques for developing high quality software [1]. The project manager aims at four activities such as problem recognition, problem sequence, problem controlling, problem evalution [2] that ensures development of high qulaity software. The role of a project manger includes managing his expert team, adopting the mangement and engineering processes, planning, scheduling, handling resources etc [4]. Scope creep has proven to be one of the most influential factors on project success. The study made by authors in [3] indicates that occurrences of scope creep is due to the lacunae in various process such as change request, work break down structure, documentation, scope management plan etc. The projects management process is equipped with several tools, techniques and metrics which assure the quality software development. Hence, industry targets setting of quality standards and compliances from the point of conception till the maintenance and support phase of software development cycle. It is evident that quality is a continuous process which is always quantified and rectified immediately to stop the trivial penalty. Further, quality is dependent on comprehensiveness and volatility of requirements which are well communicated and well elicited by both customers and developers during software development cycle [5]. Additionally, it is also quite apparent that quality is not perceived from process of development but is also viewed in various dimensions such as the factors which modulate the level of project success and the models used to develop these projects. However, factors such as time, cost, scope creep, availability of resources are deemed to be some of the important quality influencing factors, this research aimed at investigating the significance of scope creep upon the two popular approaches of software development namely traditional approach using perspective models and agile approach using agile models. According to [6], scope, time and cost are integrated. Function points are measurement units of the project. They are used as the comparison factor between projects for the quantitative analysis. The points of measurement include data functions and transaction functions. Data functions are internal logical files and external interface files. Transaction functions include external inputs, external outputs and external inquiries [7] [8].

3 Research Methodology

In order to understand the dependency of function points and scope creep on project success that are developed using conventional process models this research comprises of a deep study carried out at various software industries. These software industries are either product based or service based software developing centres which are holding CMMI Level 4 and Level 5 standard certification. The empirical data are collected from data repository databases and from industry personnel in the model of interviews, questionnaires, mails and face to face communication. The sampled projects that are collected include projects developed for non critical applications such as Enterprise Resource Planning and Financial projects developed since 2007 onwards up to 2012. These projects are intentionally sampled so that there is no variation either with technology or process or maturity level of companies. They are developed in Microsoft operating system using .Net as programming language. Data analysis comprises of comparison of influence of scope creep on project modulating factors such as function points, Number of use cases, time, cost, personnel, experience level of these personnel in the projects and project success. The comparison is made between projects developed for ERP domain and financial domain. Data analysis infers that function points and scope creep are dependent and has high influence on project success developed using traditional process models.

4 Case Study

This case study incorporates the study conducted on different software organizations which are domain specific and product based companies. However, this research restricted towards companies which are CMMI Level 5 standards. Further, hhypothesis considered for this research has led to the specific analysis and inferences drawn in this research

Hypothesis 1: All projects are developed in similar environment and using similar programming language.

To resolve the intrinsic complexity of software projects, this investigation constrained itself to study of those projects which are developed in common environment namely Microsoft Operating System and using .Net language. Analysis is carried out on projects developed using waterfall approach of traditional process models.

From the investigation, in order to further resolve complexities involved such as customer satisfaction index levels varying from industry to industry and from project to project, this research broadly classifies project success to fall under three major umbrellas such as successful, challenged and failure projects depending on the customer satisfaction index levels. Figure 1 depicts the Classification of project success sequence.

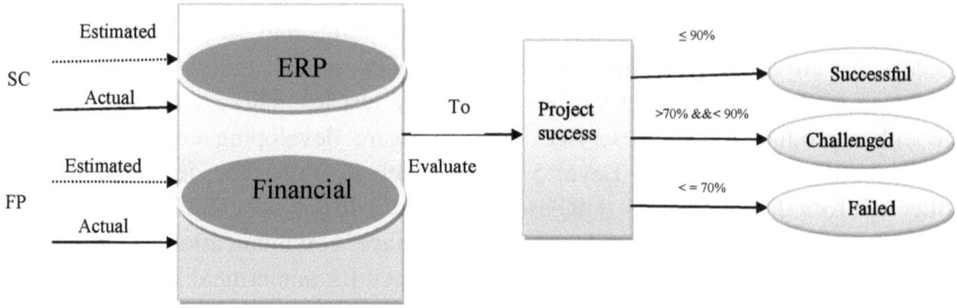

Fig. 1. Classification of Projects

Majority of the software industries therefore follow functional point analysis as one of the popularly used complexity measurement for large number of their application developments. This focused our research to further narrow down with the formulation of hypothesis 2.

Hypothesis 2: Projects are classified as small, medium and large project based on function point analysis.

Hence, the projects are considered to be as small projects having fewer complexity when the estimated function points are below 700, while projects having function points varying between 700 and 5000 are considered as medium sized projects, and those of the projects whose function points are above 5000 is always considered as large projects. Table 1 depicts 5 projects which are sampled from one company of ERP domain and Table 2 depicts 5 projects which are sampled from a company of financial domain.

Table 1 and Table 2 below infers that there exist huge variations between estimated and actual resources utilized for the development of project. The sampled projects are arranged in increasing order of function point complexity. However, the conventional belief in industry is that estimation and allocation of resources should not be more than 10% in variation. From the table it is quite apparent that these randomly selected sampled projects have undesirable variations between estimation and actual function points which may influence also the resources such as cost, time, and number of developers which is a noticeable alert to the project managers.

As mentioned above quality is a measurable unit, the success of the project is always estimated to be evaluated in terms of total customer satisfaction which is measured in percentage. Hence, any variation from this unit of measurement leads to overheads and reduce customer satisfaction. It is apparent that the projects are developed having various success classifications. Table1 infers the project success is high when the function point's difference is less between expected and actual.

Table 1. Function Point Variations on ERP Projects

Company A (ERP)										
	Medium Sized Projects						Large Projects			
Parameters	Project 1		Project 2		Project 3		Project 4		Project 5	
	Expected	Actual	Expected	Actual	Expected	Actual	Expected	Actual	Expected	Actual
Function Point (FP)	1350	1550	2125	2875	4125	4500	6250	8125	20000	23750
Difference of FP		14.81		35.29		9.09		30		18.75
Project Success (%)	100	85	100	64	100	90	100	70	100	82
Success Level Classification		Challenged		Failure		Successful	Failure			Challenged

Table 2. Function Point Variations on Financial Projects

Company B(Financial)										
	Small Projects				Medium Projects				Large Projects	
Parameters	Project 1		Project 2		Project 3		Project 4		Project 5	
	Expected	Actual	Expected	Actual	Expected	Actual	Expected	Actual	Expected	Actual
Function Point (FP)	600	700	700	1125	1625	2375	3000	3625	9500	12875
Difference of FP		16.67		60.7		46.15		20.83		35.53
Project Success (%)	100	83	100	40	100	55	100	79	100	65
Success Level Classification		Challenged		Failure		Failure	Challenged			Failure

Figure 2 depicts the graph of sampled projects which are presented in this paper which are collected from software industries as mentioned in the research methodology of this paper. Figure illustrates the percentage of variations observed between the change of function point and project success. The observation has resulted in inference that whenever the function points are added more than 30% the projects are always failure. If the change in function between actual and expected are between 10 % – 30 % then the projects are challenged and when the function points are added below 10% the project are successful.

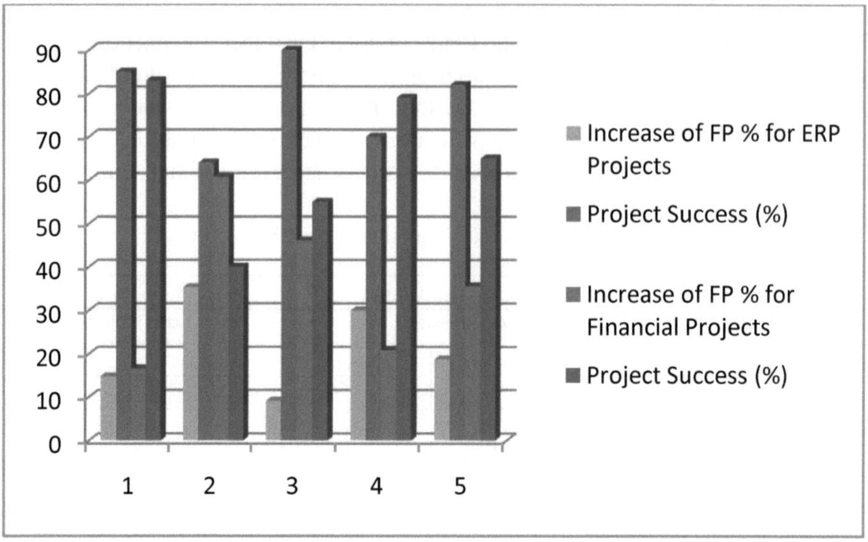

Fig. 2. Success Variation Graph basing on Function Points

5 Conclusion

Sustainability of software industries depends upon the total customer satisfaction which in turn depends upon the quality of software projects that are generated. In order to achieve above said objective, it is essential to manage the software projects effectively during software development process.

Software organizations follow various strategies towards effective project management. Nevertheless, there still prevails challenge of generation of successful projects. Scope creep and function points are deemed to be one of the factors that modulates project management.

This paper therefore presents a case study upon several projects developed in two software industries in order to gain insight on the existence of relation between scope creep, function points and project success. The study has revealed the transitivity relation between them which throws light for project managers to formulate effective tactics towards production of successful projects.

Acknowledgement. The authors are extremely thankful for all the industry personnel who have helped in carrying our investigation under the framework of non disclosure agreement.

References

1. Duncan, W.R.: A Guide to Project Management Body of Knowledge. Project Management Institute, Pennsylvania (2000)
2. Chang, C.K., Christensen, M.: A Net Practice for Software Project Management. Researchers Corner. IEEE Software (November/December 1999)

3. Kazemipoor, H., Shirazi, F.: A Methodology for Preventing and Managing Scope Creep in Projects (Case Study in Mapna-Iran). Research Journal of International Studies (23) (March 2012)
4. Gopalakrishnan Nair, T.R., Suma, V., Shashi Kumar, N.R.: Significance of Project Manager in Effective Defect Management in Software Development Process. In: The 5th Malaysian Software Engineering Conference (MySEC 2011), Johor Bahru, Malaysia, December 13-14 (2011)
5. Suma, V., Shubha Mangala, B.R., Manjunatha Rao, L.: Impact Analysis of Volatility and Security on Requirements during Software Development Process. In: International Conference on Software Engineering and Mobile Application Modeling and Development (ICSEMA), Chennai, India, December 19-21 (2012)
6. Staub, S., Fischer, M.: The Practical Needs of Integrating Scope Cost and Time. In: Lacasse, M.A., Vanier, D.J. (eds.) Institute for Research in Construction, Ottawa ON, K1A 0R6, Canada, pp. 2888–2898. National Research Council Canada (1999)
7. Fundamentals of Function Point Analysis, Knowledge@SoftwareMetrics.Com, Longstreet Consulting Inc., Blue Springs (1992), `http://www.softwaremetrics.com/fpafund.htm`
8. Garmus, D., Harron, D.: Function point Analysis: Measurement Practices for Successful Software Projects. Addison Wesley Professional (2000) ISBN-10: 0201699443, ISBN-13: 978-0201699449
9. Sutherland, J., Schwaber, K.: The Crisis in Software: The Wrong Process Produces the Wrong Results, pp. 3–16, `http://www.controlchaos.com/storage/S3D%20First%20Chapter.pdf`
10. Horvath, D.: Controlling Project Scope with Function Point Analysis. A Publication for Information Technology Professionals Q/P Management Group, Inc. Copyright 2008-2009

Prediction of Human Performance Capability during Software Development Using Classification

Sangita Gupta[1] and V. Suma [2]

[1] Jain University, Bangalore Dept. of CSE,Bangalore, India
sgjain.res@gmail.com
[2] RIIC, Dayanada Sagar Institute, Bangalore, India
sumavdsce@gmail.com

Abstract. The quality of human capital is crucial for software companies to maintain competitive advantages in knowledge economy era. Software companies recognize superior talent as a business advantage. They increasingly recognize the critical linkage between effective talent and business success. However, software companies suffering from high turnover rates often find it hard to recruit the right talents. There is an urgent need to develop a personnel selection mechanism to find the talents who are the most suitable for their software projects. Data mining techniques assures exploring the information from the historical projects depending on which the project manager can make decisions for producing high quality software. This study aims to fill the gap by developing a data mining framework based on decision tree and association rules to refocus on criteria for personnel selection. An empirical study was conducted in a software company to support their hiring decision for project members. The results demonstrated that there is a need to refocus on selection criteria for quality objectives. Better selection criteria was identified by patterns obtained from data mining models by integrating knowledge from software project database and authors research techniques.

Keywords: software projects, data mining, selection criteria, performance.

1 Introduction

Human aspect of software engineering has become one of the main concerns in software companies to achieve quality objectives. Software industries are now paying attention to select the right talent who can perform consistently throughout all generic framework activities and execute the process properly. Software quality depends on people and process quality during development[10]. Hence, this paper proposes the use of data mining algorithms that can exploit the patterns in the historical data and predict the performance based on project personnel attributes and thereby enhance the process and quality of software .Data mining is a new and promising field for knowledge discovery. Data mining is the process of extracting knowledge from data [8]. It uses a combination of an explicit knowledge base, sophisticated analytical skills, domain knowledge to uncover hidden trends and patterns. These trends and

DOI: 10.1007/978-3-319-03095-1_50, © Springer International Publishing Switzerland 2014

patterns can be extracted on by using various data mining algorithms. To create a model, the algorithm first analyzes a large set of data and finds specific patterns. The algorithm then uses the results of the analysis to define the parameters of the mining model. These parameters are then applied across the entire data set. Subsequently, patterns and detailed statistics can be extracted. Through classification, one can identify association rules. Categorization uses rule induction algorithms to handle categorical outcomes, such as good, average and poor as in this study. There are a wide range of available algorithms for such purpose. Many of them are implemented in WEKA [7]. WEKA is a cluster of machine learning algorithms for data mining tasks. The algorithms can be applied directly to a dataset. WEKA contains large variety of tools for data pre-processing, classification, regression, clustering, association rules, and visualization. The paper is organized as follows: Section 2 provides the related work and background of data mining Algorithms. Section 3 presents the research methodology to derive at a conclusion while Section 4 depicts the obtained result. Section 5 discusses the summary of this paper and its future scope.

2 Related Work and Background

The growing complexities of software and increasing demand of software projects have led to the progress of continual research in the areas of effective project management. Data mining has proven as one of the established techniques for effective project management recently. Data mining methodologies are developed for several applications including various aspects of software development. It works on large quantities of data to discover meaningful patterns and rules. Authors in [4] surveyed different data mining algorithms used for defect prediction in software and also discuss the performance and effectiveness of data mining algorithms. Authors of [5] made a comparative analysis of performance of Density-Based Spatial Clustering of Applications with Noise (DBSCAN) and Neuro-Fuzzy System for prediction of level of severity of faults present in Java based object oriented software system. Data which comprises of project personnel data provides a rich resource for knowledge discovery and decision support[9]. Data mining results in decision through methods and not assumptions. In [13] authors have done an empirical study for selection criteria for software industry by introducing a knowledge based decision tree algorithm. Authors in [2] have worked on the improvement of employee selection, by building a model, using data mining techniques. Depending on few selected attributes, the model could predict their performance. Some of these attributes are personal characteristics, educational and professional attributes. They specified age, gender, marital status, experience, education, major subjects and school tires as potential factors that might affect the performance. As a result for their study, they found that employee performance is highly affected by education degree, the school tire, and the job experience. The authors in [3] searched on certain factors that affect the job performance. They reviewed previous studies, experience, salary, training, working conditions and job satisfaction on the performance. As a result of their research, it was found that several factors affected the employee's performance such as position of the employee in the company, working conditions and environment. Highly educated and qualified employees showed dissatisfaction of bad working conditions and thus affected their performance negatively. Employees of low qualifications, on the other hand, showed high performance in spite of the bad conditions. Experience showed positive relationship in most cases, while education did not yield clear relationship with the performance. Data mining thus supports various techniques including statistics, decision tree, genetic algorithm, bayes classification and visualization techniques for analyses and prediction. It further deals with association, clustering and classification [1].

This part of the research therefore involves applying data in WEKA tool and derives a classification model for selection criteria. Some of the data mining algorithms are ID3, C4.5 and CART [12]. Decision trees are a hierarchical structure with leaves and stems. The hierarchical structure of decision trees represents different levels of attributes. Every leaf reveals the classification of an attribute, while the stems indicate the conditions of the attributes. Given training set, a decision tree can be constructed depending on various methods to provide valuable information about the attributes and their patterns [6]. The authors of [6] have developed ID3 (Iterative Dichotomise 3) which is based on Hunts algorithm. The tree is constructed in two phases namely tree building and pruning. Authors have thus developed C4.5 which is a successor to ID3 and is based on Hunt's algorithm [11]. Classification And Regression Trees (CART) is yet another popularly available algorithm for WEKA users. CART was introduced by Breiman. It handles both categorical and continuous attributes to build a decision tree in addition to handle missing values. ID3 and C4.5 algorithms have multiple branches, however CART produces binary splits and thereby binary tree. WEKA contains tools for regression, classification, clustering, association rules and visualization. The classify panel enables the user to apply classification algorithms to the resulting dataset, estimate the accuracy of the resulting predictive model, visualize erroneous predictions and the model itself as shown in section 4 of this study. The further section will discuss more about research methodology and results obtained from ID3, CART and C4.5.

3 Research Methodology

The main objective of the study is to build a performance model of an employee based on his/her attributes. This investigation focused upon the selection criteria of right project personnel that yield effective results for better software quality through a prediction technique. The objective was to concentrate on human aspect of software engineering by selecting the appropriate people.

This research therefore aimed to construct a framework for human resource data mining to explore the relationships between personnel profiles and its effect on software development. Through the proposed methodology, hidden information could be extracted from large volumes of personnel data and thus the project leaders are able to comprehend and focus on selection for the software project through the discovered knowledge. Fig. 1 shows the framework.

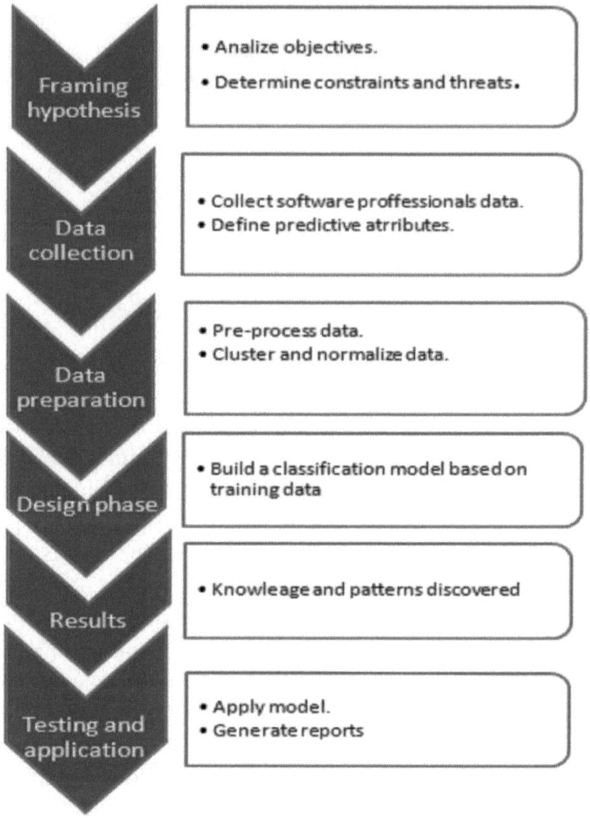

Fig. 1. Research Framework

Figure 1 indicates the methodology followed for this research. Having made a deep study of literature survey and industry based investigations, this research focused upon the formulation of hypothesis.

a. **Hypothesis** –Project personnel with certain attributes values will perform similarly.

Project personnel information is collected from those projects which are developed in similar domain using common technology and programming language. The constraint of this study is that it has dealt with short term projects of duration two to three years with windows operating system and C/C++ as programming language for web based applications.

b. Data Collection

Data was collected in collaboration with the project team. Modes of data collection included brainstorming, discussions, interviews, and argumentation. The project leaders provided goals and experts in human resources suggested criteria. The most relevant questions considered were

- How to select the relevant attributes
- How to structure them

In order to collect the required data, a questionnaire was prepared and given to both project leaders and project personnel. Information regarding various attributes was asked in the questionnaire that might predict the performance class. Attributes involving personal, educational and work related details were collected. However, this study aimed at educational, internal assessment results and work related attributes.

c. Data Preparation

Having obtained the responses for the questionnaires, the process of preparing the data was accomplished. The list of collected and selected attributes which is relevant for this research is specified below.

PS-Programming Skill. This variable was obtained through internal assessment by the company. It was mapped into three categories as good, average and poor based on the marks out of 100. Good for 75% and above, average for 60 to 75% and poor below 60%. The three categories with the above mentioned grading were taken to map the input with the output or target.

RS-Reasoning Skill. During selection company takes various assessments. This variable was one of the internal assessments, which is categorized similar to PS.

DKA-Domain Knowledge Assessment. This is yet another level of internal assessment made by the company. It was categorized and normalized similar to PS and RS.

TE-Time Efficiency. It was obtained from the project data through project leaders. It was obtained in the form of YES and NO.

GPA-General Percentile Assessment. This is obtained from the database pertaining to personal attributes of candidate. It is mapped into good (for >7.5), average (for <7.5 an >6.5) and poor for (<6.5)

CS-Communication Skills. The value for this variable is fetched from the project leaders. This is mapped into good, average and poor.

P-Performance. This is the target or output class. The value for performance variable is acquired from project team leaders in terms of good, average or poor, which is based on the quality of software developed. The company has a evaluation system every month and the consolidated results for the tenure of the project was considered. Sample instances of the training data is shown in Table 1.

Table 1. Portion of Training Data Set

S.No.	GPA	DKA	PS	TE	CS	RS	P
1	Good	Good	Good	Yes	Good	Good	Good
2	Good	Good	Average	No	Good	Good	Good
3	Good	Good	Average	No	Average	Average	Average
4	Good	Average	Good	Yes	Good	Good	Average
5	Good	Average	Average	Yes	Good	Good	Average
6	Good	Poor	Poor	No	Average	Poor	Poor

d. Implementation of Mining Model

Based on the background and related work, a training set with the attributes depicted in data preparation is selected to test against their effectiveness on the employee performance.

The algorithm used for classification in this study is ID3, C4.5 and CART. Under the "Test options", the 10-fold cross-validation is selected for the evaluation approach. Since, there is no separate evaluation data set, this option was necessary to get a reasonable idea of accuracy of the generated model. The model is generated in the form of decision tree as shown in the results section. These predictive models provide analytical way to formulate selection criteria.

4 Results

The three decision trees of predictive models obtained from the training data set by three machine learning algorithms: the ID3 decision tree algorithm, the CART decision tree algorithm and the C4.5 also called J48 in WEKA environment are shown in TABLE 2, TABLE 3 and TABLE 4.

Table 2 infers the decision tree obtained from ID3 in the tree option of classification in WEKA tools. Table 3 depicts the binary decision tree obtained in run information for CART option in tree option of classification. It shows the result with root node again being PS. Since it is a binary tree it has two branches with PS as poor and not poor. Table 4 shows the C4.5 pruned tree in J48 option of tree option in classification. Table 5 shows the accuracy of ID3, C4.5 and CART algorithms for classification applied on the above data sets using 10-fold cross validation.

Table 2 , result obtained from run information of WEKA tool, clearly indicates that if PS is good and RS is good or average, the performance is good. Also generally when PS is good then RS is either good or average and performance is good. If PS is average then performance is average. However, if PS is average but if GPA is good then there exists a possibility of performance being good. When PS is poor then his performance is poor irrespective of all other attribute values.

Table 2. ID3 Decision Tree

=== Classifier model (full training set) ===
Id3
PS = Good | RS = Good: Good | RS = good: Good | RS = Average: Good | RS = Poor: Average | RS = poor: null PS = Average | DKA = Good: Average | DKA = Average: Average | DKA = Poor | | GPA = Good: Good | | GPA = Average: Average | | GPA = Poor: Poor PS = Poor: Poor

Table 3. Cart Decision Tree

=== Classifier model (full training set) === CART Decision Tree Number of Leaf Nodes: 3 Size of the Tree: 5
PS=(Poor): Poor(13.0/0.0) PS!=(Poor) | PS=(Average)|(Poor): Average(12.0/2.0) | PS!=(Average)|(Poor): Good(12.0/1.0)

Table 3 infers that if PS is poor then the training set showed all 13 records with performance as poor. There were 13 records in the training set with PS poor and all showed poor performance. In the second level CART tree once again split PS into average and good. When PS value is average then 12 records showed average performance and two records showed poor performance. When PS is good then 12 records showed good performance and only one showed not good. Table 3 also indicates that PS is most dominating attribute. The best match is when all poor show poor and all average show average and all good show good performance. However, compared to other attributes, PS showed the best match and thereby to deem the attribute with highest importance.

Table 4. C4.5/J48 Pruned Tree

=== Classifier model (full training set) ===J48 pruned tree-- Number of Leaves: 3Size of the tree: 4
PS = Good: Good (13.0/1.0) PS = Average: Average (14.0/2.0) PS = Poor: Poor (13.0)

Table 4, run information of J48 depicts that if PS is good then out of 13 records all mapped to performance as good except 1. If PS is average then apart from 2 records all indicated an average performance. However, when PS is poor all records showed poor performance. These decision trees also provide interesting insights into hidden patterns in the project personnel performance by showing the importance of attributes in a hierarchical manner. The tree indicated that programming skills, domain knowledge and reasoning skills are more relevant than college aggregate. The tree generated using the C4.5 algorithm also indicated that the programming skills is most effective attribute for an IT company.

Table 5. Comparison of Techniques

Algorithm	Correctly Classified Instances	Incorrectly Classified Instances
ID3	90.5%	7.5%
CART	92.5%	7.5 %
C4.5	90.5%	7.5%

After building the tree, WEKA tool again cross validates the result obtained with the data set and gives the information about correctly and incorrectly classified records. Table 5 indicates that ID3 algorithm could classify 90.5% of data correctly. CART and C4.5 showed similar results by classifying 92.5% and 90.5% correctly. CART showed highest correctly classified instances.

Since all three options showed similar results we stopped our investigation for other classification model in WEKA tools.

Thus, this investigation indicates that performance of project member is good if programming skill and reasoning skills are good irrespective of his college aggregate or communication skills.

5 Conclusion

In this paper data mining is used to predict the performance of software project member on the basis of previous database or training set. This paper has focused on the human aspect of software engineering to achieve good quality of software by building a classification model for predicting employees' performance based on certain attributes. Data mining techniques could identify those attributes required in a project member which will contribute to good performance and thereby enhance software quality and success. Classification techniques like ID3, CART and C4.5 showed similar results. Performance was earlier assumed to be best for candidates having a good college aggregate.

However, this study showed that other talent attributes like programming skills and reasoning skills have proved to be more important, although software companies emphasize upon aggregate percentile and followed the same trend for many years. Due to lack of analytical method in human aspects, software companies were not selecting the right people who could perform well in the software process and thereby failed to achieve the desired quality in the time and cost constraints.

Data mining techniques have given very interesting patterns and helped in earlier identification of project members who will perform well. This study enables the managers to refocus on human capability criteria and thereby enhance the development process of software project. Just like all process within generic framework of software development is given importance, human aspect also needs a deeper investigation for effective software development. Without the right people even the best process analysis and development are bound to fail.

Further scope of this study is to investigate on several projects of different domains and take into account more attributes of project personnel and correlate it with software quality and success.

References

1. Han, J., Kamber, M.: Data Mining Concepts and Techniques, 2/e. Morgan Kaufmann Publishers, An imprint of Elsevier (2010)
2. Chien, C.F., Chen, L.F.: Data mining to improve personnel selection and enhance human capital: A case study in high-technology industry. Expert Systems and Applications 34, 280–290 (2008)
3. Jantan, H., et al.: Human Talent Prediction in HRM using C4.5 Classification Algorithm. International Journal on Computer Science and Engineering (IJCSE) 2(8), 2526–2534 (2010)
4. Suma, V., Pushpavathi, T.P., Ramaswamy, V.: An Approach to Predict Software Project Success by Data Mining Clustering. In: International Conference on Data Mining and Computer Engineering (ICDMCE 2012), pp. 185–190
5. Singh, P.: Comparing the effectiveness of machine learning algorithms for defect prediction. International Journal of Information Technology and Knowledge Management, 481–483 (2009)
6. Quinlan, J.R.: Introduction of decision tree. Journal of Machine Learning, 81–106 (1986)
7. Witten, I., Frank, E., Hall, M.: Data Mining: Practical Machine Learning Tools and Techniques, 3rd edn. Morgan Kaufmann Publishers (2011)
8. Kusiak, A., Kern, J.A., Kernstine, K.H., Tseng, B.: Autonomous decision-making: A data mining approach. IEEE Trans. Inform. Technol. Biomedicine 4(4), 274–284 (2000)
9. Chang, A.S., Leu, S.S.: Data mining model for identifying project profitability variables. International Journal of Project Management 24, 199–206 (2006)
10. Gopalakrishnan Nair, T.R., Suma, V., Tiwari, P.K.: Analysis of Test Efficiency during Software Development Process. In: 2nd Annual International Conference on Software Engineering and Applications (SEA 2011) (2011)
11. Quinlan, J.R.: C4.5: Programs for Machine Learning. Morgan Kaufmann Publishers, Inc. (1992)
12. Maimon, O., Rokach, L.: The Data Mining and Knowledge Discovery Handbook. Springer (2005)
13. Gupta, S., Suma, V.: Empirical study on selection of team members for software project- A data mining approach. International Journal of Computer Science and Informatics 3(2), 97–102 (2013) ISSN (PRINT): 2231 –5292

Defect Detection Efficiency of the Combined Approach

N. Rashmi and V. Suma

Dayananda Sagar College of Engineering,
Shavige Malleshwara Hills, Kumaraswamy Layout, Bangalore -78
rashmi_jul@yahoo.com,
sumavdsce@gmail.com

Abstract. Software Testing is the process of identifying the defects in the software and hence improving the quality of software. There are several testing techniques and combinations of these testing techniques existing. This paper discusses the defect detection efficiency of the combined approach of scripted and exploratory testing by comparing the projects which are done using only scripted testing and the combination of scripted and exploratory testing.

Keywords: Software testing, combined approach, scripted testing, exploratory testing.

1 Introduction

Delivering high quality software is a challenging task as there is always a continuing demand towards test coverage, identifying defects, learning the product assessing the risks involved with product, performance of the product from the customers [1]. High quality software is one which is error free, produces predictable results with less manageable efforts, understandable, dependable and efficient[2]. Software Testing is an important process which contributes towards the quality product by identifying defects before the product is actually delivered to the customers.

Software Testing can be manual or automated. Though there is a spectrum of automation testing tools available, automation cannot be used for all types of projects for various reasons. Also most of the serious and interesting defects are found through manual testing [3][4]. This is due to the fact that manual testing highly depends on the skills of the tester testing the product [5]. There are several manual testing techniques existing today. Software industries use these techniques depending upon several factors such as the type of the product under development, size of the project, time required to spend on testing etc. So among these testing techniques scripted testing or test case based testing is the most popular one. It is also named as specification based testing. In this type of testing the product is executed against the test cases written by testers. This type of testing emphasizes accountability and decidability of tests. But as and when the changes are made to the product, executing the same test cases may not be useful as many of the defects go undiscovered. This is where the skills of the testers play an important role. Exploratory testing is a type of testing where the skills of the testers are given high priority. Also exploratory testing sometimes results in the

S.C. Satapathy et al. (eds.), *ICT and Critical Infrastructure: Proceedings of the 48th Annual Convention of CSI - Volume II*, Advances in Intelligent Systems and Computing 249,
DOI: 10.1007/978-3-319-03095-1_51, © Springer International Publishing Switzerland 2014

identification of enormous defects which go undetected during scripted testing. In exploratory testing the tester learns, designs and executes the tests simultaneously. There is no need of prior documentation of test cases required. Hence this paper discusses the importance of performing exploratory testing in addition to scripted testing which yields even better defect detection efficiency.

2 Literature Survey

Authors of [1] performed a controlled experiment to show the significance of exploratory testing. Through the experiment they prove that exploratory testing is as effective as scripted testing. In 2011, T. R. Gopalakrishnan et al. in [2] provide an empirical investigation of several projects through a case study comprising of four software companies having various production capabilities. The aim of this investigation was to analyze the efficiency of test team during software development process. The study indicates very low-test efficiency at requirements analysis phase and even lesser test efficiency at design phase of software development. Subsequently, the study calls for a strong need to improve testing approaches using techniques such as dynamic testing of design solutions in lieu of static testing of design document. In 2002, C. Andersson et al. in [3] presented a qualitative survey of the verification and validation processes at 11 Swedish companies. The purpose was to exchange the information between the companies. It is concluded from the survey that there are substantial differences between small and large companies. In large companies, the documented process is emphasized while in small companies, single key persons have a dominating impact on the procedures. Large companies use commercial tools while small companies in-house tools or use shareware. In 2004, Juristo, N et al. in [4] analyzed the maturity level of the knowledge about testing techniques by examining existing empirical studies about these techniques. Authors of [5] made some observation in a dozen projects in the area of software testing, especially in automated testing and conclude that automation testing cannot replace manual testing. In 2000, J. A. Whittaker in [6] answers questions from developers how bugs escape from testing. Undetected bugs come from executing untested code, difference of the order of executing, combination of untested input values, and untested operating environment. A four-phase approach was described in answering to the questions. By carefully modeling the software's environment, selecting test scenarios, running and evaluating test scenarios, and measuring testing progress, the author offers testers a structure of the problems they want to solve during each phase.

3 Scripted Testing

Scripted Testing is a type of testing where the tests are designed at an early stage. These tests are executed many times in later stages of software development. Each time the test is executed the tester looks into the same things. A script specifies – the test operations, the expected results, the comparisons that human or machine should make. The problem with this approach is that the early the tests are designed the less will be known about the risk profile in the project. That means the tests remains same while the risk profile is changing. Also there may be changes due to change in the requirements/specifications, change in the environment etc. The benefits of scripted testing include careful thinking about the design of tests, optimizing the tests for power, credibility, review by stake holders, reusability, and known comprehensiveness of the set of tests [6][7][8].

4 Exploratory Testing

Exploratory Testing is a style of testing that emphasizes personal freedom and responsibility of individual tester to continually optimize the value of her work testers by treating test- related learning, test design and test execution and test result interpretation, as mutually supportive activities that run in parallel throughout the project. In contrast with scripted testing the tests are designed as needed executed at the time of design or reused later. Also in exploratory testing the tests can be varied appropriately when needed. The supporting materials such as data sets, failure mode lists, combination charts can be used during exploratory testing. The exploratory tester is always responsible for managing the value of her own time. This might include reusing old tests, creating and running new tests creating test-support artifacts such as failure mode lists, and conducting background research that can then guide test design [6][7][8].

5 Case Study

This paper discusses a case study which involved collecting data from two companies performing scripted and exploratory testing and comparing the defect detection efficiency of scripted and exploratory testing with scripted testing alone.

The companies are CMMI level 5 companies specializing in developing software products for high end printers and solutions for pharmaceutical companies. The projects here being compared are of embedded and life sciences type. The projects here are implemented using C, C++ on WINDOWS platform. All projects are non-critical in nature. All projects are of standalone type and of maintenance type [9].

Data collection is through the data centers and quality assurance departments of the above mentioned company. Data analysis is done using comparison of testing techniques using project success as a criteria measured through defect capturing capability.

The objective of this paper is to emphasize the benefit obtained by performing exploratory testing in addition to conventional scripted testing. Hence, this paper presents a case study where an investigation of several projects is carried in a leading product based software industry.

Table 1 illustrates the data collected from the company. It depicts the defect capturing capability of the testing team practicing scripted testing. The table provides information on the total development time required for the project completion which is measured in terms of person hours, the choice of process model followed, number of testers assigned in addition to the time scheduled for testing. The table further specifies the number of defects estimated and captured by the testing team.

(*) – Person Hours. The Project Development Time is expressed in person hours which are given by Project Development Time = (9 hours of work per day)*(number of personnel)*(number of months required)

Table 1. Scripted Approach

Sl.No	Parameters	P1	P2	P3	P4	P5	P6
1	Project Development Time(*)	324	243	432	162	1728	1144
2	Development life cycle model	V-Model	V-Model	V-Model	V-Model	V-Model	V-Model
3	No. of Testers	6	9	4	3	18	11
4	Scripted Test time of the project (*)	270	162	360	135	1584	1056
5	No. of defects captured by testing team	451	251	972	1022	800	550

Table 2. Combined Approach

Sl.No	Parameters	P1	P2	P3	P4	P5	P6
1	Project Development Time (*)	324	243	432	162	1728	1144
2	Development life cycle model	V-Model	V-Model	V-Model	V-Model	V-Model	V-Model
3	No. of Testers	6	9	4	3	18	11
4	Scripted Test time of the project (*)	270	162	360	135	1584	1056
5	ET test time(*)	27	16.2	36	13.5	316.8	211.2
6	No. of defects captured by testing team	602	851	1244	1644	905	612

Table 2 illustrates the defect capturing capability of the testing team practicing the combined approach. The parameters considered here are the project development time, testing time, number of defects captured, number of testers and so on

There are four projects being studied in the case study. The same projects were compared using the scripted and the combined approach. There are five parameters being used in the tables Table 1 and Table 2. They are defined as follows. First parameter is the total development time of the project which includes the coding time and the testing time of the project, second is the type of the development model used, third is the total number of testers in the project. The testers are having the relevant experience in their respective domain. Fourth parameter is the amount of time allotted for testing the product. This involves the time spent on unit, in the combined approach 10-20% of the total test time is allotted for exploratory testing.

6 Inferences

When we compare the scripted and the combined approaches we find the combined approach yielding more number of defects in the same time as the time allotted for scripted testing. This is shown in the form of a table, Table 3.

Table 3. Comparison Chart

Projects	Scripted Approach	Combined Approach	Additional defects detected using combined approach
P1	451	602	151
P2	251	851	600
P3	972	1244	272
P4	1022	1644	622
P5	800	905	105
P6	550	612	62

This can also be represented graphically as shown below.

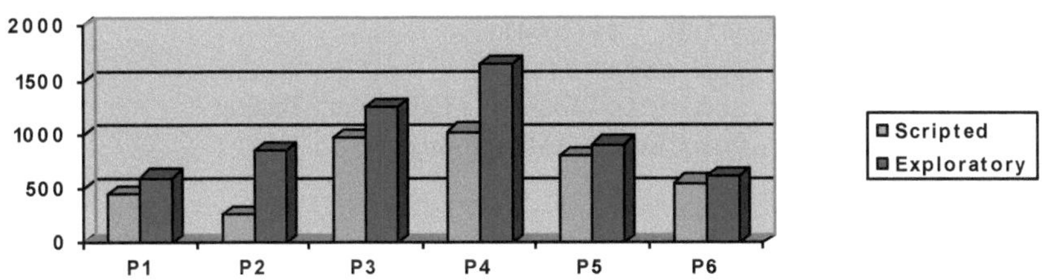

Fig. 1. Comparison Graph

Hence from the above analysis we infer that the combined approach is more efficient in terms of the number of defects found and in terms of the time spent on testing. The only concern here is that the testers performing the exploratory testing are having relevant domain experience.

It needs further investigations to be made on various other defect influencing parameters in addition to defect estimation and defect capturing capability of combined approach on varied complex projects.

7 Conclusions

Defect free product is deemed to be one of the needs for achieving high quality products. Software testing is the last opportunity for the testers to detect as many defects as possible before the product is delivered to the customers.

Despite of existence of a spectrum of testing approaches, techniques and tools, each of these approaches or the techniques and tools has their own strengths and weakness. Therefore, identifying the appropriate combination of these approaches or techniques is a challenging task. One such combination is the combined approach which is discussed in this paper. Though scripted and exploratory testing techniques individually contribute greatly towards the quality of the product when combined they increase the defect detection efficiency to a greater extent. Hence, the quality of the product is even more improved.

References

1. Prakash, V.G.: Testing efficiency exploited: Scripted versus exploratory testing. In: S. Sastra Univ., Thanjavur India 3rd International Conference on Electronics Computer Technology (ICECT) (April 2011)
2. Gopalakrishnan Nair, T.R., Suma, V., Tiwari, P.K.: Analysis of Test Efficiency during Software Development Process. In: 2nd Annual International Conference on Software Engineering and Applications (SEA 2011) (2011)
3. Andersson, C., Runeson, P.: Verification and validation in industry - a qualitative survey on the state of practice. In: Proceedings of International Symposium on Empirical Software Engineering, pp. 37–47 (2002)
4. Berner, S., Weber, R.: Observations and Lessons Learned from Automated Testing. In: Proceedings of International Conference on Software Engineering, pp. 571–579 (2005)
5. Juristo, N., Moreno, A.M.: Reviewing 25 years of Testing Technique Experiments. Empirical Software Engineering 9(1-2), 7–44 (2004)
6. http://www.kaner.com/pdfs - ETatQAI
7. http://www.kaner.com/pdfs - Exploring Exploratory Testing
8. http://www.kaner.com/pdfs - QAIExploring

Risk Measurement with CTP² Parameters in Software Development Process

Raghavi K. Bhujang[1] and V. Suma[2]

[1] Dept. of MCA, PESIT, Bangalore, Member, Research and Industry Incubation Centre, Research Scholar from Jain University, RIIC, DSI, Bangalore
raghavi@pes.edu
[2] Research and Industry Incubation Center, Dayananda Sagar Institutions Bangalore, India
sumavdsce@gmail.com

Abstract. Organizational success in software industry is completely dependent on accomplishment in terms of successful project delivery that results in client satisfaction associated with end user requirements. This victory can be sustained if the IT industry is able to cope up with all the challenges involved in software development targeted towards either product development or providing a service. One of the major challenges that can be listed as a key encounter in software development is Risk. Growth of an industry can take up well-defined criteria with respect to attainment of success if Risk management implements an efficient technique. This risk has to be measured foremost in terms of Cost, Time, People and Process (CTP2) to understand the impact of the same in the project. This paper aims to classify various Risks based on the impact analysis which is conducted in terms of CTP2 . This mode of classification enables one to prioritize the risk in order to mitigate or deal with it effectively.

Keywords: Risk management, Risk analysis, Risk Factors, Software Engineering, Software Development Process

1 Introduction

Software Development Process is said to be successful if IT industry has a well-defined plan to deal with the challenges and complexities involved in the development process. This can be attained by dexterous experts in the team assigned to bring up best practices in the process. However, risk management is one of the major challenges in the software development process. Furthermore, type of the risk needs to be identified before comprehending the impact of the same on the development process. Currently, in IT industries, there are different types of scenarios that a software development team comes across in the engineering process that are related to various factors like project schedule, budget, resource allocation and bearing with advancement in technology. Authors of [7] have made a Graphical Analysis on Risk Management where in Risk needs to be identified and assessed with

S.C. Satapathy et al. (eds.), *ICT and Critical Infrastructure: Proceedings of the 48th Annual Convention of CSI - Volume II*, Advances in Intelligent Systems and Computing 249,
DOI: 10.1007/978-3-319-03095-1_52,

Probability and frequency and finally to bring out a management technique or mitigation strategy. By considering the different projects across the platforms and domains, risk can be further classified into different types as follows: Communication Risk, Quality Risk, Technological Risk, System Configuration Risk and Estimation Risk.

The organization of this paper is as follows, section 2 of this paper provides details of the related work in the domain of risk measurement and management. Section 3 focuses on identifying the different types of risks i.e by a sample of data collected from across the projects and various domains. Section 4 emphasizes on risk measurement in terms of CTP^2 parameters and section 5of this paper provides the summery of this investigation.

2 Literature Survey

Authors of [1] have investigated the impact of specific risk management strategies and residual performance risk on objective performance measures such as cost and schedule overrun. They have also investigated the impact of two alternative conceptualization of software development risk on both objective performance and subjective performance.

Authors of [2] have presented a six-step metrics-based methodology for assessing the risks that are bound to happen and the resources required to implement- the requirements contained within a software requirements specification (SRS). The method seeks to eliminate the use of subjective probability assessments in models of risk exposure (RE) and risk reduction leverage (RRL).

Authors of [3] have provided a view towards how software risk enters into the enterprise and shown a data model of risk identification with the overview of risk assessment and mitigation process. They have also shown how they can apply insurance to software. The authors have also presented a solution to managing software risks.

According to authors of paper [4], software development team can be aware of different kinds of risks involved in the development process through risk management process. Hence, it is always appreciable to understand the probability of occurrence of risk to know its potential reach on negative impact.

Authors of [5] have carried out a literature survey on Risk Modeling and Assessment and they have suggested that analyzing risk cost would be an innovative solution in overcoming the short comes of risk assessment. Apart from the cost, there are many more critical factors which need to be managed being the core fundamentals for a risk free completion of the project.

Authors of [6] aimed to investigate an approach for the assessment of risks in globally distributed software projects. This research proposes to apply stochastic simulation technique to analyze project data and identify factors that are likely to impact team productivity and that could affect the team's ability to meet its schedule objective.

According to authors of [7], risk assessment factors like analyzing the impact of risk and assigning the priorities helps the organization in accurate prediction of success rate of project.

The analysis made by author [8] infers that occurrence of risk is always high irrespective of long term or short term project. However, the impact of risk on long term projects is higher when compared to the impact of risk on short term projects which is validated against CTP^2 (Cost, Time, People, and Process).

According to authors of [9], software risk management is considered to be an integral part of project or the business management. One of the ways in which the risk factor can be solved is by using automated tool. It further leads towards immediate call for risk resolving strategies towards development of sustainable and high quality products.

As per the authors of [10], a risk management plan is created that identifies control actions that will reduce the probability of the risk occurring and/or reduce the impact of the risk that can turn into a problem. They have also introduced a tool that identifies the risks, in terms of cost and time that can be resolved by rescheduling of the work through resource management.

3 Risk Identification

Risk is the chance of failure that might result in the damage or disaster in software development that needs to be identified at the right time to avoid the same. It further helps in successful delivery of the project. In the present scenario of project development in IT industries, risks can be classified as related to Technology, Cost, Estimation, Scope, etc. Once the risks are identified and classified, their impact can be analyzed with the evaluation of CTP2 parameters.

3.1 Case Study

This case-study involves analysis of numerous projects developed in different platforms from various domains which are collected from few of the leading software industries and across few start-up companies that are product based and service oriented IT industries.

Table 1 indicates the identified nature of risks.

Table 1. Nature of Risk Table

Nature of Risk	Description of Nature
Catastrophic	Work Stopper
High	Deployment / Syntax error
Moderate	Recoverable failure
Low	Cosmetic

Below are the risks seen in various software development projects across different domains and different industries

Table 2. Table of Risks

Risk	Risk Description	Risk Occurring Phase	Nature of Risk	Classification of Risk
R1	**Teams speak a different language**	**General**	High - it occurs all the phases	Communications
R2	**Pushing for Development to start** :the emphasis is on getting to the coding / development at the earliest.	Requirement Gathering	High - it goes up to Coding Phase	Principles and Practices
R3	**Delaying the documentation -** Sometimes people want to cut corners and leave the documentation to be done later. If it is all in a person's head there are bound to be mistakes.	**General**	Low - Can occur at any phase of the project	Principles and Practices
R4	**Keeping requirements Feasible and Relevant** :You also need to keep the requirements are in line with the rules and regulations of the company, regulatory authorities, government etc. Examples – You might specify a date format of mm/dd/yyyy in your requirements; it will be valid in the USA. Is your software only for use in the USA? If yes, it is fine but if it is for global use this format may not be valid in the country of use. The same applies to currency.	Requirement Gathering	High – can continue up to implementation	Compliance with Organisational Standards
R5	**Product Integration** : Connectivity cannot be established between each of the components due to variation in development and interfaces.	Testing	Catastrophic	System Configuration
R6	**Deployment Errors :** The deliverable/component that can be deployed and tested on local environment cannot do the same on client environment due to configuration problems	Testing	High	System Configuration

Table 2. (*continued*)

R8	**Change in Requirements :** Change that demand the creation of new algorithms, input or output technology	Requirements	Catastrophic	Technology Risk
R9	**Usage of Unproven Software Products:** the software to be built interface with vendor supplied software products that are unproven	Design	High	System Configuration, Technology Risk
R10	**Take-over of client firm :** Client Firm being taken over by another company	**General**	Catastrophic / low	Compliance with Organisational Standards
R11	**Variation in size estimate of the product :** Size estimation of the product is done according to LOC for the first time, and done according to FP for the second time	General	Moderate	Inaccurate Project Estimation
R12	**Inaccurate progress tracking :** Inaccurate progress tracking results in not knowing the project is behind schedule until late in the project	Requirements /design/build/test	High	Inaccurate Project Estimation

As given in Nature of Risk Table, risks which act as work-stoppers have the nature that until and unless such risks are resolved, the project cannot move ahead. They are further identified as Catastrophic. Apart from them, all those risks which cause the run time errors are said to be of High nature. Also, the risk that causes failures that can be recovered easily, are said to be of Moderate nature. Finally, the risks that cause very less loss in the project are said to be of Cosmetic nature.Refer Table 2 Table of Risks.The above table contains a sample of the data on risks collected from the real time projects across the domains like financial, service based, product based and few start-up companies. According to the sample data, the type of the risk can be any of the individual types mentioned above such as either Risk1 – Communication type or Risk2 – Principles and Practices type or it can be a mix of two different types such as Risk 9 – System Configuration type or Technology type as well.

An analysis on these parameters in terms of impact on the same is made by considering this classification of Communication Risk, Quality Risk, Technological Risk, System Configuration Risk and Estimation Risk taken into reflection. According to the impact analysis made on these five different types, risks can be prioritized depending on their occurrence and affected parameter.

4 Risk Measurement

Considering the above table, it follows that Communication Risk has high impact or the parameter that is affected foremost is the People component of CTP^2, followed with the Process, and subsequently Time and finally Cost.

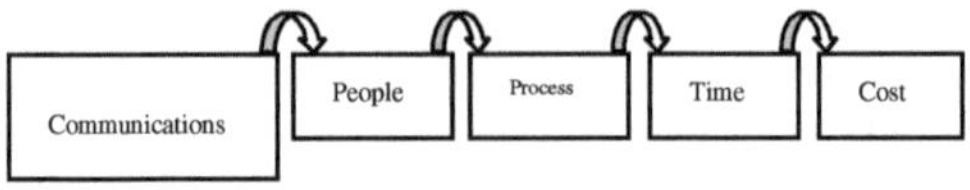

Fig. 1. Impact of Communication Risk on CTP2

infers that in the Communication Risk, the first component to be modulated is People gets highly impacted first, which affects the process component in turn, that results in delay in time due to rework and loss in budget. According to next classification, it follows that Quality Risk has high impact on the Process component of CTP^2, followed with impacting People parameter, then Time and finally Cost.

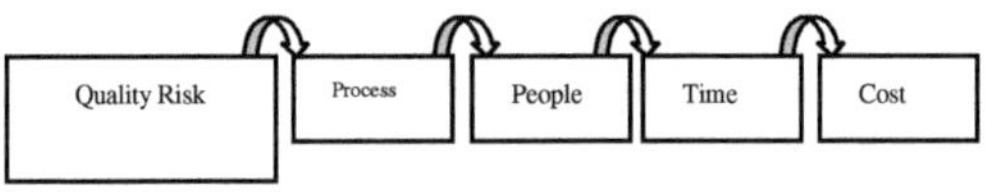

Fig. 2. Impact of Quality Risk on CTP 2

In Quality Risk, the main Process component in turn affects the development methodology followed the people that results in delay in time and loss in budget.

Yet another classification type is Technological Risk, and it follows that Technological Risk has high impact on Process component of CTP^2, followed with impacting People, then Time and finally Cost.

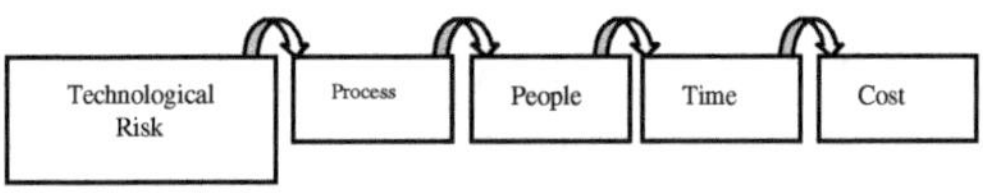

Fig. 3. Impact of Technological Risk on CTP 2

In Technological Risk, the directly impacted component will be Process again which impacts the resource which has the consequences on delay in delivery that beats up the expected cost. While following the next type in risk classification is System Configuration Risk, it trails that System Configuration Risk has high on Process component of CTP^2, followed impacting on the Time, then People and finally Cost.

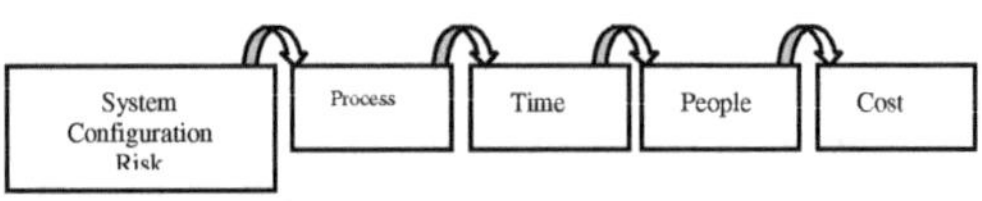

Fig. 4. Impact of System Configuration Risk on CTP 2

Finally, risk is further classified as Estimation Risk has high impact on Time component of CTP^2, followed with next parameter being impacted as Process, then People and finally Cost.

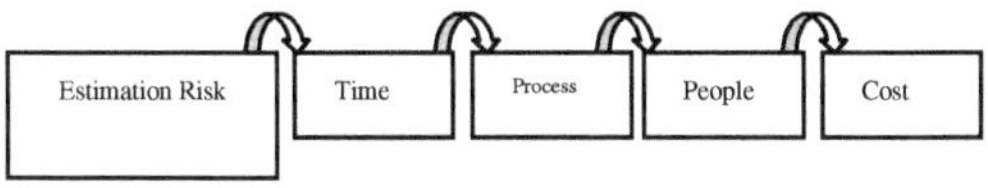

Fig. 5. Impact of Estimation Risk on CTP 2

In Estimation Risk, foremost impacted component will be Time, bringing up its effect on process, which results in risk in people component and cost component. The below bar chart represents the impact analysis of these classified risks.

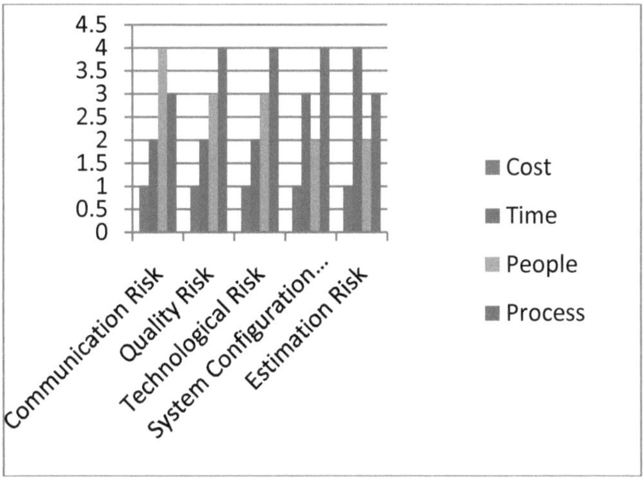

Fig. 6. Risk chart representing the impact of classified risks on CTP2

By analyzing the Risk chart, it can be clearly seen that the process component (Purple bar) in the CTP^2 parameters is the more affected parameter due to the impact of risks. Process being the core component of Software Project development from the technical aspect plays a major role and hence needs acute care in the development. This means that, however the impact remains on any of Time or Process or People components, all these will finally affect the Cost component of CTP^2 parameter.

5 Conclusion

Software Development being the hard core process to be followed in IT industries, maintaining the quality and reaching a successful delivery is a must in the organization to get client satisfaction. In terms of maintaining a good quality Risk stands as one of the key parameters to be assessed so that there would be less damage caused while reaching the successful delivery stage. Henceforth, it is obligatory for each organization to deal with risk of any type in software development process.

This paper presents a high level classification of the risks that occur in software development process as Communication Risk, Quality Risk, Technological Risk, System Configuration Risk and Estimation Risk. These classified risks are further measured in terms of impact with CTP^2 parameters. By analyzing the Risk Chart, it is observed that the process component gets greater impact of all the risks. Also, irrespective of the impact of risk on any of the component in CTP^2, the final affected component will always remain to be Cost. Further, research focuses on the risk analysis through each of the phases of software development life cycle and understanding the impact in terms of propagated Risk.

References

1. Na, K.-S., Simpson, J.T., Li, X., Singh, T., Kim, K.-Y.: Software development risk and project performance measurement: Evidence in Korea. The Journal of Systems and Software, 596–605 (2007)
2. Moores, T., Champion, R.E.M.: A Methodology for Measuring the Risk Associated with a Software Requirements Specification
3. MuradChowdhury, A.A., Arefeen, S.: Software Risk Management: Importance and Practices. IJCIT 2 (2011) ISSN 2218- 5224
4. Research 2 (2012) ISSN PP: 2277-7970
5. Taroun, A., Yang, J.B., Lowe, D.: Construction Risk Modelling and Assessment: Insights from a Literature Review. The Built & Human Environment Review 4(1) (2011)
6. Lima, A.M.: Risk Assessment on Distributed Software Projects. In: ACM/IEEE 32nd International Conference on Software Engineering, Cape Town, South Africa, May 2-8, pp. 349–350 (2010)
7. Bhujang, R.K., Suma, V.: Graphical Visualization of Risk Assessment for Effective Risk Management during Software Development Process. In: International Joint Conference on Emerging Intelligent Sustainable Technologies (EISTCON 2012), Bangalore, May 3-4, pp. 978–993 (2012) ISBN : 978-93-81693-76-6; Journal Reference - Cite as : arXiv:1210.1291v1 [cs.SE] also received the 'Best Paper Award' (2012)
8. Bhujang, R.K., Suma, V.: A Study of Risk and CTP2 during Software Development Process. To be Indexed in IEEE (2013)
9. Raju, P., Nirmala, R., Shruthi, R., Ravindra, A., Bhujang, R.K., Suma, V.: Risk Identification Tool for Cost and Time Parameters in Software Development. In: International Conference on Computer Science and Engineering, Article No:28, pp. 128–132. IRNet (2013) ISBN NO:978-93-83060-05-4
10. Bhujang, R.K., Raju, P., Nirmala, R., Shruthi, R., Ravindra, A., Suma, V.: Risk Prevention Technique In Software Development. In: International Conference on Electrical, Electronics and Computer Engineering, Article No:19, pp. 86–90. ASAR: Asian Society For Academic Research (2013) ISBN NO:978-81-927147-3-8

A Cryptographic Privacy Preserving Approach over Classification

G. Nageswara Rao, M. Sweta Harini, and Ch. Ravi Kishore

Department of Information Technology,
Aditya Institute of Technology & Management, Tekkali, A.P., India
{gnraoaitam,swetaharini35,cauchy9}@gmail.com

Abstract. We introduce a cryptographic based approach that will ensure the protection of data sets, which are used by third parties for constructing decision tree models using classification techniques, specifically ID3 algorithm. There is no necessity to increase the data sets size through perturbation or sanitize the samples before forwarding the data sets to third parties for further processing. The suggested method does not affect the accuracy of the data mining results. Cryptography techniques are applied after the collection of the entire data. This ensures privacy protection as the data sets are encrypted before they are sent to third parties preventing inadvertent disclosure or theft. This would prevent hackers/people who would like to misuse the data as the information is in encrypted form. We propose to use ID3 algorithm for classification, which is used extensively in machine learning/data mining, in construction of decision tree models

Keywords: Cryptography, Datasets, ID3 Algorithm, Decision tree models, Perturbation, Data mining.

1 Introduction

Advances in storage technologies, has resulted in explosive growth of data through the creation of ultra large databases. With increase in the volume of data, privacy concerns have become very important as there is possibility of information misuse [1], [3]. The problem is further complicated where the access to World Wide Web (WWW) has made it easy for people to upload personal information on to the net [7], [8].

As data mining and machine learning are receiving undue recognization and are being treated by computer professionals as panacea for the ills that are being faced in handling large or ultra large data sets. Data mining promises to find valuable relationships from hidden/non-obvious information contained in large databases. These relationships are the cause for data misuse as the relationships can be used or converted into commercial transactions. The concepts developed in data mining are applied to create models that deal with aggregate data without actually bothering about the precise information in individual data records.

S.C. Satapathy et al. (eds.), *ICT and Critical Infrastructure: Proceedings of the 48th Annual Convention of CSI - Volume II*, Advances in Intelligent Systems and Computing 249,
DOI: 10.1007/978-3-319-03095-1_53, © Springer International Publishing Switzerland 2014

2 Related Work

Sameer Ajmani and others [1] talked about Trusted Execution Platform (TEP) which supports shared computation between mutually distrusting parties. In this they suggested using cryptographic protocols to ensure a) The program communicates with the specified participant, b) Each participant knows what program is being done/executed. They also talked about isolation, which is nothing but communication over channels is encrypted and authenticated to prevent theft, impersonation or eves dropping. They have used secure session protocols such as SSL (Secure Socket Layer). The paper concludes by suggesting that optimizing the cryptography and channel protocol can greatly improve performance.

Sanjay Keer and Prof. Anju Singh [2] in their work have found that security of large databases that contain crucial information against unauthorized access. They have applied association rule hiding algorithm to get/obtain efficient performance for protecting confidential and crucial information. It also goes on to say that current technical challenge is the development of techniques that incorporate security and privacy issues. They concluded that sanitized database is as useful as the original database. They also discussed data distortion techniques which would result in sub-Optimal solution. In cryptography based approaches, they suggested encryption of original database instead of distorting it before sharing it with other parties.

Rakesh Aggarwal and Ramakrishnan Srikant [3] propose building of a decision tree classifier from training data in which the individual records are perturbed. They proposed a novel reconstruction procedure to accurately estimate the distribution of original data values from the perturbed dataset. They were able to built classifiers whose accuracy is comparable to the accuracy of classifiers built with original data. They built decision tree classifiers [9], [10]. Decision tree classifiers are relatively fast and yield comprehensive models. In the technique, they proposed was query restriction and data perturbation. Perturbation which they proposed is swapping values between records [11], [12].

It is well known that discretization is the method used for hiding individual values. Here the intervals need not be of equal width. They provide the interval in which the true value of the attribute lies. They have found that the difference between two randomized distributions is not a reliable indicator of the difference between the original and reconstructed distributions.

A decision tree is a class discriminator that recursively partitions the training set until each partition consists entirely or dominantly of examples from the same class [13], [14].The tree is built by recursively partitioning the data until each partition contains members belonging to the same class. It is pruned to generalize the tree eliminating dependence on noise or variation. The objective/goal of each node is to determine the split point that best divides the training records. Gini-is used to determine the goodness of the split [9].Decision trees using perturbed train data, we need to modify two key operations in the tree growth phase a) Determining a split point, b) Partitioning the data. Since we partition into intervals, the reconstruction procedure gives us an estimate of the number of points in each interval. Reconstruction is done by three methods a) global, b) class and c) local. The goal of a classification model is to understand different classes in the target population. The classification model is shipped to the user and applied there. XOR's which are known to be troublesome for decision tree classifiers. The concept of privacy preserving data mining was introduced by R.aggarwal and Srikant [3].

This concept has been relegated as it has been proved that the original data can be recovered from the perturbed data by any intruder. The technique of adding noise turned out to be flawed. There are some ideas that are still good, like, the owner can release only the perturbed data. This facilitates development of decision trees for original data and perturbed data and both are quite similar and accurate. This idea is very useful in many real-time applications. Here data perturbation is done through random substitution. Random substitution perturbation is done on one attribute. It is immune to attacks as explained in [15], [16]. A reconstructed dataset is not the original dataset. Once we say that the original and reconstructed datasets are not the same. Then one has to explore and estimate the error and find ways of their elimination or factoring it, in knowledge creation, using data tree mining.

Decision trees learned from re-constructed datasets are less accurate than those learned from the original datasets. Data mining is used widely in business application development. It is important to find the patterns, the data exhibits, to help in business decision making. Many people have worked on the protection techniques of sensitive information before allowing access to third parties for processing the information/data. It is suggested that decision tree can be built from sanitized data so that the original need not be reconstructed. There are different data modification techniques/methods – one such is k-anonymity [17].

Perturbation based approaches attempt to achieve privacy protection by distorting original information in the dataset. Most cryptographic techniques are derived for Secure Multiparty Computation (SMC). Building meaningful decision trees needs encrypted data to be decrypted or interpreted in its encrypted form [18]. This approach does not work well for discrete-valued attributes.

Generally ID3 algorithm builds a decision tree by calling choose-attribute recursively. With the smallest entropy and maximum information gain [19]. One can obtain enhanced protection by adding dummy values for any attribute thereby expanding the size of the datasets. This generally has no impact on the sample dataset other algorithms such as C4.5, C5.0.

3 Proposed Work

Having studied the related work done on datasets using trusted third party computation service, hiding sensitive and confidential information, using perturbation as a method for preservation of privacy in data mining, construction of decision trees using random substitutions using machine learning algorithms to create decision trees, creating decision tree learning through perturbation on unrealized datasets, using sanitized data for construction of data trees and finally using cryptography techniques, we have found that most of these approaches are yielding only partial results. We propose in this paper, a cryptographic classification approach. The below architecture describes its layout and working.

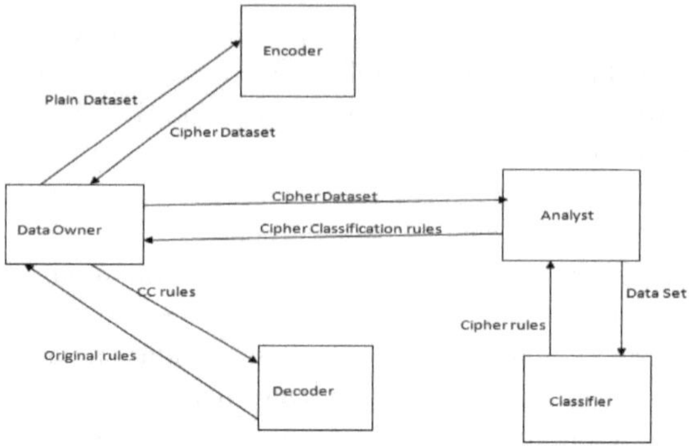

Fig. 1. Privacy preserving Architecture

Initialize Training Datasets for Data Mining

Datasets are the collection of tuples with different attributes and possible values for each attribute and with class labels, this data is given for the classification process to develop a decision tree.

Unrealized Dataset Creation

Usually data can be passed to third party analysts for the data mining purpose, but there is a privacy preserving issue regarding the confidentiality of the information. So in this paper we introduce AES (Advanced Encryption Standard) algorithm for the privacy issue. After applying this mechanism dataset can be constructed as unrealized dataset. i.e cipher dataset, this can be passed to the analyst for classification instead of plain, sensitive or confidential information.

Classification with ID3

ID3 is one of the efficient Machine learning approaches for implementing the decision trees. Decision trees are used for classification purpose. Tree can be constructed based on the attribute based entropy or information gain values. We can efficiently analyze the classification rules by sending the testing data on to the training datasets.

ID3 Algorithm. ID3 (Examples, Target_attribute, Attributes). Examples are the training examples, Target_attribute is the attribute whose value is to be predicted by the tree. Attributes is a list of other attributes that may be tested by the learned decision tree. Returns a decision tree that correctly classifies the given examples.

- Create a Root node for the tree
- If all Examples are positive, Return the single node tree Root, with label= “+”
- If all Examples are negative, Return the single node tree Root, with label= “-“
- If Attributes is empty, Return the single-node tree Root, with label = most common value of Target_attribute in Examples
- Otherwise Begin

- A← the attribute from Attributes that best* classifies Examples
- The decision attribute for Root← A
- For each possible value, v_i, of A,
- Add a new tree branch below Root, corresponding to the test A = v_i
- Let Examples v_i be the subset of Examples that have value v_i for A
- If examples v_i is empty
- Then below this new branch add leaf node with label = most common value of Target_attribute in Examples
- Else below this new branch add the sub tree
  ID3 (Examples v_i, Target_attribute
  Attributes – {A})
- End
- Return Root

Retrieval of Original Classified Results

After generating the classification results, results can be passed to the Data owner, the administrator can perform attribute oriented decryption for the resulted set. Original data set can be reconstructed by the decoder and classified rules can be obtained finally at the data owner end.

4 Experimental Analysis

Play tennis original data set:

Day	Outlook	Temperature	Humidity	Wind	Play Tennis
D1	Sunny	Hot	High	Weak	No
D2	Sunny	Hot	High	Strong	No
D3	Overcast	Hot	High	Weak	Yes
D4	Rain	Mild	High	Weak	Yes
D5	Rain	Cool	Normal	Weak	Yes
D6	Rain	Cool	Normal	Strong	No
D7	Overcast	Cool	Normal	Strong	Yes
D8	Sunny	Mild	High	Weak	No
D9	Sunny	Cool	Normal	Weak	Yes
D10	Rain	Mild	Normal	Weak	Yes
D11	Sunny	Mild	Normal	Strong	Yes
D12	Overcast	Mild	High	Strong	Yes
D13	Overcast	Hot	Normal	Weak	Yes
D14	Rain	Mild	High	Strong	No

The encrypted data to be sent to the third parties (Analyst) for classification:

[B@d70d7a	[B@b5f53a	[B@1f6f0bf	[B@137c60d	[B@ab853b	[B@b82368
437e3eb8cc074a2b38...	f3788018d287b7010c...	2b764d262c56b7a29a...	d4a3811ab7d6d0813e...	73f46018235c6faea4f...	f3be8507c605cf6e11c...
b0be4247563e390918...	f3788018d287b7010c...	2b764d262c56b7a29a...	d4a3811ab7d6d0813e...	58cc58fe96c1d87a859...	f3be8507c605cf6e11c...
414b0f1e6282aa3fa03...	1741947bade359f3a9...	2b764d262c56b7a29a...	d4a3811ab7d6d0813e...	73f46018235c6faea4f...	9f6c245816ef8eb8a6a...
3365353624f5f67ef8fc...	e75a3f848c49ee3a13...	63ac96c506bb40c38c...	d4a3811ab7d6d0813e...	73f46018235c6faea4f...	9f6c245816ef8eb8a6a...
ae41c9ad62b117b8ed...	e75a3f848c49ee3a13...	2c6b6848968f76bb0df...	eddd7a4ea2fc48ec25...	73f46018235c6faea4f...	9f6c245816ef8eb8a6a...
0858f3d11e4dba6668...	e75a3f848c49ee3a13...	2c6b6848968f76bb0df...	eddd7a4ea2fc48ec25...	58cc58fe96c1d87a859...	f3be8507c605cf6e11c...
94c65c1106743dd31e...	1741947bade359f3a9...	2c6b6848968f76bb0df...	eddd7a4ea2fc48ec25...	58cc58fe96c1d87a859...	9f6c245816ef8eb8a6a...
6689e5fff92b5614c59...	f3788018d287b7010c...	63ac96c506bb40c38c...	d4a3811ab7d6d0813e...	73f46018235c6faea4f...	f3be8507c605cf6e11c...
9b513e20c5f542bd5b...	f3788018d287b7010c...	2c6b6848968f76bb0df...	eddd7a4ea2fc48ec25...	73f46018235c6faea4f...	9f6c245816ef8eb8a6a...
c29c8ffd40f8b0f77c13...	e75a3f848c49ee3a13...	63ac96c506bb40c38c...	eddd7a4ea2fc48ec25...	73f46018235c6faea4f...	9f6c245816ef8eb8a6a...
48c9ae91fd29e5ec47...	f3788018d287b7010c...	63ac96c506bb40c38c...	eddd7a4ea2fc48ec25...	58cc58fe96c1d87a859...	9f6c245816ef8eb8a6a...
3cec75faf35ccdc192a...	1741947bade359f3a9...	63ac96c506bb40c38c...	d4a3811ab7d6d0813e...	58cc58fe96c1d87a859...	9f6c245816ef8eb8a6a...
2a134249e24b262856...	1741947bade359f3a9...	2b764d262c56b7a29a...	eddd7a4ea2fc48ec25...	73f46018235c6faea4f...	9f6c245816ef8eb8a6a...
83076ecc814e976d9b...	e75a3f848c49ee3a13...	63ac96c506bb40c38c...	d4a3811ab7d6d0813e...	58cc58fe96c1d87a859...	f3be8507c605cf6e11c...

The constructed decision tree from the encrypted data:

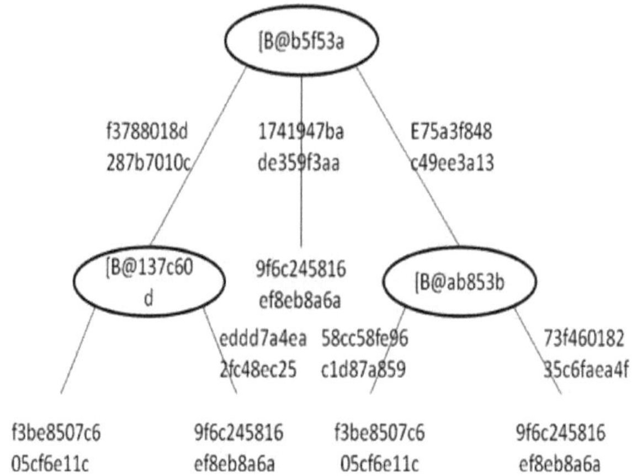

The decrypted decision tree:

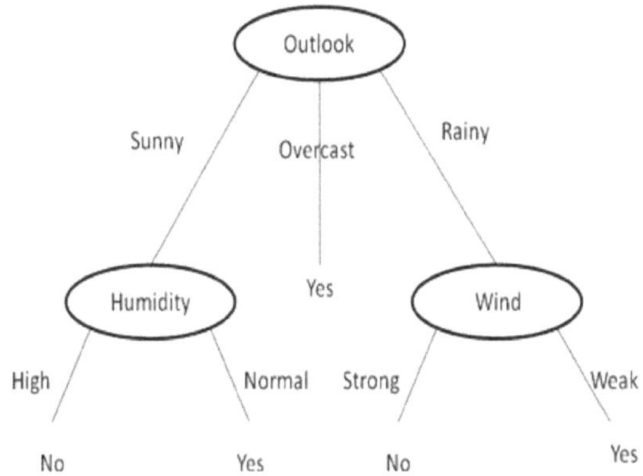

The decision tree from the original dataset:

Here, we can see that the decision tree formed from the original dataset is same as that of the encrypted decision tree formed from the encrypted data.

5 Conclusion

In this paper we propose an efficient privacy preservation technique during classification of unrealized datasets. It prevents the data owner from falling prey to unauthorized access and privacy issues, our proposed approach works efficiently without violating the classification properties. Meanwhile, an accurate decision tree can be built directly from the unrealized data sets. Finally the results are accurate even though classification is applied on the cipher dataset.

6 Proposed Future Work

1. During the process of encryption, the encrypted version of the attribute values is in the form of a 32-bit string. We proposed to explore ways of reducing the string length and at the same time retaining the high level security features provided by AES.

2. During the construction phase of a decision tree, entropy and information gain have to be calculated/determined for each attribute. For the purpose of calculation, the data set has to be scanned. This cause complications and time constraints due to the large string length of the encrypted data. We propose to develop simpler and efficient methods for reducing the scanning time.

References

1. Ajmani, S., Morris, R., Liskov, B.: A Trusted Third-Party Computation Service. Technical Report MIT-LCS-TR-847, MIT (2001)
2. Keer, S., Singh, A.: Hiding Sensitive Association Rules Using Clusters of Sensitive Association Rule. IJCSN 1(3) (June 2012)
3. Agrawal, R., Srikant, R.: Privacy Preserving Data Mining. In: Proc. ACM SIGMOD Conf. Management of Data (SIGMOD 2000), pp. 439–450 (May 2000)
4. Dowd, J., Xu, S., Zhang, W.: Privacy-Preserving Decision Tree Mining Based on Random Substitutions. In: Müller, G. (ed.) ETRICS 2006. LNCS, vol. 3995, pp. 145–159. Springer, Heidelberg (2006)
5. Lindell, Y., Pinkas, B.: Privacy preserving data mining. In: Bellare, M. (ed.) CRYPTO 2000. LNCS, vol. 1880, pp. 36–54. Springer, Heidelberg (2000)
6. Fong, P.K., Weber-Jahnke, J.H.: Privacy Preserving Decision Tree Learning Using Unrealized Data Sets. IEEE Transactions on Knowledge and Data Engineering 24(2) (2012)
7. Ma, Q., Deng, P.: Secure Multi-Party Protocols for Privacy Preserving Data Mining. In: Li, Y., Huynh, D.T., Das, S.K., Du, D.-Z. (eds.) WASA 2008. LNCS, vol. 5258, pp. 526–537. Springer, Heidelberg (2008)
8. Lomas, N.: Data on 84,000 United Kingdom Prisoners is Lost (August 2008), `http://news.cnet.com/8301-1009_3-10024550-83.html` (retrieved September 12, 2008)
9. Breiman, L., Friedman, J.H., Olshen, R.A., Stone, C.J.: Classification and Regression Trees. Wadsworth, Belmont (1984)
10. Ross Quinlan, J.: C4.5: Programs for Machine Learning. Morgan Kaufman (1993)
11. Denning, D.E.: Cryptography and Data Security. Addison-Wesley (1982)
12. Estivill-Castro, V., Brankovic, L.: Data swapping: Balancing privacy against Precision in mining for logic rules. In: Mohania, M., Tjoa, A.M. (eds.) DaWaK 1999. LNCS, vol. 1676, pp. 389–398. Springer, Heidelberg (1999)
13. Mehta, M., Agrawal, R., Rissanen, J.: SLIQ: A fast scalable Classifier for data mining. In: Apers, P.M.G., Bouzeghoub, M., Gardarin, G. (eds.) EDBT 1996. LNCS, vol. 1057, pp. 18–32. Springer, Heidelberg (1996)
14. Shafer, J., Agrawal, R., Mehta, M.: SPRINT: A scalable parallel Classifier for data mining. In: Proc. of the 22nd Int'l Conference on Very Large Databases, Bombay, India (September 1996)
15. Kargupta, H., Datta, S., Wang, Q., Sivakumar, K.: On the privacy preserving properties of Random data perturbation techniques. In: IEEE International Conference on Data Mining (2003)
16. Agrawal, D., Aggrawal, C.C.: On the design and quantification of Privacy preserving data mining algorithms. In: ACM Symposium on Principles of Database Systems (2001)

17. Sweeney, L.: k-Anonymity: A Model for Protecting Privacy. Int'l J. Uncertainty, Fuzziness and Knowledge-based Systems 10, 557–570 (2002)
18. Bu, S., Lakshmanan, L., Ng, R., Ramesh, G.: Preservation of Patterns and Input-Output Privacy. In: Proc. IEEE 23rd Int'l Conf. Data Eng., pp. 696–705 (April 2007)
19. Russell, S., Peter, N.: Artificial Intelligence. A Modern Approach 2/E. Prentice-Hall (2002)

An Advanced Authentication System Using Rotational Cryptographic Algorithm

Sk. Shabbeer Hussain[1], Ch. Rupa[1], P.S. Avadhani[2], and E. Srinivasa Reddy[3]

[1] Department of Computer Science and Engineering
Vasireddy Venkatadri Institute of Technology, Nambur, Guntur (Dt)
[2] Andhra University, Visakhapatnam
[3] Nagarjua University, Guntur (Dt)
{ssh.shabbeer,rupamtech}@gmail.com,
{psavadhani,edara_67}@yahoo.com

Abstract. The rapid extensive growth of the technology and ever increasing availability of computing resources are existed in the IT field. Furthermore, users have unauthorized access to information due to improper security measures. In this paper we proposed a new security approach to protect the information as well as to overcome the limitations which are existed in earlier approach. Here the algorithm generates Base Index, Numeric & Alpha Numeric Indexing tables and should apply rotational shifts on these tables by the user. The Indexing tables should use 6-bit character indexing with 6 time rotational shifts to eliminate duplicated values. This paper allows the Encryption and Decryption process to Alphanumeric with space and comma (,) characters.

Keywords: Rotational Shifts, Base Index, Numeric Indexing, Alphanumeric Indexing, 6-bit Character Indexing, Key less Algorithm.

1 Introduction

Nowadays, Internet is responsible to establish an essential communication between the people in globally. It is also being progressively used as a tool for commerce; as a result the security becomes a significant issue. One important aspect for secure communication is that of cryptography [7]. Cryptography is the process of creating secured messages by encoding them, and to make them unreadable. It is the combination of both mathematics and computer science and it has some relationship with information theory. Security algorithm is a technique to protect the Sensitive information in digital form. Steganography techniques involve the concealment of information within a text or images and transmit the same to the receiver with minimum distortion. This is an upcoming area of research. These techniques will have a significant effect in the areas of defense, business, copyright protection etc. where information needs to be preserved at all cost from the attackers.

Many Cryptography and Steganography techniques have been devised mainly focusing on the invisibility of the original data and robustness against various signal manipulation techniques and hostile attacks.

S.C. Satapathy et al. (eds.), *ICT and Critical Infrastructure: Proceedings of the 48th Annual Convention of CSI - Volume II*, Advances in Intelligent Systems and Computing 249,
DOI: 10.1007/978-3-319-03095-1_54, © Springer International Publishing Switzerland 2014

The existing system is responsible for a mechanism to transmit the encrypted data i.e. cipher text without any external key and it uses simpler operations like transmission [5]. This leads to saving of certain amount of time consumed to transmit the key; however it has some drawbacks which are effect on the message security that are shown in section 2. In this paper, we proposed a security technique that helps to overcome the limitations as shown in section 3. This approach consists of three main phases, which are generation of base index table, Encryption and Decryption.

The rest of the paper is organized by the following way. Section 2 consists of Existing System and its drawbacks. The proposed method is described in section 3, which holds the solutions to the drawbacks of the existing system. Section 4 holds the results and its study.

2 Existing System

In this system, the algorithm used a method to get the cipher text which requires some components like reference table, Transpose operations, etc. [5]. This System consists of three modules that are Reference table creation, creation of three character and index Blocks with the size of 3x3 and creation of Transpose blocks. Here, Encryption and decryption of an original text is done based on these three modules. This process is shown in the Figure 1.

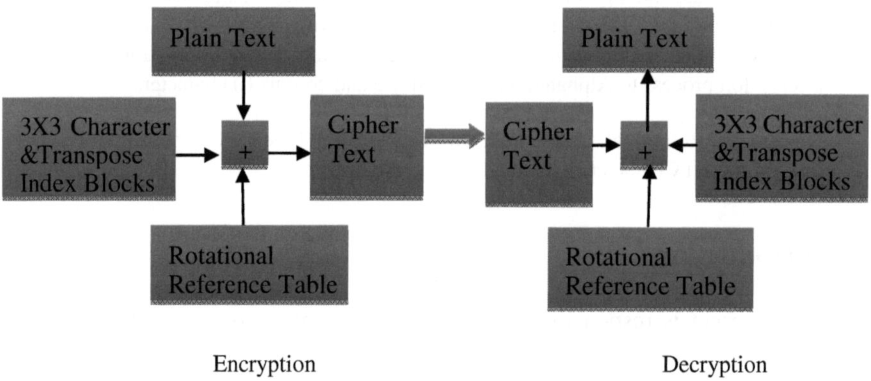

Fig. 1. Architecture of Existing System

2.1 Drawbacks of Existing System

The existing system 'Rotational shifts and Building blocks based security cipher' [5] proposed by using some simpler algebraic operations like TRANSPOSE, SHIFT. These operations are found to require lesser time complexity as compared to number generation [1] or any other logarithmic [2, 3] or exponential operations [4,6] as in other encryption techniques. As well as it has some limitations, those are explained below:

- The algorithm uses 5- Bits to represent the character, so that the table requires 2^5 =32 characters to represent the text. But here considered only 27 characters that includes ‘A’ to ‘Z’ and space.
- In the conversion process Duplicate characters are generated because it performs one rotation. To overcome this problem, it requires 5 rotations.
- The algorithm creates 3 different 3X3 blocks, in which there is no collaboration between those blocks. It effects on efficiency with respect to the time complexity, memory and lack of optimization.
- Diagonal values are not changing after transpose operation.

3 Proposed System

In this work we have proposed some limitations to the existing system and proposed a new security method to overcome the drawbacks and it helps to improve the efficiency of the existing system. Initially, this method depends on Base Index table which is used to encrypt and decrypt the message. The Figure 2 describes “Encryption process” and the Figure 3 describes the “Decryption Process”. Generation of Base Index Table is shown in section 3.1.

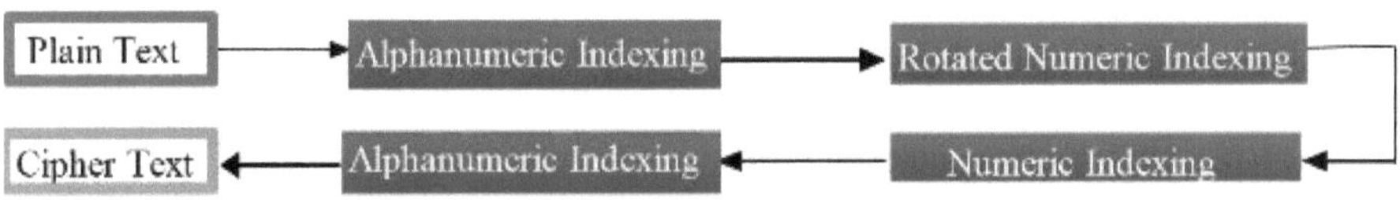

Fig. 2. Encryption Process

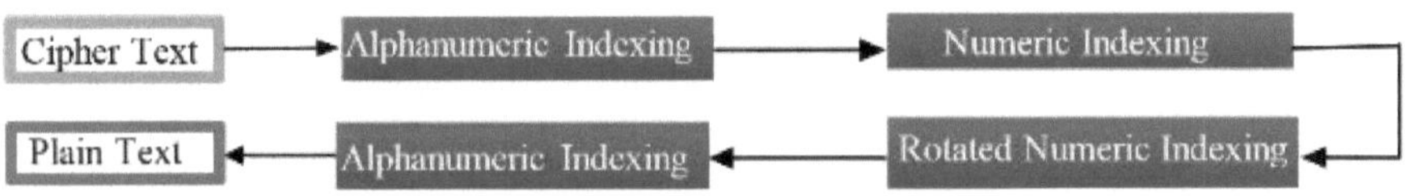

Fig. 3. Decryption Process

3.1 Base Index Table Algorithm

Step 1: Initialize "base index" with alphanumeric (A-Z, a-z, 0-9) characters, index number, and corresponding binary number. It will be stored in "**Index[64][8]**"array variable. Index[][] variable contains values shown in figure.

Index	*Binary*	*Alphanumeric*
0	*000000*	*A*
1	*000001*	*B*
2	*000010*	*C ……..*

Step 2: Rotational Shift Process:
Rotate the binary values (column wise, from 2nd to 7th column) from top-bottom if the column has odd index, bottom-top if the column has even index. The rotation should be for 6 times each column. The "rotated index" values will be stored in "Index [64] [8]" variable. Here binary values contain a column of binary values from 2nd to 7th column.

Index	*Binary*	*Alphanumeric*
46	*101110*	*A*
47	*101111*	*B*
40	*101000*	*C ……..*

Step 3: Copy Index values of Base Index (Index[row][0] column) into 2d array of string variable **"NumericIndexing[11][6]"**,as shown in Table 1.

Step 4: Apply Rotational Shift Process to the "NumericIndexing" variable, which is shown in Table 1. It should be applied to each column of "NumericIndexing" variable and should be rotated 6times. The result will be copied into the array **"Rotated NumericIndexing [11][6]"** variable as shown in Table 2.

Table 1. Numeric Indexing (NI)

46	47	40	41	42	43
4	5	6	7	16	17
18	19	28	29	30	31
24	25	26	27	20	21
22	23	0	1	2	3
12	13	14	15	8	9
10	11	36	37	38	39
48	49	50	51	60	61
62	63	56	57	58	59
52	53	54	55	32	33
34	35	44	45	Null	Null

Table 2. Rotated Numeric Indexing (RNI)

12	11	14	37	2	9
10	49	36	51	8	39
48	63	50	57	38	61
62	53	56	55	60	59
52	35	54	45	58	33
34	47	44	41	32	43
46	5	40	7	42	17
4	19	6	29	16	31
18	25	28	27	30	21
24	23	26	1	20	3
22	13	0	15	Null	Null

Table 3. Alphanumeric Indexing (AI)

A	B	C	D	E	F
G	H	I	J	K	L
M	N	O	P	Q	R
S	T	U	V	W	X
Y	Z	A	B	c	D
e	F	G	H	i	J
K	L	m	N	o	P
q	R	S	T	u	V
w	X	y	Z	0	1
2	3	4	5	6	7
8	9	,		Null	Null

Step 5: Copy the "Index[row][7]" (last column of index) into Alphanumeric Indexing[11][6]" String variable as shown in Table 3.

3.2 Encryption Algorithm

The algorithm creates cipher text from the original data by using the Encryption algorithm and with the help of Base Index table, it is shown below steps.

Input: Plain Text (Text Data) = SHABBEER
Output: Cipher text (Text Cypher) = wrellccy

Step 1: Extract each character from TextData& find the row index and col index of that character in AlphanumericIndexing variable.
Eg: Consider 1st character "S". The index of "S" in Alphanumeric Indexingis AlphanumericIndexing[3][0]. In the same way Index of "H" is AlphanumericIndexing[1][1]. Etc.....

Step 2: Now find the numeric value of each character in "RotatedNumericIndexing" variable with the help of "AlphanumericIndexing" index.
Eg: Consider numeric value of "S" is "RotatedNumericIndexing[3][0] = 62" Numeric value of "H" is "RotatedNumericIndexing[1][1] = 49". i.e., The Numeric values of "SHABBEER" is "62, 49, 12,11,11,2,2, and 61".

Step 3: Store these values into "EncryptNum" String variable.

Step4: Find each Numeric value of each value of "EncryptNum" in "Numeric Indexing" variable store its index values. Eg: The index of 62 is "NumericIndexing[8][0], The index of 49 is "NumericIndexing[7][1] Etc....

Step 5: Using the index of each value of "EncryptNum" extract Alpha numeric values.
Eg: AlphanumericIndexing[8][0] = w, AlphanumericIndexing[7][1] = r Etc.

Step 6: Store these values into TextCypher. The Cipher text of "SHABBEER" is "wrellccv".

3.3 Decryption Algorithm

The algorithm creates plain text from the cipher text by using the Decryption algorithm and with the help of Base Index table, it is shown below steps.

Input: Cipher Text = wrellccy
Output: Plain Text = SHABBEER

Step 1: Extract each character from TextCypher& find the row index and col index of that character in AlphanumericIndexing variable.
Eg: Consider 1st character "w". The index of "w" in Alphanumeric Indexingis AlphanumericIndexing[8][0]. In the same way, Index of "r" is AlphanumericIndexing[7][1]. Etc.....

Step 2: Now find the numeric value of each character in "NumericIndexing" variable with the help of "AlphanumericIndexing" index.
Eg: Consider numeric value of "w" is "NumericIndexing[8][0] = 62" Numeric value of "r" is "NumericIndexing[7][1] = 49". Etc...
i.e., The Numeric values of "wrellccv" is "62, 49, 12, 11, 11, 2, 2, and 61".

Step 3: Store these values into "DecryptNum" String variable.

Step 4: Find each Numeric value of each value of "DecryptNum" in "Rotated NumericIndexing" variable store its index values.

Eg: The index of 62 is "RotatedNumericIndexing[3][0] index of 49 is "RotatedNumericIndexing[1][1] Etc.... Step 5: Using the index of each value of "DecryptNum" extract Alphanumeric values.

Eg: AlphanumericIndexing[3][0] = S, AlphanumericIndexing[1][1] = H Etc. The Decrypted textof "wrellccv" is "SHABBEER".

3.4 Solutions to the Drawbacks of Existing System

This system gives the solutions to the drawbacks of the existing system [5]. The proposed solutions are as follows.

- The algorithm uses 6- Bit character representation, so that $64(2^6)$ characters are considered. This method strengthens the cipher text. It includes A-Z, a-z, and 0-9 along with special characters space and comma (,).
- The algorithm performs 6 rotations to reduce the duplicated characters in the cipher text. This process improves the security level.
- The algorithm produces a single &2-dimensional array with the size of (11X6) instead of 3 different 3x3 blocks to improve the optimization, efficiency and to reduce the time complexity and memory size.

4 Results and Analysis

The Figure 4 shows the Java Application for Encryption or Decryption process, which is proposed in this paper. The Application's Ribbon shows the tabs (Encryption / Decryption), contains the 4 commands to open text/cipher text, to

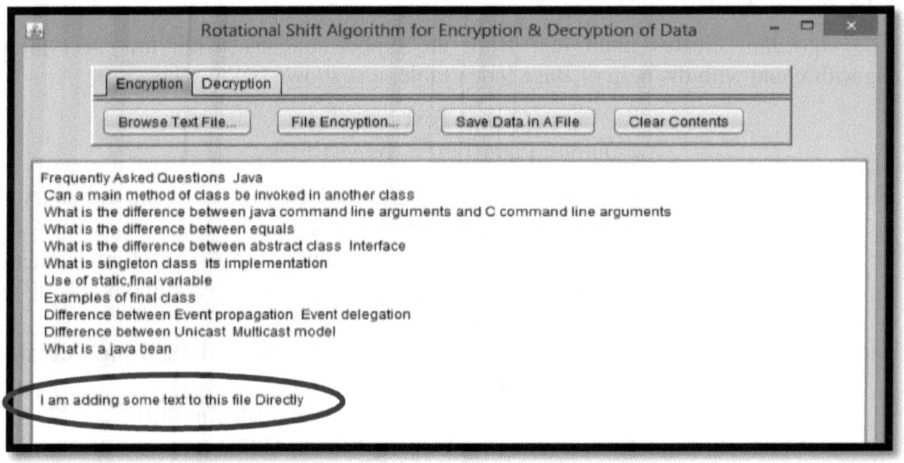

Fig. 4. Application's Ribbon with tabs and commands

encrypt or decrypt the data, to save the result into hard disk, to clear the data from application. You can also append the data by typing into the application.

Figure 5 and Figure 6 show the results of the encryption and decryption using proposed technique for security.

Fig. 5. Generating Cipher text by clicking Encryption command

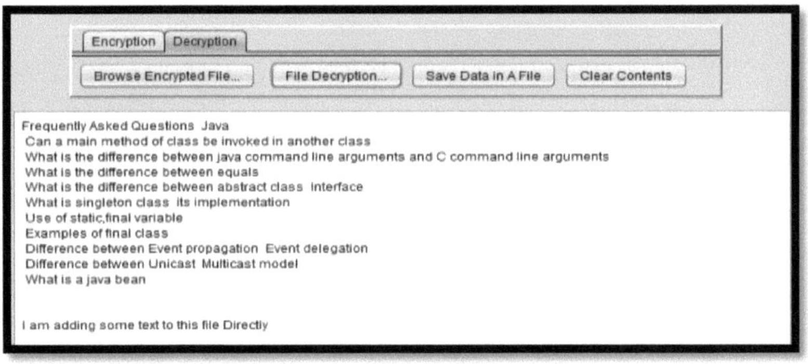

Fig. 6. Generating Plain text by clicking Decryption command

5 Conclusion

Attacks on the information are a serious threat. Many Cryptography and Steganography techniques have been devised, mainly focusing on the invisibility of the original data and robustness against various signal manipulation techniques and hostile attacks. A design of effective and efficient detection and response strategy is must. This proposed security approach solutions are successfully controls the drawbacks of the existing system. Compared to existing method it improve the optimization, efficiency and reduce the time complexity, duplication in the cipher text and memory size Further research on security services is needed to refine the system.

References

1. Zibideh, W.Y.: Modified-DES encryption algorithm with improved BER performance in wireless communication. In: Radio and Wireless Symposium (RWS), pp. 219–222. IEEE (2011)
2. Godavarty, V.K.: UsingQuasigroups for Generating Pseudorandom Numbers, Arxiv preprint arXiv:1112.1048 - arxiv.org (2011)
3. Ayushi: A Symmetric Key Cryptographic Algorithm. Int. Journal of Computer Applications 1(5) (2010)
4. Pointcheval, D.: Asymmetric cryptography ad practical security. Journal of Telecommunication and Information Technology, 42–56 (2002)
5. Rupa, C., Sudha Kishore, R., Avadhani, P.S.: An Advanced Authentication using Rotational Shifts. Int. Journal of Advanced Engineering and Technology 5 (2012)
6. Rubesh Anand, P.M., Bajpai, G., Bhaskar, V.: Real-Time Symmetric cryptography using Quaternion Julia Set. International Journal of Computer Science and network Security 9(3), 20–26 (2009)
7. Xiang, L., et al.: A Secure Steganographic Method via Multiple Choice Questions. Information Journal 10(5), 992–1000 (2011)

Privacy Preserving Data Mining

D. Aruna Kumari[1], K. Rajasekhara Rao[2], and M.Suman[1]

[1] Department of Electronics and Computer Engineering,
CSI Life Member K.L.University, Vaddeswaram, Guntur
{aruna_D,suman.maloji}@kluniversity.in
[2] Department of CSE, CSI life member
K.L.University, Vaddeswaram, Guntur
krr@kluniversity.in

Abstract. Now a day's detailed personal data from large data bases is regularly collected and analyzed by many applications with data mining, some times sharing of these data is beneficial to the application users. On one hand it is an important asset to business organizations and governments for decision making at the same time analysing such data opens treats to privacy if not done properly. This work aims to reveal the information by protecting sensitive data. Various methods including Randomization, k-anonymity and data hiding have been suggested for the same. In this work, a novel technique is suggested that makes use of LBG design algorithm to preserve the privacy of data along with compression of data. Quantization will be performed on training data it will produce transformed data set. It provides individual privacy while allowing extraction of useful knowledge from data, Hence privacy is preserved. Bit Error rate and Distortion measures are used to analyze the accuracy of compressed data.

Keywords: Privacy, Vector quantization, lbg.

1 Introduction

Data Mining is a technique that extracts hidden predictive information from large volumes of data bases. Data mining tools predict future trends and behaviours, allowing businesses to make proactive, knowledge-driven decisions. With the amount of data doubling each year, more data is gathered and data mining is becoming an increasingly important tool to transform this data into information[1][2].

Huge volumes of detailed personal data are regularly collected and analyzed by applications. Such Data include shopping habits, criminal records, medical history, credit records, among others. Analyzing such data opens new threats to privacy .Privacy preserving data mining (PPDM) is one of the important area of data mining that aims to provide security for secret information from unsolicited or unsanctioned disclosure. Data mining techniques analyzes and predicts useful information. The concept of privacy preserving data mining is primarily concerned with protecting secret data against unsolicited access. It is important because now a day's Treat to privacy is becoming real since data mining techniques are able to predict high sensitive knowledge from huge volumes of data [1].

S.C. Satapathy et al. (eds.), *ICT and Critical Infrastructure: Proceedings of the 48th Annual Convention of CSI - Volume II*, Advances in Intelligent Systems and Computing 249,
DOI: 10.1007/978-3-319-03095-1_55, © Springer International Publishing Switzerland 2014

- What data mining causes is social and ethical problem by revealing the data which should require privacy?
- Providing security to sensitive data against unauthorized access has been a long term goal for the database security research community and for the government statistical agencies.
- Hence, the security issue has become, recently, a much more important area of research in data mining.

Therefore, in recent years, privacy-preserving data mining has been studied extensively

This work provides one of the solution for privacy preserving data mining (central data warehouse not on distributed databases) Performance is measured in terms of accuracy of data mining result and privacy of sensitive data.

2 Literature Review

The term "privacy preserving data mining" was first introduced in papers by Agrawal & Srikant,2000, they worked on Randomization. Lindell and Pinkas (2000) introduced a cryptographic protocol for decision tree construction over a dataset horizontally partitioned between two parties.A Protocol for secure association rules (Kantarcioğlu and Clifton, 2004a), k-means clustering (Lin *et al.*, 2005), k-nn classifiers (Kantarcioğlu and Clifton, 2004b).Again, secure protocols for the vertically partitioned case have been developed for mining association rules (Vaidya and Clifton, 2002), decision trees (Du and Zhan, 2002) and k-means clusters (Jagannathan and Wright, 2005).

Other areas that influence the development of PPDM include cryptography and secure multiparty computation (Goldreich, 2004) (Stinson, 2006), database query auditing for disclosure detection and prevention (Kleinberg et al. 2000) (Dinur & Nissim, 2003) (Kenthapadi et al. 2005), database privacy and policy enforcement (Agrawal et al. 2002) (Aggarwal et al. 2004), database security (Castano et al. 1995), and of course, specific application domains[3][4].

Now a day's privacy preserving data mining is becoming one of the focusing area because data mining predicts more valuable information that may be beneficial to the business, education systems, medical field, political ,…etc.

3 Methodology

3.1 Vector Quantization:

The Ancient and best example of Quantization is rounding off, It was first introduced and implemented by Sheppard for many applications.Number 'S' can be rounded off to the nearest integer, say Q(s), with quantization error e=Q(s)-S.In reality, Quantization used for data compression. The key point of quantization is to divide large set of points (vectors) into groups(or regions) having approximately the equal number of points closest to them. Each group(or region) is represented by its centroid, as in k-means clustering and some other clustering algorithms.

Vector Quantization (VQ) is an efficient and simple approach for data compression. Since it is simple and easy to implement, VQ has been widely used in different applications, such as pattern recognition, image compression, speech recognition, face detection and so on [11].

Vector quantization (VQ) is generally used for data compression. In previous days, the design methodology of a vector quantizer (VQ) is treated as a big problem in terms of the need for multi-dimensional integration. Linde, Buzo, and Gray (LBG) Introduced an algorithm for Vector quantization design based on training sequence. A VQ that is designed based on this algorithm are referred as LBG-VQ.

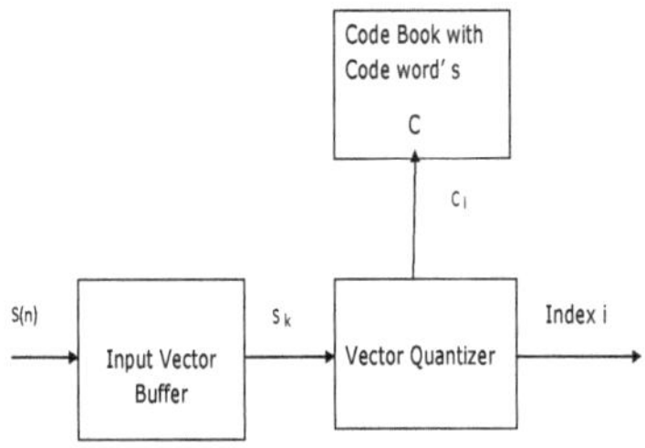

Fig. 1. Block diagram of Vector Quantizer

The central component of a Vector Quantizer (VQ) is a codebook C of size N x k, which maps the k-dimensional space R^k onto the reproduction vectors (also called code vectors or code words):

$$Q : R^K \rightarrow C, \quad C=(Y_1, Y_2, \ldots\ldots Y_N)^T, \quad Y_i \in R^K$$

The codebook can be thought of as a finite list of vectors, y_i: i = 1,2…N.

The codebook vectors are preselected through a clustering or training process to represent the training data. In the standard approach to VQ, the encoder minimizes the distortion D to give the optimal estimated vector $\widehat{X}_t$;

$$\widehat{X}_t = \min_{Y_i \in C} D(X_t, Y_i)$$

This is referred to as nearest neighbor encoding. The code rate or simply the rate of a vector quantizer in bits per component is thus

$$r = \frac{\log_2 N}{K}$$

This measures the number of bits per vector component used to represent the input vector and gives an indication of the accuracy or precision that is achievable with the vector quantizer if the codebook is well designed. Since $N = 2^{rk}$, and thus both the encoding search complexity and codebook storage size grow exponentially with dimension k and rate r. A practical limitation is that codebook design algorithms, such as the Generalized Lloyd Algorithm (GLA) yield only locally optimized codebooks. More recent methods, such as deterministic annealing and genetic optimization promise to overcome this drawback at the expense of greater computational requirements[12][13].

A VQ is nothing more than an approximator. The idea is similar to that of`rounding-off" (say to the nearest integer).

An example of a 1-dimensional VQ is shown below

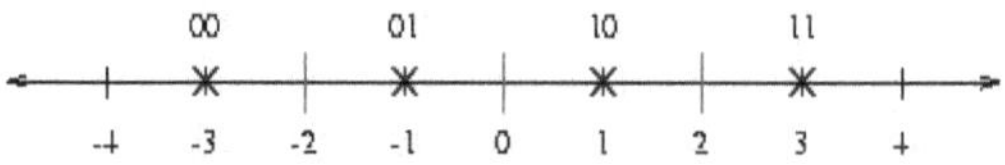

Fig. 2. 1-dimensional Vector Quantizer

numbers less than -2 are rounded (i.e, approximated) to -3. Numbers between -2 and 0 are rounded to –1. Numbers between 0 and 2 are rounded to +1 and numbers which are greater than 2 is replaced by +3. The approximate values are uniquely represented by 2 bits.

This is a one -dimensional, 2-bit VQ. The rate is 2 bits/dimension.

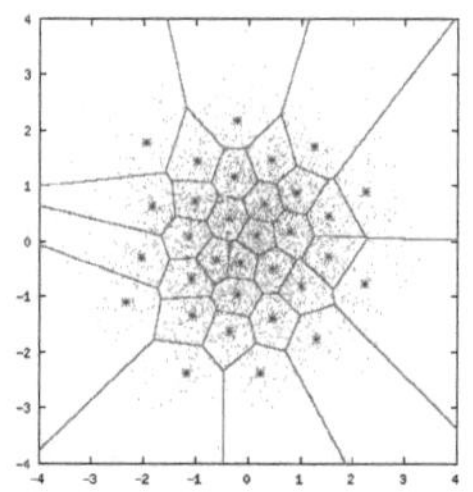

Fig. 3. Two-dimensional Vector Quantizer

An example of a Two-dimensional VQ is shown above .Here, pair of numbers falling in a particular region is grouped by a red star associated with that region. We have 16 regions and 16 red stars. Each of the regions and red stars can be uniquely represented by 4 bits. Thus, this is a two-dimensional, 4-bit VQ. The rate also known as 2 bits/dimension. In the two examples mentioned here, the stars are called code vectors or code words, and the blue border regions are called encoding regions. The codebook is the set of all code vectors. It contains a set of code words, also referred to as centroids of all clusters. The partition of the space is the set of all encoding regions.

Optimality Criteria

If C and P are a solution to the above minimization problem, then it must satisfy the following two criteria.

Nearest Neighbor Condition

$$S_n = \{X : \|X\text{-}C_n\|^2 \le \|X\text{-}C_n^{'}\|^2 \ \forall n^{'} = 1, 2, ..., N\} \qquad (1)$$

The condition says that the encoding region S_n should consist of all vectors that are closer to C_n than any of the other code vectors. For those vectors lying on the boundary any tie breaking procedure will do.

Centroid Condition

$$C_n = \frac{\sum_{x_m \in s_n} X_m}{\sum_{x_m \in s_n} 1} \tag{2}$$

This condition says that the code vector C_n should be average of all those training vectors that are in encoding region S_n . In implementation, one should ensure that at least one training vector belongs to each encoding region (so that the denominator in the above equation is never 0).

4 Code Book Generation Using LBG

1. Initially the codebook generation requires a Training sequence which is the input to LBG algorithm. The training sequence is obtained from UCI Data repositories water plant treatment data set[23]
2. Let „R" be the region of the training sequence.
3. Generate an initial codebook from the training sequece , now it will be the centroid or mean of the training dataset and let the initial codebook be „". C
4. Split the initial code book in to C_n^- and C_n^+ Where $C_n^+ = C(1+\in)$ And $C_n^- = C(1-\in)$ $\in$ =0.01 is the minimum error to be obtained between old and new codewords.
5. Compute the difference between the training sequence and each of the codeword"s C_n^- and C_n^+ and let the difference be D^1
6. Split the training sequence into two regions R1 and R2 depending on the difference „D" between the training sequence and the codeword"s C_n^- and C_n^+ . The training vectors closer to C_n^+ falls in the region R1 and the training vectors closer to C_n^- falls in the region R2.
7. Let the training vectors falling in the region R1 be TV1 and the training sequence vectors falling in the region R2 be TV2.
8. Obtain the new centroid or mean for TV1 and TV2. Let the new centroids be CR1 and CR2.
9. Replace the old centroids C_n^+ and C_n^- by the new centroids CR1 and CR2
10. Compute the difference between the training sequence and the new centroids CR1 and CR2 and let the difference be D 1 .

11. Repeat steps 5 to 10 until $\frac{D^1 - D}{D} < \in$

12. Repeat steps 4 to 11 till the required number of codewords in the codebook are obtained. where $N=2^b$ represents the number of codewords in the codebook and „ b“ represents the number of bits used for codebook generation, D represents the difference between the training sequence and the old codewords and D^1 represents the difference between the training sequence and the new codewords[12][13].

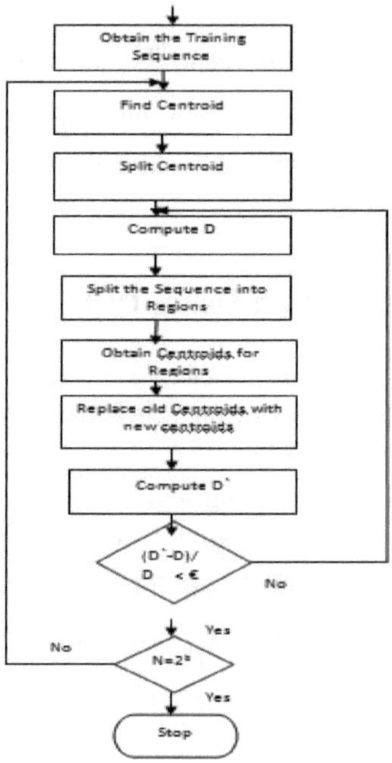

Fig. 4. Flow chart representation of lbg

4.1 Steps Involved in PPDM Using Vector Quantization

As stated in [11][12][7] the design of a Vector Quantization-based system mainly consists of three steps:

1. Constructing a codebook from a set of training samples;
2. Encoding the original data with the indices of the nearest code vectors in the codebook;
3. Using an index representation to reconstruct the signal by looking up in the codebook.

Our privacy preserving data mining requires first two steps. Because original data should not be revealed once the encoding is performed, transformed data set represents approximate data not exact and original data thus preserving privacy.

Step 1: Construct code book using lbg by taking data from training sequence call it as D

Step2 :Code book contains centroids of all k-clusters, call it as D'

Step3: Quantization performed i.e, Data set D is mapped into new data set D' by replacing each of the point with the point which fall nearest to it in its codebook. That is the point is replaced by the cluster centroid in which it falls.

Step4: clustering is performed on both data sets and results are compared.

5 Results

We have implemented above LBG algorithm using Matlab Software, and tested the results. In the output screen shots Blue line represents original data and red line represents Codebook that is compressed form of original data , hence it does not reveal the complete original information and it will reveal only cluster centroids

Results for Random Data

Input 1 :- [10 20 30 22 30 12 32 13 11 40 34 45 35]
Input 2:- Dataset [1:0.5:100]

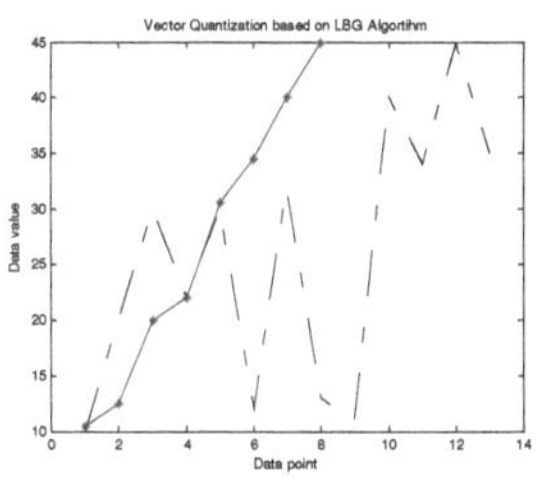

Fig. 5. Output for Input 1

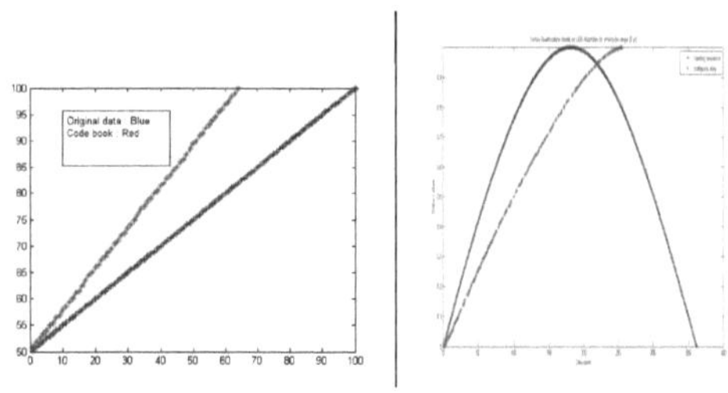

Fig. 6. Output for Input 2

Comparison Between Original data and Code book for Input1
mean_codebook = 24.3958; var_codebook = 93.1176; sd_codebook = 9.6497
mean_input = 24.1538; var_input = 97.2071; sd_input = 9.8594

6 Conclusions and Future Work

This Work gives a different approach of using vector quantization for privacy preserving data mining. This work shows analytically and experimentally that Privacy-Preserving data mining is to some extent possible using vector quantization approach. To support this work, water treatment dataset available on UCI Machine Learning Repository was taken and performed experiments on it. Performance is also evaluated by taking into account two important parameter: distortion and Fmeasure (quality of data mining results).

As future work new and effective quantization method can be used rather than LBG approach that we have used. K nearest neighbor approach is one of the approach which can give better result.

References

1. Agrawal, R., Srikant, R.: Privacy Preserving Data Mining. In: Proc. of ACM SIGMOD Conference on Management of Data (SIGMOD 2000), Dallas, TX (2000)
2. Evfimievski, A., Grandison, T.: Privacy Preserving Data Mining. IBM Almaden Research Center 650 Harry Road, San Jose, California 95120, USA
3. Agarwal Charu, C., Yu Philip, S.: Privacy Preserving Data Mining: Models and Algorithms. Springer, New York (2008)
4. Oliveira, S.R.M., Zaiane Osmar, R.: A Privacy-Preserving Clustering Approach Toward Secure and Effective Data Analysis for Business Collaboration. In: Proceedings of the International Workshop on Privacy and Security Aspects of Data Mining in Conjunction with ICDM 2004, Brighton, UK (November 2004)
5. UCI Repository of machine learning databases, University of California, Irvine, `http://archive.ics.uci.edu/ml/`
6. Wikipedia. Data mining, `http://en.wikipedia.org/wiki/Datamining`
7. Sinha, B.K.: Privacy preserving clustering in data mining
8. Tsai, C.W., Lee, C.Y., Chiang, M.C., Yang, C.S.: A Fast VQ Codebook Generation Algorithm via Pattern Reduction. Pattern Recognition Letters 30, 653–660 (2009)
9. Somasundaram, K., Vimala, S.: A Novel Codebook Initialization Technique for Generalized Lloyd Algorithm using Cluster Density. International Journal on Computer Science and Engineering 2(5), 1807–1809 (2010)
10. Somasundaram, K., Vimala, S.: Codebook Generation for Vector Quantization with Edge Features. CiiT International Journal of Digital Image Processing 2(7), 194–198 (2010)
11. Verykios, V.S., Bertino, E., Fovino, I.N.: State-of-the-art in Privacy Preserving Data Mining. SIGMOD Record 33(1) (March 2004)
12. Madhavi Latha, M., Satya Sai Ram, M., Siddaiah, P.: Multi Switched Split Vector Quantizer. International Journal of Computer, Information and Systems Science and Engineering 2(1)
13. Madhavi Latha, M., Satya Sai Ram, M., Siddaiah, P.: Multi Switched Split Vector Quantization. Proceedings of World Academy of Science, Engineering and Technology 27 (February 2008) ISSN 1307-6884
14. Agarwal Charu, C., Yu Philip, S.: Privacy Preserving Data Mining: Models and Algorithms. Springer, New York (2008)
15. Atallah, M., Elmagarmid, A., Ibrahim, M., Bertino, E., Verykios, V.: Disclosure limitation of sensitive rules. In: Workshop on Knowledge and Data Engineering Exchange (1999)

Enhanced Trusted Third Party for Cyber Security in Multi Cloud Storage

Naresh Sammeta[1], R. Jagadeesh Kannan[2], and Latha Parthiban[3]

[1] Computer Science & Engg, RMKCET, Chennai
Tamilnadu, India
samnaresh@gmail.com
[2] Computer Science & Engg, RMKEC, Chennai
Tamilnadu, India
rjk1979@yahoo.com
[3] Computer Science, Pondicherry University Community College,
Pondicherry, India
lathaparthiban@yahoo.com

Abstract. Cloud Computing offers an business model and it is tempting for companies to delegate their IT services, as well as data, to the Cloud. But in Cloud environment, lacking of cyber security users may suffer a serious data loss without any compensation for they have lost all their control on their data. Cyber security is the body of technologies and it is designed to protect networks, computers, programs and data from attack, damage or unauthorized access. Security audit is an important solution enabling trace back and analysis of any activities including data accesses, security breaches, application activities, and so on. Provable data possession (PDP) is an audit technique for ensuring the security of data in storage outsourcing. However, this existing audit schemes have focused on static data and the fact that users no longer have physical possession of the possibly large size of outsourced data makes the data integrity protection is very challenging task. For the cyber security we present a novel way implementation of a Trust Enhanced Third Party Auditor (TETPA), a trusted and easy-to- use auditor for Cloud environment. TETPA enables the Cloud Service Providers' accountability, and protects the Cloud users' benefits. Moreover our audit service is using for dynamic integrity verification in multi cloud storage. This scheme is based on the techniques, fragment structure, random sampling and index-hash table, Zero-Knowledge supporting provable updates to outsourced data and timely anomaly detection.

Keywords: Storage Security, Provable Data Possession, Audit Service, Zero-Knowledge.

1 Introduction

In recent years, cloud storage service has become a faster profit growth point by providing a comparably low-cost, scalable, position-independent platform for clients' data. Since cloud computing environment is constructed based on open architectures

S.C. Satapathy et al. (eds.), *ICT and Critical Infrastructure: Proceedings of the 48th Annual Convention of CSI - Volume II*, Advances in Intelligent Systems and Computing 249,
DOI: 10.1007/978-3-319-03095-1_56, © Springer International Publishing Switzerland 2014

and interfaces, it has the capability to incorporate multiple internal and/or external cloud services together to provide high interoperability. We call such a distributed cloud environment as a *multi-Cloud* (or *hybrid cloud*). Often, by using virtual infrastructure management (VIM) [1], a multi-cloud allows clients to easily access his/her resources remotely through interfaces such as Web services provided by Amazon EC2. Outsourcing storage prompts a number of interesting challenges. One problem is to verify that the server continually and faithfully stores the entire file. The server is untrusted in terms of both security and reliability. These security risks come from the following reasons: first, the cloud infrastructures are much more powerful and reliable than personal computing devices, but they are still susceptible to internal threats (e.g., via virtual machine) and external threats (e.g., via system holes) that can damage data integrity. Second, for the benefits of possession, there exist various motivations for cloud service providers (CSP) to behave unfaithfully towards the cloud users. Furthermore, disputes occasionally suffer from the lack of trust on CSP since the data changes may not be timely known by the cloud users, even if these disputes may result from the users' own improper operations. Therefore, it is necessary for cloud service providers to offer an efficient audit service to check the integrity and availability of stored data. In this paper, we introduce a dynamic audit service for integrity verification of untrusted and outsourced storages. Constructed on interactive proof system (IPS) with the zero-knowledge property, our audit service can provide public auditability without downloading raw data and protect privacy of the data. Also, our audit system can support dynamic data operations and timely anomaly detection with the help of several effective techniques, such as fragment structure, random sampling, and index-hash table. We also propose an efficient approach based on probabilistic query and periodic verification for improving the performance of audit services. A proof-of concept prototype is also implemented to evaluate the feasibility and viability of our proposed approaches. Another major concern is the security issue of dynamic data operations for public audit services. In clouds, one of the core design principles is to provide dynamic scalability for various applications. This means that remotely stored data might be not only accessed by the clients, but also dynamically updated by them, for instance, through block operations such as modification, deletion and insertion. However, these operations may raise security issues in most of existing schemes, e.g., the forgery of the verification metadata (called as tags) generated by data owners and the leakage of the user's secret key. Hence, it is crucial to develop a more efficient and secure mechanism for dynamic audit services, in which a potential adversary's advantage through dynamic data operations should be prohibited.

1.1 Existing Techniques

The traditional cryptographic technologies for data integrity and availability, based on Hash functions and signature schemes cannot support the outsourced data without a local copy of data. It is evidently impractical for a cloud storage service to download the whole data for data validation due to the expensiveness of communication, especially, for large-size files. Recently, several PDP(Provable Data Possession) schemes are proposed to address this issue. In fact, PDP is essentially an interactive proof between a CSP and a client because the client makes a false/true decision for data possession without downloading data. Existing PDP schemes mainly focus on integrity verification issues at untrusted stores in public clouds, but these schemes are not suitable for a hybrid cloud environment since they were originally constructed based on a two-party interactive proof system. For a hybrid cloud, these schemes can only be used in a trivial way: clients must invoke them repeatedly to check the integrity of data stored in each single cloud. This means that clients must know the exact position of each data block in outsourced data. Moreover, this process will consume higher communication bandwidth and computation costs at client sides. Thus, it is of utmost necessary to construct an efficient verification scheme with collaborative features for hybrid clouds. Solving the problems will help improve the quality of PDP services, which can not only timely detect abnormality, but also take up less resources, or rationally allocate resources. Hence, a new PDP scheme is desirable to accommodate these application requirements from hybrid clouds. Even though existing PDP schemes have addressed various aspects such as public verifiability, dynamics, scalability, and privacy preservation, we still need a careful consideration to the following attacks in data leakage attack and tag forgery attack.

2 Proposed Architecture and Techniques

The proposed architecture TEPTA is divided into two domains, the public domain and the inner security domain. These two domains construct the TETPA's perimeter, and all log files and information should never go out of this perimeter unless there are requests for evidence seeking.

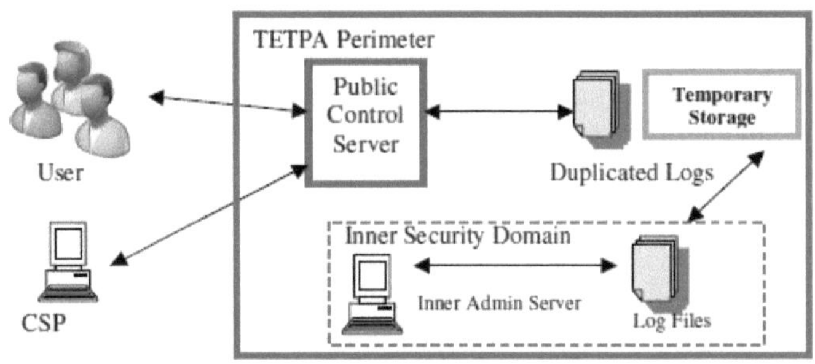

Fig. 1. Architecture of TEPTA

2.1 Public Domain

The public domain processes basic audit operations. It contains two main components.

Public Control Server: In this server is equipped a Trusted Platform Module (TPM) chip t51 or other security chips which implement the TPM specification for remote attestation. With TPM, users and CSPs can be protected from cheating attacks. Key and certification management mechanisms have been included in the public control server. These Mechanisms insure that the uses' behavior information would not be intercepted, and the data integrity can also be guaranteed. *Temporary Storage:* In this storage space there are the duplications of the log files. These duplications are the data that can be access and modified from the outside of TETPA's perimeter. The messengers and duplications are all encrypted and signed so the TPA itself cannot read and modify it neither. Otherwise, they will be discarded and notify the CSPs or the users for potential violation.

2.2 Inner Security Domain

The inner security domain maintains the unique and effect log files. Log files are created according to the user-CSP relationships. Namely, when a new user contract with a CSPs for services and want to use the TETPA for benefits, TETPA *would* create a new log file to record the audited information. After the log file is created all the behaviors between user and CSP would use the ever-existed file.

2.3 Achieving Integrity Verification

Although existing techniques offer a publicly accessible remote interface for checking and managing the tremendous amount of data, the majority of existing PDP schemes are incapable to satisfy the inherent requirements from multiple clouds in terms of communication and computation costs. To address this problem, we consider a multi-cloud storage service as illustrated in Fig.2. In this architecture, we consider that a data storage service involves four entities: data owner (DO), who has a large amount of data to be stored in the cloud; cloud service provider (CSP), who provides data storage service and has enough storage space and computation resources; third party auditor (TPA), who has capabilities to manage or monitor the outsourced data under the delegation of data owner; and authorized applications (AA), who have the right to access and manipulate the stored data. Finally, application users can enjoy various cloud application services via these authorized applications.

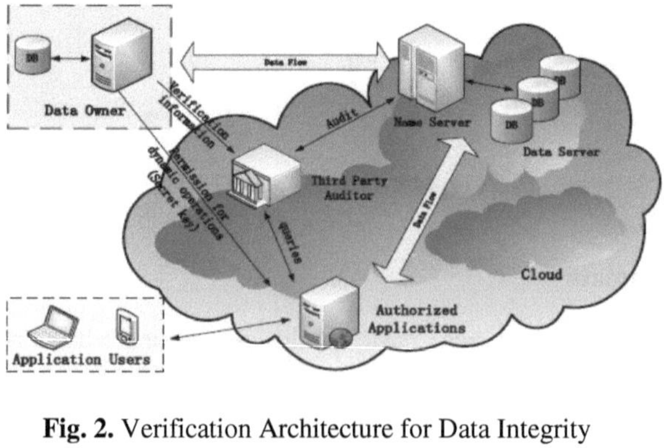

Fig. 2. Verification Architecture for Data Integrity

2.4 Fragment Structure and Secure Tags

To maximize the storage efficiency and audit performance, our audit system introduces a general fragment structure for outsourced storages.

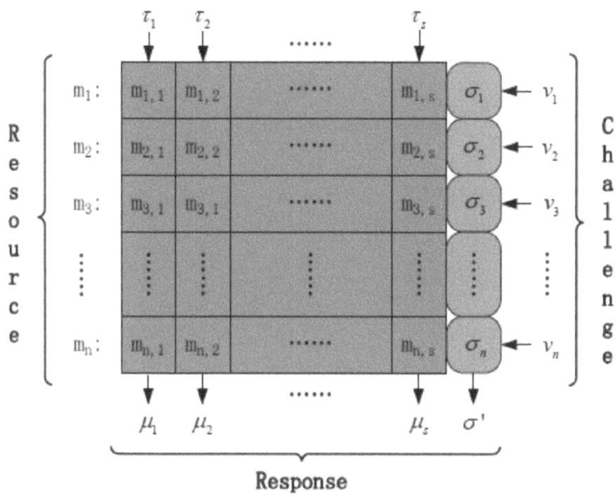

Fig. 3. Fragment Structure and Sampling Audit

An instance for this framework which is used in our approach. An outsourced file F is split into n blocks $\{m_1, m_2, \cdots, m_n\}$, and each block m_i is split into s sectors $\{m_i,1, m_i,2, \cdots, m_i,s\}$. The fragment framework in Fig.3. consists of n block-tag pair $(mi, \sigma i)$, where σi is a signature tag of a block mi generated by some secrets $\tau = (\tau_1, \tau_2, \cdots, \tau_s)$. We can use such tags and corresponding data to construct a response in terms of the TPA's challenges in the verification protocol, such that this response can be verified without raw data. If a tag is unforgeable by anyone except the original signer, we call it a *secure tag*. Finally, these block-tag pairs are stored in CSP and the encrypted secrets τ are in TPA. These schemes, built from collision-resistance hash functions and a random oracle model, support the features of scalability, performance and security.

2.5 Index-Hash Table

In order to support dynamic data operations, we introduce a simple index-hash table to record the changes of file blocks, as well as generate the hash value of each block in the verification process. The structure of our index-hash table in fig.4. is similar to that of file block allocation table in file systems. Generally, the index-hash table χ consists of serial number, block number, version number, and random integer. Note that we must assure all records in the index-hash table differ from one another to prevent the forgery of data blocks and tags. In addition to recording data changes, each record χ_i in the table is used to generate a unique hash value, which in turn is used for the construction of a signature tag σi by the secret key s_k. The relationship between χ_i and σ_i must be cryptographically secure, and we make use of it to design our verification protocol. Although the index-hash table may increase the complexity of an audit system, it provides a higher assurance to monitor the behavior of an untrusted CSP, as well as valuable evidence for computer forensics, due to the reason that anyone cannot forge the valid χ_i (in TPA) and σ_i (in CSP) without the secret key s_k.

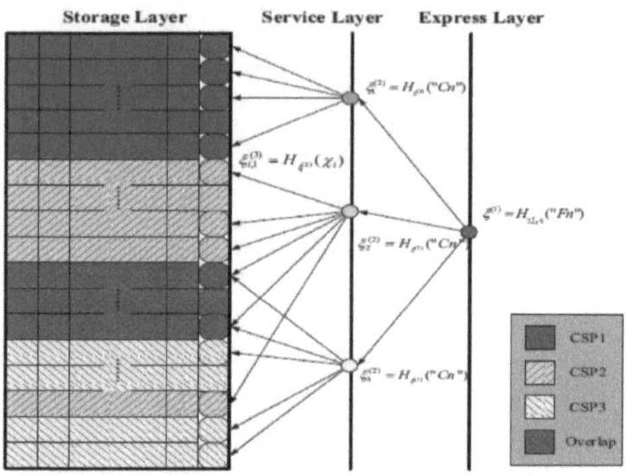

Fig. 4. Index-Hash Hierarchy for CPDP Model

2.6 Proposed Dynamic Audit Scheme

According to the dynamic audit scheme architecture in fig.5. four different network entities can be identified as follows: the verifier (V), third party auditor (TPA), the organizer (Q), and some cloud service providers (CSP's). The organizer is an entity that directly contacts with the verifier. Moreover it can initiate and organize the verification process. Often, the organizer is an independent server or a certain CSP in P. In our scheme, the verification is performed by a 5-move interactive proof protocol as follows:

- the organizer initiates the protocol and sends a commitment to the verifier;
- the verifier returns a challenge set of random index-coefficient pairs Q to the organizer;
- the organizer relays them into each P_i in P according to the exact position of each data block;
- each P_i returns its response of challenge to the organizer;
- the organizer synthesizes a final response from these responses and sends it to the verifier.

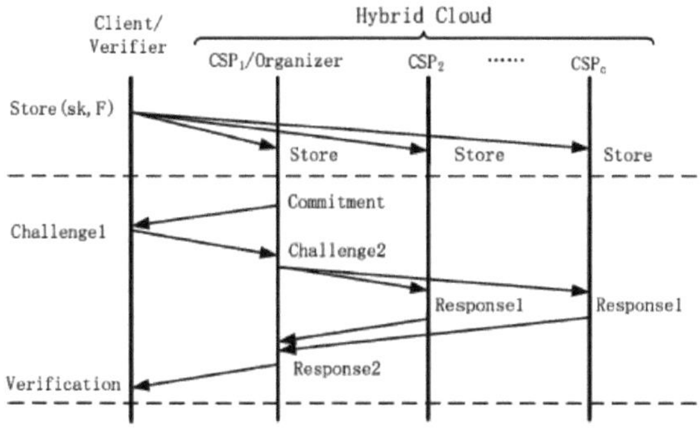

Fig. 5. Dynamic Audit Scheme

The above process would guarantee that the verifier accesses files without knowing on which CSP or in what geographical locations their files reside.

Setup Phase

We start with a database D divided into y blocks, such as : $D = D_{1,} D_{2,} D_y$. We want to be able to challenge storage *SER* t times. We make use of a pseudo-random function(*PRF*), f, and a pseudo-random permutation (*PRP*) g with the following parameters: In our example, $l = \log y$ since we use g to exchange index. The output of f is used to generate the key for g and $c = \log t$ (t as a challenge to the number of *SER*) We note that using a standard block cipher can be f and g generated, such as *AES* . In this case, $L = 128$. In setup phase, we will use the pseudo-random function f and two k -bit master secret keys W and Z. The key W is used to generate session permutation keys while Z is used to generate the current challenges

$$f:\{0,1\}^c \times \{0,1\}^k \rightarrow \{0,1\}^L$$

$$g:\{0,1\}^l \times \{0,1\}^L \rightarrow \{0,1\}^l$$

Audit Phase

When owners want to save i_{th} the verification server get proof of, the first owner to recalculate the tag key k_i and c_i .For example, setup phase for step 1, from this, Only need to encryption save files owner public key p_k and private key s_k and a verification of the master key W , Z , K and the current tags i . Then the owner send k_i and c_i to the *SER* (such as the verification phase of the algorithm, step 2.. When the server receives the owner of the message, then the calculated z value:

$$z = \{H(c_i, M[g_{k_i}(1)], \ldots, M[g_{k_i}(r)])\} \bmod N$$

Zero-Knowledge Property of Verification

The CPDP construction is in essence a Multi-Prover Zero-knowledge Proof (MP-ZKP) system [10], which can be considered as an extension of the notion of an interactive proof system (IPS). Roughly speaking in the scenario of MP-ZKP, a polynomial-time bounded verifier interacts with several provers whose computational powers are unlimited. In which every cheating verifier has a simulator that can produce a transcript that "looks like" an interaction between a honest prover and a cheating verifier, we can prove our CPDP construction has Zero-knowledge property.

The verification protocol $Proof(\mathcal{P}, V)$ in CPDP scheme is a computational zero-knowledge system under a simulator model, that is, for every probabilistic polynomial-time interactive. Zero-knowledge is a property that achieves the CSPs' robustness against attempts to gain knowledge by interacting with them.

3 Conclusion

In this paper, we proposed TETPA, a case for trusted and practical auditor in Cloud environment. Our TETPA enables a trustworthy auditing to the users and CSPs', behaviors, and make the CSPs' more accountable. With the TETPA, users can be protected from non-sense data losses. We also presented an efficient method for periodic sampling audit to enhance the performance of third party auditors and storage service providers. Our experiments showed that our solution has a small, constant amount of overhead, which minimizes computation and communication costs. Furthermore, we optimized the probabilistic query and periodic verification to improve the audit performance. As part of future work, we would extend our work to explore more effective CPDP constructions. First, from our experiments we found that the performance of CPDP scheme, especially for large files, is affected by the bilinear mapping operations due to its high complexity. To solve this problem, RSA based constructions may be a better choice, but this is still a challenging task because the existing RSA based schemes have too many restrictions on the performance and security.

References

1. Sotomayor, B., Montero, R.S., Llorente, I.M., Foster, I.T.: Virtual infrastructure management in private and hybrid clouds. IEEE Internet Computing 13(5), 14–22 (2009)
2. Ateniese, G., Burns, R.C., Curtmola, R., Herring, J., Kissner, L., Peterson, Z.N.J., Song, D.X.: Provable data possession at untrusted stores. In: Ning, P., di Vimercati, S.D.C., Syverson, P.F. (eds.) ACM, pp. 598–609. ACM (2007)
3. Juels, A., K. Jr., B.S.: Pors: proofs of retrievability for large files. In: Ning, P., di Vimercati, S.D.C., Syverson, P.F. (eds.) ACM Conference on Computer and Communications Security, pp. 584–597. ACM (2007)
4. Ateniese, G., Pietro, R.D., Mancini, L.V., Tsudik, G.: Scalable and efficient provable data possession. In: Proceedings of the 4th International Conference on Security and Privacy in Communication Netowrks, SecureComm, pp. 1–10 (2008)

5. Erway, C.C., Kupc¸, A., Papamanthou, C., Tamassia, R.: Dynamic provable data possession. In: Al-Shaer, E., Jha, S., Keromytis, A.D. (eds.) ACM Conference on Computer and Communications Security, pp. 213–222. ACM (2009)
6. Shacham, H., Waters, B.: Compact proofs of retrievability. In: Pieprzyk, J. (ed.) ASIACRYPT 2008. LNCS, vol. 5350, pp. 90–107. Springer, Heidelberg (2008)
7. Wang, Q., Wang, C., Li, J., Ren, K., Lou, W.: Enabling public verifiability and data dynamics for storage security in cloud computing. In: Backes, M., Ning, P. (eds.) ESORICS 2009. LNCS, vol. 5789, pp. 355–370. Springer, Heidelberg (2009)
8. Zhu, Y., Wang, H., Hu, Z., Ahn, G.J., Hu, H., Yau, S.S.: Dynamic audit services for integrity verification of outsourced storages in clouds. In: Chu, W.C., Wong, W.E., Palakal, M.J., Hung, C.C. (eds.) SAC, pp. 1550–1557. ACM (2011)
9. Fortnow, L., Rompel, J., Sipser, M.: On the power of multiprover interactive protocols. Theoretical Computer Science, 156–161 (1988)
10. Zhu, Y., Hu, H., Ahn, G.J., Han, Y., Chen, S.: Collaborative integrity verification in hybrid clouds. In: IEEE Conference on the 7th International Conference on Collaborative Computing: Networking, Applications and Worksharing, CollaborateCom, Orlando, Florida, USA, October 15-18, pp. 197–206. IEEE (2011)

Performance Analysis of Multi-class Steganographic Methods Based on Multi-Level Re-stegnography

Rajesh Duvvuru[1], P. Jagdeeswar Rao[2], Sunil Kumar Singh[1], Rajiv R. Suman[1], Shiva Nand Singh[3], and Pradeep Mahato[4]

[1] Department of Computer Science and Engineering, National Institute of Technology, Jamshedpur, Jharkhand, India
{rajeshduvvuru.cse,sunilkrsingh.cse,rrsuman.cse}@nitjsr.ac.in
[2] Department of Geo-Engineering, Andhra University College of Engineering, Visakhapatnam, A.P, India
pjr_geoin@rediffmail.com
[3] Dept. of Electronics and Communications Engineering, National Institute of Technology, Jamshedpur, Jharkhand, India
snsingh.ece@nitjsr.ac.in,
[4] Konylabs IT Service Pvt. Ltd, Hyderabad, A.P, India
pradeepmahato007@gmail.com

Abstract. The rapid advances of network technologies and digital devices make information exchange fast and easy. However, distributing digital data over public networks such as the Internet is not really secure due to copy violation, counterfeiting, forgery, and fraud. The paper presents a study and analysis on Multi-Level Steganography (MLS) which defines a new concept for hidden communication in telecommunication networks. In MLS, at least two steganographic methods are utilized simultaneously, in such a way that one method (called the upper-level) serves as a carrier for the second one (called the lower-level). In our work we have used two steganographic algorithms. This paper presents a steganographic algorithm based on the spatial domain, Selected Least Significant Bits (SLSB) and Least Significant Bits (LSB). This method is further exploited to do a 3-level re-steganography to enhance security. We have performed some simulations by using MATLAB software on different possible combination of the SLSB and LSB. Finally in our analysis we have concluded the LSB–LSB –LSB combination result best than rest of the combinations.

Keywords: Multi-Level Steganography, Selected Least Significant Bits, Least Significant Bit.

1 Introduction

Steganography is now more important due to the rapid growth and confidential communication of possible computer users on the internet. Steganography literally means 'covered writing'. However, the digital media formats in use for data exchange and communication today provide abundant hosts for Stenographic communication. Hence the interest in this practice has increased. Coupling this fact with the multitude

S.C. Satapathy et al. (eds.), *ICT and Critical Infrastructure: Proceedings of the 48th Annual Convention of CSI - Volume II*, Advances in Intelligent Systems and Computing 249,
DOI: 10.1007/978-3-319-03095-1_57, © Springer International Publishing Switzerland 2014

of the freely available easy to use stenographic tools available on the internet, the ability to exchange secret information without detection is available to anyone who wants to do so. A multi-class classification method that focuses on classifying unseen instances to their specific embedding method (class) [1]. This paper mainly focuses on the multi-class JPEG image classification using the re-steganography.Re-stegnography is the concept of can be used to detect itself, and are other alike stegnography algorithms have related recognition possibility? The react is "yes". According to the finale above, we can use the re-stegnography scheme to accomplish widespread recognition and multiclass classification. In literature it is proved that the re-stegnography technology has good results in these stegnography algorithms for specific detection, such as OutGuess, Steghide, Jsteg, and F5 [2].

This paper comprises two major contributions (1) Simulator of Multi-Level Re-stegnography algorithm was implemented and tested. (2) Performance evolution of different combinations of different combinations of LSB and SLB. This paper is organized in the fallowing way: section 2 describes how the multi-level re-stegnography is achieved. Section 3 explains about the analysis of the Re-stegnography. Section 4 discusses the simulation results. Lastly we will conclude and assert the future scope for this work in section 5.

2 Related Works

A simple and well known approach is directly hiding secret data into the least-significant bit (LSB) of each pixel in an image. Then based on the LSB technique, a genetic algorithm of optimal LSB substitution is now also available to improve the stego-image quality of the simple LSB method. (Wang et al., 2001) [3]. In addition, Chang et al. (2003) have also presented a fast and efficient optimal LSB method based on the dynamic programming strategy that improves the computation time of Wang et al.'s scheme [4]. A novel simple LSB technique based on optimal pixel adjustment was presented to achieve the goal of improving the stego-image quality (Lou, D.C., Hu, M.C., Liu, J.L., 2009)[6]. Besides, Then and Lin also presented a simple LSB scheme based on the modulus function to improve the stego-image quality (Juan José Roque). Wu et al. have also presented a combination scheme on the basis of pixel-value differencing and LSB replacement with a view to improving the hiding capacity while maintaining acceptable stego-image quality [5]. Lou and Liu (2002) proposed a LSB-based stenographic method that can resist the common-cover-carrier attack by embedding variable-size secret data and redundant Gaussian noise. The paper presented by WojciechFrączek, WojciechMazurczyk, Krzysztof Szczypiorski on Multi-Level Steganography (MLS), which defines a new concept for hidden communication in telecommunication networks [7]. In which MLS, at least two steganographic methods are utilized simultaneously, in such a way that one method (called the upper-level) serves as a carrier for the second one (called the lower-level).

2.1 Least Significant Bit

Least significant bit (LSB) technique works well for image steganography [8]. LSB based steganographic techniques either change the pixel value by ±1 or leave them unchanged. This is dependent both on the nature of the hidden bit and the LSB of the corresponding pixel value. Let I = { xi, i ϵ Ω } where Ω is an index set denote the mean subtracted cover image. The set Ω can be partitioned into three subsets A1, A2 and A3, where, Ω={ A1,A2,A3 }. Then, the pixel values in a LSB based stego-image, Is = { yi, I ϵ Ω } can be represented as

$$y_i = \begin{cases} x_i + 1 & \text{if } i \in \Lambda_1 \\ x_i - 1 & \text{if } i \in \Lambda_2 \\ x_i & \text{if } i \in \Lambda_3 \end{cases}$$

2.2 Selected Least Significant Bit

Selected Least Significant bit (SLSB) works with the LSB color of one pixel components in the image and altering them in accordance with the message's bits to conceal. The remaining bits in the color pixels are selected and also changed in order to achieve the nearest similar color of the original image [9].

2.3 Multi-Level Steganography

In typical single-method network steganography, overt communication traffic is used as a carrier for secret data. By influencing the carrier, a certain steganographic bandwidth (BS), which is defined as the amount of the steganogram transmitted using a particular method in one second ([b/s]), is achieved. However, the utilization of BS may result in a certain steganographic cost (CS) that expresses an impact (degradation) of a hidden data carrier due to steganographic procedure operations (see Section 1). The higher B S for given steganographic method we want to utilise the higher CS (the steganographic method has a greater impact on a hidden data carrier). If CS is excessive, then the detection of the method can be straightforward. Thus, a trade-off between BS and CS is always necessary. As mentioned in Section 1, MLS is based on at least two steganographic methods. First, the upper-level method uses overt traffic as a secret data carrier. The second, the lower-level method, uses the way the upper-level method operates as a carrier. The indirect carriers for lower-level methods are still packets from overt communication, but the direct carrier is another (upper-level) method [7].

3 Analysis of Multi-Level Stegnography

A multi-level stegnography is implemented so that the security is increased while transmitting the message. Even if, the image is decoded by, any hacker, he/she only receives a decoy message or an incomplete message (depends upon the implementation). Figure No. 1 and 2 shows, the MLS consists of 3 levels, so we use the terms Level 1 (or 2 or 3) steganographic method rather than upper- or lower-level to refer to each of them. Level 0 is considered as the overt channel. Of course, on each level, more than one steganographic method may be utilized; however, it may quickly degrade the carrier quality and thus make easy detection possible. The construction of an MLS has certain benefits compared to the scenario in which two (or more) unrelated steganographic methods are simultaneously utilized on the same carrier (Fig. 2, left):

- In general, the total steganographic cost of the MLS can be lower (for a given number of levels) than for the same number of methods used simultaneously on the same carrier (especially for the case where $CSk0 \approx 0$ where $k>1$ is a number of levels in MLS).
- The detection of MLS is harder to perform because only the discovery of the higher level method can lead to the detection of the lower level methods.
- There is a direct relationship between the steganographic methods used for MLS construction. If some additional data are carried in lower level methods, this is a direct indication that it can be used for the benefit of the higher level method.

A schematic representation of the used algorithm is as follow:-

3.1 Encoding

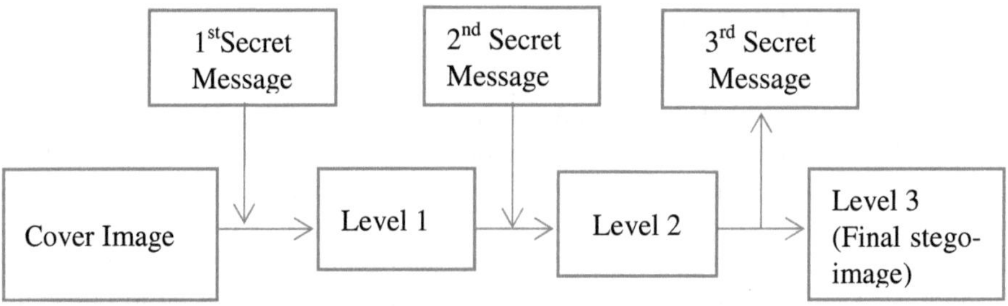

Fig. 1. Embedding of message in the Cover image

3.2 Decoding

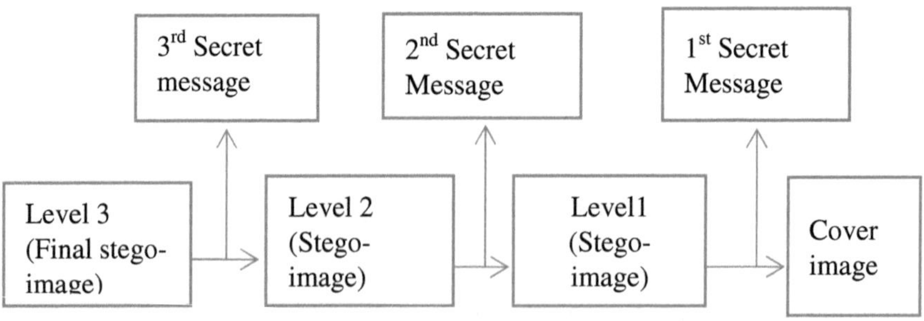

Fig. 2. Dig up of message from Cover image

4 Simulation and Results

We have used the Matlab7.5.0.342 for our simulations. Here, we have make use of two algorithms namely LSB and SLSB. These two algorithms further applied on the different level of stenographic image with different combinations like LSB - LSB – LSB, LSB – LSB – SLSB, LSB – SLSB – LSB, SLSB – LSB – LSB, LSB – SLSB – SLSB. SLSB – LSB – SLSB and SLSB – SLSB – LSB. Here, we have calculated the Peak Signal-to-noise ratio (PSNR) at each level of stegnography and also shown using histogram equalization.

4.1 Simulation of a 3-Level Re-stegnography

We have taken Standard Lina Image (512 x 512 pixels with resolution of 96 dpi) as cover image and secrete message taken in different combinations for our simulations. The messages are as fallows

Message 1 (All Alpha character)

She found Jack Crawford alone in the cluttered suite of offices. He was standing at someone else's desk talking on the telephone and she had a chance to look him over for the first time in a year. What she saw disturbed her.

Message 2 (All Digits)

3.14159265358979323846264338327950288419716939937510582097494459230781640628620899862803482534211706798214808651328230664709384460955058223172535940812848111745028410270193852110555964462294895493038196442881097566593344612847564823378678316527120190 91

Message 3 (Combination of Alphabets and Digits)

S1M1L4RLY, Y0UR M1ND 15 R34D1NG 7H15 4U70M471C4LLY W17H0U7 3V3N 7H1NK1NG 4B0U7

Encoding Message by using the combination of LSB-LSB-SLB for an JPEG image.

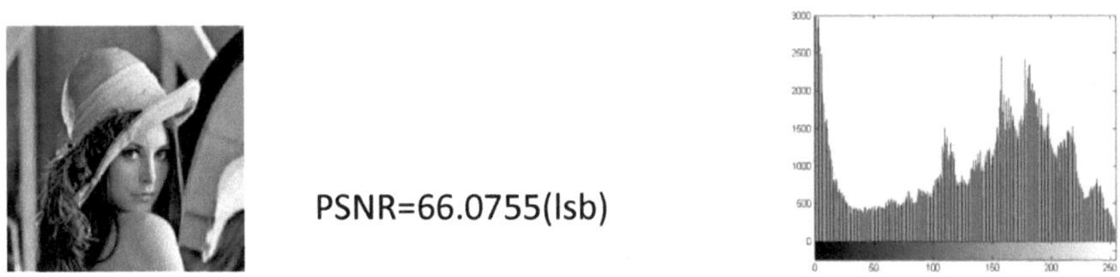

Fig. 3(a)

PSNR=64.1752(lsb)

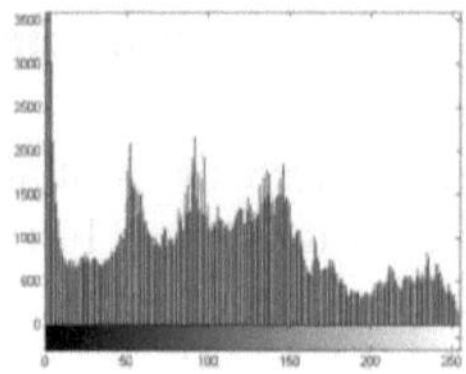

Fig. 3(b)

PSNR=62.3485(slsb)

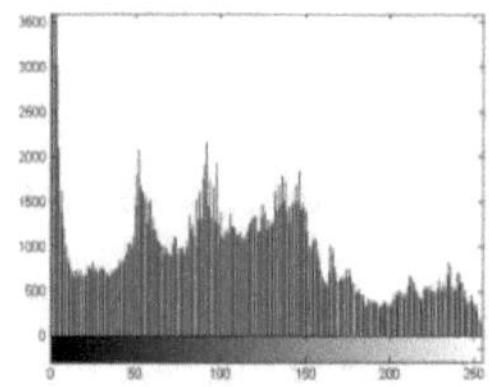

Fig. 3(c)

Fig. 3(a), Fig. 3(b), Fig. 3(c). Simulation result of the combination LSB-LSB-SLB using re-stegnography. Each figure represents each steganographic level. Here the simulation is made for 3-level.

The corresponding secrete messages are shown in figure 4, that is extracted at different level during the Multi-level Re-steganography with slight change in PSNR.

Embedded Secrete message extracted at third level:

Embedded Secrete message at extracted Second level:

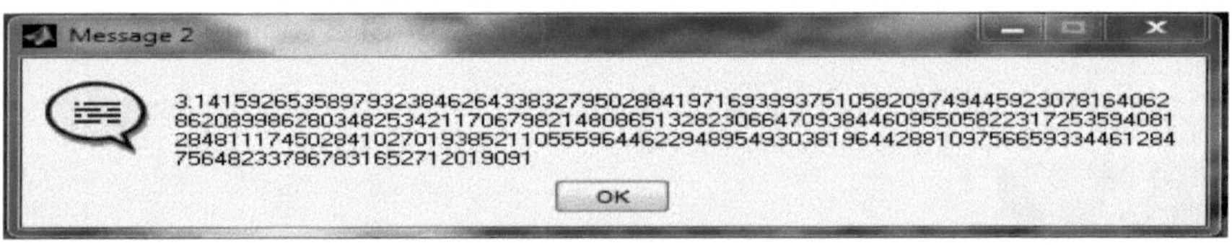

Embedded Secrete message at extracted First level:

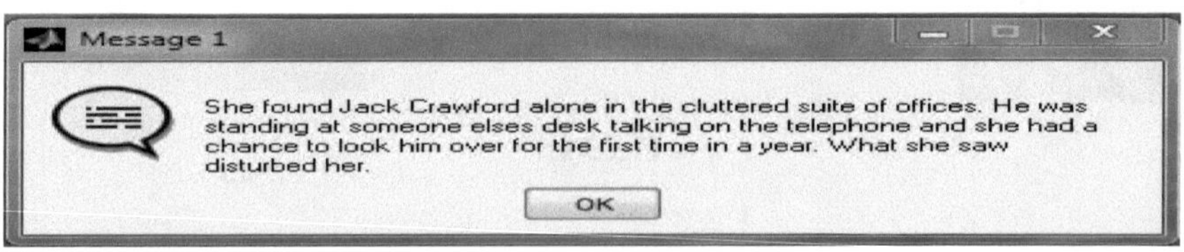

Fig. 3. Extraction of embedded messages at different level of three-level re-steganography

According to the algorithm we have embedded the first secrete message into the cover image. But it is extracted at the last steganographic image. The message extraction is fallowing the order Last-In-First-Out (LIFO).In our work we have carried out a 3-level multi-level steganography. Image of Lena has been used as a cover image. Also three different sets of messages were used with a maximum message length of 1000 characters. First message consists of only alphabets, second message consists of only digits and the third message consists of combinations of digits and alphabets.

Table 1. Detailed Analysis of Multilevel- Stegnography using LSB and SLSB

Combination	PSNR level-1	PSNR level-2	PSNR level-3
LSB - LSB - LSB	66.0745	63.1999	61.5809
LSB – LSB – SLSB	66.0755	64.1752	62.3485
LSB – SLSB – LSB	66.0745	63.5096	61.7918
SLSB – LSB – LSB	66.5046	63.4165	61.7289
LSB – SLSB – SLSB	66.4855	63.7325	62.0564
SLSB – LSB – SLSB	66.5058	63.6345	61.9895
SLSB – SLSB – LSB	66.5046	63.7426	61.9473
SLSB – SLSB – SLSB	66.4708	63.7247	62.0511

Two different algorithms were used in the experiment namely LSB (Least Significant Bit) and SLSB (Selective Least Significant Bit). Different combinations of algorithms were used. Since there exist three-level and also there were two different algorithms, so a total combination of 8 set would be formed. Peak-Signal-to-Noise-Ratio is noted at each level. Thus a data set of 24 experimental values is obtained. By observing the table no. 1 the following observations are made:-

1) SLSB generates larger PSNR difference (combination 1 & combination 2, combination 3 & combination 5)
2) LSB to LSB PSNR difference is not always same (combination 1, combination1 & combination2)
3) SLSB to SLSB PSNR difference is not always same (combination 8, combination 5 & combination7)

Combination 8 (SLSB – SLSB – SLSB) was encoded with a valid PSNR ratio but the hidden messages gets corrupt but this not always the case. Consider the following case where message 1 is 'a', message 2 is 'b' and message 3 is 'c'. The combination (SLSB – SLSB – SLSB) successfully encodes the messages into the carrier file i.e cover image.

5 Conclusion and Future Scope

In our work we have carried out two different algorithms at three levels. Least Significant Bit is the oldest and most insecure algorithm but it is very fast. Whereas Selective Least Significant Bit is slower comparatively but is more secure with respect to LSB. The best combination among all the eight combination is SLSB – SLSB – SLSB, but it is successful only for short messages. When messages are long enough, this combination even thou encode without any error and with a valid PSNR value, but the messages get corrupt. LSB- LSB- LSB is the least strong among the whole possible combinations. Choosing of a particular combination basically depends on the user and the fact that how much secure do they want the messages to be. Combination of two SLSB and one LSB is the strongest except combination 8. This work will be extended in future with inclusion of F5 and StegHide algorithms which may result in better information hiding and security.

References

1. Rodriguez, B., Petersons, G.: Detecting steganography using multi-class classification. In: Craiger, P., Shenoi, S. (eds.) Advances in Digital Forensics III. IFIP, vol. 242, pp. 193–204. Springer, Boston (2007)
2. Pan, X., Yan, B., Niu, K.: Multiclass Detect of Current Steganographic Methods for JPEG Format Based Re-stegnography. In: 2nd International Conference on Advanced Computer Control (ICACC), Shenyang, pp. 79–82. IEEE (2007)
3. Wang, R.-Z., Lin, C.-F., Lina, J.-C.: Image hiding by optimal LSB substitution and genetic algorithm. Pattern Recognition 34(3), 671–683 (2001)
4. Chang, C.C., Hsiao, J.Y., Chan, C.S.: Finding optimal least-significant-bit substitution in image hiding by dynamic programming strategy. Pattern Recognition 36, 1583–1595 (2003)
5. Wang, C.-M., Wu, N.-I., Tsai, C.-S., Hwang, M.-S.: A high quality steganographic method with pixel-value differencing and modulus function. Journal of Systems and Software, 1–8 (2007)
6. Lou, D.C., Hu, M.C., Liu, J.L.: Multiple layer data hiding scheme for medical images. Computer Standards and Interfaces 31(2), 329–335 (2009)
7. Fraczek, W., Mazurczyk, W., Szczypiorski, K.: Multi-Level Steganography: Improving Hidden Communication in Networks. In: arXiv:1101.4789v3 [cs.CR], CUL, pp. 1–8 (2012)
8. Neeta, D., Kesselman, K.S., Jacobs, D.: Implementation of LSB Steganography and Its Evaluation for Various Bits. In: 1st International Conference on Digital Information Management, Banglore, pp. 173–178. IEEE (2006)
9. Roque, J.J., Minguet, J.M.: SLSB: Improving the Steganographic Algorithm LSB. In: WOSIS, pp. 57–66. INSTICC Press (2009)

A Stylometric Investigation Tool for Authorship Attribution in E-Mail Forensics

Sridhar Neralla, D. Lalitha Bhaskari, and P.S. Avadhani

Dept. of CS&SE, Andhra University, Visakhapatnam, Andhra Pradesh, India
{neralla_sridhar,psavadhani}@yahoo.com,
lalithabhaskari@yahoo.co.in

Abstract. E-Mail forensics is one of the several cyber forensics approaches for identifying the cyber crimes that are happened through e-mails. This paper focuses on stylometric approach for finding accurate author of an e-mail. Stylometry is used to identify unique writing styles of an author; we used parameter minimization approach to reduce the overhead. In this paper we introduced java-based Stylometric Investigation Tool that is based on minimum number of parameters for stylometric approach.

Keywords: Authorship, Cyber Crimes, E-Mail Forensics, Stylometry.

1 Introduction

Identification of accurate author of e-mail messages is becoming intricate task due to increase in the use of e-mail for illegal intention. Good research is going in this direction with different paradigms; one of such paradigm is stylometric analysis. This paper is organized into four sections. Section 2 focuses on e-mail forensics and current research trends in e-mail forensics. Section 3 focuses on Stylometric analysis. Section 4 and Section 5 discusses about algorithm for stylometric analysis and proposed tool for stylometric investigation respectively.

2 E-Mail Forensics

E-Mail is one of the common ways to establish communication between people. More e-mail is generated than phone conversations. Most of the times all office based communications are going through e-mails. Authentication is one of the reasons for usage of e-mail. Cyber crimes are also increasing at alarming rate by using computer as a source/target. Computer forensics analysis finds traces about cyber crimes happened through digital devices. Forensics analysis experts focusing on e-mail communications to trace out the behavior of suspects.

2.1 Process of E-Mail Forensics

E-Mail forensics involves identification and extraction of evidence. The first step in an e-mail examination is to identify the sources of e-mail. E-mail clients and servers

S.C. Satapathy et al. (eds.), *ICT and Critical Infrastructure: Proceedings of the 48th Annual Convention of CSI - Volume II*, Advances in Intelligent Systems and Computing 249,
DOI: 10.1007/978-3-319-03095-1_58,

have expanded into packed databases, document repositories, contact managers etc. Irrespective e-mail forensics paradigm, computer forensics starts with collection of evidence in the form e-mail. Cyber criminals always try to remove the traces of their crimes, so they delete e-mails. These criminal believe that once they delete that the mail can't be recovered. Nowadays computer forensics is able to explore evidence from the damaged disks or crashed systems too. E-mails contain header information about sender and recipient; sometimes multiple recipients. E-mails reside on servers; criminals can't have any access to those servers and also sometimes e-mails stored on backup tapes.

Cyber forensics programs are able to retrieve deleted e-mails from either e-mail clients or e-mail servers. New evidences can be created with the help of recovered mails by correlating e-mails by data available on header like date, subject, recipient or sender, content properties etc. Forensics analysis plays vital role for providing these correlations. Necessary care should be taken related to spam mails which mislead the forensics experts to move in different direction. The bulk spam email messages are generated by dumb botnets using certain predefined templates [1]. Alex et.al, discussed about robust approach of detecting spam in social media [2].

2.2 E-Mail Forensics Approach

Sridhar et al [3] proposal related to inverted pyramid approach confer about three types of e-mail forensics scenarios e-mail header analysis, stylometric approach and timestamp analysis. In this paper we presented a tool related to stylometric investigation for authorship attribution in e-mail forensics. E-mail header analysis explores the header portion of e-mail and tries to trace out the originator of the e-mail. Timestamp analysis uses log files and recent file viewers to find out the system usage and file usage respectively.

Several tools are available to perform e-mail header analysis and timestamp analysis. Fig. 1 show PC on/off time that indicates usage of system and Fig. 2 shows recent file viewer indicates the files that are opened recently. Schatz et.al [4] pointed out the reasons affecting timestamp accuracy are unstable crystal oscillators, region

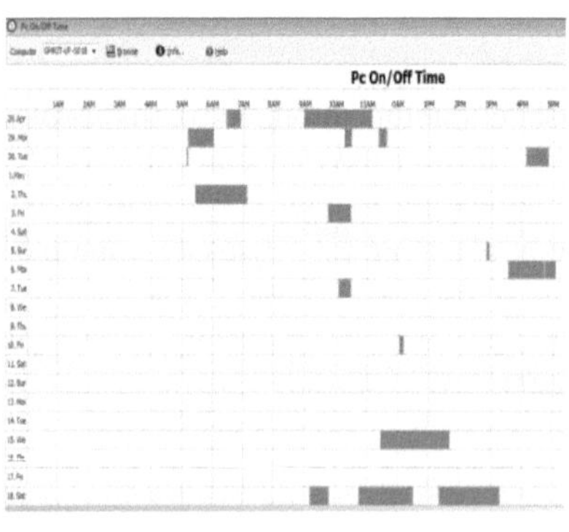

Fig. 1. PC on/off time

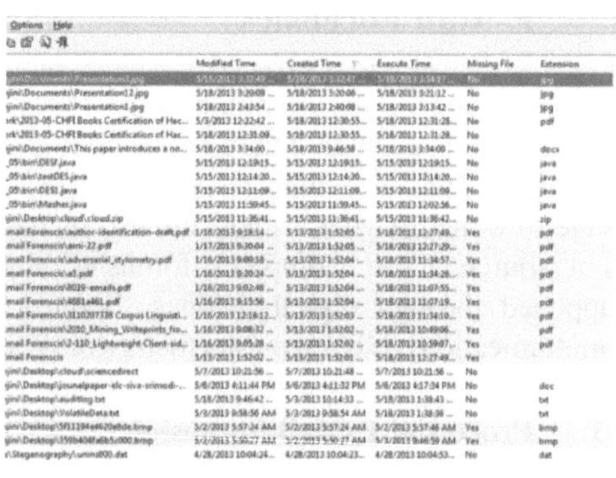

Fig. 2. Recent File Viewer

specific time zones, non-cryptographic authentication causes protocol based attacks, and software bugs. The limitations of the system clock for accurate timekeeping were given by Weil [5] who proposed dynamic timestamp analysis. So, we proposed that individually e-mail forensics or timestamps could not be stand as evidence in court of law but their combination strengthens the evidence.

3 Stylometric Analysis

Stylometric analysis is a combination of methods for author identification which is based on large volumes of data. This model utilizes statistical techniques on features within the texts. There are several methods in stylometric analysis. Khmelev and Tweedie [6] used Markov models based on letter coordination that uses first-order chain of letters and spaces. Tweedie et al used non-linear neural network models for stylometric analysis [7]. Chandrasekaran et al had shown the use of generalized regression neural network in Authorship Attribution [8]. Functional word approach is one of the best suited approaches for stylometric approach that uses multivariate statistics. Recent studies also had shown the importance of functional word approach. Paramjit et al [9] explored stylometric analyses using Dirichlet process mixture models.

3.1 Our Approach

In this paper we are using function word approach for building stylometric investigation tool. This approach first select text in the e-mail and break text into blocks/tokens. Next counts specific function words within each block, and then performs stylometric analysis on the resultant data. Anderson et al [10] studies related with number of parameters such as text size, feature sets and the number of documents by the author. Basic techniques used in stylometric approach are neural network and basic feature set [7], synonym based approach [11] and support vector machine and write-prints approach [12]. There are several studies related to number of parameters required for stylometric approach. We developed our tool based on the parameter minimization approach. This tool is Java-based tool that provides graphical comparison of various features of different authors. Widespread study of e-mails related to authorship attribution done by De Vel et al. [13], Koppel & Schler [14].

Parameter minimization approach is based on the number of parameters required for forensic. These parameters depend on type of corpus that is used for testing. As we are using e-mail as our corpus, we reduced number of parameters in word based features and character based features.

3.2 Methodology

We collected various e-mails that are available from Internet, classified all the mail based on authors, after that by using our Build Profiles procedure to build authorship profiles for each author. Then we assumed two different mails, one is from same author and other one is from different author. Finally we applied our Stylometric Investigation algorithm to find whether the mail is related to suspect or not.

4 Stylometric Investigation Algorithms

```
Algorithm: Stylometric Investigation
INPUT: Anonymous E-Mail E,
Set of E-Mails of Suspect Author {EA1,EA2,…EAn}
Various Other E-Mails belongs to other
suspects [Optional] {OS1,OS2,..OSm}
OUTPUT: Boolean Result with Visual Representation
of Authorship Profiles
(TRUE value confirms suspect author as actual author,
 FALSE value confirms suspect author is not)

BuildAuthorProfile for Emails of Suspect Author
    P={ PEA1, PEA2, PEA3,….}
BuildOthersProfile for Other Suspects
    OP={POS1,POS2,…}
Read E
BuildProfiles( PEAs)
BuildProfiles(POSs)
Draw Visual Representation for P,OP and PE
If StyloFeatures(E)==Average(StyloFeatures(P))
Then
   Return True
Else
  Return False
```

Fig. 3. Algorithm for Stylometric Investigation

```
PROCEDURE BuildProfiles
INPUT: Set of E-Mails {E1,E2,…..,En}
OUTPUT: Authorship Profile
Foreach E-mail in {E1,E2,….,En} as X
   Foreach token in X as T
    Foreach character in T as C
       Compute Vowels
       Compute Special Characters
    End for
    Find Frequency of Words
    Locate n-grams both fixed and variable
    Build Vocabulary Features
    Build Structural Features
   Build Syntactic Features
  End For
EndFor
Return Profile
```

Fig. 4. Procedural Algorithm for Build Profile

5 Stylometry Investigation Tool

We developed a Java based authorship attribution tool that takes the entire authors list from a specified directory along with their documents. Then we can supply the suspect document to the list and our tool generates a graphical output which can show individual feature comparison. Syntactical feature analysis is shown in Fig. 5. This shows that even most features are relevant, visualization techniques used for finding the one suspect profile is different from the other. Character Feature Analysis is shown Fig. 6 and Word based features are shown in Fig. 7. The problem with word based features can be shown clearly some of the features are common for most of the suspects. So, after combining all these features we can identify the accurate author of an e-mail.

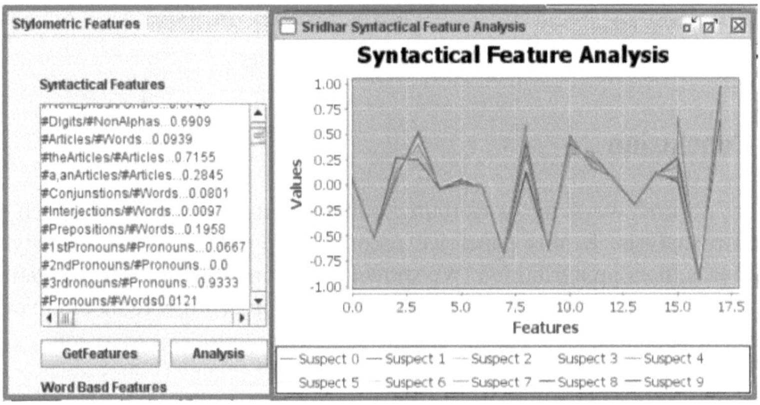

Fig. 5. Syntactical Features

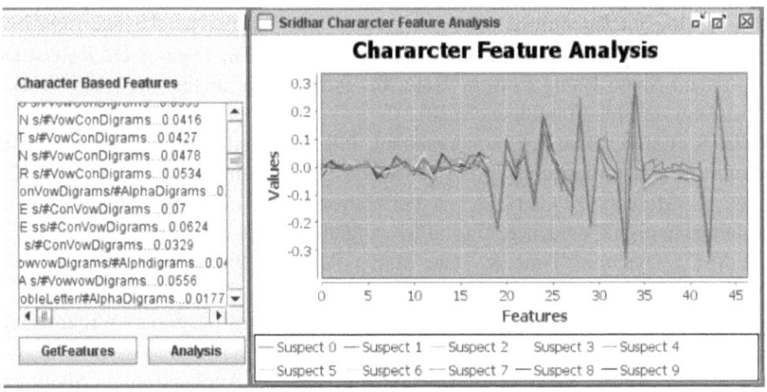

Fig. 6. Character Feature Analysis

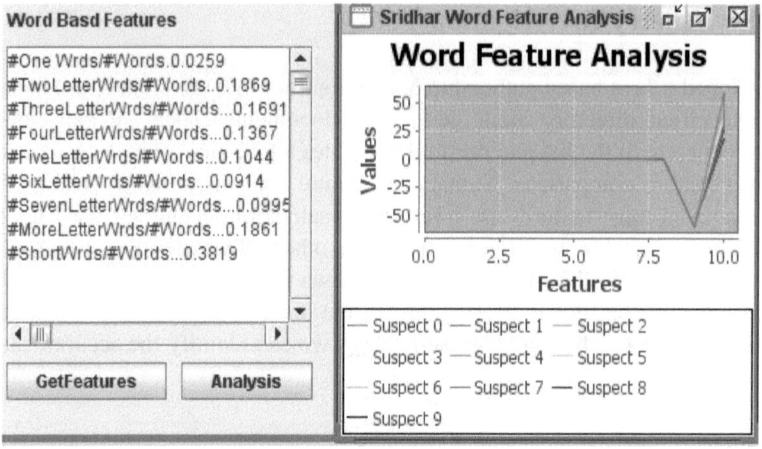

Fig. 7. Word Based Features

6 Conclusion

There are several ways for performing e-mail forensics, where one such way is stylometric analysis. In this paper we focused on e-mail forensics and stylometry features for authorship attribution. We shown results in Stylometry Investigation tool, and finally concluded the parameter minimization approach that reduces the overhead process of analysis.

References

1. Ramachandran, A., Feamster, N.: Understanding the network-level behavior of spammers. In: Proceedings of Sigcomm (2006)
2. Kantchelian, A., Ma, J., Huang, L., Afroz, S., Joseph, A.D., Tygar, J.D.: Robust Detection of Comment Spam Using Entropy Rate. In: Proceedings of the 5th ACM Workshop on Security and Artificial Intelligence (2012)
3. Neralla, S., Lalitha Bhaskari, D., Avadhani, P.S.: Inverted Pyramid Approach for E-mail forensics using heterogeneous forensics tools. CSI Communications, 20–22 (July 2013)
4. Schatz, B., Mohay, G., Clark, A.: A correlation method for establishing provenance of timestamps in digital evidence. Digital Investigation 3, 98–107 (2006)
5. Weil, M.C.: Dynamic Time & Date Stamp Analysis. International Journal of Digital Evidence (2002)
6. Khmelev, D., Tweedie, W.: Using Markov Chains for Identification of Writer. Literary and Linguistic Computing 16(4), 299–307 (2001)
7. Tweedie, F.J., Singh, S., Holmes, D.I.: Neural Network Applications in Stylometry. The Federalist Papers. Computers and the Humanities 39(1), 1–10 (1996)
8. Chandrasekaran, R., Manimannan, G.: Use of Generalized Regression Neural Network in Authorship Attribution. International Journal of Computer Applications (0975 – 8887) 62(4) (January 2013)

9. Gill, P.S., Swartz, T.B.: Stylometric analyses using Dirichlet process mixture models. Journal of Statistical Planning and Inference 141, 3665–3674 (2011)
10. Anderson, A., Corney, M., de Vel, O., Mohay, G.: Identifying the Authors of Suspect E-mail. Communications of the ACM (2001)
11. Clark, J.H., Hannon, C.J.: A classifier system for author recognition using synonym-based features. In: Gelbukh, A., Kuri Morales, Á.F. (eds.) MICAI 2007. LNCS (LNAI), vol. 4827, pp. 839–849. Springer, Heidelberg (2007)
12. Abbasi, A., Chen, H.: Writeprints: a stylometric approach to identity level identification and similarity detection in cyberspace. ACM Transactions on Information Systems 26(2) (March 2008)
13. de Vel, O., Anderson, A., Corney, M., Mohay, G.: Mining E-mail Content for Author Identification Forensics. News Letter ACM SIGMOD Record 30(4), 55–64 (2001)
14. Koppel, M., Schler, J.: Exploiting Stylistic Idiosyncrasies for Authorship Attribution. In: Proceedings of IJCAI 2003 Workshop on Computational Approaches to Style Analysis and Synthesis, pp. 69–72 (2003)

Privacy Preserving in Association Rule Mining by Data Distortion Using PSO

Janakiramaiah Bonam[1], A. Ramamohan Reddy[2], and G. Kalyani[1]

[1] Department of Computer Science and Engineering,
DVR and Dr. HS MIC College of Technology, Vijayawada
[2] Department of Computer Science and Engineering, S.V. University, Tirupathi

Abstract. Association Rule Mining is one of the core data mining tasks that is used to show the associations between data items. The distribution of data from association rule mining can bring lot of advantages for research, and business teamwork.However, huge repositories of data contain secret data and sensitive patterns that must be confined before being published. We address this problem of privacy preserving association rule mining by applying data sanitization to avoid the confession of sensitive rules while maintaining data effectiveness. Particle Swarm Optimization is an artificial intelligence technique, proficient of optimizing a non-linear and multidimensional problem which typically reaches high-quality solutions efficiently while requiring negligible parametrization. To recognize the most sensitive transactions for hiding given sensitive association rules we are with Particle Swarm Optimization technique. The performance of the algorithm is validated against representative synthetic and real datasets with some performance measures.

Keywords: Association rule mining, Sensitive rules,Particle Swarm Optimization, Data distortion, Data Sanitization, Privacy.

1 Introduction

Association rule mining extracts novel, hidden and useful patterns from huge repositories of data. These patterns are useful for effective analysis and decision making in telecommunication network, marketing, business, medical analysis, website linkages, financial transactions, advertising and other applications. On demand to various mismatched requirements of data sharing, privacy preserving and knowledge discovery, Privacy Preserving Data Mining(PPDM) has become a research hotspot in data mining. Simply, the association rule hiding problem is to hide secret, sensitive patterns contained in data from being discovered, while without losing non sensitive at the same time. The problem of frequent association rules hiding motivated many authors [4,3], and proposed different approaches.The majority of the proposed approaches can be classified along two principal research directions: (i)Data hiding approaches and (ii)Knowledge hiding approaches.

S.C. Satapathy et al. (eds.), *ICT and Critical Infrastructure: Proceedings of the 48th Annual Convention of CSI - Volume II*, Advances in Intelligent Systems and Computing 249,
DOI: 10.1007/978-3-319-03095-1_59,

2 Particle Swarm Optimization

Swarm Intelligence (SI) is a pioneering distributed intelligent paradigm for solving optimization problems. Particle Swarm Optimization (PSO) incorporates swarming behaviors observed in flocks of birds, schools of fish, or swarms of bees, and even human social behavior [2,5]. For applying PSO successfully, one of the key issues is finding how to map the problem solution into the PSO particle.The original PSO formulae define each particle as potential solution to a problem in D-dimensional space. The position of particle i is represented as $X_i = (x_{i1}, x_{i2}, x_{i3}, ...x_{iD})$. Each particle also maintains a memory of its previous best position, represented as $Pbest_i$. A particle in a swarm is moving; hence, it has a velocity, which can be represented as $V_i = (v_{i1}, v_{i2}, v_{i3}, , v_{iD})$. Each particle knows its best value so far (pbest) and its position. Moreover, each particle knows the best value so far in the group (gbest) among pbests. Each particle tries to modify its position. Velocity of each particle can be modified by the following equation (1) in Inertia Weight Approach (IWA)

$$v_i = w * v_i + c1 * r1 * (Pbest_i - X_i) + c2 * r2 * (gbest - X_i) \tag{1}$$

where, v_i velocity of particle, X_i current position of particle, $c1$ determine the relative influence of the cognitive component, $c2$ determine the relative influence of the social component, $Pbest_i$ best position of particle i, $gbest$ global best of the group. W ,the inertia factor controls the influence of previous velocity on the new velocity, r1 and r2 are the random numbers, which are used to maintain the diversity of the population, and are uniformly distributed in the interval $[0, 1]$. From equation (1), a particle decides where to move next, considering its own experience, which is the memory of its best past position, and the experience of its most successful particle in the swarm. The following equation(2)for inertia factor is usually utilized.

$$W = W_{max} - \frac{W_{max} - W_{min}}{iter_{max}} * iter \tag{2}$$

where, W_{max} - initial weight, W_{min} - final weight, $iter_{max}$ - maximum iteration number, iter - current iteration number.
The current position (searching point in the solution space) can be modified by means of the equation (3):

$$X_i = X_i + V_i \tag{3}$$

3 Problem Definition

The problem of protecting sensitive knowledge in association rule mining can be stated as, Given a data set D to be released, a set of association rules R mined from D, and a set of sensitive rules $R_S \subseteq$ R to be hidden. How can we get a new data set D^1, such that the rules in R_S cannot be mined from D^1, while the rules in $R - R_S$can still be mined as many as possible. In this case, D^1 becomes the released database.

4 Algorithm: RSIF-PSOW(Reducing Sensitive Item Frequency Using Particle Swarm Optimization)

```
Input:
    1. A Transactional Dataset D
    2. Minimum Support Threshold MST
    3. A set of Sensitive rules Rs
Output:
     A Sanitized Dataset
Method:
begin
repeat
{
 Step 1: call initialization function;
 Step 2: X <- initial population position vector;
 Step 3: V <- initial velocity vector;
 Step 4: call fitness function with X;
 Step 5: pbest <- X ; gbest <- pbest[maximum fitness position];
         iteration <- 0;
 Step 6:
  while(iteration < iteration_max)
  {
   W <- Wmax- (Wmax-Wmin) * iteration / iteration_max;
   for i= 1 to size of population
   {
   Vi <-((W * Vi +c * r1 *(pbest (i)- Xi))+ c2*r2*(gbest - Xi));
   Xi <- Xi + Vi;
   }
   call fitness function with X;
   check lower and upper limits of X,V and adjust to with in range;
   for i=1 to size of population
   {
    Compare current and previous fitness value of position vector,
    and whichever is best store that as pbest of that particle;
   }
   current gbest  <- maximum value in pbest of particles;
   current gbest fitness <- fitness of current gbest value;
   if (current gbest fitness < previous gbest fitness value)
   {
    current gbest fitness <- previous gbest fitness;
    current gbest <-  previous gbest;
   }
   iteration <- iteration + 1;
  }
 Step 7:  Identify the particle with highest fitness value
          which also existed in the database;
 Step 8:  Calculate the frequencies of items in sensitive rules Rs
```

```
          and choose the item with highest frequency;
 Step 9:  Change the value of selected item as 0 from 1
          in the selected transaction;
 Step 10: Update the support of rules in Rs
          and discard rules with support less than MST;
 }until (Rs is empty);
end
function  initialization()
begin
     c1 <- 2; c2 <- 2; r1 <- random(1,1); r2 <- random(1,1);
     Wmax <- 0.9; Wmin <- 0.4; W <- 1;  n <- 10;
     iterationmax <- n; lowerlimit <- 0;
     upperlimit <- 1023; vmax <-(upperlimit-lowerlimit) /10;
     vlowerlimit <- -1* vmax;  vupperlimit <- 1*vmax;
end
function fitness value
input  : position vector  X ; output : fitness value
begin
 1. items <- items of the database D;
 2. Sensitive items  <- items in Rs;
 3. Calculate the SIF of each sensitive item with respect to
     particle as follows: SIFij = Fij/Tj
    Where  Fi,j= no.of items of ith sensitive item that are
    present in jth transaction, Tj=No of items in jth transaction;
 4. Calculate the DF of each item with the following steps.
    4.1. Calculate the DC of each item with respect to
            the each sensitive item as Ci -MST *n + 1
         where Ci is occurrence frequency of Ith sensitive item
         in new particles, and n is no of particles;
    4.2. Calculate the MDC of each Item;
    4.3. Calculate the DF of each item as:
          DF = log( n / ( MDC-support)) Where support is
          no of times the item is in the particles;
 5. Calculate the DF of each particle with respect to each
    sensitive item using Equation (4);
```

$$\sum_{k=1}^{p} \log \frac{n}{MDC - Support} \tag{4}$$

```
 6. Calculate the Fitness value(FF) of each particle
    using Equation (5);
```

$$FF = \sum_{i=1}^{m} \frac{F_{ij}}{T_j} * \sum_{k=1}^{p} \log \frac{n}{MDC - Support} \tag{5}$$

```
end
```

5 Example

In this section an example is given to demonstrate the proposed RSIF-PSOW algorithm.Assume a database shown in Table.1. It consists of 20 transactions and 10 items denoted a to j. Assume the set of user specified sensitive rules are $R_s = \{(a, g), (b, h), (g, l), (c, g, h)\}$. Also Assume the User specified MST is set at 30% which indicates that the minimum count is 0.3*20 which is approximately 6.The proposed approach to hide the sensitive rules proceeds as follows.

Table 1. Transactional Datadase

tid	j	i	h	g	f	e	d	c	b	a	tid	j	i	h	g	f	e	d	c	b	a
T1	1	0	0	1	0	0	0	0	0	0	T11	0	0	0	1	1	1	0	1	0	0
T2	0	1	1	1	0	1	1	1	1	1	T12	1	0	0	1	1	0	0	0	0	1
T3	1	1	1	1	1	1	1	1	0	0	T13	1	0	1	1	0	0	1	1	0	0
T4	0	0	1	0	1	0	0	0	0	1	T14	1	0	0	1	0	0	0	1	1	1
T5	1	0	1	0	0	1	1	1	1	0	T15	0	1	1	0	0	0	0	0	1	0
T6	0	0	1	1	1	0	1	0	1	0	T16	0	0	1	0	0	0	0	0	1	0
T7	1	1	0	0	1	0	1	0	1	1	T17	0	1	0	1	1	1	0	1	0	1
T8	1	0	1	1	1	0	0	1	0	1	T18	0	1	1	1	1	1	1	0	1	0
T9	1	0	1	0	0	0	0	1	0	1	T19	1	1	1	1	1	0	0	1	1	1
T10	0	0	1	1	1	1	0	0	1	1	T20	1	1	1	1	1	0	1	1	1	1

Step 1: Initialization values are indicated in initialization function.
Step 2: The results of 2 to 5 steps are shown in the following Table 2.
Step 6: Maximum number of iterations are here taken as 10. After first iteration the values are shown in the following Table.3 . Results after 10 iterations are shown in the following Table 4.
Step 7: The highest fitness value 9 and its position vector 1111111111 is not present in the data base .So select the transaction with TID 3 because another position vector with fitness value 9 is existed in the database.
Step 8: The support count values of the rules in Rs are (ag:8,bh:9,gi:6,cgh:6).The frequencies of the items in Rs are a:1,b:1,c:1,g:3,h:2,i:1.Because g has highest frequency it will be selected .

Table 2. Result of step 2 to 5 of the RSIF-PSOW algorithm

Iteration No	Position vector X	Equivalent decimal number	Velocity vector V	Fitness value	pbest	gbest
0	0010100001	161	15	3	161	243
	0011101010	234	50	6	234	
	0011110011	243	80	7	243	
	1011001100	716	95	6	716	
	0010000010	130	-25	3	130	
	0111111010	506	-90	7	506	

Table 3. Result of step 6 after 1 iteration

Iteration No	Position vector X	Equivalent decimal number	Velocity vector V	Fitness value	pbest	gbest
	0100010000	272	102.3	1	161	
	0100011111	287	53	4	234	
1	0100110111	311	68	4	243	479
	0111011111	479	2.61	9	479	
	0101001101	333	102.3	6	333	
	0000000000	0	-102.3	0	506	

Table 4. Result after 10 iterations

Iteration No	Position vector X	Equivalent decimal number	Velocity vector V	Fitness value	pbest	gbest
	1111111111	1023	2.59	9	1023	
	1111111111	1023	4.93	9	1023	
10	1111111111	1023	3.45	9	1023	1023
	1111111111	1023	4.93	9	1023	
	1111111111	1023	4.93	9	1023	
	1111111111	1023	2.59	9	1023	

Step 9: The value of g in TID:3 will be changed from 1 to 0. The modified data base is shown in Table 5.
Step 10: update the Support counts of Rs. The values are (ag:7,bh:8,gi:5,cgh:5). so g $\rightarrow i$ and $c \rightarrow g, h$ were hidden because the support value is less than MST . Repeat the above procedure until Rs is empty i.e. all the sensitive rules were hidden. The final sanitized data base D^1 is shown in Table.6.

Table 5. Modified Data Base after step 9

tid	j	i	h	g	f	e	d	c	b	a	tid	j	i	h	g	f	e	d	c	b	a
T1	1	0	0	1	0	0	0	0	0	0	T11	0	0	0	1	1	1	0	1	0	0
T2	0	1	1	1	0	1	1	1	1	1	T12	1	0	0	1	1	0	0	0	0	1
T3	1	1	1	0	1	1	1	1	0	0	T13	1	0	1	1	0	0	1	1	0	0
T4	0	0	1	0	1	0	0	0	0	1	T14	1	0	0	1	0	0	0	1	1	1
T5	1	0	1	0	0	1	1	1	1	0	T15	0	1	1	0	0	0	0	0	1	0
T6	0	0	1	1	1	0	1	0	1	0	T16	0	0	1	0	0	0	0	0	1	0
T7	1	1	0	0	1	0	1	0	1	1	T17	0	1	0	1	1	1	0	1	0	1
T8	1	0	1	1	1	0	0	1	0	1	T18	0	1	1	1	1	1	1	0	1	0
T9	1	0	1	0	0	0	0	1	0	1	T19	1	1	1	1	1	0	0	1	1	1
T10	0	0	1	1	1	1	0	0	1	1	T20	1	1	1	1	1	0	1	1	1	1

Table 6. The final sanitized database D^1

tid	j	i	h	g	f	e	d	c	b	a	tid	j	i	h	g	f	e	d	c	b	a
T1	1	0	0	1	0	0	0	0	0	0	T11	0	0	0	1	1	1	0	1	0	0
T2	0	1	1	1	0	1	1	1	0	1	T12	1	0	0	1	1	0	0	0	0	1
T3	1	1	1	0	1	1	1	1	0	0	T13	1	0	1	1	0	0	1	1	0	0
T4	0	0	1	0	1	0	0	0	0	1	T14	1	0	0	1	0	0	0	1	1	1
T5	1	0	0	0	0	1	1	1	1	0	T15	0	1	1	0	0	0	0	0	1	0
T6	0	0	1	1	1	0	1	0	1	0	T16	0	0	1	0	0	0	0	0	1	0
T7	1	1	0	0	1	0	1	0	1	1	T17	0	1	0	1	1	1	0	1	0	1
T8	1	0	1	1	1	0	0	1	0	1	T18	0	1	1	1	1	1	1	0	1	0
T9	1	0	1	0	0	0	0	1	0	1	T19	1	1	1	0	1	0	0	1	1	1
T10	0	0	0	1	1	1	0	0	1	1	T20	1	1	1	1	1	0	1	1	1	0

6 Performance Measures

Hiding Failure(HF): When some restrictive rules are discovered from D^1, we call this problem as Hiding Failure, and it is measured in terms of the percentage of restrictive rules that are discovered from D^1. The hiding failure is measured by$HF = \frac{\sharp R_s(D^1)}{\sharp R_s(D)}$ where$\sharp R_s(D^1)$ denotes the number of sensitive rules discovered from sanitized database(D^1), and $\sharp R_s(D)$ denotes the number of sensitive rules discovered from original database(D).

Misses Cost(MC): Some non-sensitive rules can be hidden by mining algorithms accidentally. This happens when some non-sensitive rules lose support in the database due to the sanitization process. We call this problem as Misses Cost, and it is measured in terms of the percentage of legitimate patterns that are not discovered from D^1. The misses cost is calculated as follows: $MC = \frac{\sharp\sim R_S(D)-\sharp\sim R_S(D^1)}{\sharp\sim R_S(D)}$ where$\sharp \sim R_S(D)$ denotes the number of non-sensitive rules discovered from original database D, and $\sharp \sim R_S(D^1)$ denotes the number of non-sensitive rules discovered from sanitized databaseD^1.

7 Experimental Results

All the experiments were conducted on PC, Intel i5 CPU @ 2.50 GHz and 4 GB of RAM running on a windows 7 ,64-bit operating system. To measure the effectiveness of the algorithm, we used a dataset generated by the IBM synthetic data generator and FIMI Repository [1]. The Characteristics of the dataset

Table 7. Database Characteristics

Database	No.of Items	Avg.Length	No.of Transaction
bms1	497	2.5	59,602
retail	16,469	103	88,162
T10I4D10K	1000	10	10000

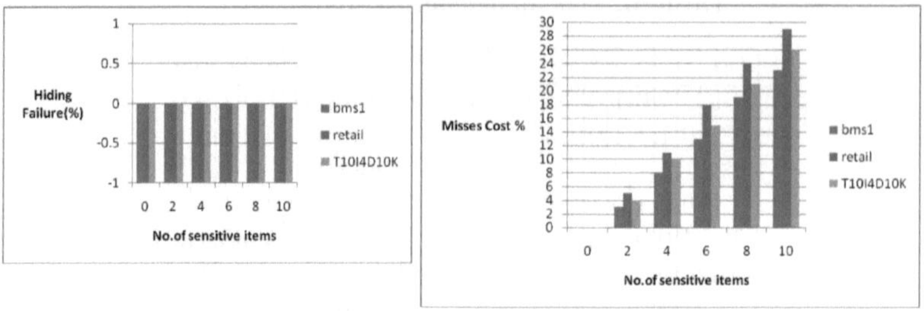

Fig. 1. (a): Hiding Failure (b): Misses Cost

were shown in Table.7. Fig. 1(a) shows efficiency of the proposed algorithm in the Hiding Failure. Accordingly, the RSIF-PSOW algorithm will not produce any sensitive rules from D^1 , when hiding any number of sensitive rules. Fig. 1(b) shows the efficiency of the proposed algorithms in the Misses cost minimization. Accordingly, the RSIF-PSOW algorithm achieved better results in reducing Misses cost.The time needed by sanitization algorithm, increases proportional to size of D, R_S and also depends on MST.

8 Conclusion

We have introduced an efficient implementation of Reducing Sensitive Item Frequency using Particle Swarm Optimization(RSIF-PSOW)for hiding sensitive rules from transactions and generating a sanitized database D^1. This Sanitization algorithm preserves privacy for sensitive rules and the non-sensitive rules that are found when mining this original database can still be mined from its sanitized database. Our further research will focus on integrating other soft computing techniques to get better performance of the proposed approach.

References

1. Goethals, B.: The fime repository. In: FIME 2003 (2003)
2. Kennedy Clerc, M.: The particle swarm explosion stability and convergence in a multidimensional complex space. IEEE Transactions on Evolutionary Computation (2002)
3. Dasseni, E., Verykios, V.S., Elmagarmid, A.K., Bertino, E.: Hiding association rules by using confidence and support. In: Moskowitz, I.S. (ed.) IH 2001. LNCS, vol. 2137, pp. 369–383. Springer, Heidelberg (2001)
4. Gkoulalas-Divanis, A., Verykios, V.S.: Association rule hiding for data mining, pp. 1386–2944. Springer (2010) ISSN 1386-2944
5. Kennedy, J., Eberhart, R.: Swarm intelligence. Morgan Kaufmann Publishers, Inc., San Francisco (2001)

Inline Block Level Data De-duplication Technique for EXT4 File System

Rahul Shinde, Vinay Patil, Akshay Bhargava, Atul Phatak, and Amar More

MIT Academy of Engineering, Pune, India
{rahul.shindeat,vinay18.patil,bhargava.akshay14,
atul.phatak5,amarmore2006}@gmail.com
http://www.mitaoe.ac.in

Abstract. Day by day data centers are growing and also their data. Data is key part of their organization and hence backed up after a regular interval. Due to huge data size, to improve utilization and life span of the disks, data de-duplication techniques are followed. In data de-duplication single copy of the data is stored on the disk by finding and eliminating the redundant copies. Now a days EXT4 has become a popular file system as it supports increased file system size and improved performance. So EXT4 file system can be used to store the backups and the data de-duplication could still increase the disk capacity virtually and could reduce the number of disk writes. In this paper we present a data de-duplication algorithm for EXT4 file system. Using this algorithm the duplicate data is eliminated before it is actually written to the disk and the extents in the EXT4 file system are arranged accordingly.

Keywords: File system, EXT4 file system, Data de-duplication, Data backup.

1 Introduction

For every datacenter, storage efficiency is one of the crucial factors since the data is backed up regularly, which results in huge amount of data to be stored on disk and hence there arises a need for data de-duplication. Data de-duplication [1] makes disk more affordable by eliminating redundant data from the disk thereby increasing the efficiency. In this method only unique copy is stored on the disk and the rest of the copies which are same as the one which is already stored will only be a reference to that unique copy and no extra storage space will be allocatd to them.

Consider an example of backup server which takes backup on weekly basis. Suppose in the first week the data stored is 50GB and in the next week the data is increased to 70GB out of which 50GB is same as the first week, then the total data stored during first and second week will be 50GB + 70GB = 120GB. But if de-duplication is applied on the backup server, then the actual data that would be stored at the end of second week is just 70GB since 50GB of it was same as that of first week i.e. only modified data gets saved.

S.C. Satapathy et al. (eds.), *ICT and Critical Infrastructure: Proceedings of the 48th Annual Convention of CSI - Volume II*, Advances in Intelligent Systems and Computing 249,
DOI: 10.1007/978-3-319-03095-1_60,

In EXT4 file system [13], the physical block number is of 48 bits and the size of each block is of 4KB or 4096 bytes (2^{12}). Hence the maximum file system size could reach up to (2^{60}) i.e. (2^{48}) * (2^{12}) or 1EB. Due to this large file system capacity, EXT4 file system could be used as a file system for back up servers and by applying de-duplication, the large amount of data could be saved on the disk. As compared to previous EXT file systems which has small file system capacity, EXT4 has larger file system capacity which makes it more efficient to handle various applications. Also block allocation and inode allocation strategies [11] in EXT4 file system are improved compared to previous versions of EXT file systems which makes EXT4 file system more efficient for backup applications.

In Section 2 we mention the related work done in the area of data de-duplication, Section 3 provides information about extents and extent structure in EXT4 file system. Section 4 shows overall design of our system and Section 5 gives the implementation details of our system. Results are discussed in Section 6 followed by the conclusion.

2 Related Work

Data de-duplication has received lot of interest in storage research and industry. If we review related work in the area of de-duplication, we can say that most of the work has been carried out in the context of various types and levels of de-duplication. Microsoft storage server [2], EMC's Centra [3] use file level de-duplication. Venti [4] perform de-duplication with respect to a fixed block size. The NetApp de-duplication [5] for file servers makes use of hashes to find duplicate data blocks. By using byte by byte comparison, hash collisions are resolved in this system. This process runs in the background, therefore it is a post-process de-duplication system. Inline de-duplication at block level [6, 7] is implemented in ext3 file system. SDFS is a file system [8] for Windows and Linux designed to support the unique needs Virtual Environments and supports enhanced functionality for VMWare, Xen, and KVM. Extreme Binning [9], proposed a scalable de-duplication technique for chunk based [10] file backup. We have used inline data de-duplication approach. which is applied Our layer added to the EXT4 file system is implemented for inline block level data de-duplication.

3 Structure of Extents in EXT4 File System

Extents [11–14] are an indivisible part of EXT4 file system. They were introduced in order to improve the throughput of the file system through sequential read and write operations.

Fig. 1. EXT4 extent structure

The features of extents are delayed allocation and persistent pre-allocation. Pre- allocation deals with allocating space of specific size at the time of file creation. Delayed allocation deals with allocation of blocks after page is flushed. Extent consists of three parts starting logical number that the extent covers, length i.e. total number of blocks stored inside the extent and starting physical block number (Higher 16 bits and lower 32 bits) as shown in Fig 1.

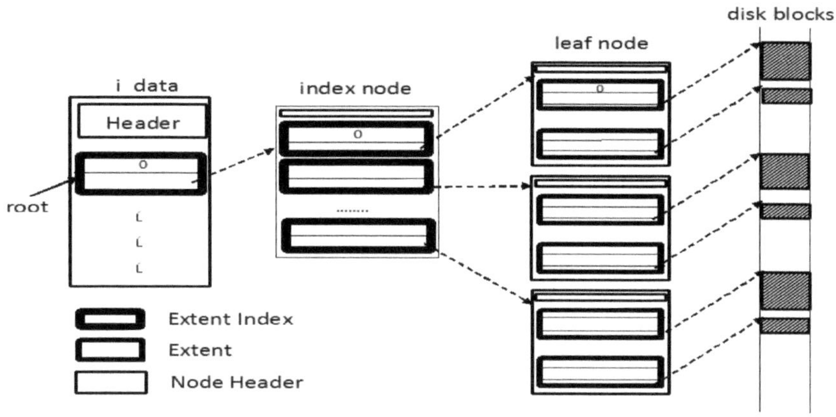

Fig. 2. Extent tree for sparse file in EXT4 file system

An inode contains a maximum of 4 extents but in case of huge, sparse files the extents are stored on the disk in the form of H-tree [13, 14] as shown in Fig. 2. So, an improved technique of accessing blocks for sequential read/write is seen in EXT4 file system as compared to its descendants.

4 System Design

Our system deals with eliminating redundant data from disk. User application collects required data and it is sent to kernel for storage. Kernel invokes write system call which is intercepted in the VFS layer. At VFS layer the data is held in the buffer. Further buffer is divided into 4KB chunks [1] which is a standard block size in EXT4 file system. For each block, hash value is calculated and the hash table is maintained where this hash value is stored. So, when each time 4KB of chunk arrives, its hash value is calculated and is compared with the previous hash values from hash table.

As shown in Fig. 3 the algorithm has to deal with two possibilities:

1. The hash value obtained is different from the hash value which is already stored in hash table
2. The hash value already exists in hash table

Also the system deals with handling of extents by verifying whether the block is contiguous with the previously encountered duplicate block. This is further explained in Section 5.

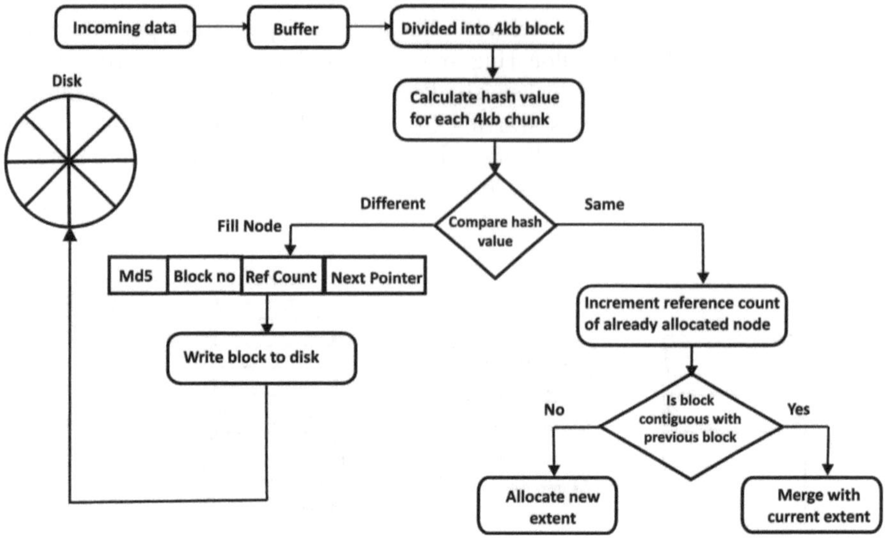

Fig. 3. EXT4 de-duplication system design

5 Implementation Details

Our de-duplication algorithm deals with eliminating redundant data from the buffer itself before it is written to the disk and also handles extents which provide sequential read/write operations on blocks. At the generic level the data that is held in the buffer is divided into chunks of 4KB each. For each block, hash value is calculated and is stored in the hash table. For this purpose we have used two hash functions viz. MD5 - Message Digest 5 and FNV - Fowler Nollvo hash.

For calculating the hash value of 4KB data chunk MD5 algorithm is used which returns 128 bit unique value and is used to check the duplicate data. The main issue here is how to organize the hash table in order to lookup the hash values efficiently. If we store 128 bit value sequentially, then it would take O(n) time to search the required hash value and hence we have furhter applied FNV algorithm to construct the hash table. This algorithm returns a 32 bit hash value of integer type. But if complete 32 bits are used to build our data structure it would require $2^{32} = 4$GB of memory. In order to optimize the memory, 32 bits are truncated to 21 bits [7] which reduced memory requirements to $2^{21} = 2$MB. So with this 21 bit value, a total of $2^{21} = 2097152$ indices can be used to build the hash table.

At every index a singly linked list is maintained. The node structure of this list is as shown in the Fig. 4. There are four fields in the node structure which comprises of hash value, block number which is of 6 bytes since the size of block number in EXT4 is 48 bits [13], reference count which shows how many blocks are referring to the unique stored copy and the next pointer, which points to the next node. Corresponding to the correct hash index this node structure is

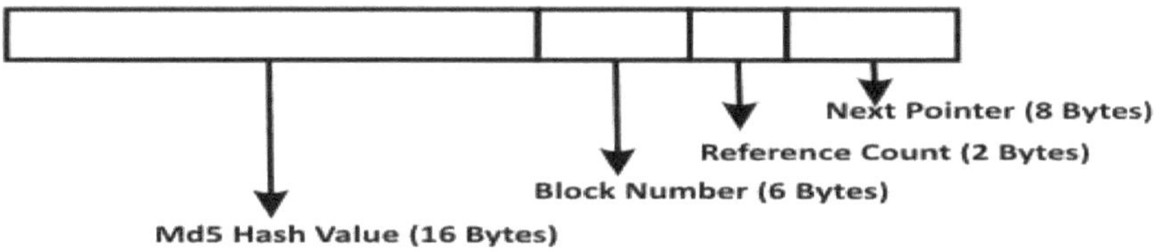

Fig. 4. Node Structure for the de-duplication database

filled. So, when each time 4KB of chunk arrives its hash value is calculated and is compared with the previous hash values stored in this node structure. As shown in Fig. 3 the two possibilities may arise as follows:

1. The value obtained is different from already stored hash value in table
2. The value is already existing in hash table

Considering the first case, each time hash value is calculated for 4KB chunk and if it is not there in the hash table then the corresponding hash value is stored in the node structure, normal write operation is performed and the block is written to the disk. However when the second case occurs, firstly the reference count is incremented to denote that there is a block which is referring to the original block which is already saved. Then the block is checked to verify if it is contiguous with the previous duplicate block. If true, it means its block number is one greater than the previous block, so the block is merged in the current extent and the block is not written to the disk. Otherwise a new extent is allocated for the duplicate block. In both the cases the metadata i.e. inode table and bitmaps are updated.

Our system deals with the following scenarios for blocks in a file:

1. The non-duplicate blocks present before the first duplicate block
2. The duplicate block or multiple duplicate blocks are present
3. The number of blocks are present after the duplicate block/s

Dealing with first case, the number of extents allocated will be 1 till block count reaches up to 32768 since one extent can contain 32768 blocks [13], or else the next extent is allocated so that a sequential read/write can be performed on blocks that are contained in a single extent. Second and third cases are explained as shown in Fig. 5.

Suppose there are 2 files say a.txt and b.txt, each file containing 7 blocks of 4KB each. The hash value of each block of a.txt is calculated and stored in hash table. On the other side, when second file is to be stored, the hash value of each 4KB block is calculated and is compared with already stored hash values in hash table. If match is found, the same hash value will not be stored again in hash table instead the reference count field in the node structure of already allocated node with similar hash value will be incremented. If match does not occur, the

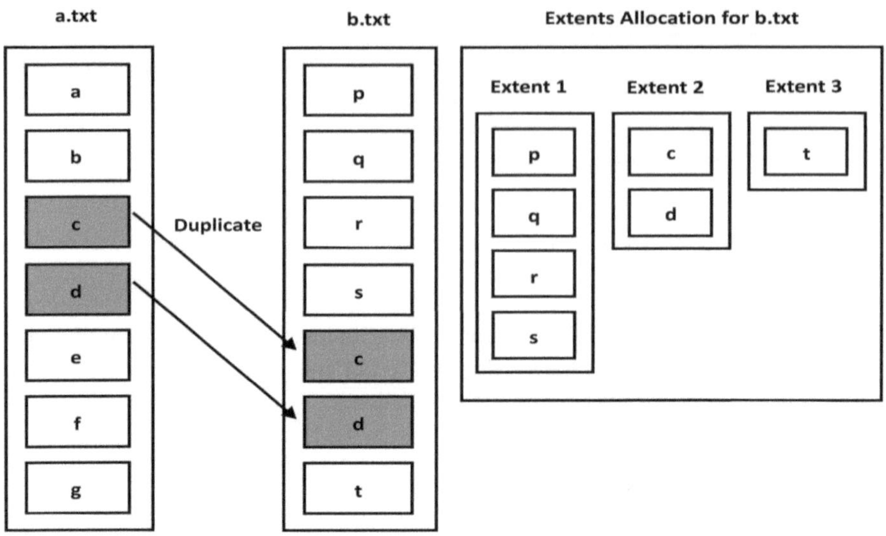

Fig. 5. Node Structure for the de-duplication database

hash value is stored in the hash table and the same process is repeated for all other blocks of file. Let's assume that the starting block number of a.txt is 1000 and that of b.txt is 2000. As shown in the Fig. 5 block c and d of file b.txt contain same data, so blocks p, q, r, s are stored in the first extent with length 4 and with starting block number 2000.For blocks c and d a separate extent is allocated with length 2 and starting block number as 1002 and for t a separate extent is allocated with length 1 and block number 2004.

6 Results

We have deployed the algorithm by modifying the source code of EXT4 file system in the Linux Kernel. In order to get the lower level details about the extents and block allocation by EXT4 file system we have used ghex [12] followed by istat, fsstat and blkcat commands [15].

```
73 30 00 A4 81 00 00 00 A0 00 00 A7 43 3C 51 A7 43 3C 51 A7 43 3C 51 00 00 00 00 00 00 01 00
50 00 00 00 00 00 08 00 01 00 00 00 0A F3 01 00 04 00 00 00 00 00 00 00 00 00 00 00 0A 00 00
00 2C 86 00 00 00 00 00 00 00 00 00 00 00 00 00 00 00 00 00 00 00 00 00 00 00 00 00 00 00 00
00 00 00 00 00 00 00 00 00 00 84 C1 D9 5F 00 00 00 00 00 00 00 00 00 00 00 00 00 00 00 00 00
00 00 00 00 00 00 00 1C 00 00 00 C4 CA FA 93 C4 CA FA 93 60 AF 43 93 A7 43 3C 51 60 AF 43 93
```

Fig. 6. Node Information of a.txt

```
3A 66 69 6C 65 5F 74 3A 73 30 00 A4 81 00 00 00 A0 00 00 C4 43 3C 51 C4 43 3C 51 C4 43 3C 51
00 00 00 00 00 00 01 00 40 00 00 00 00 00 08 00 01 00 00 00 0A F3 03 00 04 00 00 00 00 00 00
00 00 00 00 00 06 00 00 00 40 86 00 00 06 00 00 00 02 00 00 00 32 86 00 00 08 00 00 00 02 00
00 00 46 86 00 00 00 00 00 00 00 00 00 00 00 00 00 00 88 C1 D9 5F 00 00 00 00 00 00 00 00 00
00 00 00 00 00 00 00 00 00 00 00 00 00 00 00 1C 00 00 00 DC C0 68 27 DC C0 68 27 C4 A1 B1 26
```


Fig. 7. Node Information of b.txt

Each inode entry starts with A481 value as shown in Fig. 6 and Fig. 7. Bytes 40- 41 represent the magic number which is statically defined to 0xF30A, bytes 42-43 represent the number of extents in our file, bytes 52-55 represent the logical block number, bytes 56-57 represent number of blocks in extent, bytes 58-59 give the upper 16 bits of physical block address and bytes 60-63 give the lower 32 bits of physical block address.

We have gathered these results by keeping the EXT4 partition size to 10GB with default 4KB block size. Assume that initially there are no files on the disk, we sotred the first file say a.txt having 10 blocks in it. In the second file b.txt, assume that 7th and 8th blocks are same as that of a.txt. Looking at the inode information of a.txt the number of extents allocated is 1 (bytes 42-43) and the number of blocks in that extent is 0x000A => 10. The starting block number of this extent (bytes 60-63) is 0x0000862C => 34348, so for the 7th and 8th block of a.txt block numbers are 34354 and 34355 respectively.

Let's analyze b.txt to check whether we got the same block numbers. Observing the inode information of b.txt it is seen that 3 extents (bytes 42-43) are allocated for b.txt, whereas the first extent covers 6 blocks (bytes 56-57) and the starting block number of this extent is 0x00008640 => 34368. The next 4 bytes represents the logical block number of the 7th block. The shaded portion (0x0002) next to it represents that this is the next extent and there are 2 blocks in this extent. The next shaded portion 0x00008632 => 34354 represents the starting block number of this extent and the next block number is 34355.

We got the same block numbers as we have got in a.txt for 7th and 8th logical blocks. If these two blocks were not duplicate then 7th and 8th block number starting from 34374 i.e. 34374 and 34375 would not have been assigned for these blocks. But since they are duplicate blocks, 34374 and 34375 block numbers are assigned for the remaining two blocks, as can be seen in the shaded portions 0x0002 and 0x00008646=>34374 respectively.

Now if we calculate the size of a.txt using "ls -s"command which gives the allocated size of each file in blocks it is 40KB, and the size of b.txt is 32 KB since two blocks were duplicate. This proves the working of de-duplication algorithm.

7 Conclusion

The outcome of our de-duplication approach for EXT4 file system is inspiring and it shows how our approach is able to save substantial amount of disk space by actually avoiding writing duplicate data to the disk. Our approach has successfully handled the allocation of extents for contiguous and non-contiguous duplicate block/s. Since the duplicate blocks are identified before they are actually written to the disk, our algorithm also helps in reducing the number of writes to disks which may be helpful in SSDs where the number of writes are limited.

References

1. El-Shimi, A., Kalach, R., Kumar, A., Oltean, A., Li, J., Sengupta, S.: Primary Data Deduplication-Large Scale Study and System Design. In: Proc. USENIX ATC, Boston, MA (2012)
2. Windows Storage Server, `http://technet.microsoft.com/en-us/library/gg232683WS.10.aspx`
3. EMC Corporation: EMC Centera: Content Addresses Storage System, Data Sheet (2002)
4. Quinlan, S., Dorward, S.: Venti: a new approach to archival storage. In: The First USENIX Conference on File and Storage Technologies (Fast 2002), vol. 2, pp. 89–101 (2002)
5. Alvarez, C.: NetApp de-duplication for FAS and V-Series deployment and implementation guide. Technical ReportTR-3505 (2011)
6. Brown, A.: Kristopher Kosmatka: Block-level Inline Data de-duplication in EXT3. In: University of Wisconsin - Madison Department of Computer Sciences (2010)
7. More, A., Shaikh, Z., Salve, V.: DEXT3 Block Level Inline De-duplication using EXT3 File System. In: Linux Symposium, p. 87 (2012)
8. Larabel, M.: SDFS: A File-System With Inline De-Duplication (2011)
9. Bhagwat, D., Eshghi, K., Long, D.D., Lillibridge, M.: Extreme Binning: Scalable, Parallel de-duplication for Chunk-based File Backup. In: IEEE International Symposium on Modeling, Analysis & Simulation of Computer and Telecommunication Systems, MASCOTS 2009, pp. 1–9. IEEE (2009)
10. Zhu, B., Li, K., Hugo Patterson, R.: Avoiding the disk bottleneck in the data domain de- duplication file system. In: Fast, vol. 8, pp. 269–282 (2008)
11. Cao, M., Santos, J.R., Dilger, A.: EXT4 Block and Inode Allocator Improvements. In: Linux Symposium, p. 263 (2008)
12. Fairbanks, K.D.: An analysis of EXT4 for digital forensics. Digital Investigation 9, S118–S130 (2012)
13. Avantika, M., Cao, M., Bhattacharya, S., Dilger, A., Tomas, A., Vivier, L.: The new ext4 filesystem: current status and future plans. In: Proceedings of the Linux Symposium, vol. 2, pp. 21–33 (2007)
14. Kadekodi., S., et al.: Taking Linux Filesystems to the Space Age: Space Maps in EXT4. In: Linux Symposium (2010)
15. `http://computer-forensics.sans.org/blog/2010/12/20/digital-forensics-understanding-ext4-part-1-extents#part1-5`

Unique Key Based Authentication of Song Signal through DCT Transform (UKASDT)

Uttam Kr. Mondal[1], and J.K. Mandal[2]

[1] Dept. of CSE & IT,
College of Engg. & Management, Kolaghat
Midnapur W.B., India
uttam_ku_82@yahoo.co.in
[2] Dept. of CSE, University of Kalyani
Nadia W.B., India
jkm.cse@gmail.com

Abstract. Authentication of song signal is one of the assessments to detect the originality in ease of alteration of its content. In this paper, DCT transform has been applied to song signal to extract a unique key for a particular song and which itself represents the characteristics of whole song signal. Comparing song signal with computed and extracted unique keyword unauthorized ownership can be verified. A comparative study has been made with similar existing techniques to compare its characteristics which show better performances. Computed characteristics are also supported through mathematical formula based on Microsoft WAVE (".wav") stereo sound file.

Keywords: Audio song authentication, DCT, song security, protection of intellectual property, unique identification of song signal.

1 Introduction

Unique identification is one of the basic ways to determine a product/object. In most cases, the unique feature is extracted from the characteristics of product/object. Changing the characteristics need to change the unique identification. Therefore, the unique identification will reveal itself the uniqueness in features of the particular object/ product. Applying similar technique in the case of song signal, a unique feature may be extracted from song signal which represents the characteristics of whole song signal. If any change occurs, it will be differed from its quantifying value [4, 8, 9, 10].

Another important aspect of identification technique is to assess human readability and practical handling of the unique feature. Therefore, we represent the unique feature in text form, then a text context or label will determine a particular song signal and it also maps to its unique features those are differed from other song signal [1, 3, 5].

In this paper, DCT (Discrete Cosine Transform) has been deployed for extracting the unique characteristics from the sampled values of song signal. DCT will extract a set of component and selecting some components along with all positions of the

S.C. Satapathy et al. (eds.), *ICT and Critical Infrastructure: Proceedings of the 48th Annual Convention of CSI - Volume II*, Advances in Intelligent Systems and Computing 249,
DOI: 10.1007/978-3-319-03095-1_61, © Springer International Publishing Switzerland 2014

selected coefficients of transform components – the unique text is generated which will vary if the DCT component changes.

Organization of the paper is as follows. Section 2.1 of the paper deals with generation of secret key and technique of embedding secret key is described in section 2.2. The authentication procedure has been depicted in section 2.3. Experimental results are given in section 3. Conclusions are drawn in section 4. References are outlined at end.

2 The Technique

The scheme fabricates a unique key generated through Discrete Cosine Transform followed by embedding authenticating code. Extracting unique features of song signal and represent those into human readable format are done for generating secret key. Embedding secret key (of 8 characters) into song signal is also designed as such a way which will be strongly authenticated the song signal without compromising its audible quality. The details of these two processes are described in section 2.1 and section 2.2 respectively.

2.1 Unique Key Generation (UKASDT-UKG)

The Discrete Cosine Transform (DCT) is used to separate the signal into parts (or spectral sub-bands) of different importance (with respect to the song's audible quality). The general equation for 2D (N by M) DCT is expressed by equation (1), whereas equation (2) is depicting the inverse DCT [12].

$$F(u,v)=\frac{2}{N}C(u)C(v)\sum_{x=0}^{N-1}\sum_{y=0}^{N-1}f(x,y)\cos\left[\frac{\pi(2x+1)u}{2N}\right]\cos\left[\frac{\pi(2y+1)v}{2N}\right] \tag{1}$$

for $u=0,\ldots,N-1$ and $v=0,\ldots,N-1$

where $N=8$ and $C(k)=\begin{cases}1/\sqrt{2} & \text{for } k=0\\ 1 & \text{otherwise}\end{cases}$

$$f(x,y)=\frac{2}{N}\sum_{u=0}^{N-1}\sum_{v=0}^{N-1}C(u)C(v)F(u,v)\cos\left[\frac{\pi(2x+1)u}{2N}\right]\cos\left[\frac{\pi(2y+1)v}{2N}\right] \tag{2}$$

for $x=0,\ldots,N-1$ and $y=0,\ldots,N-1$ where $N=8$

Applying DCT to sampled values of song signal and representing produced transform eight consecutive values into binary pattern for observing most common portion among them where the alternation of bit pattern occurred frequently, then select the region as most sensitive bit portion. The variation of bit pattern of coefficient values is determined using Hamming distance as described in algorithm 1.

Algorithm 1

Input : DCT coefficients of song signal.

Output : A window of minimum. Hamming distance from fractional part of DCT value

Method: A binary window (8X8 matrix) is selected using following steps.

Step 1: Select eight consecutive coefficient values from the DCT transform of song signal.

Step 2: Represent each individual digit of fraction part of the selected values into equivalent bit stream of 8 bits consecutively (considering up to maximum 24 bits).

Step 3: Select a window 8 X 8 and fill each cell value along row (8 positions consecutively) from 1st position of binary representation of selected values of step 2 sequentially i.e. first 8 bits of each from selected values.

Step 4: Find maximum Hamming distance taking any combination of rows and this will be considered as Hamming distance value for this particular position of the window.

Step 5: Shift the window a bit position towards the last bit and apply step 3 to 4 for determining Hamming distance for present shifted position.

Step 6: Repeat step 5 up to the last bit position and determine Hamming distance for each shifted window position.

Step 7: Find maximum Hamming distance values among all shifted positions of the window (step 6) and determine its each cell value at this particular position. If more than one positions having same Hamming distance select only 1st position.

Applying above steps 1-7, a particular 8 X 8 bits combination will be found which is graphically shown in figure 1 where the DCT of sampled values of song signal for a channel are given as follows: 0.2022, 0.4395, 0.3841, 0.0014, 0.1628, 0.2045, 0.4652, 0.1512 with respective bit patterns (considering 20 bits for fraction part) are shown in figure 1.

Decimal Part	Fraction part
0 0	0 0 0 0 0 0 1 0 0 0 0 0 0 0 0 0 0 0 0 0
0 0	0 0 0 0 0 1 0 0 0 0 0 0 0 0 1 1 0 0 0 0
0 0	0 0 0 0 0 0 1 1 0 0 0 0 0 0 0 0 0 0 0 0
0 0	0 0 0 0 0 0 0 0 0 0 0 0 0 0 0 0 0 0 0 0
0 0	0 0 0 0 0 0 0 1 0 0 0 0 0 1 1 0 0 0 0 0
0 0	0 0 0 0 0 0 1 0 0 0 0 0 0 0 0 0 0 0 0 0
0 0	0 0 0 0 0 1 0 0 0 0 0 0 0 1 1 0 0 0 0 0
0 0	0 0 0 0 0 0 0 1 0 0 0 0 0 1 0 1 0 0 0 0

Fig. 1. Bit pattern and searched window in fractional part of DCT values

The window (darken double line box) is representing a shifted position where Hamming distance is 2 (max. Hamming distance considering any combination of 2 from 8 rows).

Therefore, applying above algorithm a particular 8 X 8 bits combination (a window) is obtained from a selected region (of 8 coefficient values.) Selection (a block of 8 coefficient values), is done by applying algorithm 2.

Algorithm 2
Input : DCT coefficient of song signal.
Output : A collection of blocks (of 8X8 matrix).
Method: Representing DCT values song signal into a series of blocks is done as follows.

Step 1: Divide all coefficient values of song signal for a particular channel into groups of 8 coefficient values sequentially as a series of blocks having 8 consecutive coefficient values.

Step 2: Determine a particular block number based on the Fibonacci series i.e., 1, 2, 3, 5, 8, 13, etc. for finding a selective window as described above.

On obtaining a unique block or unique character set is generated using algorithm 3.

Algorithm 3
Input : DCT coefficients of song signal.
Output : A set of 8 integer numbers (ASCII values) for generating secret key
Method: Creating secret key with selected windows is obtained as follows.

Step1: Determine a set of selected regions (of 8 coefficients block) up to N^{st} term of Fibonacci series (say, N is 20) as described above.

Step2: Find 8 windows from step1 according to their Hamming distances in descending order from each of the block selected from step 1.

Step3: Choose i^{st} column from i^{st} selected window from step2 (where 1<=i<=8) and find corresponding integer value of constituted 8 bits values of each column.

Representing the 8 integers into equivalent ASCII values respectively and placing them side by side will generate a secret key of 8 characters. In case of stereo type song, all above methods can be used alternately for both channels.

2.2 Embedding Secret Key (UKASDT-ESK)

Embedding a secure key (generated by section 2.1) with a secure code in the DCT form of sampled values carries another level of authentication in the song signal without affecting its audible quality. Embedding a secure key in the DCT form of signal is done using algorithm 4.

Algorithm 4

Input : DCT coefficients of song signal and chosen 8 characters key

Output : A set of 8 integer numbers (ASCII values) for generating secret key

Method: Creating secret key with selected windows is obtained as follows.

Step 1: Apply DCT on a particular channel of stereo type song signal and arrange the generated transform values in form of n matrices (frames) of 8X8 size as describe in section 2.1. Choose any 8 frames based on Fibonacii series as described 2.1 excluding the key generated frames.

Step 2: Convert the secret key into equivalent ASCII value of each character and represent them also a set of lower magnitude values by converting into fractional values [let, A is one of the character in secret key, equivalent ASCII value is 65, therefore, converted fraction value is 0.65]

Step 3: Select a particular frame and replace the coefficient value of (8, 1) position[row =8, column=1] by the average value of 8th row of the frame.

Step 4: Choose same frame (step 3) and replace the coefficient value of (1, 8) [row =1, column=8] by the 1st lower magnitude value (step 2). Again, compute the average value of 8th column (after adding secret character's value) and replace the coefficient value of (8, 8) [row =8, column=8] by the new average value.

Step 5: Continue the step 4 for 7 times more for adding remaining lower magnitude values of secret key (step 2).

Step 6: Replace the modified frame (step 5) back to original position of song signal.

Therefore, justifying 3 positions [(8, 1), (1, 8), (8, 8)]of each embedded frame, easily modified portion or changes in song signal as well as key value can be determined.

2.3 Authentication

As discuss in the section 2.1**,** each song signal will produce a unique character set (of 8 characters) which itself represent the song signal characteristic, i.e., each original song will associated a key and in any stage of processing, if necessary, need to apply same technique to extent song signal characteristic and comparing extracted character set and original character set, originality of song signal can be verified. The secret frame is also used to detect the original song signal by finding secret key from the specified position of it.

Therefore, if any change occurs during processing of signal by any alternation, it can be easily detected by computing secret key from the modified song and comparing it with hidden key as described in section 2.2. It will also be applied for determine the portion of song signal which has been altered during processing by observing changing character set pattern.

3 Experimental Results

Encoding and decoding technique have been applied over 1 minute recorded song, the song is represented by complete procedure along with results in each intermediate step has been outlined in subsections 3.1.1 to 3.1.3. The results are discussed in two sections out of which 3.1 deals with result associated with UKASDT and that of 3.2 gives a comparison with existing techniques.

3.1 Results

For experimental observation, a strip of 1 minute song ('One day in your life', sung by Michael Jackson) has been taken. Figure 2 shows amplitude-time graph of the original signal. UKASDT is applied on this signal and the output generated in the process is shown in figure 3. Figure 4 shows the difference of amplitude values before and after modification of original song. From figure 4 it is seen that the deviation of the modified song signal is very less, i.e., its audible quality will not be affected at all.

3.1.1 Original Recorded Song Signal (1 Minute)

The graphical representation of the original song, considering sampled values (2646000) of x(n,2) [stereo type song] is given in the figure 2.

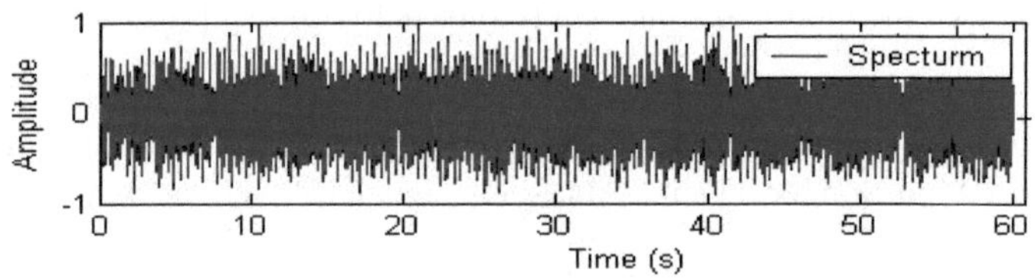

Fig. 2. Original song ('One day in your life', sung by Michael Jackson)

3.1.2 Modified Song after Applying DCT with Secret Frame (1 Minute)

The graphical representation of the modified authenticated song signal is shown in the figure 3.

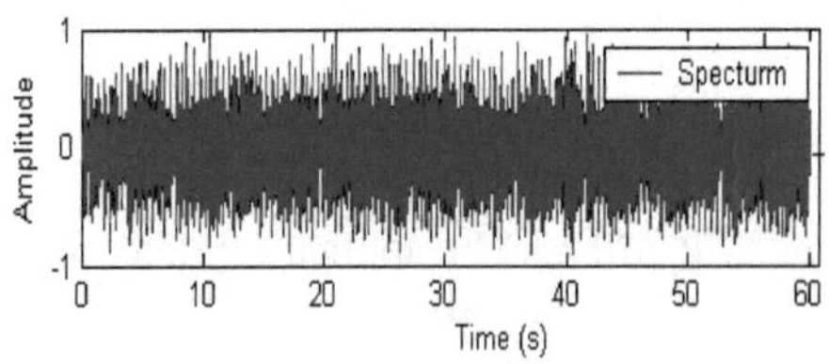

Fig. 3. Modified song after adding authenticated code

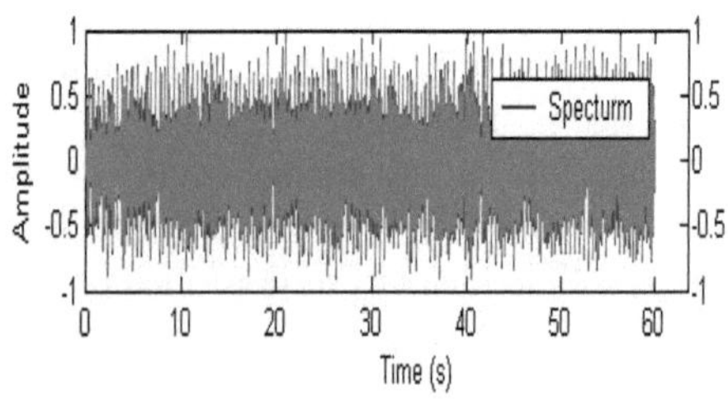

Fig. 4. The difference of sample values between signals (figure 2 and figure 3)

3.1.3 Difference of Magnitude Values between Original and Modified Songs

The graphical representation of the difference of magnitude values of original and modified songs is shown in the figure 4.

3.2 Comparison with Existing Systems

Various algorithms [6, 7] are available for embedding information with audio signals, but we are enforcing our authentication technique without changing the quality of song. A comparison study of properties of our proposed method with Multi-Channel Audio Information Hiding (MCAIH) [2] before and after embedding secret message/modifying parts of signal (16-bit stereo audio signals sampled at 44.1 kHz) is given in table 1, table 2 and table 3. Average absolute difference (AD) is used as the dissimilarity measurement between original song and modified song to justify the modified song. Whereas a lower value of AD signifies lesser error in the modified song. Normalized average absolute difference (NAD) is quantization error is to measure normalized distance to a range between 0 and 1. The higher the PSNR represents the better the quality of the modified song. Thus from our experimental results of benchmarking parameters (NAD, MSE, NMSE, SNR and PSNR) in proposed method obtain better performances without affecting the audio quality of song. The Table 4 shows PSNR, SNR, BER (Bit Error Rate) and MOS (Mean opinion score) values for the proposed algorithm. Figure 5 summarizes the results of this experimental test. This quality rating (Mean opinion score) is computed by using equation (3).

$$Quality = \frac{5}{1 + N * SNR} \tag{3}$$

Where N is a normalization constant and SNR is the measured signal to noise ratio. The ITU-R Rec. 500 quality rating is perfectly suited for this task, as it gives a quality rating on a scale of 1 to 5 [11]. Table 5 shows the rating scale, along with the quality level being represented.

Table 1. Metric for different distortions

Sl No	**Statistical parameters for differential distortion**	**Value using UKASDT**	**Value using MCAIH**
1	MD	3.0518e-005	4.8828e-004
2	AD	4.7093e-006	5.6543e-005
3	NAD	5.0315e-005	4.6591e-004
4	MSE	1.4372e-010	1.3976e-008
5	NMSE	1.1222e+008	1.8257e+006

Table 2. SNR and PSNR between original and modified song

Sl No	**Statistical parameters for differential distortion**	**Value using UKASDT**	**Value using MCAIH**
1	Signal to Noise Ratio (SNR)	52.5007	62.6143
2	Peak Signal to Noise Ratio (PSNR)	97.461	78.3738

Table 3. NC and QC between original and modified song

Sl No	Statistical parameters for correlation distortion	Value using UKASDT	Value using MCAIH
1	Normalised Cross-Correlation(NC)	1	1
2	Correlation Quality (QC)	-0.0011	-0.1137

Table 4. SNR, PSNR BER, MOS for finding consisting of different characteristics

Audio (1s)	SNR	PSNR	BER	MOS
Song1	52.5007	97.461	0	5
Song2	62.8319	90.5186	0	5
Song3	59.2604	77.8709	0	5
Song4	65.2257	94.7453	0	5
Song5	66.6017	84.6538	0	5

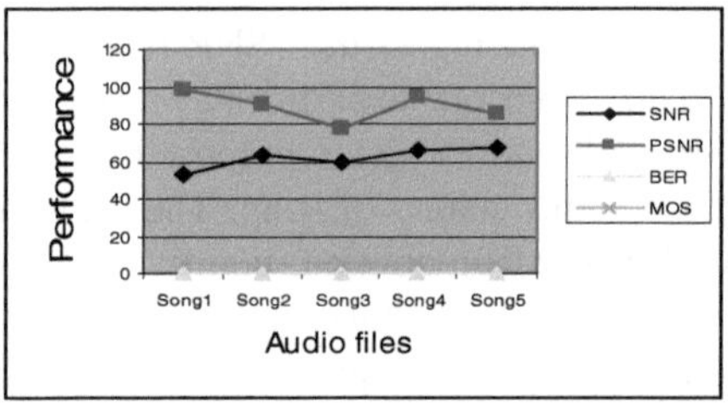

Fig. 5. Performance in terms of SNR, PSNR, BER and MOS for different audio signals

Table 5. The ITU-R Rec. 500 quality rating for audio signal

Rating	Impairment	Quality
5	Imperceptible	Excellent
4	Perceptible, not annoying	Good
3	Slightly Annoying	Fair
2	Annoying	Poor
1	Very annoying	Bad

4 Conclusion and Future Work

In this paper, an algorithm for extracting an authenticating code from original song signal for uniquely identifying it with the help of DCT in the specified portion as well as passing secret information has been proposed which will not affect the song quality but it will detect the distortion of song signal characteristics without effecting audible quality of song signal.

This technique is developed based on the observation of characteristics of different songs but the mathematical model for representing the variation of those characteristics after modification may be formulated in future. It also can be extended to embed an image into an audio signal instead of text message. The perfect estimation of percentage of threshold numbers of sample data of song that can be allow to change for a normal conditions will be done in future with all possibilities of errors.

References

1. Mondal, U.K., Mandal, J.K.: A Novel Technique to Protect Piracy of Quality Songs through Amplitude Manipulation (PPAM). In: International Symposium on Electronic System Design (ISED 2010), pp. 246–250 (2010)
2. Blackledge, J.M., et al.: Multi-Channel Audio Information Hiding. In: Proc. of the 15th Int. Conference on Digital Audio Effects (DAFx 2012), York, UK, September 17-21, pp. 1–8 (2012)
3. Erten, G., Salam, F.: Voice Output Extraction by Signal Separation. In: ISCAS 1998, vol. 3, pp. 5–8 (1998) ISBN 07803-4455-3
4. Dong, X., Bocko, M.F., Ignjatovic, Z.: Data Hiding Via Phase Manipulation of Audio Signals. In: Proc. IEEE International Conference on Acoustics, Speech, and Signal Processing (ICASSP 2004), vol. 5, pp. 377–380 (2004) ISBN 0-7803-8484-9
5. Mondal, U.K., Mandal, J.K.: Song Authentication Technique Through Concealment of Secret Song (SATCon). Presented at IEEE-International Conference on Recent Trends in Information Technology, ICRTIT 2011, June 3-5, pp. 145–150. MIT, Anna University, Chennai (2011)
6. Petitcolas, F.A.P., Anderson, R.J., Kuhn, M.G.: Information Hiding –A Survey. Proc. the IEEE 87(7), 1062–1078 (1999)
7. Anderson, R.J., Petitcolas, F.A.P.: On the Limits of Steganography. Proc. IEEE Journal of Selected Areas in Communications 16(4), 474–481 (1998)
8. Doddington, G.R.: Speaker Recognition- Identifying People by their Voices. Proc. of the IEEE 73(11), 1651–1665 (1985)
9. Suzuki, H., Zen, H., Nunkuku, Y., Miyajima, C., Tokuda, K., Kitumuru, I.: Proc. Speech Recognition Using Voice- Characteristic Dependent Acoustic Models, ICASSP 2003, vol. 3, pp. 740–743 (2003)
10. Lin, K.S., Frantz, G.A.: Proc. Voice Characteristics Conversion, IEEE Transactions on Consumer Electronics CE-30(4), 598–603 (1984)
11. Arnold, M.: Audio watermarking: Features, applications and algorithms. In: IEEE International Conference on Multimedia and Expo (2000)
12. https://en.wikipedia.org/wiki/Discrete_cosine_transform#DCT-II (last accessed on July 30, 2013)

References

DCT-PCA Based Method for Copy-Move Forgery Detection

Kumar Sunil[1], Desai Jagan[1], and Mukherjee Shaktidev[2]

[1] Faculty of Engineering & Technology,
Mody Institute of Technology & Science, Laxmangarh, India
skvasistha@ieee.org, jagandesai@yahoo.com
[2] Moradabad Institute of Technology, Moradabad, India
mukherjee.shaktidev@gmail.com

Abstract. Copy move forgery detection is emerging as one of the hot research topic among researchers in the area of image forensics. There are many techniques suggested to detect such type of tampering with the original image but, many issues still remained either unsolved or there is a lot of scope for performance improvement. The most commonly used algorithm to detect such type of tampering is block matching algorithm. Robustness against post processing operations and the time taken by the detection techniques are few of the major challenges. Change of intensity of the copy moved part is one of the post processing operations that may be employed by the attacker to evade the image forgery detection methods. This is successfully addressed in the proposed algorithm. Discrete cosine transform and principal component analysis have been used to represent and compress the feature vector of overlapping blocks respectively. Features , invariant to local change of intensity are created using down sampling of low frequency DCT coefficients.

Keywords: Blind forgery detection techniques, image forensics, intensity invariant forgery detection.

1 Introduction

Image forgery now a days is a challenge for the authorities relying on the visual information. In the early days, when computers were not available easily for everybody, there was a tendency to believe what we see. But manipulation of the digital images has become very easy with the availability of fastly improving hardware and sophisticated software. The visual document has to be authenticated before drawing any conclusion based upon it. So, detecting the manipulations in digital images is not only important but also necessary. Image forgery detection techniques may be classified as per Figure 1. A few such techniques are listed in [1].Active forgery detection methods like watermarks are already being used to protect digital images. But using such techniques has their limitations, as the source of capturing the images may not be controlled and authenticated always. Blind forgery detection is used to detect manipulations in the absence of active techniques. Digital image forgery may

S.C. Satapathy et al. (eds.), *ICT and Critical Infrastructure: Proceedings of the 48th Annual Convention of CSI - Volume II*, Advances in Intelligent Systems and Computing 249,
DOI: 10.1007/978-3-319-03095-1_62,

be performed in many different ways [2].Out of these the commonly used method is copy move forgery, where parts of the same image are used either to hide some other parts of image or to amplify some fact. In case of natural images, existence of two same regions is not common. So, the similarity created as a result of copy move attack is exploited to detect copy move forgery. One of the most frequently used method for detecting such type of forgery is to use block matching algorithm [3][4].In the block matching algorithm the image is divided into overlapping blocks and the blocks are matched to find the duplicated regions. Many people have used it to find duplication of the region with different features representing a block of image [5][6]. There may be some post processes performed, like edge smoothing, blurring, noise adding and change of intensity, but even after the post processing operations almost identical regions created in the manipulated image.

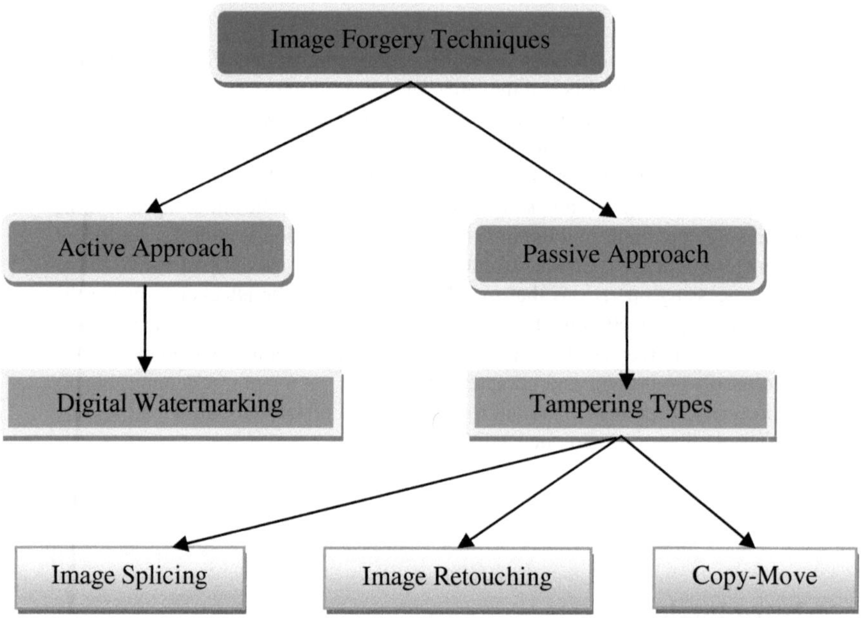

Fig. 1. Classification of image forgery

2 Related Work

One of the landmark method for copy move forgery detection was suggested by Fridrich[3]. He suggested a block matching forgery detection method based on discrete cosine transform (DCT). Popescu proposed a similar method [4], which used principal component analysis (PCA) instead of DCT. DCT is supposed to be a good feature for digital images and has been utilized by [7] [8] [9].In [7], the authors proposed a method to make the similarity criteria more robust by calculating the component ratio for matching the feature vectors. In [8], effort has been made to reduce time complexity by applying DWT. In [9], both low and high frequency bands of DyWT are used to achieve robustness. When the length of feature vector is large, handling the DCT coefficients become more difficult. In such cases the lower frequency coefficients are only retained to curtail the number of coefficients. This process has been done manually and it is difficult to suggest a threshold for faithful representation of feature vector. PCA is good at reducing the dimensionality of the data and has been utilized in [4].The method using PCA is quite fast but lack robustness. There are some methods that can detect more sophisticated post-processing, like Mahdian's work [10] can detect the duplicated region even the copy-move region been blurred or added noise. However, some issues have been reported by the researchers [10][11] in the algorithms, like the time taken by the algorithm to detect the matching blocks. In the present paper efforts have been made to address the issue of robustness in the form of intensity invariance with good time complexity and the experimental results have shown that the performance of the method is good with improved robustness.

3 Proposed Method

The proposed method addresses a particular type of robustness against change of intensity of the copy moved region. The manipulator may increase or decrease the brightness of the copy moved region to fit in some other part of the image. In such cases the DC component of the DCT feature vector will be significantly different even for the copy moved regions and evade the existing DCT based forgery detection techniques. However other components in the feature vector will be same. So the low frequency coefficients have been down sampled using very large quantization factor. The updated feature vector is invariant to the intensity variations of the copy moved part. However, it is assumed that the intensity variation will be uniform at least over the single block. It has been proved that PCA can be used directly in DCT domain [12]. So PCA is applied to reduce the dimensionality of the feature vector. To further reduce the time taken by the algorithm DCT coefficients are calculated in parallel for the overlapping blocks. Also to make the final decision process efficient morphological operations are used to remove the isolated pixels and only retain the significant connected components. The algorithm structure is as follows:

(1) The input image is a gray scale image I of the size m x n. If it is a color image, it can be converted to a grayscale image using the standard formula I = 0.299R + 0.587G + 0.114B.
(2) A fixed-sized b x b window is slided one pixel along from the upper left corner to the button right, dividing I into (m-b +1)(n-b +1) blocks.
(3) For each block, apply DCT and reshape the b x b coefficient matrix to a row vector in zigzag order.
(4) Down sample the low frequency coefficients by using large quantization factor such that the DC component is very close to zero.
(4) Do PCA to the array of row vector to reduce the dimensionality and result a m-b +1)(n-b +1) x qb^2 matrix A.

(4) Use lexicographical sorting on A to sort the row vectors according to their similarity.

(5) For each row a_i in A, test its neighboring rows a_j which satisfy the threshold condition of minimum distance between duplicated rows (N_n).

(6) If the distance between similar blocks (N_d) is greater than block size, then a shift vector 's' is calculated and normalized.
$s= (s_1, s_2) = (i_1-j_1, i_2-j_2)$, where (i_1, j_1) and $(i_{2,,} j_2)$ are similar block coordinates . Then the shift vector's existing frequency is increased by one.

(7) The frequency matrix is sorted and third value is taken to be threshold frequency (N_f). Also it should be more than b x b.

(8) For all the blocks having shift value greater than the threshold mark (N_f) the block points in the dark image are registered as white points.

(9) Finally apply morphological operations to remove the isolated points and show the binary image as output representing copy moves regions with white regions.

4 Experimental Setup and Results

To check the performance of the proposed algorithm, different sized images are taken. The image dataset contains 90 images with varying texture and intensity distributions. The dataset includes some images created specifically for the experiment and some from the dataset [13] with intensity variation of the copy moved part ranging from -50 to +80. The intensity of copy move part is varied using Photoshop. The algorithm is implemented in MATLAB 2012a on the machine equipped with Intel i5 1.8 GHz Core 2 duo processor and 8GB DDR3RAM. N_n (number of rows selected for potential similar block vector) is selected 30.The number of components retained in the feature vector after applying PCA are ¼ of the original components (q=0.25). The value of q is set experimentally. Lower values have shown adverse effect on the efficiency of the algorithm.

4.1 Visual Results

Fig. 2(a). Original image barrier.bmp

Fig. 2(b). Tampered image without any intensity variation of copy moved region

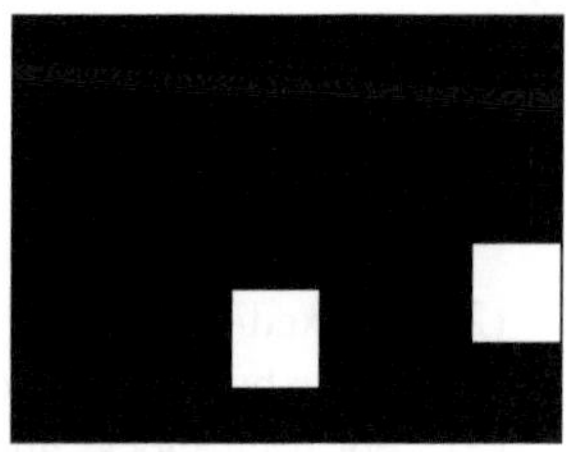

Fig. 2(c). Detection result of proposed method

Fig. 3(a). Original image barrier.bmp

Fig. 3(b). Tampered image with intensity variation of (+10) in copy moved region

Fig. 3(c). Detection result of proposed method

Fig. 4(a). Original image logo.bmp

Fig. 4(b). Tampered image with intensity variation of (+80) in copy moved region

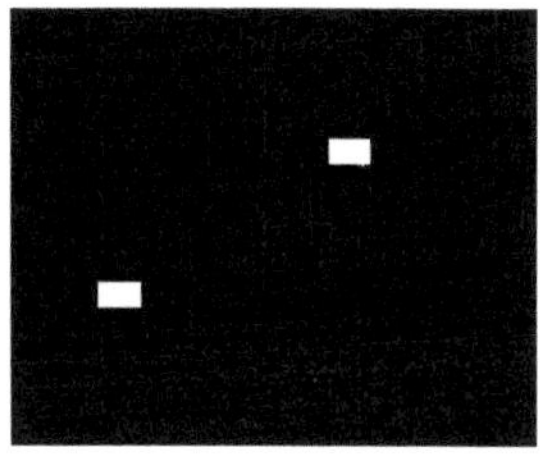

Fig. 4(c). Detection result of proposed method

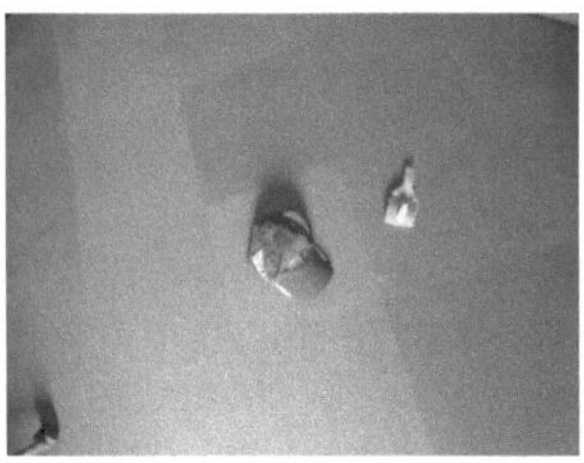

Fig. 5(a). Original image cap.bmp

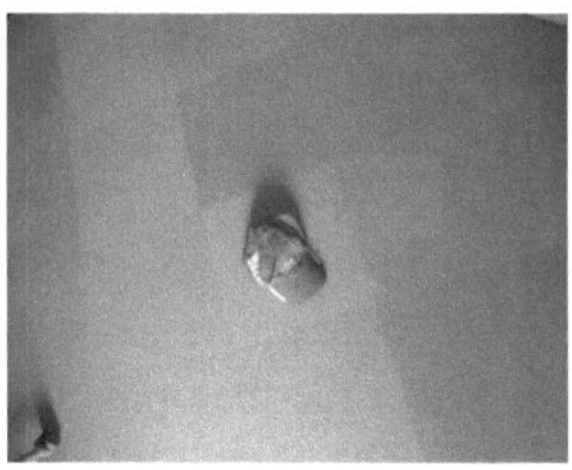

Fig. 5(b). Tampered image with intensity variation of (-50) in copy moved region

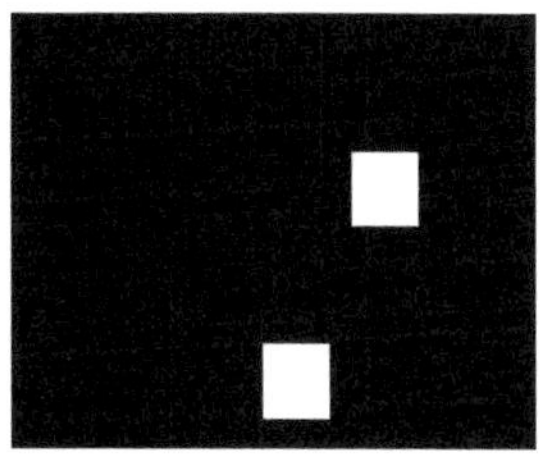

Fig. 5(c). Detection result of proposed method

Fig. 6(a). Original image pen.bmp

Fig. 6(b). Tampered image with intensity variation of (-2) in copy moved region

Fig. 6(c). Detection result of proposed method

4.2 Comparison with the Method Suggested in [3]

Table 1. Performance comparision

Image name/Image size	Intensity variation in the copy moved part	Block size	Execution Time in Seconds		Copy Move attack detected successfully	
			Method in [3]	Proposed method	Method in [3]	Proposed method
barrier.bmp (800x600)	0	8x8	92.3	71.72	yes	yes
	+10	8x8	96.03	79.85	no	no
	+10	4x4	93.98	72.4	no	yes
logo.bmp (640x480)	+50	8x8	58.86	52.11	no	yes
	+80	8x8	59.65	53.44	no	yes
	-10	8x8	59.12	52.71	no	yes
cap.bmp (640x480)	-50	8x8	61.23	53.43	no	yes
	+20	8x8	59.29	51.9	no	yes
pen.bmp (640x480)	-2	8x8	59.24	53.65	no	yes

4.3 Observations

Figure 2-6 show the visual results of the proposed algorithm on some selected images from the dataset. Table I provides comparative analysis for the images shown in figure 2-6.The method in [3] fails to detect the manipulation even with slight change in the intensity of the copied moved part. The default block size of 8x8 works well in 95% of

trials, however when the copy moved region is significantly smaller, then the block size has to be changed to 4x4 as shown in Figure 3. The proposed method has successfully detected the copy moved part with intensity changes. Also the time taken by the method is less comparable to [3].

5 Conclusion and Future Work

Another post processing technique has been introduced as change of intensity of the copy moved region. The proposed method successfully detects copy move attack in the presence of such post processing technique. The attacker may perform this operation to fit in some other part of the image to deceive the available DCT based methods. However, the feature vector may be enhanced to include the other types of invariance in a single method in future.

References

[1] Lin, E., Podilchuk, C., Delp, E.: Detection of image alterations using semi-fragile watermarks. In: Proc. SPIE, Security and Watermarking of Multimedia Contens II, vol. 3971, pp. 52–163 (2000)
[2] Birajdar, G.K., Mankar, V.H.: Digital image forgery detection using passive techniques: A survey. Digital Investigation, 1–20 (2013)
[3] Fridrich, J., Soukalm, D., Lukáš, J.: Detection of copy-move forgery in digital images. In: Digital Forensic Research Workshop, Cleveland, OH, pp. 19–23 (2003)
[4] Popescu, A.C., Farid, H.: Exposing Digital Forgeries By Detecting Duplicated Image Regions. Tech. Rep. TR2004-515, Dartmouth College (2004)
[5] Kumar, S., Das, P.K., Mukherjee, S.: Copy-Move Forgery Detection in Digital Images: Progress and Challenges. International Journal on Computer Science and Engineering 3(2), 653–663 (2011)
[6] Al-Qershi, O.M., Khoo, B.E.: Passive detection of copy-move forgery in digital images: State-of-the-art. Forensic Science International 231(1), 284–295 (2013)
[7] Huang, Y., Lu, W., Sun, W., Long, D.: Improved DCT-based detection of copy-move forgery in images. Forensic Science International 206(1), 178–184 (2011)
[8] Ghorbani, M., Firouzmand, M., Faraahi, A.: DWT-DCT (QCD) based copy-move image forgery detection. In: 2011 18th International Conference on Systems, Signals and Image Processing (IWSSIP), pp. 1–4 (2011)
[9] Muhammad, G., Hussain, M., Mirza, A.M., Bebis, G.: Dyadic wavelets and dct based blind copy-move image forgery detection. In: IET Conference on Image Processing (IPR 2012), pp. 1–6 (2012)
[10] Mahdian, B., Saic, S.: Detection of copy-move forgery using a method based on blur moment invariants. Forensic Science International 171(2), 180–189 (2007)
[11] Bacchuwar, K.S., Ramakrishnan, K.: A Jump Patch-Block Match Algorithm for Multiple Forgery Detection. In: International Multi-Conference on Automation, Computing, Communication, Control and Compressed Sensing, pp. 723–728 (2013)
[12] Chen, W., Er, M.J., Wu, S.: PCA and LDA in DCT domain. Pattern Recognition Letters 26(15), 2474–2482 (2005)
[13] Christlein, V., Riess, C., Jordan, J., Riess, C., Angelopoulou, E.: An Evaluation of Popular Copy-Move Forgery Detection Approaches. IEEE Transactions on Information Forensics and Security 7(6), 1841–1854 (2012)

Online Hybrid Model for Fraud Prevention (OHM-P): Implementation and Performance Evaluation

Ankit Mundra and Nitin Rakesh

Department of Computer Science and Engineering,
Jaypee University of Information Technology, Wakhnaghat, Distt. Solan,
Himachal Pradesh, India–173234
{ankitmundra8891,nitin.rakesh}@gmail.com

Abstract. Online Hybrid Model (OHM) approach effectively prevents, detects and eliminates the online frauds. OHM consists of two approaches: i) OHM-P which is for prevention of online frauds; ii) OHM-D which is for detection of online frauds and eliminates the detected frauds. OHM works in three layer infrastructure which comprises user, OHM systems, and web-server. Thus, an OHM system provides the secure interface for the user and web-server interaction. In this paper we have implemented the OHM-P approach using JAVA modules which provides registration interface for both user and web-server. We have evaluated our OHM-P approach on a 5-user nodes, 2-web-server, and 1-OHM system based testbed, and analyzed the OHM-P approach using security and robustness as the performance parameters.

Keywords: OHM-D, Fraud, Prevention, Detection.

1 Introduction

The advent of World Wide Web (WWW) emerges a new mode of business which is based on internet known as e-commerce. The trend of e-commerce becomes popular day by day [1]. Today, several organizations like Amazon, E-bay, Flipkart, Snapdeal *etc*. provides a platform to buy-sell goods as per consumers need. Traditional online business model provides the facility to buy-sell the product or services [2]. In that model seller sets the price for the product and if buyer feels a desired deal then he/she buys that product or service. But, nowadays several other services are also provided by e-commerce organizations like online auction, online chit-funds, online charity *etc*. [3].

As the usability of these services has increased the possibility of fraudulent cases are also increased [4]. In [5] we have shown that several types of online frauds are stirring such as online auction fraud, identity theft fraud, credit card fraud, non-delivery/merchandise fraud, online charity fraud and online investment scheme frauds. In order to restrict these online frauds we have proposed the OHM approach [5]. OHM approach is built upon two main approaches i.e. OHM-P which is stands for OHM for prevention and OHM-D which is stands for OHM for detection. In [5] we have described the three layer infrastructure of OHM approach. In that OHM

S.C. Satapathy et al. (eds.), *ICT and Critical Infrastructure: Proceedings of the 48th Annual Convention of CSI - Volume II*, Advances in Intelligent Systems and Computing 249,
DOI: 10.1007/978-3-319-03095-1_63, © Springer International Publishing Switzerland 2014

resides in the middle layer which provides interface between the users (at top layer) and web-server (at bottom layer) and regularly monitors the interaction between these two layers.

In this paper we have proposed a testbed based on OHM-P module implementation over JAVA platform. We have designed web-based interface for user and web-server registration process of OHM. Further we have evaluated OHM-P approach by deploying a five user, two web-servers and one OHM system based network environment in our university campus. To analyze the proposed approach we have used security and robustness as the performance matrices of OHM-P.

The layout of this paper is as follows. In first section we have introduce the trends of several online frauds and the OHM approach. In section second we have particularized the related work to overcome online from fraudulent cases. In section third we have shown the implementation of OHM-P approach based on JAVA modules. In section fourth we evaluate the OHM-P approach by deploying a testbed in our university campus. Further in section fifth we have concluded this paper and publicized the future work on OHM approach.

2 Background and Related Work

This section elaborates the OHM approach in the field of online fraud prevention and detection. In our previous research work we have discussed and analyzed the comparative study of several online frauds and to hamper the analyzed frauds we have proposed the OHM-P approach [5]. OHM is embedded in between the interaction process of user and web-server.

OHM-P is the prevention module of OHM approach. This module explicitly prevents the frauds (such as online auction fraud, identity theft fraud, credit card fraud, non-delivery/merchandise fraud, online charity fraud and online investment scheme frauds) during initial interaction between user and web-server. OHM-P provides major advantages for user authentication and web-server authentication. This approach limits major part of fraud occurrence by verification of user and web-server then issues OHM certificate (*OC*) as validation. This certificate will always be considered at each interaction step. Further we have analyzed the robustness of the OHM-P approach against credit card frauds, identity theft fraud, online auction fraud and non-delivery/merchandise fraud.

In [6] we have the performance of both user and web-server authentication algorithms. We have also examined the lifecycle of OHM (*OHMLC*) and each steps of OHMLC using three layer infrastructure of OHM. OHMLC eliminates the chances of online fraud during the registration process. In [6] we have divided the OHM-P approach in two parts: i) user registration which contains four interrelated modules; ii) web-server registration which also consist four interrelated modules. Then we have performed requirement engineering for both user registration modules and web-server registration modules. Further in [6] we have shown the operational interaction among user, OHM systems and web-server and have shown the effectiveness of OHM-P approach.

3 Implementation of OHM-P

In this section we have shown the implementation of OHM-P approach. As we have already discussed that OHM-P provides registration mechanism for both user and web-server so that they cannot hide their identities on internet. Thus, we have developed JAVA based interfaces for the registration of user and web-server. To implement these interfaces we have used following software specifications (shown in Table 1):

Table 1. Software specification for OHM-P

Sr. No.	Specification	Description
1	Platform	JDK 1.6.0
2	Programming Language	J2SE, JDBC, JAVA swing
3	Development Tool	Netbeans IDE 7.0
4	Operating System	Windows 7,XP
5.	Database Tool	Microsoft SQL server 2008

Fig. 1 shows the home page of the OHM system. This provides multiple services such as user registration, web-server registration, user account management, web-server account management. When a user or web-server has already registered on OHM system than OHM provides direct login to them for accessing their accounts. Whereas for a new user or web-server OHM system provides the sign-up functionality where they can register with the OHM and get their respective OHM id and password.

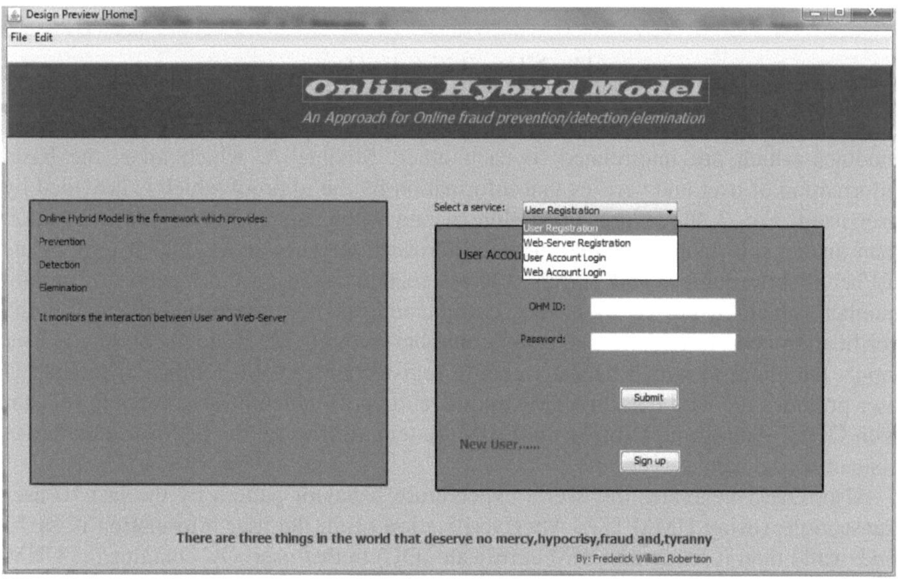

Fig. 1. OHM home page

3.1 OHM User Registration Module

Here we have shown the interface for the user registration process. As deliberated in [5] that for each user it is necessary to register with the OHM in order to accessing the services of the web-server.

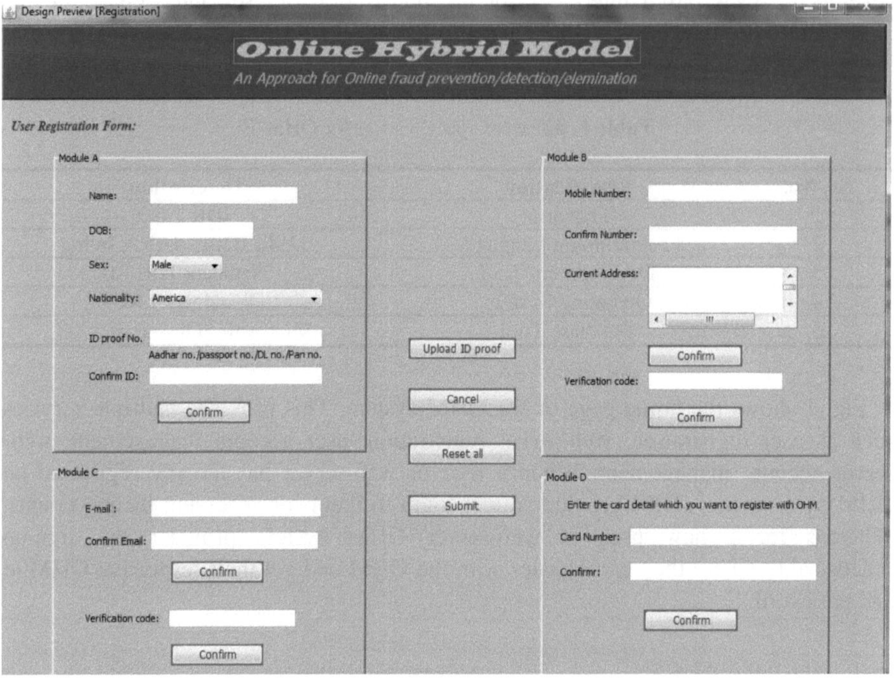

Fig. 2. User Registration form

As discussed in [6] that OHM user registration module consists of four sub-modules which are interrelated to each other. Module A, which takes the basic information of user and verifies that information by the id proof which is provided by user itself. Fig. 2 shows that in Module A and when user clicks on confirm button than his/her information is submitted for verification. Then in Module B user inputs his/her mobile number and current address so that based on the location of mobile number (obtain using GPS) his/her current address is verified and asking of the verification code to register that mobile number with OHM. Now, in Module C user inputs the email id which he/she wants to register with OHM. Further in module D user provides the card detail (ATM/Credit/Debit card) which he/she wants to register with OHM. And through this detail OHM system verifies all the previous details via contacting the bank server.

Also, OHM generates the user's expenditure behavior pattern by the last 10 user transactions (using HMM [7]). After verification of all the user information if OHM finds valid than it issues an OHM certificate (OC) to that user. OC contains the OHM id and password for the user and also having the time validity. We have discussed the complete format of OC in [6].

3.2 OHM User Registration Module

Now, we are describing the web-server registration module. For preventing the legitimate users from the fraudulent web-server organization it is necessary for the web-server to register with OHM.

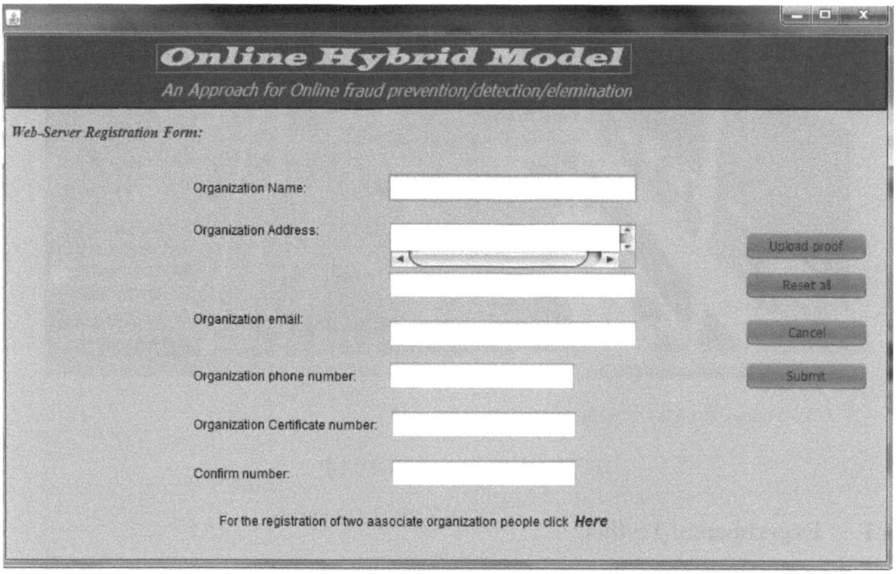

Fig. 3. Web-Server Registration form

Fig. 3 shows the interface for web-server registration. In this OHM needs the organization name, address, contact information (email, phone number) and most important is the organization certificate which is issued by government to that organization. Also, the organization has to submit the proof of government registration. Further OHM needs registration of two peoples who represent the organization. The registration process is same as user registration. For this user has to click on '*Here*' link (at the bottom of page) then it redirects the user to the user registration page.

Then the verification process takes some time and after validates all the information OHM issues an OC to the web-server organization. It is mandatory for an organization to make visible this OC to its user so that a user can rely on the legitimacy of that web-server. The format and field of OC has been discussed in [6].

4 Performance Evaluation of OHM-P

This section illustrates the experiment done by us to analyze the performance of OHM-P approach. Our experiment uses measurements from a five user, two web-servers and an OHM system based testbed to study the performance of OHM-P approach on the metrics of security and robustness.

Fig. 4. Testbed for OHM-P

4.1 Experimental Testbed

1. **Characteristics:** For deploying the testbed we have taken 5 user, 2 web-server and OHM system spans over two floors in our university campus. They are connected with the Wi-Fi technology. The 5 user nodes are distributed at different locations and separate from each other as shown in fig. 4. The 2 web –server nodes are also distant apart from each other shown in fig. 4. The OHM system is located at the center location of University campus which is also shown in Fig. 4.
2. **Software:** In our testbed all the nodes are run on windows platform. In which OHM system has implemented by JAVA modules (J2SE and JAVA swing) and run on glassfish 3.0 servers. Further user nodes interact with web-server by the internet medium. For accessing the web-server services user nodes have used the web-browser Mozilla Firefox 15.0. Whereas web-servers (having web-site interface) developed on Google sites.
3. **Hardware:** To implement the five user nodes we have used laptop computers, each having the Intel CORE 2 Duo processor, 4 GB RAM and 500 GB Hard-disk. Whereas web-server nodes have implemented on desktop computers which are having Intel i3 processor, 4 GB RAM, and 500 GB Hard-disk.
4. **Interaction process:** The interaction process has took placed as follows:

   a. First both the web-server *A* and *B* registered themselves on OHM server through the interface shown in section 3.2. And after validated the information OHM issued the OC to the web-servers.

b. Now, among the five users three users want to access web-server *A*. And because they are interacting first to the web-server, they have redirected to the OHM server. Similarly remaining two users how want to access web-server *B*

c. Then, each user registers themselves on OHM server through the interface shown in 3.1.

d. After verification of each user information, OHM issues OC to the users that contains the information shown in [2].

e. Thereafter, users are allowed to access the services of the respective web-servers by logged-in with OHM id and password.

4.2 Performance Matrices

To show the performance of the OHM-P approach we have considered the two performance matrices i.e. i) Security, which refers to the authenticity of the user information and how secure the data (user and web registration) is; ii) Robustness, which refers to the capability of OHM system to respond under the fraudulent information provide by user or web-server.

1. **Security:** OHM insures the authenticity of the user information as well as web-server information. For user authentication process OHM verified the information provided by users at two levels i) form the id proof (which is of passport id, aadhar id (only for India), pan id, driving license id); ii) from the card detail (obtain by the corresponding bank server). Further for web-server authentication OHM validates the authenticity of web-server organization through organization registration proof and demands the registration of two associating peoples so that web-server can't endeavor any online fraud.
2. **Robustness:** For the user registration process at any module, user provides false information than OHM stops the user registration process and sends a message to user about the false information. For web server registration process if the web-server organization does not have the organization id (registration from corresponding authority) then OHM stops the registration process. For validate the information of both user and web-server OHM takes some time and after validation it issues OC to the registering party.

5 Conclusion and Future Work

In this paper we have implemented the interfaces for OHM-P approach and deliver the platform for user and web-server registration. The interfaces are designed using J2SE and JAVA swing platform. The performance of OHM-P approach has been evaluated by deploying a testbed in our university campus. We have deliberated the characteristics and functioning of our testbed. Further, using the testbed we have analyzed the OHM-P approach on the performance matrices, security and robustness. We have shown that OHM-P provides highly secure registration environment and robust against the fraudulent situations. In future work we will implement the OHM-D approach by creating java modules and will study this approach on real time data.

References

1. Prasad, B.: Intelligent Techniques for E-Commerce. Journal of Electronic Commerce Research 4(2), 65–71 (2003)
2. Zhang, L., Yang, J., Tseng, B.: Online Modeling of Proactive Moderation System for Auction Fraud Detection. In: International World Wide Web Conference Committee (IW3C2), pp. 669–678 (2012)
3. U.S. Commerce Department, Forrester Research, Internet Retailer, ComScore, http://www.statisticbrain.com/total-online-sales/
4. Internet Crime Complain Center, Internet Crime Report (2004-2011), http://www.ic3.gov/media/annualreports.aspx
5. Mundra, A., Rakesh, N.: Online Hybrid Model for Online Fraud Prevention and Detection. AISC. Springer (in press, 2013)
6. Mundra, A., Rakesh, N.: Empirical Study of Online Hybrid Model for Internet Fraud Prevention and Detection. Accepted in IEEE International Conference on Human Computer Interactions (ICHCI 2013) (2013)
7. Srivastava, A., Kundu, A., Sural, S., Majumdar, A.K.: Credit Card Fraud Detection Using Hidden Markov Model. IEEE Transactions on Dependable and Secure Computing 5(1), 1062–1066 (2008)

Cyber Crime Investigations in India: Rendering Knowledge from the Past to Address the Future

V.K. Agarwal[1], Sharvan Kumar Garg[2], Manoj Kapil[3], and Deepak Sinha[1]

[1] IIMT Mgt. College, Meerut
[2] Dept. of Computer Sc., S.D. College of Management Studies, Muzaffarnagar
[3] IIMT Engineering College, Meerut

Abstract. Cyber Crime and ensuing victimization is not individual incidence. It is conjointly hampered or inspired by the group of people within which it is located. Are group of people characteristics relevant for victimization online? This paper examines the cyber crime activities within the perspective of augmentation. Our methodology analyses historical information and its relationship with structural characteristics of the communities that are exposed to cyber crime. We discover that cyber crimes are increasing in context of years, however targeted towards specific age group. The ensuing policy insight is for creating public awareness campaigns in upcoming years

Keywords: Cybercrime, Campaigns, Investigations, Policy, Governance.

1 Introduction

Cyber crime is a subcategory of computer crime and it refers to criminal offenses committed using the internet or another computer network as a component of the crime [3].Cyber-crime has a strong transnational character and high technical level As opposed to other type of crimes, in which investigations could take months, cyber crimes must be rapidly investigated due to the fact that traces and evidence in cyber space can easily disappear. Real time co-operation and understanding of legal issues Informal exchange of information Formal exchange of evidence. Criminal investigation has been a topic of study for academics and practitioners alike, and is defined as 'the process of legally gathering evidence of a crime that has been or is being committed'. Most of the Internet users and perpetrators of Cyber crime in India are young [4]. The number of cyber crime incidents is constantly growing. For this there is a need for an automated system which is capable to support a computer forensic investigation [5].

In regard of the above, through this paper we have calculated the expectancy of cyber crime in coming years with the help of some Statistical tools, which are used to analyze the data for the purpose of finding the results and making pragmatic recommendations to Cyber Crime controlling authorities in India, for the purpose of controlling cyber crime & to competently fulfill their intended objectives and also to

S.C. Satapathy et al. (eds.), *ICT and Critical Infrastructure: Proceedings of the 48th Annual Convention of CSI - Volume II*, Advances in Intelligent Systems and Computing 249,
DOI: 10.1007/978-3-319-03095-1_64, © Springer International Publishing Switzerland 2014

implement salient positive changes, which would enhance the routine operations of their organization. This is presented via a hypothetical example of a specialized unit.

2 Objective

- To find out the number of persons arrested under different cyber crimes are significantly increasing during year 2009-2012 under IT Act.
- To find out whether the cyber crime is conducted more by a particular age-group or equally by all age-groups.
- To find out the expectancy of cyber crime in coming years.

3 Methodology

The data used in this paper is secondary, collected from National Crime Records Bureau [1]. Statistical tools used to analyze the data for the purpose of finding the results and making recommendations are Chi Square Test and Time Series analysis (Least Square Method) [2]. Test is analytical in nature.

4 Scope of Paper

This paper can be used as referral material by Cyber Crime controlling authorities in India, to control cyber crime. The cyber user of a particular age group, involved in cyber crime, can be controlled.

Basically, the scope of this paper is to highlight the principles of scientific management as a controlling tool in the field of Cyber Crime.

5 Data Analysis

Table 1. Cyber crimes/cases registered and persons arrested under IT Act during 2009 – 2012 [1]

S.No.	Crime heads	Cases registered				Persons arrested			
		2009	**2010**	**2011**	**2012**	**2009**	**2010**	**2011**	**2012**
1	Tampering computer source documents	21	64	94	161	6	79	66	104
2	Hacking with computer system	233	510	983	1,875	107	294	552	749
3	Obscene publication/transmission in electronic form	139	328	496	589	141	361	443	497
4	Failure of compliance/orders of certifying authority & To assist in decrypting the information intercepted by govt. agency	3	2	9	9	6	5	4	7

Table 1. (*continued*)

5	Un-authorised access/attempt to access to protected computer system	7	3	5	3	16	6	15	1
6	Obtaining licence or digital signature certificate by misrepresentation/suppression of fact	1	9	6	6	1	4	0	5
7	Publishing false digital signature certificate	1	2	3	1	0	2	1	0
8	Fraud digital signature certificate	4	3	12	10	6	4	8	3
9	Breach of confidentiality/privacy	10	15	26	46	5	27	27	22
10	Other	1	30	157	176	0	17	68	134
	Total	**420**	**966**	**1791**	**2876**	**288**	**799**	**1184**	**1522**

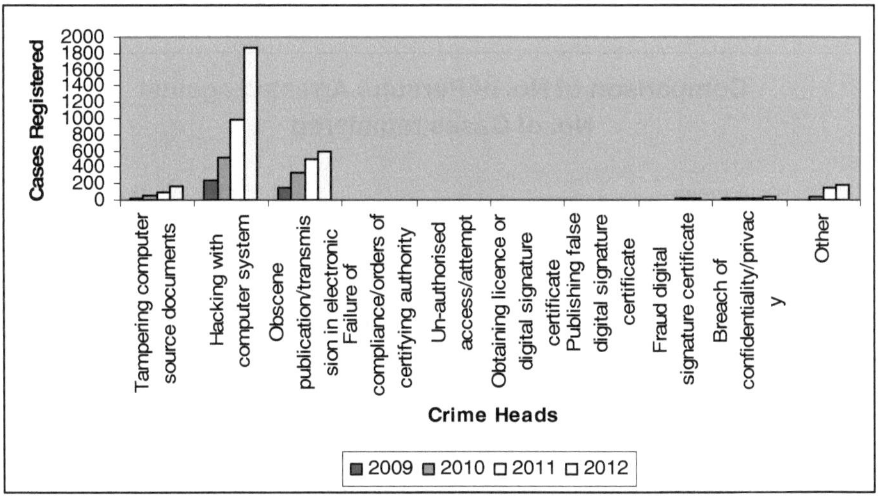

Fig. 1. Number of cases registered for different cyber crimes under IT Act

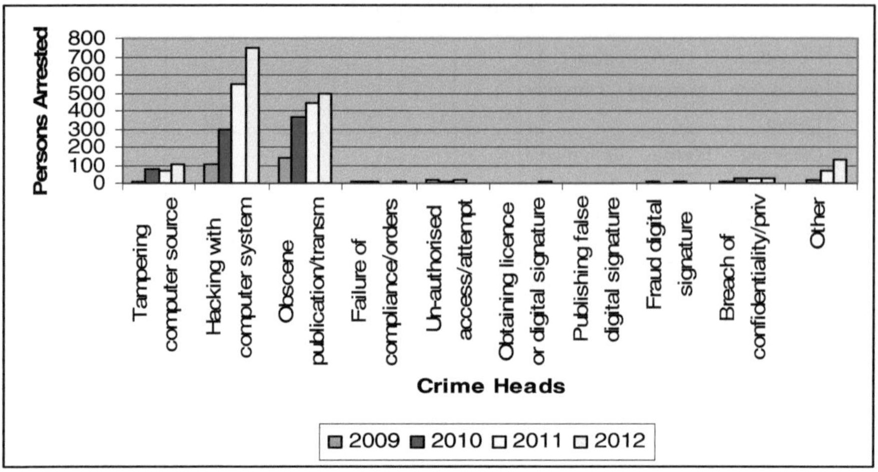

Fig. 2. Number of persons arrested for different cyber crimes under IT Act

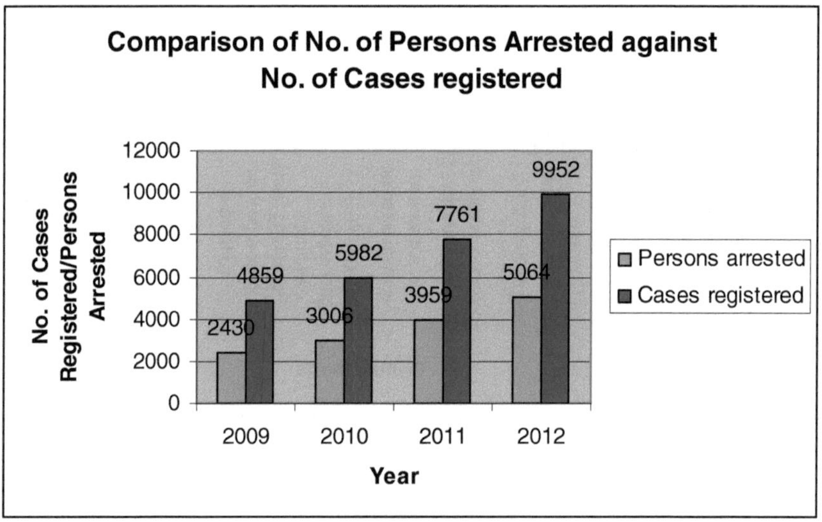

Fig. 3. No. of persons arrested/cases registered for cyber crimes under IT Act during 2009 – 2012

Table 2. Demographic details of persons arrested for Cyber Crime under IT Act during 2009 – 2012 [1]

S.No.	Crime heads	Persons arrested				
		Below 18 Years	**18-30 Years**	**30-45 Years**	**45-60 Years**	**Above 60 Years**
1	Tampering computer source documents	3	150	89	10	3
2	Hacking with computer system	36	989	575	97	5
3	Obscene publication/transmission in electronic form	70	871	413	83	5
4	Failure of compliance/orders of certifying authority	0	14	5	3	0
5	Un-authorised access/attempt	0	21	13	4	0
6	Obtaining licence or digital signature certificate	0	5	4	1	0
7	Publishing false digital signature certificate	0	2	1	0	0
8	Fraud digital signature certificate	0	15	6	0	0
9	Breach of confidentiality/privacy	1	41	27	10	2
10	Other	6	134	60	18	1
	Total	116	2242	1193	226	16

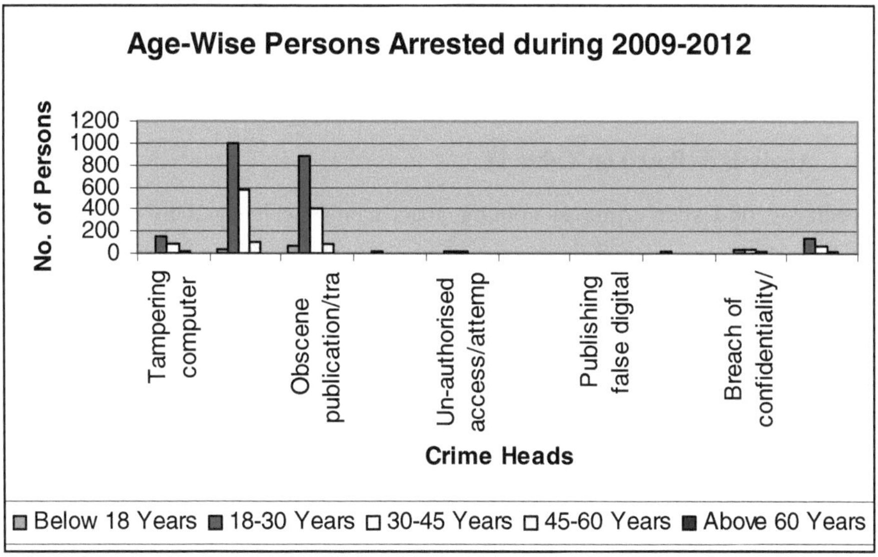

Fig. 4. Demographic details of persons arrested for Cyber Crime under IT Act during 2009 – 2012 [1]

5.1 Analysis 1 (Based on Table 1)

5.1.1. Null Hypothesis: Number of persons arrested under different cyber crimes is not significantly increasing during the year 2009-2012 under IT Act.

5.1.2. Alternative Hypothesis: Number of persons arrested under different cyber crimes is significantly increasing during the year 2009-2012 under IT Act

5.1.3. Test Statistic: $\chi^2 = 258.985$

5.1.4. Degree of Freedom: (10-1) (4-1) =27

5.1.5. Critical Value: Critical value of χ^2 at 27 degree of freedom is 40.113

5.2 Analysis 2 (Based on Table 2)

5.2.1. Null Hypothesis: Cyber crime is conducted equally by persons of all age-groups.

5.2.2. Alternative Hypothesis: Cyber crime is conducted more by persons of particular age-group.

5.2.3. Test Statistic: $\chi^2 = 66.803$

5.2.4. Degree of Freedom: (10-1) (5-1) =36

5.2.5. Critical Value: Critical value of χ^2 at 36 degree of freedom is 50.998

5.3 Analysis 3 (Based on Table 1)

Expectancy of Cyber crime in coming years can be calculated by the following formula

$$\mathbf{Y=928.25 + 214.35\ X} \tag{1}$$

where,

X= (Year-2010.5)*2 and

Y=No. of Persons Arrested.

6 Interpretation

6.1 From Analysis 1

Since the calculated value of χ^2 = 258.985 is greater than the critical value of χ^2 = 40.113, it falls in the rejection region. Hence, our Null Hypothesis is rejected, and it may be concluded that Number of persons arrested under different cyber crimes is significantly increasing during the year 2009-2012 under IT Act.

6.2 From Analysis 2

Since the calculated value of $\chi^2 = 66.803$ is greater than the critical value of $\chi^2 = 50.998$, it falls in the rejection region. Hence, our Null Hypothesis is rejected, and it may be concluded that Cyber crime is conducted more by persons of particular age-group.

6.3 From Analysis 3

Based on the formula given by analysis 3, we can predict the number of persons arrested under IT Act during 2013 – 2016 as shown in the table 3.

Table 3. Expected number of persons arrested under IT Act during 2013 – 2016

Year	**2013**	**2014**	**2015**	**2016**
No. of Persons	2000	2429	2857	3286

7 Findings

1. Number of persons arrested under different cyber crimes is significantly increasing during the year 2009-2012 under IT Act.
2. Cyber crime is conducted more by persons of particular age-group.
3. Expected number of persons involved and arrested in cyber crime will be increasing year by year.

8 Recommendations

1. As shown in Table 1, there is a large gap between the number of cases registered and the number of persons arrested under IT Act during 2009 – 2012. This shows loopholes in the current cyber laws and/or unavailability of appropriate tools and techniques, to prove the crime against the person registered for conducting cyber crime. Hence, cyber laws need to be strengthened and advance tools and techniques need to be developed.
2. Since, there are loopholes as discussed earlier, it is suggested that the Cyber crime cell in India should be equipped with appropriate Human Resource which is well-equipped to IT.
3. As shown in Table 2, the age group of 18-30 years is involved in most of the Cyber Crimes; an academic module/awareness workshop should be introduced at the school/college level to make the cyber users aware about the current cyber laws and the punishment against its offence.
4. As shown in Table 3, the expected number of persons arrested under IT Act will be increased year by year. To reduce this number, advanced Internet Security Systems should be introduced and use of Virtual keyboard should be popularized. Phishing equipment should be licensed and their sale should be registered to de-motivate the persons to involve in cyber crime.

9 Conclusion

We find that Number of persons arrested under different cyber crimes is significantly increasing during the previous years. Cyber crime is conducted more by persons of particular age-group. There is a large gap between the number of cases registered and the number of persons arrested under IT Act. We have also provided the anticipatory figure of the persons who will be intended to involve for the purpose of cyber crime in the coming years. Local government initiatives, such as public awareness campaigns and education efforts should take this vulnerability into account. This study is also limited to urban cities of India. Thus, the results are likely not generalizable.

References

1. National Crime Records Bureau, `http://www.ncrb.nic.in/`
2. Sancheti, D.C., Kapoor, V.K.: Statistics Theory Methods & Applications. Sultan Chand & Sons, New Delhi (2009)
3. Shinder, D.L.: Scene of the Cybercrime: Computer Forensics Handbook (2002)
4. Shrivastav, A.K., Ekata Dr.: ICT Penetration and Cybercrime in India: A Review. International Journal of Advanced Research in Computer Science and Software Engineering (2013)
5. Bielecki, M., Quirchmayr, G.: A prototype for support of computer forensic analysis combined with the expected knowledge level of an attacker to more efficiently achieve investigation results. In: International Conference on Availability (2010)

A Novel SVD and GEP Based Image Watermarking

Swanirbhar Majumder[1], Monjul Saikia[1], Souvik Sarkar[2], and Subir Kumar Sarkar[3]

[1] NERIST, (Deemed University), Itanagar, Arunachal Pradesh, India
swanirbhar@ieee.org, monjuls@gmail.com
[2] IBM, Hyderabad, Andhra Pradesh, India
[3] Dept. of ETCE, Jadvpur University, Kolkata, West Bengal, India
sksarkar@etce.jdvu.ac.in

Abstract. In this age of cloud computing, androids and smart phones the popularity of digital media has reached heights that have never been imagined. This is due to the efficient and omnipresent internet connectivity. So the copyright protection of intellectual properties and multimedia data has become a necessity for prevention of illegal copying and content integrity verification. Thus latest digital watermarking techniques that satisfy the requirements of imperceptibility, robustness, capacity, and security are being developed time to time. That's why everyday newer techniques are being employed for the same. Here we present a novel method of digital image watermarking using singular value decomposition (SVD) and Gene Expression Programming (GEP). The popular wavelet based watermarking techniques have been coupled with the GEP which helps in providing a robust watermarking scheme.

Keywords: Watermarking, singular value decomposition (SVD), gene expression programming (GEP).

1 Introduction

Copyright protection of multimedia data via digital image watermarking is one of the most popular techniques used over the years [1-4]. Imperceptibility, robustness and trustworthiness of the watermarking scheme are the main areas where we normally focus for implementation [5]. The two major ways of watermarking are, the spatial domain and the popular transform domain embedding of watermark. Due to implementation of the robust singular value decomposition (SVD), this method is mainly transform domain based.

This work is more in the lines of our previous work where artificial neural network was included in watermarking along with the SVD technique [6] [7]. Here the trained neural network system has been replaced by gene expression programming (GEP) [8-10].

1.1 Singular Value Decomposition

The SVD technique is a generalization of the eigen-value decomposition, used to analyze rectangular matrices. This mathematical technique has been used in various

S.C. Satapathy et al. (eds.), *ICT and Critical Infrastructure: Proceedings of the 48th Annual Convention of CSI - Volume II*, Advances in Intelligent Systems and Computing 249,
DOI: 10.1007/978-3-319-03095-1_65,

fields of image processing and its counterparts. SVD technique mainly decomposes any rectangular matrix into three simple matrices (two orthogonal matrices and one diagonal matrix). It has been widely studied and used for watermarking by researchers for long [11-13].

When SVD is undergone on an image (**I**$_{MxN}$) matrix it produces 3 matrices (**U**$_{MxM}$, **S**$_{MxN}$ and **V**$_{NxN}$) as per equation 1. The main image characteristics are in the square diagonal matrix **S**. High frequencies are at the lower end corner diagonal of **S** and low frequencies at the upper end corner diagonal. **U** and **V** contain the finer details respective to the Eigen values at **S**. So the most important characteristics of the image are distributed from the left corner diagonally and the importance of these values decrease as it proceeds towards the lower corner at the right of the matrix. Thus by using any rank R, the **U**$_{MxM}$ becomes **U**$_{MxR}$ and **S**$_{MxN}$ becomes **S**$_{RxR}$ and **V**$_{NxN}$ becomes **V**$_{NxR}$. Their resultant operation is image **I'**$_{MxN}$, where **I'** is the image generated from **U**$_{MxR}$, **S**$_{RxR}$ and **V**$_{NxR}$ as from equation 2. This **I'** does have approximately similar features as **I** for optimum value of R as shown in Fig 1 [6].

$$\mathbf{I}_{M\times N} = \mathbf{U}_{M\times M} \times \mathbf{S}_{M\times N} \times \mathbf{V}^{T}_{N\times N} \tag{1}$$

$$\mathbf{I}'_{M\times N} = \mathbf{U}_{M\times R} \times \mathbf{S}_{R\times R} \times \mathbf{V}^{T}_{N\times R} \tag{2}$$

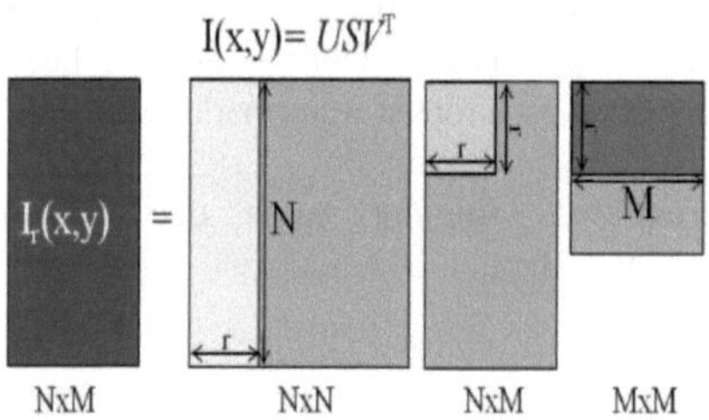

Fig. 1. SVD operation on an image I (x,y)

Here SVD is used to hide the logo for watermarking, in its Eigen values. To improve the robustness error control coding is applied, for which the convolution encoder is used with Vitrebi decoder as per a particular polynomial and code trellis has been implemented. The logo image, to be embedded is taken in binary form, converted to one dimensional array and passed through the convolution encoder to get the encoded single dimensional logo. The embedding logo in the carrier image is done at the Eigen value corner of the image via zig-zag scanning to concentrate the encoded logo predominantly at the highest information location. This is done deliberately in order to make it a necessity that the attack destroys the host image, in order to remove the embedded watermark [6].

1.2 Gene Expression Programming

Genetic Algorithm (GA) [14] is the best known algorithm from the Evolutionary Algorithm (EA) class. In the conventional version, chromosomes are representing as a fixed length binary string. Another version of GA is Genetic Programming (GP) [9], where chromosomes are represented as a LISP expression translated graphically into a tree. A newly introduced version of the Evolutionary Algorithms, called GEP was proposed by Candida Ferreira in 2001. Her motivation was from biological evolution as this method overcame certain limitations of GA and GP by working with two elements, the chromosome and the expression tree. The chromosome is the encoder of the candidate solution which is then translated into an expression tree. GEP is an example of a full-fledged replicator/phenotype system where the chromosome /expression trees form a truly functional, indivisible whole [10]. That's why GEP is a big breakthrough in evolutionary computation, and it continuously attracting more and more researcher attentions recently, especially in the areas of data mining. It should be noted that GEP chromosomes are multigenic. It encodes multiple expression trees or sub-programs, later on which can be structured into a much more complex program. Because of this, as like the DNA/protein system of life on Earth, the gene/tree system of GEP not only explores all the crannies and paths of the solution space but it has also the scope to explore sophisticated levels of organization.

A GEP gene and its responses expression tree (ET) can be illustrated considering the algebraic expression of equation 3. It can be represented as a diagram or an expression tree as shown in Fig 2. Here "Q" represents the square root function [8].

$$\sqrt{(a - b)(c + d)} \tag{3}$$

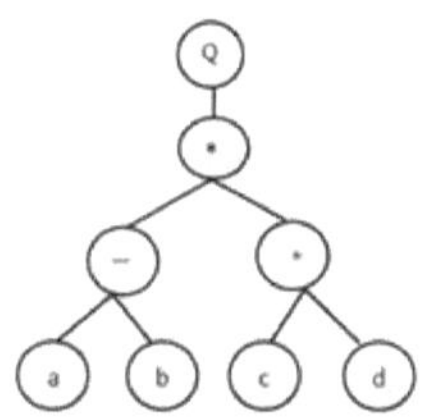

Fig. 2. Gene Expression tree of Equation 3

This way of diagram depiction is called as the phenotype in GEP. And then the genotype can easily be derived from the phenotype as follows:

0	1	2	3	4	5	6	7
Q	*	-	+	a	b	c	d

(4)

It is just the direct reading of the expression tree from left to right and from top to bottom; as like we read a text page. This is an ORF expression, beginning from "Q" (location 0) and ending at "d" (location 7). These were named K-expressions from Karva notation.

2 The Watermarking Scheme

The whole watermarking scheme is mainly divided in two parts:

2.1 Watermark Embedding

The whole process of embedding the LOGO in the cover image is as shown in Fig 3. After generation of the code trellis, the watermark is passed through an error control convolution encoder to obtain an encoded logo data stream as in [6][7][15]. The SVD of host image is performed to obtain the matrices **U**, **S**, and **V**. The **S** matrix is arranged as one dimension via zig zag scan in order to add the logo near the most prolific diagonal Eigen values. This leads to a matrix **S'**.

$$\mathbf{S}'_{1D} = \mathbf{S}_{zigzag1D} + key \times Logo_{1D} \tag{5}$$

Since the number of bits in **S** is greater than that of the logo, number of bits in **S'** is same as that of **S**. Moreover, to reduce the intensity of the logo, it is multiplied with a number '*key*', which is less than 1. This reduces the intensity of logo in the **S** matrix and does not degrade the host image significantly.

This one dimensional **S'** is then converted back to two dimensional (2D) form using the anti-zigzag algorithm. On having the 2D **S'** the SVD operation is applied on it for the second time to have **S1**, **U1** and **V1** as output of the SVD operation on **S'**. The **S1** of second SVD operation along with the earlier extracted **U** and **V** from the first operation are incorporated to obtain the watermarked image $\mathbf{I_W}$ from equation below:

$$\mathbf{I}_W = \mathbf{U} \times \mathbf{S1} \times \mathbf{V}^{\mathrm{T}} \tag{6}$$

Now using the leftover matrices **U1**, **S** and **V1** we get the key image $\mathbf{I_K}$ given by:

$$\mathbf{I}_K = \mathbf{U1} \times \mathbf{S} \times \mathbf{V1}^{\mathrm{T}} \tag{7}$$

To implement convolution encoding, the trellis structure used is a feed-forward convolution encoder with input as a vector of length k (here k=2). Rate is n/k (here 3/2) code, the encoder output is a vector of length n (here n=3). So the constraint length taken is 1-by-k vector specifying the delay for each of the k input bit streams represented by matrix cl; and the code-generator is a k-by-n matrix of octal numbers specifying the n output connections for each of the k inputs represented by matrix cg[15].

$$cl = \begin{vmatrix} 4 & 3 \end{vmatrix} \qquad cg = \begin{vmatrix} 4 & 5 & 17 \\ 7 & 4 & 2 \end{vmatrix} \tag{8}$$

Thus the trellis generated has the following structure:

number of input symbols	$= 2^k = 4$;
number of output symbols	$= 2^n = 8$;
number of states	$= 2^k \text{ x } 2^n = 32$;
next state matrix	= [32x4]
output matrix	= [32x4]

Based on the above trellis structure the convolution encoder is used to encode the subjective logo having 256 bits in order to obtain an encrypted message with 384 bits.

In case of convolution decoding, the same trellis is used with the Viterbi algorithm. The encoder is assumed to have started at the all-zeros state and the decoder traces back from the state with the best metric, expecting binary input values. Here the trace back depth for the particular *cg* and *cl* is assumed to be 2 for the coding [6][7][15].

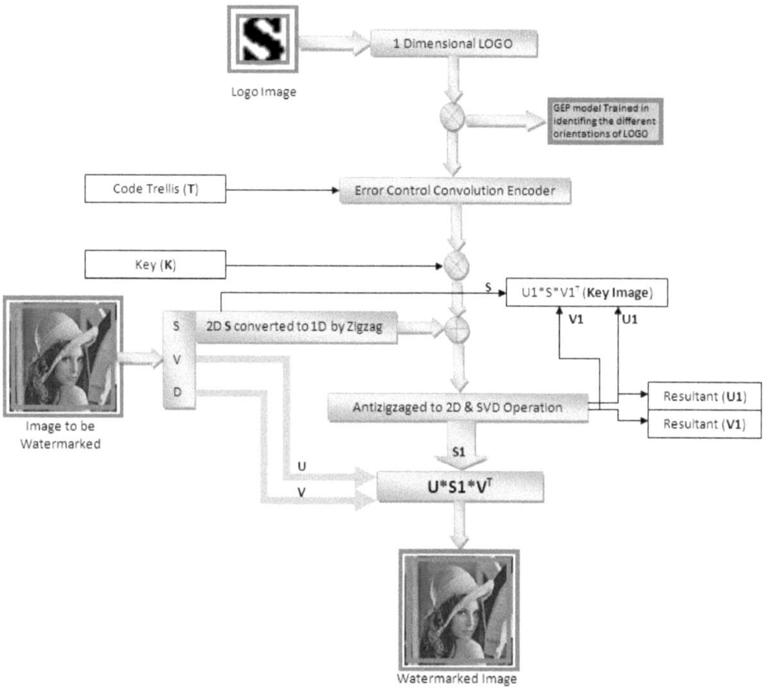

Fig. 3. Watermark embedding steps

Further the 16x16 logo (watermark) is arranged to the orientation of the zeros '0's and ones '1's and all of their possible permutations and combinations are generated. The whole data set is used to optimize a gene expression model with the help of GeneXpro Tools 4 by GepSoft. This model provides 'true'('1') output for any input which has near inter relationship of zeros and ones matching to the same as the logo. Else for any other input totally uncorrelated to the logo it provides 'false'('0') output. Thereby with some amount of pixel alteration the model can still identify the logo (that it has been trained to recognize).

2.2 Watermark Extraction

The detection of the watermarked logo from the stego image is just the opposite of the embedding method as shown in Fig 4. This is of non-oblivious type, as the *key* and Key image $\mathbf{I_K}$ are to be available at the receiver end, where the stego image $\mathbf{I'_W}$ (due to malicious attacks $\mathbf{I_W}$ turns to $\mathbf{I'_W}$) is received instead of $\mathbf{I_W}$. Therefore, SVD is applied on the key image $\mathbf{I_K}$ to obtain to obtain **U1**, **S1** and **V1** and the distorted watermarked image $\mathbf{I'_W}$ to obtain **U2**, **S2** and **V2**.

Here **S1** is replaced by **S2** and the inverse SVD is performed to obtain **D** where:

$$\mathbf{D} = \mathbf{U1} \times \mathbf{S2} \times \mathbf{V1}^{\mathrm{T}} \tag{9}$$

Now, this **D** and the previously obtained **S1** from key image are reshaped to one dimension by zigzag operation. Then at the receiver end, encoded logo code **C** is estimated.

$$\mathbf{C} = \frac{1}{key} \times (\mathbf{D} - \mathbf{S1}) \tag{10}$$

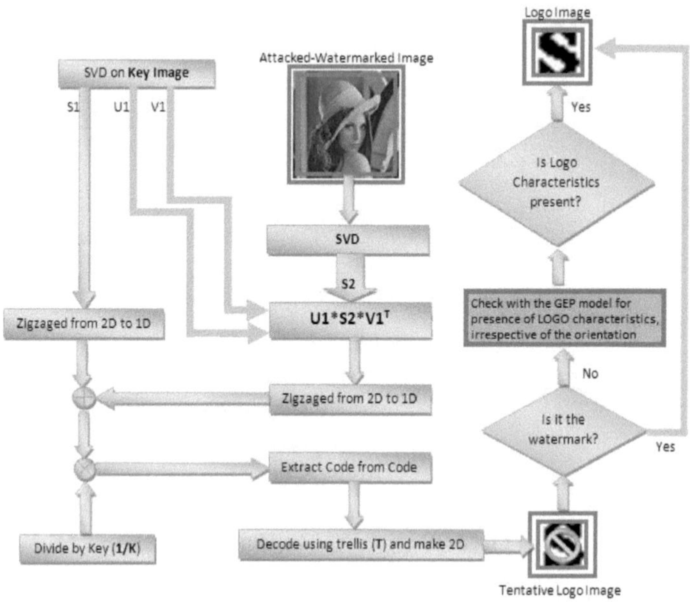

Fig. 4. Watermark extraction Steps

The difference of the two arrays is multiplied by the reciprocal of *key* to reinstate the lost intensity of code in the decoder. This extracted code is further decoded using a hard Viterbi decoder with the same trellis structure used during encoding.

But the problem is that sometimes due to heavy duty attacks; even the presence of error control coding cannot help the recognition of the logo. In these conditions the tentative logo is checked via the GEP model trained during the embedding process. In case nominal logo characteristics are present in the tentative logo it is recognized, as in the Fig 5 below. Else it is considered to be a malicious logo with no characteristics of the logo that was embedded.

3 Results and Discussion

This Algorithm has been simulated on the Checkmark 1.2 developed by Shelby Pereira of University of Geneva, Vision Group on their 'Logo' application. On application of the MATLAB based attack program on the watermarked image, 77 corrupted output images were generated; these 77 images were due to application of 77 different signal impairments which are subsets of 11 main signal degradations. All of these attacks are standard attacks as recognized by the Watermarking World community[15].

The attacked image initially undergoes the '*bicubic*' interpolation incase the size of the attacked watermarked image is not as per the required size for which the algorithm is designed. This application initiated to reduce the complexity of the decoding algorithm issues a further attack on the watermark image.

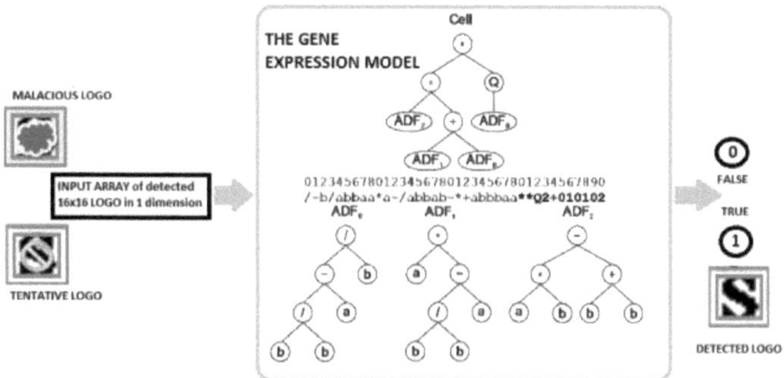

Fig. 5. GEP model detecting tentative logo and rejecting malicious ones

Table 1. Number of checkmark 1.2 'logo' application attacks sustained by the algorithm

Major Attack Type	No of Sub-Attacks	Detection via our algorithm
Aspect ratio	35	35
Crop	7	7
JPEG	7	7
Scale	6	6
MAP	6	6
Up-Down Sample	4	4
Re-modulation	4	4
Filtering	3	3
Bending	2	2
Wavelet	2	2
Copy	1	0

Except for the one '*copy*' attack, all the other 76 attacks were detected when the GEP model was used along with the SVD and error control coding based detection algorithm. This is around 98% of success. This too without any allowable percentages of error, as the network either provides full detection with a *one* ('1') or no detection with a *zero* ('0'). The result of this comparison is as tabulated in Table 1.

4 Conclusion

In this paper a novel method of watermarking logo using the SVD technique with GEP is proposed. This scheme has been enhanced with the usage of error control coding. The encoding and detection method has been simulated with the standard Checkmark 1.2 attacks, for the 'Logo' application. Around 98% attack detection was achieved. This was particularly for 'bicubic' interpolation method followed by the detection using SVD and error control coding and finally application in the GEP model. These results are all self generated as per the inbuilt programs of Checkmark 1.2. So for these standard attacks the robustness of the watermarking method may thereby be judged.

References

1. Macq, B.R., Pitas, I.: Special issue on water making. Signal Processing 66(3), 281–282 (1998)
2. Swanson, M.D., Kobayashi, M., Tewfik, A.H.: Multimedia data embedding and watermarking technologies. Proc. IEEE 86, 1064–1087 (1998)
3. Acken, J.M.: How watermarking adds value to digital content. Commun. ACM 41(7), 74–77 (1998)
4. Low, S.H., Maxemchuk, N.F., Lapone, A.M.: Document identification for copyright protection using centroid detection. IEEE Trans. Commun. 46, 372–383 (1998)
5. Liu, R., Tan, T.: A SVD-based watermarking scheme for protecting rightful ownership. IEEE Transactions on Multimedia 4, 121–128 (2002)
6. Majumder, S., Das, T.S., Mankar, V.H., Sarkar, S.K.: SVD and Neural Network based Watermarking Scheme. In: Das, V.V., et al. (eds.) BAIP 2010. CCIS, vol. 70, pp. 1–5. Springer, Heidelberg (2010)
7. Majumder, S., Das, T.S., Sarkar, S.K.: BPNN and SVD based Watermarking Scheme. International Journal of Recent Trends in Engineering 4(1), 44–47 (2011), doi:01.IJRTET 04.01.3
8. Ferreira, C.: Gene Expression Programming: A New Adaptive Algorithm for Solving Problems. Complex Systems 13(2), 87–129 (2001)
9. Koza, J.R.: Genetic Programming: On the Programming of Computers by Means of Natural Selection. MIT Press (1992)
10. Ferreira, C.: Gene Expression Programming: Mathematical Modelling by an Artificial Intelligence, 2nd edn. Springer, Heidelberg (2006)
11. Abdi, H.: Singular Value Decomposition (SVD) and Generalized Singular Value Decomposition (GSVD). Encyclopaedia of Measurement and Statistics (2007)
12. Goldrick, C.S.M., Dowling, W.J., Bury, A.: Image coding using the singular value decomposition and vector quantization. In: Image Processing and Its Applications, pp. 296–300. IEE (1995)
13. Yang, J.F., Lu, C.L.: Combined Techniques of Singular Value Decomposition and Vector Quantization for Image Coding. IEEE Transactions on Image Processing 4(8), 1141–1146 (1999)
14. Holland, J.H.: Adaptation in Natural and Artificial Systems. University of Michigan Press, Ann Arbor (1975)
15. Majumder, S., et al.: SVD and Error Control Coding based Digital Image Watermarking. In: Proceedings of ACT 2009, pp. 60–63. IEEE CSI, India (2009) ISBN 978-0-7695-3915-7

Complete Binary Tree Architecture Based Triple Layer Perceptron Synchronized Group Session Key Exchange and Authentication in Wireless Communication (CBTLP)

Arindam Sarkar and J.K. Mandal

Department of Computer Science & Engineering,
University of Kalyani, Kalyani-741235,
Nadia, West Bengal, India
{arindam.vb,jkm.cse}@gmail.com

Abstract. In this paper, Triple Layer Perceptron harmonized one time session key exchange and authentication (CBTLP) has been proposed based on the structural design of complete binary tree. In this proposed technique 3 hidden layers are used in the architecture of the TLP to enhance the security. This proposed CBTLP scheme offers structure of complete binary tree. Group of parties can participate in TLP synchronization and key switch over process. In CBTLP scheme only O($\log_2$ N) synchronization is needed for swap over the session key among N parties.

Keywords: Triple Layer Perceptron (TLP), Session Key, Wireless Communication.

1 Introduction

These days a variety of techniques are available to secure data and information from eavesdroppers [1, 5]. Every algorithm has its own advantages and disadvantages. Security of the encrypted text entirely depends on the key used for encryption. The most important hazard of cryptography is how to firmly swap over the shared secrets between the parties. As a result, key exchange protocols are mandatory for transferring keys in a protected manner. As the same time as key exchange protocols are developed for exchanging key between two parties, many applications do necessitate the need of swapping over a secret key among group of parties. A lot of proposals have been proposed to accomplish this goal. In this proposed technique a key swap over by synchronization among cluster of TLPs has been proposed for this purpose which is a fresh addition to the field of cryptography. This proposed scheme implements the key swap over algorithm with the help of complete binary tree which make the algorithm scales logarithmically with the number of parties participating in the protocol. Two parties can swap over a common key using synchronization between their own perceptrons [2, 3, 4]. But the problem crop up when group of N parties desire to swap over a key. Since in this case each communicating party has to synchronize with other for swapping over the key. So, if there are N parties then total number of synchronizations needed before swapping over the actual key is O(N^2).

S.C. Satapathy et al. (eds.), *ICT and Critical Infrastructure: Proceedings of the 48th Annual Convention of CSI - Volume II*, Advances in Intelligent Systems and Computing 249,
DOI: 10.1007/978-3-319-03095-1_66, © Springer International Publishing Switzerland 2014

This proposed scheme offers a novel technique in which complete binary tree structure is follows for key swapping over. Using proposed algorithm a set of N parties can be able to share a common key with only $O(\log_2 N)$ synchronization.

The organization of this paper is as follows. Proposed Triple Layer Perceptron (TLP) architecture has been discussed in section 2. Section 3 deals with the proposed CBTLP based group session key authentication. Section 4 deals with the complexity measurement of proposed CBTLP. Experimental results of this technique are given in section 5. Analysis regarding various aspects of the technique & results has been presented in section 6. Conclusions & future scopes are drawn in section 8 and that of references at end.

2 Architecture of Triple Layer Perceptron

The architecture of triple layer perceptron consists of one input layer, one output layer and three hidden layers. Addition of these extra hidden layers improves security of the CBTLP technique by making the attackers life difficult. Here, the parameter K is being divided into K1, K2 and K3 value. K3 represents number on hidden neurons adjacent to the output layer. For each K3 neuron there are K2 number hidden neurons, i.e. the middle -hidden layer 2 i.e.between the hidden layer 1 and 3. Each of K2×K3 numbers of hidden neurons there are K1 number of hidden neurons at the first hidden layer adjacent to the input layer. So, hidden layer 1 has K1×K2×K3neurons. Layer 1 now for each K1×K2×K3 neurons there are N inputs possible. So, finally it can be stated that, the input layer has K1×K2×K3×N input neurons and this number represents the size of the triple layer peceptron . Each hidden layer number 1 (i.e. with K1×K2×K3 neurons) neuron produces σ^1_i values, each hidden layer number 2 neurons (i.e. with K2×K3 neurons) produces σ^2_i value. Each hidden layer number 3 neurons (i.e. with K3 neurons) produce σ^3_i value. These can be represented as –

$$\sigma^1_i = \text{sgn}\left(\sum_{j=1}^{N} W_{i,j} X_{i,j}\right)$$

$$\sigma^2_i = \text{sgn}\left(\sum_{j=1}^{N} \sigma_i^1\right)$$

$$\sigma^3_i = \text{sgn}\left(\sum_{j=1}^{N} \sigma_i^2\right)$$

Sgn is a function, which returns -1, 0 or 1:

$$\text{sgn} = \begin{cases} -1 \ \ if\ x < 0 \\ 0 \ \ if\ x = 0 \\ 1 \ \ if\ x > 0 \end{cases}$$

The output of neural network is then computed as the multiplication of all values produced by hidden elements:

$$\tau = \prod_{i=1}^{K2} \sigma_i^3$$

The, basic difference between these double layer perceptron and triple layer perceptron is, double layer TPMs are calculating the σ^i_i where i ={1,2,3} and j ={1,2…,K1×K2×K3×N } value for two times. So, this CBTLP ends up with 1 set of weight vector and 3 sets of σ values.

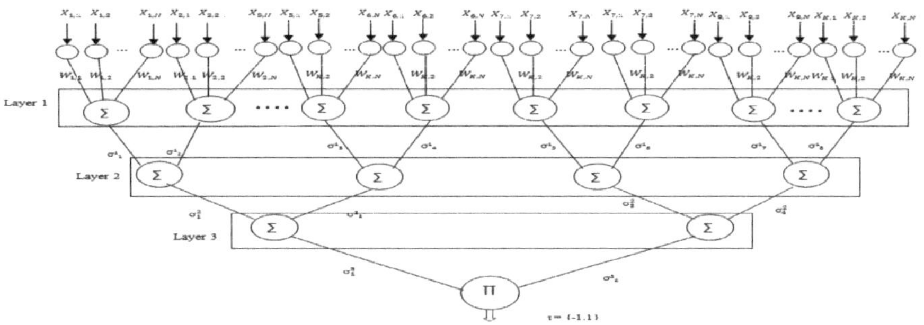

Fig. 1. A Triple Layer Perceptron with 3 Hidden Layers

The figure 1 shows a perceptron with 3 hidden neuron layers. Here the K3 =2, K2=2 and K1=2 as well. So, the first hidden neuron from the top has K1×K2×K3 = 2×2×2 = 8 hidden neurons. The second hidden layer contains 2 neurons. The number of inputs = N×K, where N is the number of inputs per hidden layer 1 neuron and K= K1×K2×K3. The output is τ, which is either -1 or 1. Proposed group key swap over technique offers two novel procedures for exchanging group key among different TLP. Both procedures are based on the structure of complete binary tree. In the TLP group key exchange algorithm, *N* number of TLP need to synchronize together and they are represented by an M number of leaves of a complete binary tree where M is defined as $M = 2^{(\log 2\ N)}$.

2.1 Synchronization Technique

In this method, the mutual learning algorithm is take place between every two parties having the same parent in the binary tree structure. Let the $max\ depth$ is the depth of the complete binary tree and $cur\ depth$ is the current depth where the algorithm is functioning. Starting from a $cur\ depth = max\ depth{-}1$, apply the mutual learning algorithm between each pair of leaves having the same parent. Following the synchronization, one level up is marked.

($cur\ depth = cur\ depth - 1$) and a exchange method is applied between the right leaves of both right and left branches for all subtrees in that $cur\ depth$. Once the $cur\ depth$ becomes equal to zero, all leaves will be synchronized together. For sake of simplicity the group of parties will be represented as vector with indices {0, 1, . . . , M−1}. Figure 2 shows the scenario of synchronization. Figure 2a shows the preliminary configuration of unsynchronized parties. In figure 2b, pairs of parties are synchronized together, {(0, 1), (2, 3), (4, 5), (6, 7)}. Then, the exchange operation is performed, {(0, 2), (1, 3), (4, 6), (6, 7)}, and the mutual learning is applied again. This results in synchronization of two groups each with four parties, {(0, 1, 2, 3), (4, 5, 6, 7)}, as shown in figure 2c. After that, the exchange operation is applied again and the vector takes the form {(0, 4), (1, 5), (2, 6), (3, 7)}. The algorithm terminates when pairs in the new vector apply mutual learning that produces full synchronization between all parties figure2d.

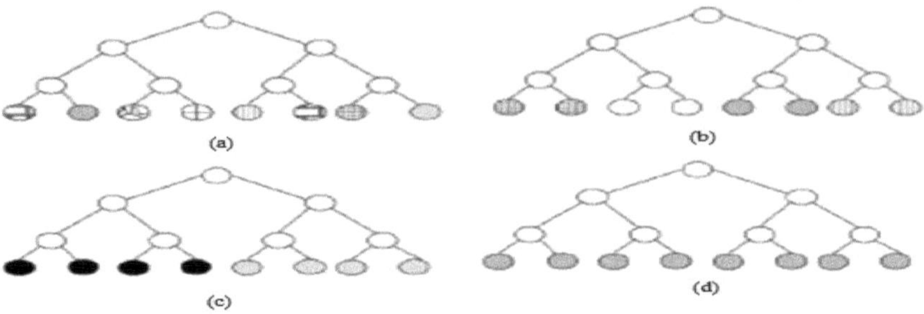

Fig. 2. (a) Shows the preliminary configuration of unsynchronized parties. (b) Pairs of parties are synchronized together, {(0, 1), (2, 3), (4, 5), (6, 7)}. (c) Synchronization of two groups each with four parties, {(0, 1, 2, 3), (4, 5, 6, 7)}. (d) After that, the exchange operation is applied again and the vector takes the form {(0, 4), (1, 5), (2, 6), (3, 7)}.

At end of full weight synchronization process, weight vectors of hidden layers of both TLP systems become identical. This synchronized weight vector of both the TLP's is used to construct the secret session key. This session key is not get transmitted over public channel because receiver TLP has same identical weight vector. At the same time as the proposed key exchange protocol is scalable; it remains susceptible to active attacks. An attacker can take part in the protocol and synchronize with the group and finally obtain the shared key which endangers the secret communication between the groups later. As a result, it is compulsory to build up an authenticated key exchange protocol to permit only certified users to get hold of the mutual secret.

3 Group Session Key Authentication

In double layer perceptron based key generation mechanism if two parties' do not have the identical input vectors i.e. $\forall t: x^A(t) \neq x^B(t)$ then synchronization is not achievable between them. If the inputs are identical for both parties' then only two parties can be trained using each other outputs. Given diverse inputs, the two parties are trying to learn totally dissimilar relations between inputs $x^{A/B}(t)$ and output $\tau^{A/B}(t)$ as result synchronization is not possible and thus in turn prevent the generation of time-dependent equal weights. The development of normalized sum of absolute differences $diff(w^A(t), w^B(t)) \in [0,1]$ over time for different offsets $\forall t: x^A(t) = x^B(t+\varphi), \varphi \in N$ in the input vector and for completely different input vector.

Case1: Case 1 represents a circumstance where attackers have random number generator with a dissimilar initialization process.

Case2: In Case 2 attackers are deals with completely different set of inputs. It is observed that the distance between two parties that do not acquire the same inputs remains fluctuating within a certain limited range around 0.4 and never decreases towards zero.

Case3: Two parties with entirely diverse inputs illustrate the same qualitative performance. Taking into consideration the number of repulsive and attractive steps, it can be observed that on average there must be as many repulsive as attractive steps for such performance. Two parties having the same inputs (offset zero) soon decrease their distance and synchronies.

Case4: In case 4 both parties uses identical inputs but a certain proportion of uniformly scattered 'noise' has been imposed on the transmitted outputs of either party. Despite of presence of noise in a certain time, the system would synchronies with a delay of approximately the duration of the noisy period plus the time used up for unproductive synchronization before the noisy period.

Now, ${w_{kj}}^A(t)$ and ${w_{kj}}^B(t)$ get a dissimilar random element $x_{kj}\ (t)$ of their input vectors. The distance between the elements is therefore not going to in turn condensed to zero after each bounding action and the two parties deviate. So, no common inputs lead to the non-synchronization. For this reason common input of both patties i.e. $x^{A/B}(t)$ kept secret between the two parties in addition to their own arbitrarily assigned secret initial weights $w^{A/B}(t)$.There are, which is a large enough practical amount for the parameters that makes Brute force attacks become computationally very costly because of $(2^{KN} - 1)$ computations are needed for finding out possible common inputs. By this authentication scheme attack likes Man-In-The-Middle attack and all other known attacks can be prevented.

4 Complexity of CBTLP Technique

O(N) computational steps are desired to create a key of length N. The average synchronization time up to N=1000 asymptotically one expects an increase like O (log N).

5 Experimental Results

In this section the results of implementation of the proposed technique has been presented in terms of encryption decryption time, Chi-Square test, degree of freedom. The results are also compared with existing RSA & TDES [1] technique. Figure 3 shows memory heap of TLP at run time and Figure 4 shows memory Gantt chart during execution of TLP.

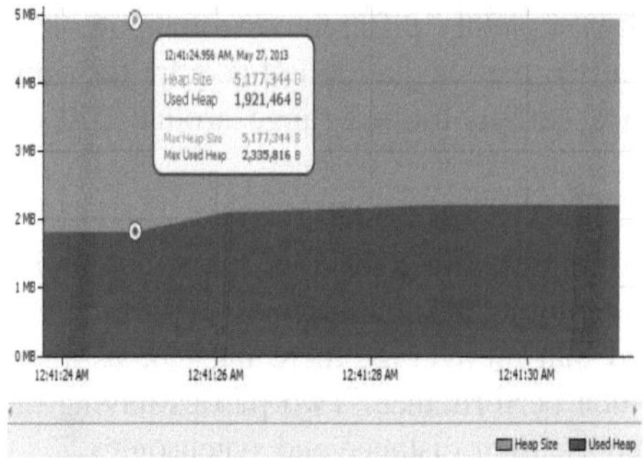

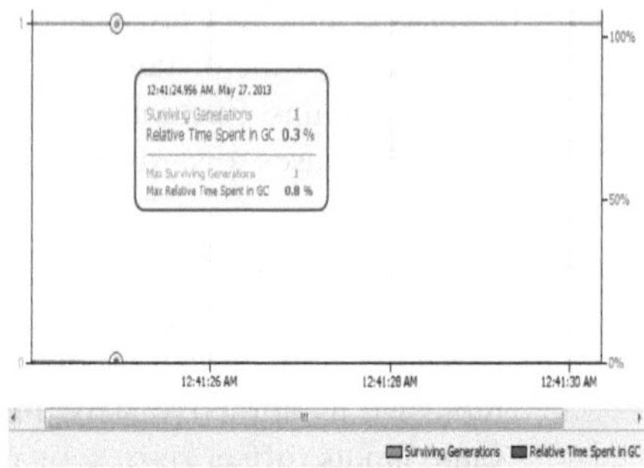

Fig. 3. Memory Heap of TLP **Fig. 4.** Shows Memory Gantt Chart during execution

Figure 5 shows the graph of TLP Synchronization Vs TLP size graph. X axis represents the size of the network and Y-axis represents time in Microseconds. All three learning rules were applied while examining the TLP. From the above graph its evident that Anti-Hebbain rule takes more time to mutually synchronize both TLP. Random walk takes less time among other two learning rules. While Hebbian learning rule takes more time than random walk but less than Anti-Hebbian rule.

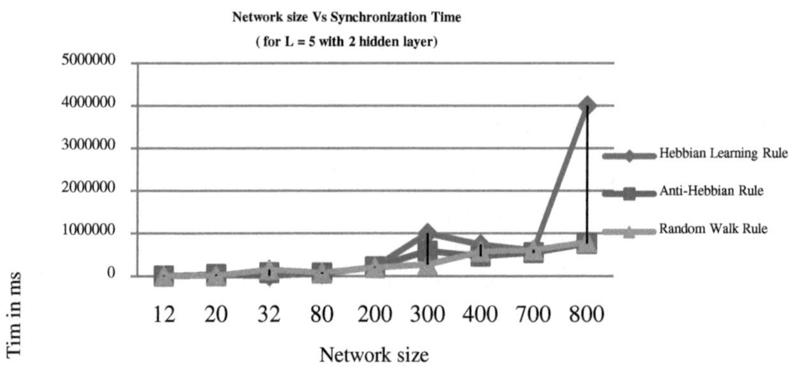

Fig. 5. Shows Synchronization time Vs. Network size of TLP

Table 1. Shows Encryption Decryption Time, Chi-Square Value and degree of Freedom Comapred to TDES and RSA

Source File	Source Size (In Bytes)	Encryption Time (In Seconds)	Decryption Time (In Seconds)	Chi Square Value (CBTLP)	Degree of Freedom (CBTLP)	Chi Square Value (TDES)	Chi Square Value (RSA)
POSTFIX.CPP	32150	13.7185	13.7068	128493	90	82946	149273
PFPICK.DLL	58368	25.0172	25.0105	13518	255	10348	18437
MAKER.EXE	59398	24.6149	24.5938	21936	255	15910	29087
MODE.COM	29271	13.1038	13.0971	5093	255	2056	5186
HIMEM.SYS	33191	14.1532	14.1487	10362	255	6182	11047

6 Analysis of Results

In this paper, CBTLP technique has shown Chi-Square value of proposed technique is higher than TDES and also quite comparable with RSA. So, its confirmed degree of freedom & the characters are well distributed in the range of 0 to 255 parameter value. From experimental results it is clear that the proposed technique may achieve optimal performances. In this case, the two partners A and B do not have to transmit a common secret but use their indistinguishable weights as a secret key needed for encryption [5].

7 Future Scope and Conclusion

This paper presents a novel approach for generation of secret key proposed algorithm using TLP synchronization. This technique enhances the security features of the key exchange algorithm by increasing the synaptic depth L of the TLP. In this case, the two partners A and B do not have to exchange a common secret key over a public channel but use their indistinguishable weights as a secret key needed for encryption or decryption. So likelihood of attack proposed technique is much lesser than the simple key exchange algorithm.

Future scope of this technique is that this TLP model can be deploy in wireless communication for key distribution & authentication purpose. And also this technique can be used in password authentication system in E-commerce. Some evolutionary algorithm can be incorporated with this TLP model to get well distributed weight vector.

Acknowledgment. The author expressed deep sense of gratitude to the Department of Science & Technology (DST) , Govt. of India, for financial assistance through INSPIRE Fellowship leading for a PhD work under which this work has been carried out, at the department of Computer Science & Engineering, University of Kalyani.

References

1. Kahate, A.: Cryptography and Network Security. Tata McGraw-Hill Publishing Company Limited (2003) (Eighth reprint 2006)
2. Mislovaty, R., Perchenok, Y., Kanter, I., Kinzel, W.: Secure key-exchange protocol with an absence of injective functions. Phys. Rev. E 66, 066102 (2002)
3. Ruttor, A., Kinzel, W., Naeh, R., Kanter, I.: Genetic attack on neural cryptography. Phys. Rev. E 73(3), 036121 (2006)
4. Engel, A., Van den Broeck, C.: Statistical Mechanics of Learning. Cambridge University Press, Cambridge (2001)
5. Mandal, J.K., Arindam, S.: An Adaptive Genetic Key Based Neural Encryption For Online Wireless Communication (AGKNE). In: International Conference on Recent Trends In Information Systems (RETIS 2011), Jadavpur University, Kolkata, India, December 21-23. IEEE (2011) ISBN 978-1-4577-0791-9

Color Image Authentication through Visible Patterns (CAV)

Madhumita Sengupta and J.K. Mandal

Department of Computer Science & Engineering, University of Kalyani
{madhumita.sngpt,jkm.cse}@gmail.com

Abstract. In this paper a copyright protection technique based on visual patterns (CAV) has been proposed for color image. This technique manipulates bits of color image to hide the hash of secret without embedding secret directly. Three layers of random noise, generated by R, G and B through hash function, when fall upon a base noise can able to generate imprint of secret in the form of visual patterns. Same process on receiver end authenticates the originality of image and protects ownership. CAV also optimized the intensity value of pixel after embedding by comparing it with original pixel value. Proposed CAV technique has been compared with existing Wu-Tsai's Method, H.C. Wu Method, SAWT and STMDF techniques, where proposed technique shows better performance in terms of MSE, PSNR and fidelity of the stego images.

Keywords: Steganography, authentication, copyright protection, Visual steganography, Mean Square Error (MSE), Peak Signal to Noise Ratio (PSNR), Image fidelity (IF) and Universal Quality Image (UQI).

1 Introduction

Steganography is a technique of hiding secret in cover signals in such a manner that no one apart from the sender and the intended receiver can able to identify the presence of secret message. To hide such secret some feature of the cover image needs to be extracted through hash function, slight change over those feature solve our purpose. Many such steganographic technique are already in existance LSB replacement embedding technique is one of such very popular technique and easy to implement where LSB is replaced by secret message bits directly[1]. On the other hand the fundamental goal of steganalysis is to detect message in cover medium or break the secret of steganography. Eavesdroppers without knowing any fact about secret can damage the secret message or try various permutation and combination to come out the secret. But in the present proposal secret is not embedded in the original image rather a special noise is generated and embedded. This paper introduces CAV technique on digital color images, where randomly generated noise is embedded rather than original secret into the cover signal. Thus any steganalysis software if able to break the hash, may not be able to decode the secret without the base noise. And the base noise is generated by mutual understanding without sharing through mesh network.

S.C. Satapathy et al. (eds.), *ICT and Critical Infrastructure: Proceedings of the 48th Annual Convention of CSI - Volume II*, Advances in Intelligent Systems and Computing 249,
DOI: 10.1007/978-3-319-03095-1_67, © Springer International Publishing Switzerland 2014

Various parametric tests are performed on original image with reconstructed stego-image are compared with most recent existing techniques such as PVD [1], LSB [2], Li's Method [3], Region-Based[4], and STMDF [5], based on Mean Square Error (MSE), Peak Signal to Noise Ratio (PSNR), Standard Deviation (SD), and Image Fidelity (IF) analysis [6] to show a consistent relationship with the recent existing techniques.

Section 2 of this paper deals with the scheme, section 3 elaborates the entire technique with four sub sections, 3.1. base pattern generation, 3.2. embedding secret information with adjustment, 3.3. noise creation for R G and B color band and 3.4. authentications. Results and discussions are outlined in section 4, conclusions are drawn in section 5 and references are cited at end.

2 Proposed Scheme

On transmitting the image any eavesdropper can claim his false ownership. To avoid this problem CAV manipulate bits at sender side to authenticate the image at recipient on clam. Traditional authentication techniques embed bits to insert a secret message/image. Whereas CAV authentication technique flip bits not to insert message but to generate a special noise in a manner that the noise when complied with base noise computed in both end revels the secret in human visual form.

In a digital system the color image I_{NxN} is organized in a form of three primary color red, green and blue represented in three matrixes of intensity value ranges from 0 to 255 of same dimension labelled as R_{NxN}, G_{NxN} and B_{NxN} as shown in fig. 1 and fig. 3. Three secret tiles are used to shape three separate patterns R_P, G_P and B_P of dimension N/2 x N/2 used for three separate color bands.

A single base pattern B_N generates through non sharable computation on both side of communication. Based on the three secret pattern and single base pattern B_N shares of secret are generated. Those shares of secrets N_R, N_G and N_B are embedded through hash function on matrixes of color band R, G and B respectively. On embedding N_R, N_G and N_B on R, G and B the embedded R_e, G_e and B_e generates stego-Image. The overall computation of CAV technique at sender side is shown in fig. 1.

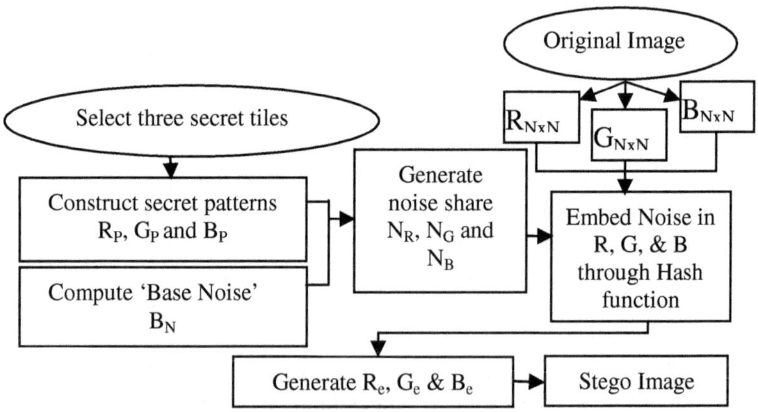

Fig. 1. The schematic representation of CAV technique (sender side)

For authentication the stego-image received at the destination needs to generate three separate bands of color R_{eNxN}, G_{eNxN} and B_{eNxN}. Seed for hash function and noise through hash function is extracted. Generate three noise patterns $N`_R$, $N`_G$ and $N`_B$. When base pattern B_N compile with any of the three extracted noise it regenerate visual patterns of the secret tiles as shown in fig. 10. On comparing the visual patterns with R_P, G_P and B_P authentication is done. The overall schematic representation of the procedure of authentication is shown in fig. 2. Step wise elaboration of the technique has been done in section 3.

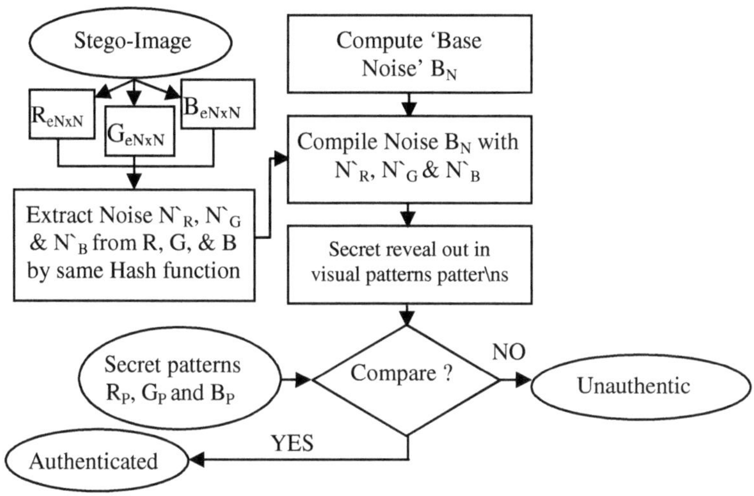

Fig. 2. The schematic representation of CAV technique (receiver side)

3 CAV Technique

The proposed CAV technique is a visual steganographic technique, where authentication is done through human judgment. And the secret information is passively embedded into the cover image. The four steps of CAV algorithm has been discussed in subsequent subsections. The algorithm starts with separation of the three color bands R, G and B of the image taken as cover image for embedding. The color bands are shown in figure 3 for Baboon and Monalisa image respectively.

Three secret tiles are taken as input to generate secret patterns R_P, G_P and B_P as shown in figure 4 to authenticate the whole image. These secret patterns R_P, G_P and B_P are purely having black and white pixel intensity (Binary image) no gray value is allowed through preprocessing by aplying thresholding of 127, that is average of minimum intensity plus maximum intensity allowed in color image, that is ((0 + 255) / 2).

Fig. 3. Three color band of Baboon & Monalisa image of dimension 512 x 512

(i) Three secret tiles for authentication

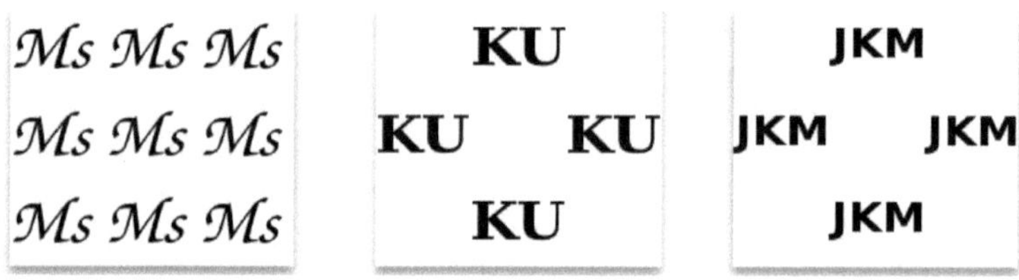

secret pattern R_p secret pattern G_p secret pattern B_p

(ii) Three patterns of secret for authentication of color image (Binary Image) 256 x 256 in dimension

Fig. 4. Pattern of secret tiles for three color bands red, green, and blue as R_P, G_P and B_P respectively

3.1 Base Pattern Generation

The base pattern is common information from the secret patterns which is generated through a hash function and the seed of the hash function is also embedded with the share of secret message in cover image. The same seed extracted from stego-image at receiver end and able to generate the same base pattern B_N. Base pattern is the major share of the secret which need not to be sent through transmission channel to ensure more security to the CAV technique. Four cases are used in this base pattern construction, based on the row plus column mod 4 hash function. Cases are labelled as case 0 to case 3 respectively for horizontal, vertical, front diagonal and back diagonal as shown in figure 5, and that of base pattern generated for the baboon image is shown in figure 6.

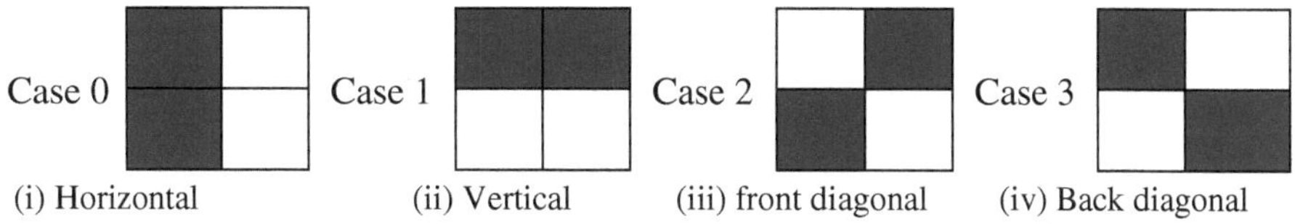

Fig. 5. Base pattern B_N generation rule

Fig. 6. Base noise B_N for 512x512 color Baboon image in gray scale

3.2 Embedding Secret Information with Adjustment

After secret and base pattern construction the next step is to embed the secret into the cover image. In CAV technique the embedding is done by single comparison and incrementing the pixel value. This technique of embedding is directly done on the pixel intensity value of three color bands based on 2 x 2 window in a row major order. As per the rule of binary and digital number system that every even number in digital number system is having LSB '0' (zero) in binary conversion and that of odd numbers are having LSB '1' (one). Here embedding in LSB is just adjusting the odd and even number in digital number system.

Input: Three distinct secret logo RP, GP and BP of dimension 256 x 256 in binary image format shown in figure 4.ii and base pattern BN shown in figure 6.

Output: Stego-image 512 x 512 in dimension

Process: Single share of secret pattern generated from secret binary logo and the base patterns will embed in the original image to generate stego-image.

Algorithm:

Step 1: For every black pixel of secret logo RP and the 2 x 2 window in row major order of base pattern BN the original color bands red 'RNxN' manipulates based on four cases of BN.

Step 2: If the 2 x 2 window of BN followed case 0 then manipulation will be based on case 0`, for case 1 case 1` will be followed and so on as shown in figure 7. Where E|O symbol indicates even or odd both the number are allowed that means no change required, whereas O symbolise odd number only.

Case 0`

E|O	O
E|O	O

(i) Horizontal

Case 1`

E|O	E|O
O	O

(ii) Vertical

Case 2`

O	E|O
E|O	O

(iii) front diagonal

Case 3`

E|O	O
O	E|O

(iv) Back diagonal

Fig. 7. Original pixel of 512 x 512 red color band of Baboon image

Step3: Find the position where to make the number odd, check for even number with mod 2 function and increment the value to make it odd number with LSB 1.

Step 4: For color band green and blue same procedure repeats with secret logo G_P and B_P respectively.

Adjustment: In case the original Image pixel intensity value is 255 and on comparison it needs to increment to become even, that is 256, then special adjustment is required to make the pixel intensity value in range and even too by subtracting digit 2 from the embedded number.

3.3 Noise Creation from R G and B Color Band

Few bits from every pixel of color intensity R, G and B is captured separately for three color bands of cover image based on hash function, to generate three layers of random noise. As shown in figure 8.

(i) Original Red band Noise N_R (ii) Original Green Band Noise N_G (iii) Original Blue Band Noise N_B

Fig. 8. Noise creation from three color band for Baboon Image at sender side

On generation of base noise pattern B_N, the next step is to calculate the secret share and manipulate the LSB with even and odd number parameters. After manipulation, the combination of three colors generates the stego-image. The stego-image at receiver side again passes through a step to extract noise. This extracted noise looks similar to figure 9. Which when compiled with base noise can revel back the secret as shown in figure 10 as a step of authentication.

(i) Modified Red band Noise $N`_R$ (ii) Modified Green Band Noise $N`_G$ (iii) Modified Blue Band Noise $N`_B$

Fig. 9. Noise creation from three color band for stego-Baboon Image at receiver end

3.4 Authentications

For authentication process of CAV, at recipient the stego-image needs to pass through noise generation step to generate three layers of noise $N`_R$, $N`_G$ and $N`_B$. The black pixels of three noise when fall upon the common base noise B_N respectively through any image editing software three secret patterns taken as input at sender side revels out at receiver side. The Human visual patterns can easily identify the authenticity and protect the copyright of any document/image.

Fig. 10. Black noise pixel when compiled with Base noise B_N of Baboon Image512 x 512 in dimension at receiver end generates the secret

4 Results and Discussion

This section deals with the results of computation after embedding single share of hidden data. Ten PPM [7] images have been taken and CAV is applied on each of them. All cover images are 512 x 512 in dimension with secret patterns of 256 x 256 binary image as authenticating image. All the ten images are shown in figure 11. Table 1 shows the computation results of ten color images in PPM format. The results of average of MSE are 0.036203 and PSNR is 62.543495in dB and image fidelity is 0.999997 and that of UQI is 0.999992.

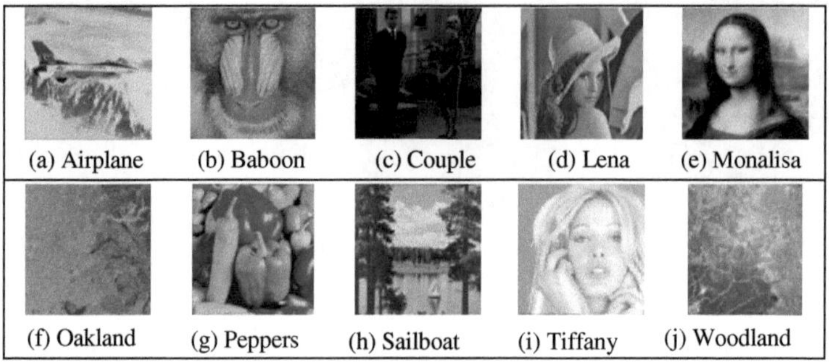

Fig. 11. Color Cover image of dimension 512 x 512 in PPM format

Table 1. The statistical calculation for the MSE, PSNR and IF with UQI

Cover Image	MSE	PSNR	IF	UQI
Airplane	0.036316	62.529833	0.999999	0.999991
Baboon	0.036196	62.544151	0.999998	0.999994
Couple	0.035465	62.632775	0.999984	0.999983
Lena	0.036316	62.529833	0.999998	0.999995
Monalisa	0.036354	62.525274	0.999998	0.999997
Oakland	0.036242	62.538662	0.999998	0.999991
Peppers	0.036369	62.523451	0.999998	0.999996
Sailboat	0.036179	62.546287	0.999998	0.999996
Tiffany	0.036294	62.532419	0.999999	0.999990
Woodland	0.036296	62.532267	0.999999	0.999991
Average	**0.036203**	**62.543495**	**0.999997**	**0.999992**

On comparison with other existing techniques, proposed CAV technique shows better performance as per the statistical results shown in table ii. The minimum requirement for CAV in terms of bytes or bits to authenticate a color image of 512 x 512 as compared with other techniques is lesser with better PSNR in dB which again enhances the image fidelity as compared with other existing techniques.

Table 2. Comparison between various techniques with proposed CAV

Technique	Hiding Capacity (bytes)	Size of cover image	bpB (Bits per byte)	PSNR (dB)
Li's Method[3]	1089	257 x 257	0.13	28.68
H.C. Wu Method[2]	95355	512 x 512	0.97	35.75
Wu-Tsai's Method[1]	51611	512 x 512	0.525	38.66
Region-Based[4]	16384	512 x 512	0.50	40.79
STMDF[5]	50700	512 x 512	0.515	42.70
CAV	196608 bits	512 x 512	0.25	62.54

5 Conclusions

In this paper the authentication of the image is done through CAV algorithm, where the embedding information is not the direct secret message but rather it is a noise. That noise is a share of original secret and the base information. That base information is common information in all the three color bands for three separate secret patterns. This technique helps to authenticate the image as well as provide copyright on clam with minimum degradation of the original image.

Acknowledgment. The authors express deep sense of gratuity towards the Dept of CSE University of Kalyani where the computational resources are used for the work.

References

1. Wu, D.C., Tsai, W.H.: A steganographic method for images by pixel-value differencing. Pattern Recognit. Lett. 24(9), 1613–1626 (2003)
2. Wu, H.-C., Wu, N.-I., Tsai, C.-S., Hwang, M.-S.: Image steganographic scheme based on pixel-value differencing and LSB replacement methods. In: IEE Proceedings of the Vision, Image and Signal Processing, vol. 152(5), pp. 611–615 (2005)
3. Yuancheng, L., Wang, X.: A watermarking method combined with Radon transformand 2D-wavelet transform. In: IEEE Proceedings of the 7th World Congress on Intelligent Control and Automation, Chongqing, China, June 25-27 (2008)
4. Nikolaidis, I.P.: Region-Based Image Watermarking. IEEE Transactions on Image Processing 10(11), 1721–1740 (2001)
5. Mandal, J.K., Sengupta, M.: Steganographic Technique Based on Minimum Deviation of Fidelity (STMDF). In: IEEE Second International Conference on Emerging Applications of Information Technology (EAIT 2011), February 19-20, pp. 298–301 (2011) Print ISBN: 978-1-4244-9683-9, doi:10.1109/EAIT.2011.24
6. Kutter, M., Petitcolas, F.A.P.: A fair benchmark for image watermarking systems. In: Security and Watermarking of Multimedia Contents, Electronic Imaging 1999, vol. 3657. The International Society for Optical Engineering, Sans Jose (1999), `http://www.petitcolas.net/fabien/publications/ei99-benchmark.pdf` (last accessed on February 12, 2013)
7. Weber, A.G.: The USC-SIPI Image Database: Version 5, Original release. In: Signal and Image Processing Institute. Department of Electrical Engineering, University of Southern California (1997), `http://sipi.usc.edu/database/` (last accessed on January 25, 2013)

[illegible] the three color bands [illegible] separate [illegible]. The [illegible] helps to authenticate the image [illegible] the original image.

[illegible]

References

[illegible]

Smart Infrastructure at Home Using Internet of Things

D. Christy Sujatha[1], A. Satheesh[1], D. Kumar[2], and S. Manjula[1]

[1] Department of Software Engineering,
[2] Department of Electronics and Communication Engineering
Periyar Maniammai University
Thanjavur, Tamilnadu, India
{christy_se,asatheesh,kumar_durai,manujula_se}@pmu.edu

Abstract. The Internet of things can be defined as a world-wide network of interconnected objects, which are uniquely addressable. Smart home is one of the applications of the Internet of things in the development of networking technologies. In a smart home, all the home appliances like Lights, Fans, Computer, etc. can be interconnected with a Wireless Sensor Network and can be manipulated by a smart mobile from inside and outside the door through gateway. In the traditional smart home architecture, the Memory and CPU of the Sensor Nodes are idle for most of the time, and these resources were not used in an efficient manner. Using the Virtualization concept of Cloud computing technology, in this paper, we have proposed an energy and cost effective smart home architecture, which connect the Wireless Sensor Nodes through the Virtual Sensor Network (VSN). Virtual Sensor Networking is a developing approach which enables the dynamic collection of a subset of sensor nodes. Our simulation result shows that there is a falling of CPU utilization by 70%, and it also increases the efficient usage of memory capacity.

Keywords: Virtualization, Wireless Sensor Network, ZigBee protocol, Set Top Box, Smart Phone.

1 Introduction

The Internet of things can be defined as a world-wide network of interconnected objects, which are uniquely addressable, based on standard communication protocols. The next revolution will be the interconnection between objects to create a smart environment. Currently, there are 9 billion interconnected devices, and it is expected to reach 24 billion devices by 2020. Smart home is one of the applications of the Internet of things in the direction of the development and promotion of networking industry. In smart home [1] all the devices and appliances at home like Washing Machines, Refrigerators, Electricity Meters, Television, Air Conditioner, Lights, Fans, Computer and other things are interconnected through the Wireless Sensor Networks, so they can communicate with each other and with the residents. The home devices and equipments can be controlled using the latest high-tech smart phones or PDAs.

S.C. Satapathy et al. (eds.), *ICT and Critical Infrastructure: Proceedings of the 48th Annual Convention of CSI - Volume II*, Advances in Intelligent Systems and Computing 249,
DOI: 10.1007/978-3-319-03095-1_68, © Springer International Publishing Switzerland 2014

In the traditional smart home architecture[2] shown in Fig.1., all the home appliances were connected through the Wireless Sensor network which were considered to perform a very specific tasks. All sensor nodes in the network perform more or less as equal nodes to achieve the goal of deploying sensor nodes. The communication protocols of sensor networks were also very simple. Most of the sensor nodes remain idle for the maximum periods of its lifetime. So the resources like memory and CPU of the sensor nodes were not utilized in an efficient manner.

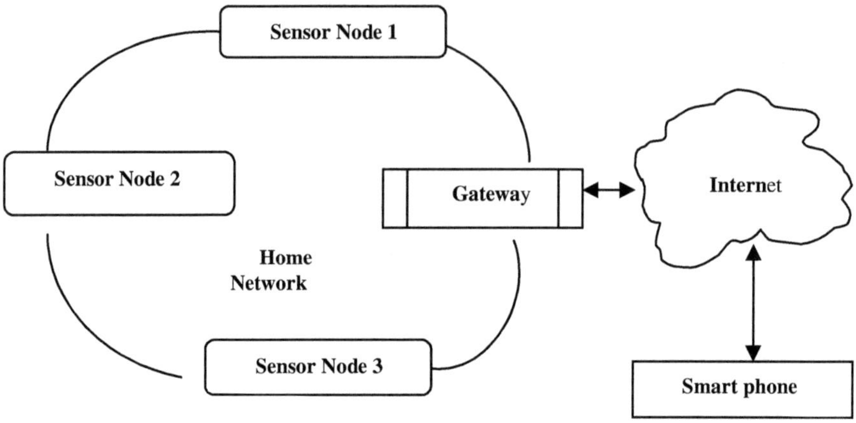

Fig. 1. Traditional Smart home Architecture

Our proposed smart home architecture has a set-top box that acts as a home control box which connects all the electronic equipments at home. The home control box once connected to the Internet, all these home appliances can be reached with any mobile web device or smart phone from anywhere in the world. This set-top box routes, all the home appliances and it also performs monitoring and management systems through Virtual Sensor Network. Virtual Sensor Networking [3, 4, 8] is a growing technique which contains the dynamic collection of a subset of sensor nodes. VSN is one of the best ways to utilize the resources and services in an effective manner, and it provides a platform upon which traditional sensor network architectures can be built, experimented, and evaluated.

The Virtual Sensor Networks are connected through the Set-Top Box that supports ZigBee protocol [5] which is the standard based IPv6 specification for wireless sensor networks. We have chosen the ZigBee protocol since it is specifically designed to operate on low-cost, low-power devices and support the requirements of smart home. We have also carried out simulation based assessment using Imote2 sensor node and VMware technology [6]. The result shows that, our proposed architecture reduces 70% of CPU utilization and increases the efficient usage of memory capacity.

The rest of this paper is organized as follows: Section 2 denotes the proposed smart home architecture. Section 3 presents the Simulation result and discussion and Section 4 gives the conclusion.

2 Proposed Architecture

The proposed architecture implements the sensing technology and a cloud-based virtualization technology. Besides turning home appliances on and off in an intelligent manner using the proposed smart home architecture, an efficient energy consumption can be achieved by continuously monitoring every electricity point within a house and using this information one can modify the way the electricity is consumed.

2.1 Hard Ware Design

The hardware design for the proposed Smart Home architecture is based on the integration of the emerging technologies like Set-top Box, Virtual sensor Ntwork service Provider and Virtual Sensor Network.

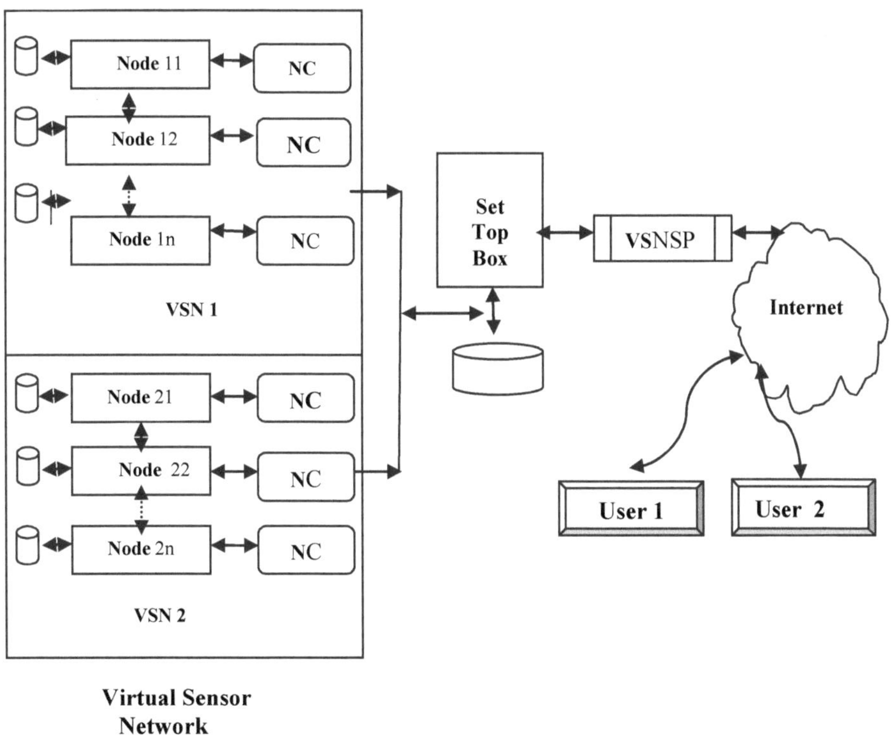

Fig. 2. Smart Home Architecture

2.1.1 Set Top Box

The set-top has a central database which acts as a storage point for all data in the smart-home environment. Any device can be registered with the storage whenever connection is needed. Whenever a new device is connected it creates new records, describing all supported commands, functions, controls, and other information. All the connected devices have a built-in local database (LDB) which is used for storing current status, scheduled tasks and a list of commands in its queue. The Smart-Home central database is coordinated with the device's LDB, by sending and receiving commands. During registration, it loads the information into the control box database from the local database. Any new integrated device is synchronized upon connection, and can be removed when not in use. This is done automatically by sending data to the control box upon initial connection.

2.1.2 Virtual Sensor Network (VSN)

For the smart home automation and control system, Virtual Sensor Network offers a wide range of services: local or remote access from the Internet not only to monitor the home (temperature, humidity, activation of remote video surveillance, status of the doors (locked or open) *etc.* but also for home control (activate the air conditioning/heating, door locks, sprinkler systems, *etc.*).

The resources in the traditional Wireless Sensor Network remain idle for most of the time. Sensor network virtualisation is one of the best ways to utilize the physical sensor node which can provide cost-effective and green technology solutions to design smart houses. In a traditional sensor network, all the nodes in the network perform more or less equal task or application (e.g. Sensing gas leakage) to achieve the goal of deploying sensor nodes.

Virtual Sensor Network consists of collaborative wireless sensor network, which is formed by a subset of sensor nodes of a wireless sensor network, with the subset being dedicated to a certain task or an application at a given time. A virtual sensor network is formed by providing logical connectivity among collaborative sensor nodes. The nodes that do not sense the particular event / application could be part of the VSN which can be used to communicate messages through the sensors used. There may exist more VSN formation at the same time on the physical network. The members of any VSN may be changing over time.

The proposed Virtual Sensor Network Architecture contains

i. End Users
ii. Sensor Node Controller (NC)
iii. Virtual Sensor Network Service Provider (VSNSP)

2.1.2.1 End User

The end user is an actor who needs the smart services of virtual sensor network. The End Users can register their Identification, Name, Descriptions and type of services, they are privileged with. The user information is stored in the VSN Service Provider log record. A user interface is available to submit their details, send the queries and get the responses. The remote device with a user interface may be a Smart phone, PDA or Lap top.

2.1.2.2 Node Controller (NC)

Every smart sensor node is connected with each home appliance such as Washing Machines, Refrigerators, Electricity Meters, Television, Air Conditioner, Lights, Fans, and Computer, etc. The smart sensor nodes can run autonomously and translate data information to the other nodes which are deployed in the Wireless Sensor Network (WSN). This Virtual sensor network is connected with the outside world through the Set-Top Box.

The Node Controller manages the services requested by the user, and the services provided by the sensor node. It stores the details of the sensor nodes like Sensor Id, Sensor Name, Description, Memory and RAM size, Unoccupied Memory and Ram size, Remaining energy, Queue Status, Chanel's availability and usage, etc. It also maintains an association between the QoS requirements imposed on the delivery of the services like Reliability and Delay and it maps different service and resource request to different processes.

2.1.2.3 Virtual Sensor Network Service Provider (VSNSVP)

When the end user requests for any service to the VSN Service Provider, It checks the authentication of the end user and starts initiating an instance of Virtual Sensor Network if the end user is a valid user. Then the connection is established between the end user and the VSN and starts the required execution. The execution of the service is taken care by the Service Provider automatically and transparently to the user. The execution time may be very short or long lasting. When the required service is completed the service is terminated, and the established connection is disconnected and all the resources used by the VSN are released. The service providers also ensure isolation between coexisting VSNs to improve fault-tolerance, security, and privacy. Isolation allows logical separation of the VSNs although they coexist on the same physical substrate sensor network.

2.2 Software Design

The gateway once connected to the Internet, all the home appliances can be reached with any smart mobile phone from anywhere in the world. Any user can access the home appliance from inside or outside the home through Smart User Interface. When the user selects a device, its current status and energy consumption by the device is displayed on the smart phone. Fig.3. shows the proposed Software Layered Architecture. The bottom of the layer is formed by the Hardware *Components* that we have used for the proposed smart home-like Sensor Nodes, Set- Top Box, and Gateway, etc. In order to allow low energy and optimal usage of existing home communication infrastructures; the proposed smart-home control box supports ZigBee protocol as the *Communication Interface.* ZigBee protocol is the standard based IPv6 specification for wireless sensor networks, and it is a software standard that sits on top of the IEEE802.15.4 low data rate wireless standard. *Remote Service*

Interface layer provides various remote network services like Control providers and Content providers. The *Monitoring Software* is written in Dot NET language with ACCESS database. Since it is built on Dot NET software platform, this software is independence of platforms, and it has tremendous expandability. *Application Layer* denotes the type of request we have chosen like Energy Monitoring, Energy Control, Energy Meter and display of on / off status of the device, etc.

Application Layer
Monitoring Software
Remote Service Interface
Communication Interface
Hardware Components

Fig. 3. Software Layered Architecture

3 Implementation Details

3.1 Experimental Environment

In this section, we discuss the simulation environment and evaluation results. We have implemented and evaluated the Virtual Sensor Network environment using the Imote2 Sensor node network which consists of a set of 100 sensor nodes programmed with a Linux kernel. VMware is installed in all the sensor nodes. The sensor nodes can communicate with each other using 2.4 GHz frequency band. The Imote2 sensor node has more memory size and high speed processor 64 GB RAM and Quad 4.0 GHz CPUs. VMware supports synchronized application execution and dynamic application operation, and it also support dynamically loadable modules.

3.2 Experimental Result and Discussion

3.2.1 CPU Utilization

We have plotted a graph to find out the CPU usage for our proposed VSN approach and the traditional approach. In Fig. 4., we have plotted the number of applications in X axis and CPU Usage in Y axis. In the proposed VSN approaches, the sensor node is not restricted to perform specific application; it uses CPU resources more efficiently than the traditional approach. The evaluation result shows that the proposed VSN approach reduces 70% of average CPU utilization than the traditional approach.

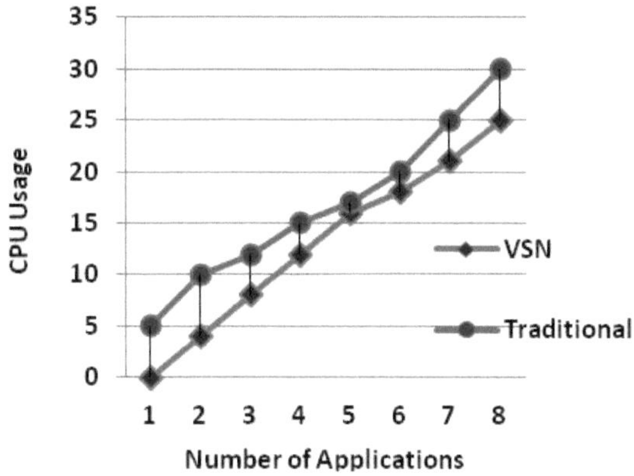

Fig. 4. Comparison of CPU Utilization

3.2.2 Memory Utilization

In Fig.5. we have plotted the number of applications in X- axis and the Memory usage by different applications in traditional Wireless Network approach and the proposed Virtual Sensor Network approach. We illustrate the memory usage of different applications such as temperature, humidity, smoke and light detections. Since the memory size in the Imote2 sensor node more (64 MB), it can provide the required memory to run the application. In all the applications, the proposed VSN approach uses less memory when compared with the traditional approach.

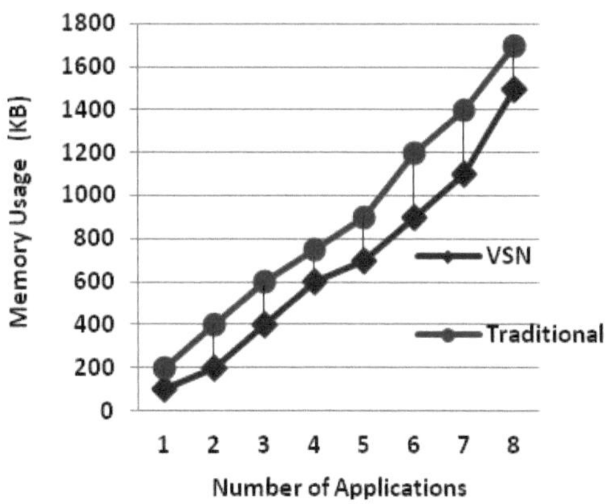

Fig. 5. Comparison of memory usage

4 Conclusion

In this paper, we proposed the Virtualization approach in Wireless Sensor Network in smart home. In the traditional smart home architecture, the memory and CPU of the sensor nodes are idle for most of the time, and these resources were not used in an efficient manner. Our proposed approach allows multiple heterogeneous nodes in different sensor network architecture deployed on a logically shared virtual sensor network. So the resources like memory and CPU of the sensor nodes were utilized in an efficient manner. We have also performed simulation using the Imote2 sensor nodes and VMware technology to evaluate the CPU and memory usage. The result shows that our proposed architecture reduces 70% of CPU utilization and increases the usage of memory capacity effectively.

References

1. Bregman, D., Korman, A.: A Universal Implementation Model for the Smart Home. International Journal of Smart Home 3, 3–8 (2009)
2. Ronnie, D.: Caytiles1 and Byungjoo Park: Mobile IP-Based Architecture for Smart Homes. International Journal of Smart Home 6, 1–8 (2012)
3. Motaharul Islam, M., Huh, E.-N.: Virtualization of Wireless Sensor Network. JNW 7(3), 412–418 (2012)
4. Virtual Sensor Network, http://en.wikipedia.org/wiki/Virtual_sensor_network
5. Zhang, L., Wang, Z.: Integration of RFID into Wireless Sensor Networks: Architectures, Opportunities and Challenging Problems. In: Fifth IEEE International Conference on Grid and Cooperative Computing Workshops (GCCW 2006) (2006)
6. http://VMware.Wikipedia.freeencyclopedia.htm
7. Zhang, Y., Zhang, J., Zhang, W.: Discussion of a smart house solution basing cloud computing. In: Communications and Intelligence Information Security, ICCIIS (2010)
8. Sarakis, L., Zahariadis, T., Leligou, H.-C., Dohler, M.: A framework for service provisioning in virtual sensor networks. EURASIP Journal on Wireless Communications and Networking (2012)

A Novel Approach for Ipv6 Address

S. Deepthi, G. Prashanti, and K. Sandhya Rani

Department of Computer Science & Engineering,
Vignan's Lara Institute of Technology and Science,
Guntur, A.P, India
deepthi.rachakonda@gmail.com

Abstract. One of the biggest challenges is IPv4 address nearing exhaustion, so planning for the adoption of IPv6. As IPv6 deployment increases, so many services running in Cloud computing will face problems associated with IPv6 addressing. In IPv6 address the notation is too long 39 bytes, there are too many variants of a single IPv6 address and a potential conflict may exist with conventional http_URL notation caused by the use of the colon (:) With that in mind, this paper explores a new scheme to represent an IPv6 address with a shorter, more compact notation 28 bytes, and 26 Bytes without variants or conflicts with http_URL. The proposal uses Base64 and the well-known period as a group delimiter instead of the colon and non symbolic usage of the Base 64.The paper mainly concentrated on two ways of representing the IPv6 Address one is reducing the no of segments in the IPv6 address to 5 segments and second is 4 segments which are compact and user-friendly textual representation of IPv6 address.

Keywords: IPv6 address, Cloud computing, Non Symbolic Base64, Colon hexadecimal, Unicast address, Text Conversions.

1 Introduction

Cloud computing allows dynamic, on-demand, and elastic generation of both virtual infrastructure, and virtual machines running over the infrastructure. The ongoing expansive growth of the Internet and the need to provide IP addresses to accommodate it—including addresses for virtualized machines and resources in the cloud—is accelerating the emergent use of IPv6. IPv6 with its robust architecture was designed to support increasing numbers of new users, computer networks, and Internet-enabled devices, applications for collaboration and communication, and virtualized resources. As they increase in number, applications and services within clouds render the need for transition to IPv6 even more immediate.

From the Fig. 1 one third of the cloud service provider's surveyed are available natively via IPv6 [10]. Adoption of IPv6 by Google users is trending upwards Fig .2, indicating that user IPv6 connectivity is gaining slowly but is still nowhere near widespread adoption.

S.C. Satapathy et al. (eds.), *ICT and Critical Infrastructure: Proceedings of the 48th Annual Convention of CSI - Volume II*, Advances in Intelligent Systems and Computing 249,
DOI: 10.1007/978-3-319-03095-1_69, © Springer International Publishing Switzerland 2014

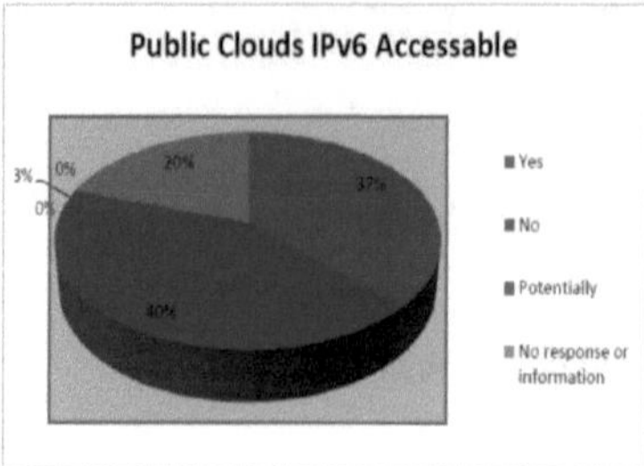

Fig. 1. Cloud Service Providers Natively Available via IPv6.

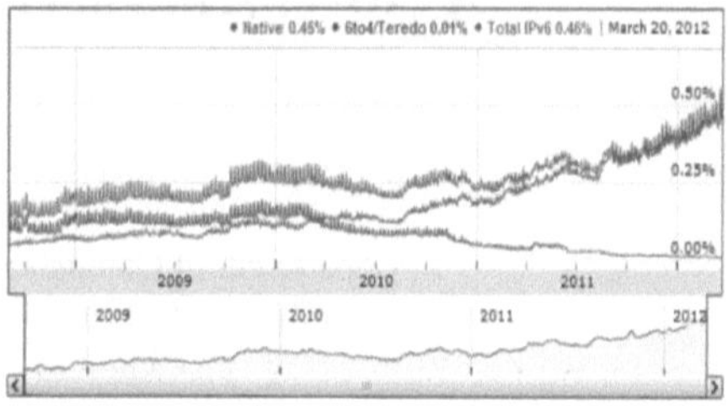

Fig. 2. IPv6 Connectivity of Google Users

The demand for new IP addresses is continuously increasing and it is speculated that after the depletion of IPv4 as shown in Fig.3 [11]. The increasing number of Internet users, systems, and the convergence of services into common infrastructure will drive the demand for IPv6. Percentage of IPv6-enabled Autonomous Systems registered in each RIR is showed in Fig.4.

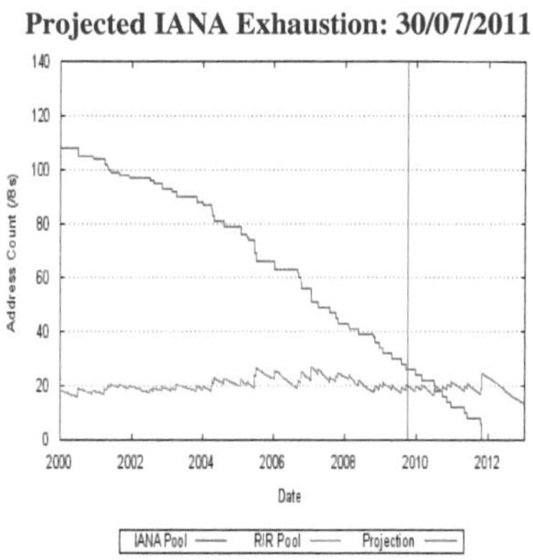

Fig. 3. IPv4 Consumption: Projection

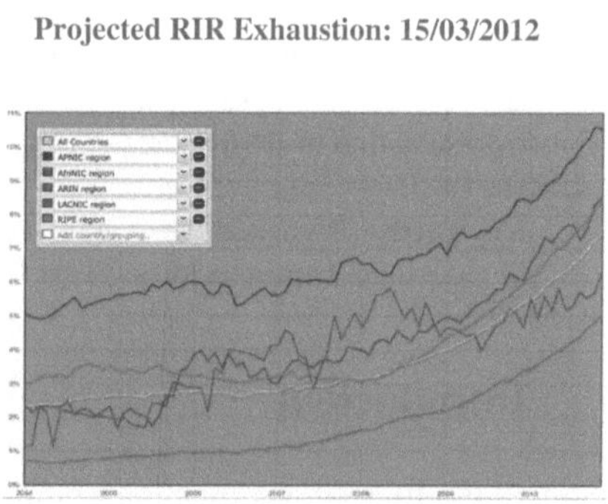

Fig. 4. Percentage of IPv6-enabled Autonomous Systems registered in each RIR

The wireless market requires a low latency, always on, auto-roaming always reachable IP service. Percentage of IPv6-enabled Autonomous Systems registered in each RIR is showed in Fig.4. Some of the benefits of IPv6 are128 bits in length address that the pool of addresses will be large enough to serve all the present and future hosts on the internet. Optimized for next-generation networks: Getting rid of NAT re-enables the peer-to-peer model and helps with deploying new applications (e.g., communications and mobility solutions, such as VoIP.) [3]. IP sec is mandatorily supported, an aspect that can solve a lot of the security issues that arise with careless users. QoS handling through flow label field in the header which opens new possibilities in how to manage communications and application traffic[6].

2 Literature Survey

In order to present our work we made a lot of study regarding existing methods. Those are having some difficulties on representation and remembrance. Our paper has made modifications which overcome them. The models we studied are described below.

2.1 RFC 4648

This describes the commonly used base 64, base 32, and base 16 encoding schemes. It also discusses the use of line-feeds in encoded data, use of padding in encoded data, use of non-alphabet characters in encoded data [14].

In Padding of encoded data the use of padding ("=") in base-encoded data is not required or used. In the general case, when assumptions about the size of transported data cannot be made, padding is required to yield correct decoded data. Implementations MUST include appropriate pad characters at the end of encoded data unless the specification referring to this document explicitly states.

Choosing the Alphabet also plays very important role where "0" and "O" are easily confused, as are "1", "l", and "I". For base 64, the non-alphanumeric characters (in particular, "/") may be problematic in file names and URLs. Certain characters, notably "+" and "/" in the base 64 alphabet, are treated as word-breaks by legacy text search/index tools. There is no universally accepted alphabet that fulfills all the requirements.

Some other ways: For example, consider the address

1080:0:0:0:8:800:200C:417A

In decimal, considered as a 128 bit number, that is

21932261930451111902915077091070067066.

As we divide that successively by 85 the following remainders emerge [4]:

51, 34, 65, 57, 58, 0, 75, 53, 37, 4, 19, 61, 31, 63, 12, 66, 46, 70, 68, 4.

Thus in base85 the address is:

4-68-70-46-66-12-63-31-61-19-4-37-53-75-0-58-57-65-34-51.

Then, when encoded as specified above, this becomes:

4)+k&C#VzJ4br>0wv%Yp

➢ Proposals to resolve colon-related conflicts Extra square brackets in domain part of http_URL:IPv6 addresses are transcribed as a hostname or sub domain name within this name space, in the following fashion[11]

2aa1: da8 : 65b3 : 8d3 : 1369 : 0a8e : 570 : 7448

Is written as 2aa1 - da8 - 65b3 - 8d3- 1369- 0a8e- 570- 7448 literal: net:

IPv6 addresses are 128-bit identifiers for interfaces and sets of interfaces. There are three types of addresses: Unicast, Anycast , and Multicast.

3 Proposed Method

Whilst we believe that IPv6 will begin a new and improved communications era for the whole IT industry, we also accept that IPv6 itself is not perfect. First, it is obvious that with such a large address space (3.4 × 1038 or 340 undecillion addresses) a significant number of characters will be required to represent any single address. The following example shows the IPv6 address format

AB80:0000:0000:0000:0560:97FF:EE8F:64BB

Here a full IPv6 address consists of 32 bytes or a string of 39 characters (including 7 delimiters) in human readable form which is both challenging to remember and prone to mistakes when read, written or deployed. A longer notation means more buffer space is needed when saving, there is an increased cost in bandwidth and latency time during transit, and more computing power is used when reading/writing, searching/parsing, etc.

Second, the current IPv6 notation of "colon hexadecimal" [5] has another issue that there are too many variants of text representation for a single IPv6 address [6]. With such a degree of flexibility in representing an address, it might become prone to misinterpretation in both human and computer environments (searching, parsing and modifying, logging and operating).

Third, the use of the colon (:) separator in place of the dot (.) presents both a potential ambiguity with current http_URL/Windows UNC and the annoyance of being a "two-key" entry on most. It is unpredictable that how many systems and applications will be affected by this incompatibility [5].

Bearing these issues in mind and considering the increasing demands of cloud computing, this paper introduces two methods one is Non symbolic Representation of IPv6 Address which consist of 5 segments and second one is enhancement of first method where the IPv6 address consists of 4 segments. These two approaches present an IPv6 address in non symbolic Base64 with period (or "dot") delimiters as used in IPv4.

Table 1. Base 64 Representation

value	encoding	value	encoding	value	encoding	value	encoding
0	A	17	R	34	i	51	z
1	B	18	S	35	j	52	0
2	C	19	T	36	k	53	1
3	D	20	U	37	l	54	2
4	E	21	V	38	m	55	3
5	F	22	W	39	n	56	4
6	G	23	X	40	o	57	5
7	H	24	Y	41	p	58	6
8	I	25	Z	42	q	59	7
9	J	26	a	43	r	60	8
10	K	27	b	44	s	61	9
11	L	28	c	45	t	62	+
12	M	29	d	46	u	63	/
13	N	30	e	47	v	pad	=
14	O	31	f	48	w		
15	P	32	g	49	x		
16	Q	33	h	50	y		

Table 2. Non Symbolic Base 64 Representation

value	encoding	value	encoding	value	encoding	value	encoding
0	0	17	H	34	Y	51	p
1	1	18	I	35	Z	52	q
2	2	19	J	36	a	53	r
3	3	20	K	37	b	54	s
4	4	21	L	38	c	55	t
5	5	22	M	39	d	56	u
6	6	23	N	40	e	57	v
7	7	24	O	41	f	58	w
8	8	25	P	42	g	59	x
9	9	26	Q	43	h	60	y
10	A	27	R	44	i	61	z
11	B	28	S	45	j	62	A[1]
12	C	29	T	46	k	63	B[1]
13	D	30	U	47	l		
14	E	31	V	48	m		
15	F	32	W	49	n		
16	G	33	X	50	o		

3.1 Non Symbolic Base64notation of IPv6 Address with 5 Segments

Non Symbolic Base64 was first described in a paper [16] where it is proposed as an alternative approach to Base64 for non-alphanumeric characters and is intended to be an improved implementation of Base64.The differences between the original Base64 and non Symbolic Base64 can be seen in Table1 and Table2. In the new scheme, the symbols "+", "/" and "=" are not used. Instead, the character "A^1" is a special tag and subsequently A^1 represents number 62[14], B^1 for 63. As a result, the new alphabet series is 0–9, A-Z, a-z and A^1, B^1. Since there is no symbol used in Non Symbolic Base 64 index table, it shortens the length of IPv6 address without adversely affecting readability, one of the important requirements of the proposed IPv6 address notation. Below given shows hexadecimal representations of an IPv6 address of 128 bits represented along with hex format

100 1111 1010 1010 0001 1010 1110 0011
0000 0000 1010 0111 0110 0001 0000 0111
0000 0000 0000 0000 0001 1011 1111 1111
0000 0000 0010 0010 0111 1000 1001 1000
Represented in Hex (4 bit) 4FAA:1BE2:00A7:6107:0000:1BFF:0022:7898

This long address is commonly depicted as eight pairs of bytes, but it can also be considered in three sections as shown in Figure 5. The general format for IPv6 global unicast addresses is as follows [1]:

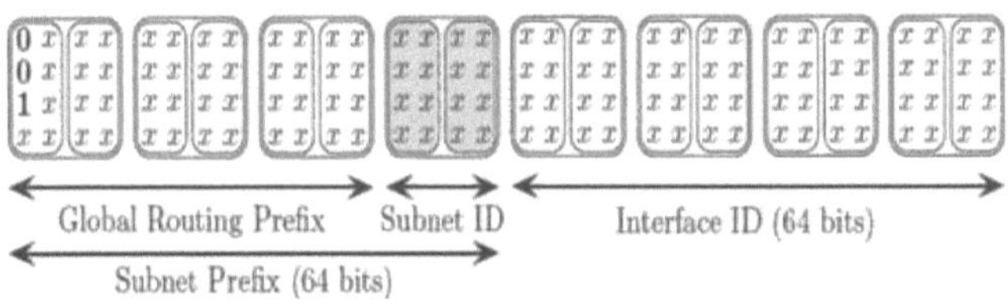

Fig. 5. Global Unicast Addresses

The first half of the address is a 64-bit subnet prefix comprising of a six byte (48 bits) Global Routing Prefix and a two byte (16 bits) Subnet ID. The second part of the address is another 64 bits known as the Interface ID and is used mainly in a unicast addressing. For the purpose of this paper, IPv6 addressing could be described using

The format is having following features:

- Encoded in Non Symbolic Base64.
- Dot-separated six segments
- Prototype length: 24 codes + 4 dots = 28 characters
- Character range: 0–9, A-Z, a-z
- Case-sensitive

Conversions to/from Non Symbolic Base64

The process of converting an IPv6 address into Non Symbolic Base64 can be summarized as these steps:

- ➢ S1. Divide the given 16-byte address into 5 segments as 4:2:2:4:4
yyyy:yy:yy:yyyy:yyyy (4:2:2:4:4)
- ➢ S2. Convert each segment into Non symbolic Base64
- ➢ S3. Separate the Non symbolic Base64 encoded string into 6:4:4:6:6 as
xxxxxx:xxxx:xxxx:xxxxxx:xxxxxx (6:4:4:6:6)

Here we will see some example for this approach

0011011010010101110000001000101100000000000
1000100101011001111010010000010101111100000
0001010110011111111111111111100110011010110 0

The above there is a string is the Binary representation of IPv6 address

3695:C08B:0011:2B3D:20AF:00AC:FFFF:99AC

Firstly, divide the binary string into 5 segments by the proportions of 4:2:2:4:4

0011 0110 100101 011100 000010 001011 (4 Bytes,32 Bits)
0000 000000 010001 (2 Bytes, 16 Bits)
0010 101100 111101 (2 Bytes, 16 Bits)
0010 0000 101011 110000 000010 101100 (4 Bytes,32 Bits)
1111 1111 111111 111001 100110 101100(4 Bytes,32 Bits)

Secondly, encode each segment using 6-bitNon Symbolic Base64,

36bS2B 00H 2iz 20hm2i ffB^{1}vci

Thirdly, add the period (or dot) as a delimiter,

36bS2B. 00H .2iz .20hm2i .ffB^{1}vci

Using the steps listed, a conversion program was written to automate the process of converting an IPv6 address from base 16 to Non Symbolic Base64

3.2 Non Symbolic Base64notation of IPv6 Address with 4 Segments

According to Non Symbolic Base64 Method where we have 5 segments. Further we can shorten these 5 segments into a IPv6 address which consists of only 4 segments and these are constructed according the global unicast addressing as shown in Figuer.5.This method which reduces the 28 Bytes(5 segments) to 26 Bytes (4 segments) definitely a better representation of IPv6 address. This method is similar to that of previous method except that the prototype length: 23 codes + 3 dots = 26 characters

Conversions to/from Non Symbolic Base64 with 4 segments

The process of converting an IPv6 address into Non Symbolic Base64 can be summarized as these steps:

- ➢ S1. Divide the given 16-byte address into 4 segments as 4:2:2:8
- ➢ S2. Convert each segment into Non symbolic Base64
- ➢ S3. Separate the Non symbolic Base64 encoded string into 6:4:4:11 as xxxxxx:xxxx:xxxx:xxxxxxxxxxx (6:4:4:11)

Example: 3695:C08B:0011:2B3D:20AF:00AC:FFFF:99AC

The Binary representation of above IPv6 address is

0011011010010101110000000100010110000000000
1000100101011001111010010000001010111100000
0001010110011111111111111111001100110101100

Firstly, divide the binary string into 4 segments by the proportions of 4:2:2:8,

0011 0110 100101 011100 000010 001011 (4 Bytes,32 Bits)
0000 000000 010001 (2 Bytes, 16 Bits)
0010 101100 111101 (2 Bytes, 16 Bits)
0010 000010 101111 000000 001010 110011 111111 111111 111001 100110 101100(8 Bytes, 64 Bits)

Secondly, encode each segment using 6-bitNon Symbolic Base64,

36bS2B 00H 2iz 22i0ApB^{1}B^{1}vci

Thirdly, add the period (or dot) as a delimiter,

36bS2B. 00H .2iz .22i0ApB^{1}B^{1}vci

Using the steps listed, a conversion program was written to automate the process of converting an IPv6 address from base 16 to Non Symbolic Base64

4 Conclusion

The original objective of this study was to find a shorter textual representation for IPv6 addressing. The length of an IPv6 address encoded in Non Symbolic Base64 with 5 segments has a theoretical reduction in length of (39–28)/39 = 28.8% where as the Non Symbolic Base64 with 4 segment has a reduction in length of(39-26)/39=33.3% .Even though some method having 27 bytes of representation but it used five dot representations. Here we got one byte extra but as a human being remembering 5 segments or 4 segments are better than remembering 6 segments. So definitely it has an advantage.

Acknowledgment. We take this opportunity to acknowledge those who have been great support and inspiration through the research work.

References

1. Liu, Z., Liu, L., Hardy, J., Anjum, A., Hill, R., Antonopoulos, N.: Dot-base62x: building a compact and user-friendly text presentation scheme of ipv6 addresses for cloud computing
2. Liu, Z., Lallie, H.S., Liu, L.: A Hash-based Secure Interface on Plain Connection. In: Proceedings of CHINACOM 2011. ICST.OTG & IEEE Press, Harbin, China (2011)
3. Davies, J.: Understanding IPv6, 2nd edn., pp. 43-45, 50, 92. Microsoft Press, Redmond (2008) ISBN-10: 0735624461, 978–0735624467
4. RFC 1924 A Compact Representation of IPv6 Addresses, http://tools.ietf.org/html/rfc1924
5. RFC 5952 Recommendation for IPv6 Address Text Representation, http://tools.ietf.org/pdf/rfc5952.pdf
6. Understanding IPv6 Addressing - Technical Documentation - Support - Juniper Networks
7. Illustrating the Impediments for Widespread Deployment of IPv6 ALA HAMARSHEH and MARNIX GOOSSENS
8. RFC 3513 Internet Protocol Version 6 (IPv6) Addressing Architecture, http://www.ietf.org/rfc/rfc3513.txt
9. Grayeli, P., Sarkani, S., Mazzuchi, T.: Performance Analysis of IPv6 Transition Mechanisms over MPLS
10. Vail, J.: Cloud Providers that Support IPv6. East Carolina University, Deeepartment of Technology Systems
11. Huston, G.: IPv4 Address Report, daily generated (2011), http://www.potaroo.net/tools/ipv4/index.html (retrieved in August 2011)
12. Stockebrand, B.: IPv6 in Practice—A Unixer's Guide to the Next Generation Internet. Springer, Heidelberg (2007) ISBN 3-540-24524-3, 978-3-540-24524-7
13. Hinden, R., Deering, S.: IP Version 6 Addressing Architecture, IETF RFC 4291 (2005), http://www.rfc.net/rfc4291.html (retrieved in August 2011)
14. RFC 4648 Base-N Encodings, http://www.ietf.org/rfc/rfc4648.txt

Secured Internet Voting System Based on Combined DSA and Multiple DES Algorithms

K. Sujatha[1], A. Arjuna Rao[1], L.V. Rajesh[1], V. Vivek Raja[1], and P.V. Nageswara Rao[2]

[1] Miracle Educational Society Group of Institutions, Vizianagram
[2] GITAM University, Visakhapatnam

Abstract. Internet Voting with the widespread use of Internet is becoming increasingly appealing to groups in place of paper based elections or vote-by-mail elections to geographically distributed voters, as more people are gaining access to the Internet. However, I-Voting systems should be designed carefully or otherwise they may corrupt results or violate voters privacy. This paper proposes the concept of Internet Voting using combined Digital Signature Algorithm and Multiple Data Encryption Standard Algorithm that supports every phase of the electoral process to ensure its security, privacy and transparency. The idea of using security algorithms is to provide a secure method to encrypt and sign data. First the Digital Signature Algorithm is used to generate digital signature that is used as a key for Data Encryption Standard Algorithm. The random number and Digital Signature are send to authorized person using a secure channel. Then Data Encryption Algorithm is applied random number of times for obtaining Multiple Data Encryption Standard. In this work, the aspects and risks involved in Internet Voting architectures are considered and a solution is derived for better usage. The Secured Internet Voting System is simulated and the test results show that the proposed system is secured and flexible. The function of this system is to maintain the most basic principles of an elector process, such as secrecy, correctness, anonymity, non-coercion.

Keywords: Internet Voting(I-Voting), Digital Signature Algorithm(DSA), Multiple Data Encryption Algorithm(MDES), Data Encryption Standard Algorithm (DES), The Secured Internet Voting System(SIVS).

1 Introduction

Information Technology has colossal influence in our daily lives. Election mechanisms are no exception. Electronic Voting Machines(EVMs) are replacing paper based and mechanical balloting system. Electronic voting technology can speed the counting of ballots and can provide improved accessibility for disabled voters. But some concerns with these machines are trustworthiness of both their hardware and their software. Generally public and political parties have raised their doubts that whether presently used EVMs are developed without any scope for tampering. Voting systems use to ensure that votes were cast correctly to detect possible fraud or

S.C. Satapathy et al. (eds.), *ICT and Critical Infrastructure: Proceedings of the 48th Annual Convention of CSI - Volume II*, Advances in Intelligent Systems and Computing 249,
DOI: 10.1007/978-3-319-03095-1_70, © Springer International Publishing Switzerland 2014

malfunction provides a means to audit the original machine. Hence it is required to increase confidence in the EVM-based election process. Trustworthiness is a complex concept which is difficult to quantify and hard to achieve but is a necessary condition for the legitimacy of the electoral process[1]. Election commission of India is also seriously thinking about this problem and recently conducted all party meeting for possible solutions. They are even thinking about providing a printed receipt to the voter which increases manual overheads with handling paper slips in this electronic generation.

The most fundamental problem with electronic machine based voting system is that the entire election concentrates on the robustness, correctness and security of the software within the voting terminal. Hence EVMs cannot be considered to be alternative to paper voting. Paperless voting systems should be software dependent, so that a change in the software or an undetected error could cause an undetectable change in the election outcome. Internet voting can use remote locations or can use traditional polling locations with voting booths consisting of Internet connected voting systems with secure channel. Corporations and organizations routinely use Internet voting to elect officers and Board members and for other proxy elections. Internet Voting is the technical evolution of ballot paper voting and therefore new approaches of software will make electronic elections as secure as remote banking[2]. Internet Voting is one form of Electronic voting and offers many advantages over traditional systems as it has the ability to easily handle multiple languages and by meeting the needs of voters with disabilities and also eliminates problems such as over voting and other voter intent issues. In this proposed Secured Internet Voting people from various remote locations can access the voting system from their computer and the results are encrypted and stored in database as shown in figure 1[3].

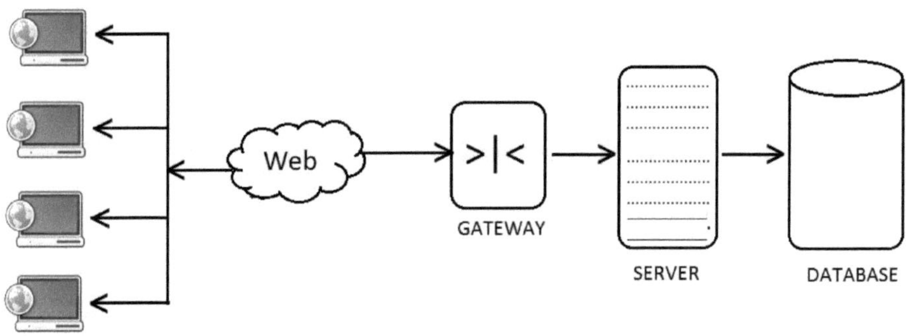

Fig. 1. Secured Internet Voting System

2 Secured Internet Voting System

Internet elections have the prospective of being cheaper and less time consuming. Voting requires certain principles like directness, freedom, equality, publicness which are difficult to attain by using the traditional voting methods but possible with Internet based Voting. Proposed Secured Internet Voting System proposed maintains all the principles and provides the authentication and encryption features by using proved algorithms in a unique process.

2.1 Internet Voting Process

The Internet Voting Process involves handling the ballots and voter specifications and the security concerns in handling the voter's choice in network as shown in figure 2. The Steps involved in this process are listed below[4].

Step 1: Setup Ballot: The election manager will set up the Internet ballot.
Step 2: Start Election: Ballots officially open based on time settings.
Step 3: Voters Vote: The Internet voting system is anonymous. This means that voters identity is separated from the choices they make. The voters can easily vote sitting at their desktops at home or office.
Step 4: Split Voter Identity and Votes: In order to preserve the voter privacy the voter identity and Votes are separated and the identity is directly placed in database and votes are send for encryption. In addition to the encrypted timestamps this phase of the process takes security one step further. Once an election is “finalized”, the encrypted timestamps are completely deleted, so there is no physical way to connect choices made to the voter.
Step 5: Encrypt and Transfer: As ballots are passed in the database, timestamps will be encrypted. To the naked eye, it will be impossible to match-up two timestamps and match a voter and choices where even the Technicians cannot manipulate any votes or results. Then the voter choice is encrypted by using a combined DSA and Multiple DES Algorithms.
Step 6: Store Values in Secure Database: The Internet voting system will have its own database. The timestamps and votes will be stored in encrypted format in secure databases.
Step 7: End Election: Once the election is complete, the administrators must “Finalize” it to view the results. Prior to finalizing the election the reports and results are not available to administrators which restricts administrator access to the system while the election is running. Since no reports are available while an election is in progress there is no chance of manipulation by administrators.
Step 8: Decrypt Votes: Decryption of votes placed in encrypted format in database is done by taking the signatures through secure channel.
Step 9: Start Counting: Counting the votes is done after applying decryption.
Step 10: Tabulate Results: The system will tabulate the votes and generate results.
Step 11: Display Reports: Results will be provided various formats for downloads and printing purpose.
Step 12: Download/Print : This download option is provided for distribution of results and printing is for displaying results in notice boards and serves filing purpose.

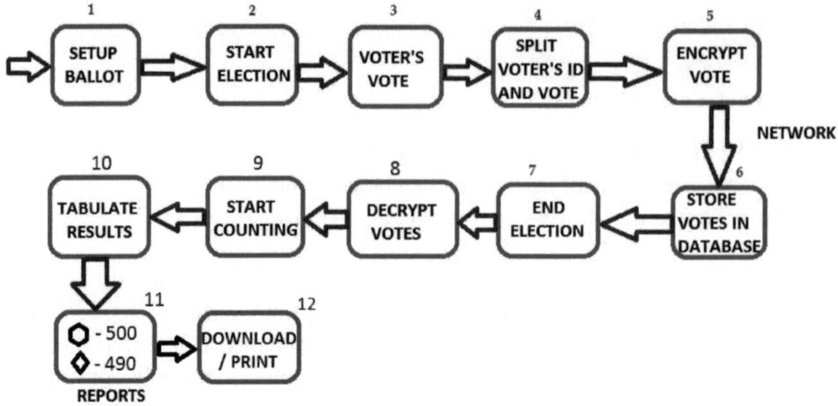

Fig. 2. Internet Voting Process

2.2 Digital Signature Algorithm

Digital Signature Algorithm[5] enables the computation of a public key for an entity when only some general scheme parameters and a string identifying the entity are given. The Digital Signature Algorithm (DSA) designed by NIST & NSA in early 90's and is used with SHA hash algorithm. DSA creates a 320 bit signature, but with 512-1024 bit security which again rests on difficulty of computing discrete logarithms which has been widely accepted. This involves Key Generation and Signature Creation and Verification and is the signature scheme that has advantages, being both smaller and faster than RSA[7,8].

2.3 Multiple Data Encryption Standard (MDES) Algorithm

Present authors propose the MDES Algorithm based on standard DES. DES is an unclassified crypt algorithm adopted by the National Bureau of Standards for public use. The DES, which was approved by the National Institute of Standards and Technology(NIST) which is intended for public and government use. The algorithm is designed to encipher and decipher blocks of data consisting of 64 bits under control of a 64-bit key1[6,9]. In Multiple Data Encryption Standard(MDES) the DES is applied random number of times. The equations for encryption and decryption are as given in (1) and (2).

$$C = DES_{kn}\,\{\,DES_{kn-1}\,\langle \ldots DES_{k1}(M)\rangle\} \tag{1}$$

$$M = DES_{k1}^{-1}\,\{\ldots\,[DES_{kn-1}^{-1}\,\langle DES_{kn}^{-1}(C)\rangle]\} \tag{2}$$

The multiple DES encryption and decryption processes are as shown in figure 3 and figure 4. The first key, k_1 is the signature which is generated by DSA. Then the subsequent keys, $k_{2,}$,,, k_n are obtained by left rotating the previous key. Similarly in decryption right rotation is used for obtaining keys.

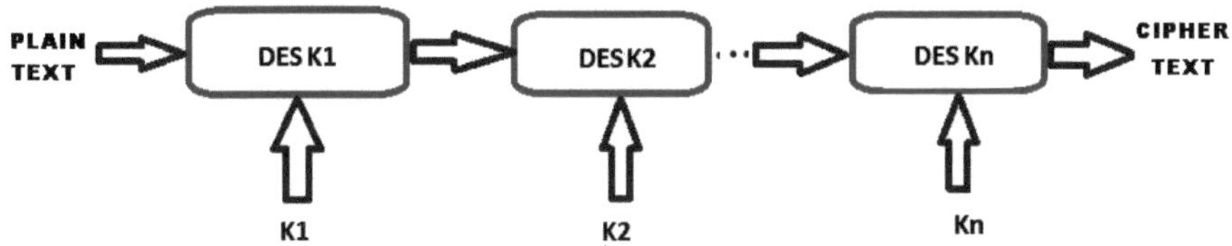

Fig. 3. Multiple Data Encryption Standard Algorithm (MDES) Encryption

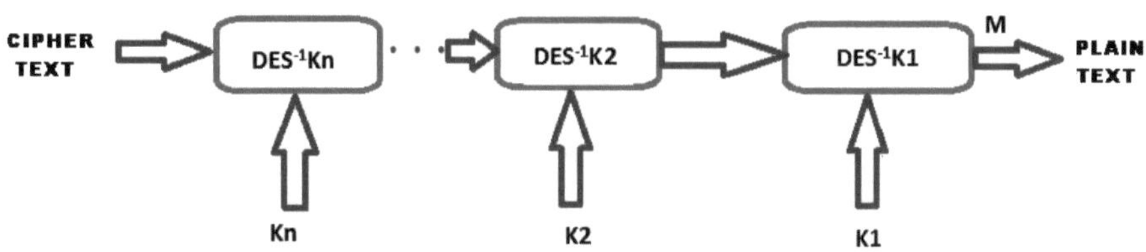

Fig. 4. Multiple Data Encryption Standard Algorithm (MDES) Decryption

2.4 Combined DSA and MDES Algorithms

The Combined DSA and MDES Algorithms is the proposed technique that will encrypt vote timestamps making it impossible to link a voter and his/her choices by mere observation and further transmit them in secure format on a public channel. The encryption technique shown in figure 5 serves the purpose.

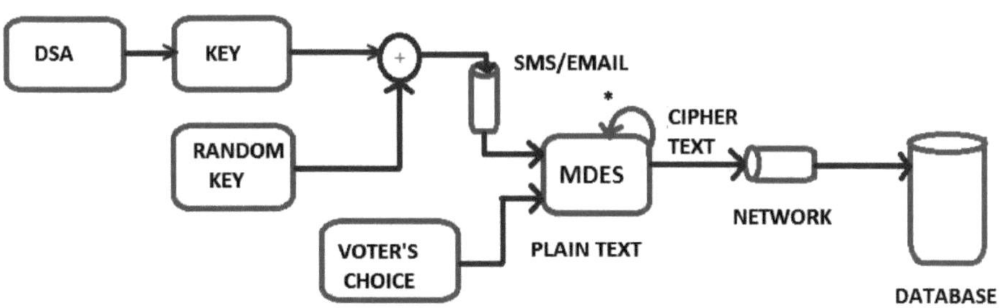

Fig. 5. Encryption Process with aggregation of DSA and MDES Algorithms

Step 1: Digital signature is generated and transmitted secure channel such as email or an SMS alert.
Step 2: Voter downloads the signature and the random number to run MDEs.
Step 3: Voter invokes a locally stored application program and enters signature that acts as a key to MDES Algorithm.
Step 4: MDES Algorithm is applied random number of times from the random number is also passed to voter signed with Digital Signature.

3 Results

An online webbased application is simulated and is tested by in a college environment for electing class representatives by using students database and their mobile phones. The application involves two actors administrator and voters. Administrator manages Ballots and counting encrypted votes as shown in figure 6.

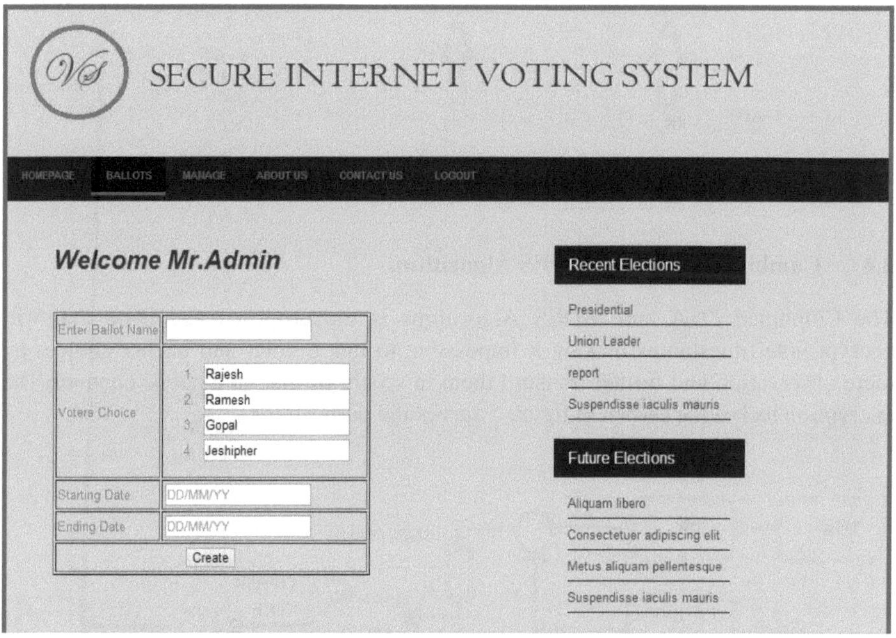

Fig. 6. Encryption Process with aggregation of DSA and MDES Algorithms

Voters are given SMS the DSA encrypted signature and they have to enter that in the relevant screens along with their voter's choice as shown in Fig 7. The random number for running the MDES algorithm is also embedded in the signature given in Security Password.

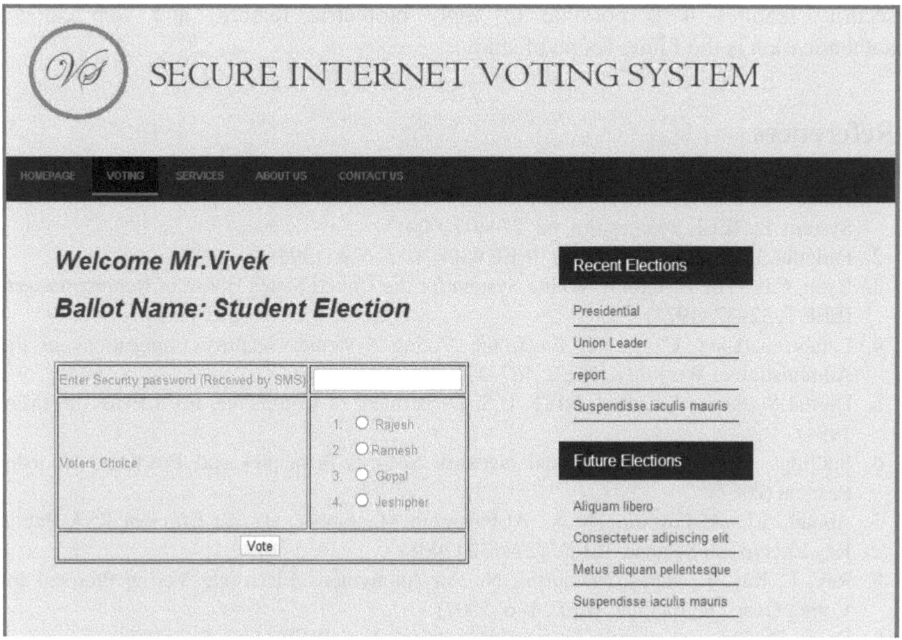

Fig. 7. Encryption Process with aggregation of DSA and MDES Algorithms

This Secured Internet voting with Combined Digital Signature Algorithm and Multiple Data Encryption Standard Algorithms is found to be effective and offers all the advantages mentioned in this context. This application can be used in a variety of applications which require privacy, authentication and reliability with other secure features those are the basic mechanisms of security [10]. Mathematical strength of the algorithm also increases and it is not easy for any hacker to detect the content even when the able to view the cipher text. This also provides authentication and the votes not signed can be considered to be invalid. Hence the application developed in the present scenario can be utilized by the Election committee who can easily manage to conduct trade union elections, College level elections, Governing body elections which reduces manual overheads and increases security and authentications levels[11,12].

4 Conclusion

Internet Voting System provide certain advantages like Correctness, Privacy, Verifiability, Robustness, Coercion-Resistance. These ensure that the Final Election result should be the exact voters choice and only eligible voters can vote, and each eligible voter can cast at most one vote that counts. Also this assures that votes cannot be altered, deleted, or substituted after casting and all valid votes are counted, invalid votes are not counted. The secure encryption process applied here with an aggregation of DSA and MDES supports all the principles of correctness. To provide more security features it is possible to apply biometric features and web camera authentication is the future scope of study.

References

1. Kohno, T., Stubblefield, A., Rubin, A.D., Wallach, D.S.: Analysis of an Electronic Voting System. In: IEEE Symposium, pp. 27–40 (2004)
2. Parhami, B.: Voting Algorithms. IEEE 43(4), 617–629 (1994)
3. Ryan, F.B.: The Electronic Voting System for the United States House of Representatives. IEEE 5, 32–37 (1972)
4. Lambrinoudakis, C., et al.: Electronic Voting Systems: Security Implications of the Administrative Workflow. IEEE, 467–471 (2003)
5. Digital Signature Standard, NIST, U.S. Department of Commerce, FIPS PUB 186 (May 1994)
6. Stallings, W.: Cryptography and Network Security-Principles and Practices, 4th edn. Pearson (2007)
7. Aboud, S.J., Al-Fayoumi, M.A., Al-Fayoumi, M., Jabbar, H.: An Efficient RSA Public Key Encryption Scheme. IEEE, 127–130 (2008)
8. Ray, I., Ray, I., Narasimhamurthi, N.: An Anonymous Electronic Voting Protocol for Voting Over The Internet. IEEE, 1–6 (2001)
9. Davis, R.: The data encryption standard in perspective. IEEE 16(6), 5–9 (1978)
10. Barbara, D., Garcia-Molina, H.: The Reliability of Voting Mechanisms. IEEE 36, 1197–1208 (1987)
11. Centinkoya, O.: Analysis of Security Requirement of Cryptographic Voting Protocols (Extended Abstract). IEEE, 1451–1456 (2008)
12. Yacoub, S., Lin, X., Simske, S., Burns, J.: Automating the Analysis of Voting Systems. IEEE, 203–214 (2003)

Defending Approach against Forceful Browsing in Web Applications

K. Padmaja[1], K. Nageswara Rao[2], and J.V.R. Murthy[3]

[1] Dr. B.R.A GMR Polytechnic, Rajahmundry, India
padma920@yahoo.com
[2] PSCMR College of Engineering & Technology, Vijayawada, India
knrao@ieee.org
[3] Dept. Of CSE, JNTU, Kakinada, India
mjonnalagedda@gmail.com

Abstract. Web Applications have become crucial components in providing services on the internet. But at the same time, vulnerabilities are effecting the functioning of web applications severely. Web Applications may expose organizations to significant risk if they are not properly protected. Several hardware dependent solutions are there.But hardware maintenance is big problem. This paper proposes a new approach of locking the restricted or confidential pages with authentication page. This would prevent the unauthorised direct access of the restricted pages by the malicious user, then by keeping the confidential information secure.

Keywords: Web application, Forceful browsing, WAF, Security, Lattice.

1 Introduction

E-Businesses have increased the amount and the sensitivity of corporate information that can be accessed through the Web. More Web based enterprise applications deal with financial and medical data [1].Providing a secure Web environment has become a high priority for companies as divulging of any confidential information of the company or the stakeholder of the company could be devastating. Rahul Telang and Sunil Wattal [11], state that even the announcement about Web application vulnerability can have vast impact on the stock markets. Forceful Browsing is one of such vulnerabilities which can be very harmful to the organizations largely depending on e-commerce. EjikeOfuonye et al. [23][24] state that the e-commerce Websites can hardly be trusted due to vulnerabilities. This paper proposes a new concept of locking the restricted pages containing the vital information of an individual or the organization with the authentication pages. This would prevent the direct access of the restricted pages and keep the project environment secure.

Generally, the restricted pages are accessible only after appropriate authentication. However, these pages are sometimes accessed by direct access by the malicious user by giving the direct predicted address of the Web page that may have broken links [36]. Currently many Web applications are using WAFs to counter this vulnerability

S.C. Satapathy et al. (eds.), *ICT and Critical Infrastructure: Proceedings of the 48th Annual Convention of CSI - Volume II*, Advances in Intelligent Systems and Computing 249,
DOI: 10.1007/978-3-319-03095-1_71, © Springer International Publishing Switzerland 2014

according to LievenDesmet et al. [10]. This requires installation of hardware and separate software comprising of user-defined protocols. This increases the implementation cost and requires regular monitoring of different approaches of forceful browsing, so as to update the protocols to counter them.

2 Web Application and Browsing

Web Application or webapp [34] is an application that is accessed via Web browser over a network such as the Internet or an Intranet. Information exchange is the new lifeblood of the 21st century. Today Bank balances, order books, hotel reservations and airline tickets purchasing are all done online. However, Web applications are still highly insecure. Common reasons for Web Application Vulnerabilities include Design Flaws, Programming flaws, and Malicious Users, continuously and persistently trying to attack the Web Applications for their illegal benefits[13][14]. J.D. Meier [22] states that engineering of Web application security is one of the top priority activities while developing the Web application.

2.1 Forceful Browsing

Forceful browsing means making several requests to the web server with the URL patterns of typical web application components such as CGI programs [9].The process of personalizing the *Websites* [7 – 8] involves link analysis. The broken links can easily be identified during this analysis. By guessing URLs, an attacker can get access to those files. An attacker, in this act forcefully browses through several parts of a website via direct *URL* [21] entry. These parts are otherwise inaccessible, but a skilled hacker with good experience can easily find his way through them by implementing *Web directory search* [3] or *Web services discovery* [4] or simple *Web search* [2]. Forceful browsing can prove devastating to any website. It leads to information leakage that may diminish the goodwill of a website.

3 Problem Definition

3.1 Broken Session Data Dependency

Breaking data dependencies is a general hazard in composing data-centred applications. Data-centred Web compositions are vulnerable to broken data dependencies. Bypassing the intended application flow in a Web application can generally lead to unauthorized access to resources or unexpected application behaviour [11]. Firewalls (hardware) associated with the web servers are specifically re-designed to counter the problem [15]. *WAFs* [31] are defined by the consortium as an intermediary device, sitting between a Web Client and a Web Server, analysing OSI Layer 7 messages for violations in the programmed security policy. A WAF does not require modification of source code. A WAF can use a proxy-based architecture, a deep packet inspection-based architecture or both.

3.2 Lattice

Lattice ,a mathematical tool used in software analysis, reengineering [12] to address a frame work like structure. The lattice properties permit concise formulations of the security requirements of different existing systems and facilitate the construction of mechanisms that enforce security [10].Lattice is used as tool in securing computer application [38]. Lattice-based access control (LBAC)[6] is a complex access control model based on the interaction between any combination of objects (such as resources, computers, and applications) and subjects (such as individuals, groups or organizations) in computer security[16][17]. Lattice is a partially ordered set(poset) $(L, \leq)$ for any two elements a and b of L, the set $\{a, b\}$ has a join: a V b (also known as the least upper bound, or the supremum).Existence of binary meets For any two elements a and b of L, the set $\{a, b\}$ has a meet: a Λ b (also known as the greatest lower bound, or the infimum).

3.3 Access Control Matrix

Access control matrix is a model of system resource's protection, proposed by Butler W. Lampson [39], an American computer scientist, in 1971. The protection schemes in this model do not allow unauthorized users or subjects to use system resources. Access control matrix [40] consists of triple parts such as subject, object, and access operation. A subject is an active entity in a computer system such as user, program, process, and thread. An object is a passive entity or system resource such as file, directory, database record and printer. In access control matrix's schema, the subjects and objects are placed in a table. Each row represents a subject and each column represents an object. The data inside the table are set of access operations such as read, write, and execute. The access operations are responsible for interactions between subjects and objects.

3.4 Random Number Generator

Random number has application in cryptography [41].Random numbers are a critical part of computer and Internet security. They allow websites and browsers to encrypt the data sent between them using a session key [42]. *Fechner and Osterloh* explain that a good random number in computer binary would usually comprise discrete and uniformly distributed ones and zeroes.

4 Proposed Approach

In this section we specified a framework approach with the help of lattice and also random number generator for protecting web application against vulnerabilities. Our approach proposes to build a banking application with lattice frame structure and Lock number .User will login mainpage.when he login with username and password, user page will open for access, and then he can send or view transactions. But if unauthorized user crawls with URL and forceful get into web application with the help of session id, then security problem arises. A malicious user is prohibited to access data by locking mechanism. According to Denning's [5], lattice based access control can be used for information flow.

Procedure

1. Build a Banking application. It's having :

- H-Home Page/Main Page
- P_A-Authentication page
- P_B -Restricted page
- RNB_C -Random number generator
- D -Database to store random numbers

First directed graph is generated between web pages and the graph is as follows:

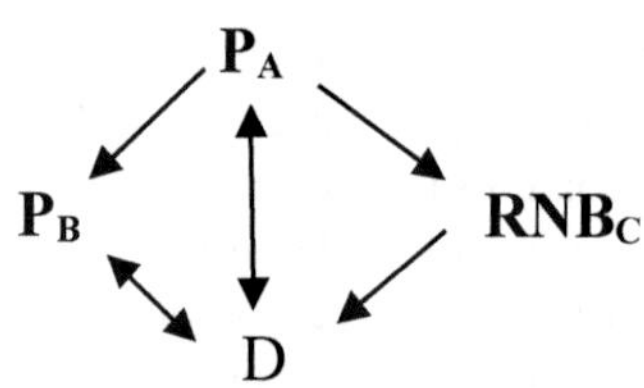

Fig. 1. Directed graph

Now generate access control matrix form given graph as follows:

	P_A	P_B	RNB_C	D
P_A	1	1	1	1
P_B	0	1	0	1
RNB(C)	0	0	1	1
D	1	1	0	1

From this construct hasse diagram as follows.

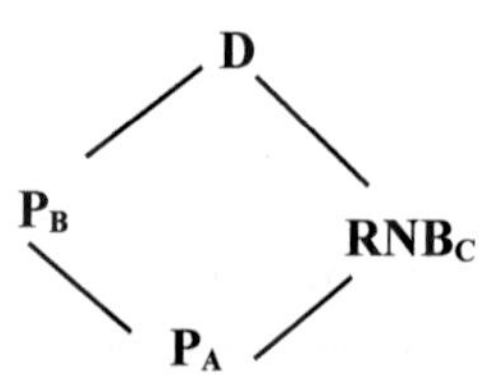

Fig. 2. Hasse diagram

Now given hasse diagram is as bounded lattice with $\mathbf{P_A}$ and D boundaries. According to Denning axioms [5], information flow is designed.

True random number generator can protect systems against third-party snooping [33].The measure of randomness is called entropy. Entropy measures how uncertainty about a value. Entropy can be as the average number of bits needs to specify the value to use an ideal compression algorithm. Mathematically entropy (measure of randomness),H(X) ,for a variable X is :

$$H(X) = -\sum P(X{=}x)\log_2 P(X{=}x)$$

Where, P(X=x) is the probability that the variable X takes on the Value x. For a value consisting of a sequence of 8 bytes or 64 bits (8 digits x 8 bits/digit=64 bits) entropy is 64 bits (Security strength).Thus the amount of work required to break the security is 2^{64} operations. This is computationally infeasible within the valid session. Thus, this random number would minimize the chances to nil for malicious user to predict not only the session id but also the random number required for accessing the restricted web page. This can be shown in figure.

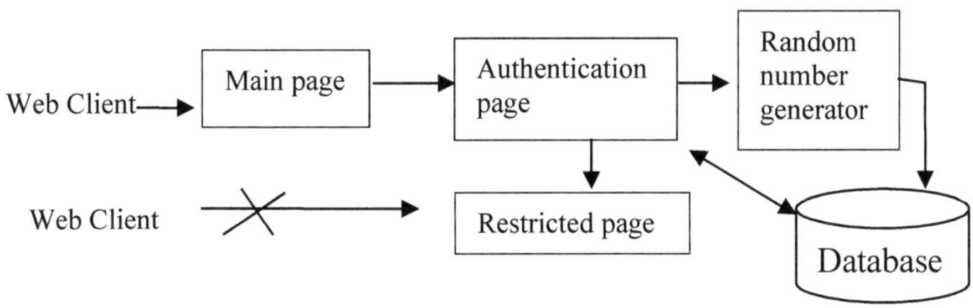

Fig. 3. Locking of the restricted web page

4.1 Encryption of Lock Numbers

The 8 digit Random Number that is generated during *authentication* [12 – 13] is first encrypted and then stored in the database. This encryption is done to ensure better security. i.e. even if the malicious user is successful in tracing the Random number that is generated, still he / she would not be able to access the restricted pages using that random number alone. For successful Forceful Browsing, the malicious user needs to know the 64 bits of the Lock Number that is generated using this Random Number. This encryption takes place as shown below in figure 4.

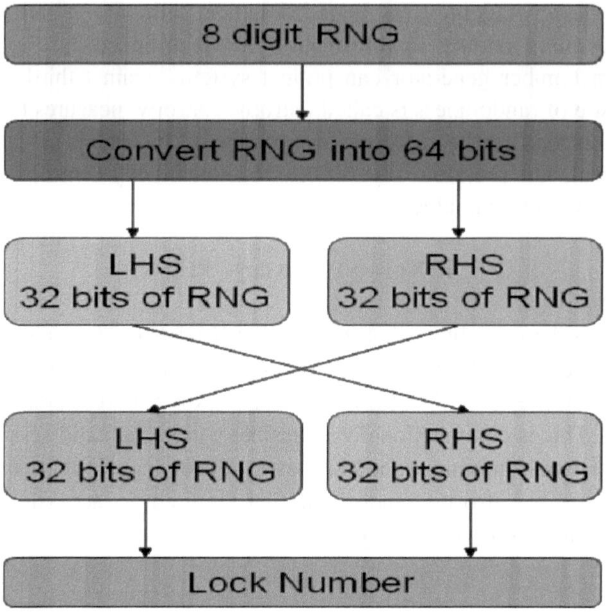

Fig. 4. Lock Number Generation Using 8 digit Random Number

As shown in the figure 5, the 8 digit random number is first converted into its binary equivalent. Hence, 64 bits binary number containing only zeros and ones is obtained. This 64 bits binary number is then divided into two equal halves of 32 bits each. Then, the two binary number obtained by dividing into 32 bits are interchanged and kept together to obtain a 64 bit binary number. This 64 bit binary number is the lock number.

4.2 Algorithms Used

Web servers are extended by application frameworks, such as J2EE or ASP.NET, implementing a state management scheme, tying individual user's requests into a *session* by tying a cryptographically unique random value stored in a cookie against state held on the server, giving users the appearance of a stateful application. The ability to restrict and maintain user actions within unique sessions is critical to web security. With Web applications, the web server serves up pages in response to client requests. By design, the web server is free to forget everything about pages it has rendered in the past, as there is no explicit state. The use of these session states that includes the session id along with the Random number generation after appropriate authentication guarantees total security to the restricted pages.

Java has a rich toolkit for generating random numbers, in a class named *Random*. The best way to think of class *Random* is that its instances are random number generator objects, i.e. Objects that go around spitting out random numbers of various sorts in response to messages from their clients. *Random* is defined in the *java.util* library package. Hence any Java source file that uses *Random* shall begin with a line of the form:

`import java.util.Random;` or `import java.util.*;`

4.2.1 Algorithms

The algorithm for generating Random numbers can be given as below in algorithm 1:

```
Step 1.Private String get64BitNumber (String rno)
Step 2.{  String lockBitNum="";
Step 3.for (int i = 0; i <= rno.length(); i++)
Step 4.{ lockBitNum=lockBitNum+getBinarynumber
         (rno.substring(i, i+1)); }
Step 7.return lockBitNum;
Step 8.          }
```

4.2.2 Algorithm for Authenticating

1. P_A ←Authentication page
2. P_B ←restricted page
3. IF (RNB(P_B) ==RNB(P_A)) && ((LUB((B ,C))=D) &(GLB((B,C))=A))
4. B←R(B)
5. ELSE
6. B←"UNAUTHORISED ACCESS"

4.3 Session Termination

Generally session will expires with in fixed time. So we should access the pages only within the limited time, otherwise the session will expired.

```
<Session-config>
<Session-timeout>40000</Session-timeout>
..
</Session-config>
```

This will provide the security. The no-broken-data dependencies property is examined in every possible step of execution within user's session. This proposed approach ensures protection against vulnerabilities in a given application.

5 Conclusion

In this paper, we have proposed an approach to prevent web application vulnerability, forceful browsing. This could provide more protection and security for web applications. In particular, we have proposed a solution to prevent access to unauthorised users with the help of mathematical tools lattice and random number. This improves Web application security by providing appropriate solution in order to reduce effect of forceful browsing.

References

[1] Benjamin Livshits, V., Lam, M.S.: Stanford University. Finding Vulnerabilities in Java Application with Static analysis

[2] Joachims, T., Granka, L., Pan, B., Hembrooke, H., Radlinski, F., Gay, G.: Evaluating the accuracy of implicit feedback from clicks and query reformulations in Web search. ACM Transactions on Information Systems 25(2), 1–27 (2007)

[3] Gerstel, O., Kutten, S., SanyLaber, E., Matichin, R., Peleg, D., Souza, C.: Reducing Human Interactions in Web Directory Searches. ACM Transactions on Information Systems 25(4), Article 20, 20–27 (2007)

[4] Shehab, M., Bhattacharya, K., Watson, T.J., Ghafoor, A.: Web Services Discovery In Secure Collaboration Environments. ACM Transactions on Internet Technology 8(1), Article 5, 5–22 (2007)

[5] Web Application Security, Wikipedia

[6] Lattice-based acess control, Wikipedia

[7] Eirinaki, M., Vazirgiannis, M.: Web Site Personalization Based On Link Analysis And Navigational Patterns. ACM Transactions on Internet Technology 7(4), Article 21, 21–27 (2007)

[8] Anand, S.S., Kearney, P., Shapcott, M.: Generating Semantically Enriched User Profiles for Web Personalization. ACM Transactions on Internet Technology 7(4), Article 22, 22–26 (2007)

[9] Auronen, L.: Helsinki University of Technology, Telecommunications Software and Multimedia Laboratory. Tool-Based Approach to Assessing Web Application Security

[10] Desmet, L., Verbaeten, P., Joosen, W., Piessens, F.: Provable Protection Against Web Application Vulnerabilities Related To Session Data Dependencies. IEEE Transactions on Software Engineering 34(1), 50–64 (2008)

[11] webSSARI, Wikipedia

[12] Snelting, G.: Software Engineering based on concept of lattices

[13] Song, H.-G., Kim, Y., Doh, K.-G.: Automatic Detection of Access Control vulnerabilities in Web applications by URL Crawling and Forced Browsing

[14] Song, H.-G., Kim, Y., Doh, K.-G.: Automatic Detection of Access Control vulnerabilities in Web applications by URL Crawling and Forced Browsing

[15] Kanchi, P., Heera Singh, B., Nageswara Rao, K., Murthy, J.V.R.: Hardware Independent Protection against Vulnerabilities in Web Applications. In: NCATS 2012 (2012)

[16] Denning, D.E.: A lattice model of secure information flow. ACM, New York (1976)

[17] Sandhu, R.S.: Lattice-based access control models, vol. 26. IEEE computer (1993)

[18] Coates, R.F.W., Janacek, G.J., Lever, K.V.: Monte Carlo Simulation and Random Number Generation. IEEE Journal on Selected Areas in Communications 6(1), 58–66 (1988)

[19] Han, T.S., Hoshi, M.: Interval Algorithm for Random Number Generation. IEEE Transactions on Information Theory 43(2), 599–611 (1997)

[20] Clewett, J.: Random Numbers. Numberphile. Brady Haran

[21] Khare, R.: Anatomy of a URL. IEEE Internet Computing, 78–81(September/October 1999)
[22] Meier, J.D.: Web Application Security Engineering. IEEE Security & Privacy, 16–24 (2006)
[23] Ofuonye, E., Beatty, P., Reay, I., Dick, S., Miller, J.: How Do We Build Trust Into E-Commerce Web Sites? IEEE Software, 7–9 (2008)
[24] Humeau, P., Jung, M. In: depth benchmark of 12 ecommerce solutions (June 21, 2013)
[25] Rossi, G., Schwabe, D.: Object-Oriented Design Structures in Web Application Models. In: Annals of Software Engineering, vol. 13, pp. 97–110. Kluwer Academic Publishers (2002)
[26] Ricca, F., Tonella, P.: Testing Processes Of Web Applications. In: Annals of Software Engineering, vol. 14, pp. 93–114. Kluwer Academic Publishers (2002)
[27] Läufer, K.: A Hike Through Post-EJB J2EE Web Application Architecture. In: IEEE Computing in Science & Engineering, pp. 80–88 (2005)
[28] Farrell, S.: Password Policy Purgatory. IEEE Internet Computing, 84–87 (2008)
[29] Bellovin, S.M., Cheswick, W.R.: Network Firewalls. IEEE Communications Magazine, 50–57 (September 1994)
[30] Kanchi, P., Nageswara Rao, K.: A Preventive Approch against vulnerabilities in Web Applications. In: ARIES 2012 (2012)
[31] Desmet, L., Piessens, F., Joosen, W., Verbaeten, P.: Bridging the Gap between Web Application Firewalls and Web Applications. In: Proceedings of The Fourth ACM Workshop On Formal Methods in Security, pp. 67–77 (2006)
[32] Bayross, I.: SQL, PL/SQL – The programming language of Oracle. BPB Publications (2006)
[33] Random Number generators, Science Daily
[34] Web Applications (November 17, 2008), `http://en.wikipedia.org/Wiki/web_application`
[35] Web Application Firewalls (January 28, 2009), `http://www.owasp.org/index.php/web_application_firewall`
[36] Broken links of Web pages (February 6, 2009), `http://www.sigchi.org/web/chi97testing/ricknote.html`
[37] Random number generation (February 24, 2009), `http://www.random.org/integers`
[38] Meadows, C.: Applications of Lattices To Computer Security, `http://chacs.nrl.navy.mil`
[39] Access Control Matrix, `http://en.wikipedia.org/wiki/Access_Control_Matrix` (accessed February 2009)
[40] Stamp, M.: Information Security Principles and Practice. John Wiley & Sons Inc., NJ (2006)
[41] Random number, Wikipedia
[42] New Approach to Generating Truly Random Numbers May Improve Internet Security, Weather Forecasts, Science Daily

Effect of Indexing on High-Dimensional Databases Using Query Workloads

S. Rajesh[1], Karthik Jilla[1], K. Rajiv[1], and T.V.K.P. Prasad[2]

[1] Nalla Narasimha Reddy Educational Society's Group of Institutions, Hyderabad, AP, India
[2] Dept. of CSE, SRKR Engineering College, Bhimavaram, AP, India

Abstract. High-dimensional indexes do not work because of the often-cited "curse of dimensionality." However, users are usually interested in querying data over a relatively small subset of the entire attribute set at a time. A potential solution is to use lower dimensional indexes that accurately represent the user access patterns. To address these issues, in this paper we propose a parameterizable technique to recommend indexes based on index types.

Keywords: Object Oriented, Database Model, High dimensional indexes, indexing.

1 Introduction

Query pattern evolution over time presents a challenging problem. Researchers have proposed workload- based index recommendation techniques. Their long term effectiveness is dependent on the stability of the query workload. However, query access patterns may change over time, becoming completely dissimilar from the patterns on which the index set was originally determined. There are many common reasons that query patterns change. A pattern change could be the result of periodic time variation, a change in the focus of user knowledge discovery, a change in the popularity of a search attribute, or simply the random variation of query attributes. When the current query patterns are substantially different from the query patterns used to recommend the database indexes, the system performance will drastically degrade, since incoming queries do not benefit from the existing indexes. To make this approach practical in the presence of a query pattern change, the index set should evolve with the query patterns. For this reason, a dynamic mechanism is introduced to detect when the access patterns have changed enough that the introduction of a new index, the replacement of an existing index, or the construction of an entirely new index set is beneficial[1-10].

Because of the need to proactively monitor query patterns and query performance quickly, the index selection technique that we have developed uses an abstract representation of the query workload and the data set that can be adjusted to yields a faster analysis. An abstract representation of the query workload is generated by mining patterns in the workload. The query workload representation consists of a set of attribute sets that frequently occur over the entire query set that has nonempty

S.C. Satapathy et al. (eds.), *ICT and Critical Infrastructure: Proceedings of the 48th Annual Convention of CSI - Volume II*, Advances in Intelligent Systems and Computing 249,
DOI: 10.1007/978-3-319-03095-1_72, © Springer International Publishing Switzerland 2014

intersections with the attributes of the query for each query. To estimate the query cost, the data set is represented by a multidimensional histogram, where each unique value represents an approximation of data and contains a count of the number of records that match that approximation [1-10].

Initial index selection occurs by traversing the query workload representation and determining which frequently occurring attribute set results in the greatest benefit over the entire query set. This process is iterated until an indexing constraint is met or no further improvement is achieved by adding additional indexes. Analysis speed and granularity are affected by tuning the resolution of the abstract representations. The number of potential indexes considered is affected by adjusting the data mining support level. The size of the multidimensional histogram affects the accuracy of the cost estimates associated with using an index for a query[1-10].

2 Literature

The work done here differs from the related index selection work in that an index selection framework that can be tuned for speed or accuracy. This technique is optimized to take advantage of the multidimensional pruning offered by multidimensional index structures. It takes both data and query characteristics into consideration, and it can be applied to perform real-time index recommendations for evolving query patterns[1-10].

i. High-Dimensional Indexing

A number of techniques have been introduced to address the high-dimensional indexing problem such as the X-tree and the GC-tree. Although these index structures have been shown to increase the range of effective dimensionality, they still suffer performance degradation at higher index dimensionality.

ii. The X-tree

Stefan Berchtold et.al. proposed a new method for indexing large amounts of point and spatial data in high dimensional space. Their analysis showed that index structures such as the R*-tree are not adequate for indexing high–dimensional data sets. The major problem of R-tree-based index structures is the overlap of bounding boxes in the directory, which increases with growing dimensions. They proposed a solution in order to avoid growing dimension problem is that a new organization of the directory which uses a split algorithm minimizing overlap and additionally utilizes the concept of super nodes[11].

iii. GC-tree

Guang-Ho cha et.al., proposed a new dynamic index structure called the grid cell tree for efficient similarity search in image databases. The GC-tree which they proposed is based on a special subspace partitioning strategy which is a clustered high-dimensional image dataset. They proposed the basic idea into three-fold: the first fold is that they adaptively partition the data space based on a density function that identifies the dense and sparse regions in a data space. The second fold they proposed is that they concentrate on the dense regions and the objects in the sparse regions of a certain partition level are treated as if they lie within a single region. The third fold is they dynamically construct an index structure that corresponds to the space partition hierarchy[12].

iv. Feature Selection

Avrim L.Blum et.al. proposed at a conceptual level one can divide the task of concept learning into two subtasks deciding which features to use in describing the concept and deciding how to combine those features In this view the selection of relevant features and the elimination of irrelevant ones is one of the central problems in machine learning and many induction algorithms incorporate some approach to addressing it. Feature selection techniques are a subset of dimensionality reduction targeted at finding a set of untransformed attributes that best represent the overall data set. These techniques are also focused on maximizing data energy or classification accuracy rather than query response. As a result, selected features may have no overlap with queried attributes[13]

v. Index Selection

Surajit chaudary et.al. described a novel technique that make it possible to build an industrial-strength tool for automating the choice of indexes in the physical design of a SQL database. The tool takes as input a workload of SQL queries, and suggests a set of suitable indexes. They ensure that the indexes chosen are effective in reducing the cost of the workload by keeping the index selection tool and the query optimizer "in steps". The number of index sets that must be evaluated to find the optimal configuration is very large. They reduce the complexity of this problem using three techniques[14].

vi. Automatic Index Selection

The ideas of having a database that can tune itself by automatically creating new indexes as the queries arrive have been proposed. In, a cost model it is used [15] to identify beneficial indexes and decide when to create or drop an index at runtime. Costa and Lifschitz propose agent-based database architecture to deal with an automatic index creation. Microsoft Research has proposed a physical-design alerter to identify when a modification to the physical design could result [16] in improved performance.

3 System Specifications

Software Requirement Specification is the starting point of the software developing activity. As system grew more complex it became evident that the goal of the entire system cannot be easily comprehended. Hence the needs for the requirement phase arise. The software project is initiated by the client needs. The Software Requirement Specification is the means of translating the ideas of the minds of clients (the input) into a formal document (the output of the requirement phase.)

Best application for this approach is to apply the more time consuming no-constraint analysis in order to determine an initial index set and then apply a lightweight and low control sensitivity analysis for the online query pattern change detection in order to avoid or make the user aware of situations where the index set is not at all effective for the new incoming queries.

Objective: Introduction of a flexible index selection technique designed for high-dimensional data sets, which uses an abstract representation of the data set and query workload (the resolution of the abstract representation can be tuned to achieve either a high ratio of index-covered queries for a static index selection or a fast index selection to facilitate online index selection).

In The Current System Query response does not perform well if query patterns change, because it uses static query workload. Its performance may degrade if the database size gets increased. Tradition feature selection technique may offer less or no data pruning capability given query attributes. For this here we propose a flexible index selection frame work is developed to achieve index selection for high dimensional data. A control feedback technique is introduced for measuring the performance. Through this a database could benefit from an index change. The index selection minimizes the cost of the queries in the work load. Online index selection is designed in the motivation if the query pattern changes over time. By monitoring the query workload and detecting when there is a change on the query pattern, able to evolve good performance as query patterns evolve. Advantages:

- By creating index one can minimize the searching time.
- Index will automatically adjust itself based on the query workloads over time.

Problem Definition: The definition of the problem of index selection for a multidimensional space by using a query workload is given below. A query workload W consists of a set of queries that select objects within a specified subspace in that data domain. A workload W is a tuple W= (D, DS,Q),Where D is the domain, DS $\subseteq$ D is a finite subset(the data set),and the Q(the query set) is a set of subsets of DS. The problem can be defined as finding a set of indexes I, given a multidimensional data set DS, a query workload W, an optional indexing constraint C, and an optional analysis time constraint t_{a},, that provides the best estimated cost over W.

4 System Structure

Index Selection: The goal of the index selection is to minimize the cost of the queries in the workload, given certain constraints. Given a query workload, a data set, the indexing constraints, and several analysis parameters, this framework produces a set of suggested indexes as an output.

Fig.1. shows a flow diagram of the index selection framework. Table.1. provides a list of the notations used in the descriptions. Three major components in the index selection framework were identified: the initialization of the abstract representations, the query cost computation, and the index selection loop. These components and the data flow between them are discussed below. The goal of the index selection is to minimize the cost of the queries in the workload, given certain constraints. Given a query workload, a data set, the indexing constraints, and several analysis parameters, this framework produces a set of suggested indexes as an output. Figure 1 shows a flow diagram of the index selection framework.

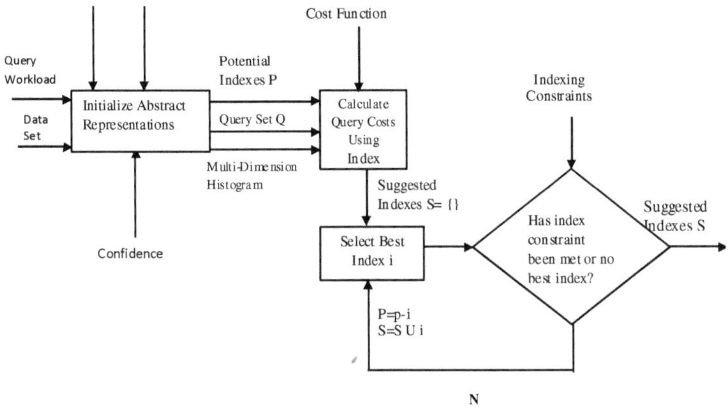

Fig. 1. Index selection Flowchart

Online Index Selection: Figure 2 represents the implementation of dynamic index selection. System input is a set of indexes and a set of incoming queries. The system simulates and estimates costs for the execution of incoming queries. System output is the ratio of the potential system performance to the actual system performance in terms of database page accesses to answer the most recent queries. Two controls feedback loops were implemented. One is for fine-grained control and is used to recommend minor inexpensive changes to the index set. The other loop is for coarse control and is used to avoid very poor system performance by recommending major index set changes. Each control feedback loop has decision logic associated with it.

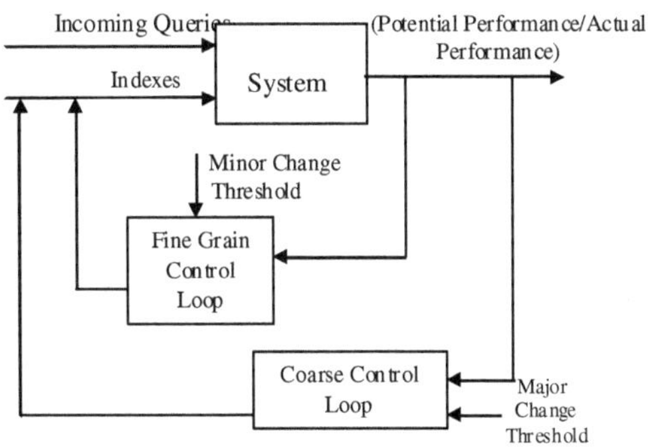

Fig. 2. Dynamic index analysis frameworks

System Input: The system input is made up of new incoming queries and the current set of indexes I, which is initialized to be the suggested indexes S from the output of the initial index selection algorithm.

System: The system simulates query execution over a number of incoming queries, that is, the abstract representation of the last w queries stored as *W*, where *w* is an adjustable window size parameter. W is used to estimate the performance of a hypothetical set of indexes I_{new} against the current index set *I*. This representation is similar to the one kept for query set *Q* in the static index selection. In this case, when a new query *q* arrives, this tool determines which of the current indexes in I most efficiently answers this query and replace the oldest query in W with the abstract representation of *q*. It also incrementally computes the attribute sets that meet the input *support* and *confidence* over the last w queries. This information is used in the control-feedback-loop decision logic. The system also keeps track of the current potential indexes P and the current multidimensional histogram *H*.

System Output: In order to monitor the performance of the system, the query performance using the current set of indexes *I* to the performance using a hypothetical set of indexes I_{new} were compared. The query performance using *I* is the summation of the costs of queries using the best index from *I* for the given query. Consider the possible new indexes P_{new} to be the set of attribute sets that currently meet the input support and confidence over the last w queries.

5 Results

The system input is a set of indexes and a set of incoming queries, with theses two inputs the system simulates and estimates costs for the execution of incoming queries. The system output is the ratio of potential system performance to the actual system performance in terms of database page accesses to answer the most recent queries. Query performance is the obvious parameter to monitor as shown in fig 3. However lower query performance could be related to other aspects rather than the index set.

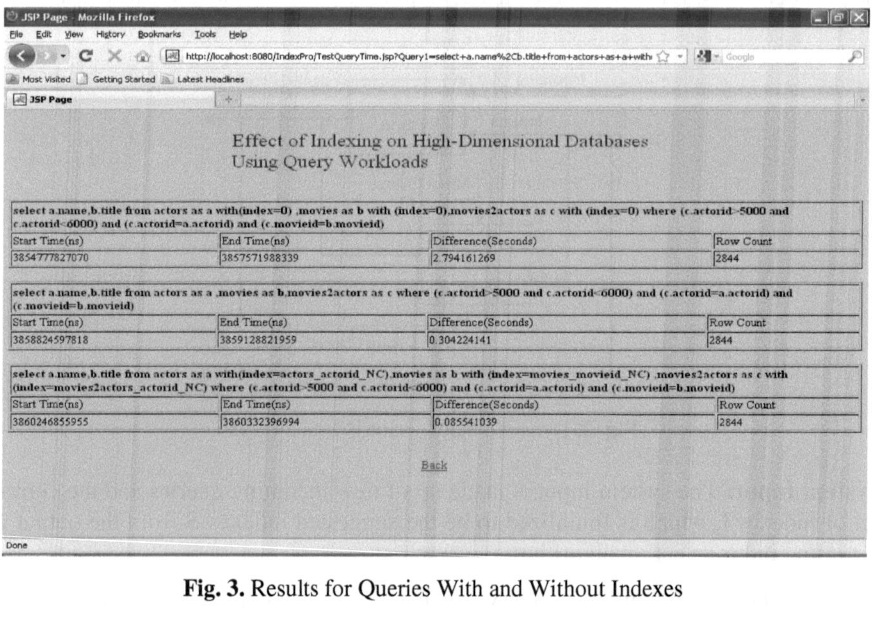

Fig. 3. Results for Queries With and Without Indexes

A workload using select statement is applied to the three tables Movies, Actors, Movies Actors which consists of more than seven lakh records by giving index as zero. The given query is input to the system and the system simulates and estimates costs for the executing the query. The system performance is measured in terms of database page accesses to answer the query. The result is as shown in figure 4.

select a.name,b.title from actors as a with(index=0) ,movies as b with (index=0),movies2actors as c with (index=0) where (c.actorid>5000 and c.actorid<6000) and (c.actorid=a.actorid) and (c.movieid=b.movieid)			
Start Time(ns)	End Time(ns)	Difference(Seconds)	Row Count
3854777827070	3857571988339	2.794161269	2844

Fig. 4. Result for query with index value 0

A workload using select statement is applied to the three tables Movies, Actors, Movies Actors which consists of more than seven lakh records without giving indexes to the columns. The given query is input to the system and the system simulates and estimates costs for the executing the query. The system performance is measured in terms of database page accesses to answer the query. The result is as shown in figure 5.

select a.name,b.title from actors as a ,movies as b,movies2actors as c where (c.actorid>5000 and c.actorid<6000) and (c.actorid=a.actorid) and (c.movieid=b.movieid)			
Start Time(ns)	End Time(ns)	Difference(Seconds)	Row Count
3858824597818	3859128821959	0.304224141	2844

Fig. 5. Result for query without index

A workload using select statement is applied to the three tables Movies, Actors, Movies Actors which consists of more than seven lakh records by giving indexes to the columns. The given query along with the indexes is input to the system and the system simulates and estimates costs for the executing the query. The system performance is measured in terms of database page accesses to answer the query. The result is as shown in figure 6.

select a.name,b.title from actors as a with(index=actors_actorid_NC),movies as b with (index=movies_movieid_NC) ,movies2actors as c with (index=movies2actors_actorid_NC) where (c.actorid>5000 and c.actorid<6000) and (c.actorid=a.actorid) and (c.movieid=b.movieid)			
Start Time(ns)	End Time(ns)	Difference(Seconds)	Row Count
3860246855955	3860332396994	0.085541039	2844

Fig. 6. Result for query with indexes

6 Conclusions

The proposed technique affords the opportunity to adjust indexes to new query patterns. From the initial experimental results, it seems that the best application for this approach is to apply the more time consuming no-constraint analysis in order to determine an initial index set and then apply a lightweight and low control sensitivity analysis for the online query pattern change detection in order to avoid or make the user aware of situations where the index set is not at all effective for the new incoming queries. A limitation of the proposed approach is that if index set changes are not responsive enough to query pattern changes, then the control feedback may not affect positive system changes. As a future present two system enhancements that provide further robustness and scalability to the framework

References

[1] Goldstein, J., Platt, J.C., Burges, C.J.C.: Indexing High Dimensional Rectangles for Fast Multimedia Identification. Technical Report MSR-TR-2003-38 (2003)
[2] Bohm, C., Berchtold, S., Keim, D.A.: Searching in High-Dimensional Spaces—Index Structures for Improving the Performance of Multimedia Databases. ACM Computing Surveys 33(3), 322–373 (2001)
[3] Hand, D., Mannila, H., Smyth, P.: Principles of Data Mining. The MIT Press (2001)
[4] Han, J., Kamber, M.: Data Mining: Concepts and Techniques. Morgan Kaufmann Publishers (2000)
[5] Lawder, J.K., King, P.J.H.: Using Space-filling Curves for Multi-dimensional Indexing (June 2000)
[6] Heesch, D., Rueger, S.: NNk networks for content-based image retrieval. In: Proceedings of the 26th European Conference on Information Retrieval (ECIR), Sunderland, UK (April 2004)
[7] Bingham, E., Mannila, H.: Random projection in dimensionality reduction: Applications for image and text data. In: KDD 2001, San Francisco, CA (2001)
[8] Chakrabarti, K., Mehrotra, S.: Local Dimensionality Reduction: A New Approach to Indexing High Dimensional Spaces. In: VLDB Conference Proceedings (2000)
[9] Yu, C., Ooi, B.C., Tan, K.L., Jagadish, H.: Indexing the distance: an efficient method to knn processing. In: Proc. 27th International Conference on Very Large Data Bases, pp. 421–430 (2001)
[10] Valentin, G., Zuliani, M., Zilio, D., Lohman, G., Skelley, A.: DB2 Advisor: An Optimizer Smart Enough to Recommend Its Own Indexes. In: Proc. 16th Int'l Conf. Data Eng., ICDE 2000 (2000)
[11] Han, J., Pei, J., Yin, Y.: Mining Frequent Patterns without Candidate Generation. In: Chen, W., Naughton, J., Bernstein, P.A. (eds.) Proc. ACM SIGMOD Int'l Conf. Management of Data (SIGMOD 2000), pp. 1–12 (2000)
[12] Berchtold, S., Keim, D., Kriegel, H.: The X-Tree: An Index Structure for High-Dimensional Data. In: Proc. 22nd Int'l Conf. Very Large Data Bases (VLDB 1996), pp. 28–39 (1996)
[13] Chung, C.-W., Cha, G.-H.: The GC-Tree: A High-Dimensional Index Structure for Similarity Search in Image Databases. IEEE Trans. Multimedia 4(2), 235–247 (2002)

[14] Blum, Langley, P.: Selection of Relevant Features and Examples in Machine Learning. Artificial Intelligence (1997)
[15] Chaudhuri, S., Narasayya, V.R.: An Efficient Cost-Driven Index Selection Tool for Microsoft SQL Server. VLDB J., 146–155 (1997)
[16] Kai-Uwe, S., Schallehn, E., Geist, I.: Autonomous Query- Driven Index Tuning. In: Fourth Int'l Database Eng. And Applications Symp., IDEAS 2004 (2004)
[17] Costa, R.L.D.C., Lifschitz, S.: Index Self-Tuning with Agent- Based Databases. In: Proc. 28th Latin-Am. Conf. Informatics, CLEI 2002 (2002)

A Novel Architecture for Dynamic Invocation of Web Services

Venkataramani Korupala[1], Amarendra Kothalanka[1], and Satyanarayana Gandi[2]

[1] Department of Computer Science & Engineering, Dadi Institute of Engineering & Technology, Anakapalle-531002, A.P., India
[2] Department of Information Technology, Dadi Institute of Engineering & Technology, Anakapalle-531002, A.P., India
{ramanikorupala,amarendradiet,satyanarayanagandi}@gmail.com

Abstract. Now a day's usage of mobile phones is higher when compared with the usage of laptops and desktops. Researchers are interested in invocation of functionalities to the user with minimal over head by using dynamic invocation and with language interoperability approach. Even though various traditional approaches introduced in the traditional mechanisms those are not optimum and used for only few simple query or minimal parameters. For dynamic invocation we introduced a novel Proxy based approach and for the language interoperability we introduced WSDL files.

Keywords: Web services, Mobile computing, UDDI, SOAP, WSDL.

1 Introduction

A web service is a collection of open protocols and standards used for exchanging data between applications or systems [10]. Software applications written in various programming languages and running on various platforms can use web services to exchange data over computer networks like the Internet in a manner similar to inter-process communication on a single computer. This interoperability (e.g., between Java and Python, or Windows and Linux applications) is due to the use of open standards. Web services standards enable applications to actively participate in various forms of Internet transactions without a Web browser, such as receiving up-to-date stock quote information, obtain flight status information, being notified with calendar events, and even perform a Google search using the Google API. The basic Web services platform is XML + HTTP. All the standard Web Services [3] [7] works using following components [16].

- SOAP (Simple Object Access Protocol)
- UDDI (Universal Description, Discovery and Integration)
- WSDL (Web Services Description Language)

S.C. Satapathy et al. (eds.), *ICT and Critical Infrastructure: Proceedings of the 48th Annual Convention of CSI - Volume II*, Advances in Intelligent Systems and Computing 249,
DOI: 10.1007/978-3-319-03095-1_73, © Springer International Publishing Switzerland 2014

The Web services architecture model is shown below. This layered architecture defines the levels at which the Web services protocols are supposed to be used.

The transport protocol layer is at the bottom of the layered architecture model. The firewall-friendly HTTP (Hyper Text Transfer Protocol) and encrypted HTTPS are commonly used Ease of Use over TCP/IP to invoke Web services over a network. HTTP/1.1 supports request-response style message exchanges. Recent Web services standards aim to support a variety of message exchange patterns.

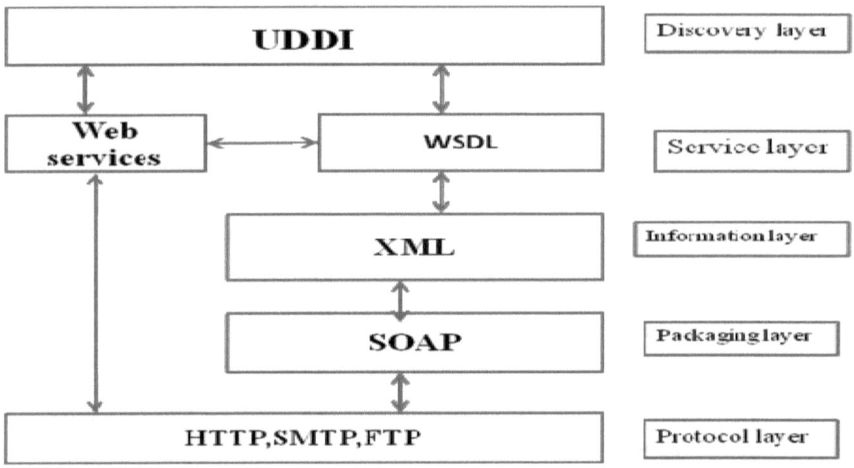

Fig. 1. Layered Architecture Model

The packaging layer uses SOAP, which defines a modular packaging model and the encoding mechanisms for encoding data within XML-based modules.

The information layer carries XML-formatted data, which may consist of arbitrary XML documents, but more commonly transports XML-encoded application data that is transferred with remote procedure calls and responses.

The services layer provides meta-data on the interface to Web services as defined by WSDL. A WSDL document is a description of a Web service that promotes reusability by defining its functionality and access mechanisms.

The discovery layer offers a way to publish information about web services, as well as provide a mechanism to discover what web services are available through the Universal Description, Discovery, and Integration (UDDI) specification.

The general use of the World Wide Web is for effective access of applications. Most of the cases the access is done by users through web browsers and other interactive front-end systems. Without common framework, coordination of each component produces complexity and minimizes interoperability. The framework allows flexible interaction.

Framework: From online technical dictionary "The web service is defined as a standardized way of integrating Web-based applications using the XML, SOAP, WSDL, and UDDI open standards over an Internet protocol backbone"[8]. Specific standards that could be used for performing binding and for interacting with a Web service are mentioned here as

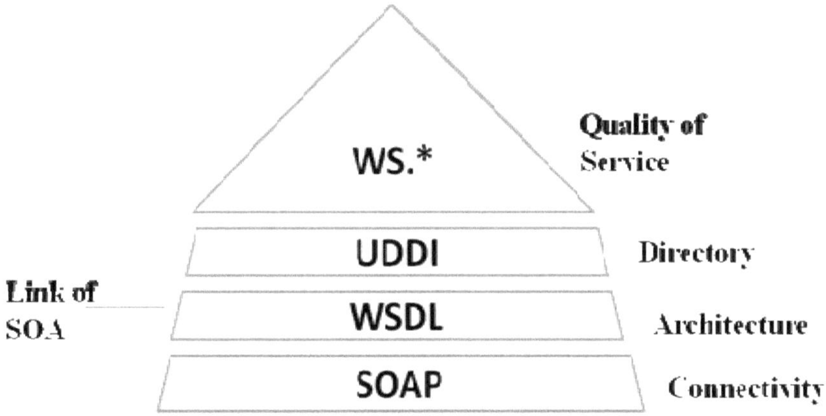

XML is used to tag the data, SOAP is used to transfer the data, WSDL is used for describing the services available, and UDDI is used for listing what services are available".

The above figure includes standards in Web services. Core Web services standards [4] [5] [14] include: Simple Object Access Protocol (SOAP), Web Services Description Language (WSDL) Universal Description, Discovery, and Integration XML.

UDDI: Universal Description Discovery and Integration (UDDI) is a platform-independent, Extensible Markup Language (XML)-based registry by which businesses worldwide can list themselves on the Internet, and a mechanism to register and locate web service applications [11]. It is a directory for storing information about web services described by WSDL.

WSDL: WSDL stands for Web Services Description Language. WSDL stands for Web Services Description Language which is an XML document used to describe Web services. It is also used to locate Web services which are recommended by W3C [8] [13]. The WSDL describes services as collections of network endpoints, or ports. The WSDL specification provides an XML format for documents for this purpose.

2 Existing System

The particularities and limitations of mobile devices and the Mobile environment pose great challenges for consuming web services. In the traditional Mobile invocation of web services those services are not optimal. To begin with, when using UDDI registries for service discovery, multiple costly network round trips are needed, and frequent unavailability of the wireless network may cause failures in the service discovery process additionally, there are several issues and challenges that emerge from the fact that mobile devices have lower processing power, limited bandwidth, less memory, and finite battery power when compared to desktops. Because of that the architecture which includes mobile devices should reduce network interactions and maintain less resource consumption [1].

To access the web services dynamically, a solution is its WSDL files to be parsed by the mobile device application for direct interaction of service. WSDL stands for Web Services Description Language. WSDL is a document written in XML. The document describes a Web service. It specifies the location of the service and the operations (or methods) the service exposes. There are two different techniques for accessing web services: SOAP and REST [1]. SOAP is a simple XML-based protocol [15] that allows applications to exchange information over HTTP. A better way to communicate between applications is over HTTP, because HTTP is supported by all Internet browsers and servers. SOAP was created to accomplish this [9].

A SOAP message is an XML document information item that contains three elements: **<Envelope>**, **<Header>**and **<Body>**.The Envelope is the root element of the SOAP message and contains an optional Header element and a mandatory Body element [12]. REST, on the other side, abbreviated as Representational State Transfer and is an architectural style.

REST style architectures conventionally consist of clients and servers [1]. REST has been applied to describe the desired web architecture, to help identify existing problems, to compare alternative solutions, and to ensure that protocol extensions would not violate the core constraints that make the Web successful.

3 Proposed System

In the proposed approach we introduced a dynamic invocation of web services from the mobile client by implementing the client side proxy (cache) [2], man in the middle server implementation with implementation of soap protocol [6]. The proposed system can be easily understood using below figure. In our approach the data communication can be done in the WSDL so users need not bother about the client side tools, because of language interoperability and dynamic invocation through web services.

3.1 Modules

This project is divided into 4 Modules.

1. Business Logic: Service contains the business logic of the operations, which can be accessed by the mobile users dynamically and there are registered in the server and these business language uses the web service description language (WSDL) for the communication and uses the protocol soap(Simple access protocol),Everything is transmitted in the form of soap objects.

2. Mobile Users: Mobile users requests the specific string or operation sends a request to the server(i.e. web service), in terms of soap messages after that it performs text matching approach and provides the description in terms of WSDL. This mobile application is developed using the android for efficient data access from the service.

3. Proxy Generation and Dynamic Invocation: We are generating client side proxy, it caches the previously accessed information from the server, whenever user makes a request again for the same information, and user need not access the information of WSDL files from the server whenever the data updated in the server.
4. Consuming Services: User consumes the service by creating the soap object make a call to the service with the specified name space and operation ,service process the request sends the response in the format of web service description language, at receiver end it can be converted to native language.

3.2 General Architecture

The process of web service invocation starts when the client-side proxy encapsulated the user's request into SOAP message and sends it to the service, which extracts the call from the received message, executes the call to produce the results, wraps the results into a SOAP message and sends it to the client. Upon receiving the message, the same proxy extracts the results and hands them over to the calling client application. Before an application can begin communicating with a service, it must first discover it and get its specifics and then generate its proxy. The detailed process flow of a web service is represented as

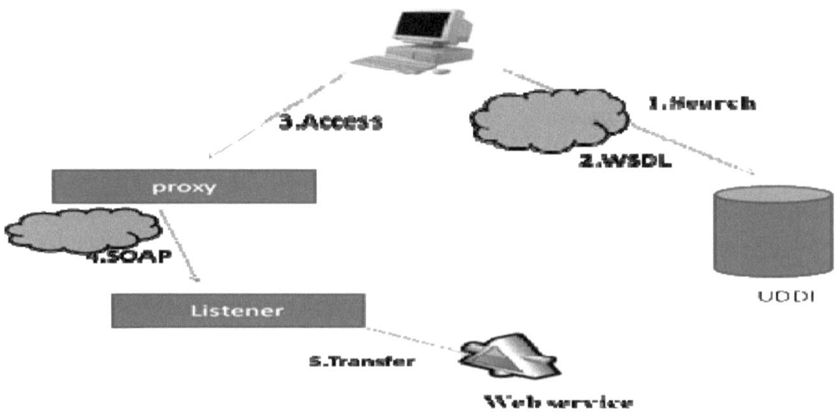

The proxy allows a client application to make method calls as though it were calling a local function. The proxy is created by first generating a source file from the service's Web Service Description Language file, which describes the web service, how to access it, the operations it performs, the types of parameters to be passed to each of the supported methods, and the types of returned results. After the source file is generated, it is compiled into a proxy class that is finally registered with the client application.

4 Implementation

The diagram depicts the general architecture of the proposed system. The user will send a query then the text will be compared. The text will be changed with respect to the architecture of web server that is wsdl urls' will be created for interoperability, in return resultant information for the query in the form of wsdl files will be sent to a proxy. It will generate a source file and keeps a file for future use. In this manner the effective dynamic services will be invoked.

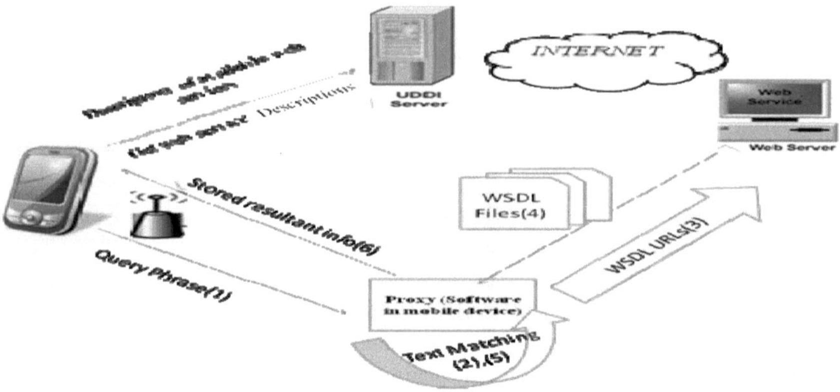

The web services client on the mobile device was implemented using the windows .NET framework. Here the IDE is Eclipse Indigo, Visual studio dot net. Android plug-in, Android SDK is used. Proxy is taken as software and 1 GB of memory. The **Android** is a Linux-based operating system designed primarily for touch screen mobile devices such as smart phones and tablet computers. Android is open source. Android application development is considered to be the best for application integration.

Also refer to the cross-coupled connection or basically run Android products, is considered to be very easy. It effectively means more than satisfied with only the Android user's overall user satisfaction. Allows one to obtain input data for one application and configuration data to other applications. First of all, most of the times the application is smart enough to hire the data required by other installed application.

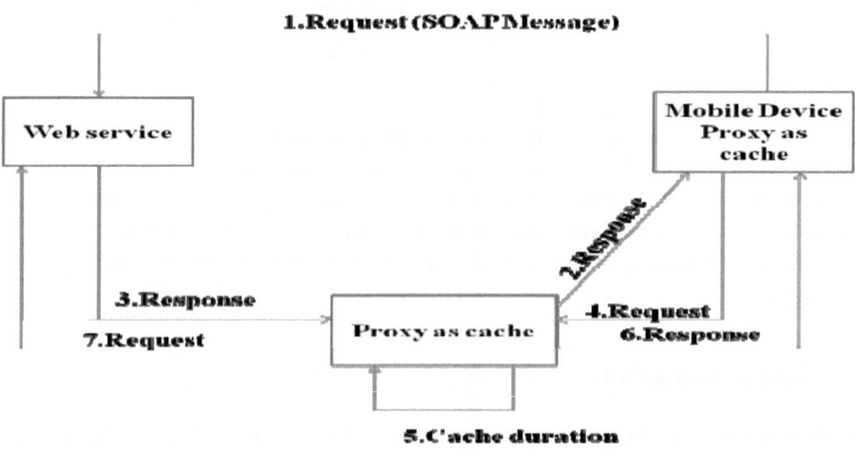

The complete procedure in proposed system is described as Mobile Host is a light weight web service provider built for resource constrained devices like cellular phones and it has been developed as a web service handler built on top of a normal Web server and SOAP based web service requests sent by HTTP tunneling and handled by the web service handler component. Android mobile is used as mobile client, Web services are implemented in .net, it efficiently shows the language interoperability.

Business logic runs at specific location that can be identified by the url and which is used by the mobile client to access the business logic by forwarding the input parameters to respective url in the form of soap object, implicitly it serializes the object and converts into the web service description language again at the receiver end it converts the WSDL into receiver native language by using the deserializer, that shows the language interoperability due to WSDL. For implementation purpose we designed an application which uses the business logic as bank common manipulations and result analysis if any recent accessed information is there, it is available in cache otherwise forward the request to the server, during the reply before displaying to the user place the accessed content in cache for next time access. We can reduce the load on client by maintaining the time duration for storing the copy of information in cache.

5 Analysis and Results

In this we provide analysis about the characteristics of proposed system.

Battery Power Saving: This architecture will speed up the experiments .Once the query is executed by the user then resultant information is stored in client side proxy. When the same request is given it will not search in main web server. In this way fast access is provided.

Scalability: By creating proxy at client side, the load on the server is reduced and allows it to serve more users.

Effectiveness: The services provided to the user basing on their needs. Effectively the request and responses are correlated.

Speed: Once the server application starts, it downloads available services and one copy is stored in cache. So accessing the query again will be done very fast because it will not go to the web server first it searches in proxy.

Suitability to Emerging and Current Platforms: Because of the usage of Android, the current and emerging platform suitability is high.

Adaptability to REST: We can do dynamic invocation of web services using SOAP messages but some groups prefer REST. The similar efforts will lead to standardize models for publishing REST web services that is analogous to UDDI registries.

6 Conclusion

The present architecture will give possibility to mobile device users to invoke web service methods dynamically that meet their needs. The implemented solution overcomes technical limitations and also saves device battery power by providing a proxy thus leading to better access in wireless network. Thus the project will provide a better service for the applications running on mobile devices.

References

1. Artail, H., Fawaz, K., Ghandour, A.: A proxy-Based Architecture for Dynamic Discovery and Invocation of Web Services from Mobile Devices
2. Aggarwal, C., Wolf, J., Yu, P.: Caching on the World Wide Web. IEEE Trans. Knowledge and Data Eng. 11(1), 94–107 (1999)
3. Li, L., Li, M., Cui, X.: The Study on Mobile Phone-Oriented Application Integration Technology of Web Services. In: Proc. Int'l Conf. Grid and Cooperative Computing (GCC) (April 2004)
4. Halteren, A., Pawar, P.: Mobile Service Platform: A Middleware for Nomadic MobileService Provisioning. In: Proc. IEEE Int'l Conf. Wireless and Mobile Computing, Networking and Comm. (WIMOB) (2006)
5. Sadhukhan, P., Das, P.K., Sen, R., Chatterjee, N., Das, A.: A Middleware-Based Approach to Mobile Web Services p. Sadhukhan, Centre for Mobile Computing and Communication (CMCC), Jadavpur University, Kolkata- 700032
6. Tere, G.M., Jadhav, B.T., Mudholkar, R.R.: Dynamic invocation of web services (1). Department of Computer Science, Shivaji University, Kolhapur, Maharashtra. 2. Department of Electronics & Computer Science, Y.C. Institute of Science, Satara, Maharashtra - 4, India. 3. Department of Electronics, Shivaji University, Kolhapur, Maharashtra - 416004, India
7. Laukkanen, M., Helin, H.: Web services in wireless networks: What happened to the performance. In: Proceedings of the Int. Conf. on Web Services (ICWS 2003), pp. 278–284. CSREA Press (2003)
8. Web Services: Usage and Challenges in mobile phones (computers) W3c seminar, Paris, France (March 6, 2006)
9. Robert, S., Khaled, K., Tharam, D.: Mobile Web Services Discovery and Invocation Through Auto- Generation of Abstract Multimodal Interface. In: Proceeding of Third International Conference on Information Technology: Coding and Computing, ITCC 2005 (2005)
10. Forouzan, B.A.: Data Communications and Networking
11. http://xml.coverpages.org/uddi.html
12. http://www.w3.org/TR/soap/
13. http://www.w3.org/TR/wsdl
14. Park, Kim, Bae, Kim Sok, Kang: An Automated WSDL Generation and Enhanced SOAP Message Processing System for Mobile Web Services. In: Proceeding of Third International Conference on Information Technology: New Generation, ITNG 2006 (2006)
15. Curbera, F., Duftler, M., Khalaf, R., Nagy, W., Mukhi, N., Weerawarana, S.: Unraveling the Web Services Web: An Introduction to SOAP, WSDL, and UDDI. IEEE Internet Computing 6(2), 86–93 (2002)
16. IBM-WSTK Toolkit (2009), http://alphaworks.ibm.com/tech/webservicestoolkit

Development of Web-Based Application for Generating and Publishing Groundwater Quality Maps Using RS/GIS Technology and P. Mapper in Sattenapalle, Mandal, Guntur District, Andhra Pradesh

Aswini Kumar Das[1], Prathapani Prakash[1], C.V.S. Sandilya[2], and Shaik Subhani[3]

[1] A.P. State Remote Sensing Applications Centre
[2] Geosciences Division, A.P. State Remote Sensing Applications Centre
[3] Nagarjuna University, Guntur
{aswini.das81,prakashmhbd}@gmail.com, cvs_sandilya@yahoo.co.in, subbu_buddu@ymail.com

Abstract. Groundwater is an essential source of drinking water for many Indian habitats. Large number of people consumes ground water as it is free from pathogenic bacteria and it is easily available through open well / bore well / tube well. The quality of water plays a prominent role in promoting human health. The paper discusses the usage of P-mapper for publishing the maps on to web portal. Element wise and integrated ground water quality maps are prepared for the elements pH, Total Alkalinity, Total Hardness, Total Dissolved Solids, Chloride and Fluoride. The ground water quality maps are studied with respect to different thematic maps which were prepared by using Remote Sensing and GIS techniques. The main aim and objectives of the current study is in building up of open source web GIS based application which has been developed using open sources like Postgress SQL, Map Guide and Pmapper. This application can be used for bringing the generated ground water quality maps to the authority and or the community and to alert the local community about the steps to be taken about the contaminated water and decision making. In conclusion web based GIS application is a useful tool especially in formulating policy related to water quality using P-mapper as a media.

Keywords: GIS, Ground Water quality, Map server, P-mapper, Remote Sensing, Web GIS.

1 Introduction and Background

Water has become a scarce resource all over the world. Water resources management has often focused on satisfying increasing demands for water without adequately accounting for the need to protect water quality [1]. Rapidly growing cities and industries, expansion of the mining industry, and the increasing use of chemicals in agriculture have undermined the quality of many rivers, lakes, and aquifers [2]. If pollution makes the water unfit for human use, degraded surface and groundwater

S.C. Satapathy et al. (eds.), *ICT and Critical Infrastructure: Proceedings of the 48th Annual Convention of CSI - Volume II*, Advances in Intelligent Systems and Computing 249,
DOI: 10.1007/978-3-319-03095-1_74, © Springer International Publishing Switzerland 2014

quality can even add to water shortages in water-scarce regions. Even though water quality deterioration is often not as visible as water scarcity, its impacts can be just as serious with significant economic consequences. Health hazards, agricultural production losses, and losses of ecological function and biodiversity have long-term effects that are costly to remediate and impose real suffering on those affected [3].

In view of this a study has been attempted here to understand the significance of remote sensing and GIS techniques in groundwater potential and quality assessment [4]. It can easily adhere that in which district/mandal/village/habitation and durinal season the ground water quality is depleted or deviated from the average. The software helps the users not only an overall picture of the region with regards to the condition of the ground water quality over the seasons but also the site specific locations for preparedness.

2 Study Area

The study area i.e. Sattenapalle mandal (Figure 1) falls in the Guntur district of Andhra Pradesh, in between latitudes 16.318^0 to 16.521^0 N and longitudes 80.059^0 to 80.281^0 E and covering the topographic sheet no. 65D/2, 65D/3, 65D/7. The study area covers an area of 238.204 sq. km. It has a population of 123,697 and the average rainfall of the area is 1200 mm. The climate is comfortable around the year.

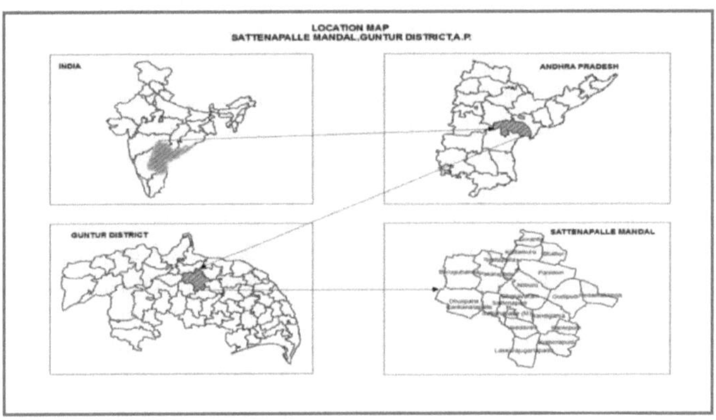

Fig. 1. Image Showing Index Map of Study Area

3 Aim and Objective

The main aim and objective of the current study is to generate open source web based application using the satellite based thematic maps on ground water quality. In the process of generating the ground water quality map, the following thematic layers / tables are generated [5].

- Excel table providing classified input ground water quality data and its database.
- Ground water sample layer showing location of ground water samples.
- Element-wise and integrated Ground water quality layer.

4 Data Sets

In the present project, the following datasets are used for the study:

1. Base map
2. Hydrogeomorphology map
3. Ground Water quality maps
   3.1 Groundwater quality sample layer
   3.2 Element-wise Groundwater quality layer
   3.3 Integrated ground water quality map in Pre and Post-Monsoon Season
   3.4 Ground Water Quality map

5 Methodology

The spatial distribution of the quality of ground water has been generated by considering element-wise average values of a habitation. After preparing element-wise ground water quality maps, the integrated ground water quality maps are prepared by combining all the element-wise ground water quality maps. Element-wise ground water quality maps are generated by using the interpolation technique namely Inverse Distance Weightage (IDW) [6]. The entire process of ground water quality mapping considered for the present purpose, broadly can be divided in to three major parts [7].

Part-1: The input data required for generating the ground water quality map is to be processed and to be organized in to a GIS data base.

Part-2: Element-wise ground water quality layers are to be created from the GIS database using interpolation technique.

Part-3: The final ground water quality map is to be derived based on the integration of the element-wise ground water quality layers.

A schematic diagram showing various components of the process and their flow in realizing final ground water quality map is given in following (Figure: 2).

The methodology for Groundwater quality can be summarized as follows:

The legacy data from line department RWS&S (Rural Water Supply & Sanitation Department, Govt. of Andhra Pradesh) is used for the study. It consists of the ground water quality data pertaining to the elements Total Dissolved Solids (TDS), Total Hardness (TH), Fluoride (F), Total Alkalinity (TA), Chloride (Cl) & pH. There are 31 Habitations which are included in to 19 villages / 26 Gram Panchayats in the Mandal. The following table 1 shows element-wise ranges of values for pre and post monsoon seasons.

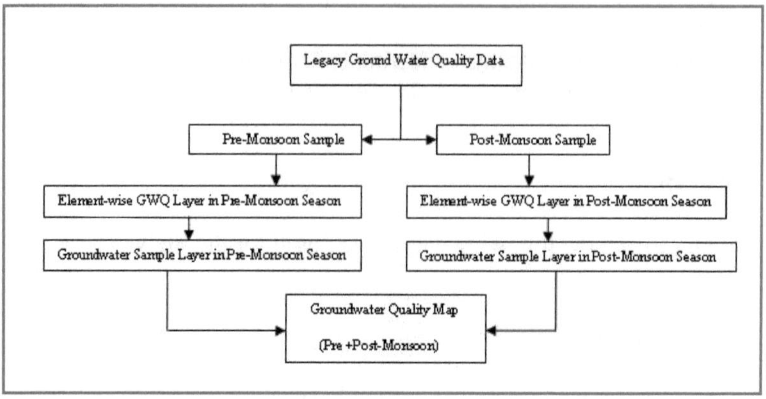

Fig. 2. Methodology for Generating Groundwater Quality Map

Table 1. Element-wise ranges of values for pre and post-monsoon seasons

Pre -Monsoon	pH	TDS	TH	TA	F	Cl
Minimum	6.53	77	40	120	0.2	20
Maximum	8.64	9779	3540	442	4.2	3040

Post -Monsoon	pH	TDS	TH	TA	F	Cl
Minimum	6.10	320	100	108	0	39
Maximum	8.89	18988	2048	442	4.8	1750

From the above table it is observed that Total Dissolved Solids (TDS), Total Hardness (TH), Fluoride (F) & Chloride (Cl) is above the permissible limits in the Mandal. After element-wise interpolation is completed for pre and post monsoon seasons, integrated ground water quality map is prepared (Figure 3) for pre and post monsoon season by combining (union) all the elements in a season.

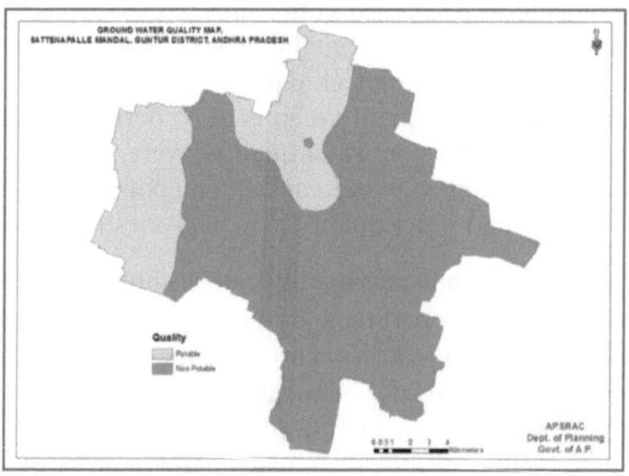

Fig. 3. Final Ground Water Quality Map (Pre+Post)

6 Thematic Map

6.1 Geology

The total area of mandal is underlain by Peninsular Gneissic Complex which may be divided in to two rock groups namely 1) Banded biotite-hornblende gneiss with migmatite patches and younger 2) Grey / pink granite gneiss. The boundary of two rock groups is separated by a thrust in North Western part of the mandal. In general if we look at the ground water quality map overlaid on geology map, it is revealed that the ground water quality in Grey / pink granite is good whereas the other formation it is of bad quality (most of the area).

6.2 Geomorphology

The major geomorphic units identified in this study area are pediplain and pediment. The pediplain under canal command covers major portion of the study area (Figure 4).

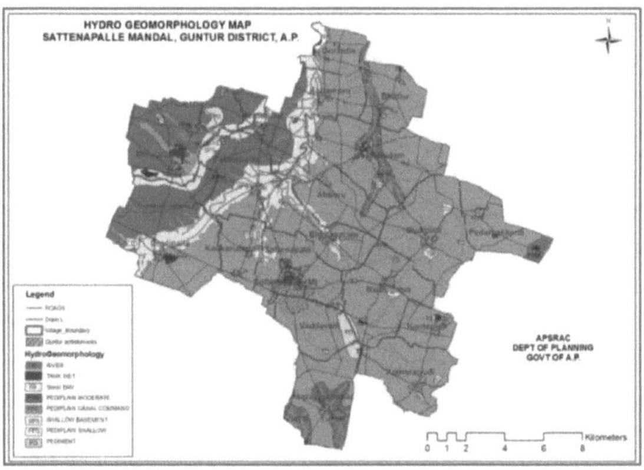

Fig. 4. Hydrogeomorphology map of the study area

7 Database and Software

Database was designed for efficient handling of raster, vector and tabular data. For this application PostgresSQL database is used for storing water quality information. For each sample season, district, mandal, village and habitation separate tables are created and designed. Here the database is designed such that there is no transitive dependency and all non-primary key attributes are mutually independent. The new data can also update without affecting entire database.

7.1 Development of P Mapper Based Web GIS

MapServer is an open source development environment for building spatially enabled internet applications [9]. It is a framework based on Map Server and PHP/Map Script.

It can run as a CGI program or via MapScript which supports several programming languages (using SWIG). It provides a good set of tools ready to use and it has a plug-in API to add functionalities [10]. It is released at no cost. It is possible to modify the source code.

When a request is sent to MapServer, it uses information passed in the request URL and the Mapfile to create an image of the requested map. The request may also return images for legends, scale bars, reference maps, and values passed as CGI variables.

8 Results and Discussion

Both natural processes and human activities can cause deterioration in water quality. Various water quality standards have been developed to assess the suitability of a water resource for particular uses. Human activities influence groundwater quality primarily through contamination. Major sources of groundwater contaminants include landfills, septic systems, abandoned water wells and excessive fertilizer use. Aquifers vary in their susceptibility to contamination. Shallow aquifers consisting of permeable sediments are extremely vulnerable, as contaminated surface water can enter them very quickly. In contrast, deep aquifers or those of less permeable materials are less vulnerable either because of the longer travel and filtering time or the overlying protective confining layers of rock preventing the downward migration of contaminants. Contamination of groundwater is difficult to detect in early stages. By the time contamination is realized, its effects are often significant and costly to clean up. If contaminated groundwater migrates to a surface water body such as a wetland or lake, those supplies can also be affected. The result of the suitability of the publication of map on the Pmapper is to adhere that in which district/mandal/village/habitation and diurnal season the ground water quality is depleted or deviated from the average. The software helps the users not only an overall picture of the region with regards to the condition of the ground water quality over the seasons but also the site specific locations for preparedness (Figure:5a & 5b).

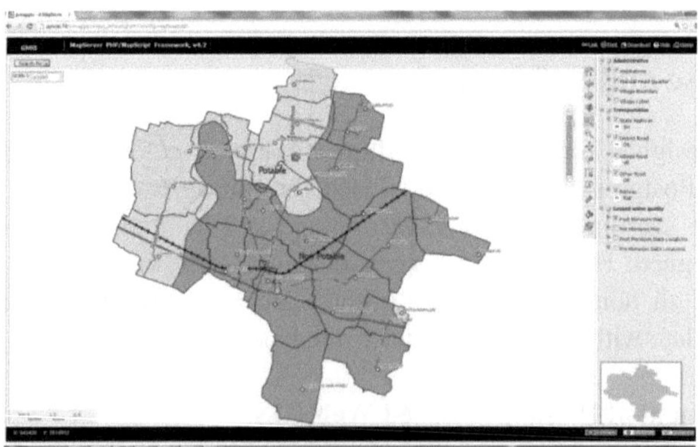

Fig. 5a. Output Window shows Post-Monsoon Layer

Fig. 5b. Output Window shows Pre-Monsoon Layer

9 Conclusion

The majority of sources of the mandal are contaminated with Fluoride (geogenic), TDS (geogenic/anthropogenic) and Total Hardness (geogenic / anthropogenic). TDS & TH are out of range for most of the sources because of extensive irrigation practised in command areas and recharge from the nutrient rich surface waters. The origin for fluoride contamination occurs geogenically. Fluoride in ground waters occurs because of fluoride bearing minerals present in the parent rock. When the fluoride is dissolved in ground waters because of chemical action with water, the concentration of fluoride is getting increased.

Preventing groundwater contamination is achievable through a combination of actions including:

- Dynamic water quality maps using p-mapper and decision making
- Public awareness programs quality affected areas and corrective measures if required
- Proper construction, maintenance and plugging of water wells.

Acknowledgement. Authors are thankful to Sri Sanjay Gupta, Director General and Dr. S.V.B. Krishna Bhagavan, Director (Technical), Andhra Pradesh State Remote Sensing Applications Centre (APSRAC), Hyderabad for giving permission to carry out this work at esteemed office. Authors are also thankful to the reviewer for critically going through the manuscript and giving valuable suggestions for the improvement of manuscript.

References

1. Winter, T.C., Harvey, J.W., Lehn Franke, O., Alley, W.M.: Ground Water and Surface Water A Single Resource. U.S. Geological Survey Circular1139, Denver, Colorado (1998)

2. Elfithri, R., Toriman, M.E.B., Mokhtar, M.B., Juahir, H.B.: Perspectives and Initiatives on Integrated River Basin Management in Malaysia: A Review. The Social Sciences 6(2), 169–176 (2011)
3. Coye, M.J.: The Health Effects of Agricultural Production: I. The Health of Agricultural Workers. Journal of Public Health Policy 6(3), 349–370 (1985)
4. Ramachandran, S.: Application of Remote Sensing and GIS
5. Lilesand, T.M.: Remote Sensing and Image Interpretation. John Wiley andsons. U.S.A. 721p. (1989)
6. Watson, D.F., Philip, G.M.: A Refinement of Inverse Distance Weighted Interpolation. Geoprocessing 2, 315–327 (1985)
7. RGNDWM Manual, Groundwater Quality Mapping for Rajiv Gandhi national drinking water mission (RGNDWM) National Remote Sensing Agency, NRSA (2011)
8. BIS 10500 (Bureau of Indian Standards), Indian Standard Drinking Water Specification, 1st edn., pp. 1-8 (1991)
9. McKenna, J.: Mapsever for Windows (MS4W), Gateway Geomatics (2012)
10. Valentini, L.: P.mapper User Manual v. 4.x, Gis course 2011, Development of a p.mapper-based webGIS, Politecnico di Milano – Polo Regionale di Como (2011)

A Reactive E-Service Framework for Dynamic Adaptation and Management of Web Services

T. Hemalatha[1], G. Athisha[2], and C. Sathya[3]

[1] Associate Professor
[2] Professor, Departement of ECE
[3] Lecturer Departement of CSE
[1,2,3] PSNA College of Engg. & Tech., Dindigul,
624622 Tamilnadu, India
hemashek@yahoo.com, {gathisha,sathi.saras}@gmail.com

Abstract. Web service is both a process and a set of protocols for finding and connecting to software exposed as service over the web. Different strategies have been suggested to implement governmental, nongovernmental and general services, which are characterized by high sharing and reuse of strategic applications. Dynamic adaptation is a technique, which enables the rule server to create rules dynamically based on the service requested by the user. The execution of such user's service may involve multiple web services obtained from various service providers with in the trusted environment in a cooperative manner. Under such circumstances it becomes necessary to manage multiple web services corresponding to many users request, which leads to multiple transactions. Hence a Web service Management System (WSMS) is proposed in this work to manage the web service resources and to control the activities that take place between different transactions which provides user-friendly interface to clients.

Keywords: Web Service, Reactive, Event Condition Action (ECA), and Extensible Markup Language XML Rules, Simple Object Access Protocol (SOAP), Information and Communication Technology (ICT).

1 Introduction

Web Services are emerged to provide a systematic and extensible framework for application-to-application interaction built on top of existing web protocols and based on open XML standards. Web Service is both a process and a set of protocols for finding and connecting to software exposed as services over the web. It provides a RPC style mechanism for accessing the resources on the web. SOAP [9] is an XML based protocol for Messaging and RPC. SOAP works on existing transports such as HTTP, SMTP. SOAP is used as a simple messaging protocol. SOAP message has a simple structure an XML element with two child elements one of which contains the header and the other body. The SOAP specification dictates how recipients should process SOAP messages since it is a lightweight protocol. Reactive Service is pushing

S.C. Satapathy et al. (eds.), *ICT and Critical Infrastructure: Proceedings of the 48th Annual Convention of CSI - Volume II*, Advances in Intelligent Systems and Computing 249,
DOI: 10.1007/978-3-319-03095-1_75, © Springer International Publishing Switzerland 2014

of information to clients in reaction to occurrence of new events. Reactive Service is implemented using Push technology. Push technology [2] is the ability of sending relevant information to clients in reaction to occurrence of new events. ECA paradigm is useful and necessary to implement the push technology in this work [10].

The government recently set out the need for reformation of Civil services to embrace the Digital age. ICT enables government to operate efficiently and keep pace with citizens desire to engage with government and access public services online [13]. The objective of this work is to bring the Government closer to the citizens through the concept of Service Oriented Architecture (SOA). SOA can bring major improvements in the delivery of government services and information provision in convenient ways. It helps to systemize and stream line the Government processes in the most convenient and flexible manner. A Service framework is proposed in this paper, which constitutes following features Security, Transaction Management and trust negotiation that plays major role in implementing most of the general, government and non-government activities as web service. For simple service only one or two services may involve, but for some complex services several services should interact with each other to process the request. Hence it is to process the request in a secured and authenticated environment. Since such environment is essential for many of the organizations to provide their service as an E-Service, which was lagging, that prevents many of the organizations to provide their services in web. Hence in this paper a service framework is proposed that consists of trusted domains and allows only trusted domain web servers to interact with each other in a secured way that provides co-operative and reactive web services.

2 Related Work

In order to provide electronic service delivery several activities involving different public agencies need to be related and carried out in coordinated manner, thus resulting in a cooperative process" by Mariangela Contenti [2]. A repository of workflow components for co-operative e-Applications which was described by Massimo Mecella explains the integration of different e-services to support centralized federated and virtual enterprises [1]. Co-operative e-applications require the development of a complex framework in which the dynamic interchangeability of different e-services is possible in a semiautomatic way. Pushing reactive services to XML repositories using Active Rules which was described by Angela Bonifati [3] explains implementation of reactive technology using XML query languages and explained the negotiation protocol and the interchange mechanism between the rule broker and XML repository. SOAP protocols and Envelopes were used for this protocol. Angela Bonifati proposed Rules using XML supporting technologies [11]. Bailey proposed Rules using ECA paradigm [12]. But to write an efficient rule it may be necessary to know the schema of XML rules. In order to overcome this problem rules are installed by the trusted domains through regulating authority. The Infrastructure for E-Government Web Services, which was described by Brahim Medjahed , proposed that the framework for automatically composing e-government services is based on set of rules that check composability of services [5]. SOAP supports RPC and message passing [10] [8]. Hence SOAP plays major role in interaction between web services and clients. Model driven trust negotiation for web services which was described by Halvard Skogsrud explains "trust negotiation as an approach to access control whereby access is granted based on trust established in a negotiation between the service requester and the service provider" [4]. Trust builder is intended [6] for use in any situation where two entities from different security domains need to establish trust between business-to-business and retail interactions and so on is explained by Marianne & Ting Yu in Negotiating Trust on the web [6]. The need for web service in a co-operative environment was described by Massimo Mecella et. Al.[1] in which the interaction was the core issue. But in a co-operative environment the reactive behavior is more important. A web service Interaction models which was described by Steve Vinoski explains about the various factors that occurred when exposing the object as a web service [7]. Hence in this paper, we proposed a reactive service framework for dynamic adaptation and management of web services. This framework consists of a service agent, service scheduler, a rule engine, a rule tracker, and a Web Service Management System [WSMS] and a regulating authority.

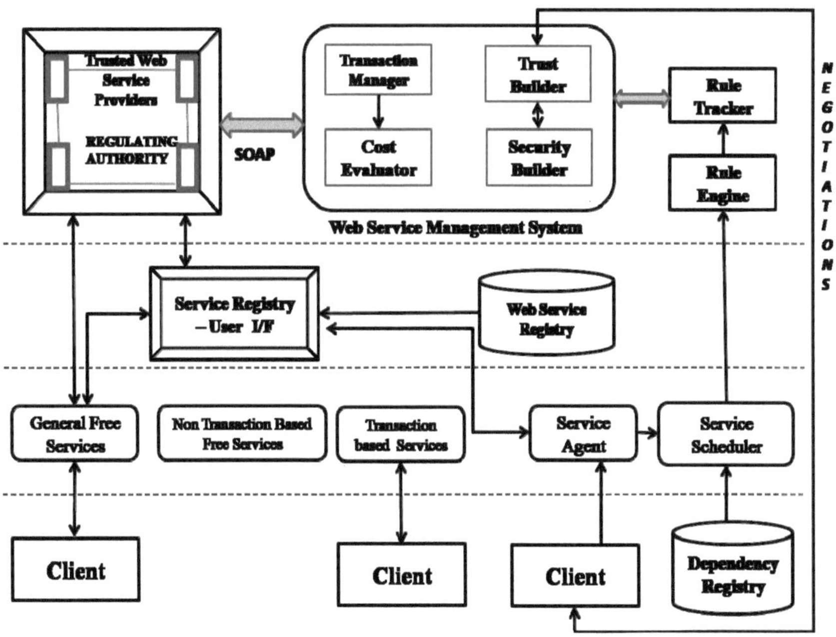

Fig. 1. System Architecture – *Regulating Authority and Web Service Management System*

3 System Architecture

The proposed architecture of the service framework is shown in Fig. 1. The system consists of service agent, service scheduler, rule engine, rule tracker and WSMS.

WSMS consists of Transaction Manager, Trust builder, Security builder and cost evaluator. A regulating authority is required in order to build trusted domains. The service provider has to register with regulating authority in order to provide service to user in a secured and trusted environment. The service provider has to register its rule and dependencies which are used further by the rule engine and service scheduler. Service agent acts as a proxy between the service consumer and the service provider. The service agent provides the list of services to the clients, which are fetched from the repository of regulating authority. Then the client selects the service that is needed. The service agent sends a login form to be filled by the client. The client should give a username and a password. The agent will provide a valid token to the client on receiving the filled form, which will be used with the request for identity or authenticity. The token is included in the SOAP message for validity and security. Then based on the selected service the service agent will send the requested service and its respective method name with the token to the service scheduler.

3.1 Service Scheduler

The service scheduler process the service requested by the client by accessing the dependency information. The dependency information is stored in a XML data store. If the service requested is a simple service, which involves with a single service provider, it checks the rule registered by the service provider in regulating authority. If the rule does not contain any dependency details it immediately forwards the newly created schedule to the rule engine else it will access the dependency information and frames the rule with the token and create a new schedule. It does the same if requested service is a complex service. Then with the dependency information it will forward the scheduled request to the rule engine with the token.

3.2 Rule Engine

The rule engine receives the request from the service scheduler and frames a rule with a unique rule id and with all the constraints in the XML file format. The structure of the rule is given below. The rule template for the requested service is accessed from the rule repository. After framing the rule, the rule is forwarded to the Rule tracker. The rule tracker receives the rule with a unique rule id. Then it parses the XML rule. It takes care of executing the request by constructing the SOAP message with the proper parameters to invoke the requested service. Once the service is invoked it uses the push technology to notify the client about the reaction of event thereby provides reactive service. If the rule has any dependency then the trust builder will be invoked. On invocation the negotiator module sends a SOAP request to the client. Then the client will supply the requested credentials to the trust builder only if the message contains the token that is similar to the token possessed by client. Then if it is identical it will provide the requested credentials to the negotiator. Then the negotiator will check the credential supplied by the client. If the credentials are true and valid and if the response message contains the same token it will proceed further. If the set of credentials are disclosed between the service provider and consumer the negotiator will send a response SOAP message to both the ends and thereby the trust is built. During negotiations if any credential is found to be invalid then the trust builder will send a message to service provider thereby it aborts the entire transaction.

3.3 Web Service Management System

The transaction manager keeps track of all the transactions that are happening between the service provider and the service consumer or between two service providers. It uses the Two Phase Commit Protocol to maintain consistency in databases. If the transaction aborts in the middle, then all the changes that are done so far in the data sources are rolled back else it commits if the whole transaction takes place smoothly. The security builder uses symmetric algorithm AES (Advanced Encryption standard) to provide confidentiality while disclosing credentials during negotiation between two parties. The cost evaluator evaluates the total cost involved in particular transaction and it pushes the cost information to the client. Thus WSMS provides various functionalities to handle the transaction in secured manner in trusted environment.

4 Implementation

The major functions of the system are as follows. Initially a service provider has to register in the regulating authority if it wants to provide service to the users through a trusted environment in a secured manner. While registering in regulating authority it has to provide its dependency details and the default rule schema. Various services provided by different public agencies (service providers) that are registered in regulating authority are stored in XML data source. In Fig. 1, General free services are freely available without disclosing any credentials. It does not require any authentication from users. General free services are directly accessed from Universal Dynamic Discovery Integration UDDI. E.g. Getting the address and phone number of a particular person. Non-Transaction based free services need not be kept track in order to compute the cost associated with it e.g. Enquiring about the status of flights in Airline reservation system. Service Agent provides a user-friendly interface to the end user that provides list of web services which are available in trusted domains. Service scheduler schedules the service based on the request from the clients by accessing the dependency registry. Service scheduler creates a new schedule for every new request and returns the file name which is an XML instance to the Rule Engine. The file created by service scheduler is shown below

```
A Sample instance of the XML file from Service Scheduler
  <?xml version="1.0" encoding="UTF-8" ?>
   <Scheduler>
   <MainService>
  <Name>NominationWs</Name>
  <Url>http:// msaul13/NominationWs/</Url>
```

```
 <MethodName>Register</MethodName>
</MainService>
<DependentServices>
<DependentService>
<Name>Bank Services</Name>
<Url>http:// msaul13/BankWs/</Url>
<MethodName>Verify</MethodName>
<ParamCount>1</ParamCount> </DependentService>
```

Rule Engine is a rule framer that dynamically frames the rule based on the type of request from the clients. If the request is simple it invokes the web service directly by using XML-RPC (Remote Procedure Call) request and response interactions. Otherwise if the request requires more than a simple service which is to be invoked by composing different simple services together that can be executed either sequentially or parallel based on the rule. It accesses the XML file created by the service scheduler. A part of a rule that is created by rule engine is shown below.

4.1 Structure of the XML File from Rule Engine

It is a sample rule which provides reactive service. Rule tracker will take care of execution of events in the web servers either in a serial or in a parallel fashion based

```
<!DOCTYPE Rule[<!ELEMENT Rule ANY>]>
<Rule>
<Token>dsjad</Token> <ReactiveRule>  <if>
<condition>cid&eq</condition> <value>value</value>
</if> <then> <action processCWS> <if>
<condition>cid&eq</condition> <value>value</value>
</if>  <then>  <action processPWS> <if>
<condition>cid&eq</condition> <value>value</value>
</if>  <then> <action>Commit</action>  </then> <else>
<value>exit</value>  </else>
</action> </then> <else> <value>exit</value> </else>
 </action>  </then><else>
<value>exit</value>  </else>
</ReactiveRule>
</Rule>
```

on the rule framed by the rule engine. Trust builder is an agent that performs negotiations between B2B and B2C and verifies the credentials and the policy. Security builder provides confidentiality and authenticity to the system. Transaction manager will keep track of transactions and the cost evaluator will evaluate the cost. Two-phase commit protocol is used by the transaction manager to make consistent transactions. For testing this proposed system some government and non-government services are created in distributed manner to form trusted environment. For example the following several services are created and the dependent services are also created and registered with the regulating authority. They are Election Nomination service, Law and order service, Civil service, Income tax service, Credit card service and Passport service. All these services are registered in regulating authority with its dependencies and the basic rule template. After registration is completed the service agent is updated dynamically with the newly registered services and dependency registry is also updated. The end user selects the service and the method associated with it. The service agent generates a unique token and it sends the token along with the selected service to the service scheduler. The service scheduler accesses the dependency registry to fetch the dependent details for the service requested by the end user. Then it creates a new schedule and sends the schedule to the rule engine. The rule engine creates a new rule with a unique rule id by accessing the rule repository that contains the rule template for various services. After creating the rule, the rule tracker will take care of executing the services through WSMS. On getting a response from a web service, the rule tracker pushes the status to the client. The rule tracker also initiates the negotiator if the rule contains dependencies. It invokes the method dynamically and returns the result returned by the execution of the web method. Then the rule tracker checks the dependencies to find whether all web services within that DEPENDENCY_DETAILS have been completed. This is accomplished by checking the status of the rule-completed levels. Rule tracker uses XML DOM to parse, update and insert in the rule on execution of event.

4.2 Experimentation

The proposed system is implemented by using C# for Web services and ASP .Net for Web Interfaces. For testing the framework several services as mentioned in section 4.1 are designed and implemented with different service requirements. The services are identified from the existing government and non-government operations so that when there is a need for ICT in government, this proposed system is suitable for transforming the existing setup to e-Government. The proposed system uses the Web service architecture and its standards for its flexible and interoperable features.

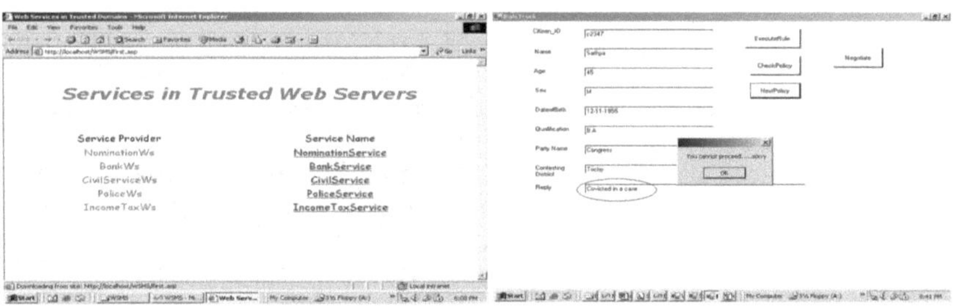

Fig. 2a. Different services registered with regulating authority (*LHS*) b. Nested invocation of different web services based on the created schedule and rule (*RHS*)

For example, when the user request to perform nominee registration for Election process the rule tracker uses civil service to check if the person is a citizen or not. If it returns true then it executes the police service to verify whether the particular citizen is convicted in any of the case. When it returns "No Case" or "Not Found" then it checks the Income Tax service and finally Credit Card service to finish registration process. If every level is completed the Nomination Web service registers the nominee details in the Election Commission database. Here the transaction manager performs transaction in consistent manner and cost evaluator will evaluate the total cost involved in one particular request. Finally credit card service is invoked to complete the process. These are all carried out in secured environment and in reliable manner. Another sample scenario is tested using this framework. For instance set of services like RTO service, Law and order service and General insurance were developed and deployed for testing. In case of any accident any personnel can access this framework to get the complete details about the driver and the vehicle owner and the corresponding status of the vehicle insurance in an adhoc manner without any complexity. Few snapshots of our proposed framework are given below in Figure 2.

5 Conclusion

Since ICT is gaining popularity and most of the Government and non-governmental organizations are coming forward to implement their service using this technology. In this paper a service framework is designed which has a repository of rules that provide services to different categories of customers in reliable and secured manner. The service framework is tested for both Government and Non-government services and the results are found to be promising and light weight without much dependency.

References

1. Mecella, M., et al.: A Repository of Workflow Components for Cooperative e-Applications. In: Proceedings of the 1st IFIP TC8 Working Conference on E-commerce/E-business, pp. 1–19 (2001)
2. Contenti, M., et al.: An e-service-based framework for inter-administration cooperation. In: Wimmer, M.A. (ed.) KMGov 2003. LNCS (LNAI), vol. 2645, pp. 13–24. Springer, Heidelberg (2003)
3. Bonifati, A., Ceri, S., Paraboschi, S.: Pushing Reactive Services to Xml repositories using Active Rules. In: Proc. 10th WWW Conf., pp. 633–641. ACM
4. Skogsrud, H., Benatallah, B., Casati, F.: The Model-Driven Trust Negotiation for Web Services. IEEE Internet Computing, 45–52 (November- December 2003)
5. Medjahed, B., Rezgui, A., Bouguettaya, A., Ouzzani, M.: Infrastructure for E-Government Web Services. IEEE Internet Computing, 58–77 (January-February 2003)
6. Winslett, M., Lu, T.: Negotiating Trust on the Web. IEEE Internet Computing, 30–37 (November-December 2002)
7. Vinoski, S.: Web Services Interaction Models – Current Practice. IEEE Internet Computing, 89 – 91 (May-June 2002)

8. Curbera, F., Duftler, M., Khalaf, R.: Unraveling the Web Services Web. IEEE Internet Computing, 86 – 93 (March–April 2002)
9. Sahai, A., Graipner, S., Kim, W.: The Unfolding of the Web Services Paradigm. HPL-2002-130, Internet Encyclopedia
10. Simple Object Access Protocol (SOAP) 1.1, `http://www.w3.org/TR/2000/NOTE-SOAP-20000508`
11. Bonifati, A., et al.: Active Rules for XML: A new paradigm for E-Services. The VLDB Journal 10, 39–47 (2001)
12. Bailey, J., Alexandra, Wood, P.T.: An ECA language for XML. In: WWW 2002, pp. 486–495. ACM (May 2002)
13. National Audit Office, UK, `http://www.nao.org.uk/wp-content/uploads/2013/03/`

Enabling Time Sensitive Information Retrieval on the Web through Real Time Search Engines Using Streams

S. Tarun and Ch. Sreenu Babu

Department of Computer Science and Engineering
GMR Institute of Technology, Rajam, AP, India
s.tarun013@gmail.com, sreenubabu.ch@gmrit.org

Abstract. Real time search engines constantly index web content originated by data streams also. This is because, the web sources like social networking sites, news, and tweets provide up to date information through streams. As new content is arrived constantly from those sources, it is very challenging job for search engines to have efficient indexing mechanisms to ensure index freshness and coverage of the index. Such updated index supports faster search whose results also include the latest content available. Latencies such as retrieval latency and indexing latency play an important role in index freshness. The former is the time taken to fetch the content after its publication while the latter is the time taken to make index on the newly fetched content. This paper presents a framework which optimizes indexing latency and also indexing coverage. The empirical results revealed that the proposed framework is capable of achieving index freshness and coverage in order to support faster processing of search queries.

Keywords: Indexing, search engines, index freshness, index coverage, information retrieval.

1 Introduction

Data stream is the flow of data that continuously arrives. In this paper, the data stream refers to a set of ordered documents that are published by a web site. In order to obtain content from stream there are two means namely Atom feeds and notification through distribution protocols that are push-based and polling from RSS. Users of late are expecting latest content from the web through Google real time search and Twitter search [1], [2]. Real time search engines work in offline for index updates. They constantly look out for new content and update their index freshness in order to enable end users to obtain latest content. This is because there are data streams that come from various sources like social media, blogs and news [3], [4]. Through subscription or monitoring data streams are made available to real time search engines [5], [6]. Moreover data streams also help search engines to improve their coverage of index for better performance. However, optimizing index freshness is a challenging job with respect to data streams. Fig. 1 shows the proposed framework for real time search engines to ensure index freshness and index coverage. As can be seen in fig. 1

DOI: 10.1007/978-3-319-03095-1_76,

there are five components in the framework architecture. The components are stream database, entry repository, indexer, stream fetcher, and crawler. The programs which are responsible to obtain new data streams from World Wide Web and updating them database with the new content. Stream fetchers are responsible to obtain new entries through polling process or notification from the publishers of streams. It also places the entries into entry repository. Entry repository is the place to store new entries which are obtained by stream fetchers. Indexer is a software component that is responsible to build indexes and update them as content is changed. The new entries are named indexing – ready entries. A policy is associated with entry repository. Capability and buffer size of the entry repository are the possible criteria present in the policy.

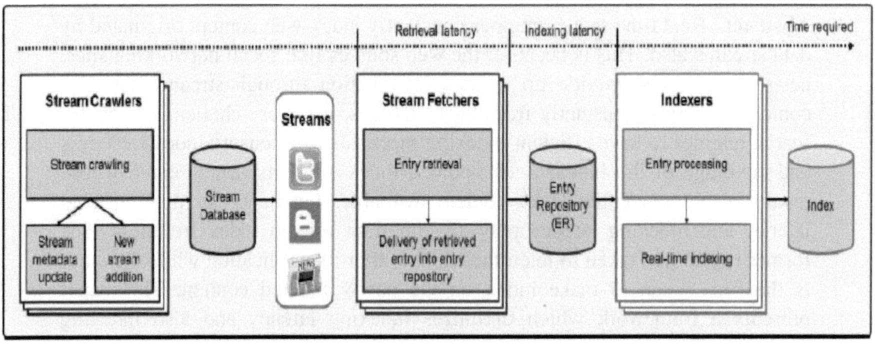

Fig. 1. Proposed architecture for real time search engines

As can be seen in fig. 1, there are two delays that can affect the freshness of indexes and index coverage. They are named retrieval latency and indexing latency respectively. The time taken by stream fetcher to fetch new entry, when it is published and adding to entry repository is known as retrieval latency. The indexing latency is the time taken to prepare index for the newly added entry into entry repository. Prior works to optimize index freshness tried to minimize retrieval latency. The index freshness is focused in [7], [8], [9], [10], [11] and [12]. Data stream aggregation also explored in [13] and [14]. As a matter of fact, the polling based protocols are replaced by PubSubHubbub [15] and XMPP (eXtensible Messaging and Presence Protocol) [16]. They are best used to reduce retrieval latency [3]. Thus the retrieval latency is effectively addressed while the indexing latency is a challenging problem. The indexing latency is increased when there are number of new items arrived constantly. It is also affected by the capacity of buffering. Large buffer size can also harm the index freshness. This is because the buffer maintains big queue of items that are yet to be indexed. There is tradeoff between the buffer size and index freshness. This problem is also surfaced in many crawling machines [17], [18] and [19]. To overcome such problems an attempt is made in [20] by making crawling policies that control queue size. However, it could not compute an optimal queue size for achieving index freshness.

In this paper we propose an optimal entry repository policy which is based on the theory of inventory control. The results revealed that the proposed mechanism successfully addressed the problem of index freshness and index coverage with respect to data streams that continually arrive from various web sources. The remainder of the paper is organized as follows. Section II presents the proposed optimization model. Section III provides details of prototype implementation. Section IV describes experiments and results while section V concludes the paper.

2 Proposed Optimization Model

The proposed optimization model is similar to the inventory theory where reorder level is optimized based on the prior experience, season, and other parameters. In the same fashion, the buffer size usage for entry repository has to be optimized to reduce indexing latency. The following assumptions are made for building the proposed optimization model. The fetching rate of stream fetchers is known and the value is a constant. The time taken for stream fetcher to deliver new items into entry repository is negligible. Data streams are visited by stream fetchers as per the pre-defined time interval. With respect to entry repository deterioration rate is constant and such items are not replaced and no indexing is made for such entries.

A) The Model

Based on the assumptions described above, the proposed model's objective function is defined by considering factors such as indexing shortage cost throughput, deterioration cost, lost entry cost, and indexing delay cost. Minimization of expected total ratio cost is computed as follows.

$$\min_{m\,M} K(m, M) = \frac{Expected\ Total\ Cost}{Expected\ cycle\ time}$$

$$= \frac{C_{ID}N_{ID}+C_{IS}N_{IS}+C_{LE}N_{LE}+C_{DE}N_{DE}}{T}$$

For different segments of repository level differential equations are as given below.

$$\frac{dL(t)}{dt} + L(t)\ .\ r_d = (r_f - r_i)\ ,\ 0 \le t \le T_1$$

$$\frac{dL(t)}{dt} + L(t)\ .\ r_d = -r_i\ ,\ T_1 \le t \le T_2$$

$$\frac{dL(t)}{dt} = -r_i\ ,\ T_2 \le t \le T$$

Finally the objective function is computed as follows.

$$L(t) = \left(\frac{r_f - r_i}{r_d}\right)(1\text{-}\exp(-r_{d.t}))$$

B) Algorithm for Revising an Optimal Policy

A heuristic has been proposed to revise optimal policy for the proposed model. The algorithm is as shown in algorithm.

1: Find an optimal policy, (m,M), by using the proposed optimization model.
2 : Calculate {(M-m)-[(M-m)/β].β}
3 : If {(M-m)-[(M-m)/β].β}>0, then revise m to m', where m' = M-[(M-m)/β].β
4 : Produce the final policy, (m,M) or (m',M).

Algorithm: Algorithm for revising an optimal policy

As can be seen in algorithm, the algorithm uses the proposed optimization model to find an optimal policy. Then it computes revised "m" in order change the optimal policy. Finally the algorithm produces final policy which serves best.

3 Prototype Application

We built a prototype application to demonstrate the proof of concept. The environment used to develop the application includes PC with 4 GB RAM and Core 2 Dual processor. Operating system is Windows 7. IDE (Integrated Development Environment) used is Net Beans. It is a web based application which makes use of tweetstream API which helps to connect to twitter and obtain live tweets. When the application is notified about the presence of new streams the following UI is shows with available stream trends.

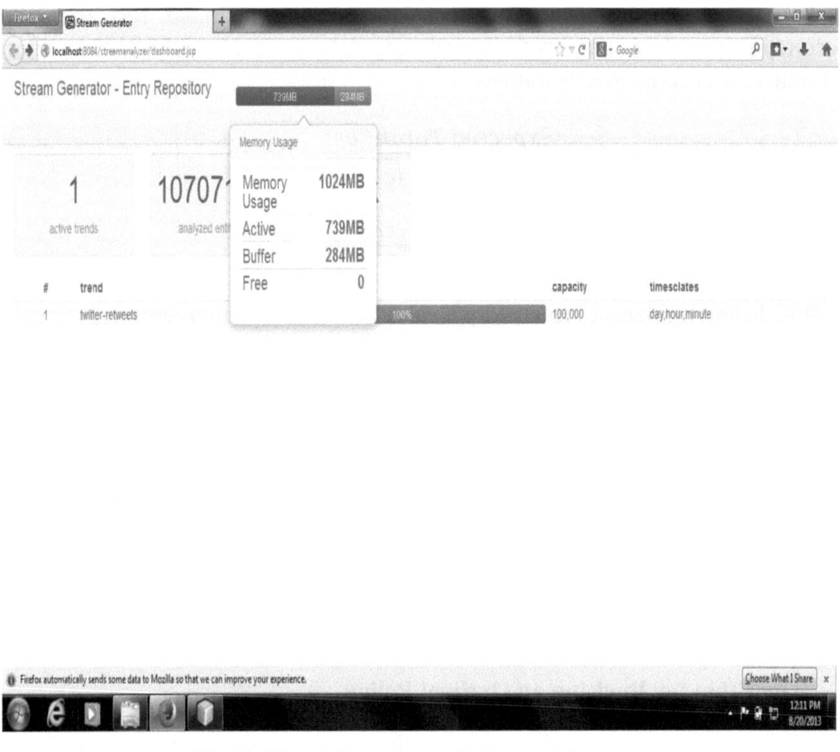

Fig. 2. UI to initiate stream fetcher and indexer

As can be seen in fig. 2, the prototype application running in local server fetches the trends into entry repository. Entry repository active memory usage is 706 MB and the buffer size is 317MB at that particular instance. The stream generator is responsible to interact with twitter server using tweetstream API and generate streams in real time. It stores the collected entries into entry repository. The entry repository is the storage place for stream entries. However, it plays an important role in performance of the overall application. The performance of the application depends on certain parameters such as buffer size of the ER. The ER works as per the settings specified. The settings should not be static. It has to adapt to the dynamic situations. For this reason it has to be optimized from time to time. Optimization involves changing buffer size other possible parameters. There is impact of ER optimization on the index freshness and index coverage. These two are the corner stone of the application as they can ensure users to gain access to latest content added to WWW through streaming. On clicking the trend "twitter-retreats" the tweets and their statistics are shown as in figure 3.

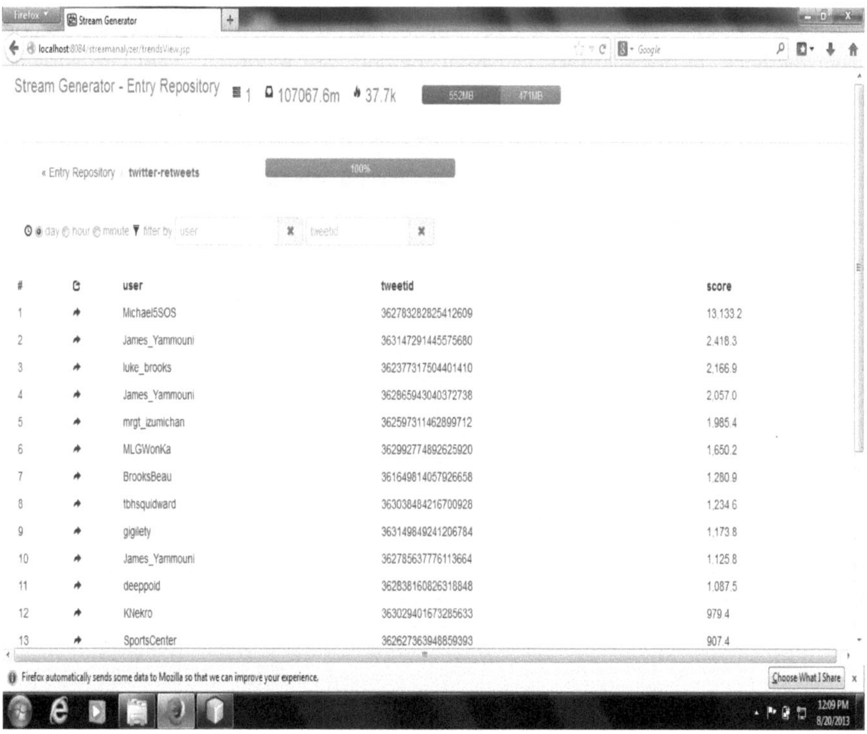

Fig. 3. The top scored tweets in entry repository

As can be seen in fig. 3, it is evident that based on the batch size given the Stream Fetcher processes the incoming steams batch wise and stores the entries into ER. As soon as new entries are found in ER, the indexer too takes them as batches and completes indexing. The resultant memory dynamics and other live changes are shown to end user.

As mentioned earlier, it is essential to optimize the ER in order to adjust its parameters like buffer size for optimal performance. The optimization of ER takes place automatically based on the experience of the application on the fly. The optimization is a continuous process. The algorithm used for optimizing ER is as given in algorithm. With optimization process carried out from time to time, the index freshness and index coverage are achieved.

4 Experimental Results

For the purpose of experiments we used the Twitter tweets obtained through tweetstream API. The streams are generated periodically by Stream Generator as shown earlier. The number of active trends, analyzed entities, updates per second, active memory, buffer memory and total memory are recorded. They are shown in table 1.

Table 1. Observations of statistics regarding buffer optimization

S no	Active trends	Analyzed entities	Updates /sec	Active memory	Buffer memory	Total Memory
1		96167.1	37.2	844	180	1024
2	1	96167.2	38.6	616	407	1023
3	1	96167.3	22.4	678	338	1016
4	1	96167.3	22.5	631	364	995
5	1	96167.5	20.6	743	254	997

As can be seen in algorithm, statistics about runtime process and buffer size optimization are presented. The following graphs show the changes in the statistics over a period of time with two seconds time interval.

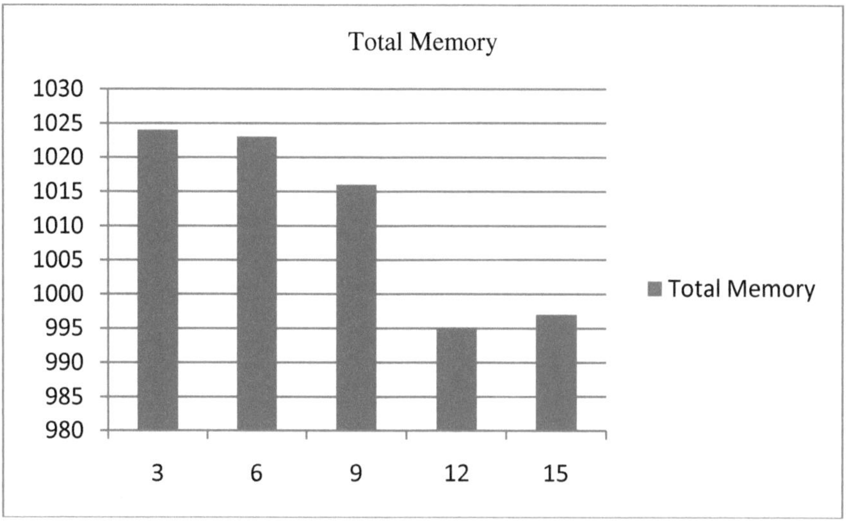

Fig. 4. Statistics about total memory

As can be seen in fig. 4, the total memory quotient varies with impact of tweet stream inputs. The maximum memory utilized is 1024MB.

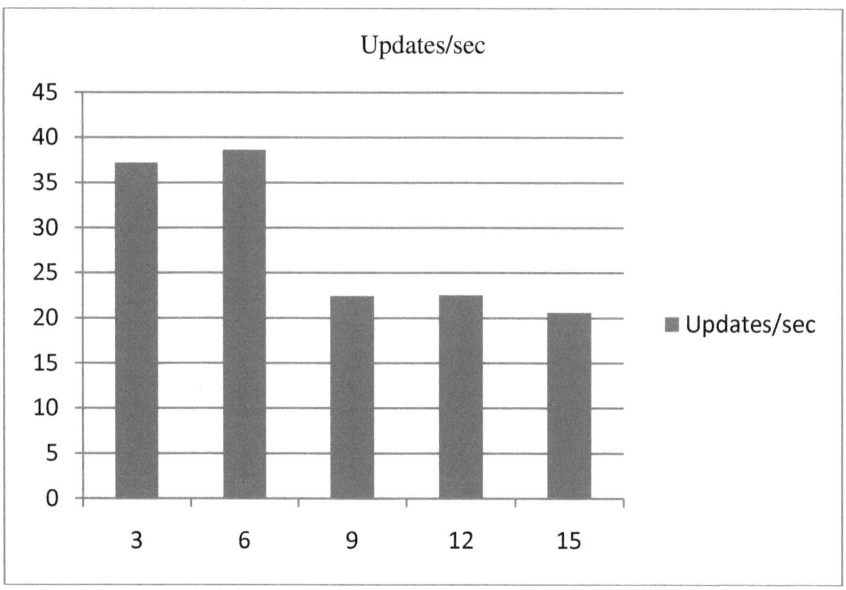

Fig. 5. Statistics about updates/sec

As can be seen in fig. 5, the updates/sec varies w.r.t. stream input flow and buffer capacity.

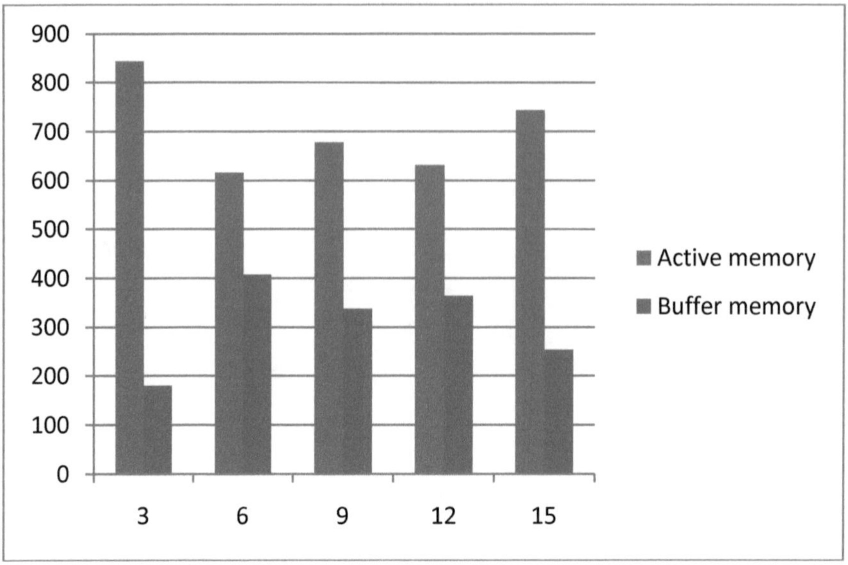

Fig. 6. Statistics about active memory and buffer memory w.r.t. Total memory

As can be seen in fig. 6, the active memory and buffer memory are parallel fluctuated w.r.t. optimization policy mentioned in algorithm and total memory. The summation of both memories is equal to its respective total memory.

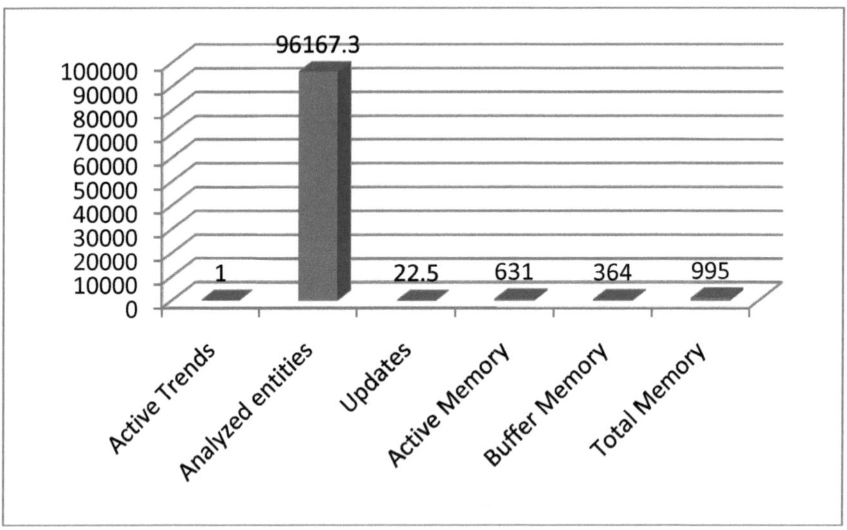

Fig. 7. Statistics about various entities w.r.t. buffer optimization

As can be seen in fig. 7, the number of the active trends is 1. There are changes in analyzed entities, updates per second, active memory, buffer memory and total memory. This reflects the dynamic and active optimization of buffer size. Based on this the index freshness is also achieved.

5 Conclusion

In this paper, we have proposed a framework for real time search engines that helps in ensuring optimizing index freshness and index coverage. This is required by real time search engines as streaming data arrives continuously. There is need for updating index to reflect the latest content in the search results. There are two delays that affect the process of index freshness. They are known as retrieval latency and indexing latency. The time gap between publication and fetching of the published document is known as retrieval latency. The indexing latency is nothing but the time gap between the arrival of entry repository and indexing of the entry. The buffer size used for entry repository where new entries are arrived also determines the index freshness. Optimal buffer size has to be computed. In this paper, we proposed an algorithm that optimizes the buffer usage thus causing index freshness. Experiments revealed that the proposed algorithm performs better in ensuring index freshness and index coverage. It also revealed that the average cost of indexing process is reduced considerably. With this the users of search engines can obtain latest information in the search results.

References

[1] Google Real-Time Search (2012), http://www.google.com/realtime
[2] Twitter Search (2012), http://search.twitter.com
[3] Geer, D.: Is It Really Time for Real-Time Search? Computer 43(3), 16–19 (2010)
[4] Gurumurthy, S., et al.: Improving Web Search Relevance andFreshness with Content Previews. In: Proc. 19th ACM Int'l Conf. Information and Knowledge Management, CIKM (2010)
[5] Jansen, B.J., Campbell, G., Gregg, M.: Real Time Search UserBehavior. In: Proc. 28th ACM Conf. Human Factors in ComputingSystems, CHI (2010)
[6] Gurler, U., Ozkaya, B.Y.: Analysis of the (s, S) Policy forPerishables with a Random Shelf Life. IIE Trans. 40, 759–781 (2008)
[7] Cho, J., Garcia-Molina, H.: Synchronizing a Database toImprove Freshness. In: Proc. ACM SIGMOD Int'l Conf. Managementof Data (2000)
[8] Cho, J., Garcia-Molina, H.: Effective Page Refresh Policies forWeb Crawlers. ACM Trans. Database Systems 28(4), 390–426 (2003)
[9] Coffman Jr., E.G., Liu, Z., Webber, R.R.: Optimal RobotScheduling for Web Search Engines. J. Scheduling 1(1), 15–29 (1998)
[10] Edwards, J., McCurley, K., Tomlin, J.: An Adaptive Model ofOptimizing Performance of an Incremental Web Crawler. In: Proc. Ninth Int'l World Wide Web Conf., WWW (2000)
[11] Pandey, S., Olston, C.: User-Centric Web Crawling. In: Proc. 14th Int'l World Wide Web Conf., WWW (2005)

[12] Wolf, J.L., et al.: Optimal Crawling Strategies for Web SearchEngines. In: Proc. 11th Int'l World Wide Web Conf., WWW (2002)
[13] Chmielewski, D., Hu, G.: A Distributed Platform for Archiving and Retrieving RSS Feeds. In: Proc. Fourth ACIS Int'l Conf. Computer and Information Science, pp. 215–220 (2005)
[14] Sia, K.C., Cho, J., Cho, H.: Efficient Monitoring Algorithm forFast News Alerts. IEEE Trans. Knowledge and Data Eng. 19(7), 950–961 (2007)
[15] Fitzpatrick, B., et al.: PubSubHubbub Core 0.3 (2012), http://pubsubhubbub.googlecode.com/svn/trunk/pubsubhubbubcore-0.3.html
[16] Saint-Andre, P.: Extensible Messaging and Presence Protocol(XMPP): Core (2012), http://tools.ietf.org/html/draft-ietf-xmpp-3920bis-05
[17] Arasu, A., et al.: Searching the Web. ACM Trans. Internet Technology 1(1), 2–43 (2001)
[18] Heydon, A., Najork, M.: Mercator: A Scalable, Extensible WebCrawler. World Wide Web 2, 219–229 (1999)
[19] Pant, G., Srinivasan, P., Menczer, F.: Crawling the Web. In: Web Dynamics: Adapting to Change in Content, Size. Topology and Use. Springer (2004)
[20] Castillo, C., Nelli, A., Panconesi, A.: Crawling the Web WithLimited Memory. In: Proc. Web Intelligence Conf. (2006)

New Architecture for Flip Flops Using Quantum-Dot Cellular Automata

Paramartha Dutta[1] and Debarka Mukhopadhyay[2]

[1] Department of Computer and System Sciences, Visva Bharati University, Santiniketan, W.B., India
[2] Department of Computer Science and Engineering, Bengal Institute of Technology and Management, Santiniketan, W.B., India
{paramartha.dutta,debarka.mukhopadhyay}@gmail.com

Abstract. This article proposes a thorough design and analysis of Quantum-dot Cellular Automata (QCA) flip flops. QCA is an emerging technology which provides implementation of digital technology at the nano-scale level. These circuits have the advantages of higher switching speed, smaller size and less power consumption compared to CMOS circuit. Compared to the previous designs this circuits have less number of cells and less area. All the proposed designs are simulated using the QCADesigner and the analysis of the result proves the validity of the circuits.

Keywords: QCA, Flip Flop, Emerging technology, Nano-Technology.

1 Introduction

One of the interesting and promising fields of research has grown to be Quantum-Dot Cellular Automata [4, 5, 6] and it is an implementation of digital concept at the nano-scale level[8,9]. In QCA logic states are not determined by the voltage level but by the position of the electrons. The two electrons are there within each cell and are always located in the opposite corners owing to coulomb repulsion. The two different positions of each electron determine two different states. Now if we assign binary values to these two different positions of each electron, then positions of cell electrons correspond to single bit '0' or '1'(Fig. 1a). QCA inverter and MV gate are the two primitive gates in QCA logic. MV gate is equivalent to logic function F (A,B,C) = AB + BC + CA. Cell "out" is the output cell that is latched based majority polarization of the input cells (Fig. 1b).

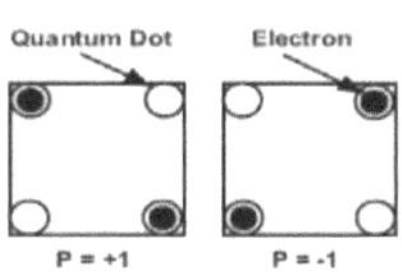

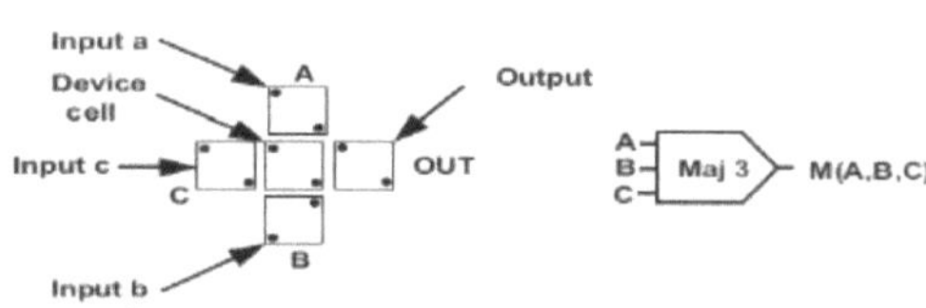

Fig. 1a. Two possible states of basic QCA cells **b.** MV gate

S.C. Satapathy et al. (eds.), *ICT and Critical Infrastructure: Proceedings of the 48th Annual Convention of CSI - Volume II*, Advances in Intelligent Systems and Computing 249,
DOI: 10.1007/978-3-319-03095-1_77,

Nowadays increase in power consumption has become a critical issue for modern VLSI circuits. Thus flip flops[7,10] has been proposed for low power CMOS technology[1,2] having advantage of low power consumption. In this article we are reporting new design of Quantum-Dot Cellular Automata Flip Flops. These architectures are aiming at building unique nanostructure circuit

2 QCA Review

In this section QCA wires, QCA gates and QCA clocking will be considered.

2.1 QCA Wires

This is an array of QCA cells. The electrostatic repulsion between neighboring cells electrons determine the polarization of the cells in the array. In QCA concept there are two types of wires: binary wire and inverter chain. All cells in a binary wire have the same polarization (Fig. 2a). In case of inverter chain polarization of cells change alternatively (Fig. 2b).

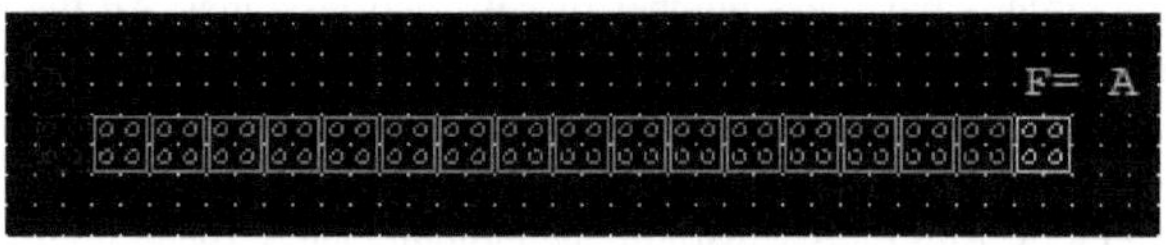

Fig. 2a. QCA binary wire

2.2 QCA Gates

Primary gates are known to be MV and NOT gates. MV gate with three inputs supports the logic function F (A, B, C) = AB + BC + CA. When one of the inputs is assigned to '1', this will be equivalent to an AND gate. When one of the inputs is set to '1', OR gate will be achieved. NOT gate is like conventional classical inverter gate.

2.3 QCA Clocking

CMOS clocking and QCA clocking has difference in concept. In CMOS, clocking is used for synchronization of sequential circuit. But in QCA clocking is used mainly for controlling the data flow and supply of power to the week signal which has been lost during the process of flow. QCA clock consists of four clocking phases, Switch, Hold, Release and Relax as in Fig. 3. During the first phase i.e. switch phase, the inter dot barrier is slowly and linearly raised and electrons are pushed to the corner dots following the influence of its neighbors. In hold phase the inter dot barrier is very high and cells retains polarity and acts as inputs to the neighboring cells.

During release phase the barrier is slowly lowered and the electrons slowly start to be delocalized. During final phase or relax phase the electrons are completely delocalized and lose its polarization

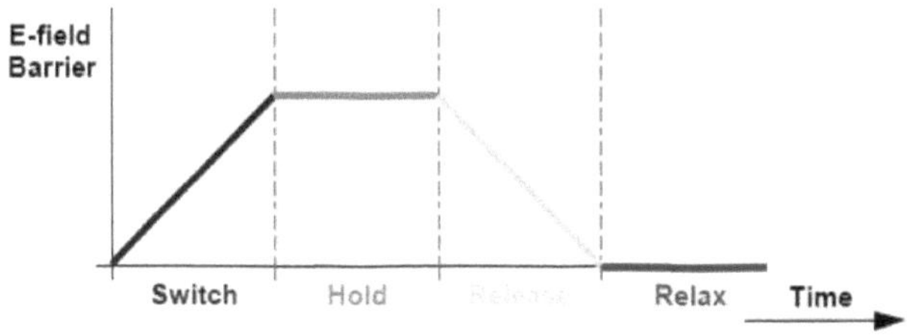

Fig. 3. Inter-Dot barrier potential during four clock phases

2.4 QCA Designer Simulator

For the proposed circuit layout and functionality checking, a simulation tool for QCA circuits QCADesigner [3] version 2.0.3, is used. The following parameters are used for a bitable approximation: cell size=18 nm, clock high = 9.800000e-022 J, clock low=3.800000e-023 J, Dot diameter= 5 nm. Most of the mentioned parameters are default values in QCADesigner.

3 Flip Flops and the Proposed Architecture

In this section SR-FF, JK-FF, D-FF and T-FF should be discussed.

3.1 SR-Flip Flop

Memory use flip flops as the basic elements. As yet no static memory has been developed in QCA; loops are created to keep memory in motion. The SR-Flip Flop is constructed from 18 cells. If there is need of Q output only, 15 cells are enough to construct the flip flop.

S-R Flip Flop is a sequential circuit that maintains a stable output even after the inputs are turned off. This simple Flip Flop has a Set(S) input and a Reset(R) input. This design has 18 cells which is much less figure than any of the previous designs [10].

Result Analysis : This QCA S-R Flip Flop holds the previous output when both inputs are set. So this design provides no ambiguous results. The output is displayed at phase 2 of first clock.

3.2 J-K Flip Flop

The J-K Flip Flop is the most versatile of the basic Flip Flop. The R-S Flip Flop is connected with two AND gates at the S and R line interface. Feedback from Qbar is connected with J and Feedback from Q is connected with K. This design consists of 54 cells, which is much less than any of the previous designs [10].

Result Analysis: The result in Fig. 5b is matching with the conventional classical Flip Flop. The Result is displayed at '0' phase of 2nd clock.

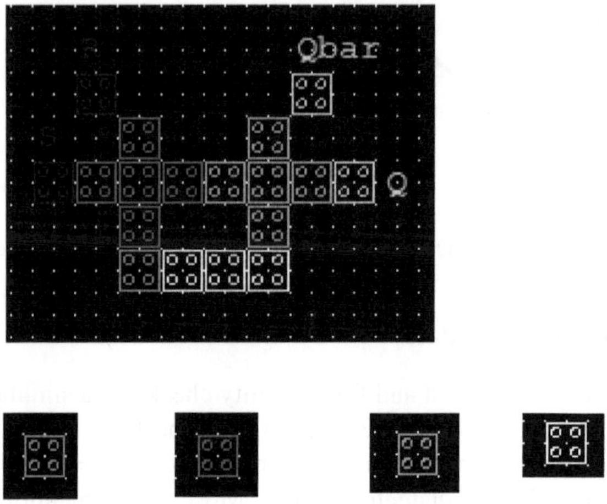

'0' Phase '1' Phase '2' Phase '3' Phase

Fig. 4a. Design of SR Flip flop

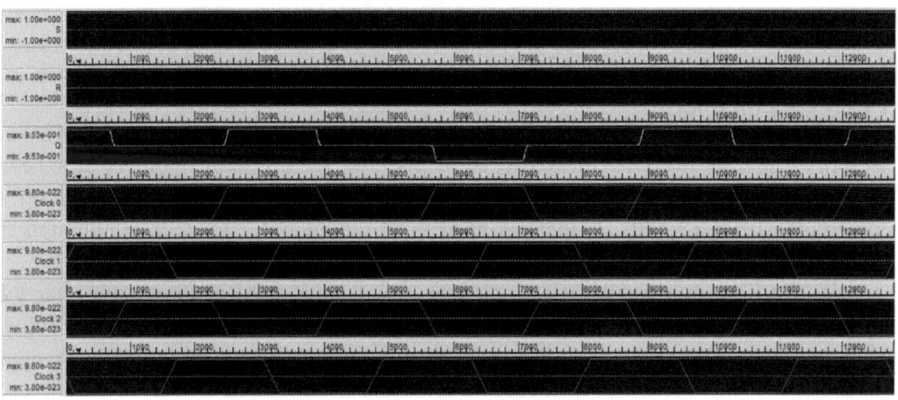

Fig. 4b. Simulation Result for SR- Flip Flop

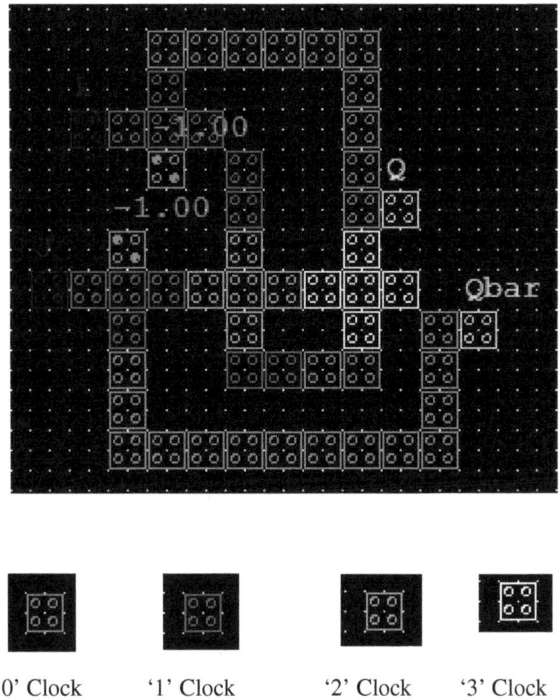

Fig. 5a. Design of J-K Flip Flop

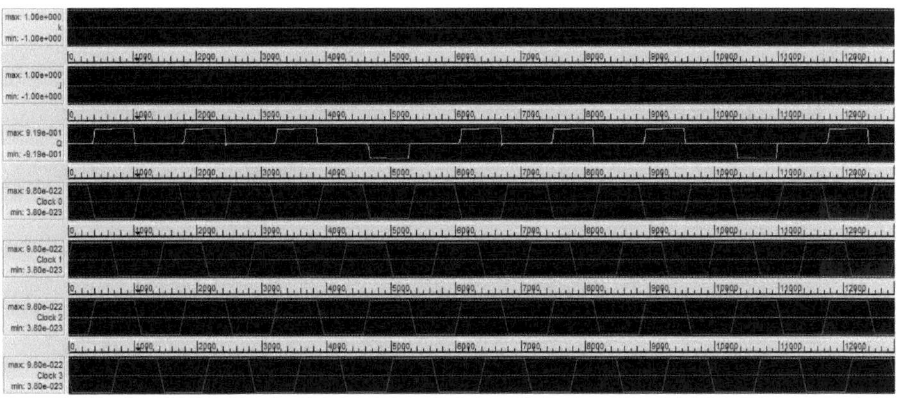

Fig. 5b. Simulation result of J-K Flip Flop

3.3 D-Flip Flop

The D Flip Flop is constructed by connecting the two inputs of S-R Flip Flop through a NOT gate. This proposed design is consisting of 21 cells which is less than any of the previous designs [10].

Result Analysis: The output is matching with the conventional classical D-Flip Flop. The output is displayed at Phase 2 .of the first clock cycle.

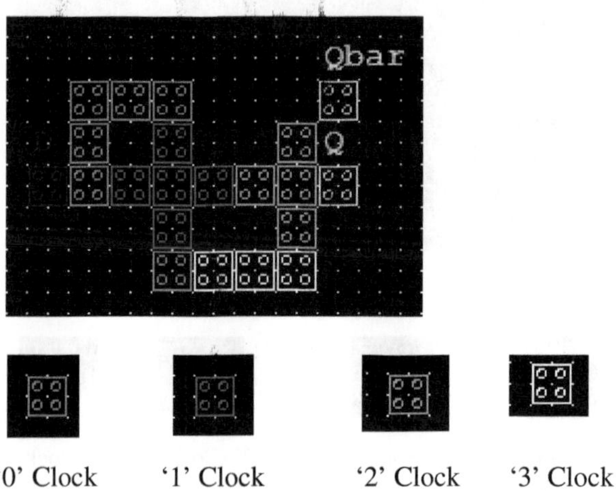

Fig. 6a. Design of D Flip Flop

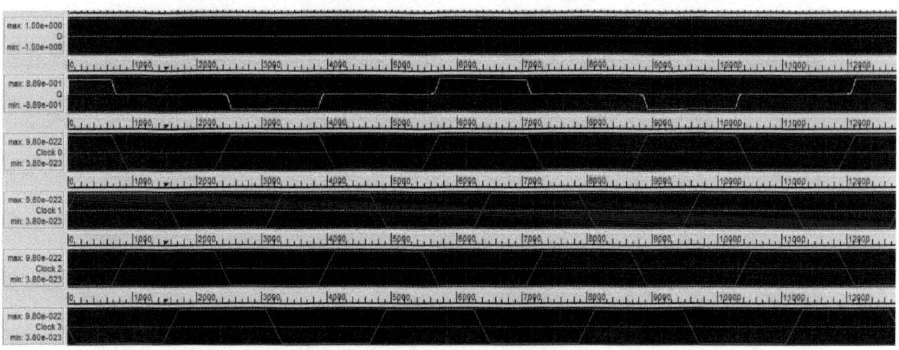

Fig. 6b. Simulation result of D Flip Flop

3.4 T Flip Flop

The T Flip Flop is constructed with shorting the J and K end of J-K Flip Flop in Fig.5a. This design has 58 cells which is much less than it's nearest competitor[10].

Result Analysis : The result is matching with the conventional classical T Flip Flop. It is available at the '0' phase of the 2nd clock in Fig. 7b.

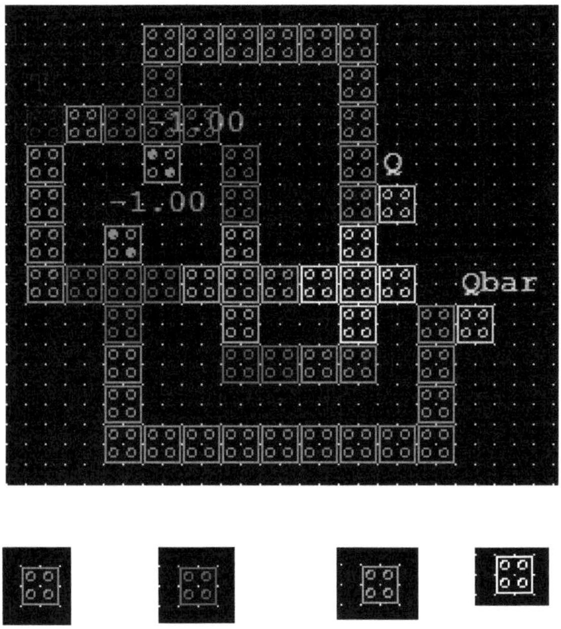

'0' Clock '1' Clock '2' Clock '3' Clock

Fig. 7a. Design of T Flip Flop

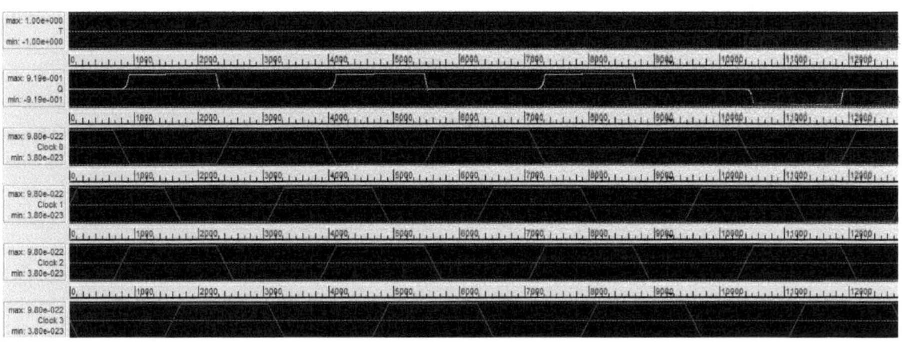

Fig. 7b. Simulation result of T Flip Flop

4 Future Work

All these designs are much more reduced than its only predecessor available on the web [10]. In the future work these Flip flops can be utilized for efficient designs of register, counter, memory etc.

5 Conclusion

In this article we have described the QCA computing paradigm and proposed layout of all the flip-flops based on this emerging technology. The proposed designs are clearly smaller than all the previously proposed designs. Using the unique features of QCA we are able to design all the flip flops within a single layer eliminating the requirement of complex interconnects found in CMOS circuits.

Acknowledgements. The authors acknowledge the support from Bengal Institute of Technology and Management, Santiniketan and Visva-Bharati University, Santiniketan.

References

1. Vetteth, A., Walus, K., Jullien, G.A., Dimitrov, V.S.: RAM Design Using Quantum-Dot Cellular Automata. In: 2003 Nanotechnology Conference and Trade Show (2003)
2. Timler, J., Lent, C.S.: Power gain and dissipation in quantum-dot cellular automata. J. Appl. Phys.,American Institute of Physics 91(2), 823 (2002)
3. Walus, K.: QCADesigner Homepage. ATIPS Laboratory, University of Calgary, Calgary, AB (2002), `http://www.atips.ca/~walus`
4. Mukhopadhyay, D., Dutta, P.: Quantum Cellular Automata based Novel Unit 2:1 Multiplexer. International Journal of Computer Applications 43(2), 22–25 (2012)
5. Mukhopadhyay, D., Dinda, S., Dutta, P.: International Journal of Computer Applications 25, 21–24 (2011)
6. Mukhopadhyay, D., Dutta, P.: QCA Based Novel Unit Reversible Multiplexer. Adv. Sci. Lett. 5, 1–6 (2012)
7. Yang, X., Cai, L., Zhao, X.: Low power dual-edge triggered flip flop structure in quantum dot cellular automata. Electronic Letters 46(12) (2010)
8. Oya, T., Asai, T., Fukui, T., Amemiya, Y.: IEEE Trans. Nanotechnol. 2, 15 (2003)
9. Orlov, A.O., Amlani, I., Bernstein, G.H., Lent, C.S., Snider, G.L.: Science 277, 928 (1997)
10. Vetteth, A., Walus, K., Dimitrov, V.S., Jullien, G.A.: Quantum- Dot Cellulur Automata of Flip Flop. ATIPS Laboratory, 2500 University Drive, N.W., Calgory, Alberta, Canada, T2N 1N4

Emerging ICT Tools for Virtual Supply Chain Management: Evidences from Progressive Companies

Prashant R. Nair

Vice-Chairman – IT, Amrita School of Engineering, Amrita Univesity, Amrita Nagar, Coimbatore, 641112 India
prashant@amrita.edu

Abstract. Industry trends like liberalization, privatization and globalization have resulted in increasing competition & pricing pressure and thereby compelled enterprises to adopt robust supply chain management technologies and policies. To survive in these times, companies will find that their conventional supply chain integration will have to be expanded beyond their boundaries so as to integrate all stakeholders. Adoption of ICT tools is vital for such efforts. ICT tools are great enablers, enhancers, levelers and facilitators of enterprise operations. This paper discusses the role of emerging ICT tools like Cloud Computing, RFID and Decision Support Systems (DSS) as an enabler in virtual supply chain management and also highlights the adoption of these technologies in various contexts. Successful ICT implementations of SCM of progressive companies are also described.

Keywords: Supply Chain Management (SCM), RFID, Information and Communication Technology (ICT), Decision Support Systems (DSS), Cloud Computing, virtual.

1 Introduction

Rapid technology advances and dynamic market forces have altered the business landscape as also fundamental altered existing business models. Information and Communication Technology (ICT) usage has opened the doors for companies to compete in any marketplace. Even companies, which have been benefiting from protectionist policies by governments, are now exposed to the perils of increased competition due to liberalization, privatization and globalization. Accessing information in a timely and actionable manner as also negotiating and managing relationships within and between organizations has become a source of competitive advantage for businesses.

ICT tools are great enablers, enhancers, levelers and facilitators of enterprise operations. Access to information is critical to expanding social and economic opportunities in any context.

S.C. Satapathy et al. (eds.), *ICT and Critical Infrastructure: Proceedings of the 48th Annual Convention of CSI - Volume II*, Advances in Intelligent Systems and Computing 249,
DOI: 10.1007/978-3-319-03095-1_78, © Springer International Publishing Switzerland 2014

1.1 ICT Tools for SCM

ICT adoption and usage across the supply chain has become a performance enabler and determinant of competitive advantage for many corporations. Extensive usages of systems and tools like ERP, Supply Chain Management (SCM) Software packages and solutions, EDI, bar-codes, inventory management systems, warehouse management systems and transportation & fleet management systems have resulted in better agility, collaboration, coordination, decision making, transaction processing and visibility both within and among enterprises.

ICT adoption and usage are critical to improving a firm's performance irrespective of the context in which the ICT tools have been placed. A study on Electronic Data Interchange (EDI) usage by 193 suppliers on Just in Time (JIT) shipments in the automobile industry in 1994 yielded good results. EDI usage sharply reduced the shipment errors and resulted in considerable saving for the enterprises which adopted this technology [1]. The performance measurement of a supply chain depends upon the efficiency and effectiveness of the internal processes of the chain. ICT enhances collaboration and communication between various processes, links, partners and stakeholders in the supply chain. Successful supply chain management relies on visibility, transparency and unfettered access to information, which can be made possible by effective integration of information resources.

Effective implementations of Supply Chain Management can be attained in multiple ways using ICT tools. For example, bar coding and RFID usage inventory tracking and control, fleet management systems to optimize truck routing, Web services and EDI for communication with supply chain partners in the extranet and MRP and ERP solutions to integrate various links in the supply chain [2].

1.2 Benefits of Using ICT Tools

Cisco had reported savings of $500 million by reengineering its internal operations and process integration with customers and suppliers with the help of web-services. Wal-Mart shares point of sale (POS) information from its many retail outlets directly with Proctor & Gamble (P&G) and other major suppliers. This information sharing has resulted in a win-win situation for both companies [3]. Intel was able to replace hundreds of their order clerks using online ordering applications. Activities like planning and forecasting, sourcing and procurement, logistics and service and spare parts management are the first to move to the cloud [4].

The tremendous potential of e-business applications like e-procurement and e-commerce can help companies respond quickly to customer demand as also help in efficient procurement of raw materials. Storage, Warehousing and transportation operations can also be effectively managed and optimized.

2 Emerging ICT Tools for SCM

Fundamental changes have occurred in today's global economy. These changes alter the relationship that we have with our stakeholders, our customers, our suppliers, our channel partners, and our internal operations. Emerging ICT tools like software agents, RFID, web services, virtual supply chains, electronic commerce, cloud computing and decision support systems are being deployed to aid various operations for supply chain planning and execution.

Application areas in the supply chain domain with tangible benefits of some emerging ICT tools like RFID, Decision Support Systems (DSS) and Cloud computing are studied with focus on existing configurations, available applications, and deployments in progressive companies. ICT as an enabler of Virtual SCM is highlighted by addressing enterprise solutions in a variety of supply chain settings. The rapid adoption of the Internet for communication with all stakeholders seems to reflect the potential of the new-age communication media. It has also been observed that several progressive companies are extensively using emerging tools like virtual supply chains, web services, RFID, and electronic commerce to shore up their supply chain operations. However, usage of tools like software agents, and decision support systems for supply chain management is limited [5].

Cloud computing and associated technologies like virtualization, software as a service etc is touted as the next 'big' thing and game changer for enterprises. Companies have started using Software as a Service (SAAS) application for managing their supply chains, which are part of larger frameworks called public clouds. A recent survey conducted by E2open in collaboration with SCM World, found that cloud led to significant improvements in metrics such as inventory days [6].

3 Cloud Application Areas in SCM

3.1 Demand Forecasting

Cloud platforms are being used for demand forecasting by coupling and coordinating various links in the supply chain like retailers, suppliers and distributors. Cloud-based tools are available for capturing and analyzing sales data and executing statistical demand forecasts [4].

3.2 Demand Planning

Order and Demand planning can be facilitating the cloud network with accurate forecasts.

3.3 E-procurement

Cloud platforms are inherently collaborative in nature. These tools can negotiate through an array of suppliers and get the best e-procurement results. Companies will be able to choose the best suppliers as per their needs and specification. Moreover, cloud-based tools enable companies and suppliers to mutually develop contracts and thereby drastically improving contract management [4].

3.4 Inventory, Warehouse and Transportation Management

Cloud computing tools are also available for inventory, warehouse and transportation management. HighJump Software is an example of a warehouse management system cloud provider which takes care of tracking shipments and inventory for the warehouse operations. Precisio Business Solutions has developed inventory management software on the cloud using salesforce.com platform. Likewise cloud apps for transportation and fleet management are also available.

4 RFID Application Areas in SCM

4.1 Inventory Tracking and Management

The most popular and widespread RFID application area in SCM is inventory management and tracking. RFID is initially used to manage and track the identification of large lots of goods at the unit, pallet, case and carton levels. Accurate information about inventory availability and speed of transportation also adds to supply chain performance [7].

4.2 Vendor Managed Inventory (VMI)

Retailers can reduce their out-of-stocks, reduce labor and warehouse costs. The supplier has total control with access to Point of Sale and Inventory information of the retailer. Inventory management can thereby be properly regulated at the supplier end. This would reduce inventory holding cost, reduce out-of-stocks and subsequent loss of sales. Data obtained from RFID can eliminate inaccuracies in data due to human error or absence of data [8].

4.3 Customer Relationship Management

RFID usage in SCM aids the engagement with customers in many ways. Retailers are able to reduce their out-of-stocks using RFID. Consumer behavior studies have shown us the adverse impact on retail establishments not having adequate stocks. Both supplier and retailer are synchronized at all times with regard to the dynamics of demand and supply. Customer order fulfillment can be realized in a better fashion as the chances of sending orders to wrong destinations is minimized due to the RFID usage [9]. Such process changes will reduce the cost of operations and also lead to reduced labor. Suppliers are able to handle product recalls and return of faulty and defective items in an effective manner using RFID.

4.4 Production and Manufacturing Workflow

RFID can be used in the production and manufacturing workflow & process automation in the production line. This will increase the velocity and visibility of products in the supply chain as also help to achieve the zero inventory paradigm. Tags can also monitor things like pilferage, tamper and environmental parameters like temperature & bacterial levels.

5 DSS Application Areas in SCM

5.1 Demand Planning

Demand forecasts play an important role in many supply chain decisions developing accurate forecasts is critical for the efficiency of the entire supply chain. Forecasts are made using various statistical techniques that take into account the history of the item, stability of demand, and other product-specific data.

There are two processes here [10]:

- Demand forecast: a process in which historical demand data are used to develop long-term estimates of expected demand
- Demand shaping: a process in which the firm determines the impact of various marketing plans such as promotion, pricing discounts, rebates, new product introduction, and product withdrawal on demand forecasts

5.2 Inventory Management and Deployment

An inventory management DSS uses transportation and holding cost information, along with lead times and projected demand, to propose inventory policies that help the decision maker achieve some combination of low cost and high customer service. The objective is to use transportation and inventory holding costs, demand forecasts and forecast error, and service levels to determine the levels of inventory, in particular, safety stock levels, to keep in each location in each period [10]. Even when the firm does not wish to modify its logistics network, decisions must be made about what inventory to keep in which warehouses and at what times. This is the inventory deployment decision. Here, transportation costs, demand forecasts, and inventory holding are used to determine the levels of inventory to keep in each location in each period. DSS may use optimal or heuristic algorithms to generate suggested policies.

5.3 Material and Distribution Resource Planning

Material Resource Planning systems use a bill of materials, inventory positions and lead times to plan when manufacturing of a particular product should begin. DSS can serve as a good example of why the decision maker should use only the output of a DSS as a possible problem solution. Often MRP systems propose impossible schedules because they typically do not take production capacities into account. It is up to the decision maker to modify the plan in such a way that it becomes a feasible schedule without becoming too expensive.

Optimal routes and inventory polices for a set of warehouses and retailers can also be determined. Given warehouse and retailer locations, transportation costs, and demand forecasts for each retail outlet, these DSS utilize analytical techniques to determine policies that will achieve high levels of customer service at minimal cost [5].

5.4 Fleet Planning

Fleet planning typically involves not only the dispatching of a company's own fleet but also decisions regarding selection of a commercial carrier on certain routes. Since rate structures can often be very complex, and speed and reliability may a difficult problem. In addition, input data such as rate structures need to be frequently updated.

5.5 Workforce and Production Scheduling

Given a series of products to make, information about their production processes, and due dated for the product, production scheduling DSS propose manufacturing sequences and schedules. A production scheduling DSS can use artificial intelligence and mathematical and simulation techniques to develop schedules.

Given production (service) schedule, information on labour costs, and a set of work rules, a workforce scheduling DSS proposes a number of possible employee schedule to ensure that the necessary labour is available at all times and at the lowest possible cost. Often these systems have to take complicated union rules into account [5].

6 Successful Enterprise Deployments

- Data from a shipping bill from Sony manufacturing plant in Penang (Malaysia) and near real-time location information from GE VeriWise automatic identification system used by the logistics provider, Schneider Corporation, for track and trace purposes, may offer an estimated arrival time for the goods from Sony to its US distributor Merisol after US customs clearance in Long Beach, California. This information may generate an alert or advice to the retail store that the Sony Playstation Portable 3 (PSP3) devices may not be on the shelves of the Circuit City store in Watertown, Massachusetts for the fourth of July sales event.
- Baan, a leading ERP vendor unveiled an application, Baan Enterprise Decision Manager for aiding corporate decision-making. Major retailers like Walmart, Sara Lee and Roebuck have increasingly started using Collaborative Forecasting and Replenishment (CFAR) which uses DSS for jointly developing forecasts. GAF Materials Corp, the largest manufacturer of asphalt-based roofing materials in the US, uses a freight-management DSS [11].
- Automobile major, Mahindra and Mahindra is another early adopter, using RFID for scheduling and logistics management [12].

- P&G estimates the cost saving of up to $200 million in inventory carrying costs with its RFID implementation [8]
- Amcor uses RFID for managing its warehouses [7]
- The US army is working on RFID tags with sensors to monitor temperature in areas where there is massive transfer of goods and services [7]
- Frito-Lay, Inc uses DSS for Price, advertising, and promotion selection
- Burlington Coat Factory uses DSS for Store location and inventory mix
- Keycorp uses DSS for targeting direct mail marketing customers
- National Gypsum uses DSS for corporate planning and forecasting
- Texas Oil and Gas Corporation uses DSS for evaluation of potential drilling sites
- United Airlines uses DSS for Flight scheduling, passenger demand forecasting [5]
- Vendors like IBM, Mercury Gate, JDA, Amber Road, Deposco, eBIZnet, and Ariba, are offering public cloud deployment models for various supply chain operations and activities
- FedEx has a private cloud deployed in 2011 with CloudX [13] as service-provider. Sales processing and customer relationship management are the activities on the cloud.
- COSCO Logistics, the largest 3PL company of China and the world's second largest ocean shipping company is using SaaS service and integrating all stakeholders like customers, subsidiaries and distributors in order all of them to use the same logistics management software
- In India, Retail giants like ITC, Wills Lifestyle, Madura Garments, Big Bazaar, and Total Mall have already implemented RFID at their retail outlets for product tracking and warehouse management [9]. As such there are no evidences of Indian companies and enterprises using Cloud Computing or Decision Support Systems for SCM.
- In India, Future Group, which manages the retail establishments like Pantaloon Retail, Big Bazaar and Food Bazaar have tied up with Cisco for their RFID deployment. This investment is worth more than Rs. 200 crores and is dubbed as one of the largest in Asia. It is expected that there would be more than 30 million scans per day.

7 Conclusion

ICT adoption and usage across the supply chain has become a determinant of competitive advantage for many corporations. Extensive usages of systems and tools have resulted in better agility, collaboration, coordination, decision making, transaction processing and visibility both within and among enterprises. ICT interventions in supply chain planning and execution using Cloud Computing, RFID and Decision Support Systems with evidences of deployment of several progressive companies has been showcased. The role of these technologies as an enabler of virtual supply chain management and its adoption in various contexts is highlighted. Considering the fact that several enterprises are adopting e-commerce strategies and slowly transitioning to an e-business, virtual supply chain management with the usage of these ICT tools will only aid in process re-alignment to the e-business framework.

References

1. Srinivasan, K., Kekre, S., Mukhopadhyay, T.: Impact of Electronic Data Interchange Technology on JIT Shipments. Mgmt. Sci. 40, 1291–1304 (1994)
2. Nair, P.R., Raju, V., Anbuudayashankar, S.P.: Overview of Information Technology Tools for Supply Chain Management. CSI Comm. 33(9), 20–27 (2009)
3. Anderson, D.L., Britt, F.E., Favre, D.J.: The Seven Principles of Supply Chain Management. Supply Chain Mgmt Rev., 31–41 (Spring 1997)
4. Schramm, T., Nogueira, S., Jones, D.: Cloud Computing and Supply Chain: A Natural Fit for the Future: Logistics Mgmt. 3, 9–11 (2011)
5. Nair, P.R., Balasubramaniam, O.A.: IT Enabled Supply Chain Management using Decision Support Systems. CSI Comm. 34(2), 34–40 (2010)
6. Jha, V.: Impact of Cloud Computing on Supply Chain Management. IIM Indore Mgmt. Canvas (2013)
7. Michael, K., McCathie, L.: The Pros and Cons of RFID in Supply Chain Management. In: International Conference on Mobile Business, pp. 623–629 (2005)
8. Sabbaghi, A., Vaidyanathan, G.: Effectiveness and Efficiency of RFID Technology in SCM - Strategic Values and Challenges. J. of Theo and Appl Electron Comm Res. 3(2), 71–81 (2008)
9. Nair, P.R.: RFID for Supply Chain Management. CSI Comm. 36(8), 14–18 (2012)
10. Simchi-levi, D., Kaminsky, P., Simchi-levi, E.: Managing the Supply Chain: The Definitive Guide for the Business Professional. McGraw Hill, Columbus (2003)
11. Lee, C., Lee, K.C., Han, J.H.: A Web-based Decision Support System for Logistics Decision-Making. Proc of the Acad. of Info. and Mgmt. Sci. 3(1) (1999)
12. Channel World Information, `http://www.channelworld.in/specialreports/index.jsp/artId=5013515`
13. Information Technology Research Institute Information, `http://rfid.uark.edu/research-papers.asp`

ICT to Renovate the Present Life Line Systems from Fossil Fuels to Green Energy

Yashodhara Manduva and K. Rajasekhara Rao

KL University
yashukishan@yahoo.com,
rajasekhar.kurra@klce.ac.in

Abstract. Most of our present life-line systems such as cooking, Domestic electricity supply, fuels for transportation etc. are based on the fossil fuels. Coal, oil and natural gas are the three different forms of fossil fuels that are widely used. Large-scale use of fossil fuels started since the Industrial Revolution. Today, these are the most cheap sources of energy available for the use of both personal as well as commercial purposes.

- **Petroleum & Natural Gas are used to fuel our vehicles.**
- **Natural gas & Fire-wood are used for cooking**
- **Coal & Natural Gas are used to produce Electricity**

In today's climate of growing energy needs and increasing environmental concern, alternatives to the use of non-renewable and polluting fossil fuels have to be investigated. Increase in usage of solar power based technologies results in….

- **Rreducing green house gas emissions**
- **Reducing dependency on exhaustible natural resources**
- **Eenergy saving**
- **Better bright lighting at low prices.**

Keywords: Fossil Fuels, Green House gasses, Renewable & Non-Renewable Energy sources, PWMs, WLED Lighting, MPPT Techniques.

1 Introduction

1.1 Usage of Fossil Fuels, and Drastic Effects on Environment

Green House Gases—About 300 years ago, humans began to burn coal and oil to produce energy and goods. By doing every activities like cooking, turning on the lights, we are taking millions of years of worth of carbon, stored beneath the earth as fossil fuels and releasing it into the air.

At the same time we are changing the way we use our land, i.e cutting down trees and tilling our farmland, which also adds co_2 to our atmosphere. It is the major green house gas that is causing global warming. and global warming is causing climate change.

S.C. Satapathy et al. (eds.), *ICT and Critical Infrastructure: Proceedings of the 48th Annual Convention of CSI - Volume II*, Advances in Intelligent Systems and Computing 249,
DOI: 10.1007/978-3-319-03095-1_79, © Springer International Publishing Switzerland 2014

Fig. 1. Impact of Global Warming & Climate Change

1.2 Climate Change Impacts

Rise of co_2 in our atmosphere is having an effect much faster and more severely than scientists once predicted. a few examples of impacts we are already seeing:

- Glaciers are melting
- See levels are rising
- Oceans are acidifying
- Weather is more severe, i.e Hurricanes, Typhoons, and droughts are becoming more frequent, harsher and unpredictable.

2 Need to Be Changed from Non-renewables to Renewables

Fossil fuels are non-renewable, they draw on finite resources that will eventually dwindle, becoming too expensive, also too environmentally damaging to retrieve.

In contrast, **Renewable energy** is energy that comes from natural resources such as **sunlight, wind, rain, tides, waves** and **geothermal heat, bio-mass**, etc., which are renewable because they are naturally replenished at a constant rate, also environmentally friendly. Use of solar cookers and biogas for cooking, Bio-mass to produce fuel, Solar Panels to produce electricity must be encouraged.

Sunlight, or **solar energy**, can be used directly for heating and lighting homes and other buildings, for generating electricity, and for hot water, solar cooling, and a variety of commercial and industrial uses.

Other technologies also exist to extract sun light, such as sterling engine dishes which use a Stirling cycle engine to power a generator.

3 Aim and Objective

Renovation of present life-line systems from fossil fuels to Renewable Energy & Increase in usage of renewable sources of energy and also Increase in usage of solar power based technologies.

ICT is being considered as" survival science" Environmentalists and academicians feel that ICT can be utilized to provide renewable energy technologies,.To provide high efficiency lighting to communities that do not have access to appropriate and affordable energy solutions.

- Increase in Implementing and usage of WLED lighting,
- Increase in use of modern technologies like microcontrollers PWMs, MPPTs in solar power based applications, benefits in power optimization.

4 New Trends in Alternative Energy

4.1 Artificial Photo-Synthesis Using……

- Hydrogen catalysts
  - Water-oxidizing catalysts
  - Photosensitizers
  - Carbon dioxide reduction catalysts
  - Bio-Fuel using Bio-mass

4.2 Light-Driven Methodologies under Development

- Photo electrochemical cells
- Photo catalytic water splitting in homogeneous systems
- Hydrogen-producing artificial systems
- NADP+/NADPH coenzyme-inspired catalyst
- Photo biological production of fuels by manipulation of photosynthetic microorganisms such as
  - **Micro-algae** and cyanobacteria
  - **Algae biofuels** such as **butanol** and **methanol**

Artificial Photosynthesis–Artificial photosynthesis is a chemical process that replicates the natural process of photosynthesis, a process that converts sunlight, water and carbon dioxide into carbohydrates and oxygen. The term is commonly used to refer to any scheme for capturing and storing the energy from sunlight in the chemical bonds of a fuel (solar fuel).

Photo Catalytic Water Splitting converts water into protons (and eventually hydrogen) and oxygen, and is a main research area in artificial photosynthesis.

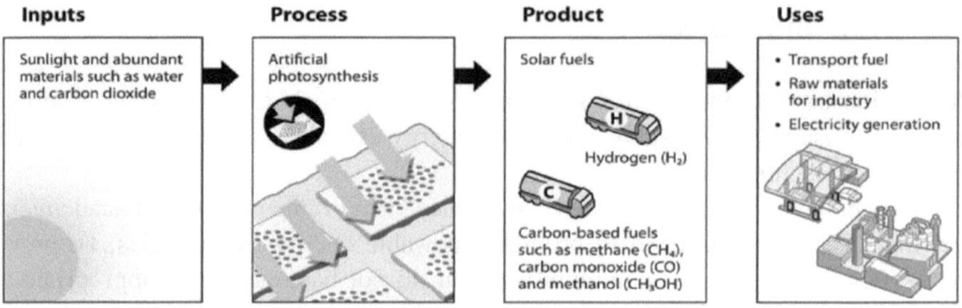

Fig. 2. Artificial Photosynthesis pathway from sunlight to fuel

4.3 Photosynthesis

During natural photosynthesis the energy of excited electrons, obtained from photons, generates a reducing power in the form of pyridine nucleotide cofactor, NAD(P)H. This photo chemically regenerated NAD(P)H is consumed by redox enzymes during the reduction of CO_2 to organic compounds.

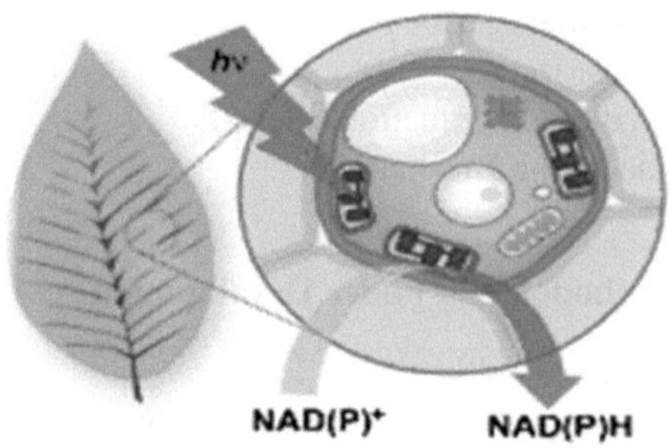

Fig. 3. The Process of Photosynthesis

5 Sun, The Source of All Kinds of Renewable-Energies

5.1 Most Renewable Energy Comes Either Directly or Indirectly from the Sun

Hydro-Electric Power—The sun's heat drives the winds, whose energy is captured with wind turbines. Then, the winds and the sun's heat cause water to evaporate, When this water vapour turns into rain or snow and flows downhill into rivers or streams, its energy can be captured using Hydroelectricity.

Bio-Energy—Along with the rain and snow, sunlight causes plants to grow. The organic matter that makes up those plants is known as biomass. Biomass can be used to produce electricity, transportation fuels, or chemicals. The use of biomass for any of these purposes is called Bioenergy.

Ocean_Energy—driven by both the tides and the winds, also The sun warms the surface of the ocean more than the ocean depths, creating a temperature difference, that can be used as an energy source. All these forms of ocean energy can be used to produce electricity.

5.2 Total World Energy Consumption

About 16% of global energy consumption comes from renewables.

With 10% coming from traditional biomass, which is mainly used for heating. and 3.4% from hydroelectricity. New renewables (small hydro, modern biomass, wind, solar, geothermal, and biofuels) accounted for another 3% and are growing very rapidly.

The share of renewables in electricity generation is around 19%, with 16% of global electricity coming from hydroelectricity and 3% from new renewables.

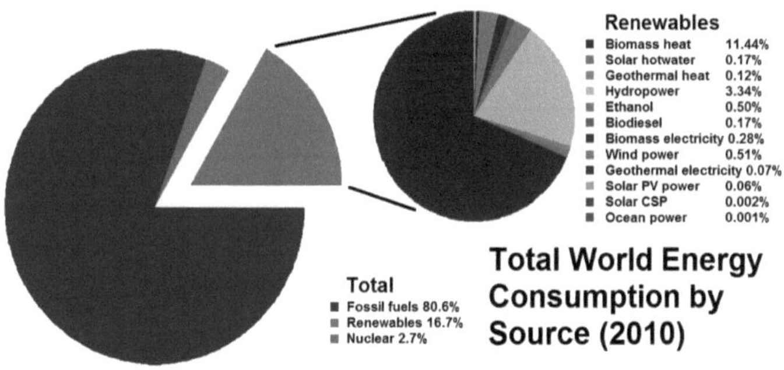

Fig. 4. Global Energy Consumption by Source

6 Cost of Electricity Production by Source

Table 1. Cost of Electricity Production by Various Sources

Plant Type	**Capacity Factor** (%)	Valise Capital Cost	Fixed O&M	Variable O & M	Transmission Investment	Total System Levelized Cost
Conventi-onal Coal	85	65.8	4.0	28.6	1.2	99.6
Advance-d Coal	85	75.2	6.6	29.2	1.2	112.2
Advanced coal with ccs	85	98.3	9.3	36.8	1.2	140.7
Natural Gas Fired:---						
Conventional combined cycle	87	17.5	1.9	48.0	1.2	68.6
Advanced combined cycle	87	17.9	1.9	44.0	1.2	65.5

Table 1. *(continued)*

Advanced CC with CCS	87	34.9	4.0	52.7	1.2	92.8
Conventional combustion turbine	30	46.0	2.7	79.9	3.6	132.0
Advanced combustion turbine	30	31.7	2.6	67.5	3.6	105.3
Advanced Nuclear	90	88.8	11.3	11.6	1.1	112.7
Geothermal	92	76.6	11.9	9.6	1.5	99.6
Biomass	83	56.8	13.8	48.3	1.3	120.2
Wind	34	83.3	9.7	0,0	3.7	96.8
Wind-Offshore	27	300.6	22.4	0.0	7.7	330.6
Solar PV	25	144.9	7.7	0.0	4.2	156.9
Solar Thermal	20	204.7	40.1	0.0	6.2	251.0
Hydro	50	76.9	4.0	6.0	2.1	89.9

6.1 Solar Power Generation (AC & DC Applications)

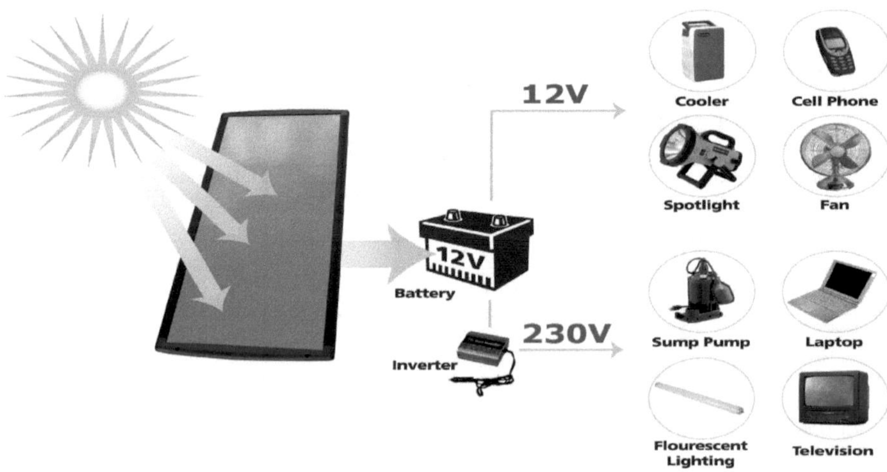

Fig. 5. Various Electrical Appliances based on solar energy

6.2 Economics of Solar-Power

Despite the overwhelming availability of solar power, prior to 2012, less was installed, compared to other power generation, due to the high installation costs. This cost has declined as more systems have been installed, As of 2011, the cost of PV has fallen well below that of nuclear power and is set to fall further.

Photovoltaic systems use no fuel and modules typically last 25 to 40 years. Installation cost is measured in \$/watt or €/watt. The electricity generated is sold for ¢/kWh. 1 watt of installed photovoltaic generates roughly 1 to 2 kWh/year, as a result of the local insolation..

The International Conference on Solar Photovoltaic Investments, organized by EPIA. , has estimated that PV systems will pay back their investors in 8 to 12 years. As a result, since 2006 it has been economical for investors to install photovoltaic for free in return for a long term power purchase agreement. Fifty percept of commercial systems were installed in this manner in 2007 and over 90% by 2009.

The cost of installation is almost the only cost, as there is very little maintenance required. **Hence solar power is more sustainable and economic.**

7 The Principle of Solar Energy

7.1 Solar Power

Solar Power is the conversion of sunlight into electricity, either directly using

- **Photo-voltaics (PV), or**
- **Indirectly using Concentrated Solar Power (CSP).**

Concentrated solar power systems use lenses or mirrors and tracking systems to focus a large area of sunlight into a small beam. The concentrated heat is then used as a heat source for a conventional power plant

Photovoltaic convert light into electric current using the **Photoelectric effect.** A **solar cel**l, or photovoltaic cell (PV), is a device that converts light into electric current using the **Photoelectric effect.**

In the **Photoelectric effect**, electrons are emitted from matter (metals and non-metallic solids, liquids, or gases) as a consequence of their absorption of energy from electromagnetic radiation of very short wavelength and high frequency, such as Ultraviolet radiation. Electrons emitted in this manner may be referred to as photoelectrons.

7.2 Solar Power Generation

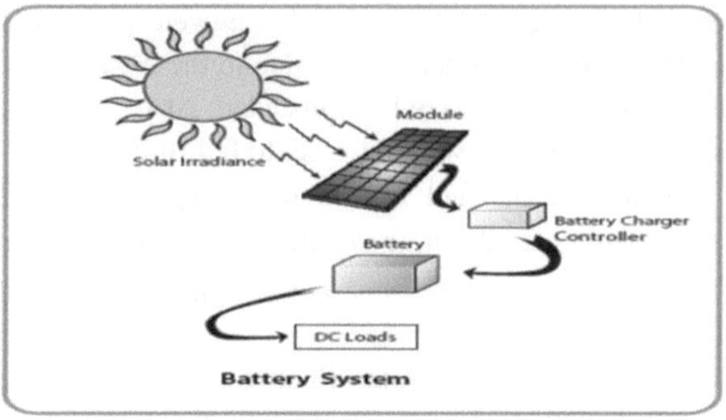

Fig. 6. Solar Power Generation

7.3 Solar Power Generation System

Solar Cells produce direct current (DC) power which fluctuates with the sunlight's intensity. For practical use this usually requires conversion to certain desired voltages or alternating current (AC), through the use of inverters.

Multiple solar cells are connected inside modules. Modules are wired together to form arrays, then tied to an inverter, which produces power at the desired voltage, and for AC, the desired frequency/phase.

7.4 Solar Power Generation Connected Back to Power Grid

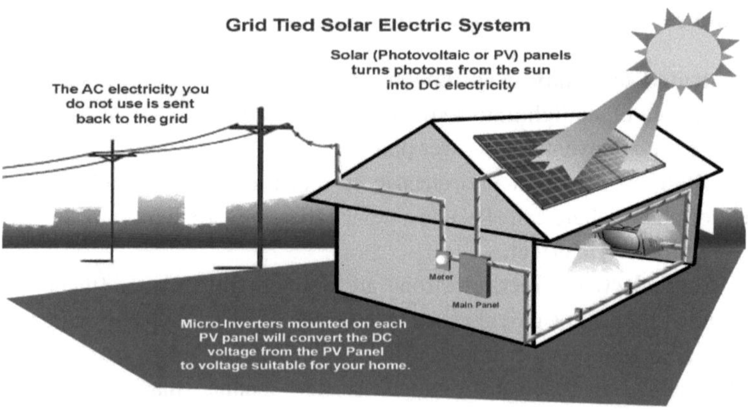

Fig. 7. Solar Power Generation connected back to power grid

8 Need of Solar Charge Controller

A solar charge controller is needed in virtually all solar power systems that utilise batteries. The job of the solar charge controller is to regulate the power going from the solar panels to the batteries. Overcharging batteries will significantly reduce battery life and at worst, damage the batteries to the point that they are unusable.

The most basic charge controller simply monitors the battery voltage and opens the circuit, stopping the charging, when the battery voltage rises to a certain level. Older charge controllers used a mechanical relay to open or close the circuit, stopping or starting power going to the batteries.

More modern charge controllers use Pulse Width Modulation (PWM) to slowly lower the amount of power applied to the batteries as the batteries get closer and closer to fully charged. This type of controller allows the batteries to be more fully charged with less stress on the battery, extending battery life. It can also keep batteries in a fully charged state (called "float") indefinitely. But PWM is more complex, also does not have any mechanical connections to break.

8.1 Using MPPT as Charge Controller

The most recent and best type of solar charge controller is called Maximum Power Point Tracking or MPPT. MPPT controllers are basically able to convert excess voltage into amperage. This has advantages in a couple of different areas.

Most solar power systems use 12 volt batteries, like we find in cars. Solar panels can deliver far more voltage than is required to charge the batteries. In essence, converting the excess voltage into amps, the charge voltage can be kept at an optimal level while the time required to fully charge the batteries is reduced. This allows the solar power system to operate optimally at all times.

Another area that is enhanced by an MPPT charge controller is power loss. Lower voltage in the wires running from the solar panels to the charge controller results in higher energy loss in the wires than higher voltage.

With a PWM charge controller used with 12v batteries, the voltage from the solar panel to the charge controller typically has to be 18v. Where as Using an MPPT controller allows much higher voltages in the wires from the panels to the solar charge controller. The MPPT controller then converts the excess voltage into additional amps. By running higher voltage in the wires from the solar panels to the charge controller, power loss in the wires is reduced significantly.

8.2 What Is Maximum Power Point Tracking?

Panel Tracking(Mechanical) - This is where the panels are on a mount that follows the sun. These optimize output by following the sun across the sky for maximum sunlight. These typically give us about a 15% increase in winter and up to a 35% increase in summer.

This is just the opposite of the seasonal variation for MPPT controllers. Since panel temperatures are much lower in winter, they put out more power. And winter is usually when we need the most power from your solar panels due to shorter days.
Maximum Power Point Tracking is electronic tracking - usually digital. The charge controller looks at the output of the panels, and compares it to the battery voltage. It then figures out what is the best power that the panel can put out to charge the battery.

It takes this and converts it to best voltage to get maximum AMPS into the battery. Most modern MPPT's are around 93-97% efficient in the conversion. We typically get a 20 to 45% power gain in winter and 10-15% in summer. Actual gain can vary widely depending weather, temperature, battery state of charge, and other factors.

8.3 Maximizing the Solar Energy, Using MPPT

Photovoltaic cells have a complex relationship between their operating environment and the maximum power they can produce. They can not function at their maximum, efficiently during cold weather, on cloudy or hazy days, or when the battery is deeply discharged.

- **Cold weather** - solar panels work better at cold temperatures, but without a MPPT we loose most of that. Cold weather is most likely in winter - the time when sun hours are low and we need the power to recharge batteries the most.
- **Low battery charge** - the lower the state of charge in the battery, the more current a MPPT puts into them - another time when the extra power is needed the most. we can have both of these conditions at the same time.
- **Long wire runs** - If we are charging a 12 volt battery, and panels are 100 feet away, the voltage drop and power loss can be considerable unless we use very large wire. That can be very expensive. But if we have four 12 volt panels wired in series for 48 volts, the power loss is much less, and the controller will convert that high voltage to 12 volts at the battery.
- MPPT's (Maximum Power Point Tracking Systems) are most effective under all these conditions.

8.4 P-V Panel I-V Curve

Maximum power point tracking (MPPT) is a technique that grid tie inverters, solar battery chargers and similar devices use to get the maximum possible power from one or more solar panels. Solar cells have a complex relationship between solar irradiation, temperature and total resistance that produces a non-linear output efficiency known as the I-V Curve. It is the purpose of the MPPT system to sample the output of the cells and apply the proper resistance (load) to obtain maximum power for any given environmental conditions.

Photovoltaic Cells have a complex relationship between their operating environment and the maximum power they can produce. The Fill Factor fd, is a parameter which characterizes the non-linear electrical behaviour of the solar cell. Fill factor is defined as the ratio of the maximum power from the solar cell to the product of V_{oc} and I_{sc}. It is often used to estimate the maximum power that a cell can provide with an optimal load under given conditions, $P = FF*V_{oc}*I_{sc}$. For most purposes, FF, V_{oc}, and I_{sc} are enough information to give a useful approximate model of the electrical behaviour of a photovoltaic cell under typical conditions.

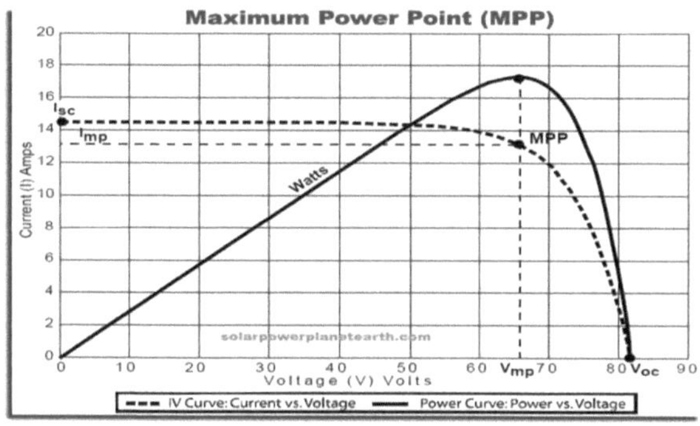

Fig. 8. P-V Panel I-V Curve

9 MPPT Methods or Algorithms

Maximum Power Point Tracking Methods

- Many MPPT techniques have been reported in the literature, but there are three main methods most widely used
    - Perturb and Observe (P&O)
    - Incremental Conductance (INC)
    - Constant Voltage (CV)

9.1 Perturb and Observe (or) Hill Climbing

In this method, the controller adjusts the voltage by a small amount from the array and measures power; if the power increases, further adjustments in that direction are tried until power no longer increases.

This is called the perturb and observe method and is most common, although this method can result in oscillations of power output. It is also referred to as a ***hill climbing*** method, because it depends on the rise of the curve of power against voltage below the maximum power point, and the fall above that point.. This method may result in top-level efficiency, provided that a proper predictive and adaptive hill climbing strategy is adopted.

9.2 Incremental Conductance

In the incremental conductance method, the controller measures incremental changes in array current and voltage to predict the effect of a voltage change.

Like the P&O algorithm, it can produce oscillations in power output. This method utilizes the incremental conductance (dI/dV) of the photovoltaic array to compute the sign of the change in power with respect to voltage (dP/dV). The incremental conductance method computes the maximum power point by comparison of the incremental conductance (ΔI/ΔV) to the array conductance (I/V). When the incremental conductance is zero, the output voltage is the MPP voltage. The controller maintains this voltage until the irradiation changes and the process is repeated.

9.3 Constant Voltage

In the constant voltage method, the power delivered to the load is momentarily interrupted and the open-circuit voltage with zero current is measured. The controller then resumes operation with the voltage controlled at a fixed ratio, such as 0.76, of the open-circuit voltage, which has empirically been determined as the estimated maximum power point.

The operating point of the PV array is kept near the MPP by regulating the array voltage and matching it to a fixed reference voltage Vref. The Vref value is set equal to the maximum power point voltage of the characteristic PV module or to another calculated best fixed voltage.

One of the approximations of this method is that variations of individual panels are not considered. The constant reference voltage can be considered as the maximum power point voltage. The data for this method varies with geographical location and has to be processed differently for different geographical locations. The CV method does not require any input.

10 General Configuration of the MPPT Solar Charge Controller

- Advanced microprocessor control with Buck regulator wide input range
- Maximum Power Point Tracking (MPPT) as solar charge controller with DC load control
- Reverse polarities protection of PV and battery with battery overcharge and over discharge protection
- Temperature compensation (-3 to -7mV/Cell/Celsius) & Lighting surge protection (TVSS)
- 3-step charging to provide quick and safe charging for battery
- Automatic cooling fan (outside enclosure)
- 7 modes timer control (ON/OFF DC load) selectable

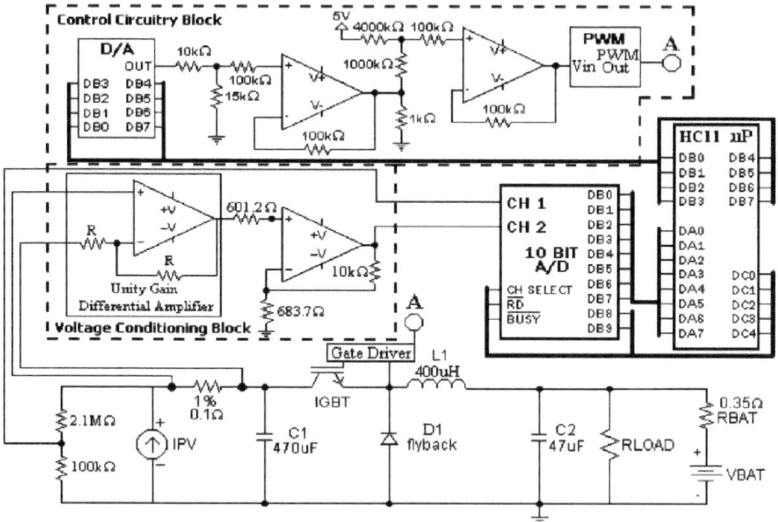

Fig. 9. Component level Diagram Of MPPT Controller

11 Comparative Statement of MPPT Algorithms

PV arrays under constant uniform irradiance has a current–voltage (I–V) characteristic like that shown in the below Figure. There is a unique point on the curve, called the maximum power point (MPP), at which the array operates with maximum efficiency and produces maximum output power.

11.1 Photovoltaic Array Current –Voltage Relationship

When a PV array is directly connected to a load (a so-called 'direct-coupled' system), the system's operating point will be at the intersection of the I–V curve of the PV array and load line shown in the below Figure.

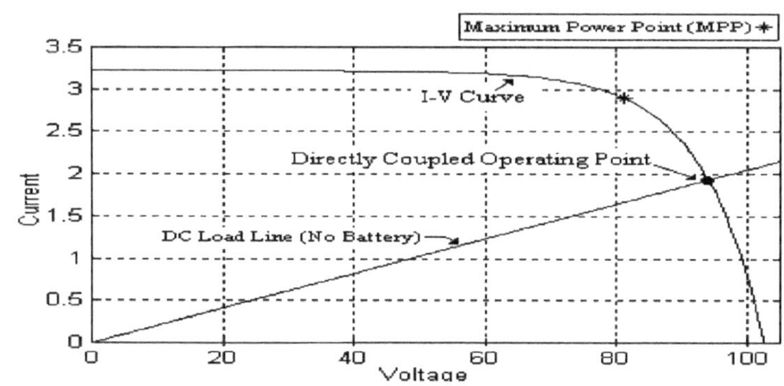

Fig. 10. P-V Array Current-Voltage Characteristics

In a direct-coupled system, the PV array must usually be oversized to ensure that the load's power requirements can be supplied. This leads to an overly expensive system. To overcome this problem, a switch-mode power converter, called a maximum power point tracker (MPPT), can be used to maintain the PV array's operating point at the MPP. The MPPT does this by controlling the PV array's voltage or current independently of those of the load.

However, the location of the MPP in the I–V plane is not known a priori. It must be located, either through model calculations or by a search algorithm. The situation is further complicated by the fact that the MPP depends in a nonlinear way on irradiance and temperature, as illustrated in the below Figure. Fig. 11 shows a family of PV I–V curves under increasing irradiance, but at constant temperature.

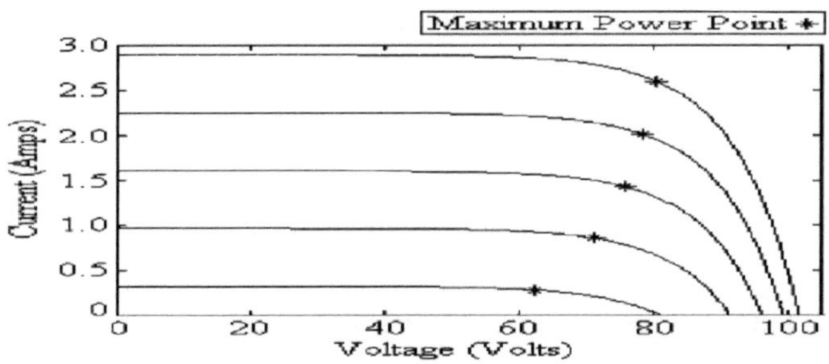

Fig. 11. PV Array Voltage – Current at 40^0 C at different Irradiance Levels

11.2 Comparison of MPP Tracking Efficiencies

Table 2. Comparative statement of MPP Methods

Sky conditions	P&O		Inc		CV	
PV array	Days of data	η_{MPPT}	Days of data	η_{MPPT}	Days of data	η_{MPPT}
Clear	20	98·7	17	98·7	20	90·4
Partly cloudy	14	96·5	11	97·0	10	90·1
Cloudy	9	98·1	11	96·7	6	93·1
Overall	43	97·8	39	97·4	36	91·2
Simulator						
Overall	99·3		99·4		93·1	

12 Various Photovoltaic System Configurations

PV Modules Generate DC Current and Voltage. However, to feed the electricity to the grid, AC current and voltage are needed. Inverters are the equipment used to convert DC to AC. In addition, they can be in charge of keeping the operating point of the PV array at the MPP. This is usually done with computational MPP tracking algorithms. There are different inverter configurations depending on how the PV modules are connected to the inverter. The decision on what configuration should be used has to be made for each case depending on the environmental and financial requirements.

- Central inverter
- String inverter
- Multi-string inverter
- Module integrated inverter

12.1 Central Inverter PV-Module Configuration

It is the simpler configuration: PV strings, consisting of series connected PV panels, are connected in parallel to obtain the desired output power. The resulting PV array is connected to a single inverter. In this configuration all PV strings operate at the same voltage, which may not be the MPP voltage for all of them.

The problem of this configuration is the possible mismatches among the different PV modules. If they are receiving different irradiation (shading or other problems), the true MPP is difficult to find and consequently there are power losses and the PV modules are underutilized.

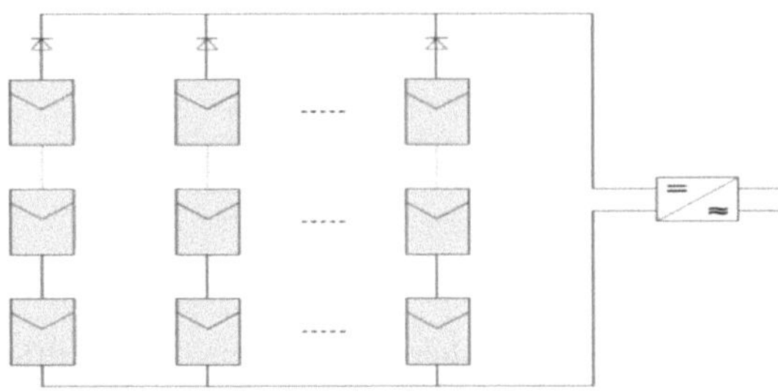

12.2 String Inverter Configuration

In this configuration, every string of PV panels connected in series is connected to a different inverter. This can improve the MPP tracking in case of mismatches or shading, because each string can operate at a different MPP, if necessary. Whereas in the central inverter there is only one operating point which may not be the MPP for each string, thus leading to power losses.

On the other hand, the number of components of the system increases as well as the installation cost, as an inverter is used for each string.

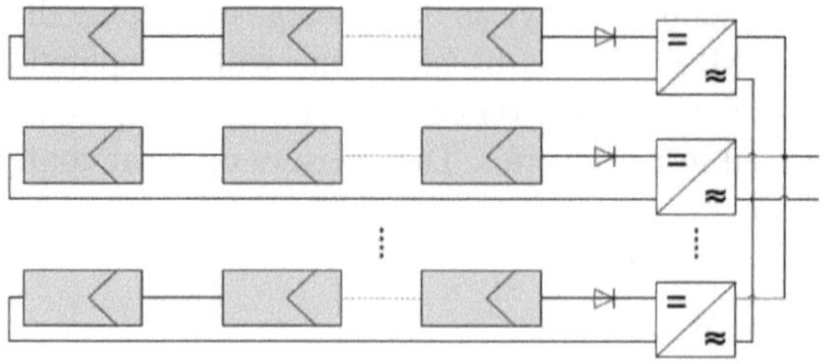

12.3 Multi-string Inverter Configuration

In this case each string is connected to a different DC-DC converter, which is in charge of the MPP tracking of the string, and the converters are connected to a single inverter. The advantages related to MPP tracking are the same as in the string configuration; each string can have a different MPP.

The disadvantages, an increase in the price compared to the central inverter, as a converter is used for each string.

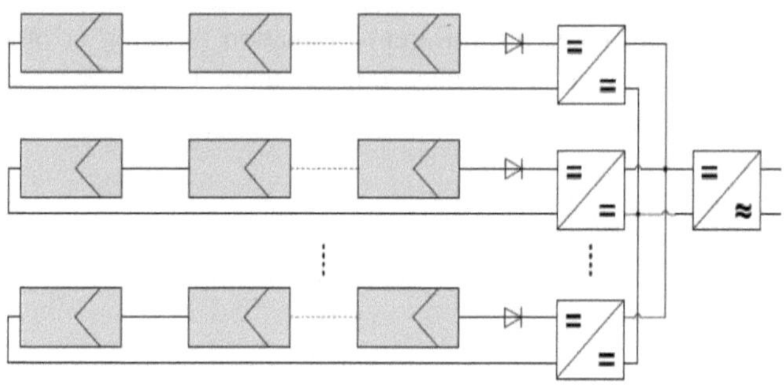

12.4 Module Integrated Inverter Configuration

In this configuration,, each PV module is connected to a different inverter and consequently the maximum power is obtained from each panel as the individual MPP is tracked by each inverter.

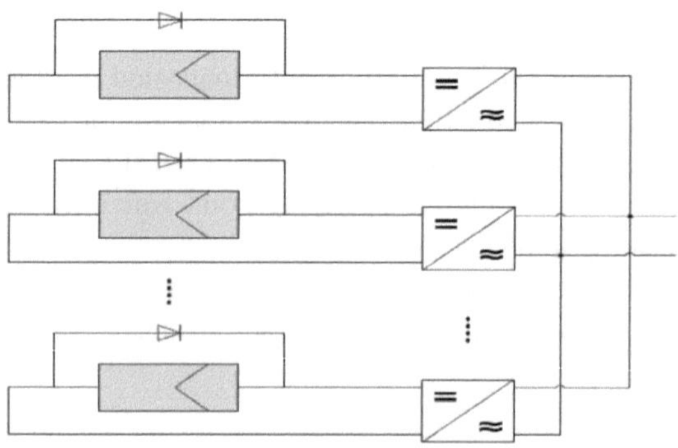

This configuration can be used when the differences in the operating point of the different modules are large. However, it is more expensive because each panel has its own inverter.

13 Analysis

We observed at the outset of this study that we would find that the perturb-and-observe and incremental conductance algorithms should have very similar overall efficiencies, but that incremental conductance should be slightly better. However, the results of this study indicate that, to within the accuracy available, the MPPT efficiencies of the incremental conductance and perturb-and-observe MPPT algorithms are essentially the same.

When optimized for the particular MPPT hardware in use, P&O and INC had the same performance under clear sky conditions, indicating that the penalty in efficiency caused by the oscillation about the MPP inherent in P&O under steady-state conditions was insignificant for the optimized algorithms. Incremental conductance outperformed P&O under partly cloudy conditions, as expected, but the difference was very small. Also, interestingly, P&O had a significantly higher efficiency than incremental conductance under cloudy skies. The reason for this can be understood through the array P–V curves in above figures.

In the future solar energy will be very important energy source. More than 45% of necessary energy in the world will be generated by photovoltaic array.. In order to reach this aspect, it is important to note that the output characteristic of a photovoltaic array is nonlinear and changes with solar irradiation and the temperature. Therefore a maximum power point tracking (MPPT) technique is needed to draw peak power from the solar array in order to maximize the produced energy. This paper presents a comparative study of widely-adopted MPPT algorithms; their performance is evaluated using the simulation tool. In particular, this study compares the behaviours of each technique in presence of solar irradiation variations.

14 Conclusions

As the conventional energy sources are rapidly depleting, the importance of solar photovoltaic (PV) energy has been emerging as replaceable energy resources to human being. Since it is clean, pollution-free, and inexhaustible, researches on the PV power generation system have received much attention, particularly, on many terrestrial applications. Furthermore, due to the continuing decrease in PV arrays cost and the increase in their efficiency, PV power generation system could be one of comparable candidates as energy sources for mankind in near future

The above objectives and proposed model ise effectively implemented by the use of solar power based technologies and with the further use of modern technology like microcontrollers and technique like PWMs and MPPT Systems and Algorithms in solar power based applications. This Objective is aimed at bridging the gap between our knowledge of Natures energy conversion and storage factor. By operating at the solar panel's maximum power point (MPP) and by intelligently drawing the power from the panel, energy can be successfully harnessed to power a pulsed load. This model presents a simple and cost effective solution for maximum-power-point tracking for use in solar based power generation systems.

References

1. Park, J., Ahn, J., Cho, B., Yu, G.: Dual-Module-Based Maximum Power Point Tracking Control of Photovoltaic Systems. Transactions on Industrial Electronics 53(4) (August 2006)
2. Lynn, P.A.: Electricity from Sunlight: An Introduction to Photovoltaics, p. 238. John Wiley & Sons (2010)
3. Markvart, T.: Solar electricity, p. 280. Wiley (2000)
4. Overall efficiency of grid connected photovoltaic inverters, European Standard EN 50530 (2010)
5. Esram, T., Chapman, P.L.: Comparison of Photovoltaic Array Maximum Power Point Tracking Techniques. IEEE Transactions on Energy Conversion 22(2), 439–449 (2007)
6. Yuvarajan, S., Xu, S.: Photo-voltaic power converter with a simple maximum-powerpoint-tracker. In: Proc. International Symposium on Circuits and Systems, vol. 3, pp. 399–402 (2003)
7. Solodovnik, E.V., Liu, S., Dougal, R.A.: Power Controller Design for Maximum Power Tracking in Solar Installations. IEEE Transactions in Power Electronics 19, 1295–1304 (2004)
8. Kobayashi, K., Takano, I., Sawada, Y.: A study on a two stage maximum power point tracking control of a photovoltaic system under partially shaded insolation conditions. In: Power Engineering Society General Meeting, July 13-17, vol. 4. IEEE (2003)
9. Blankenship, R.E.: Molecular Mechanisms of Photosynthesis. Blackwell Science (2002)

A Secure and Reliable Mobile Banking Framework

Shaik Shakeel Ahamad[1], V.N. Sastry[2], Siba K. Udgata[3], and Madhusoodhnan Nair[1]

[1] K.G. Reddy College of Engineering and Technology,
Chilkur Village, Moinabad Mandal, Ranga Reddy District-501504, India
ahamadss786@gmail.com, principal@kgr.ac.in
[2] Institute for Development and Research in Banking Technology (IDRBT),
Castle Hills, Masab Tank, Hyderabad-57, India
vnsastry@idrbt.ac.in
[3] School of Computer and Information Sciences,
University of Hyderabad, Hyderabad, India
udgatacs@uohyd.ernet.in

Abstract. In this paper we propose a secure mobile banking framework which ensures reliable end to end communication channel and end to end application security from the UICC to the Remote Bank Server via Mobile Equipment. SSL/TLS ensures secure connection from the UICC to the Remote Bank Server, TCP provides end to end reliable communication and Bearer Independent Protocol (BIP) provides and manages the link layer in achieving end to end reliable communications between the UICC and the Remote Bank Server. All the digital signatures are generated in a tamper proof hardware i.e. UICC at the client side and Hardware Security Module at the Bank side. So all the signatures generated in the framework are qualified signatures. Bank server is supported by Communication Manager, Synchronization Manager, Security Manager, Concurrency Manager, Backup Manager, Archives Manager and Error and Exception Handling Manager in order to ensure end to end security at the communication layer and at the application layer.

Keywords: Mobile Banking (MB), UICC, Bearer Independent Protocol (BIP), SSL/ TLS, TCP.

1 Introduction

As wireless telecommunication and hardware technology becoming more advanced the mobile phone/handset is evolving into a powerful computing and communication platform. With the advancement in wireless telecommunication and hardware technology customers are demanding more Value Added Services (VAS) such as Mobile Banking and Mobile Commerce. Mobile banking is a term used for performing balance checks, account transactions, payments, credit applications and other banking transactions through a mobile device such as a mobile phone or Personal Digital Assistant (PDA). Banks offer mobile banking due to the following reasons Lower operating costs, Greater geographic diversification, Improved or sustained competitive position, increased customer demand for services and new

S.C. Satapathy et al. (eds.), *ICT and Critical Infrastructure: Proceedings of the 48th Annual Convention of CSI - Volume II*, Advances in Intelligent Systems and Computing 249,
DOI: 10.1007/978-3-319-03095-1_80,

revenue opportunities. Benefits to customers are Increased Convenience, Reduced theft, fraud and mismanagement, Ability to maintain electronic records, Access to other Value Added Services and Time Saving. In order to realize customer acceptance for mobile banking, Banks must seek a solution that not only solves security concerns but also it must be simple to use and easy to deploy for all customers. Security is paramount. An insecure payment system will not be accepted by both merchants and customers. So Mobile Banking should ensure both end to end security and reliable communication which are very important. Transaction level security must ensure end-to-end security with message integrity, confidentiality and non-repudiation. In order to ensure this Wireless PKI is used and is normally implemented on the mobile phones but implementing WPKI in mobile phones has serious limitations such as secret keys stored in the memory of Mobile Phone could be infected by viruses or can be maliciously replaced. So we propose WPKI on the Secure Element (SE) of the mobile phone, SE is a UICC which is a smart card and is a well trusted and tamper proof device thus UICC card can be used for security critical applications such as Mobile Banking/Commerce. So UICC card will have WPKI functionality which can generate Private keys with OBKG procedure on the card and can store both the Private keys and Certificates securely. Existing mobile banking solutions neither ensure communication security nor application security; these solutions claim to achieve non repudiation property using private keys stored in the memory of mobile phone which is not tamper resistant so these signatures cannot be considered as qualified signatures. Universal Integrated Circuit Card (UICC), which hosts Subscriber Identification Module (SIM) and is supported by the GSMA, has many benefits. UICC smartcards are expected to be globally the most widely distributed Secure Element and the UICC has been well standardized, with well-proven enrollment processes in practice [13]. New service channels and concepts have created new requirements where cross-industry collaboration is needed for the creation of successful business models of MFS. MNOs and Banks should collaborate with each other thereby offering cost-effective and compelling services to the customers. Although UICC is the property of Mobile Network Operator there are several alternatives for ownership, issuance and management of a shared UICC based secure element. Shared use of the SE has been compared with the property world. Different property management models have different agreements between parties. They are Hotel Concept, Rental Building Concept and Ownership Concept as given in [13]. We adopt Rental Building Concept in which the Bank will hire space on the UICC from MNO to install Mobile Payment Applications. Certifying Authority will act as a Trusted Service Manager (TSM). The rest of the paper is organized as follows. Section 2 presents recent related works related to our research. In Sect. 3, we present the proposed Secure and Reliable Mobile Banking (SRMB) Framework. Section 4 presents Security Analysis; Section 5 presents the comparative analysis of our proposed framework with related works. Section 6 presents conclusion and future works.

2 Related Work

Author of [1] proposes a Mobile Payment Framework Based on 3G Network and Authors of [2] proposes an Integrated Mobile Phone Payment System Based on 3G Network by combining IC chip, mobile phone and mobile internet along with some improvement measures for current settlement mechanisms in the network transaction, which is entirely around payment channel, payment carrier, security authentication but their solution has some limitations such as private key is not installed in tamper resistant hardware and CA installs private key and in addition to these limitations the solutions does not ensure communication security. Authors of [3] proposes a new approach on secure mobile banking using public key infrastructure but there is no contribution from the author in this work it looks more like a white paper rather than a research paper. Authors of [4] proposes a Lightweight Architecture for Secure Two-Party Mobile Payment between the client and bank server but the solution is proposed in the memory of mobile phone which is not a tamper proof hardware (like UICC at the client side) and it does not ensure communication security. Authors of [5] propose a GPRS Mobile Payment System based on RFID which is an optimization design of the GPRS mobile payment system as a transitional solution before the prevalence of the NFC (near field communication). So Current Mobile Banking/ Mobile Payment solutions have the following limitations User Credentials are generated and stored in the memory of Mobile Phones, Existing mobile payment solutions does not ensure communication security and application security and Mobile Payment Applications cannot be personalized by Banks without the intervention of MNO's.

3 Proposed Framework

A. Entities Involved in the Framework

Detailed explanation of Mobile Network Operator (MNO), Banks, WPKI and UICC can be found in [9] and [10].

B. Proposed Architecture

Using TCP as a Transport Layer Protocol

In order to provide reliable end to end communication between UICC and remote bank server on a network without requiring implementation of a special purpose protocol at the remote bank server we propose a architecture shown in figure 1 in which communications at the communication layer are carried out at SSL/TLS and TCP channels are opened from the CAT in UICC card to the remote bank server. The communication module accepts data from the applications and encapsulates the data in TCP packets for the TCP layer. The TCP packets are then encapsulated by the communication module to IP data grams which the communications module transmits to the mobile terminal over the link layer. BIP is one possible link layer for communication between UICC and ME. Both UICC and ME implement BIP in their respective communications modules. This transmits IP data grams from the IP layer to the ME using BIP frames. BIP is used to monitor errors on a link. BIP is defined in [14] and [21] standards in order to allow the UICC to interact and operate with the terminal that supports specific mechanism required by UICC card applications. More precisely, BIP is a mechanism by which the terminal provides the UICC with access of the data bearers supported by the terminal and the operator's back-end. The BIP at the UICC side receives application data and delivers them to the upper layers. UICC cards have both USIM and network smart card functionalities. This enables USIM to have end to end reliable communication with the remote bank server over the internet. The internet security protocol, such as SSL/TLS, further secures this reliable communication between UICC and remote banking server via ME, SSL/TLS lies between the Application layer and Transport layer. Reliable end-to-end communication is ensured between UICC and remote banking server without requiring a implementation of a special purpose protocols (i.e. CAT_TP) at the remote banking server by using TCP as a Transport Layer Protocol.

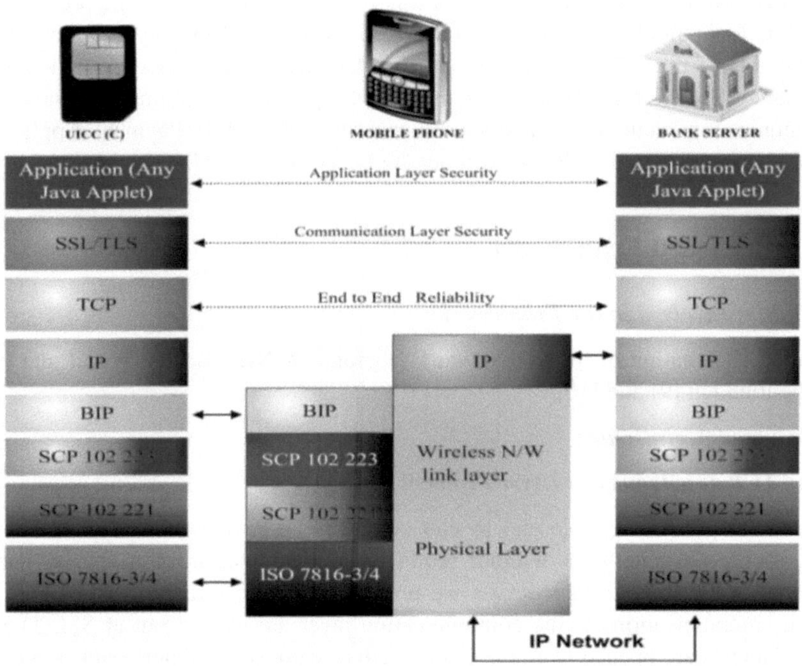

Fig. 1. Interface between UICC and Bank Server

C. Main Components at the Bank Server

Following are the main components at the Bank Server

i) **Communication Manager (ComM)**

Communication Manager (ComM) establishes the connection with the Bank Server. It allows the exchange of information between the communicating parties. The CM maintains the account for source and destination of communication, information about communicators, the time at which the communication is enabled, etc. Other communication issues are performance, security and privacy. It establishes a secure and reliable communication channel using SSL/TLS and TCP from the client to the Bank Server. It ensures reliable end to end communication security.

ii) **Synchronization Manager (SyM)**

SyM's main responsibilities are sending and receiving the data to and from the client and Bank Server, Personalizing (agreeing for a shared symmetric key) mobile payment application (which is in UICC) and updating (which is in UICC). Data is exchanged in an encrypted form between client and server.

iii) **Security Manager (SeM)**

The Security Manager is responsible for authentication (using password & Biometrics), signing the message, certificate management, secret key management and distribution. SeM encrypts and decrypts the messages, generates and verifies digital signatures, generates shared symmetric keys for all the clients, maintains password and biometric data in its database. Certificate Manager maintains all the certificates of the client. Bank generates and keeps its private key and shared symmetric keys in Tamper Resistant HSM (Hardware Security Module). All the clients Mobile numbers are mapped to Certificates and Account numbers.

iv) **Concurrency Manager (ConM)**

The Concurrency Manager handles the multiple requests that come from the various mobile clients. The concurrency manager keeps the technique of parallelism, which improves quick response time and reduces the latency period

v) **Backup Manager (BM)**

The Backup Manager supports atomic transaction in case of network disconnection. When the network is disconnected, the failed transaction is picked up from the restore point and resumes the data, instead of restarting it again.

vi) **Archives Manager (AM)**

Bank Server securely archives digital signatures received and sent by it in its archives, logs all the changes in the certificate status i.e. activation, suspension, termination of suspension and revocation) and all validity confirmations given by OCSP and CRL responder. Log records are cryptographically linked to prevent insider attacks and forging log records (for example backdating the log record).

Once a month, cryptographic hash is printed in newspaper. Log record database is backed up on timely basis with three copies stored in different locations. This kind of secure log works for dual purpose:

a) **Auditability** – in case of disputes there are means to prove that some validity confirmation was (not) issued, was issued before/after some other event, etc.
b) **Long term validity preservation of digital signatures** – even after complete breakdown of all keys and RSA algorithm one can prove that document signed at past is valid – by matching confirmation data with log data. Secure log is completely invisible to end-users with one exception – every ID-card

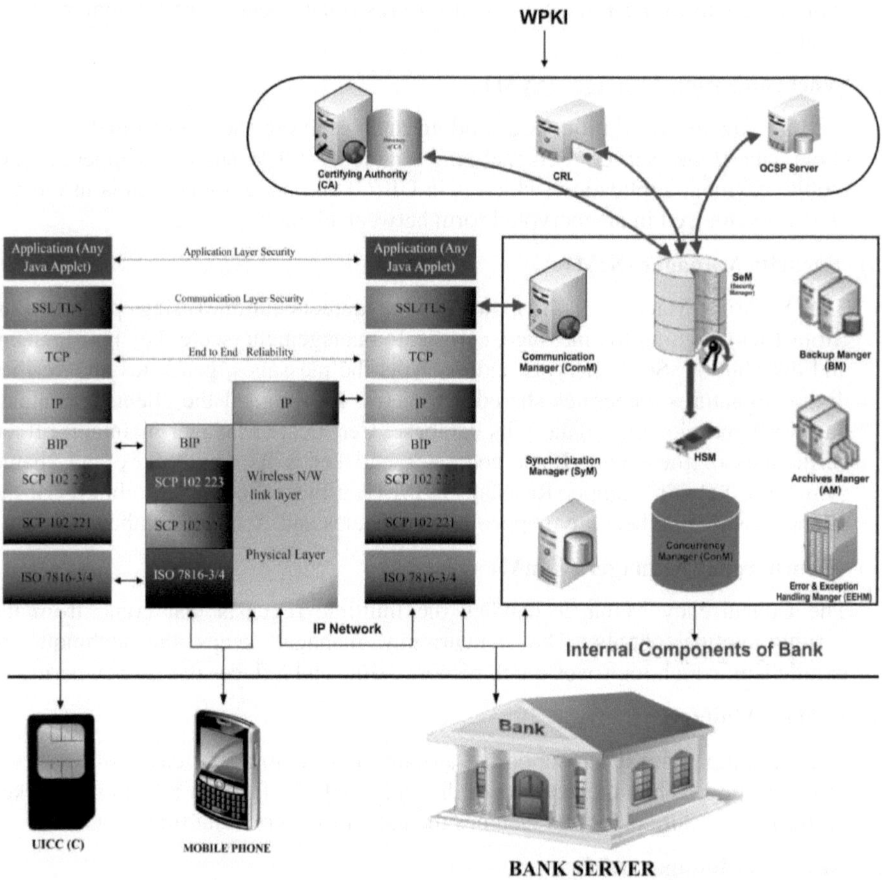

Fig. 2. Entities involved in Mobile Banking Framework

c) holder can check their personal records (certificate history and issued validity confirmations).

vii) Error and Exception Handling Manager (EEHM)

Mobile banking application server must be able to properly handle exceptions and reporting errors. It may be noted that if error reporting and exception handling are not properly managed, they can reveal information that can be misused to perform illegitimate queries. A thorough testing of application by financial Institutions or service provider or third parties should be done in this regard to make sure that application is handling exceptions and reporting errors properly.

4 Comparitative Analysis with Related Work

PROTOCOLS / FEATURES	[1]	[2]	[3]	[4]	[5]	SRMB
Authentication	NR (Not Referred)	NR (Not Referred)	NR (Not Referred)	NR (Not Referred)	NR (Not Referred)	YES
Confidentiality	NR (Not Referred)	NR (Not Referred)	NR (Not Referred)	NR (Not Referred)	NR (Not Referred)	YES
Integrity	NR (Not Referred)	NR (Not Referred)	NR (Not Referred)	NR (Not Referred)	NR (Not Referred)	YES
Non- Repudiation	NR (Not Referred)	NR (Not Referred)	NR (Not Referred)	NR (Not Referred)	NR (Not Referred)	YES
Key pairs are generated and stored in Tamper resistant device	NO	NO	NO	NO	NO	YES
Are the Signatures generated in "Secure Signature Creation Device (SSCD)"	NO	NO	NO	NO	NO	YES
Ensures Communication Security	NO	NO	NO	NO	NO	YES
Ensures Application Security	NO	NO	NO	NO	NO	YES
Withstands Replay, Impersonation & MITM Attacks	NO	NO	NO	NO	NO	YES

5 Conclusions and Future Work

This paper proposes a secure mobile banking framework which ensures reliable end to end communication channel and application security from the UICC to the Remote Bank Server via Mobile Equipment. All the digital signatures are generated in a tamper proof hardware i.e. UICC at the client side and Hardware Security Module at the Bank side. So all the signatures generated in the framework are qualified signatures. Bank server is supported by Communication Manager, Synchronization Manager, Security Manager, Concurrency Manager, Backup Manager, Archives Manager and Error and Exception Handling Manager in order to ensure end to end security at the communication layer and at the application layer. The network is assumed to be hostile as it contains intruders with the capabilities to encrypt, decrypt, copy, forward, delete, and so forth. Several examples show how carefully designed protocols were later found out to have security breaches [7] (Muhammad et al., 2006). So formal verification of security protocols is essential as it can detect flaws that lead to protocol failure. So we plan to propose a mobile payment protocol and verify our proposed protocol using BAN logic, AVISPA and Scyther Tool in the future.

References

1. Wu, H., Li, X., Dai, W., Zhao, W.: Mobile Payment Framework Based on 3G Network. In: Proceedings of the Third International Symposium on Electronic Commerce and Security Workshops(ISECS 2010), Guangzhou, P. R. China, July 29-31, pp. 172–175 (2010)
2. Dai, W., Cai, X., Wu, H., Zhao, W., Li, X.: An Integrated Mobile Phone Payment System Based on 3G Network. Journal of Networks 6(9), 1329–1336 (2011), doi:10.4304/jnw
3. Narendiran, C.: A new approach on secure mobile banking using public key infrastructure. International Journal of Computing Technology and Information Security 1(1), 40–46 (2011)
4. Zhu, Y., Rice, J.E.: A Lightweight Architecture for Secure Two-Party Mobile Payment. Computational Science and Engineering 2, 326–333 (2009)
5. Wei, L., Chenglin, Z., Wei, Z., Zheng, Z.: The GPRS Mobile Payment System Based on RFID. Comminication Technology, 1–4 (2006)
6. Manvi, S.S., Bhajantri, L.B., Vijayakumar, M.A.: Secure Mobile Payment System in Wireless Environment payment system. In: Proceedings of the Second International Conference on Mobile Technology, Applications and Systems, pp. 113–119 (2005)
7. Muhammad, S., Furqan, Z., Guha, R.K.: Understanding the intruder through attacks on cryptographic protocols. In: Proceedings of the 44th ACM Southeast Conference (ACMSE 2006), pp. 667–672 (March 2006)
8. Kumar, S.B.R., Raj, A.A.G., Rabara, S.A.: A framework for mobile payment consortia system. Computer Science and Software Engineering 2, 43–47 (2008)
9. Ahamad, S.S., Sastry, V.N., Udgata, S.K.: Secure Mobile Payment Framework based on UICC with Formal Verification. Special Issue on 'Future Trends in Security Issues in Internet and Web Applications' Int. J. Computational Science and Engineering (accepted) (in press)
10. Ahamad, S.S., Sastry, V.N., Udgata, S.K.: A secure and optimized mobile payment framework with formal verification. In: SECURIT 2012, pp. 27–35 (2012)

Challenges towards Implementation of e-Government Project in West Bengal, India: A Fishbone Analysis in Order to Find Out the Root Causes of Challenges

Manas Kumar Sanyal, Sudhangsu Das, and Sajal Bhadra

Department of Business Administration,
Kalyani University
Manas_sanyal@rediffmail.com,
{iamsud,Sajal.bhadra}@gmail.com

Abstract. E-governance is gaining momentum with various Government initiatives in India to ensure Government services delivery in a smarter way, more smoothly, more transparently and faster way. Government of India (GI) is committed to transfer its services from manual to electronic since 2006. With the past experiences, it has been anticipated that there are several challenges in India to roll-out e-Governance projects. The remedy and mitigation of those challenges could be possible if and only if the Government can address the root causes of all the challenges. The focus of this study is to identify the key challenges and to explore the root causes of the challenges using very well-known methodology, Fishbone Analysis. The analysis has been performed based on the survey data only, and the data have been collected through the interviews and questionnaires. The survey was conducted randomly among the different stake holders like citizens, Government officials and IT vendors, related with the e-Governance projects in West Bengal, a state in India.

Keywords: e-Governance, Fishbone, e-Governance Challenges, CSC.

1 Introduction

E-governance refers to Government service delivery in the form of electronic format. The service delivery may be any of the following categories: i) Government to Government (G to G), ii) Government to Citizens (G to C), iii) Government to Business (G to B) and iv) Government to Employee (G to E). The main objective of the e-Governance is not only to do transformation of Government service delivery from manual to electronic and administrative process re-engineering, but also to bring more transparency, accuracy, speed and cost effectiveness in Government service delivery.

The Government of India (GI) also has taken massive initiatives under National E-Governance Plan (NeGP) to roll out various e-Governance projects including 27 mission mode projects and 8 components from May 18, 2006 to automate various state and central Government department [8]. The core infrastructure projects are already in place to promote all other citizen centric e-Governance project like State

S.C. Satapathy et al. (eds.), *ICT and Critical Infrastructure: Proceedings of the 48th Annual Convention of CSI - Volume II*, Advances in Intelligent Systems and Computing 249,
DOI: 10.1007/978-3-319-03095-1_81, © Springer International Publishing Switzerland 2014

Data Center (SDC), State Wide Area Network (SWAN) and Common Service Center (CSC). Apart from various core infrastructure projects, National e-Governance Service Delivery Gateway (NSDG), State e-Governance Service Delivery Gateway (SSDG), and Mobile e-Governance Service Delivery Gateway (MSDG) have been established to facilitate the middle ware functionality [8].

With the other states of India, the Government of West Bengal (GoWB) is also committed to make the Government department fully ICT compliance across the state. For this, GoWB has made IT policy in the year 2003 and gradually implementing the following e-Governance projects throughout the state [9], which are shown in Fig-1.

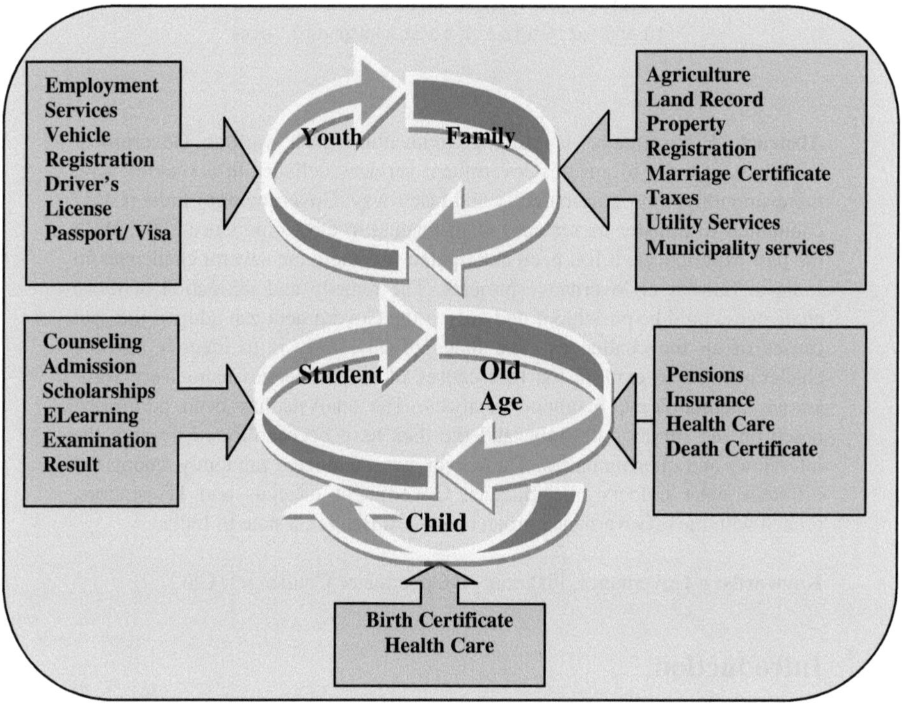

Fig. 1. Various e-Governance initiatives in West Bengal, a state of India

The e-Government projects in West Bengal are facing a lot of challenges in various forms like operational, strategic, environmental, technical, capability and planning. The success of e-Governance projects can only be possible if and only if the Government can identify the root causes of those challenges and take some corrective actions to mitigate those identified root causes. In this study, an attempt has been made to find out the key challenges those are affecting e-Governance projects in West Bengal and finally to explore the root causes behind the identified challenges with the help of Fish Bone analysis.

The Fishbone analysis is very popular methodology to prevent quality defect and to find out the root causes of any defects or problems and to figure out the relative importance of various causes [5].The fishbone diagram is also known as Ishikawa diagram or herringbone diagram or Cause-Defect diagram which is invented by Kaoru Ishikawa in the year 1960, who was the pioneer in the quality management process [4]. The fishbone analysis was first utilized in the quality circle in the year 1960's and it was treated one of the seven basic tools for the quality control. Though it was initially designed and targeted for the management field, but in the modern research it is widely used in other fields as well, like Medicine, Engineering, Manufacturing, Computer Science, Information Technology, Telecom Industry, Food Industry etc. [5]

Here, after survey work, authors have brainstormed six key challenges as problem for e-Governance initiative in West Bengal and done some rigorous survey in different districts in West Bengal to get the root causes of those challenges. Finally, the fish of fishbone analysis has been designed in order to visualize the challenges and to find the root causes of those challenges to roll out e-Governance projects in West Bengal.

2 Literature Review

The root cause analysis helps a lot to take corrective actions for any kind of fatal problems faced by any organization. The Fishbone Diagram is one of the popular approaches to do root cause analysis because it is structured and systematics in nature. Sensuse D. I. , Lusa S. (2011) has explored and mapped the different e-Governance implementation problem in Indonesia with the root causes of those problems. Authors have depicted it by the Fig-2 with the help of Fishbone Diagram.

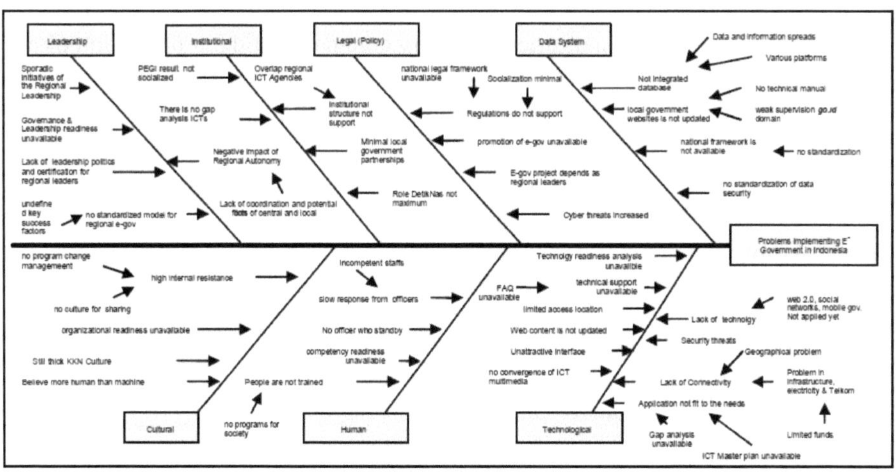

Fig. 2. Mapping Problems E-Government in Indonesia using Fishbone analysis

The Federation of Indian Chambers of Commerce and Industry (FICCI), West Bengal State Council (2012) have been used Fishbone Diagram to do root cause analysis for finding different challenges, faced to roll out e-Governance projects in the district Bankura, West Bengal. In their study, they have concluded that Document/ Guideline, Process/ Procedure, People, Technology, Delay in service delivery are the main challenges experienced by them to implement e-District project.

Adibi1 H., Khalesi N., Ravaghi H., Jafari M., and Jeddian A. R. (2012) have emphasized the different root causes for wrong transfusion medicine in medical service with the help of Fishbone Diagram and proposed some corrective actions to avoid serious concern and ensured patient safety in Tehran, Iran. Adibi1 H., et. al. have mentioned the following root causes for the problem "transfusion for wrong patient" with the help of Fishbone Diagram :

- **Assignment of blood transfusion to a relief nurse:** Insufficient supervision on relief nurse performance, Uncertainty of the duties and role of the relief nurses, Shortage of responsible and skilled nurses.
- **Employment of unskilled staff in emergency ward:** Shortage of competent nurse, Lack of efficient educational and vocational training, insufficient supervision of shift manager on staff recruitment, Unwillingness of expert nurses to be employed in emergence ward.
- **Poor adherence to transfusion protocols:** Incompetency of nurses in terms of blood transfusion skills, Stress and lack of motivation of the personnel, poor communication with the patient, Defective design of the emergency for control measures.
- **Incomplete information on the blood bag label:** Understaffed blood bank for essential control measures, Lack of or inattention to phlebotomy and transfusion protocol, failure of the medical staff to provide proper feedback to blood bank.

In Information Technology industries, it is very common that the projects do fail due to improper project management. Liu S., Wu B., and Meng Q. (2012) have researched on that and identified the following different critical and sub factors. They also have depicted all those factors by Fishbone Diagram:

- **Personnel:** Lacking personnel
- **Customers' Requirement:** Indefinite requirements
- **Team Members:** Inadequate Training, Frequent personnel flow, Lacking cooperative concept, Lacking perfect motivation mechanisms
- **Project Managers:** Emphasizing technology more than management
- **Top management:** Inadequate support
- **Communication:** Improper communication
- **Other Factors:** Changes in markets, policies and laws

In the other study, Bose T. K. (2012) has explored different problems in supply chain management and operation process for St. James Hospital with the help of Fishbone Diagram. According to him, the main problem areas are lack of proper equipment, faulty process, misdirected people, and poorly materials managed, improper environment, and inefficient management.

3 Methodology

Both the quantitative and qualitative methodologies have been adopted to carry out the research and random sampling method has been chosen for the survey. The survey was conducted among common citizens and Government officials who are associated with e-Governance projects and the respondents were selected randomly. During the survey, authors have captured the Name, Sex, Caste, Different Income Group and Different Age Group of respondents along with the e-Governance services feedback to get more visibility about the collected data. The districts were selected based on the different phases of e-Governance projects. For collecting primary data, both face to face interviews and questionnaires survey have been used and explored different websites for collecting secondary data. The fishbone diagram has been drawn from the collected data to do causal analysis.

3.1 Data Collection

The data collection process was intended to find out the root causes of the key challenges in implementing e-Governance projects. Extensive literature reviews have suggested authors about the existing challenges in different e-Governance projects. In the beginning, authors have framed questionnaires based on all the existing challenges and have taken simple testing among some people selected randomly to confirm its correctness. The primary data collections have been conducted by the following process:

i) Feed Back Form

The survey was conducted among various stake holders using feedback form based on the prepared questionnaires. In this approach, the printed questionnaires were handed over to the respondents at random and requested them to answer all the questions as per their preference. There were different multiple choice and open ended questions. The questionnaires were prepared in such a way that they were very simple and easy to answer.

ii) Interview

Both face to face and telephonic interviews, based on the prepared questionnaires, were conducted for some cases where feedback form was not possible to handover like Government officials.

The primary data collection was conducted among the following stake holders-

1) Common citizens, availing different e-Governance service,
2) IT vendor engineers, providing support for different e-Governance projects,
3) CSC operators, acting as interface in between e-Governance application and common citizens,
4) Kiosk operators, helping to common citizens to avail the services using KIOSKs
5) Government officials, involved directly or indirectly in various strategic decisions for implementing e-Governance projects.

The primary data collection was conducted at the following different locations

1) Headquarters of the selected District to meet with Kiosk operators and common citizens
2) CSC (Tathya Mitra Kendro) both located at Urban and Rural areas of the selected districts to collect data from CSC operators and common citizens
3) Webel Bhavan , Salt lake, West Bengal to collect data from Government officials
4) IT company (Tata Consultancy Services) to collect data from software engineers

3.2 Analysis and Interpretation

The analysis and interpretations of this study has been done in the form of graphical representation by Fishbone diagram. From the literature, it has been observed that Fishbone diagram is one of the most useful tools to do root cause analysis for any kind of problems especially for service control. The objective of this study is to identify the important challenges and root causes behind of those challenges for the e-Governance projects in West Bengal.

4 Analysis and Findings

A random survey has been conducted among 892 e-Governance stake holders from 9 different districts in West Bengal. In the Fig-3, it has depicted that the maximum participants of survey has been considered from North 24(Prgs) district i.e. 149 whereas minimum no. of participants from Birbhum district ie.54.

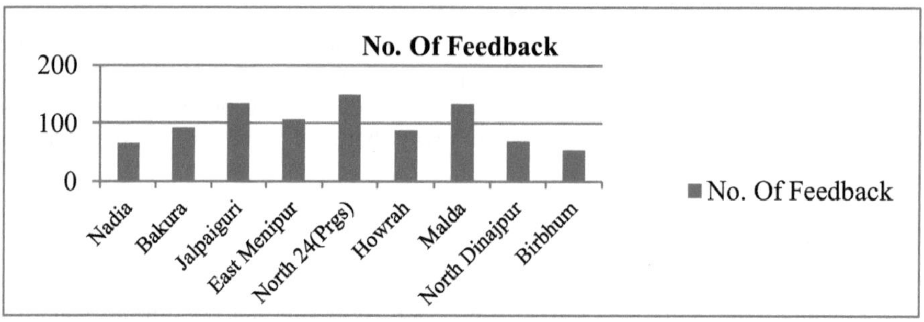

Fig. 3. Districts wise statistics

In the other way around, in Fig-4, it is showing that 67% of total survey sample has been considered from common citizens, 3% from IT Vendors Engineers, 23% from CSC operators, 6% from Kiosk Operators and rest 1% from Government officials.

The feedback has been collected in the form of different challenges to measure how different stake holders of e-Governance project feeling on the account of different challenges towards successfully roll out e-Governance projects in West Bengal.

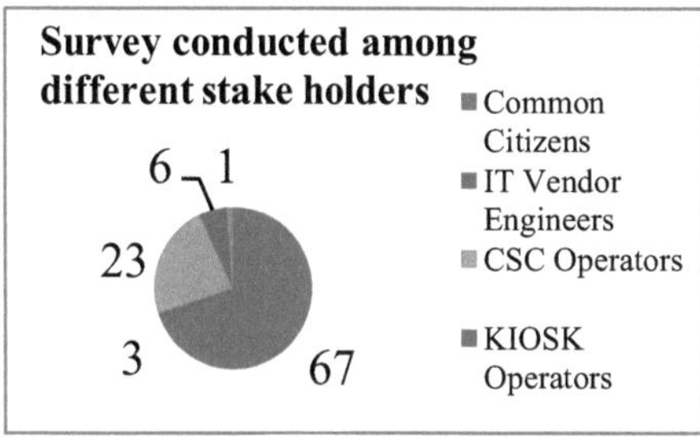

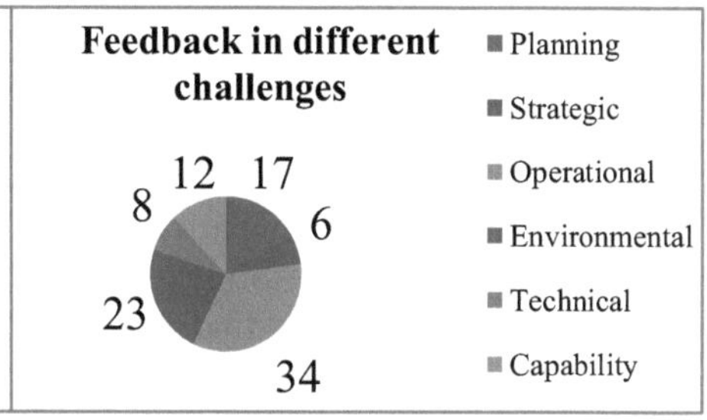

Fig. 4. Different challenge wise statistics **Fig. 5.** Different stake holders wise statistics

In Fig-5, the result of survey is showing that majority percent (34%) of people would have been said that Operational challenges effecting a lot, followed by Environmental challenges (23%) , Planning challenges (17%), Capability challenges (12%), Technical challenges (8%) and Strategic challenges (6%).The survey was intended not only to find out challenges, but the intention was to dig into more and more depth for identifying root causes of the challenges as well. In Fig-6, authors have entitled all the challenges and root causes of the challenges with the help of Fishbone Diagram. The challenges have been shown in the diagram by the main line/bone of the fish and the corresponding root causes have been indicated by figuring sub line/bone from the main line/bone of the fish.

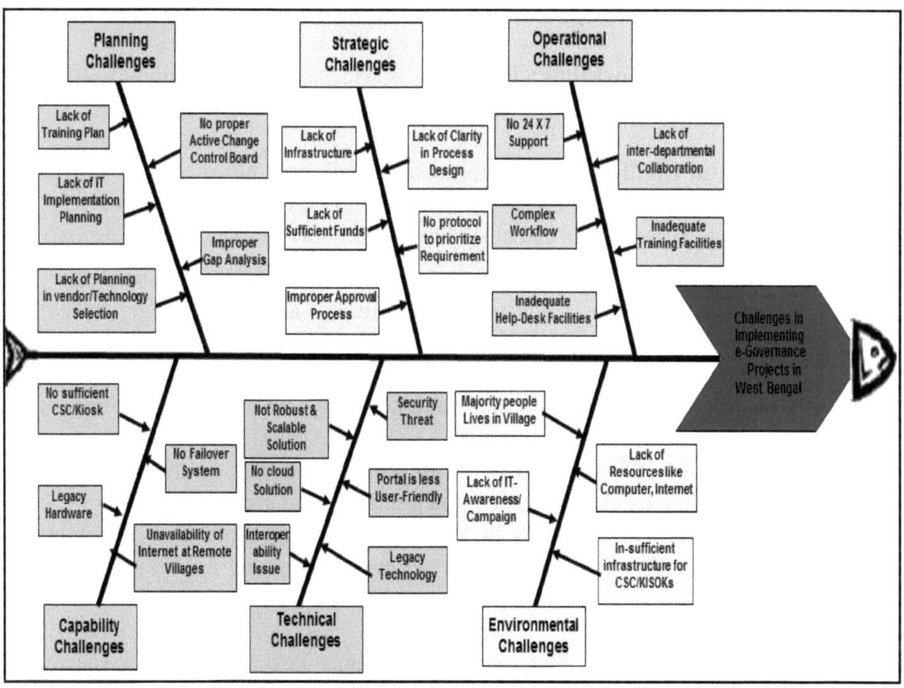

Fig. 6. The Fishbone diagram to depict the root causes of different challenges for implementing e-Governance projects in West Bengal

The following identified challenges and associated root causes of the challenges have been observed from the survey that conducted in different districts of West Bengal-

I. **Operational Challenges:** The operational challenges are typically very critical for success of any ICT initiatives. In this study, authors have been seen few operational challenges to implement e-Governance projects in West Bengal. The outcomes of this survey find the following root causes for operational challenges to roll out e-Governance projects-

- **No 24 X 7 Support:** There is no 24 X 7 support exists for e-Governance services. All the services are available during official hours i.e. 10:00 am – 17:00 pm.
- **Inadequate Help-Desk Facilities:** There is not sufficient no. of help desks to serve the huge population both in rural and urban areas.
- **Inadequate Training Facilities:** Well-trained professionals are inevitable for both the implementation and running the e-Governance projects successfully. There is lack of training facilities to groom CSC/KIOSK operators.
- **Lack of Inter-departmental Collaboration:** There is lack of collaborations among the various Government departments. Most of the departments use different software packages for their own purposes and no proper integration exist between them which make it difficult to execute certain services. Also it delays the whole service processing time.
- **Complex Workflow:** For some services, there are complex approval levels which delay the overall processing of the services.

II. **Environmental Challenges:** Environmental challenges play a vital role to make any successful roll out of e-Governance projects. Authors have found out the following root causes in this category during their survey works -

- **In-sufficient Infrastructure for CSC/KIOSKs:** Infrastructures like computer rooms, internet availabilities and supporting stuffs are adequate to run e-Governance projects.
- **Lack of Resources Like Computer, Internet:** Lack of resources like Computers, Internet is also another key factor to run e-Governance projects in rural areas.
- **Lack of IT-Awareness/ Campaign:** GoWB should increase IT awareness by continuous campaigning in rural areas.
- **Majority People Lives in Village:** Majority people live in rural areas some of them are in remote villages. They are not much computer educated.

III. **Planning Challenges:** The most important key factor for successful implementation of e-Governance projects is to do proper planning from the beginning with requirement gathering and at the end with successful deployment. In West Bengal, the major root causes behind the planning challenges are-

- **Lack of Planning in Vendor/Technology Selection:** Most of the cases there are local system provider or old-dated software are used for the solution which may become obsolete within few years.
- **Lack of IT Implementation Planning:** Improper IT implementation planning which is delaying the whole implementation e-Governance projects in West Bengal.
- **Lack of Training Plan:** Improper training plan causing long delays for CSC/KIOSK's operator to become expert.
- **Improper Gap Analysis:** In most of the cases, Gaps between actual manual processes and software processes are not being analyzed properly.
- **No proper Active Change Control Board:** No change control board. It may hamper proper enhancement/migration of the software components.

IV. **Capability Challenges:** The capability set up is another basic requirement to ensure smooth roll out of e-Governance projects. In this study, authors have found out that the implementation of e-Governance projects in West Bengal are suffering from the following root causes related to Capability challenges–

- **No Sufficient CSC/Kiosk:** Number of CSCs/KIOSKs should be increased to provide the e-Governance services to huge populations.
- **Unavailability of Internet at Remote Villages:** In some of the remote villages in West Bengal, internet facilities are yet to be activated.
- **No Failover System:** Failover strategies didn't implemented yet. There should be Disasters Recovery System (DRS) in place.
- **Legacy Hardware:** Servers, Computers and all accessories are not up to date.

V. **Technical Challenges:** The technical factors are the backbone of any IT projects because it is responsible to make the solution more robust, simple and extensible. In this survey, the following root causes have been identified as a reason of technical challenges to roll out e-Governance projects in West Bengal-

- **Not Robust and Scalable Solution:** Most of the e-Governance projects are running as pilot basis.They are not robust or highly scalable solution.
- **No Cloud Solution:** Most of the e-Governance projects follow client-server architecture or n-tier architecture. No cloud solution exists so far. Cloud solutions can reduce service price and maintenance costs.
- **Interoperability Issue:** e-Governance projects has been developed in multiple technologies and deployed in different platforms, but there is no standard interoperability solution to support cross platform solutions.
- **Security Threat:** Most of the solutions are prone to security threats. It's a big issue for e-Governance projects in West Bengal.

- **Portal Is Less User-Friendly:** Some of the portals are not user friendly at all and takes long time to expertize the solutions**.**
- **Legacy Technology:** Most of the e-Government projects are legacy in nature and they are not integrated at all.

VI. **Strategic Challenges:** Any e-Governance projects in reality can be a success story if and only if it is carried out with proper strategies. Authors have found out the following root causes related to strategic challenges -

- **Lack of Infrastructure:** Infrastructures like computer rooms, internet availabilities and supporting stuffs are adequate to run e-Governance projects.
- **Lack of Sufficient Funds:** Funds allocated to the e-Governance projects are not sufficient to run them smoothly.
- **Improper Approval Process:** There are some improper approval processes for some services which require high level simplifications in approval workflow.
- **Lack of Clarity in Process Design:** There are some processes for some services which are not clear in implementation. Clear and unambiguous processes should be defined beforehand.
- **No Protocol to Prioritize Requirement:** No standard rules/protocols in place to prioritize the existing requirement that's sometimes leads delays to implement the urgent requirements.

5 Conclusion

The outcome of this research is extensively outlined from the survey work. Literatures have assisted initially to get understand about the challenges though finally it has been formalized by doing sample survey testing. Authors have concluded that operational, strategic, environmental, technical, capability and planning are the major challenges those are extremely affecting implementation of e-Governance projects in West Bengal. Authors also do believe that the successful implementation of e-Governance projects are possible if and only if these root causes could be identified properly and Government can pay special attention to eliminate these root causes. Thus, Authors have pointed out some key root causes from survey data and finally depicted them with the help of fishbone diagram.

6 Limitation

The survey was limited to nine districts in West Bengal only and also sample size was limited to only 892. More samples should be collected to make this survey work more robust and fruitful.

References

1. Adibi, H., Khalesi, N., Ravaghi, H., Jafari, M., Jeddian, A.R.: Root-Cause Analysis of a Potentially Sentinel Transfusion Event:Lessons for Improvement of Patient Safety. Acta Medica Iranica 50(9) (2012)
2. Bose, T.K.: Application of Fishbone Analysis for Evaluating Supply Chain and Business Process-A Case Study On The ST James Hospital. International Journal of Managing Value and Supply Chains (IJMVSC) 3(2) (2012)
3. FICCI (2012), http://www.pmi.org.in/downloads/PMI_FICCI_West_Bengal_2012.pdf
4. Wikipedia, https://en.wikipedia.org/wiki/Ishikawa_diagram
5. Li, S.S., Lee, L.C.: Using fishbone analysis to improve the quality of proposals for science and technology programs. Research Evaluation 20(4), 275–282 (2011)
6. Liu, S., Wu, B., Meng, Q.: Critical Affecting Factors of IT Project Management. In: International Conference on Information Management, Innovation Management and Industrial Engineering (2012)
7. Sensuse, D.I., Lusa, S.: Socio Technology Perspective for E-Government Implementation in Indonesias, Laboratory of E-Government (2011)
8. Department of Electronics and Information Technology, http://deity.gov.in/
9. West Bengal IT Policy (2003), http://www.itwb.org/download_pdf/itpolicy_2003.pdf

Referen[illegible]

Tackling Supply Chain through Cloud Computing: Management: Opportunities, Challenges and Successful Deployments

Prashant R. Nair

Vice-Chairman, IT, Amrita School of Engineering, Amrita Univesity, Amrita Nagar, Coimbatore, 641112 India
prashant@amrita.edu

Abstract. An important requirement of supply chain management is to attain supply chain visibility among multiple stakeholders and partners. With several enterprises off shoring their manufacturing and service operations to low-cost hubs in Asia with poor infrastructure and transportation networks, visibility has become a major challenge. Various Information and Communication Technology (ICT) tools like RFID have enabled supply chain visibility and agility. Cloud computing and associated technologies like virtualization, Software as a Service (SaaS), Platform as a Service (PaaS) and Infrastructure as a Service (IaaS) have been touted as the 'next big thing' and 'game changer' for enterprises to improve their top line and bottom line. ICT interventions in supply chain planning and execution using Cloud Computing with evidences of early adoptions and deployments by pioneering enterprises have been highlighted. Application areas where cloud-based solutions are available include demand forecasting, demand planning, e-procurement, distribution, inventory, warehouse and transportation systems.

Keywords: Supply Chain Management (SCM), Cloud Computing, visibility, Software as a Service (SaaS), stakeholders, logistics, demand, transportation, warehouse, inventory.

1 Introduction

Supply chain visibility has become a source of competitive advantage for businesses. This complements the need to have access to real-time information in an actionable manner as also be able to negotiate and manage relationships within and between various stakeholders like suppliers, customers and transporters. ICT early adoption and deployment across the supply chain has become a force multiplier and determinant of competitive advantage for many enterprises [1]. Extensive use of technologies like Supply Chain Management (SCM) Software packages, both independent as well as within the ERP framework, RFID [2], bar-codes, web services, decision support systems, transportation & inventory management systems etc have resulted in better visibility, agility, collaboration and communication both within and among enterprise.

S.C. Satapathy et al. (eds.), *ICT and Critical Infrastructure: Proceedings of the 48th Annual Convention of CSI - Volume II*, Advances in Intelligent Systems and Computing 249,
DOI: 10.1007/978-3-319-03095-1_82, © Springer International Publishing Switzerland 2014

With the recent moves by enterprises in North America and Europe to move their manufacturing facilities to low-cost locations in Asia and other regions coupled with the business process outsourcing boom, it makes business sense for enterprises to make things in Asia such that the extended supply chain and additional cost was more than offset by the financial windfall stemming from reduced labor and associated cost [3]. One major off shoring challenge is poor and unreliable transportation networks in many of these hubs in Asia. Absence of warehousing and fleet data further complicates the situation resulting in poor visibility in the supply chain. The ideal situation is that these extended supply chains which span many continents and geographies need to be simple with less number of partners and suppliers with information at our finger tips. On an average some of these international shipments could have 8 to 10 different trading partners.

Cloud computing and associated technologies like virtualization; Software as a Service (SaaS), Platform as a Service (PaaS) and Infrastructure as a Service (IaaS) are touted as the next 'big' thing and game changer for enterprises. Cloud computing can integrate all partners in this increasingly global extended supply chain. These could include shippers, service providers, distributors, logistic providers, customers, sellers etc. A social network like community could be created with their participation. Typical data items that could be placed online on the cloud include information on prices, inventory, schedules, service options, contacts, announcements etc. This increases the local responsiveness of enterprises while they pursue their global strategy and plans. Timely information updates from all stakeholders renders companies to be demand-driven rather than forecast-driven [4]. Else there is a tendency to amass stockpiles of inventory based on forecasts. Major ERP vendors are also working offering cloud-based solutions.

Cloud-based supply chain management provides us the benefit of real-time pricing. Visibility of each element in the supply chain provides us with the opportunity to control costs [5]. Thereby we are in a position to monitor and control costs and thereby right price. These tools are also scalable and flexible.

Companies have started using SaaS applications for managing their supply chains, which are part of larger frameworks called public clouds. A recent survey conducted by E2open in collaboration with SCM World, found that cloud led to significant improvements in metrics such as inventory days [6].

FedEx has a private cloud deployed in 2011 with CloudX [7] as service-provider. Early cloud adopters include vendors like IBM, Mercury Gate, JDA, Amber Road and Ariba, who are offering public cloud deployment models for various supply chain operations and activities. Sales processing and CRM are the activities on the cloud.

GT Nexus is banking 'big' on the cloud by offering a cloud-based platform, which is being used by more than 15,000 organizations [8]. In addition to visibility and lower costs in terms of Total Cost of Ownership (TCO), benefits of using cloud services include less infrastructure cost and platform scalability and flexibility. One of the first cloud-based SCM solution providers is Amitive with its product Amitive Unity 5.0 targeted at large and small companies that outsource manufacturing. Orthera is another early provider [9].

However enterprises which go for deployment of cloud-based solutions will be grappled with some challenges like heterogeneity in the legacy information systems and software applications of various partners in the supply chain. Most of the data that companies need to run their supply chains resides with partners and they use their own systems and solutions, whether proprietary or open-source. Another issue that prevents companies struggle to get a unified picture of their supply chains is that most of these information systems were designed to operate within a single company, not across a network of companies [10]. There are also privacy issues in both public and private clouds as the service provider has all the data.

2 Cloud Computing

Forrester Research projects that the global cloud computing industry will grow from $40.7 billion in 2010 to more than $241 billion by 2020 [11]. Every ERP vendor is coming out with a cloud offering that reduces the Total Cost of Ownership. There is considerable interest across all sectors and geographies towards the cloud paradigm and how it can improve both the top and bottom lines for enterprises.

Cloud computing is a form of utility computing, where hardware, software, storage and platform is made available as per need and on a subscription basis. In this service model, clients can access the cloud-based application through an Internet browser. The data can be resident at a remote place also. Complementing the cloud is the usage of server farms and data centres where all applications and data can be stored, shared and accessed on demand using virtualization. Cloud is a 'green' technology as it eliminates the need for enterprises to procure and maintain large servers and associated space and infrastructure. As a form of distributed computing over a network, cloud computing makes full use of shared services and resources.

There are two deployment models for cloud, public and private cloud. Public cloud services are available to general public. In public cloud, one can access the services using the Internet. Amazon and Google have built their large public clouds. Some services like You Tube run on a cloud. Private cloud is built by enterprises for their internal operations and processes including communication with their branches and units. These could be maintained by a third party though. Of late, there is some movement to architect hybrid clouds which combine features of both private and public clouds. In a hybrid cloud, a company can maintain its own private cloud and on saturation of these resources, use the public cloud [12].

Popular cloud service models are:

- Software-as-a-service (SaaS) where application software like word processors, spreadsheets and databases can be availed on demand
- Infrastructure-as-a-service (IaaS) where hardware, servers, storage space are shared and offered on pay per use basis
- Platform-as-a-service (PaaS) where the operating system or programming language execution environment can be availed on demand

3 Cloud Application Areas in SCM

3.1 Demand Planning and Forecasting

Enterprises have always been grappling with the challenge of understanding demand requirements. Various predictive models and heuristics are in vogue to give forecasts. The 'Cloud' advantage can give enterprises access to timely and actionable information. Cloud platforms are being used for demand forecasting by coupling and coordinating various links in the supply chain like retailers, suppliers and distributors. Cloud-based tools are available for capturing and analyzing sales data and executing statistical demand forecasts [14]. Typical data items that could be placed online on the cloud include information on prices, inventory, schedules, service options, contacts, announcements etc Order and Demand planning can be facilitating the cloud network with accurate forecasts. In addition to this, insights into market segmentation, customer product preferences, integrated sales and operations planning could also be obtained [13]. Demand solutions have an offering, Demand Solutions platform in a SaaS format on the Windows Azure platform.

3.2 E-Procurement and E-Distribution

Cloud platforms are inherently collaborative in nature. These tools can negotiate through an array of suppliers and get the best e-procurement results as well as focus on customer requirements for distribution of products and services. This helps enterprises to balance supply and demand. Companies will be able to choose the best suppliers as per their needs and specifications. Moreover, cloud-based tools enable companies and suppliers to mutually develop contracts and thereby drastically improving contract management [14]. The procurement and transportation data can be analyzed and mined to achieve optimal routes and utilization of equipment as also eliminates waste [5]. Asite and Coupa are examples of SaaS providers for e-procurement. Skanska, a leading construction and facilities management company in UK is using Asite eProcurement to manage their purchasing cycle including tendering, purchasing, delivery logistics and goods receipt, and payment processes.

3.3 Inventory, Warehouse and Transportation Management

Cloud computing tools are available for inventory, warehouse and transportation management. Collaborative computing using cloud-based solutions help in inventory tracking and optimization as also improve supply chain visibility. These tools also help to integrate forward and reverse logistics in the same closed-loop supply chain and give companies the edge [15]. In particular, Third Party Logistics (3PL) providers can exploit this cloud opportunity to their advantage [10]. Real-time inventory data makes supply chains robust and well-equipped to handle surprise demand fluctuations. Jump Software is an example of a warehouse management system cloud provider which takes care of tracking shipments and inventory for the warehouse operations. Precisio Business Solutions has developed inventory management software on the cloud using salesforce.com platform. Likewise cloud apps for transportation and fleet management are also available. eBizNET is a provider of Cloud-based Warehouse Management System. Lean Logistics and JDA solutions are two vendors with cloud solutions for fleet, transportation and logistics management. These help to respond better to ever-changing customer demands as also enable timely re-ordering and replenishment of inventory. Cloud tools entail the need for various partners in the supply chain to integrate their warehouse, inventory and transportation systems. This is also an opportunity to migrate from legacy systems to a seamless online community.

4 A Conceptual Model of Cloud Supply Chain

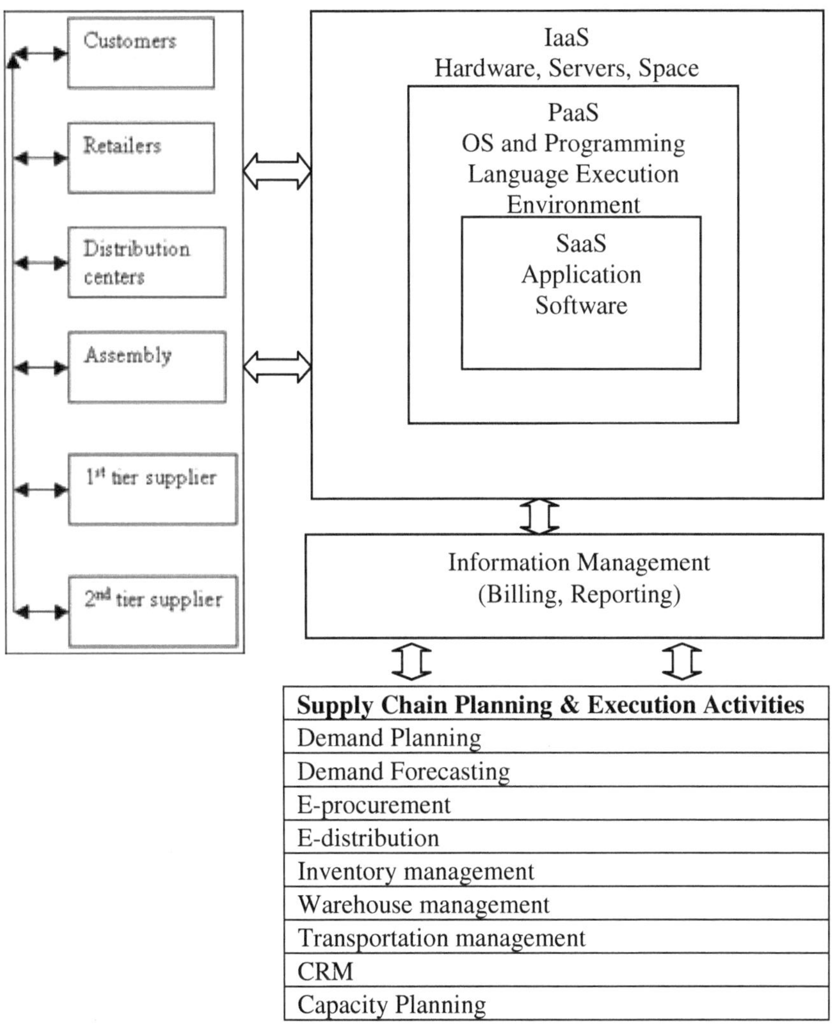

Fig. 1. A Conceptual Model of Cloud Supply Chain

5 Enterprise Deployments of Cloud Computing for SCM

- Several medium size enterprises have deployed JDA Cloud Services with spectacular results. One of them is Massdiscounters, a South African retail player in FMCG, grocery, appliances and electronics with 133 stores. JDA cloud has Massdiscounters has enabled accurate forecasting and fulfillment of demand. This cloud solution is has enhanced supply chain visibility by improving inventory positions and enabled quick response to demand fluctuations.
- NII Holdings, a company based in Virginia, which offers mobile services in Latin America to over 11 million subscribers under the Nextel brand, has also deployed JDA Cloud Services. This has improved demand shaping and inventory positions. Elimination of manual interventions in inventory tracking and control through online media and cloud infrastructure has increased accuracy.
- Joyent Cloud solutions have made a marked difference for Container and Pooling Solutions (CAPS), which is the largest container management service in North America. CAPS containers are used by all industries like automobile, grocery, bottled drinks, manufacturing etc. Joyent has migrated the CAPS tracking and management system to the cloud with tangible benefits accrued includes scalability, reliability, fault tolerance and quick response to demand fluctuations especially handling sudden spurts and surges [16].
- Intel was able to replace hundreds of their order clerks using online ordering applications. Several supply chain activities like planning and forecasting, sourcing and procurement, logistics and inventory management were migrated to the cloud [14].
- COSCO Logistics, the largest 3PL company of China and the world's second largest ocean shipping company is using SaaS service and integrating all stakeholders like customers, subsidiaries and distributors in order all of them to use the same logistics management software [7].
- Exceedra is a UK based software vendor providing supply chain solutions to leading companies like Pirelli, Revlon and Heineken. This solution is primarily used for monitoring and analytics of inventory as also tracking the reach of product promotions. Using the Microsoft Azure platform, Exceedra has unveiled cloud versions of their solutions, which go by brand names, Procast and ActNow. These applications use software agents which automatically sign on to a retailer's supplier portal to extract data. Data storage and processing are performed on the Windows Azure platform [17]. The solution includes web-based dashboards for tweaking, mining, analysis and viewing of data. One Exceedra customer, Plum Baby has spectacular results. In a few years, Plum Baby, a baby food company has graduated from being a startup to an enterprise with more than $ 40 million in revenues.

6 Conclusion

Cloud-based Supply Chain Management solutions offer visibility, agility, transparency, flexibility, scalability, simplicity, competitiveness, collaboration, cost benefit and operational efficiency to enterprises. Cloud computing can integrate all partners in this increasingly global extended supply chain into an online social network like community with real-time information on all elements in the supply chain. Several solutions are now available and many enterprises have made the shift with good results. Application areas where cloud-based solutions are available include demand forecasting, demand planning, e-procurement, distribution, inventory, warehouse and transportation systems. Enterprise case studies of successful cloud deployments for supply chain management are also showcased. One challenge that enterprises which go for deployment of cloud-based solutions will face would be heterogeneity in the legacy information systems and software applications of various partners in the supply chain

References

1. Nair, P.R., Balasubramaniam, O.A.: IT Enabled Supply Chain Management using Decision Support Systems. CSI Comm. 34(2), 34–40 (2010)
2. Nair, P.R.: RFID for Supply Chain Management. CSI Comm. 36(8), 14–18 (2012)
3. Computer World Information, http://www.computerworld.com
4. Christopher, M.: The agile supply chain: Competing in volatile markets. Ind. Mark. Mgmt. 29, 37–44 (2000)
5. Supply Chain 24/7 Information, http://www.supplychain247.com/article/7_benefits_of_cloudbased_logistics_management/cloud
6. Jha, V.: Impact of Cloud Computing on Supply Chain Management. IIM Indore Mgmt Canvas (2013)
7. Information Technology Research Institute Information, http://rfid.uark.edu/research-papers.asp
8. The CIO Information, http://www.cio.com/article/692784/The_Cloud_Solves_Those_Lingering_Supply_Chain_Problems?page=2&taxonomyId=3024
9. Ojha, J.: Distributed environment of cloud for supply chain management. Int. J. of Engg. & Inno. Tech. 1(4), 50–55 (2012)
10. Toka, A., Aivazidou, E., Antoniou, A., Arvanitopoulos-Darginis, K.: E-Logistics and E-Supply Chain Management: Applications for Evolving Business. IGI Global, Hershey (2013)
11. Supply Chain Europe Information, http://www.scemagazine.com/the-rise-of-the-cloud/
12. Sujay, R.: Hybrid Cloud: A new Era. Int. J. of Comp. Sci. & Tech. 2(2), 323–326 (2012)
13. Grimson, J.A., Pyke, D.K.: Sales and Operations Planning: An Exploratory Framework. Int. J. of Logi. Mgmt. 11, 255–274 (2007)
14. Schramm, T., Nogueira, S., Jones, D.: Cloud computing and supply chain: A natural fit for the future. Logistics Mgmt. 3, 9–11 (2011)
15. Guide, V., Harrison, T., Wassenhove, L.V.: The Challenge of Closedloop Supply Chains. Interfaces: The INFORMS. J. of Oper. Res. 33(6), 3–6 (2007)
16. Joyent Information, http://hoffmancloud.com/docs/Cloud-Case-Study-Supply-Chain-Joyent.pdf
17. Microsoft Case Studies Information, http://www.microsoft.com/casestudies/Case_Study_Detail.aspx?casestudyid=4000011040

e-Health and ICT in Insurance Solutions

Josephina Paul

Kerala Agricultural University, Thrissur
jpkktom@hotmail.com

Abstract. Health informatics is an emerging field of the day. It is the appropriate and innovative application of the concepts and technologies of the information age to improve health care and health [2]. Right from consultation, through prescription till the completion of treatment and medication can be implemented online. A network that includes the hospitals, clinics, pharmacies, diagnostic centers and insurance companies sharing the pooled resources can perform in co-ordination and work for the common goal of delivering medical claims to the stakeholders. An online software that works on Windows Server platform with SQL Server as backend and the ASP.NET programs do the purpose.

Keywords: Health Informatics, e-health, medi-claims, ASP.NET, SQL Server.

1 Introduction

Healthcare is the prime care that is intended for a human being. According to the Wikipedia, primary care is that term for healthcare services, refers to the work of health care professionals who act as a first point of consultation for all patients within the healthcare system [1]. Living healthy is every man's right. To ensure healthy life of the citizen is thus, the mandate of the state and the Government. Providing state-of-the-art healthcare network all over the country at reasonable price is every government's responsibility. Unfortunately, there have been so many constraints which pull the government machinery backwards from this very divine drive. Though the health policies of the countries across the globe are varying, there has been a general trend of welcoming the influx of insurance providers to the health care sector over the past two decades. The corporates and public limited companies are gorgeously handshake with this paradigm shift of health policy by implementing the health insurance to its employees and workers with a tie up between private or public limited insurance providers.

Health insurance is the insurance against the risk of incurring medical expenses among individuals. According to the Health International Association of America, health insurance is defined as "coverage that provides for the payments of benefits as a result of sickness or injury. It includes insurance for losses from accident medical expenses, disability or accidental death and dismemberment."

Since an accident, sickness or death makes not only the person who suffer from this invalid temporarily or permanently, but also the family of the members of the

S.C. Satapathy et al. (eds.), *ICT and Critical Infrastructure: Proceedings of the 48th Annual Convention of CSI - Volume II*, Advances in Intelligent Systems and Computing 249,
DOI: 10.1007/978-3-319-03095-1_83,

person to a great extent. The insurance sector is playing a significant role in maintaining a healthy living of the society. In this context, it is worth saying about the e-health.

The area of e-health encompasses products, systems and services, including tools for health authorities and professionals as well as personalized health systems for patients and citizens [4]. The scope of e-health includes the hospital treatment to the population health activities, which present complex information management challenges to support individualized patient care [2].

The sooner the insurance sector enact in the situation of emergency, the faster the family/society gets relief. However, there are a few limitations from their side. Unless and until they get the medical reports, the claims cannot be processed and delivered. The online servicing of medical claims is an application of health informatics comes into the scene at this juncture. Health informatics is an emerging field of the day. It is the appropriate and innovative application of the concepts and technologies of the information age to improve health care and health [2]. Health Informatics has also been defined by WHO as "an umbrella term used to encompass the rapidly evolving discipline of using computing, networking and communications – methodology and technology – to support the health related fields, such as medicine, nursing, pharmacy and dentistry" [3].

Right from consultation, through prescription till the completion of treatment and medication can be implemented online. A network that includes the hospitals, clinics, pharmacies, diagnostic centers and the insurance companies can perform in co-ordination and can work for the common goal of delivering medical claims to the stakeholders with utmost urgency.

2 Software System

An integrated online software that works on windows platform with SQL server as backend and the ASP.NET programs do the purpose. Such a system was developed and it could implement successfully with the co-operation of more than 30 clinics and hospitals, laboratories, pharmacies and insurance companies in the network.

The website with fully online ASP.NET software that runs on MS-Windows 2003 server and its Internet Information Server(IIS) showed good results. The concept of the system is that it consists of service providers – Insurance service companies and Medical service providers such as hospitals, laboratories, pharmacies and diagnostic centers- and service receivers/members ie. the individuals who insured and companies who pays the insurance premium of their employees and dependents. The system is designed in such a way that a third party/agent can act as the coordinator and use the system to serve many providers and members simultaneously.

2.1 Software Platform

The entire system works on an IIS Server that runs on Windows 2003 Server platform. The hardware is a high end IBM Server machine with dual hard disk of

500GB each with automatic back up facility and is connected to the ISP via optical fibre network. The program is written in ASP.NET on its VB.NET language and java script for efficient GUI performance. The ASP.NET program runs on dot net framework 2.0, and therefore the system is three tier with adequate inbuilt security. The back end database is SQL Server 2000/+ with multiple tables in normalized form with joins and views and stored procedures. Since stored procedures have been written for query retrieval within the program and for the reports, the smooth extension and modification of the system is ensured. The reports are written in Crystal reports 9.0 that generates customized reports according to the user's requirement but with a professional outlook. The system is online completely on a security enabled http server and available on World Wide Web. All the service providing members of the system such as hospitals, clinics and the insurance companies can access the system online with their own login Id and password. All the admin utilities like password change, back up utilities etc are incorporated in the system as well.

2.2 Software Modules

The operation of the system is divided into modules, and has mainly four modules. All the modules have been password protected and can be operated only by the user groups such as doctors, Pharmacists, Laboratory Technicians etc, who have been registered as authenticated users of the system. A report section that generates various reports into soft and hard copy, with customization facility makes the software robust and flexible to the end users.

1. Admin Window

This module is used to input and maintain the master data and images to the system with security. The users of the system are categorized into multiple groups with different levels of rights for the access and execution/use of available options /functions of the system and they are created and managed by the admin module. They are given unique user Ids and passwords with permission of password change.

Whenever a new provider hospital or a servicing company joins the system, the details are entered through the admin window providing an identification number to the company, a unique number used for the retrieval and access of the company. This window enables the administrator to modify and delete the existing details with a security password and controlled privileges. Similarly, when a new patient registers into the system, the details are entered via admin or from the doctor's diagnostic window itself, as it automatically transfers the control to the patients' details screen, where the necessary fields can be filled in. Each patient is provided with a card by the insurance company with a unique identification number and the details. A third group is doctors, whose details also can be entered through the admin window, where each doctor is attached to a hospital or a clinic.

2. Doctor's Consultation Window

In the second module, doctor's diagnosis of the patient's diseases and prescriptions of medicine, treatments, lab tests etc. are entered. This window is handled by the doctors themselves so that nobody else can operate on this option and can do any malpractices. Each doctor is given with a login id and password unique for them. As soon as the patient's id on the card is entered, a window is loaded, where the details of treatments and medicine can be entered by the doctors. Moreover, the past history of treatments if any, is loaded in the same window.

Once the prescription is entered by the doctor, the pharmacist can view the prescription online when the patient's id is entered. According to the prescription the pharmacist dispenses the medicine, and the system puts a flag on this medicine in the system showing that the medicine has been delivered so that a duplicate dispensing/billing of the same medicine cannot be possible by the pharmacists or anybody else. Nobody except the administrator has the privilege to make any changes necessitated in the course of errors if any, occurred. These kind of security measures will protect the insurance companies from financial loss due to double entries or malpractices. The functioning of laboratory and diagnostic tests options are in similar fashion as well.

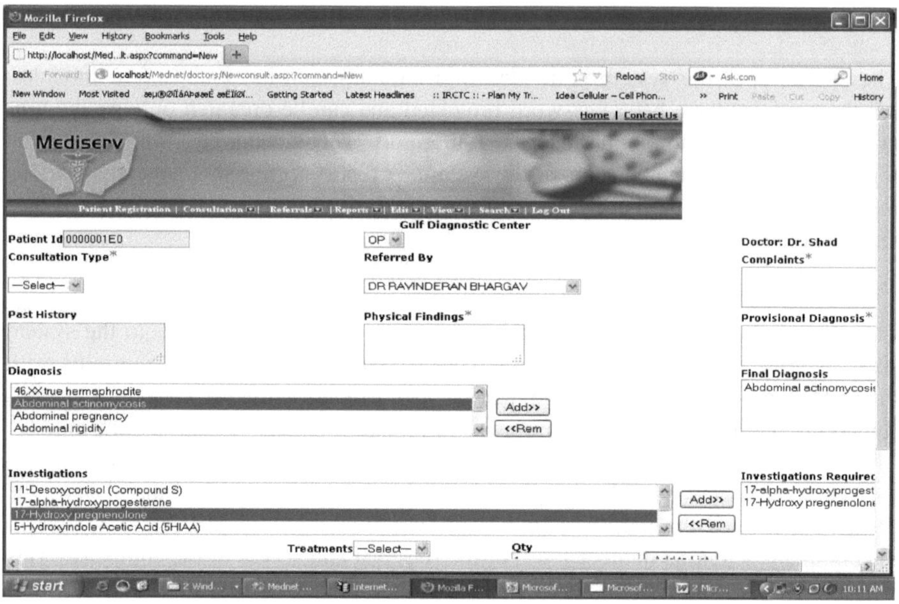

Fig. 1. Consultation and Medication window of the System

3. Report Section

The system has the capacity to generate various kinds of customized reports to meet the diversified needs of the stakeholders. The reports can be from user level to admin level and from detailed to summarize ones. The reports include general details of providers, patients, treatment details, diagnostic tests, medicines prescribed, dispensed, its cost and details of hospital where the treatments has been done with the period of treatments, doctor who treated, pharmacy and laboratory where the services has been rendered etc.

An important group of reports are account level details and statements. Reports such as monthly amount to be paid to the hospitals, clinics and other service providers, details of account statements in different level ie. transaction wise and summary wise etc can be generated. Specialized reports of transaction details can be retrieved patient wise, doctor wise, per service provider etc. All the details can be obtained customized to a specific period/beneficiaries as well. Yearly account statement is another specialty of the package. All the reports are displayed on the screen as well as the hard copy can be printed in full or as selected pages.

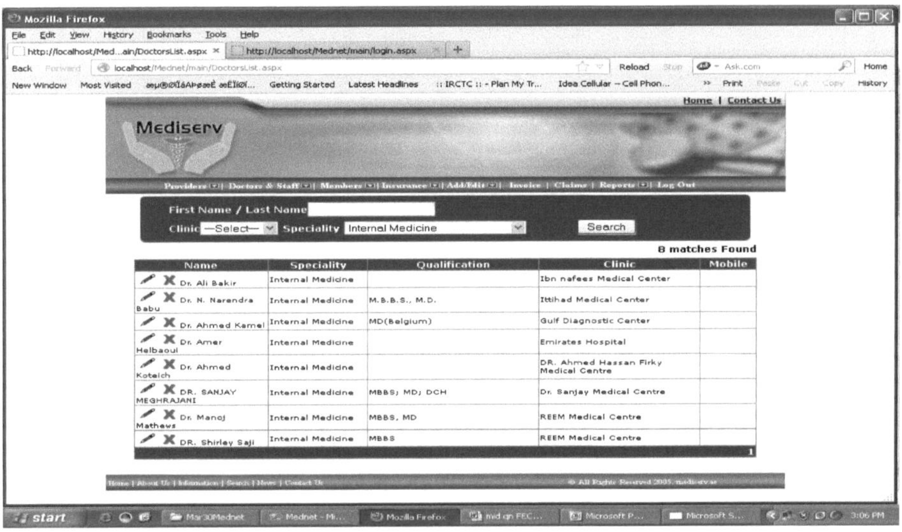

Fig. 2. Search for Doctors Listing window of the System

4. Monitoring Module

The software incorporates a module for online monitoring of the prescriptions of the doctors by the insurance company, ie. the coordinating insurance company has got a team of doctors who will monitor the treatments and doctor's prescriptions online so that a doctor cannot prescribe any medicine which is too costly, while the low cost alternative is available in the market. For necessary cases where a diversion of treatments is required, prior permission of the insurance company can be sought out and the same would be granted if the request is genuine. The double checking of the insurance company and its online monitoring do help maintain the insurance company free from bad debts and unnecessary overheads caused by fake prescriptions.

5. Website Module

The system has a website module with a good graphical outlook and interface. The website carries information such as health news, health tips, company details, link to advances in medical and allied branches of science and the like.

Fig. 3. Web site window of the System

3 Discussion

The menu driven system integrates the operations of diversified business entities to a single platform performing in coordination for a common goal by taking advantages of the pooled resources [5]. As the software is intended to support the fast delivery of the insurance claims to the incumbents, it has been designed as a web application to ensure its real time availability to a network of large number of customers spread over a wide geographical area. The system is fast and robust and at the same time leverage with multi-tier security of ASP.NET, dedicated server and password authentication. For added security, the site can be deployed on an http server without www access. The software can be utilized by the insurance providers directly or by a third party who can act as a service co-coordinator by providing the services to the insurance companies and the stake holders such as hospitals, pharmacies and the beneficiary companies. The headache of offline processing of large volumes of insurance claims and the time delay incurred at various check points can be avoided to a great extent which expedites the delivery of claims to the affected party. The online paperless claims not only save time and labor but also are environment friendly.

The best part of this package is that, the expenses incurred by each patient will be managed by the software itself without allowing going overdraft by putting a on ceil on the credit limit of the individuals. Once the expense due to a patient is exhausted, no more treatments are allowed without payment. Another advantage is that there is a module for online monitoring, ie. the insurance company has got a team of doctors who will monitor the treatments and doctor's prescriptions online so that a doctor cannot prescribe any medicine which is too costly, while the low cost alternative is available in the market. For necessary cases where a deviation of treatments is required, prior permission of the insurance company can be sought out and the same would be granted if the request is genuine. The double checking of the insurance company and its online monitoring do help maintain the insurance company free from bad debts and unnecessary overheads caused by fake prescriptions. Handling of referral cases by the specialist doctors of referral hospitals are an added advantage of the system. The system also generates various reports relevant to the hospitals and the insurance companies as well. Another feature of the software is the website associated with it. Recent developments in medical and allied fields can be accessed via website as there are links provided to the information rich sites on the web.

While discussing on the implementation aspects the system is mainly intended to cater to the needs of the medical claims of the corporates and big companies where a large number of employees and their dependents are benefited with the insurance claims. It can also be used by companies having medium to small staff strength. The software system is feasible in every aspect of social and technical nature and economically viable too.

Since the entire system is operating on a high end dedicated server running with routine backup services, the huge amount of data generated by medi-claims at physically divided systems on various locations can be stored in the machine and thus reduces the risk of loss of data at the individual points of generation. Secondly, the common resources and data have been shared by companies that differing in their nature of operations. The efficiency and scalability of the system is more as it uses the compiled aspx web application in dotnet platform with robust SQL Server database backend and its fast processing stored procedures to retrieve and execute the queries. However, the overall performance of the system may get varied with the speed and bandwidth of the network media.

4 Future Expansions

Presently, the medical claim has become an integral part of employees of public and private sector corporates and companies as they are competing to provide excellent services to their employees. Therefore, the medical insurance premium will be remitted by the companies and the employees do not have to worry about this expenditure from their purse. This software is suitable for corporate employees who have been provided with free insurance coverage for their own treatments and their family.

The system can be expanded to include any number of service members and beneficiaries. With the advent of computers and the impact of IT in the medical diagnostics, most of the hospitals are operating on enterprise software or on their own tailor made programs; this software also can be integrated with their systems so that duplication of documentation can be avoided. Those hospitals without any operating software can use the software for operating purpose as well by incorporating a few modules such as appointment fixing, inpatient routine etc. into it. Similar systems are being used in Singapore, The Middle East, Europe and western countries.

In order to adapt with the hardware/software platform updating and maintenance of the system, slight modification/conversion of the program only required without considerable recurring costs and inputs.

References

1. Srinivasan, R.: Healthcare in India- Vision 2020, Issues and prospects., http://www.planningcommission.nic.in/reports
2. http://www.e-health.standards.org.au/ABOUTIT014/WhatisHealthInformatics.aspx
3. Al-Shorbaji, N.: Health and Medical Informatics: Technical Paper WHO Cairo (2001)
4. European Commission - eHealth - making healthcare better for European citizens: an action plan for a European eHealth area (2004), http:europa.eu.int/eur-lex/pri/en/dpi/cnc/doc/2004/com2004_0356en01.doc
5. Data Center Best Practices: Managing Data with Cloud Computing, Oracle White Paper, InfoWorld, Custom Solutions Group

Modified Real-Time Advanced Inexpensive Networks for Critical Infrastructure Security and Resilience

K. Rajasekhar[1,*] and Neeraj Upadhyaya[2]

[1] NIC, Dept. of Electronics and Information Technology
MCIT, Government of India, APSC, Hyderabad,
Andhra Pradesh, India – 620 015
sekhar@nic.in
[2] J.B. Institute of Engineering & Technology
Yenkapally, Moinabad Mandal
R.R. District, Hyderabad
A.P., India-500075
drnirajup@gmail.com

Abstract. The critical Communication and Information Technology Infrastructure required for delivery of financial inclusion services of the Government to Citizen services has been considered as the Critical Infrastructure for research work by authors. On the basis of practical research work a novel, efficient, effective and sustainable system for ensuring safety, security and resilience of critical IT infrastructure has been presented in this article. The various types of attacks and threats reported in the literature were studied, the main reasons of security threats and attacks analysed. The solution hypothesis proposed. A new mode of computation, Real-time Advanced Inexpensive Network (RAIN) Computing proposed and a multi-level scalable, low cost security solution with RAIN computing to safeguard and ensure resilience of the critical IT infrastructure for e-Governance projects presented. Though focus of experimentation is critical IT & C infrastructure, security methodology presented in the paper can be extended to safeguard other types of CI also.

Keywords: Modified Real-time Advanced Inexpensive Networks, Critical Infrastructure Security, Resilience, RAIN Computing, e-Governance.

1 Introduction

Safety and security of Critical IT infrastructure has become vital for the economy of the country. This article has 9 parts. In Part 1 we gave Introduction, in part 2 the present day sophisticated organised attacks, security problems, threats, and security vulnerabilities were described. In Part 3 the reasons for existence of vulnerabilities analysed and the root cause identified. In Part 4 of this article the related work done in this area has been described based on the published research articles. In Part 5, a

* Corresponding author.

S.C. Satapathy et al. (eds.), *ICT and Critical Infrastructure: Proceedings of the 48th Annual Convention of CSI - Volume II*, Advances in Intelligent Systems and Computing 249,
DOI: 10.1007/978-3-319-03095-1_84, © Springer International Publishing Switzerland 2014

new concept of Real-time Advance Inexpensive Network computing was introduced and it can be modified to build a very sophisticated yet low cost security solution to safeguard and ensure resilience of ITC critical infrastructure. We have also given description about the prototype which we had built to safeguard CI and used it in practical real-life situation. In Part 6 we have presented the results of the experiment. In Part 7 the results observations were discussed. In Part 8 the conclusions which can be drawn from the whole pilot research project were presented. In Part 8, we have acknowledged the support received for the research project.

2 Attacks on Critical Infrastructure

The Incapacitation or destruction of Critical Infrastructure shall adversely affect national security, economy, public health or safety [1]. Organised Cyber attacks are for profit and for political reasons are increasing day by day all over the world. Cyber attacks with financial motives are resulting in financial losses to individuals and organisations. The attacks are becoming more and more sophisticated and more than 36 million euros were stolen from more than 30,000 bank accounts during 2012 through Euro grabber attack [1]. Apart from attacks, due to lack of sensitivity, awareness, organisational security policies and their enforcement, employees are sending sensitive information outside the organisation. Security Standards, Policies, People, Education, Awareness, Compliance, Auditing certification review and such measures shall avoid such in-secure practices. So finding security solutions and measures in such scenarios are outside the scope of our research work.

3 Root Causes of Security Problems

One of the objectives of our research is to find out the root cause of security problems. Most of the operating systems Windows, Linux and other platforms such as JRE, IE, Acrobat Reader, Flash Player and popular products such as Oracle, Apple are also found to be vulnerable[1][2]. Cloud technology has provided opportunities to safeguard the systems with cloud enabled knowledge based system for securing the CI. But the applications and products developed shall inherit the security vulnerabilities and limitations (if any) of the cloud based development tools. Another drawback is the hackers are also hiring hacking tools for limited period of time and able to mobilise enterprise resources to make more sophisticated attack tools to compromise CI. So such attacks are able to cause heavy damages to the CI.

So from all the previous research it is evident that, no generic purpose software is fool proof.[1][2]. The problems identified during the research work are – General purposes operating systems and system software are vulnerable to attacks and not suitable for CI. Automated security tools are also not completely reliable to tackle all types of attacks especially the zero based attacks. In such situations, the limited manpower manning CI and the limitations of their knowledge and skills set pose problems for resilience.

4 Related Work

Mostly the related work done by others highlighted the previous attacks and associated damages. Some of them offered solutions in terms of knowledge based cloud based solutions which are very expensive. [1], [5], [6]. Most of the solutions offered are aimed at fortifying and protecting the available general purpose system software such as operating systems, application servers, relational database management servers, middle tier applications, networking applications, other deployment or rendering platforms such as acrobat reader, browsers, run time environments etc., Most of the systems are used for several purposes, i.e., for computing, communicating with world, socialising, entertaining, presenting, collaborating with colleagues etc., Facility of accessibility to CI or clients connected to CI from anywhere, anytime from any device is a double edged knife. Though, the general purpose systems with several features are less expensive, their protection mechanism is very expensive and the cost for security maintenance increases exponentially. So far the assumption made by researchers who offered solutions is - it is possible to transform the human knowledge and expertise into a knowledge based system.

In reality, it is not possible to transform all the knowledge of experts into systems. It is evident from research [1][2][3], that each time the attackers are adopting a new ways and means to attack CI. So, even if systems are empowered with available human knowledge, such systems fail to tackle new types of attacks based on zero date vulnerabilities. Therefore, it is not possible to eliminate human experts in tackling the attacks and mitigate the risks to achieve resilience. In our research, we have taken into consideration this important factor in arriving at a feasible solution. In US the home security dept responsible for security of CI, is doing surveillance, of attacks as post-mortem activity [4]. Such activities shall help in analysing and working out solutions to manage risks in future. To ensure resilience, multi-agent natural immunologically inspired Artificial Immune Systems based security model was proposed safeguarding power grids [5]. However, the part of the critical system under attack when fails the recover in their model sends danger signals, these danger have to propagate till they reach an appropriate node which may offer solution. The drawback with Mavee et al., [5] approach is loss of time in linear propagation of danger signals, and uncertainty in reaching an appropriate node which can offer solution. In our research we have adopted the approach of automatic healing, and worked out solution to avoid delays and uncertainty in achieving resilience with modified real-time advance inexpensive network computing.

5 Modified Real-Time Inexpensive Network Computing and CI Security

Based on research we have worked out solution to safeguard CI. The solution has three major components. One of the components is Specialised System Software (SSS1) for CI installations. Second component is Systems Surveillance and Solution

(SSS2). Synchronized Superiors Supervision (SSS3) . To build these three components SSS1, SSS2 and SSS3, modified Real-time Inexpensive Network computing has been harnessed. In this section we describe each of the components in detail.

5.1 SSS1

It is the customized system software for CI. Hypothesis is generic purpose software is vulnerable to security threats. For CI special and customised system software is required. To test the hypothesis, we have taken two systems of similar configuration and make. On one of the machine a popular general purpose licensed software was procured and installed on the other, freely available software source course which was re-compiled with minor modifications (to file names and paths) was installed. So one system had general purpose operating system we label as A, and the other relatively proprietary OS which we shall label as B. On system A, available enterprise anti-virus software client was also installed. However, on system B, we could not directly install any of the available anti-virus solutions, (as the OS is proprietary). Both the systems were connected to internet and put to use for six months period. Both the systems were used to access same set of internet resources for six months. Besides this same set of USB memory sticks which were used in the peer group were also used to store and carry files from both systems. For six months time system A and system B were exposed to same set of security threats. The virus, malware, spam ware, adware infected by both the systems were noted. System A was infected 332 files with 5 types of deadly virus, three of which could not be healed by anti-virus installed on system A. Whereas not a single file of B was infected. The results obtained are shown in figure 1 below.

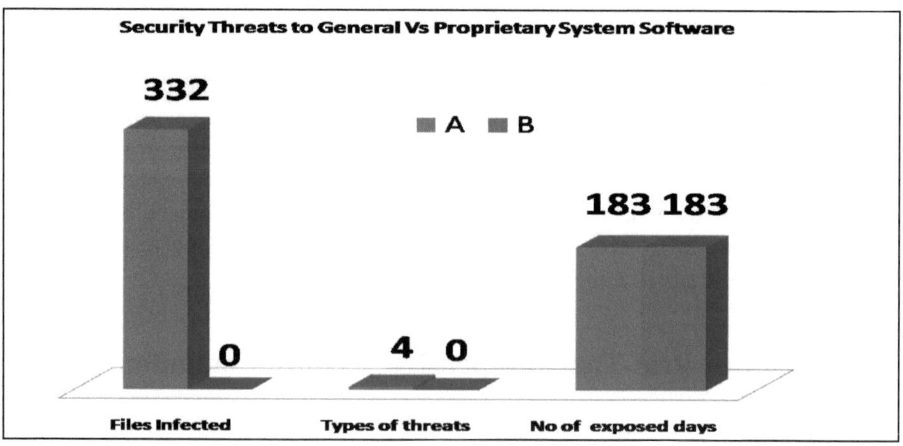

Fig. 1. Research Experimentation Results

As system B is secure and least vulnerable, it was chosen as platform for hosting critical information infrastructure. Subsequently, Critical Infrastructure for a typical Government to Citizen Services System was created by mobilising 4 servers/ virtual machines. One of which was used as Application server which also generates certificates to the Beneficiaries of welfare schemes of Government, on second machine RDBMS was installed, on third the Payment Gateway server which interacts with Banking System was installed, on fourth virtual machine the SMSC Gateway was installed for communication with mobile Telecom network which establishes the required linkage with Real-time Advanced Inexpensive Network (RAIN) by installing a RAIN node on each of the servers. The resultant infrastructure is Modified RAIN. The term modified was used, because, the OS of the systems was modified to ensure security, subsequently, the RAIN which comprises only mobile hand-held systems was connected to CI located at Data centre to ensure security as shown in the figure 2.

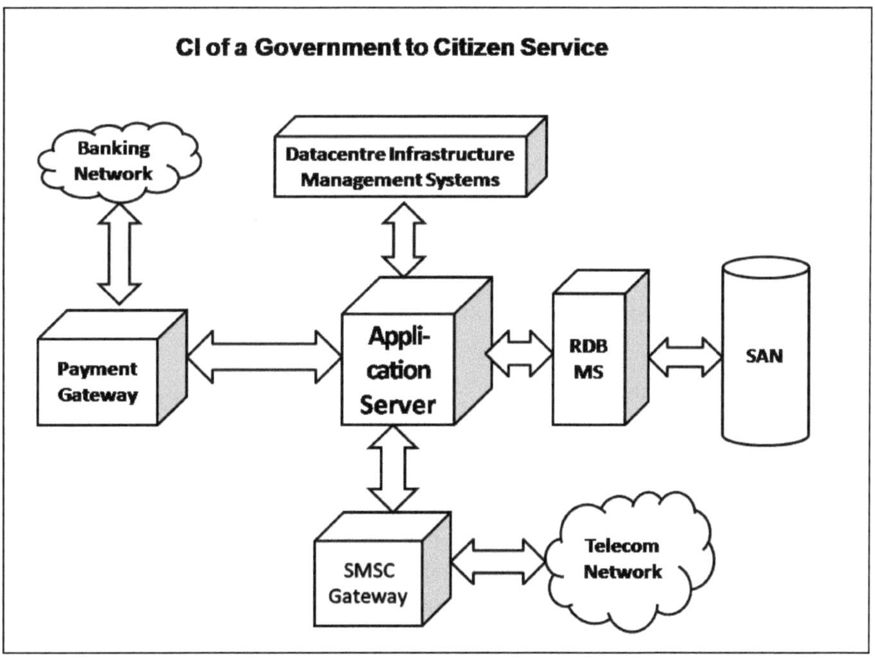

Fig. 2. CI of a Govt. to Citizen Paymnet Service

5.2 Systems Surveillance and Solution (SSS2)

As part of SSS2 research experiment the critical infrastructure been divided in to small and easily identifiable and manageable digital assets for better security surveillance based on a criteria. The default sizing criteria could be - 1.Type of CI 2. Server ID or IP number / Digital Machine Id (Digital machine can be virtual also), 3. Project ID 4. Sub-service Id, 5. Dominant stakeholder's location ID, 6. Function Id,

7.File Type, 8.Directory Id, 9. File Id. 10. Version Id. Main aim of our research is ensuring security of CI from well organised and highly sophisticated enterprise level cyber attacks. The ten parameters based criteria was defined to classify the critical digital assets to render security surveillance and security maintenance services. To ensure high level of security it is always preferable to fix a maximum limit (99) on the number of files in any directory. This type of organisation of digital assets, enable us to easily identify critical directories and the corresponding human experts who may monitor and secure the respective digital asset. The smallest unit of digital asset as per the defined criteria in cyber space is a file. We have ensured average size of 30 files in any directory of CI. Then in each directory a SSS2 script called as modified RAIN security surveillance node was installed. SSS2 script performs following system surveillance and solution functions. It checks for changes to the files in the respective directories of CI. If new files are written they are deleted immediately. If any process tries to modify existing files are modify file permissions they are restored. If any process tries to delete or modify files, the files are restored and a log is written about such incidents. So SSS2 facilitates local level surveillance and quick solution. If they fail to restore previous status of the directory of if SSS2 associated script itself is modified or deleted, automatically SMS or USSD alert goes to pre-define RAIN nodes. Which shall trigger the third level security of CI.

5.3 Synchronized Superiors Supervision (SSS3)

Finally in this step, the RAIN nodes were involved to ensure real-time involvement of experts to tackle attack situations to aid the data-centre staff. In the following section we briefly describe about RAIN.

5.4 RAIN Computing

RAIN computing is computing with intelligently networked nodes which interact in real-time. The end mobile nodes installed with RAIN computing software can be termed as RAIN- droplets. To secure the CI, the RAIN droplets can be considered as small piece of software installed in the mobile phones of the experts.The performance can be monitored by the node which initiates firing in the RAIN easily by checking the feedback from the target Droplet or set of Droplets within the admissible time. When the SSS2 level fails to provide solution, automatic alerts goes to data centre manager or the corresponding human expert to whom that particular directory security might have been assigned. This is done by SMS gateway server, which has a unit which pings the presence of SSS2 scripts in all directories regularly. If it fails to see the presence or when it receives a alert message from SSS2, it broadcasts SOS alerts to pre-defined human experts. Some times alert messages can go to more than one human expert also (if both are made responsible for a critical digital asset security). So human expert can immediately intervene and direct the few manpower manning the data centre. The type of intervention depends on the type of content. The directory listing of the CI shall be made available a priori to the human experts. So they can easily monitor the status and provide remedial measures to ensure quick resilience.

Remote command execution by experts directly to ensure resilience also can be ensured, but it is a security risk as most of the smart phones which can perform such operations themselves are vulnerable. So simple pre-defined text messages / instructions goes operators and data centre from human experts during the crisis, instructing precisely what to do. The exchange of instructions happen in small codes. At the operator end they shall be manifested as detailed step by step process for restoration/ supervision.

So with this type of organisation three tier or multi-tier and very high level of security can be ensured with modified RAIN. So with modified RAIN, a security ring consisting of three levels of security SSS1, SSS2 and SSS3 get formed around the CI. The whole experiment do not cost any additional hardware or system software. In the case of vital CI such as nuclear installations, SSS2 level scripts modified to do surveillance at VM level can be written in PROMs and hardwired into the systems. So such hardwired SSS2 security agents cannot be tampered by external attacks.

6 Results

SSS1 experiment results were shown figure 2. It was observed that 332 files in general purpose system A were affected by virus, malware, adware, spams etc., Even email password also reported to have been leaked to unknown IPs located outside the country. In the case of system Special System B, no files were affected. No threats detected. SSS2 experiment results shown that, in the lab the scripts have performed their functions well during the tests. The SSS3 experiment has shown promising results, and facilitated demarcation of roles and responsibilities clearly and provided means for multi-level security surveillance. In the simulated conditions, alarms were sent to experts. Some times SMS Gateway of service provider took some time to deliver sms messages to the RAIN nodes which are owned by the experts.

7 Discussion

The method suggested is based on Prevention, Early warning, Detection, Mitigation, Response, Recovery and Resilience. However, unlike other it is based on hypothesis of Generic system software are not suitable for CI. The results of SSS1 shows that specialised systems are very immune to attacks. However, creation of special customised unique systems can also be down with free open source systems with a little extra efforts. Even licensed softwares also can be designed to be installation specific and unique so that it will maintain uniqueness so that it becomes immune to attacks, at the same time it can perform the vital functions required for CI. Results of experiment SSS2 showed that, organisation of digital assets and in-expensive self healing solution are feasible and possible to safeguard CI. In the SSS3 experiment results shows that some times SMS message may not get delivered in time. So USSD mode is preferable for alerting SOS messages.

8 Conclusion

For prevention of attacks each CI installation platform can be made unique. For easy surveillance and building built-in risk detection and mitigation, entire CI is categorised into small domains. Each domain shall have a CI protection modified RAIN node. Such protection mechanism shall do self healing, or expert controlled self healing immediately. For threat Detection, correction, security monitoring , auditing and resilience of CI, modified RAIN computing can be successfully used.

Acknowledgements. The authors are thankful to NIC, Department of Electronics and Communication Technology, Ministry of Communication and Inform and the various Government of Andhra Pradesh Departments, for support in this research work.

References

1. Security Report (2013), `http://www.checkpoint.com/campaigns/security-report`
2. Cyber Security News, `http://ptlb.in/csrdci/?p=183`
3. Vulnerability Notes of Indian CERT, `http://www.cert-in.org.in/`
4. Electric Emergency Incident and Disturbance Report, `http://www.oe.netl.doe.gov/oe417.aspx`
5. Mavee, S.M.A., Ehlers, E.M.: A Multi-agent Immunologically-inspired Model for Critical Information Infrastructure Protection. In: 11th IEEE International Conference on Trust, Security and Privacy in Computing and Communications (TrustCom), pp. 1089–1096. IEEE Press, New York (2012)
6. Blauensteiner, P., Kampel, M., Musik, C., Vogtenhuber, S.A.: Socio-technical approach for event detection in security critical infrastructure. In: IEEE Computer Society Conference on Computer Vision and Pattern Recognition Workshops (CVPRW), pp. 23–30. IEEE Press, New York (2010)

Regression Model for Edu-data in Technical Education System: A Linear Approach

P.K. Srimani[1] and Malini M. Patil[2]

[1] R&D, Bangalore University,
Bangalore, Karnataka, India
profsrimanipk@gmail.com
[2] Dept. of ISE, JSSATE, Bangalore,
Karnataka, India
Bhartiyaar University, Coimbatore,
Tamilnadu, India
patilmalini31@yahoo.com

Abstract. Mining educational data is an emerging interdisciplinary research area that mainly deals with the development of methods to explore the data stored in educational institutions which is referred to as Edu-Data. Data mining is concerned with the analysis of data for finding patterns which are previously unknown and are presently useful for future analysis. The technique of mining Edu-data is referred to as Edu-mining. On the other hand statistics is a mathematical science concerned with the collection, analysis, interpretation or explanation, and presentation of data which plays a very important role in the process of data mining. The paper aims at developing a simple linear regression model for Edu-data using the statistical approach. The results obtained helps the management to predict the semester results and also helps in proper decision making processes in Technical Education System. It is also found that the predictions were almost nearing to the actual values. The present work is first of its kind in literature.

Keywords: Edu-data, Edu-mining, Data Mining, Regression, Prediction, Visualization.

1 Introduction

Educational Mining (Edu-mining) is a process of discovering knowledge from educational data(Edu-data), which helps the technical education system(TES) to take useful decisions for maintaining the quality of the education system. Edu-data which is a large data repository consisting of data related to educational systems. Edu-data is evolved because of huge collection of data mainly from WWW, study material available in the internet, e-learning schemes, computerization of education system, online registration schemes for admission process in the universities, student information system, examination evaluation systems etc. Recent development of such data repository not only belongs to higher education system but also to the secondary

S.C. Satapathy et al. (eds.), *ICT and Critical Infrastructure: Proceedings of the 48th Annual Convention of CSI - Volume II*, Advances in Intelligent Systems and Computing 249,
DOI: 10.1007/978-3-319-03095-1_85, © Springer International Publishing Switzerland 2014

education system. Predictive analysis is found to be one of the novel approach for proper predictions in student stakeholder of the typical Edu-mining system.

Data mining(DM) can be viewed as a confluence of many disciplines. The advancement of technology has resulted in the evolution of different techniques in the area of DM. New research findings resulted in new issues in each technique. To quote some: association rule mining, classification, clustering, SVM, SDM, data stream mining etc. DM focuses on different ideas such as sampling, estimation hypothesis testing from statistics, search algorithms, modeling techniques, machine learning theories from artificial intelligence, pattern recognition and machine learning, hi-performance computing and many more. Thus DM is a "Confluence of Many Disciplines", among which the statistical approach plays a very important role in predictions. Because of this reason both data mining and statistics are considered as two intersecting disciplines.

The paper is organized as follows: section 2 focuses on the overview of technical education system which is taken as a benchmark system for Edu-mining. Section 3 discuses the related work about statistical approach in mining education data. Section 4 focuses on Edu-Mining using linear regression analysis. Section 5 is about the implementation steps and Section 6 about the results and analysis respectively. Future enhancement of the work and conclusions are briefed at the end of the paper.

2 Overview of Technical Education System (TES)

This section mainly focuses on the Technical Education System, which is considered as a bench mark system for the study of Edu-mining. The system is organized by three main components, which are called as stakeholders shown in Fig. 1.

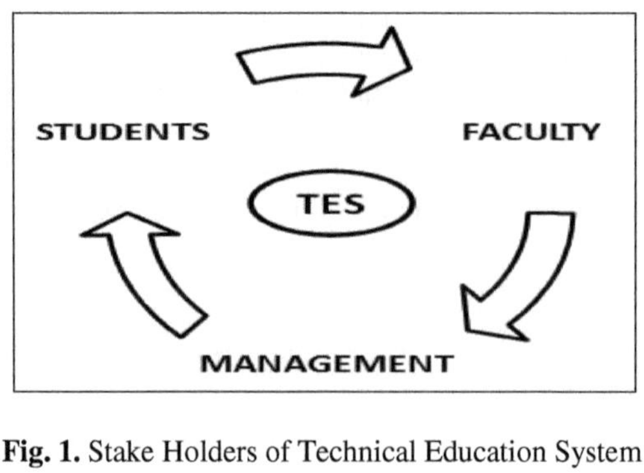

Fig. 1. Stake Holders of Technical Education System

The three important stakeholders of the system are discussed as follows: Stakeholder 1 is Management, which is the supreme authority to manage the system. Stakeholder 2 is Students who are considered as the main revenue generators in the system, who work on a give and take policy. Stakeholder 3 is Teachers who are instrumental in strengthening of the system and are in teaching and learning process. The managerial perspectives of the present analysis could be cited as: goal seeking analysis, optimization analysis, sensitivity analysis. The detailed discussion on these different approaches of analysis is based on the typical education systems approach to problem solving methods.

2.1 Goal Seeking Analysis

This analysis mainly focuses on the aims and objectives of the institution set by the management or in other words, it can be stated as the goal of the management. They are summarized as follows: The mission and vision of the management is to scale new heights and enhance the brand image of the system. It also aims at providing sophisticated infrastructure, learning platform to grow to a new height. It also aims at conducting activities to enhance the performance to achieve excellence.

2.2 Optimization Analysis

This kind of analysis in the system is mainly concerned with the qualitative measures of the system. They include standardization of policies related to administrative procedures for the students, proper faculty recruitment procedures as per the norms, designing the infrastructure, facilities to cope up with development of the institution. This is designed so as to maximize the quality output of the students and faculty.

2.3 Sensitivity Analysis

This kind of analysis deals with strengthening the important and vital factors of development. Programs like faculty development and management development programs have to be designed and developed in order to strengthen the core competence of the faculty. Similarly the student development programs should help in broadening their horizon of learning. It is imperative that each component of the system will work in harmony and the individual goals merge with the organization goals. The above three analyses will greatly act as a decision support system.

3 Related work

A thorough survey of the literature reveals that very sparse literature is available pertaining to the present work. Some amount of work in this regard has been done and is outlined briefly in this section. The authors emphasize that with regard to edu-mining the only works are [1,2,3,4] where the authors have not used the statistical approach. Therefore the present investigation is carried out to provide an excellent platform for future research. The main objective of the present investigation is to provide recommendations directly to the students, faculty and management with respect to their personalized activities. Several DM techniques have been used for this task and the most common are association-rule mining, clustering, and sequential pattern mining. But no work is available where in the regression models are studied. This paper emphasizes a linear regression model in an integrated way by considering the above aspects.

Performance, knowledge, grade, score, marks obtained are different variables which describe the student's performance. In an education system these values are used as predictive variables if a student report has to be generated. The similar work is found in [5]. It is not surprising that teachers prefer pedagogically oriented statistics that are easy to interpret [6]. Statistical analysis of educational data can tell us things such as: the most popular pages of student study material, results and analysis, comparisons of results etc. Statistical analysis is also very useful to obtain reports assessing [7] how many minutes the student has worked, how many minutes he has worked today, how many problems he has resolved, and his correct percentage, our prediction of his score, and his performance level. Visualization techniques used to understand and analyze the data [8].

4 Regression

Regression[9] is technique of DM and is used to fit an equation to a dataset. Regression modeling has many applications in trend analysis, business planning, marketing, financial forecasting, time series prediction, biomedical and drug response modeling, and environmental modeling. The simplest form of regression, linear regression, uses the formula of a straight line (y = mx + b) and determines the appropriate values for m and b to predict the value of y based upon a given value of x. A regression task begins with a data set in which the target values are known. For example, a regression model that predicts house values could be developed based on observed data for many houses over a period of time. In addition to the value, the data might track the age of the house, area, number of rooms, number of floors etc. House value would be the target, the other attributes would be the predictors, and the data for each house would constitute a case. In the model building process, a regression algorithm estimates the value of the target as a function of the predictors for each case in the built data. These relationships between predictors and target are summarized in a model, which can then be applied to a different data set in which the target values are unknown.

4.1 Methodology of Regression Analysis

Regression analysis seeks to determine the values of parameters for a function that cause the function to best fit a set of data observations that you provide. The following equation expresses these relationships in symbols. It shows that regression is the process of estimating the value of a continuous target (*y*) as a function (*F*) of one or more predictors ($\mathbf{x}^1, \mathbf{x}^2, \ldots, \mathbf{x}^n$), a set of parameters ($\theta^1, \theta^2, \ldots, \theta^n$), and a measure of error (*e*).

$$Y = F(x, \theta) + e. \tag{1}$$

The predictors can be understood as independent variables and the target as a dependent variable. The error, also called as the **residual**, is the difference between the expected and predicted value of the dependent variable. The regression parameters are also known as **regression coefficients**. The process of training a regression model involves finding the parameter values that minimize a measure of the error, for example, the sum of squared errors.

4.2 Linear Regression

Linear regression is a statistical technique that is used to learn more about the relationship between an independent (predictor) variable and a dependent (criterion) variable. A linear regression technique can be used if the relationship between the predictors and the target can be approximated with a straight line which is shown in fig. 3. and can be expressed with the following equation.

$$y = \theta^2 x + \theta^1 + e. \tag{2}$$

The **slope** of the line (θ^2) and the ***y* intercept** (θ^1) are the two parameters in simple linear regression where θ^2 is the angle between a data point and the regression line and θ^1 is the point where **x** crosses the *y* axis (**x** = 0).

4.3 The Coefficient of Determination

The **coefficient of determination** (denoted by R^2) is a key output of regression analysis. It is interpreted as the proportion of the variance in the dependent variable that is predictable from the independent variable.

- The coefficient of determination ranges from 0 to 1.
- If $R^2 = 0$, then the dependent variable cannot be predicted from the independent variable else it can be predicted without error from the independent variable.
- An R^2 between 0 and 1 indicates the extent to which the dependent variable is predictable.
- An R^2 of 0.10 means that 10 percent of the variance in *Y* is predictable from *X*; An R^2 of 0.20 means that 20 percent is predictable; and so on.

5 Experiments and Results

Regression analysis on Edu-data is performed using IBM PASW [10] Statistics 18. PASW can perform variety of data analysis and presentation functions, including the statistical analysis and graphical representation of data. The student data set in Edu-data comprises of different attributes pertaining to the student stakeholder in TES. The data set description used in the analysis is presented in table 1.

Table 1. Data set description

Attributes	Decription of attributes
USN	University Seat Number
GENDER	Either Male /Female, (0/1)
MODE	Mode of entry of a student whether diploma/regular, (0/1)
SEAT_TYPE	Type of the seat (CET/Management/ComedK), (0/1/2)
TENTH_MARKS	Marks in %
PU_MARKS	Marks in %
RANKING_NUM	Ranking Number of a student
M1	First Semester Marks in %
M2	Second Semester Marks in %
SIMPLE	Result after performing regression

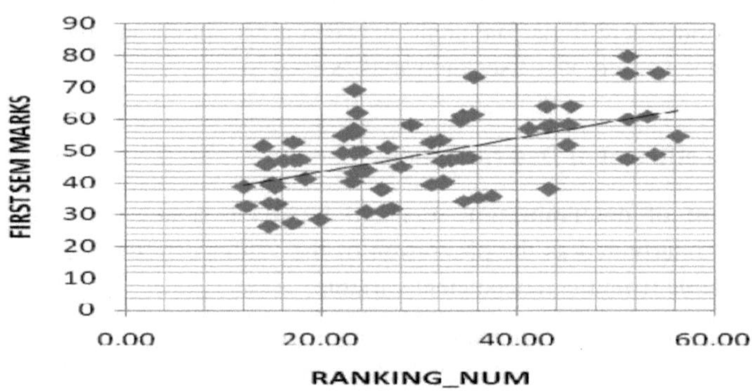

Fig. 2. Scatter plot to check the linearity

To determine the linear relationship between the variables on the edu-data it is recommended to run a Scatter plot before applying a regression analysis using PASW. If there is no linear relationship, no need to run a simple regression. From fig. 2 it is observed that points on a graph are clustered in a straight line. This clearly indicates that there is a linear relationship between the variables and the simple regression can be run in PASW. Fig. 3 depicts the results obtained by simple linear regression analysis using PASW-18 for predicting the second semester results of students.

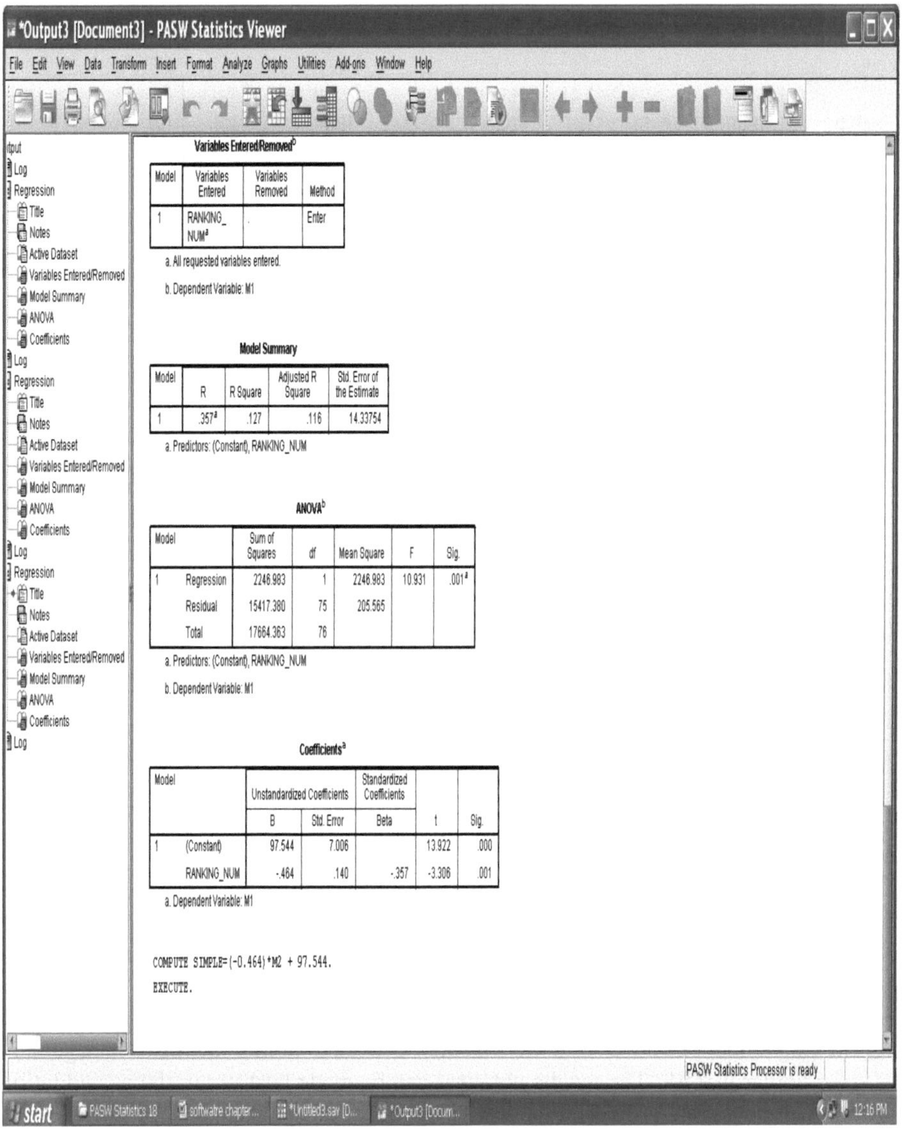

Fig. 3. Result window for simple linear regression from PASW18

Table 2 shows the results after executing the simple linear regression model. It is found that the predictions were almost nearing to the actual values.

Table 2. Simple Regression Table

USN	GENDER	MODE	SEAT_ TYPE	TENTH_ MARKS	PU_ MARKS	RANKING_ NUM	M1	M2	SIMPLE
2008_001	0	0	0	78.56	45.56	300	78.56	45.56	75.89
2008_002	1	0	1	56.67	56.23	301	56.67	56.23	76.85
2008_003	1	0	2	56.89	78.00	544	56.89	78.00	78.81
2008_004	0	0	2	56.45	78.32	234	56.45	78.32	78.84
2008_005	1	0	1	67.43	87.90	678	67.43	87.90	79.70
2008_006	0	0	2	67.00	78.45	456	67.00	78.45	78.85
2008_007	1	0	1	46.00	67.78	345	46.00	67.78	77.89
2008_008	1	0	2	56.45	98.67	456	56.45	98.67	80.67
2008_009	1	0	1	34.23	89.76	321	34.23	89.76	79.87

6 Conclusion

The authors have made the thorough analysis of Edu-data for which already the classification technique is applied. Present results are explored using statistical method and they are found to be interesting and provide the excellent platform for future scope in this regard. The results help the management to analyze the Edu- data considering student as a Stake holder, faculty as a stake holder and also management as stake holder in TES. The results obtained can be made best use in the accreditation process for the overall growth of the technical education system.

Acknowledgements. One of the authors Mrs. Malini M. Patil acknowledges J.S.S Academy of Technical education, Bangalore, Karnataka and Bhartiyaar University, Coimbatore, Tamilnadu, India for providing the facilities for carrying out the research work.

References

1. Srimani, P.K., Patil, M.M.: Edu-mining: A Machine learning approach. In: AIP. Conf. Proceedings, pp. 61–66 (2011)
2. Srimani, P.K., Patil, M.M.: A Classification Model for Edu-mining. In: PSRC-ICICS Conference Proceedings, pp. 35–40 (2012)
3. Srimani, P.K., Patil, M.M.: A Comparative Study of Classifiers for Student Module IN Technical Education System(TES). International Journal of Current Research 4(01), 249–254 (2012)
4. Srimani, P.K., Patil, M.M., Srivatsa, P.K.: Performance evaluation of Classifiers for Edu-data: An integrated approach. International Journal of Current Research 4(02), 183–190 (2012)
5. Feng, M., Heffernan, N.: Informing teachers live about student learning: Reporting in the assessment system. Technol., Instruction, Cognition, Learn. J. 3, 1–8 (2006)
6. Freedman, D., Purves, R.: Statistics 4th edn., Newyork
7. Zinn, C., Scheuer, O.: Getting to know your students in Distance learning contexts. In: Proc. 1st Eur. Conf. Technol. Enhanced Learn., pp. 437–451 (2006)
8. Mazza, R.: Introduction to Information Visualization. Springer, NewYork (2009)
9. Han, J., Kambler, M.: Data Mining Concepts and Techniques, 2nd edn. Morgan Kaufmann (2007)
10. PASW Statistics 18 Brief Guide. Prentice Hall (2007)

References

Author Index

MIX
Papier aus verantwortungsvollen Quellen
Paper from responsible sources
FSC® C105338

If you have any concerns about our products,
you can contact us on
ProductSafety@springernature.com

In case Publisher is established outside the EU,
the EU authorized representative is:
Springer Nature Customer Service Center GmbH
Europaplatz 3, 69115 Heidelberg, Germany

Printed by Libri Plureos GmbH
in Hamburg, Germany